Edited by
Gang He and Zhaoqi Sun

High-k Gate Dielectrics for CMOS Technology

Related Titles

Biswas, A. S., Mishra, V. V.

Dielectric Resonators

Analysis, Applications, and Materials

Hardcover

ISBN: 978-0-471-08613-0

Starr, G. W.

Phase Locked Loops and Clock Data Recovery Circuit Design on Nano CMOS Processes

Hardcover

ISBN: 978-0-470-04489-6

Klauk, H. (ed.)

Organic Electronics II

More Materials and Applications

2012

Hardcover

ISBN: 978-3-527-32647-1

Ramm, P., Lu, J. J.-Q., Taklo, M. M. V. (eds.)

Handbook of Wafer Bonding

2012

Hardcover

ISBN: 978-3-527-32646-4

Morkoc, H., Özgür, Ü.

Zinc Oxide

Fundamentals, Materials and Device Technology

2009

Hardcover

ISBN: 978-3-527-40813-9

Wong, B. P., Mittal, A., Starr, G. W., Zach, F., Moroz, V., Kahng, A.

Nano-CMOS Design for Manufacturability

Robust Circuit and Physical Design for Sub-65nm Technology Nodes

Hardcover

ISBN: 978-0-470-11280-9

Morkoc, H.

Handbook of Nitride Semiconductors and Devices

approx. 3000 pages in 3 volumes

Hardcover

ISBN: 978-3-527-40797-2

Garrou, P., Bower, C., Ramm, P. (eds.)

Handbook of 3D Integration

Technology and Applications of 3D Integrated Circuits

2008

Hardcover

ISBN: 978-3-527-32034-9

Baklanov, M., Maex, K., Green, M. (eds.)

Dielectric Films for Advanced Microelectronics

2007

Hardcover

ISBN: 978-0-470-01360-1

Kursun, V., Friedman, E. G.

Multi-voltage CMOS Circuit Design

2006

Hardcover

ISBN: 978-0-470-01023-5

Edited by Gang He and Zhaoqi Sun

High-k Gate Dielectrics for CMOS Technology

WILEY-VCH Verlag GmbH & Co. KGaA

The Editors

Prof. Gang He
Anhui University
School of Physics and Materials Science
Anhui Key Laboratory of Information Materials and Devices
Feixi Road 3
Hefei 230039
China

Prof. Zhaoqi Sun
Anhui University
School of Physics and Materials Science
Anhui Key Laboratory of Information Materials and Devices
Feixi Road 3
Hefei 230039
China

Library of Congress Card No.: applied for

British Library Cataloguing-in-Publication Data
A catalogue record for this book is available from the British Library.

Bibliographic information published by the Deutsche Nationalbibliothek
The Deutsche Nationalbibliothek lists this publication in the Deutsche Nationalbibliografie; detailed bibliographic data are available on the Internet at http://dnb.d-nb.de.

Print ISBN: 978-3-527-33032-4
ePDF ISBN: 978-3-527-64637-1
ePub ISBN: 978-3-527-64636-4
mobi ISBN: 978-3-527-64635-7
oBook ISBN: 978-3-527-64634-0

Cover Design Adam-Design, Weinheim, Germany
Typesetting Thomson Digital, Noida, India
Printing and Binding Markono Print Media Pte Ltd, Singapore

Contents

Preface

In 1965, Gordon Moore wrote a paper entitled "Cramming more Components onto Integrated Circuits" where he first proposed that transistor density on chips would grow exponentially. This became known as Moore's law. Remarkably, the industry has kept pace with this exponential growth for that past four decades. To keep track with the Roadmap, scaling of the gate stack has been a key to enhancing the performance of complementary metal oxide semiconductor field-effect transistors (CMOSFETs) of past technology generations. Because the rate of gate stack scaling has diminished in recent years, the motivation for alternative gate stacks or novel device structures has increased considerably. However, further scaling the FET is eventually going to be impeded by the inability to further reduce the oxide thickness without risking a breakdown of the device. New technology problems, including the dielectric thickness variation, direct tunnel current, penetration of impurities, and the reliability and lifetime of devices, arose, which cause significant concern regarding the operation of CMOS devices, particularly with regard to standby power dissipation, reliability, and lifetime. Intense research during the past decade has led to the development of high dielectric constant (k) gate stacks that match the performance of conventional SiO_2-based gate dielectrics. However, many challenges remain before alternative gate stacks can be introduced into mainstream technology. This book provides a perspective on the gate dielectric and the approaches in progress to rectify the above noted issues, that is, increasing the physical thickness of the gate dielectric to significantly reduce the power and direct tunneling current issues while enabling the continued reduction in the electrically active gate dielectric thickness by utilizing high-k dielectric constant materials. The high-k materials facilitate both an increased physical thickness and a reduction in the electrical thickness to maintain the requisite scaling methodology. This monograph, however, is envisioned to be more than just a current view of these alternative high-k gate stack technology. Rather, both previous and present directions related to scaling the gate dielectric and their impact, along with the creative directions and future challenges defining the direction of high-k gate dielectric scaling methodology, will be reviewed.

The monograph is introduced by a comprehensive review of scaling and limitation of Si-based CMOS by Gang He. In this part, a detailed discussion of current dielectrics and those proposed for future generations is included. Current beliefs regarding the limitations of silicon dioxide as the gate dielectric are reviewed.

Ultrathin oxides, including gate dielectrics below 40 Å, some of which are nitrided silicon oxides are discussed. The benefits and limitations of alternative for new high-*k* materials for possible high-performance CMOS applications are reviewed. Part 2 addresses the high-*k* deposition and material characterization with seven chapters. Initially, Hong-Liang Lu discusses the issue of high-*k* gate stacks and shows the criteria for selecting alternative high-*k* gate dielectrics. Then, Ian W. Boyd and Cheol Seong Hwang discuss the high-*k* deposition by UV-photo-CVD and ALD technology. Structure, band alignment, and electrical characteristics have been given by Tung-Ming Pan, Yi-Zhao, Elena O. Filatova1, and V. V. Afanas'ev. The challenge in interface engineering and gate electrode has been described in Part 3 by Shi-Jie Wang, Li-Qiang Zhu, and Wen-Wu Wang.

Part 4 discusses the development of non-Si-based CMOS technology. Initially, Albert Chin gives the description of metal gate/high-*k* CMOS evolution from Si to Ge platform. Then, Wei-Chao Wang reviews the progress on GaAs/high-*k* interface. Jack C. Lee continues to discuss the III–V MOSFET with ALD-derived high-*k* gate dielectrics. In Part 5, high-*k* application in novel devices has been discussed by Xu-Bin Lu and Pooi See Lee. The last chapter of this book by M. Chudzik addresses the challenges and future directions for advanced technology generations. The vision, experience, and wisdom the authors have summarized will help succeed in ensuring this monograph is a timely, relevant, interesting, and resourceful book focusing on both the fundamentals and the evolving directions to ensure the successful integration of high-*k* gate dielectrics and metal gate electrodes in future ICs as envisioned in the ITRS.

We thank Wiley-VCH for inviting us to write/edit the book and we hope to stimulate even more the incorporation of high-*k* gate technology in the near future. Last but not least, we express our thanks to Dr Martin Preuss and Dr. Ernest Kirkwood, who gave us time and resources to edit this book.

Dec. 2011
Hefei

Gang He
Zhaoqi Sun

List of Contributors

Albert Achin
National Chiao-Tung University
Electronics Engineering Department
and College of Photonics
1001 University Road
Hsinchu 300
Taiwan

Valeri V. Afanas'ev
University of Leuven
Department of Physics and Astronomy
Laboratory of Semiconductor Physics
Celestijnenlaan 200D
3001 Leuven
Belgium

Takashi Ando
IBM Semiconductor Research and
Development Center (SRDC)
IBM Systems and Technology Division
Hopewell Junction, NY 12533
USA

Ian W. Boyd
Brunel University
Experimental Techniques Center
Kingston Lane
Uxbridge, Middlesex UB8 3PH
UK

and

Irving Ian LIAW
Melbourne Materials Institute
David Caro Building
University of Melbourne
Parkville 3010, VIC
Australia

Huiming Bu
IBM Semiconductor Research and
Development Center (SRDC)
IBM Systems and Technology Division
Hopewell Junction, NY 12533
USA

Ed Cartier
IBM Semiconductor Research and
Development Center (SRDC)
IBM Systems and Technology Division
Hopewell Junction, NY 12533
USA

Mei Yin Chan
Nanyang Technological University
School of Materials Science and
Engineering
Block N4.1, 50 Nanyang Avenue
Singapore 639798
Singapore

Fu-Chien Chiu
Ming-Chuan University
Department of Electronic Engineering
Taoyuan 333
Taiwan

Kyeongjae Cho
The University of Texas at Dallas
Department of Materials Science & Engineering
800 W. Campbell Road, RL 10
Richardson, TX 75080
USA

and

The University of Texas at Dallas
Department of Physics
800 W. Campbell Road, RL 10
Richardson, TX 75080
USA

Moonju Cho
Seoul National University
Department of Materials Science and Engineering and Inter-University Semiconductor Research Center
WCU Hybrid Materials Program
Seoul 151-744
Korea

and

IMEC
75 Kapeldreef
3001 Leuven
Belgium

Michael Chudzik
IBM Semiconductor Research and Development Center (SRDC)
IBM Systems and Technology Division
Hopewell Junction, NY 12533
USA

Peter Damarwan
Nanyang Technological University
School of Materials Science and Engineering
Block N4.1, 50 Nanyang Avenue
Singapore 639798
Singapore

Yuanping Feng
National University of Singapore
Department of Physics
2 Science Drive 3
Singapore 117542
Singapore

Elena O. Filatova
Petersburg State University
St. Petersburg 198504
Russia

Kai Han
Chinese Academy of Sciences
Institute of Microelectronics
3 # BeiTuCheng West Road
Chaoyang District
Beijing 100029
China

Gang He
Anhui University
School of Physics and Materials Science
Anhui Key Laboratory of Information Materials and Devices
Feixi Road 3
Hefei 230039
China

Michel Houssa
University of Leuven
Department of Physics and Astronomy
Laboratory of Semiconductor Physics
Celestijnenlaan 200D
3001 Leuven
Belgium

Alfred C. H. Huan
Nanyang Technological University
Division of Physics and Applied Physics
School of Physical and Mathematical Sciences
21 Nanyang Link
Singapore 637371
Singapore

Cheol Seong Hwang
Seoul National University
Department of Materials Science and Engineering and Inter-University Semiconductor Research Center
WCU Hybrid Materials Program
Seoul 151-744
Korea

Hyung-Suk Jung
Seoul National University
Department of Materials Science and Engineering and Inter-University Semiconductor Research Center
WCU Hybrid Materials Program
Seoul 151-744
Korea

Mukesh Khare
IBM Semiconductor Research and Development Center (SRDC)
IBM Systems and Technology Division
Hopewell Junction, NY 12533
USA

Igor V. Kozhevnikov
Institute of Crystallography
Moscow 119333
Russia

Siddarth Krishnan
IBM Semiconductor Research and Development Center (SRDC)
IBM Systems and Technology Division
Hopewell Junction, NY 12533
USA

Unoh Kwon
IBM Semiconductor Research and Development Center (SRDC)
IBM Systems and Technology Division
Hopewell Junction, NY 12533
USA

Jack C. Lee
The University of Texas at Austin
Microelectronics Research Center
Department of Electrical and Computer Engineering
10100 Burnet Road
Austin, TX 78758-4445
USA

Pooi See Lee
Nanyang Technological University
School of Materials Science and Engineering
Block N4.1, 50 Nanyang Avenue
Singapore 639798
Singapore

Mao Liu
Chinese Academy of Sciences
Institute of Solid State Physics
Anhui Key Laboratory of Nanomaterials and Nanostructure
Key Lab of Materials Physics
Hefei 230031
China

Hong-Liang Lu
Fudan University
Department of Microelectronics
State Key Laboratory of ASIC and System
220 Handan Road
Shanghai 200433
China

Xubing Lu
South China Normal University
Higher Education Mega Center
Guangzhou
School of Physics and
Telecommunication Engineering
Institute for Advanced Materials (IAM)
Guangdong 510006
China

Somnath Mondal
Chang Gung University
Department of Electronic Engineering
Taoyuan 333
Taiwan

Vijay Narayanan
IBM Semiconductor Research and
Development Center (SRDC)
IBM Systems and Technology Division
Hopewell Junction, NY 12533
USA

Tung-Ming Pan
Chang Gung University
Department of Electronic Engineering
Taoyuan 333
Taiwan

Tae Joo Park
Seoul National University
Department of Materials Science and
Engineering and Inter-University
Semiconductor Research Center
WCU Hybrid Materials Program
Seoul 151-744
Korea

and

Hanyang University
Department of Materials Engineering
Ansan 426-791
Korea

Vamsi Paruchuri
IBM Semiconductor Research and
Development Center (SRDC)
IBM Systems and Technology Division
Hopewell Junction, NY 12533
USA

Andrey A. Sokolov
Petersburg State University
St. Petersburg 198504
Russia

Andre Stesmans
University of Leuven
Department of Physics and Astronomy
Laboratory of Semiconductor Physics
Celestijnenlaan 200D
3001 Leuven
Belgium

Zhaoqi Sun
Anhui University
School of Physics and Materials Science
Anhui Key Laboratory of Information
Materials and Devices
Feixi Road 3
Hefei 230039
China

Robert M. Wallace
The University of Texas at Dallas
Department of Materials Science &
Engineering
800 W. Campbell Road, RL 10
Richardson, TX 75080
USA

and

The University of Texas at Dallas
Department of Physics
800 W. Campbell Rd., RL10
Richardson, TX 75080
USA

Shijie Wang
A*STAR (Agency for Science, Technology and Research)
Institute of Materials Research and Engineering (IMRE)
3 Research Link
Singapore 117602
Singapore

Weichao Wang
The University of Texas at Dallas
Department of Materials Science & Engineering
800 W. Campbell Road, RL10
Richardson, TX 75080
USA

Wenwu Wang
Chinese Academy of Sciences
Institute of Microelectronics
3 # BeiTuCheng West Road
Chaoyang District
Beijing 100029
China

Xiaolei Wang
Chinese Academy of Sciences
Institute of Microelectronics
3 # BeiTuCheng West Road
Chaoyang District
Beijing 100029
China

Ka Xiong
The University of Texas at Dallas
Department of Materials Science & Engineering
800 W. Campbell Road, RL10
Richardson, TX 75080
USA

David Wei Zhang
Fudan University
Department of Microelectronics
State Key Laboratory of ASIC and System
220 Handan Road
Shanghai 200433
China

Lide Zhang
Chinese Academy of Sciences
Institute of Solid State Physics
Anhui Key Laboratory of Nanomaterials and Nanostructure
Key Lab of Materials Physics
Hefei 230031
China

Han Zhao
The University of Texas at Austin
Microelectronics Research Center
Department of Electrical and Computer Engineering
10100 Burnet Road
Austin, TX 78758-4445
USA

Yi Zhao
Nanjing University
School of Electronic Science and Engineering
Hankou Road 22
Nanjing 210093
China

Li Qiang Zhu
Chinese Academy of Sciences
Ningbo Institute of Material Technology and Engineering
519 Zhuangshi Road, Zhenhai
Ningbo 315201
China

Color Plates

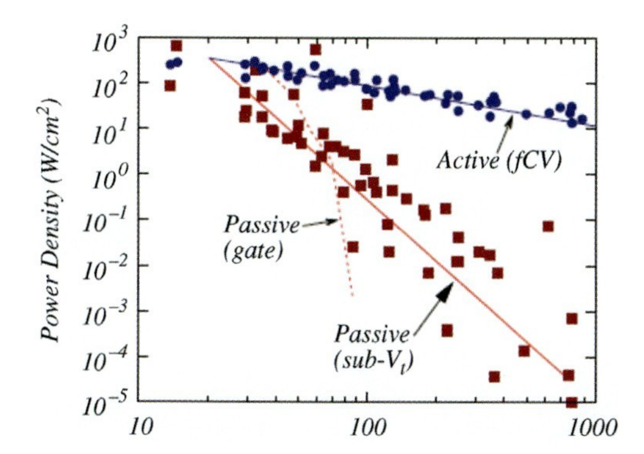

Figure 1.2 Comparison of measured active power and leakage (passive) power in devices with gate lengths ranging from 1 μm to 20 nm (symbols). Lines indicate the trend of a particular power component, as indicated. Reproduced from Ref. [11]. Copyright 2006, International Business Machines Corporation. (This figure also appears on page 6.)

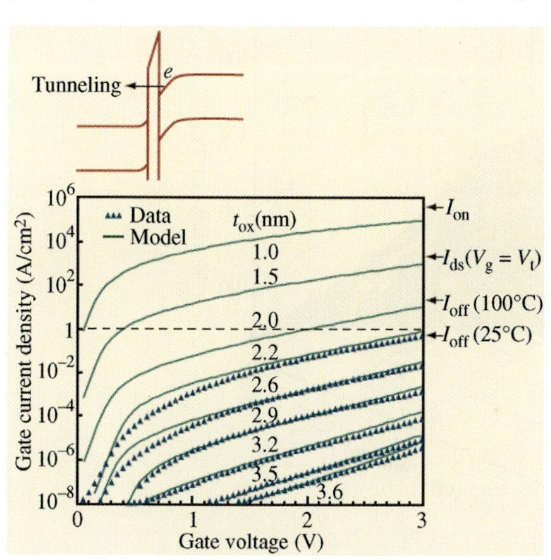

Figure 1.3 Measured and calculated oxide tunneling currents versus gate voltage for different oxide thicknesses. Reproduced from Ref. [12]. Copyright 2002, International Business Machines Corporation. (This figure also appears on page 7.)

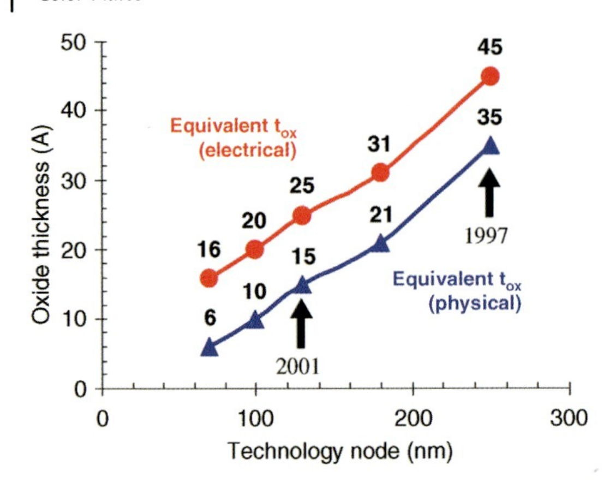

Figure 1.4 Extrapolated gate oxide scaling trend for recent CMOS technologies. Reproduced from Ref. [14]. Copyright 2000, IEEE. (This figure also appears on page 9.)

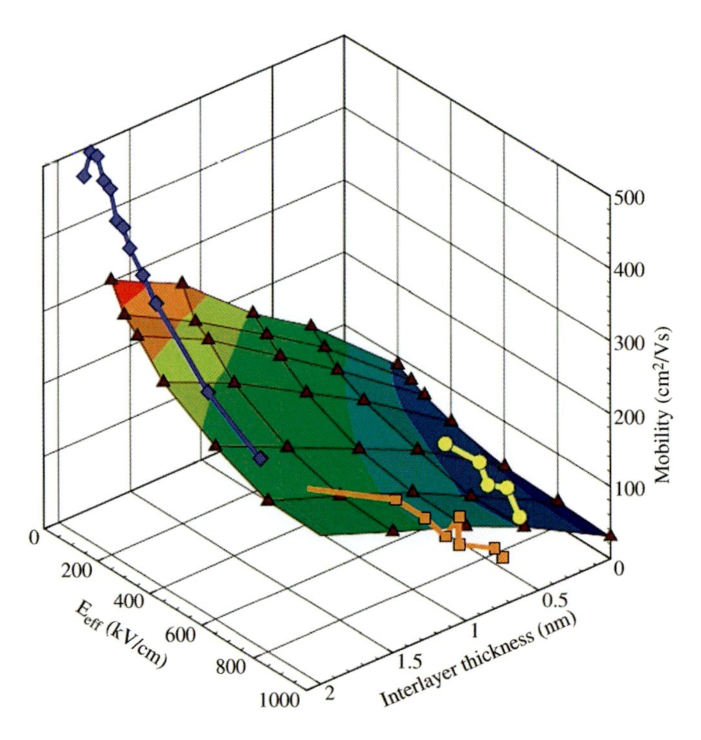

Figure 1.10 Electron mobility for HfO_2 versus effective field E_{eff} and the SiO_2 interfacial layer thickness. The mobility for pure SiO_2 and the experimental results [54] are also reported. Reproduced from Ref. [53]. Copyright 2007, Elsevier. (This figure also appears on page 18.)

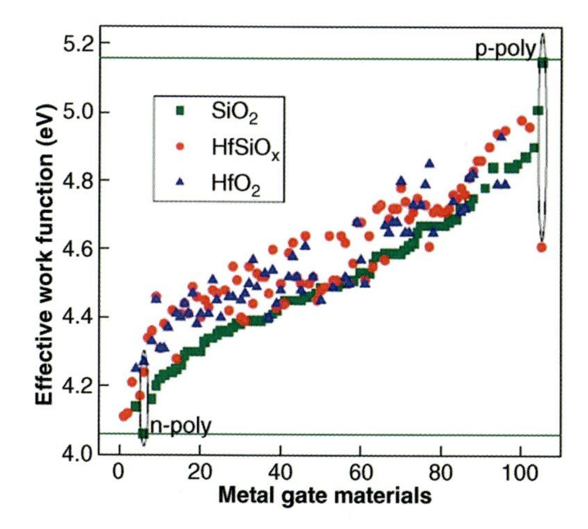

Figure 1.11 Effective work function of various metal electrode materials on SiO_2, $HfSiO_x$, and HfO_2 investigated with the terraced oxide method. Reproduced from Ref. [71]. Copyright 2008, Elsevier. (This figure also appears on page 19.)

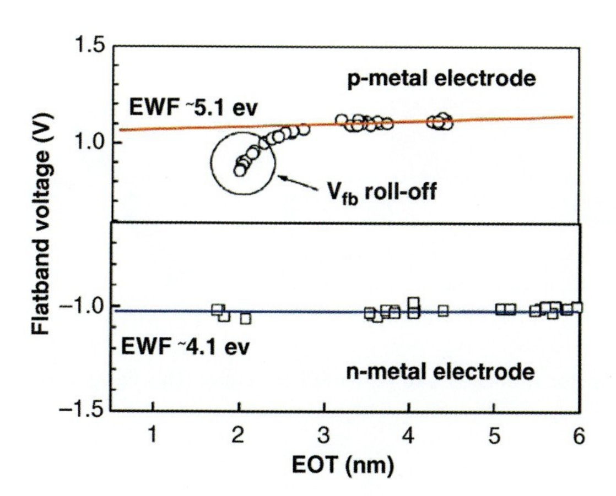

Figure 1.12 V_{fb}–EOT curve for a typical n-metal electrode and a p-metal electrode. Reproduced from Ref. [73]. Copyright 2006, Elsevier. (This figure also appears on page 20.)

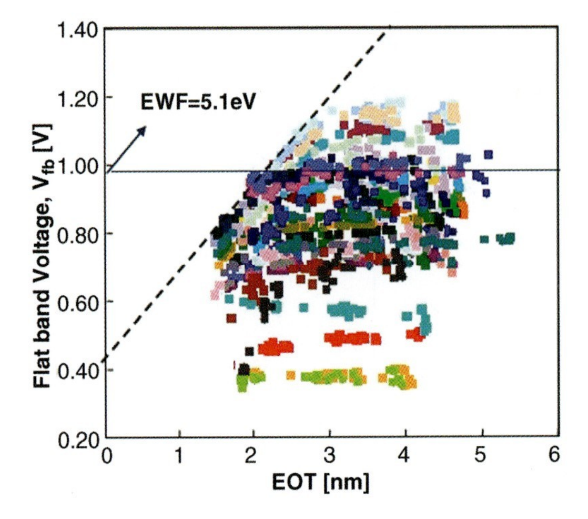

Figure 1.13 V_{fb}–EOT relationship for various materials systems investigated showing the V_{fb} roll-off for high work function metal electrodes. Reproduced from Ref. [71]. Copyright 2008, Elsevier. (This figure also appears on page 21.)

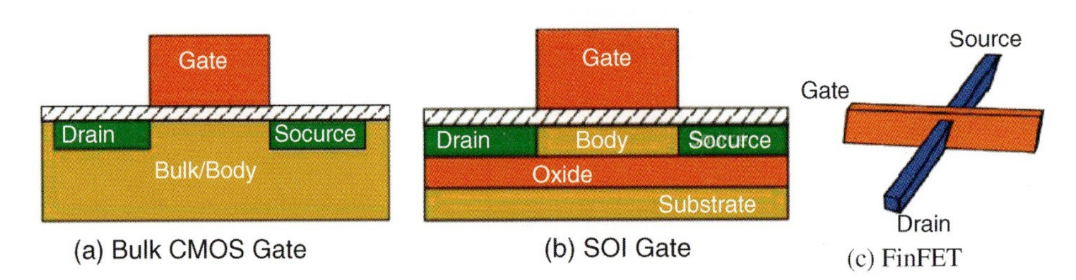

Figure 1.14 (a) Bulk, (b) SOI transistor structures, and (c) FinFET structure. (This figure also appears on page 22.)

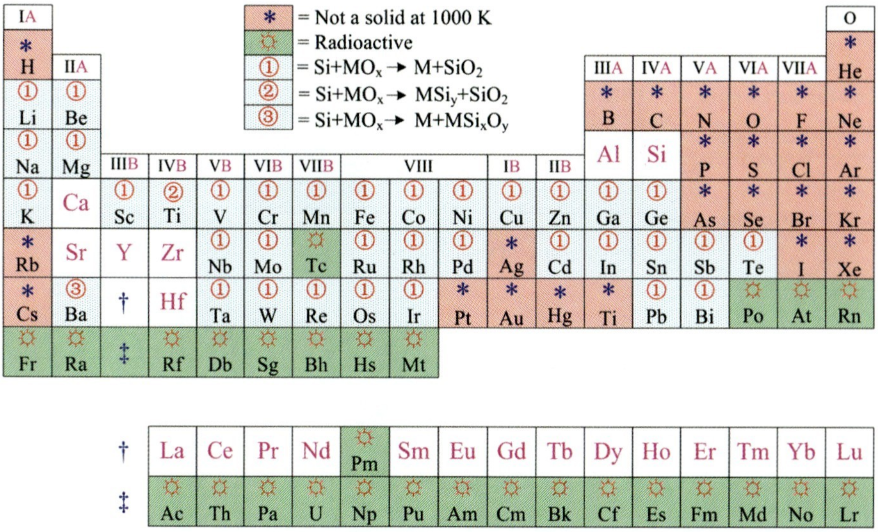

*	= Not a solid at 1000 K
☼	= Radioactive
①	= Si+MO_x → M+SiO_2
②	= Si+MO_x → MSi_y+SiO_2
③	= Si+MO_x → M+MSi_xO_y

IA																	O
* H	IIA											IIIA	IVA	VA	VIA	VIIA	* He
① Li	① Be											* B	* C	* N	* O	* F	* Ne
① Na	① Mg	IIIB	IVB	VB	VIB	VIIB	VIII			IB	IIB	Al	Si	* P	* S	* Cl	* Ar
① K	Ca	① Sc	② Ti	① V	① Cr	① Mn	① Fe	① Co	① Ni	① Cu	① Zn	① Ga	① Ge	* As	* Se	* Br	* Kr
* Rb	Sr	Y	Zr	① Nb	① Mo	☼ Tc	① Ru	① Rh	① Pd	* Ag	① Cd	① In	① Sn	① Sb	① Te	* I	* Xe
* Cs	③ Ba	†	Hf	① Ta	① W	① Re	① Os	① Ir	* Pt	* Au	* Hg	* Ti	① Pb	① Bi	☼ Po	☼ At	☼ Rn
☼ Fr	☼ Ra	‡	☼ Rf	☼ Db	☼ Sg	☼ Bh	☼ Hs	☼ Mt									

†	La	Ce	Pr	Nd	☼ Pm	Sm	Eu	Gd	Tb	Dy	Ho	Er	Tm	Yb	Lu
‡	☼ Ac	☼ Th	☼ Pa	☼ U	☼ Np	☼ Pu	☼ Am	☼ Cm	☼ Bk	☼ Cf	☼ Es	☼ Fm	☼ Md	☼ No	☼ Lr

Figure 2.2 Thermodynamic stability of binary oxides in contact with Si. (This figure also appears on page 36.)

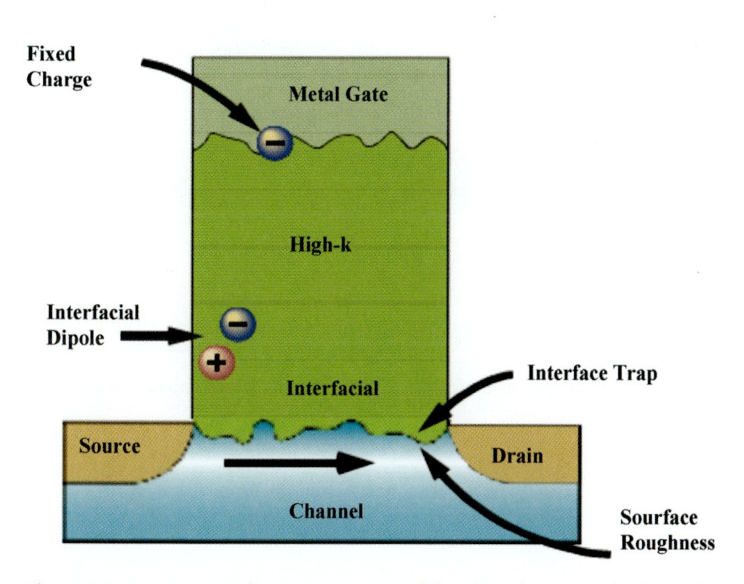

Figure 2.3 Factors contributing to carrier mobility degradation in a high-*k* oxide layer [27]. (This figure also appears on page 39.)

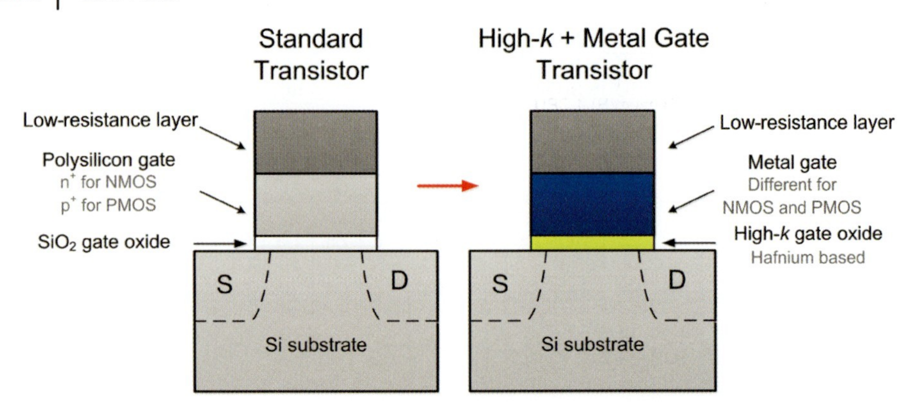

Figure 2.4 High-k + metal gate transistor. (This figure also appears on page 41.)

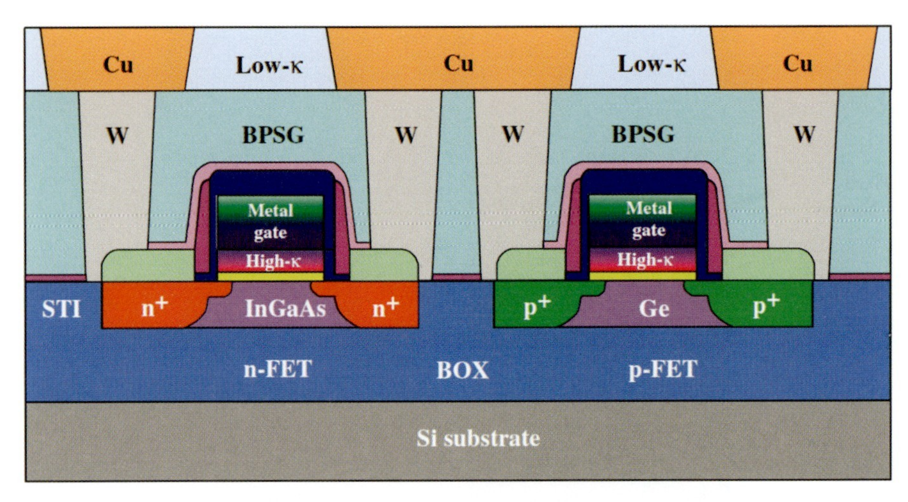

Figure 2.5 Ideal CMOS structure for high performance [87]. (This figure also appears on page 46.)

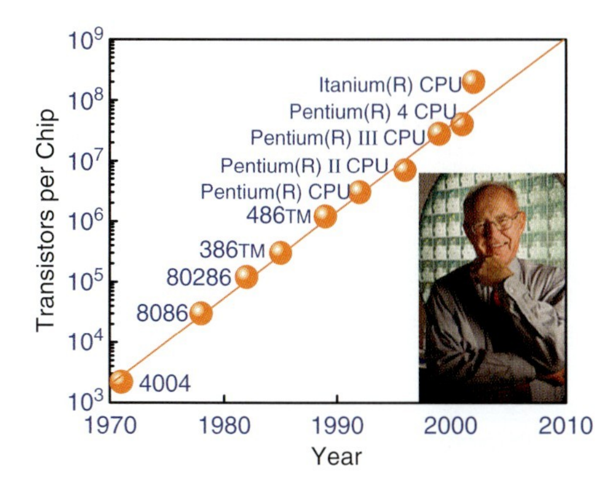

Figure 11.2 Growth of transistor counts for Intel processors (dots) and Moore's law (line). Moore's law means more performance and decreasing cost. (This figure also appears on page 357.)

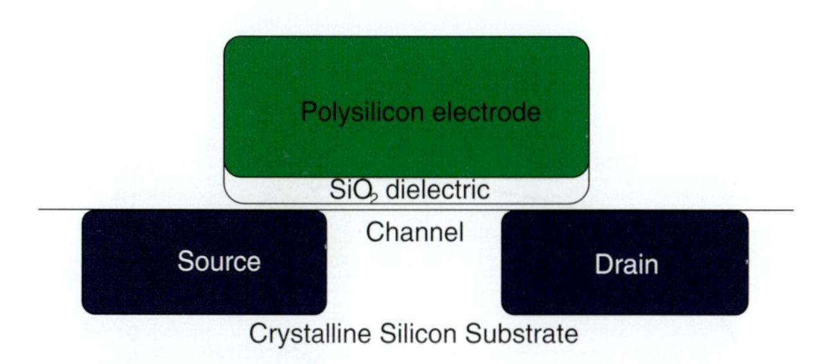

Figure 11.4 Basic CMOS transistor. Below 90 nm process node, the thickness of the SiO_2 gate dielectric will be lower than 1.2 nm (only about four atomic layers). Current leakage due to tunneling imposes significant power dissipation. (This figure also appears on page 361.)

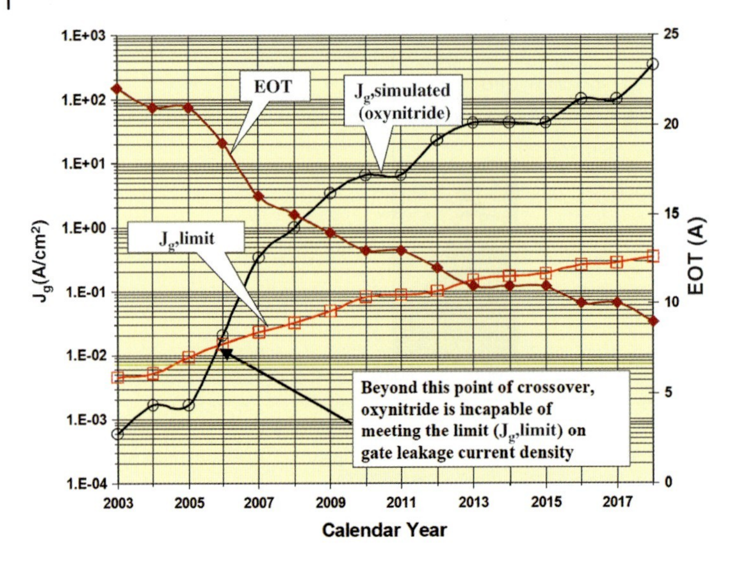

Figure 11.6 LSTP logic scaling up of gate leakage current density limit and of simulated gate leakage due to direct tunneling [12]. (This figure also appears on page 362.)

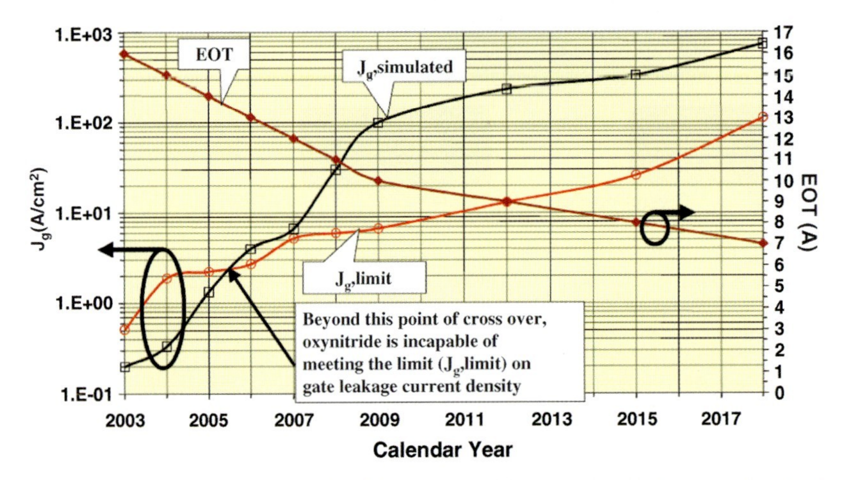

Figure 11.7 LOP logic scaling up of gate leakage current density limit and of simulated gate leakage due to direct tunneling [12]. (This figure also appears on page 362.)

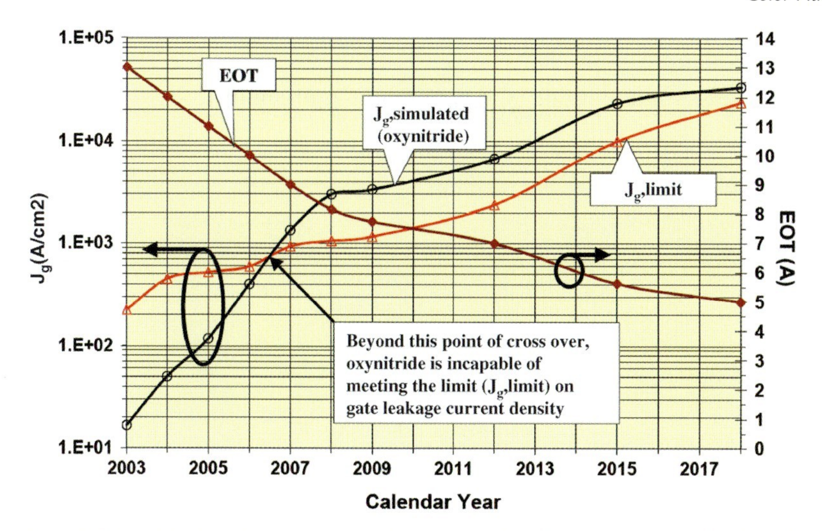

Figure 11.8 HP logic scaling up of gate leakage current density limit and of simulated gate leakage due to direct tunneling [12]. (This figure also appears on page 363.)

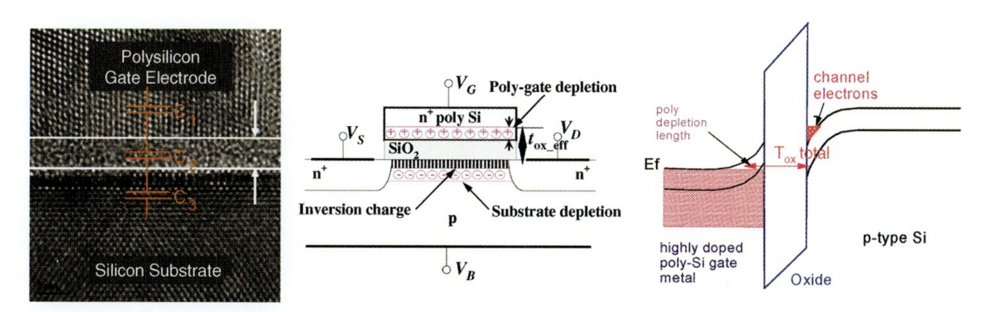

Figure 11.9 Schematic diagram of polysilicon depletion effect. (This figure also appears on page 364.)

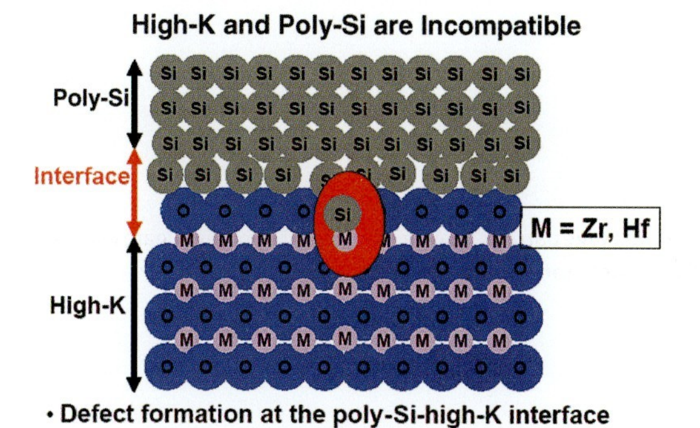

Figure 11.10 Instability of polysilicon electrode on high-κ dielectrics. (This figure also appears on page 364.)

Part One
Scaling and Challenge of Si-based CMOS

High-k Gate Dielectrics for CMOS Technology, First Edition. Edited by Gang He and Zhaoqi Sun.

1
Scaling and Limitation of Si-based CMOS

Gang He, Zhaoqi Sun, Mao Liu, and Lide Zhang

1.1
Introduction

Scaling transistor's dimensions has been the main tool to power the development of silicon integrated circuits (ICs). The more an IC is scaled, the higher its packing density and the lower its power dissipation [1]. These have been key in the evolutionary progress leading to today's computers and communication systems that offer superior performance, dramatically reduced cost per function, and much reduced physical size compared to their predecessors. However, the fundamental limits of complementary metal oxide semiconductor (CMOS) technology have been discussed, reviewed, and claimed to be at hand since the first MOS processes were developed [2, 3]. The integration of semiconductor devices has gone through different stages. At each stage of evolution, limits were reached and then subsequently surpassed, and very little has changed in the basic transistor design.

Questions about the end of CMOS scaling have been discussed, but engineering ingenuity has proven the predictions wrong. The most spectacular failures in predicting the end involved the "lithography barrier," in which it was assumed that spatial resolution smaller than the wavelength used for the lithographic process is not possible [4, 5], and the "oxide scaling barrier," in which it was claimed that the gate oxide thickness cannot be reduced below $\sim$3 nm due to gate leakage [6]. For the present and near future, it appears unlikely that lithography will limit the scaling of silicon devices. The cost of lithography tools, including that required for making masks, may, however, impede future scaling of devices. It is more likely that a fundamental limit will halt further scaling when at least one of physical dimensions of the device, be it a length, width, depth, or thickness, approaches a few silicon atoms. Manufacturing tolerance, and therefore economics, may dictate an end to the scaling of silicon devices before these fundamental limits are reached. Therefore, in this chapter on scaling of SiO_2-based gate dielectrics for MOS devices, only present perceived fundamental limits are considered.

The downscaling of an MOSFETS sets demands especially on the properties of the gate oxide, SiO_2, which nature has endowed the silicon microelectronics industry with and which has dominated as the favorite and by far most practical choice for FET

High-k Gate Dielectrics for CMOS Technology, First Edition. Edited by Gang He and Zhaoqi Sun.

gate dielectric materials since 1957. SiO_2 offers several crucial advantages and industry's acquired knowledge of its properties and processing techniques has allowed its continuous use for the past several decades. However, further scaling the FET is eventually going to be impeded by the inability to further reduce the oxide thickness without risking a breakdown of the device. Recently, however, the thickness of the gate oxide has scaled more slowly compared to its historical pace. An equivalent oxide thickness (EOT) of ~1 nm has been used for the past two to three generations because of issues such as process controllability, high leakage current, and reliability limits, signaling an end to the scaling era and the advent of a new era of material and device evolution. As we know, when it is thinned to 1 nm, which corresponds to only 4–5 atom layers, some new technology problems arose [7, 8]. A variation in thickness of only 0.1 nm could result in changes in the device operation condition, making it extremely difficult to maintain device tolerances. As one might imagine, this exponential increase in the gate dielectric leakage current has caused significant concern about the operation of CMOS devices, particularly with regard to standby power dissipation, reliability, and lifetime. This will likely be one if there is no the major contributor leading to the limited extendability of SiO_2 as the gate dielectric in the 1.5–2.0 nm regimes. Therefore, in this chapter, a detailed discussion of current dielectrics and those proposed for future generations is included. Present beliefs regarding the limitations of silicon dioxide as the gate dielectric are reviewed. Ultrathin oxides, with gate dielectrics below 40 Å, some of which are nitrided silicon oxides, are discussed. The benefits and limitations of alternative for new high-k materials for possible high-performance CMOS applications are reviewed.

1.2 Scaling and Limitation of CMOS

1.2.1 Device Scaling and Power Dissipation

The primary goal of CMOS scaling is reduction of the cost per functional power, by increasing the integration density of on-chip components. The elaboration of constant field scaling rules entails concomitant performance and power consumption improvements, which have shaped the evolution of silicon technology [1]. The concept of device scaling is illustrated in Figure 1.1. In constant field scaling, the physical dimensions of the device (gate length L_G and width W_G, oxide thickness t_{ox}, and junction depth X_j), and the supply and threshold voltages (V_{DD} and V_T, respectively), are reduced by the same factor, $\alpha > 1$, so that the two-dimensional pattern of the electric field is maintained constant, while circuit density increases by $\sim\alpha^2$. This implies that the depletion width (W_d) must also be reduced by the same amount, which is achieved by increasing the substrate doping N_B by α. Consequently, both the gate capacitance ($C = L_G W_G \varepsilon_{ox}/t_{ox}$) and the drain saturation current ($I_{D,sat}$) are scaled down by α. The saturation current determines the transistor intrinsic switching delay $\tau \sim CV_{DD}/I_{D,sat}$, which is thus reduced by α, leading to a performance

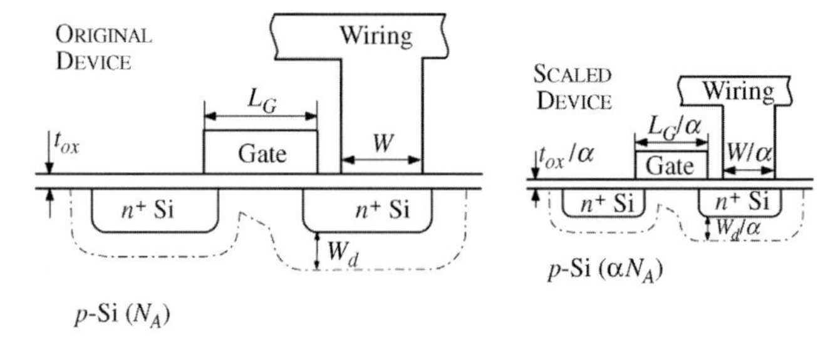

Figure 1.1 Conceptual schematic diagram of device scaling. Both device and wiring dimensions are required to scale by the same factor $1/\alpha$, in order to increase integration density by α^2. Scaling of the supply voltage by the same factor ($1/\alpha$) maintains the same 2D electric field pattern, subject to an equivalent scaling of the depletion width W_d. Reproduced from Ref. [10]. Copyright 2001, IEEE.

improvement. At the same time, the power dissipation ($P \sim I_{D,sat} V_{DD}$) is reduced by α^2, so that the power density ($P/(L_G W_G)$) remains unchanged.

Table 1.1 shows the gate length, supply voltage, and oxide thickness figures for several recent and future technology generations, as projected by the ITRS. It is obvious that present day device scaling does not adhere to the constant field scaling rules. This is because of several fundamental, nonscaling factors and practical considerations. Instead, the generalized scaling rules are followed, where the physical dimensions of the transistor are still reduced by a factor of α, providing the desired circuit density increase (α^2) and performance improvement (α), but the supply voltage is scaled by β/α, leading to an increase in the magnitude of the electrical field by $1 \leq \beta \leq \alpha$ [9]. A full list of MOSFET physical parameters, and their scaling factors, is given for constant field scaling and generalized scaling in Table 1.2. The last column in this table shows the rules for selective scaling, which relaxes one more constraint – the fixed ratio between gate length and width [10]. Such relaxation is driven mostly by the slower scaling pace of on-chip interconnect lines.

As can be seen in Table 1.2, under generalized and selective scaling scenarios, the dissipated power scales by β^2/α^2, while the power density increases by β^2. The power density is of paramount importance for chip packaging and systems design, and its increase imposes a practical limit for the exploitation of device scaling. It is worth

Table 1.1 ITRS-projected L_G, EOT and V_{DD}.

Year	L_G (nm)	EOT (nm)	V_{DD} (V)
2003	45	1.3	1.2
2005	32	1.2	1.1
2007	25	1.1	1.1
2009	20	0.9	1.0
2011	16	0.6	1.0
2013	13	0.5	0.9
2015	10	0.5	0.9

Table 1.2 Device parameters and their scaling factors. Reproduced from Ref. [10]. Copyright 2001, IEEE.

Physical parameters	Scaling rules' factors		
	Constant Field	Generalized	Selective
Gate length (L_G), oxide thickness (t_{ox})	$1/\alpha$	$1/\alpha$	$1/\alpha_D$
Wiring width, channel width (W_G)	$1/\alpha$	$1/\alpha$	$1/\alpha_W$
Voltages (V_{DD}, V_T)	$1/\alpha$	β/α	β/α_D
Substrate doping (N_B)	α	$\beta\alpha$	β/α_W
Electric field	1	β	β
Gate capacitance ($C = L_G W_G \varepsilon_{ox}/t_{ox}$)	$1/\alpha$	$1/\alpha$	$1/\alpha_W$
Drive current ($I_{D,sat}$)	$1/\alpha$	β/α	β/α_W
Intrinsic delay ($\tau \sim CV_{DD}/I_{D,sat}$)	$1/\alpha$	$1/\alpha$	$1/\alpha_D$
Area ($A \propto L_G W_G$, or $\propto W_G^2$)	$1/\alpha^2$	$1/\alpha^2$	$1/\alpha_W^2$
Power dissipation ($P \sim I_{D,sat} V_{DD}$)	$1/\alpha^2$	β^2/α^2	$\beta^2/(\alpha_W \alpha_D)$
Power density (P/A)	1	β^2	$\beta^2/(\alpha_W/\alpha_D)$

noting that in the most recent technologies, and in the projections of the ITRS, supply voltage hardly scales, that is, $\beta \approx \alpha$ [11]. This already suggests that the power density grows at the same rate as the integration density. Actually, the issue is even worse than it appears considering the scaling rules alone, because of nonscaling factors, leading to the increase and dominance of the static, leakage power. This is clearly demonstrated by the crossing lines of Figure 1.2, illustrating the trends in dynamic and leakage power densities with shrinking gate length. The actual measurements for devices with gate length between 1 μm and 65 nm are shown with symbols. The rapid escalation of leakage has ultimately led to power-constrained, application-specific

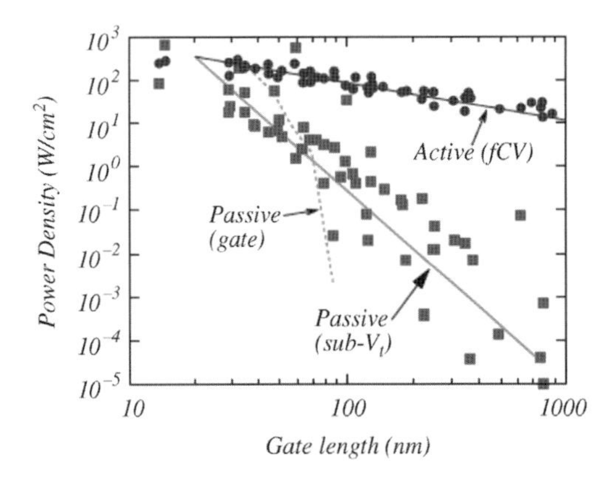

Figure 1.2 Comparison of measured active power and leakage (passive) power in devices with gate lengths ranging from 1 μm to 20 nm (symbols). Lines indicate the trend of a particular power component, as indicated. Reproduced from Ref. [11]. Copyright 2006, International Business Machines Corporation. (For a color version of this figure, please see the color plate at the beginning of this book.)

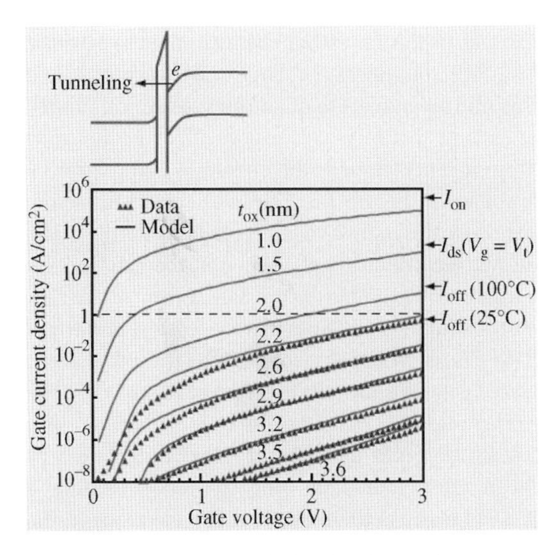

Figure 1.3 Measured and calculated oxide tunneling currents versus gate voltage for different oxide thicknesses. Reproduced from Ref. [12]. Copyright 2002, International Business Machines Corporation. (For a color version of this figure, please see the color plate at the beginning of this book.)

evaluation of scaling scenarios since different applications can tolerate different power densities.

Of the two components of consumed power, the dynamic power, dissipated during a switching between logic states, can be ameliorated by limiting the switching frequency f, to which it is proportional ($P_{dyn} \approx CV^2_{DD}f$). The other component – static, leakage power, dissipated while maintaining a logic state, is exponentially sensitive to some of the device parameters and their variability, as well as to nonscaling factors. Leakage power dissipation is time invariant and now poses the most significant scaling limit [11].

1.2.2
Gate Oxide Tunneling

When the dimensions of a MOSFET are scaled down, both the voltage level and the gate oxide thickness must also be reduced [1]. Since the electron thermal voltage, kT/q, is a constant for room-temperature electronics, the ratio between the operating voltage and the thermal voltage inevitably shrinks. This leads to higher source-to-drain leakage currents stemming from the thermal diffusion of electrons. At the same time, to keep adverse 2D electrostatic effects on threshold voltage under control, the thickness of gate oxide is reduced nearly in proportion to channel length, as shown in Figure 1.3. This is necessary in order for the gate to retain more control over the channel than the drain. For CMOS devices with channel lengths of 100 nm or less, an oxide thickness of $<$3 nm is needed. This thickness comprises only a few layers of atoms and is approaching fundamental limits.

While it is amazing that SiO_2 can take us this far without being limited by extrinsic factors such as defect density, surface roughness, or large-scale thickness, and uniformity control, oxide films this thin are subject to quantum mechanical tunneling, giving rise to a gate leakage current that increases exponentially as the oxide thickness is scaled down. Tunneling currents for oxide thicknesses ranging from 3.6 to 1.0 nm are plotted versus gate voltage in Figure 1.3 [12]. In the direct tunneling regime, the current is rather insensitive to the applied voltage or field across the oxide, so reduced voltage operation will not buy much relief. Although the gate leakage current may be at a level that is negligible compared to the on-state current of a device, it will first have an effect on the chip standby power. Note that the leakage power will be dominated by turned-on n-MOSFETs, in which electrons tunnel from the silicon inversion layer to the positively biased gate, as shown in the inset of Figure 1.3. Edge tunneling in the gate-to-drain overlap region of turned-off devices should not be a fundamental issue since one can always build up the corner oxide thickness by additional oxidation of poly-silicon after gate patterning. p-MOSFETs have a much lower leakage than n-MOSFETs because there are very few electrons in the p^+ polysilicon ("poly") gate available for tunneling to the substrate, and hole tunneling has a much lower probability. If one assumes that the total active gate area per chip is of the order of 0.1 cm^2, the maximum tolerable gate leakage current will be of the order of 10 A/cm^2. This sets a lower limit of 1.0–1.5 nm for the gate oxide thickness. Dynamic memory devices have a more stringent leakage requirement and therefore must impose a higher limit on gate oxide thickness [10].

The intolerable growth of gate tunneling leakage has triggered a radical change in the silicon technology, which aims to introduce dielectric permittivity scaling through material engineering. The goal is to scale the oxide sheet capacitance C_{ox}, which is required to maintain good electrostatic control by the gate, but avoid decreasing the physical thickness of the gate insulator. This may be achieved by increasing the gate dielectric constant of the insulator. With respect to scaling, the relevant metric is now the EOT, defined through the relation $\varepsilon_{high\,k}/t_{high\,k} = \varepsilon_{ox}/\mathrm{EOT} = C_{ox}$, where $\varepsilon_{high\,k}$ and $t_{high\,k}$ are the permittivity and physical thickness of the high-k material, respectively, and ε_{ox} is the SiO_2 permittivity. Therefore, the introduction of high-k dielectric materials is seen as the only way to realize the projections of the ITRS for effective oxide thicknesses reaching 0.5 nm.

However, the transition to high-k gate dielectric also entails a replacement of the poly-Si gate with a metal gate, which is another formidable technological challenge, in addition to the difficulties of controlled growth of ultrathin high-k oxide films [13]. These complications have delayed the adoption of alternative gate dielectric stacks and have led to the continuous relaxation of EOT scaling requirements.

1.2.3 Gate Oxide Scaling Trends

The gate oxide has been aggressively scaled in recent generations. Figure 1.4 shows extrapolated gate oxide scaling targets based on published data from recent Intel technologies [14]. The technology node refers to the smallest poly-Si gate length that

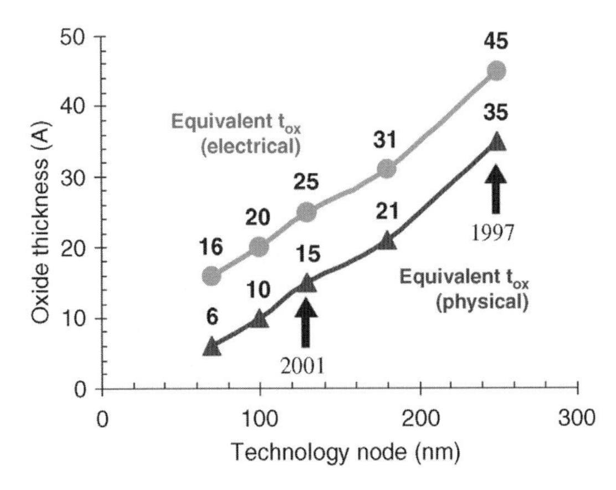

Figure 1.4 Extrapolated gate oxide scaling trend for recent CMOS technologies. Reproduced from Ref. [14]. Copyright 2000, IEEE. (For a color version of this figure, please see the color plate at the beginning of this book.)

can be defined by photolithography and roughly corresponds to the minimum channel length for a given process technology. A more complete list of projected transistor parameters is given in Table 1.3. The predictions are based on extrapolations of published state-of-the-art 180 nm technologies [15]. These projections, representative of the current targets for high-performance logic technology, aggressively outpace those compiled in the 2000 update of ITRS. The two data sets in Figure 1.4 refer to the equivalent electrical and physical thickness of the gate oxide. The EOT refers to how thin a pure SiO_2 layer would need to be in order to meet the gate capacitance requirements of a given technology. In a modern MOSFET device, the gate oxide behaves electrically as if it were 8–10 Å thicker than its physical thickness because depletion in the poly-Si gate and quantization in the inversion layer each extend the centroids of charge modulated by the gate voltage by 4–5 Å [16]. From Figure 1.4, it is clear that the physical thickness of the gate oxide is rapidly approaching atomic dimensions. The 250 nm technology, which entered volume production in 1997, used SiO_2 layer with approximately 40 Å physical t_{ox}, corresponding to approximately 20 Å monolayers of SiO_2. In contrast, the 100 and 70 nm technologies, scheduled for production in the next 5–10 years, will require gate capacitance values achievable only with SiO_2 layers as thin as 10 and 7 Å, respectively,

Table 1.3 Projected transistor parameters for future technology generations.

Generation (nm)	180	130	100	70	Scaling factor
I_{gate} (nm)	100	70	50	35	0.7×
V_{dd} (V)	1.5	1.2	1.0	0.8	0.8×
$T_{ox'}$ electrical (A)	31	25	20	16	0.8×
$T_{ox'}$ physical (A)	21	15	10	6	0.8×
I_{off} at 25 °C (nA/um)	20	40	80	160	2×

to guarantee proper device operation. A 10 Å film consists of only three–four monolayers of SiO_2.

1.2.4
Scaling and Limitation of SiO_2 Gate Dielectrics

The apparent robust nature of SiO_2 [17], coupled with industry's acquired knowledge of oxide process control, has helped the continued use of SiO_2 for over 30 years in CMOS technology. The use of amorphous, thermally grown SiO_2 as a gate dielectric provides thermodynamically and electrically stable, high-quality Si–SiO_2 interface with superior electrical isolation properties. In advanced integrated devices, the SiO_2 gate dielectrics are produced with charge densities of $10^{10}/cm^2$, mid-gap interface state densities of $10^{10}/cm^2$, and dielectric strengths of 15 MV/cm [17]. In fact, we can say that the silicon age own its existence to the superb quality of thermally grown SiO_2 as gate insulator and Si surface electrical passivator, Actually, no other known interface approaches the electrical figures quoted above the Si–SiO_2 interface. Therefore, the progress epitomized by Moore's law is best characterized as a steep but smooth development, having been achieved with no major revolution in fundamental device designs and no changes in the materials that constitute the heart of the MOSFET:Si and SiO_2.

Nevertheless, the outstanding evolution of silicon is rapidly approaching a saturation point where device fabrication can no longer be simply scaled to progressively smaller sizes. The origin of this saturation is indicated in Figure 1.5, where not only the number of transistors in an integrated circuit but also the corresponding gate dielectric thickness are plotted as a function of time. The thinning of the gate dielectric required by scaling rules, at present between 2 and 2.5 nm in fabrication, necessary for reaching the next generations of integrated devices will give origin to

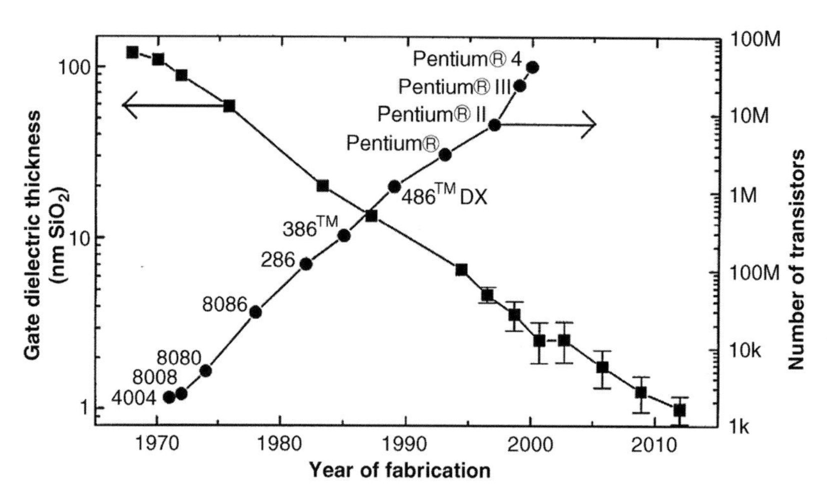

Figure 1.5 Moore's law as expressed by the number of transistors per chip. The corresponding SiO_2 gate dielectrics thickness is also shown.

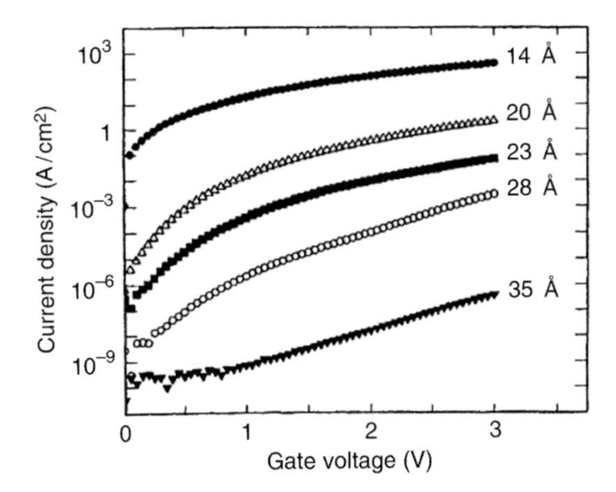

Figure 1.6 Gate current density as a function of gate voltage for MOS capacitors with different gate dielectric thickness. Reproduced from Ref. [19]. Copyright 1996, IEEE.

unacceptably high gate current arising from electron tunneling through the SiO_2 films. Gate current densities versus gate voltage are plotted in Figure 1.6 for transistor made with decreasing gate dielectric thickness below 3.5 nm, where the exponential increase in the leakage current is clearly observed. Figure 1.7 shows that progressive scaling of the gate oxide below 1.5 nm will lead to leakage current that destroys the transistor effect itself, as the ON and OFF states of the transistors might not be clearly distinguishable as necessary [18].

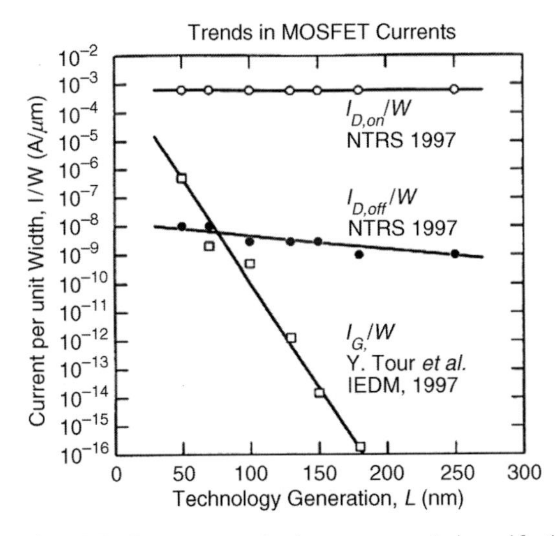

Figure 1.7 Gate current and train current per unit channel for the transistor states ON and OFF as a function of transistor channel length. Reproduced from Ref. [18]. Copyright 1997, Materials Research Society.

The inferior limit for gate oxide thickness can be extended down to 1.3 nm, owing to the characteristics of SiO_2 and the acquired knowledge of semiconductor manufacturers on oxide process control. Although high leakage current density is measured for such devices, as shown in Figure 1.6, transistors intended for high-performance microprocessor applications can sustain these currents [19]. However, further downscaling deteriorates the electrical properties rendering the devices inoperative. Several independent works showed the inconvenience of fabricated CMOSFET devices with SiO_2 gate oxides thinner than about 1.0–1.2 nm because there is no further gain in transistor drive current, setting a fundamental limit for gate oxide scaling.

The existence of a more fundamental limit to scaling SiO_2 around 1.0–1.2 nm as demonstrated at the atomic scale in a convincing experiment by Muller *et al.* from Bell Labs is reproduced in Figure 1.8 [8]. Using a scanning transmission electron microscope (STEM) probe with 2 Å resolution, they studied the chemical composition and electronic structure of oxide layers as thin as 7–12 Å using detailed electron energy loss spectroscopy (EELS) measurements. By moving the probe site-by-site through the ultrathin SiO_2 layers, they mapped the local unoccupied density of

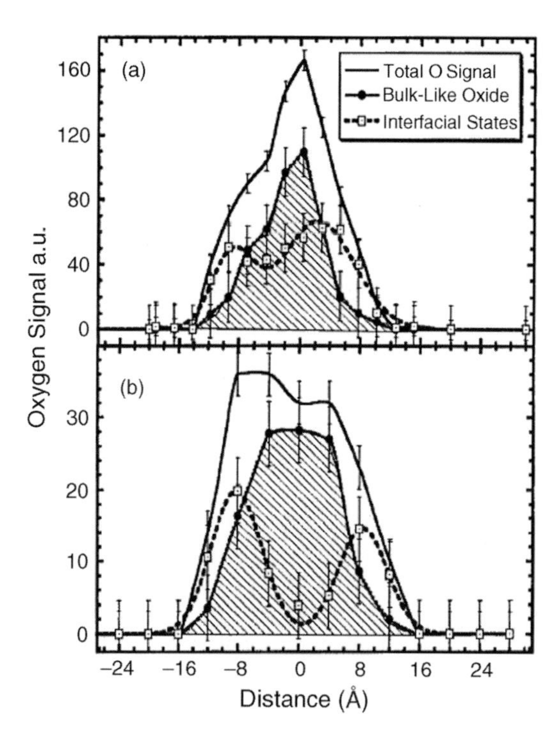

Figure 1.8 Oxygen signals in EELS performed in two different MOS capacitors made of (a) 1.3 nm and (b) 1.8 nm oxides that were analyzed in two components, one from bulk-like SiO_2 states and another from interfacial-like states. Reproduced from Ref. [8]. Copyright 1999, Nature Publishing Group.

electronic states, which provides insight into the local energy gap of the material, as a function of the probe position. In their work, the local energy gap was given by the separation between the highest occupied and lowest unoccupied states. They found that three to four monolayers of SiO_2 were needed to ensure that at least one monolayer maintained a fully bulk-like bonding environment, giving rise to the wide, insulating bandgap of SiO_2. Since the first and the last monolayers form interfaces with Si and poly-Si, respectively, they have bonding arrangements intermediate to those of bulk Si and bulk SiO_2 and hence have energy gaps smaller than that of bulk SiO_2. Based on these insights, Muller *et al.* concluded that the fundamental scaling limit of SiO_2 is likely to be in the range of 7 Å to 12 Å. Another important insight from their study was that for a 10 Å oxide, a 1 Å increase in the root mean square (RMS) interface roughness will lead to a factor of 10 increase in the gate leakage current, showing that the growth of such thin layers must be precisely controlled on atomic scales.

There has been a remarkable agreement between experiment and theory regarding the scaling limit of SiO_2. Theoretical studies by Tang *et al.* employing a Si/SiO_2 interface model based on the β-cristobalite form of SiO_2 showed that the band offset at the interface degraded substantially when the SiO_2 layer was scaled to less than three monolayers [20]. The large reduction in the band offset was attributed to a reduction in the SiO_2 bandgap and also suggested 7 Å as the scaling limit of SiO_2. A more recent study by Kaneta *et al.* using a Si/SiO_2 interface model based on quartz SiO_2 directly computed the local energy gap as a function of position through the interface [21]. While the transition from bulk Si to bulk SiO_2 in their model was structurally abrupt, it was found that the full bandgap of SiO_2 was not obtained until the second monolayer of SiO_2 was reached. Again, these calculations suggest that approximately 7 Å of SiO_2 is the minimum required for substantial band offsets to develop at the interface, indicating the formation of a large bandgap. Thus, both experiment and theory suggest that the bulk properties of SiO_2, including the wide, insulating bandgap needed to isolate the gate and channel regions, cannot be obtained for films less than 7 Å thick. Since technology roadmaps predict the need for sub-6 gate oxides in future generations, it is unlikely, from the viewpoint of both static power dissipation and fundamental materials science, which SiO_2 will scale beyond the 70 nm generation.

There are other limiting factors regarding SiO_2 gate oxide scaling. One is device reliability, which is open to debate whether the stringent 10-year reliability criterion set by industry for CMOS technology could be fulfilled by devices made of gate oxides thinner than 1.5–2 nm [22]. Indeed, dielectrics degradation due to "hot electron" irradiation of the Si–SiO_2 interface can trigger a succession of physical and chemical phenomena whose overall consequence on reliability is acceleration of dielectric breakdown of the gate oxide.

In view of the limiting factors described above, in order to achieve further scaling of integrated devices based on Si and SiO_2, the semiconductor industry has found solutions that allowed significant progress with the remarkable advantage of remaining within the Si–SiO_2 materials framework.

1.2.5 Silicon Oxynitrides

As scaling of the oxide thickness continues, it is obviously desirable to reduce the leakage current and maintain or increase the reliability of such films. The introduction of nitrogen into SiO_2 has been used to eliminate a number of concerns, although not all with equal success. The first introduction of nitrogen into SiO_2 films was for much thicker films and it was confirmed that the reliability of these films could be increased if the films were annealed in ammonia or other nitrogen-containing gas [23]. Some observations have shown that nitrogen could reduce the "defect generation rate" in these films, which can be thought to be a result of the ability of the nitrogen to getter hydrogen or reduce its diffusion [24, 25]. However, the concentration of nitrogen into the SiO_2 should be carefully controlled. Too much incorporation of nitrogen leads to large flatband or threshold-voltage shifts [26, 27]. As more experiments were performed to tailor the nitrogen profiles in oxide films, it was found that nonhydrogen nitrogen species were even more efficient at increasing device reliability and reducing defect generation rate since hydrogen was not introduced during the nitridation process. The incorporation of large amounts of nitrogen not only leads to the reduction of boron penetration but also causes threshold-voltage shifts, ΔV_t, and mobility and transconductance degradation that depend upon both the nitrogen concentration and its concentration profile, which can be due at least partially to the positive charge that results from the nitrogen incorporation into the SiO_2 matrix. The nitrogen incorporated near the Si/SiO_2 interface also reduces the mobility of the device.

The use of oxide/nitride stacks can eliminate boron penetration if the nitride film is of sufficient thickness [28–30]. With the incorporation of a nitride into the gate dielectric, the permittivity of the stack is greater than that of SiO_2. The physical thickness of the stack can be greater than that of a single SiO_2 film of equivalent thickness since the dielectric constant of silicon nitride is approximately twice that of SiO_2. Therefore, with the addition of silicon nitride to the stack, an increase in the effective thickness of approximately 30% is possible. Silicon nitride should also be able to reduce the gate leakage current [29, 30]. There are, however, many concerns involving the use of silicon nitride as a gate material, the greatest of which is the electrical stability of the material. Silicon nitride is known, at least in thicker films, to contain large amounts of positive charge; some of that charge can be unstable when a bias is applied, and this instability leads to obvious device problems. Such films are also known to contain large amounts of hydrogen, which could be a reliability concern [31, 32].

Having in hands an interim solution for reliability, the exponential increase in the leakage current with oxide thinner remains the major difficulty preventing further gate dielectric scaling. One possible solution is to consider the derive current of a MOSFET, the drain current, which in a simplified approximation can be written as

$$I_D = \frac{W}{L}\mu C\left(V_G - V_T - \frac{V_D}{2}\right)V_D$$

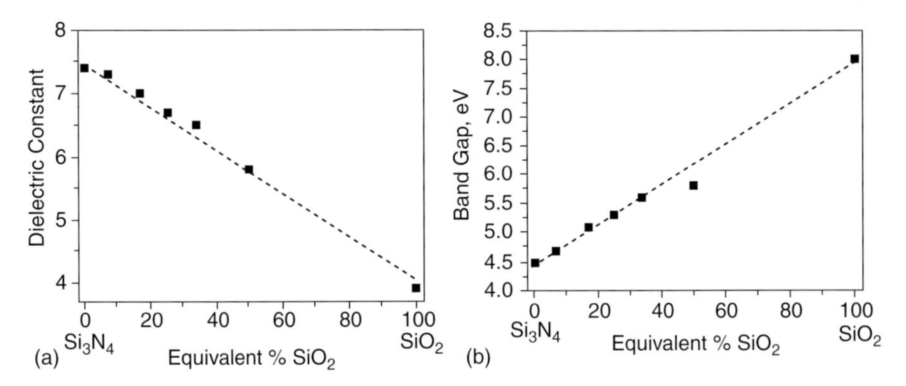

Figure 1.9 Dielectric constant (a) and bandgap (b) as a function of N content in the Si–O–N system.

where W is the width and L is the length of transistor channel, the charge carrier mobility in the channel, C the capacitance of the MOS capacitor, and V_G, V_D, and V_T are the gate, drain, and threshold voltages, respectively. I_D increase monotonically with V_D and then eventually saturates to a maximum when $V_{D.sat} = V_G - V_T$ to yield

$$I_D = \frac{W}{L}\mu C \frac{(V_G - V_T)^2}{2}$$

Since gate voltage is limited by leakage current and reliability constraints, in order to maintain enough drain current such that the transistor can operate in safe, reliable conditions, C must be increased or at least kept constant. Now

$$C = \frac{\varepsilon A}{t}$$

where ε is the permittivity of the capacitor dielectric, and A and t are the area and the thickness, respectively. According to previous references, it can be noted that progressive scaling has been performed reducing gate oxide thickness by the same factor as the horizontal dimensions determining the area A. This scaling strategy is limited, according to Figure 1.9, by leakage current originated in electron tunneling through the ultrathin oxides required. Therefore, increasing the capacitor permittivity is the only way to prevent capacitance decrease. Figure 1.9 show that the dielectrics of silicon oxynitride increases with the N content from the pure SiO_2 value up to the pure Si_3N_4 value. As known, when materials with dielectric constant higher than that of SiO_2 are used in gate dielectric, for design purposes, the relevant magnitude is not the dielectric film thickness but rather an associated quality called equivalent oxide thickness, defined as

$$\mathrm{EOT} = \frac{k_{ox}}{k_{\mathrm{high}\,k}} t_{\mathrm{high}\,k}$$

which represents the thickness of the SiO_2 layer with dielectric constant k_{ox} that would be required to achieve the same capacitance density as a given thickness $t_{\mathrm{high}\,k}$

of an alternative dielectric layer with dielectric constant $k_{\text{high }k}$. This means that it is possible to avoid electron tunneling that leads to unacceptable leakage current by simply increasing the physical thickness of the gate dielectric, without increasing its capacitance. Based on Figure 1.9, it is also shown that the bandgap decreases with increasing N concentration, which sets another limitation besides degradation of charge carrier mobility to the maximum N concentration in the oxynitride films. Thus, silicon oxynitride films are and will probably continue to be an interim solution for gate dielectrics for another few years. However, the limits of this solution are evident and therefore to keep pace with the historically steep progress of silicon technology it appears that the smoothness of this development will have to be abandoned in the near future.

1.3 Toward Alternative Gate Stacks Technology

1.3.1 Advances and Challenges in Dielectric Development

Fundamental challenge to the scaling of the gate dielectric is the exponential increase in tunnel current with reduction in film thickness. For films as thin as 20 Å, leakage currents can rise to 1–10A/cm^2. This incredibly high current can alter device performance, not to mention the difficulties associated with dissipation of such a large amount of power. Although higher power dissipation may be tolerable with some high-performance processors, it quickly leads to problems for portable machines. The reduction in gate leakage current is an important reason, if not the primary one, for replacing SiO_2-based dielectrics.

For a given technology, CMOS devices are designed with a specific gate capacitance, which is proportional to the dielectric constant and inversely proportional to the thickness of the gate material. To reduce the leakage current while maintaining the same gate capacitance, a thicker film with a higher dielectric constant is required. The gate leakage current, at least for direct quantum mechanical tunneling, exponentially depends upon the dielectric thickness, while the capacitance depends only linearly on the thickness. At first glance, this would seem to be a winning proposal since a substantial reduction in the current should be possible with only small increases in thickness. There is, however, another exponentially dependent term in the tunneling current – the barrier height between the cathode and the conduction band of the insulator. For a large number of dielectrics, the tunnel current exponentially depends upon the barrier height. Therefore, not only is a material with higher dielectric constant required but also this material must have a suitably large bandgap, and barrier height, to keep the gate leakage currents within reasonable limits.

Many materials have been suggested that could replace SiO_2 or SiON as a candidate for possible gate dielectric. Early, CeO_2 and Y_2O_3 were investigated to act as high-k dielectrics with an EOT of ~6 nm [33, 34]. However, crystallization-induced leakage currents and poor reliability in thick EOT regions prevent them to act as high-*k* candidates. In the 1990s, research on high-*k* gate dielectrics was updated. More attention has been focused on Ta_2O_5, TiO_2, and $SrTiO_3$, which were inherited from dynamic random access memory (DRAM) capacitor dielectric research [35–37]. Quickly, research on high-*k* dielectrics converged on the ZrO_2 and HfO_2 families with larger bandgaps [38, 39]. Owing to good thermal stability with Si, HfO_2 and ZrO_2 received most attention and many thorough reviews of research were also provided at that time [40–42]. Since then, Hf-based high-*k* gate oxides have emerged as promising candidates for high-*k* dielectrics [43–47]. Furthermore, improvement in mobility of CMOS devices using Hf-based gate dielectrics has been observed, almost matching that of nitride oxide with an EOT of ~1 nm [48]. The two big breakthroughs to enable mobility enhancement are to understand transient charging behaviors in high-*k* devices and scaling of high-*k* thickness below the tunneling limit to eliminate residual tunneling carriers in high-*k* layer [49–51]. By keeping the high-*k* thickness smaller than 2 nm, mobility degradation originating from transient charging effects can be eliminated. Low standby power (LSTP) applications require low leakage current. To meet the requirement of LSTP, thickness of high-*k* layer should be increased without degrading the mobility. Meanwhile, optimization of high-*k* composition should be carried out to meet these requirements in LSTP applications.

To obtain microelectronic devices with excellent performance, aggressive EOT scaling is needed [52]. What is more, to scale the EOT of high-*k* gate stack, the growth of the interfacial layer should be controlled effectively. Figure 1.10 shows the mobility dependence on the perpendicular field at different thicknesses of the SiO_2 interfacial layer [53]. Based on this figure, it can be noted that reduction in mobility degradation has been observed due to the presence of the interfacial layer. Increasing the SiO_2 interfacial thickness leads to the increase in mobility approaching that of SiO_2. So, it can be concluded that the interfacial layer is beneficial to increase the inversion layer mobility, even if it increases EOT of the gate dielectric. The increased mobility is due to the decoupling between the motion of the electrons in the inversion layer and the phonons in the HfO_2.

To avoid the balance between EOT scaling and mobility, it is important to eliminate the charge scattering source within the high-*k* layer and increase the dielectric constant of the interfacial oxide with a minimal impact on the interface state density [55–57]. Although HfSiON can be used for 32 nm generations and below, any further extension of high-*k* dielectrics will require materials with higher dielectric constant that that of HfO_2. La_2O_3-based high-*k* oxides have been proposed as high-*k* candidates, but degraded dielectric characteristics compared to Hf oxides have been observed [58]. An alternative path to future scaling is to use other channel materials other than Si. For example, formation of interfacial oxide can be controlled when high-*k* dielectric is deposited on Ge or GaAs substrates [59, 60]. The challenges in this path are to overcome the problems associated with the degraded interface and to make use of the benefits of high carrier mobility in Ge or other channel materials. If Hf-based high-*k* gate oxides can be successfully

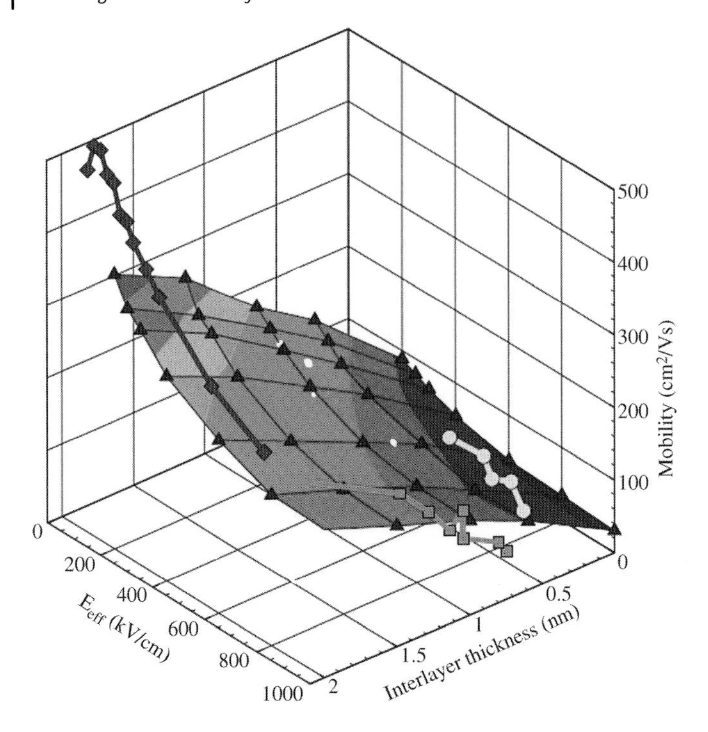

Figure 1.10 Electron mobility for HfO_2 versus effective field E_{eff} and the SiO_2 interfacial layer thickness. The mobility for pure SiO_2 and the experimental results [54] are also reported. Reproduced from Ref. [53]. Copyright 2007, Elsevier. (For a color version of this figure, please see the color plate at the beginning of this book.)

implemented with these channel materials, Hf-based oxides might then be usable beyond the 32 nm node.

Due to the thermal stability when in contact with poly-Si gate, metal electrodes have been used to integrate with high-*k* gate dielectrics. As we know, it is difficult to alter the effective work function (EWF) of a poly-Si gate when it is integrated with high-*k* dielectrics. It is attributed to the reasons that the EWF of a poly-Si/high-*k* dielectric stack is determined by Si-Hf bonding, not Fermi level of the poly-Si gate, which is called "Fermi-level pinning" [61]. In conclusion, PMOS devices with a poly-Si/high-*k* stack pose a high threshold voltage (V_{th}) that exceeds the practical range. This phenomenon has posed a serious challenge in the implementation of high-*k* dielectrics because to obtain the best performance in CMOS devices, the EWF of NMOS and PMOS should be close to the conduction (E_c) and valence bands (E_v), respectively [62]. To obtain high-performance applications, dual work function metal electrodes with EWFs approaching the conduction and valence bands are needed. However, metal electrode materials suffer from a limited EWF range when combined with high-*k* dielectrics. The following sections give more details of dielectric and electrode developments.

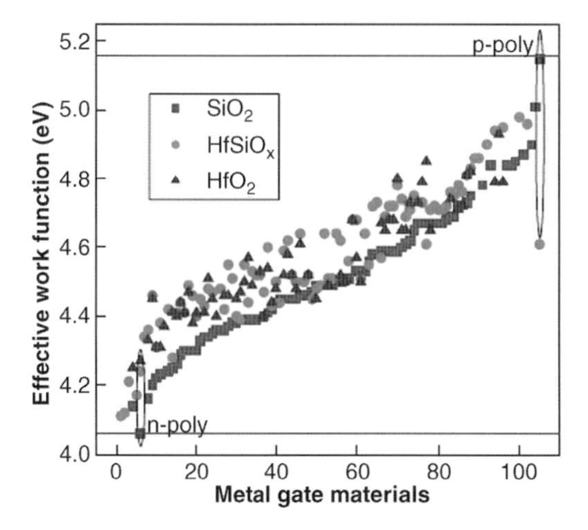

Figure 1.11 Effective work function of various metal electrode materials on SiO_2, $HfSiO_x$, and HfO_2 investigated with the terraced oxide method. Reproduced from Ref. [71]. Copyright 2008, Elsevier. (For a color version of this figure, please see the color plate at the beginning of this book.)

1.3.2 Advances and Challenges in Electrode Development

Based on the previous investigation, it can be noted that the application of poly-Si/high-k stacks are suppressed due to the V_{th} controllability, dopant penetration, and inversion oxide thickness T_{inv} scalability. Therefore, metal electrode materials with suitable EWFs close to E_c and E_v of the Si substrate should be investigated to replace poly-Si gate. The present challenge facing metal electrode is that the vacuum work function values of metals are not directly correlated with the V_{th} of devices. Thus, overall electrode processes, such as the deposition method, heat cycle, and dielectric composition, should be optimized to develop a well-defined metal electrode process. Many reported literatures suggests that experimental EWF results are consistent with the presence of a pinning effect [63, 64]; however, recently detailed studies of many of these electrode systems indicate contributions from extrinsic defect states induced by the electrode [65–67], chemical reduction of the high-k interface [68], or material characteristic changes in the electrode [69, 70] are the source of change in the EWF. Figure 1.11 shows the EWFs of vast metal systems extracted by the terraced oxide method, which were obtained using an oxide thickness series formed on a single substrate to minimize variations in interface charge [71].

The x-axis represents the material systems and three groups of data points show EWFs obtained with three different gate dielectrics. The difference in EWF between SiO_2, $HfSiO_x$, and HfO_2 is often explained by dipole formation, that is, if the EWF of the metal is smaller than the charge neutrality level (CNL) of the underlying dielectric, charge can be donated to the dielectric side and the EWF is

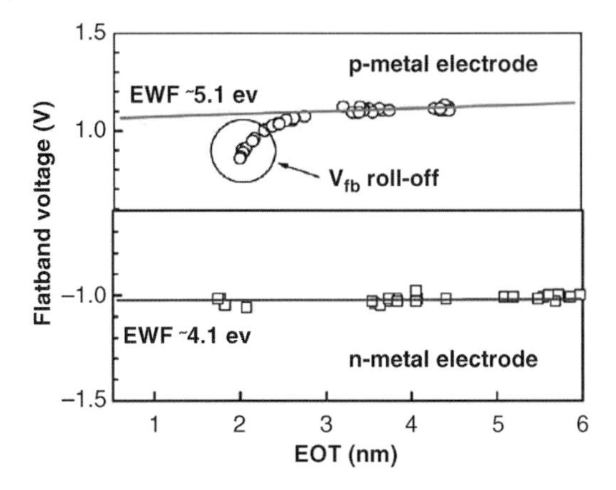

Figure 1.12 V_{fb}–EOT curve for a typical n-metal electrode and a p-metal electrode. Reproduced from Ref. [73]. Copyright 2006, Elsevier. (For a color version of this figure, please see the color plate at the beginning of this book.)

shifted toward the mid-gap value (∼4.6–4.7 eV) [72]. However, experimental data shown in Figure 1.11 indicate that the EWF shift because of dipole formation is actually unidirectional, indicating that dipole formation is more affected by metal–oxide bonding and microscopic charge transfer than the CNL of the underlying dielectric. Several metal systems show EWF values near the conduction and valence band edge of Si, even after high-temperature processing. These satisfy the requirements for gate-first CMOS integration, namely, that the EWF of the electrodes should be close to $E_{c,Si}$ for NMOS and $E_{v,Si}$ for PMOS after the CMOS heat cycle, which is typically a 1100 °C spike anneal or a 1000 °C 5 s rapid thermal anneal. The metal electrode systems are not specified in Figure 1.11 because the EWF of the metal/high-k stack can be controlled by many factors described above and, in fact, the electrode and high-k dielectric is best considered as a single material system that requires simultaneous optimization. Once thermally stable metal electrodes with band edge EWFs have been identified, the next challenge is to obtain a low V_{th} that matches the EWF values. Since the V_{th} of the actual device may be affected by additional factors, such as EOT, reactions with gate etch process gases, mixing with capping materials, and oxygen redistribution from the sidewall, it is not straightforward to predict V_{th} from the flatband voltage V_{fb}.

Figure 1.12 shows V_{fb}–EOT curve for a typical n-metal electrode and a p-metal electrode. From figure, it can be noted that high work function cannot be maintained as the device is scaled to lower EOT regimes and exhibit a "roll off"-like behavior, which can also be misinterpreted as Fermi-level pinning for p-metals. To further confirm this abnormal EWF behavior at low EOTs, a grand summary V_{fb}–EOT plot for various metal gates was presented in Figure 1.13. The V_{fb} roll-off region can be clearly found in Figure 1.14 that suggests a tradeoff in EWF in our goal of scaling the EOT. Roll off of the V_{fb} curve in the low EOT region can be contributed to several physical mechanisms including localized material interdiffusion, changes in stoichiometry,

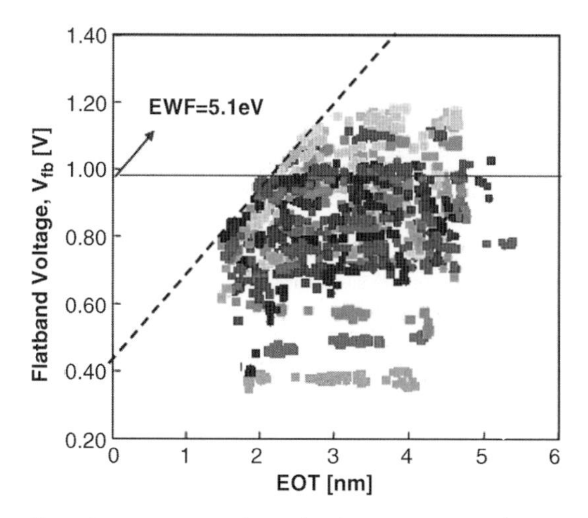

Figure 1.13 V_{fb}–EOT relationship for various materials systems investigated showing the V_{fb} roll-off for high work function metal electrodes. Reproduced from Ref. [71]. Copyright 2008, Elsevier. (For a color version of this figure, please see the color plate at the beginning of this book.)

and bulk charges of the SiO_2 bottom interface layer [65]. Based on previous analysis, it can be concluded that V_{fb} roll off should be minimized to obtain a high EWF of ~5.0 eV.

Silicides are the promising candidates for metal electrode applications. Fully silicided (FUSI) gates have been paid more attention because the poly-Si gate can be silicided with a relatively low heat cycle and minimal damage to the gate dielectric after device fabrication. $CoSi_2$ was used in the initial demonstration of a silicided gate, but the focus was shifted quickly to NiSi and $NiSi_3$ because the work function of Ni silicides can be modified with dopants in the poly-Si gate or a phase of Ni silicides [74, 75]. However, the EWF range with a silicide gate is still limited, even though the FUSI process uses only low-temperature steps. Various silicides have been studied to find a gate showing a band edge EWF. The most promising silicide gate materials are HfSi (~4.2 eV) and ErSi (~4.2 eV) for NMOS and PtSi (~4.9 eV) for PMOS [76, 77]. Even though the range of EWF is not wide enough for high-performance applications, these silicide gates can still be used in low-power applications or applications requiring quarter gap electrodes (~4.3 eV for NMOS and ~4.8 eV for PMOS), such as FinFETs or fully depleted silicon-on-insulator (FDSOI) devices. There are several potential challenges in the FUSI approach, such as pattern-dependent silicidation [78], strain control arising from volume expansion, and process complexity for a dual-silicide approach. Once electrode materials have been chosen for each application, the impact of reliability should be studied in detail. Metal electrodes can affect the reliability of gate dielectrics in many ways: diffusion and intermixing with the dielectric, oxygen, and nitrogen redistribution, and impurity contamination. None of these topics has been well investigated, primarily because the material systems have not been finalized yet.

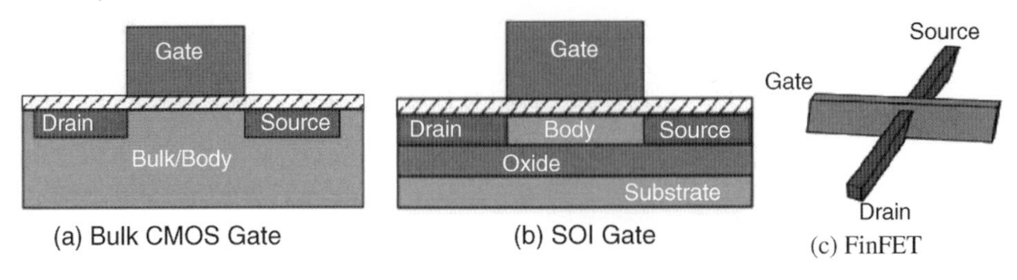

Figure 1.14 (a) Bulk, (b) SOI transistor structures, and (c) FinFET structure. (For a color version of this figure, please see the color plate at the beginning of this book.)

1.4 Improvements and Alternative to CMOS Technologies

1.4.1 Improvement to CMOS

1.4.1.1 New Materials

As analyzed in the past, the high-mobility III–V materials do not confer an advantage at the end of the scaling path since the isotropic, low-mass C valley does not provide a strong electron confinement and, furthermore, the light electron mass promotes source–drain tunneling. In addition, the smooth heterobarriers, responsible for the spectacular transport, are of low height and do not scale well. The single-transport valley also does not provide the large charge densities needed for these small devices. Materials such as Ge can provide larger carrier densities, but the low bandgap is a considerable disadvantage unless quantum confinement can be used to increase the bandgap in structures of practical dimensions.

High-k dielectrics are designed to address one particular aspect of off-state power consumptions: gate tunneling currents. It is likely that the gate dielectric thickness will be the first parameter to reach atomic dimensions. This is because the dielectric thickness indirectly controls the gate length. In general, the effective gate length needs to be 40 times the dielectric thickness to properly control short channel effects (SCE). Thus, the scaling that reduces gate length must also reduce the dielectric thickness. However, as the dielectric thickness decreases, electron tunneling through the dielectric becomes a significant issue. The proposed solution to this problem is to find a material that has a higher k value than the SiO_2 used at present as the gate dielectric. This would allow the actual thickness of the gate dielectric to be increased while still maintaining the same electric field in the channel.

While this sounds good in theory, implementation has proved to be challenging. To do this, a material must be found that meets many criteria. The material must be compatible with the surrounding silicon and the fabrication processes used. Also, it must have a breakdown time at least as long as silica. While there are many other requirements, suffice it to say that many materials have been proposed, but no good substitute has yet been found.

1.4.1.2 New Structures

It has been proposed to try and change the structure of the transistor itself. Here, two most prominent structural changes, silicon on insulator (SOI) and double-gate CMOS (DGCMOS) have been discussed. As we know, bulk CMOS structure is based on the concept that the transistor is connected to the substrate, as demonstrated in Figure 1.14a. However, for silicon on insulator, an insulating oxide is first deposited on the substrate and then the transistor is fabricated on top of that (Figure 1.14b). By doing this structure, the body is electrically isolated from its surroundings. This design leads to the improvement in performance. What is more, this new structure also lends itself to some new uses, such as using the insulating layer for a high-resistance element. However, there are also some drawbacks facing this new structure. Some techniques have been developed to address some of these issues. Based on SOI structure, IBM Company redesigned some PC line chips and performance improvement has been observed compared to bulk CMOS structure.

The second structure is DGCMOS, which is more experimental, but it may demonstrate promising application in future. In this structure, an additional gate has been added to increase coupling between the gate and the channel, also called "deal structure for scalability" [79]. However, it is difficult to design and realize this structure. Based on traditional method, second gate can be added to the body. However, the alignment issues of such a gate are troublesome. So, FinFET structure has been proposed, as shown in Figure 1.14c. This is a daunting challenge because the gate length is usually the smallest dimension that can be fabricated. There are some technologies that may address this, but more work needs to be carried out in this area.

1.5 Potential Technologies Beyond CMOS

So far, scaling has tremendous benefits that no alternative technologies can compete with the power of mainstream CMOS devices. In addition, more attention and knowledge have been focused on the investigation of MOS technology. Even if much efforts made, scaling CMOS is unquestionably approaching its limits. Although many issues have been resolved, scaling still cannot progress past the size of the molecule. The questions of what technology might surpass CMOS have come out. By now, there are many alternative devices that show promising replacement to CMOS for the future mentioned as follows.

1) Electrical-dependent nanodevices [80–81].

   Based either on ballistic transport and tunneling or on electrostatic phenomenon, these nanodevices are being investigated, such as carbon nanotubes field-effect transistors (CNTFETs), semiconductor nanowire field-effect transistors (NWFETs), resonant tunneling diodes (RTDs), and electrical quantum dot cellular automata (EQCA).

2) Magnetic-dependent nanodevices.

   Magnetic quantum dot cellular automata (MQCA) and spin field-effect transistors (spinFETs) are included in this class. Magnetostatic and spin transport are the phenomena for the operation of the devices.

3) Mechanical-dependent nanodevices.

   Compared to traditional CMOS devices, these nanodevices demonstrate some advantages. For example, CNTFETs have an extraordinary mechanical strength, low power consumption, better thermal stability, and higher resistance to electromigration [80]. The advantages of NWFETs over CMOS are similar to CNTFETs [80], plus the ability to operate at high speed produces saturated current at low bias voltage, and the potential to behave as either active or passive devices in single nanowires [81]. EQCA and MQCA exhibit a higher low-power dissipation, nonvolatility, and reconfigurability [81]. SpinFETs possess many advantages, such as small off-current, high operating speed, high-power gain, and low-power consumption compared to traditional CMOS devices. Except for the innovations already mentioned, there still exist some other proposed technologies, which may develop in parallel with CMOS allowing developers to choose the technology to satisfy their demands.

1.6 Conclusions

With the end of CMOS roadmap looming, there has been tremendous research in order to identify promising technologies to continue the historical trend of performance scaling. This chapter mainly explored the present status and challenges associated with alternative gate stack technology for future generations. Present beliefs regarding the limitations and main challenges faced by CMOS technology are reviewed. The benefits and limitations of alternative new high-*k* materials as candidate of SiO_2 and nitrided SiO_2 for possible high-performance CMOS applications are reviewed. Considering the progress of technology development, an alternative gate stack with a metal gate/high-*k* dielectric will be implemented for Si-based technology in the near future. Finally, we have also presented alternative devices that are potentially able to overcome the limitation in CMOS technology. These technologies will probably develop in parallel with CMOS allowing developers to choose the technology that best fits their needs.

Acknowledgments

The authors acknowledge the support from the National Natural Science Foundation of China (Grant Nos. 10804109 and 11104269) and Outstanding Young Scientific Foundation of Anhui University (Grant No. KJJQ1103). The authors are indebted to publishers/authors concerned for their kind permissions to reproduce their works, especially figures, used in this chapter.

References

1 Dennard, R.H., Gaensslen, F.H., Yu, H.N., Rideout, V.L., Bassous, E., and LeBlanc, A.R. (1974) Design of ion-implanted MOSFETs with very small physical dimensions. *IEEE J. Solid St. Circ.*, **SC-9**, 256.

2 Ligenza, J.R. and Spitzer, W.G. (1960) The mechanisms for silicon oxidation in steam and oxygen. *J. Phys. Chem. Solids*, **14**, 131.

3 Ligenza, J.R. and Spitzer, W.G. (1961) Effects of crystal orientation on oxidation rates of silicon in high pressure steam. *J. Phys. Chem.*, **65**, 2011.

4 Hoeneisen, B. and Mead, C.A. (1972) Fundamental limitations in microelectronics-I. MOS technology. *Solid State Electron.*, **15**, 819.

5 Wallmark, J.T. (1975) Fundamental physical limitations in integrated electronic circuits. *Inst. Phys. Conf. Ser.*, **25**, 133.

6 Stathis, J.H. and Dimaria, D.J. (1998) Reliability projection for ultra-thin oxides at low voltage. IEDM Technical Digest, p. 167.

7 Schulz, M. (1999) The end of the road for silicon. *Nature*, **399**, 729.

8 Muller, D.A., Sorsch, T., Moccio, S., Baumann, F., Evans-Lutterodt, K., and Timp, G. (1999) The electronic structure at the atomic scale of ultrathin gate oxide. *Nature*, **399**, 758.

9 Baccarani, G., Wordeman, M., and Dennard, R. (1984) Generalised scaling theory and its application to a 1/4 micrometer MOSFET design. *IEEE Trans. Electron. Dev.*, **31**, 452.

10 Frank, D., Dennard, R., Nowak, E., Solomon, P., Taur, Y., and Wong, H.-S. (2001) Device scaling limits of Si MOSFETs and their application dependencies. *Proc. IEEE*, **89**, 259.

11 Haensch, W., Nowak, E., Dennard, R., Solomon, P., Bryant, A., Dokumaci, O., Kumar, A., Wang, X., Johnson, J., and Fischetti, M. (2006) Silicon CMOS devices beyond scaling. *IBM J. Res. Dev.*, **50**, 339.

12 Taur, Y. (2002) CMOS design near the limit of scaling. *IBM J. Res. Dev.*, **46**, 213.

13 Gusev, E., Narayanan, V., and Frank, M. (2006) Advanced high-*k* dielectric stacks with poly-Si and metal gates. *IBM J. Res. Dev.*, **50**, 387.

14 Ghani, T., Mistry, K., Packan, P., Thompson, S., Stettler, M., Tyagi, S., and Bohr, M. (2000) Scaling challenges and device design requirements for high performance sub-50nm gate length planar CMOS transistors. Symposium on VLSI Technology Digest, p. 174.

15 Hargrove, M., Crowder, S., Nowak, E., Logan, R., Han, L., Ng, H., Ray, A., Sinitsky, D., Smeys, P., Guarin, F., Oberschmidt, J., Crabbe, E., Yee, D., and Su, L. (1998). High-performance sub-0.08 μm CMOS with dual gate oxide and 9.7ps inverter delay. IEDM Technical Digest, p. 627.

16 Lo, S.-H., Buchanan, D., and Taur, Y. (1999) Modeling and characterization of quantization, polysilicon depletion, and direct tunelling effects in MOSFETs with ultrathin oxides. *IBM J. Res. Dev.*, **43**, 327.

17 Green, M.L., Gusev, E.P., Degraeve, R., and Garfunkel, E.L. (2001) Ultrathin (<4 nm) SiO_2 and Si-O-N gate dielectric layers for silicon microelectronics: understanding the processing, structure, and physical and electrical limits. *J. Appl. Phys.*, **90**, 2057.

18 Massoud, H.Z., Shiely, J.P., and Shanware, A. (1999) Self-consistent MOSFET tunneling simulations? Trends in the gate and substrate currents and the drain-current turnaround effect with oxide scaling. *Mater. Res. Soc. Symp.*, **567**, 227.

19 Momose, H.S., Ono, M., Yoshitomi, T., Ohguro, T., Nakamura, S.I., Saito, M., and Iwai, H. (1996) 1.5-nm direct-tunneling gate oxide Si MOSFET's. *IEEE Trans. Electron. Dev.*, **43**, 1233.

20 Tang, S., Wallace, R., Seabaugh, A., and King-Smith, D. (1998) Evaluating the minimum thickness of gate oxide on silicon using first-principles method. *Appl. Surf. Sci.*, **135**, 137.

21 Kaneta, C., Yamasaki, T., Uchiyama, T., Uda, T., and Terakura, K. (1999) Structure and electronic property of Si

(100)/SiO_2 interface. *Microelectron. Eng.*, **48**, 117.

22 Cao, M., Voorde, P.V., Cox, M., and Greene, W. (1998) Boron diffusion and penetration in ultrathin oxide with poly-Si gate. *IEEE Electron. Device Lett.*, **19**, 291.

23 Krisch, K.S. and Sodini, C.G. (1994) Suppression of interface state generation in reoxidized nitrided oxide gate dielectrics. *J. Appl. Phys.*, **76**, 2284.

24 Cartier, E., Buchanan, D.A., and Dunn, G.J. (1994) Atomic hydrogen-induced interface degradation of reoxidized-nitrided silicon dioxide on silicon. *Appl. Phys. Lett.*, **64**, 901.

25 Buchanan, D.A., Marwick, A.D., DiMaria, D.J., and Dori, L. (1994) Hot-electron induced redistribution and defect generation in metal-oxide-semiconductor capacitors. *J. Appl. Phys.*, **76**, 3595.

26 Joshi, A.B., Ahn, J., and Kwong, D.L. (1993) Oxynitride gate dielectrics for p1 polysilicon gate MOS devices. *IEEE Electron. Device Lett.*, **14**, 560.

27 Ma, Z.L., Chen, J.C., Liu, H., Krick, J.T., Cheng, Y.C., Hu, C., and Ko, P.K. (1994) Suppression of boron penetration in p1 polysilicon gate *p*-MOSFETs using low-temperature gate oxide N_2O anneal. *IEEE Electron. Device Lett.*, **15**, 109.

28 Taur, Y., Mii, Y.-J., Frank, D.J., Wong, H.-S., Buchanan, D.A., Wind, S.J., Rishton, S.A., Sai-Halasz, G.A., and Nowak, E.J. (1995) CMOS scaling into the 21st century: 0.1mm and beyond. *IBM J. Res. Dev.*, **39**, 245.

29 Ma, T.P. (1998) Making silicon nitride film a viable gate dielectric. *IEEE Trans. Electron. Dev.*, **45**, 680.

30 Ma, T.P. (1997) Gate dielectric properties of silicon nitride films formed by jet vapor deposition. *Appl. Surf. Sci.*, **117/118**, 259.

31 Buchanan, D.A., DiMaria, D.J., Chang, C.-A., and Taur, Y. (1994) Defect generation in 3.5nm silicon dioxide films. *Appl. Phys. Lett.*, **65**, 1820.

32 Stathis, J.H. and Cartier, E. (1994) Atomic hydrogen reactions with Pb centers at the (100)Si/SiO_2 interface. *Phys. Rev. Lett.*, **72**, 2745.

33 Fukumoto, H., Imura, T., and Osaka, Y. (1989) Heteroepitaxial growth of Y_2O_3 films on silicon. *Appl. Phys. Lett.*, **55**, 360.

34 Inoue, T., Yamamoto, Y., Koyama, S., Suzuki, S., and Ueda, Y. (1990) Epitaxial growth of CeO_2 layers on silicon. *Appl. Phys. Lett.*, **56**, 1332.

35 Alers, G.B., Werder, D.J., Chabal, Y., Lu, H.C., Gusev, E.P., Garfunkel, E., Gustafsson, T., and Urgahl, R.S. (1998) Intermixing at the tantalum oxide/silicon interface in gate dielectric structures. *Appl. Phys. Lett.*, **73**, 1517.

36 Ha, H.-K., Yoshimoto, M., Koinuma, H., Moon, B.-K., and Ishiwara, H. (1996) Open air plasma chemical vapor deposition of highly dielectric amorphous TiO_2 films. *Appl. Phys. Lett.*, **68**, 2965.

37 Pallecchi, I., Grassano, G., Marre, D., Pellegrino, L., Putti, M., and Siri, A.S. (2001) $SrTiO_3$-based metal-insulator–semiconductor heterostructures. *Appl. Phys. Lett.*, **78**, 2244.

38 Wilk, G.D. and Wallace, R.M. (2000) Stable zirconium silicate gate dielectrics deposited directly on silicon. *Appl. Phys. Lett.*, **76**, 112.

39 Lee, B.H., Kang, L., Nieh, R., Qi, W.J., and Lee, J.C. (2000) Thermal stability and electrical characteristics of ultrathin hafnium oxide gate dielectric reoxidized with rapid thermal annealing. *Appl. Phys. Lett.*, **76**, 1926.

40 Wilk, G.D. and Wallace, R.M. (1999) Electrical properties of hafnium silicate gate dielectrics deposited directly on silicon. *Appl. Phys. Lett.*, **74**, 2854.

41 Qi, W.J., Nieh, R., Dharmarajan, E., Lee, B.H., Jeon, Y., Kang, L., Onishi, K., and Lee, J.C. (2000) Ultrathin zirconium silicate film with good thermal stability for alternative gate dielectric application. *Appl. Phys. Lett.*, **77**, 1704.

42 Wilk, G.D., Wallace, R.M., and Anthony, J.M. (2001) High-*k* gate dielectrics: current status and materials properties considerations. *J. Appl. Phys.*, **89**, 5243.

43 He, G., Liu, M., Zhu, L.Q., Chang, M., Fang, Q., and Zhang, L.D. (2005) Effect of postdeposition annealing on the thermal

stability and structural characteristics of sputtered HfO_2 films on Si (100). *Surf. Sci.*, **576**, 67.

44 He, G., Fang, Q., Liu, M., Zhu, L.Q., and Zhang, L.D. (2007) Structural and optical properties of nitrogen-incorporated HfO_2 gate dielectrics deposited by reactive sputtering. *Appl. Surf. Sci.*, **253**, 8483.

45 He, G., Zhang, L.D., Meng, G.W., Li, G.H., Fang, Q., and Zhang, J.P. (2007) Temperature-dependent thermal stability and optical properties of ultrathin $HfAlO_x$ films on Si (100) grown by reactive sputtering. *J. Appl. Phys.*, **102**, 094103.

46 He, G., Zhang, L.D., Meng, G.W., Li, G.H., Fei, G.T., Wang, X.J., Zhang, J.P., Liu, M., Fang, Q., and Boyd, I.W. (2008) Composition dependence of electronic structure and optical properties of $Hf_{1-x}Si_xO_y$ gate dielectrics. *J. Appl. Phys.*, **104**, 104116.

47 He, G., Fang, Q., and Zhang, L.D. (2006) High-*k* $HfSi_xO_y$ gate dielectrics grown by solid phase reaction between sputtered Hf layer and SiO_2/Si. *J. Appl. Phys.*, **100**, 083517.

48 Quevedo-Lopez, M.A., Krishnan, S.A., Kirsch, P.D., Li, H.J., Sim, J.H., and Huffman, C. (2005) High performance gate first HfSiON dielectric satisfying 45nm node requirements. IEDM Technical Digest, 2005, p. 438.

49 Lee, B.H., Young, C., Choi, R., Sim, J.H., and Bersuker, G. (2005) Transient charging and relaxation in high-*k* gate dielectrics and their implications. *Jpn. J. Appl. Phys.*, **44**, 2415.

50 Choi, R., Song, S.C., Young, C.D., and Bersuker, G., and Lee, B.H. (2005) Charge trapping and detrapping characteristics in hafnium silicate gate dielectric using an inversion pulse measurement technique. *Appl. Phys. Lett.*, **87**, 122901.

51 Sim, J.H., Song, S.C., Kirsch, P.D., Young, C.D., Choi, R., Kwong, D.L., Lee, B.H., and Bersuker, G. (2005) Effects of ALD HfO_2 thickness on charge trapping and mobility. *Microelectron. Eng.*, **80**, 218.

52 Kirsch, P.D., Quevedo-Lopez, M.A., Li, H.J., Senzaki, Y., Peterson, J.J., Song, S.C., Wang, Q., Gay, D., and Ekerdt, J.G. (2006) Nucleation and growth study of atomic layer deposited HfO_2 gate dielectrics resulting in improved scaling and electron mobility. *J. Appl. Phys.*, **99**, 023508.

53 Ferrari, G., Watling, J.R., Roy, S., Barker, J.R., and Asenov, A. (2007) Beyond SiO_2 technology: simulation of the impact of high-*k* dielectrics on mobility. *J. Non-Cryst. Solids*, **353**, 630.

54 Ragnarsson, L.A., Severi, S., Trojman, L., Brunco, D.P., Johnson, K.D., Del abie, A., Schram, T., Tsai, W., Groeseneken, G., De Meyer, K., De Gendt, S., and Heyns, M. (2005). High performing 8 angstrom EOT HfO_2/TaN low thermal-budget n-channel FETs with solid-phase epitaxially regrown (SPER) junctions. Symposium on VLSI Technology Digest, p. 234.

55 Lai, C.S., Wu, W.C., Wang, J.C., and Chao, T.S. (2005) Characterization of CF_4-plasma fluorinated HfO_2 gate dielectrics with TaN metal gate. *Appl. Phys. Lett.*, **86**, 222905.

56 Kita, K., Kyuno, K., and Toriumi, A. (2005) Permittivity increase of yttrium-doped HfO_2 through structural phase transformation. *Appl. Phys. Lett.*, **86**, 102906.

57 Osten, H.J., Liu, P., Gaworzewski, P., Bugiel, E., and Zaumseil, P. (2000) High-*k* gate dielectrics with ultra-low leakage current based on praseodymium oxide. IEDM Technical Digest, p. 653.

58 Lu, X.B., Liu, Z.G., Wang, Y.P., Wang, X.P., Zhou, H.W., and Nguyen, B.Y. (2003) Structure and dielectric properties of amorphous $LaAlO_3$ and $LaAlO_xN_y$ films as alternative gate dielectric materials. *J. Appl. Phys.*, **94**, 1229.

59 Chen, J.H. Jr., Bojarczuk, N.A., Shang, H., Copel, M., and Hannon, J.B. (2004) Ultrathin Al and HfO_2 gate dielectrics on surface-nitrided Ge. *IEEE Trans. Electron. Dev.*, **51**, 1441.

60 Frank, M.M., Wilk, G.D., Starodub, D., Gustafsson, T., Garfunkel, E., Chabal, Y.J., Grazul, J., and Muller, D.A. (2005) HfO_2 and Al_2O_3 gate dielectrics on GaAs grown by atomic layer deposition. *Appl. Phys. Lett.*, **86**, 152904.

61 Hobbs, C.C., Fonseca, L.R.C., Knizhnik, A., Dhandapani, V.,

Anderson, S.G.H., and Tobin, P.J. (2004) Fermi-level pinning at the polysilicon/metal oxide interface. *IEEE Trans. Electron. Dev.*, **51**, 971.

62 De, I., Jonri, D., Srivastava, A., and Osburn, C.M. (2000) Impact of gate work function on device performance at the 50nm technology node. *Solid State Electron.*, **44**, 1077.

63 Samavedam, S.B., La, L.B., Tobin, P.J., White, B., Hobbs, C., Fonseca, L.R.C., Demkov, A.A., Schaeffer, J., Luckowski, E., Martinez, A., Raymond, M., Triyoso, D., Roan, D., Dhandapani, V., Garcia, R., Anderson, S.G.H., Moore, K., Tseng, H.H., Capasso, C., Adetutu, O., Gilmer, D.C., Taylor, W.J., Hegde, R., and Grant, J. (2003). Fermi-level pinning with sub-monolayer MeO_x and metal gates. IEDM Technical Digest, p. 307.

64 Yu, H.Y., Ren, C., Yeo, Y.C., Kang, J.F., Wang, X.P., Ma, H.H.H., Li, M.F., Chan, D.S.H., and Kwong, A.L. (2004) Fermi pinning-induced thermal instability of metal-gate work functions. *IEEE Electron. Device Lett.*, **25**, 337.

65 Cartier, E., McFeely, F.R., Narayanan, V., Jamison, P., Linder, B.P., Copel, M., Paruchuri, V.K., Basker, V.S., Haight, R., Lim, D., Carruthers, R., Shaw, T., Steen, M., Sleight, J., Rubino, J., Deligianni, H., Guha, S., Jammy, R., and Shahidi, G. (2005). Role of oxygen vacancies in V-FB/V-t stability of pFET metals on HfO_2. Symposium on VLSI Technology Digest, p. 230.

66 Schaeffer, J.K., Fonseca, L.R.C., Samavedam, S.B., Liang, Y., Tobin, P.J., and White, B.E. (2004) Contributions to the effective work function of platinum on hafnium dioxide. *Appl. Phys. Lett.*, **85**, 1826.

67 Liang, Y., Curless, J., Tracy, C.J., Gilmer, D.C., Schaeffer, J.K., Triyoso, D.H., and Tobin, P.J. (2006) Interface dipole and effective work function of Re in $Re/HfO_2/SiO_x$/n-Si gate stack. *Appl. Phys. Lett.*, **88**, 072907.

68 Copel, M., Pezzi, R.P., Neumayer, D., and Jamison, P. (2006) Reduction of hafnium oxide and hafnium silicate by rhenium and platinum. *Appl. Phys. Lett.*, **88**, 072914.

69 Pantisano, L., Schram, T., Li, Z., Lisoni, J.G., Pourtois, G., Gendt, S.D., Brunco, D.P., Akheyar, A., Afanas'ev, V.V., Shamuilia, S., and Stesmans, A. (2006) Ruthenium gate electrodes on SiO_2 and HfO_2: sensitivity to hydrogen and oxygen ambients. *Appl. Phys. Lett.*, **88**, 243514.

70 Alshareef, H.N., Wen, H.C., Luan, H., Choi, K., Harris, H.R., Senzaki, Y., Majhi, P., Lee, B.H., Foran, B., and Lian, G. (2006) Temperature dependence of the work function of ruthenium-based gate electrodes. *Thin Solid Films*, **515**, 1294.

71 Wen, H.C., Majhi, P., Choi, K., Park, C.S., Alshareef, H.N., Harris, H.R., Luan, H., Niimi, H., Park, H.B., Bersuker, G., Lysaght, P.S., Kwong, D.L., Song, S.C., Lee, B.H., and Jammy, R. (2008). Decoupling the Fermi-level pinning effect and intrinsic limitations on p-type effective work function metal electrodes. *Microelectron. Eng.*, **85**, 2.

72 Yeo, Y.-C., Ranade, P., King, T.-J., and Hu, C. (2002) Effects of high-*k* gate dielectric materials on metal and silicon gate workfunctions. *IEEE Electron. Device Lett.*, **23**, 342.

73 Lee, B.H., Oh, J., Tseng, H.H., Jammy, R., and Huff, H. (2006) Gate stack technology for nanoscale devices. *Mater. Today*, **9**, 32.

74 Tavel, B., Skotnicki, T., Pares, G., Carriere, N., Rivoire, M., Leverd, F., Julien, C., Torres, J., and Pantel, R. (2001) Totally silicided ($CoSi_2$) polysilicon: a novel approach to very low resistive gate without metal CMP or etching. IEDM Technical Digest, p. 825.

75 Kang, C.Y., Lysaght, P., Choi, R., Lee, B.H., Rhee, S.J., Choi, C.H., Akbar, M.S., and Lee, J.C. (2005) Nickel-silicide phase effects on flatband voltage shift and equivalent oxide thickness decrease of hafnium silicon oxynitride metal-silicon-oxide capacitors. *Appl. Phys. Lett.*, **86**, 222906.

76 Park, C.S., Cho, B.J., and Kwong, D.L. (2004) Thermal stable fully

silicided Hf-silicide metal-gate electrode. *IEEE Electron. Device Lett.*, **25**, 372.

77 Nabatame, T., Kadoshima, M., Iwamoto, K., Mise, N., Migita, S., and Ohno, M. (2004) Partial silicides technology for tunable work function electrodes on high-k gate dielectrics: Fermi level pinning controlled $PtSi_x$ for HfO_x (N) pMOSFET. IEDM Technical Digest, p. 83.

78 Kedzierski, J. (2003) Issues in NiSi-gated FDSOI device integration. IEDM Technical Digest, p. 441.

79 Nowak, E.J. (2002) Maintaining the benefits of CMOS scaling when scaling bogs down. *IBM J. Res. Dev.*, **46**, 169.

80 Goser, K. (2004) *Nanoelectronics and Nanosystems: From Transistor to Molecular and Quantum Devices*, Springer.

81 Zhimov, V.V. (2005) Emerging research logic devices. *IEEE Circuits Devices Mag.*, **21**, 37.

Part Two
High-k Deposition and Materials Characterization

High-k Gate Dielectrics for CMOS Technology, First Edition. Edited by Gang He and Zhaoqi Sun.

2
Issues in High-*k* Gate Dielectrics and its Stack Interfaces

Hong-Liang Lu and David Wei Zhang

2.1
Introduction

The continual scaling of silicon dioxide (SiO_2) dielectrics has once been viewed as an effective approach to enhance transistor performance in complementary metal oxide semiconductor (CMOS) technologies as predicted by Moore's law. However, further scaling of gate oxide thickness in CMOS transistors for high performance and circuit density aggravates the problems of gate leakage current, standby power consumption, and gate oxide reliability. New materials and device structures are thus beginning to play important roles in further improving the required device performance. For example, high dielectric constant (high-*k*) dielectrics and metal gates are now replacing the conventional SiO_2 and polysilicon. The direct tunneling gate leakage current of MOS field-effect transistors (MOSFETs) with conventional SiO_2 gate dielectric can be greatly suppressed by using thicker high-*k* oxides with the same equivalent oxide thickness (EOT) [1]. Moreover, new channel materials, such as germanium (Ge) and III–V semiconductors, have at present been recognized as mandatory for future scaled CMOS, which can provide higher hole and electron mobility than Si.

However, successful integration of high-*k*, metal gate, and high-mobility channel materials into CMOS technology poses a number of challenges. The entire structure of the transistor must be considered in view of the material process constraints for fabrication. In this chapter, we first explain the various issues regarding the replacement of SiO_2 by high-*k* dielectric that can fulfill the requirements of submicron MOSFETs. Then, we describe the interfacial issues between metal gate and high-*k* dielectrics. Finally, we discuss the challenges in the combination of high-*k* dielectrics with Ge and III–V channel materials.

2.2
High-*k* Dielectrics

SiO_2 with the *k*-value of 3.9 has been used as the primary gate dielectric for over four decades due to (i) the high-quality interface between SiO_2 and Si substrate,

High-k Gate Dielectrics for CMOS Technology, First Edition. Edited by Gang He and Zhaoqi Sun.

Table 2.1 Main high-*k* materials with their parameters as tabulated by Robertson [4].

Material	*k*	E_g (eV)	CBO (eV)	VBO (eV)
SiO_2	3.9	9.0	3.2	4.7
Si_3N_4	7	5.3	2.4	1.8
Al_2O_3	9	8.8	2.8	4.9
La_2O_3	30	6.0	2.3	2.6
Y_2O_3	15	6.0	2.3	2.6
ZrO_2	25	5.8	1.5	3.2
Ta_2O_5	22	4.4	0.35	2.95
HfO_2	25	5.8	1.4	3.3
$HfSiO_4$	11	6.5	1.8	3.6
TiO_2	80	3.5	0	2.4
α-$LaAlO_3$	30	5.6	1.8	2.7
$SrTiO_3$	2000	3.2	0	2.1

(ii) chemical and thermal stability at high temperature ($\sim$1000 °C), (iii) good quality of insulation, (iv) the property of hard mask in different diffusion and doping process, and (v) high breakdown fields of 13 MV/cm. It is required to keep high capacitance density for channel formation in the submicron MOSFET with ultrathin SiO_2 layer. The recent trend shows that the high leakage current will prevent the scaling of the SiO_2 below 1 nm for future applications. Therefore, thickness reduction of SiO_2 gate layer below 1 nm is a big challenge. Again defects are formed in the gate oxide at the SiO_2/Si interface due to flow of charge carriers. If defect density reaches a certain threshold, this may cause quasi-breakdown on the gate layer. This is an important reliability issue of the transistor.

An insulator with higher *k*-value than SiO_2 can be a solution [1–3]. For CMOS application, high-*k* dielectrics are defined as those with a dielectric constant greater than that of silicon nitride (Si_3N_4, $k = 7$). There are a variety of high-*k* dielectrics being studied, such as aluminum oxide (Al_2O_3), yttrium oxide (Y_2O_3), titanium oxide (TiO_2), zirconium oxide (ZrO_2), hafnium oxide (HfO_2), tantalum oxide (Ta_2O_5), lanthanum oxide (La_2O_3), and strontium titanate ($SrTiO_3$). Table 2.1 lists the main high-*k* candidate materials with their *k* and bandgap (E_g) values as well as the conduction (valence) band offset, CBO (VBO) [4]. Moreover, there have been several excellent review papers on different aspects of high-*k* dielectrics [5, 6], but the field is progressing very rapidly and is on the verge of converging on a successful replacement for SiO_2.

2.2.1 The Criteria Required for High-k Dielectrics

As shown in Table 2.1, there are many dielectrics with much higher *k*-value than that of SiO_2. However, the choice of the "best high-*k* dielectric" is not simple. Many issues remain to be resolved in terms of process integration, as briefly illustrated in this section.

Permittivity. Selecting a gate dielectric with a higher permittivity than that of SiO_2 is clearly essential. For gate dielectric applications, the appropriate dielectric constant of the metal oxide should be over 10, preferably 20–40. However, there is a tradeoff between the *k*-value and the band offset. Generally, the *k*-value of the candidate dielectrics tends to vary inversely with the bandgap [4], as shown in Figure 2.1. For example, ferroelectrics such as $SrTiO_3$ have an extremely large *k*-value but too low bandgap. In fact, too large a *k*-value will cause undesirable strong fringing field close to source or drain electrodes and these fields can degrade short-channel performances [7].

Bandgap width and band offset. The dielectric material must be an insulator with a bandgap larger than 5 eV. The band offsets with Si must be over 1.0 eV to inhibit conduction by the Schottky emission of electrons or holes into the oxide bands [8]. As a comparison, the bandgap of SiO_2 is 9 eV, and the conduction and valence band offsets with Si are 3.1 and 4.8 eV, respectively. If the CBO values for the selected high-*k* oxides are smaller than 1.0 eV, it will likely preclude using these oxides in gate dielectric applications since electron transport would cause unacceptable high leakage currents [1].

Thermodynamic stability on silicon. For all gate dielectrics, the interface reactivity with Si substrate is very important and, in most cases, is the dominant factor in determining the overall electrical properties. Stability requires no or little reaction of the high-*k* oxide with Si to prevent formation of interfacial SiO_2

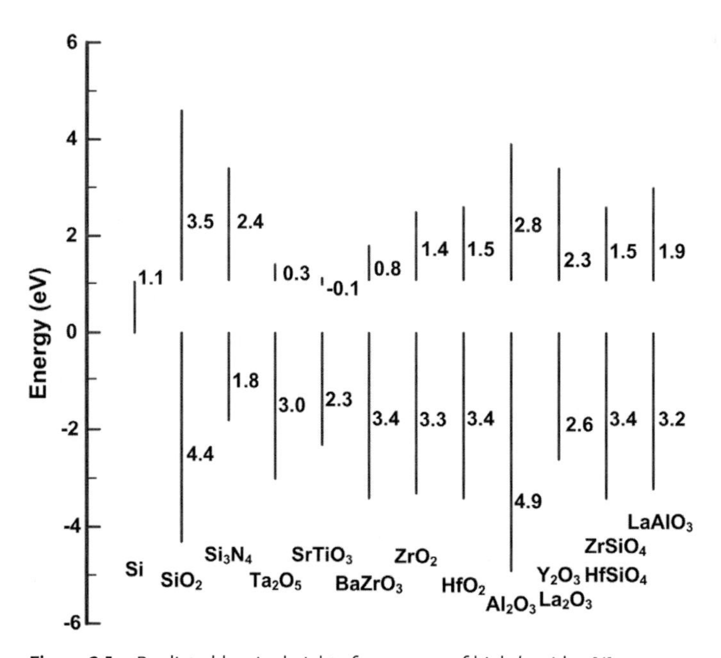

Figure 2.1 Predicted barrier heights for a range of high-*k* oxides [4].

* = Not a solid at 1000 K
☼ = Radioactive
① = Si+MO_x → M+SiO_2
② = Si+MO_x → MSi_y+SiO_2
③ = Si+MO_x → M+MSi_xO_y

IA	IIA	IIIB	IVB	VB	VIB	VIIB	VIII	VIII	VIII	IB	IIB	IIIA	IVA	VA	VIA	VIIA	O
* H																	* He
① Li	① Be											* B	* C	* N	* O	* F	* Ne
① Na	① Mg											Al	Si	* P	* S	* Cl	* Ar
① K	Ca	① Sc	② Ti	① V	① Cr	① Mn	① Fe	① Co	① Ni	① Cu	① Zn	① Ga	① Ge	* As	* Se	* Br	* Kr
* Rb	Sr	Y	Zr	① Nb	① Mo	☼ Tc	① Ru	① Rh	① Pd	* Ag	① Cd	① In	① Sn	① Sb	① Te	* I	* Xe
* Cs	③ Ba	†	Hf	① Ta	① W	① Re	① Os	① Ir	* Pt	* Au	* Hg	* Ti	① Pb	① Bi	☼ Po	☼ At	☼ Rn
☼ Fr	☼ Ra	‡	☼ Rf	☼ Db	☼ Sg	☼ Bh	☼ Hs	☼ Mt									

†	La	Ce	Pr	Nd	☼ Pm	Sm	Eu	Gd	Tb	Dy	Ho	Er	Tm	Yb	Lu
‡	☼ Ac	☼ Th	☼ Pa	☼ U	☼ Np	☼ Pu	☼ Am	☼ Cm	☼ Bk	☼ Cf	☼ Es	☼ Fm	☼ Md	☼ No	☼ Lr

Figure 2.2 Thermodynamic stability of binary oxides in contact with Si. (For a color version of this figure, please see the color plate at the beginning of this book.)

layers. Unfortunately, most of the high-*k* candidates are thermodynamically unstable at the interfaces with Si [1, 9], as shown in Figure 2.2.

Direct growth of high-*k* dielectrics on silicon is frequently accompanied by extensive interdiffusion or chemical reactions that degrade the properties of the dielectric and the underlying silicon or of both. A suitable way to avoid reactions between a high-*k* dielectric and silicon is to select a dielectric that has a large Gibbs free energy of formation to prevent reaction with Si [10]. Besides, oxygen diffusion coefficients in high-*k* materials must be low as they will also induce uncontrollable interfacial SiO_2 growth at high-temperature rapid thermal anneal (RTA) [11]. This low-*k* SiO_2 layer will compromise the total capacitance density of the gate stack and thus nullify the benefit of the high-*k* dielectric.

Film morphology. For the applications on the gate dielectrics, they must withstand a rapid thermal annealing at 1000 °C for at least 5 s [12]. This is a stringent condition for most high-*k* oxides because they usually have low crystalline temperature and can easily crystallize when subjected to RTA [1]. In particular, ZrO_2, TiO_2, and rare-earth oxides crystallize at much lower temperatures. The grain boundaries of crystallized gate dielectrics in thermal processes may serve as high leakage paths, making EOT scaling problematic and causing device failure [13]. On the other hand, it has also been found that particular phase engineering on high-*k* crystalline structures may sometimes have a positive effect on the enhancement of the dielectric constant of the film [4, 14, 15]. However, this enhancement on *k*-value due to the variations in crystalline structures and orientations may sometimes be hard to control or to reproduce, especially if the thickness of gate oxides is in the same dimension of the crystal grain size. So it is desired to select an alternative high-*k* gate dielectric material that remains amorphous throughout the necessary processing treatments.

Table 2.2 The comparison between some deposition technologies.

Method	ALD	MBE	CVD	Sputter	Evapor.	PLD
Thickness uniformity	Good	Fair	Good	Good	Fair	Fair
Film density	Good	Good	Good	Good	Poor	Good
Step coverage	Good	Poor	Varies	Poor	Poor	Poor
Interface quality	Good	Good	Varies	Poor	Good	Varies
Number of materials	Fair	Good	Poor	Good	Fair	Poor
Low-temp. deposition	Good	Good	Varies	Good	Good	Good
Deposition rate	Fair	Poor	Good	Good	Good	Good
Industrial applicability	Good	Fair	Good	Good	Good	Poor

Process compatibility. The film quality and properties are always correlated with the deposition method. The deposition process must be compatible with the CMOS processing, cost, and throughput. To obtain good electrical performance of the resulting devices, the grown dielectrics must have an excellent thickness uniformity, superior interfacial properties, and high thermal stability. The following film deposition methods have been proposed to deposit high-*k* films such as physical vapor deposition (PVD; e.g., sputtering and evaporation), pulsed laser deposition (PLD), chemical vapor deposition (CVD), atomic layer deposition (ALD), and molecular beam epitaxy (MBE). A general comparison between each deposition method is shown in Table 2.2. Among them, ALD is considered as the most promising one to deposit ultrathin high-*k* gate dielectrics due to many advantages, such as excellent thickness control, uniformity on large area, and good step coverage [16–18].

2.2.2 The Challenges of High-*k* Dielectrics

High-*k* dielectric gate oxide also faces several challenges because of the ionic nature of the chemical bonding. The major concerns are structural defects, mobility degradation, threshold voltage instability, and reliability.

2.2.2.1 Structural Defects

SiO_2 is an almost ideal insulating oxide, in that it has a low concentration of defects that give rise to states in the gap. The defects in SiO_2 are mainly due to dangling bond and low coordination. Dangling bond can be removed by rebonding the network especially at the Si/SiO_2 interface. However, the high-*k* oxides differ in that their bonding structure is ionic, and they have higher coordination number [19]. In fact, the most important contribution to the value of the dielectric constant in high-*k* materials comes from the dipole-active displacements of transition metal ions facilitated by their d-electrons [20]. The chemical bonding involving d-electrons leads to high vulnerability of the high-*k* dielectrics to the formation of structural defects during the deposition process. Therefore, the high-*k* gate dielectric has higher defect concentration than SiO_2. Most of the intrinsic defects in high-k oxides are oxygen vacancies, oxygen interstitials, or oxygen deficiency defects due to possible multiple valence of the metal [21]. Among them, the large amount of oxygen vacancies is the primary source of oxide traps. In addition, electrically active defects can also be introduced into the high-k dielectric during the gate electrode deposition or RTA process due to high diffusivity of various species in high-k materials [22]. These defects can be a source of fixed charges and electron traps, where the latter may affect not only device performance but also its reliability [23]. A more thorough understanding of defect phenomenon in high-k materials is, therefore, necessary. Much of the present work on high-k oxides is trying to reduce defect densities by processing control and annealing.

2.2.2.2 **Channel Mobility Degradation**

The degradation of carrier mobility in the channel is another major concern. The surface mobility is governed by various scattering mechanisms at the bulk silicon and at the dielectric/Si interface [24]. The major scattering mechanisms affecting the channel mobility at the SiO_2/Si interface are the Coulomb (μ_{Coul}), surface roughness (μ_{SR}), and phonon scattering (μ_{Ph}). According to Mathiessen's rule [19], the overall effective channel mobility (μ_{eff}) is given by

$$\frac{1}{\mu_{eff}} = \frac{1}{\mu_{Coul}} + \frac{1}{\mu_{SR}} + \frac{1}{\mu_{Ph}} \tag{2.1}$$

However, the channel mobility at the high-k/Si interface was reported to be greatly degraded [25, 26]. First, the surface roughness plays important roles in this degradation. The high-k/Si interface has higher degree of roughness because the metal–O and metal–Si generally have longer bond lengths that the Si–Si of the substrate. Moreover, high-k oxides have much higher oxide trap and interface trap densities than SiO_2. As a result, the Coulomb scattering would be more pronounced compared to the SiO_2 case. In addition, the soft optical phonons in the high-k metal oxide layer will also interact with the channel electrons and result in mobility degradation. All factors contributing to carrier mobility degradation in a high-k oxide layer are shown in Figure 2.3 [27]. The density of soft optical phonons is usually high in the high-k metal oxide due to the ionic bonds [28]. High-k dielectrics such as HfO_2 and ZrO_2 show high mobility degradation due to the soft optical phonons caused by the oscillation of oxygen ions [29]. It was reported that this soft phonon mechanism could be minimized either by using $HfSiO_4$ or by including SiO_2 interfacial layer to keep HfO_2 away from the channel [27, 30]. However, both methods increase the EOT. Finding a way to boost the channel mobility is, therefore, a big challenge when using high-k dielectric materials.

2.2.2.3 **Threshold Voltage Control**

Another key challenge with respect to the high-k gate dielectric system is threshold voltage (V_{th}) control. An asymmetric V_{th} shift (i.e., 0.3 V shift for n-MOSFETs and

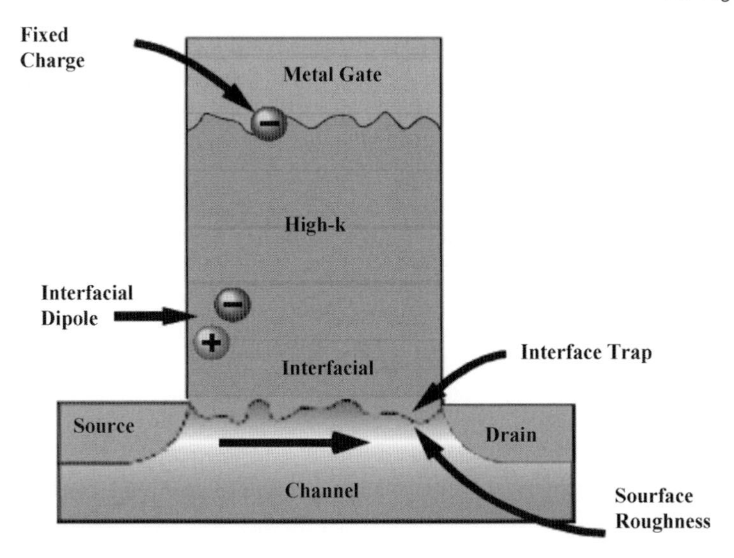

Figure 2.3 Factors contributing to carrier mobility degradation in a high-*k* oxide layer [27]. (For a color version of this figure, please see the color plate at the beginning of this book.)

0.9–1.0 V shift for p-MOSFETs) has been observed for all high-*k* materials when utilizing poly-Si gate electrodes [31]. Unlike SiO_2, high-*k* dielectric usually contains large amounts of fixed charge. The charge-trapping centers responsible for the fixed charge pose a serious issue for V_{th} control. But this is not the only reason. It has been found that Fermi-level pinning also plays an important role in V_{th} control in actual application of high-*k* materials [32]. Fermi-level pinning occurs at the poly-Si/high-*k* interface due to the defect formation through metal–Si bonding [33]. *Ab initio* [34] calculation showed that the interaction between metal and Si atoms could produce surface dipoles at the poly-Si/high-*k* interface that in effect modify the interface barrier height and then the flatband voltage (V_{fb}). The high-*k* dielectric material shifts the V_{fb} that changes V_{th} of the device. The Fermi-level pinning or interface dipole effect would result in a high threshold voltage. Diffusion of dopants (particularly boron) could be another source of the uncontrollable shift in threshold voltage [35]. Adding a relatively small amount of nitrogen to the high-*k* dielectric is expected to suppress the boron diffusion through the dielectric, as has been generally effective with current SiO_xN_y applications [36]. Moreover, replacing the poly-Si gate electrode by metal gate electrodes could be a possible solution to these issues. Metal electrode materials with work functions near the mid-gap may suffer less from this Fermi-level pinning effect [37]. Such a strategy will be discussed in the following section.

2.2.2.4 **Reliability**

The reliability of the gate insulator has always been a major concern throughout all CMOS generations. The high-*k* dielectrics tend to show two important general reliability trends: (1) the breakdown strength is lower for the high-*k* gate dielectrics versus SiO_2 while (2) the local electric field is larger [38]. Luckily, most of the models and concepts that had been developed for SiO_2 or SiON reliability could be maintained on high-*k* stacks. Similar to SiO_2, the high-*k* materials show some reliability phenomena including bias temperature instability (BTI), time-dependent dielectric breakdown (TDDB), and stress-induced leakage current (SILC) [39–41]. BTI was identified as one of the most limiting reliability issues in scaled CMOS technologies. The high-*k* dielectrics such as Hf-based dielectrics present serious instabilities for negative and positive bias, after negative bias temperature (NBT) and positive bias temperature (PBT) stresses [42]. NBT instability is caused by generation of new defects under the influence of the presence of holes at the high-*k*/Si interface, whereas PBT instability can be attributed to trapping in preexisting defects. Additional bulk traps in high-*k* dielectrics are created during positive constant voltage stress, leading to dielectric breakdown when a critical trap density is reached. The generated traps give rise to SILC, which has to be taken into account in the actual process. Further studies on the status and issues of high-*k* gate dielectric reliability can be seen in the review by Ribes *et al.* [43].

2.3 Metal Gates

Highly doped polycrystalline silicon (poly-Si) has been used as a MOSFET gate material for several decades due to its high compatibility with CMOS processing. However, as the CMOS technology is downscaling, poly-Si will encounter several inherent limitations. One of them is depletion of the poly-Si gate electrode when the gate stack is biased in inversion [44]. The depleted region (2–5 Å) is added to the dielectric thickness and the EOT is increased, which results in the additional series capacitance and reduced gate capacitance in the inversion regime. This reduction in the inversion capacitance of a poly-Si gated device can lower the drive currents significantly. This would be a serious setback to all efforts to scale down the dielectric thickness using high-*k* gate dielectrics. To reduce these problems, the active dopant density in the poly-Si gate material must be increased. This is nevertheless limited by the saturation active dopant density in p^+- or n^+-doped poly-Si [45]. In addition, increasing the dopant density in poly-Si worsens the dopant penetration problem [46].

The step from conventional SiO_2 to high-*k* materials might also induce troubles with respect to compatibility since some high-*k* materials have been shown to react with poly-Si, making it a less favorable material for continued use as a gate electrode [32, 47]. For example, the poly-Si/HfO_2 interface will lead to the so-called Fermi-level pinning effect that is believed to have been caused by Si–Hf interaction and leads to high threshold voltages of MOSFET devices [33]. Moreover, polysilicon gates are found to be thermodynamically unstable on many high-*k* materials. A viable way to circumvent the mentioned issues with poly-Si is to use a metal or metal compound as the gate electrode [48, 49], as shown in Figure 2.4. There is,

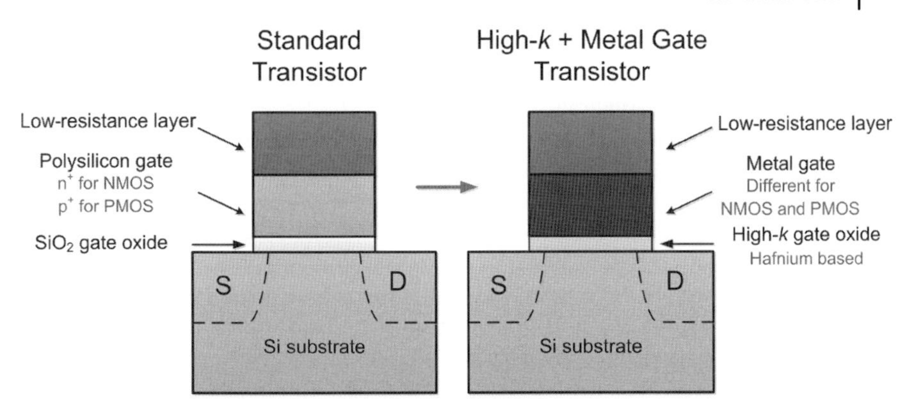

Figure 2.4 High-k + metal gate transistor. (For a color version of this figure, please see the color plate at the beginning of this book.)

therefore, immense interest in replacing conventional poly-Si gates with metal technology.

2.3.1 Basic Requirements for Metal Gates

The use of high-k/metal gate structure, however, will introduce several new challenges such as the choice of metal candidates, the development of a compatible process, and the thermal stability issue between metal and high-k gate dielectric. The metal candidate materials must be highly conductive and must have very high melting points in order to withstand the high thermal budgets commonly used in CMOS processing. The work function value of the introduced metal candidate will significantly influence threshold voltages of fabricated devices. High-performance CMOS technology generally requires two separate gate work function values for n-MOSFETs and p-MOSFETs. Providing appropriate work function values at the gate dielectric interface, one can achieve low and symmetric threshold voltages for n- and p-channel devices without high-dosage channel implantation that potentially leads to the threshold voltage nonuniformity and the carrier mobility degradation. The reported simulation results show that the required gate work function for n- and p-channel surface channel bulk devices are about 4.4–4.6 eV and 4.8–5.0 eV, respectively [50]. In addition, the chosen metal candidates should be able to possess good thermal stability with the underlying gate dielectric material. Consequently, the undesired interaction at metal/gate dielectric interface during the device fabrication process can be avoided, and the process induced work function (Φ_m) and/or the EOT variations can be suppressed.

2.3.2 Metal Gate Materials

Metal gate materials are generally classified into four major categories including pure metals [51], metallic alloys [52], metal nitrides [53], and metal silicides [54].

2.3.2.1 **Pure Metals**

The commonly used metals in IC industry are located in the group B of periodic table. The work function periodically ranges from conduction band (E_c) to valence band (E_v) or Si energy band relative to locations of elements [51]. Elements such as Ti, Y, Hf, Ta, and lanthanide series in column IIIB, IVB, and VB have the n-type work functions, while elements including Mo and W in group VIB have mid-gap work functions and metal elements including Re, Ru, Co, Rh, Ir, Ni, Pd, and Pt in group VIIB and VIIIB have p-type work functions. Since electronegativity is generally inversely related to free energy of oxide formation, and is proportional to work function, pure metals with lower work functions will exhibit problems with stability. N-type elements are chemically reactive, which would inevitably react with the gate dielectric. The formed layers can be either insulating or conducting, which will affect the EOT and the V_{fb} shift of the device. For example, metals such as Ti, Al, and Hf, which prefer reacting with the underlying dielectric, are considered inadequate for gate application in conventional process [55]. P-type elements are relatively inert and can sustain the high-temperature process. However, the challenge for them is selecting metals that exhibit good adhesion on the underlying dielectrics. The chemical inertia reflects difficulty in patterning and poorness in adhesion, which is affected by the formation of chemical bonding in the metal/dielectrics interface [56].

2.3.2.2 **Metallic Alloys**

The common metallic alloys are binary alloys consisting of an n-type metal and a p-type metal, such as Ta–Pt alloys, Ru–Ta alloys, and Hf–Mo alloys [57–59]. Metallic alloys can provide desired work function with good thermal stability through alloying low and high work function metals. The actual work function is determined by the atomic composition. Alloys with higher n-type metal content have a lower work function. On the contrary, alloys with higher p-type metal content have a higher work function. Besides the work function tuning, metallic alloys may also have thermodynamic stability due to strong metal–metal bonding. For example, Mo–Al alloys were found to be stable after rapid thermal annealing up to 900 °C [60]. Despite the advantages, the selection of the appropriate metallic alloy may be guided by the presence of stable single phases over a large composition range. It is expected that the presence of a single phase may ensure better work function uniformity. Misra *et al.* reported that binary alloys of TaRu with at. Ta% between 40 and 54% forms a single Ta_1Ru_1 phase, which has a work function of 4.3 eV with excellent thermal stability of EOT and V_{fb} [61].

2.3.2.3 **Metal Nitrides**

Metal nitrides are formed by the interaction of nitrogen 2s and 2p orbitals with the metal d orbitals [62], which are more chemically stable on dielectrics than pure metals. The metal nitride structure often consists of close packed metal atoms with nitrogen occupying octahedral interstitial positions. The commonly considered metal nitrides are tantalum nitride (TaN) [63], titanium nitride (TiN) [64], tungsten nitride (WN) [65], and hafnium nitride (HfN) [66]. Among them, TaN is considered promising for NMOS gate electrode applications due to the excellent diffusion barrier properties and high melting point. The work function can be adjusted by the nitrogen composition and the nitride phase, but the tunable range is not wide enough to be used for both n- and p-MOSFETs. Furthermore, the reported work function values of transition metal nitrides have varied widely in the literature, which may be caused by the various growth techniques or different characterization methods.

2.3.2.4 **Metal Silicides**

Metals react with silicon to form metal silicides. The commonly considered metal silicides are molybdenum silicide, tungsten silicide, nickel silicide, cobalt silicide, titanium silicide, platinum silicide, and hafnium silicide. Silicide gate is compatible with conventional CMOS process and its effective work function is adjusted to suit both n- and p-MOSFETs by the IIIA and VA impurity [67, 68]. Therefore, silicides are thought to be the mostly possible materials. Among various metal gates, full silicidation (FUSI) is highly compatible with conventional CMOS process flow and draws much attention. Recently, FUSI Ni–silicide and Co–silicide have been proposed for gate application [69, 70]. However, the compatibility with high-*k* dielectric is still a concern because unstable interface between polysilicon and high-*k* dielectric can also affect the interface between silicide and dielectric after silicidation [71].

2.3.3 Work Function

The threshold voltage of the MOSFET is directly controlled by the gate electrode through the work function difference between the gate and the channel. The work function can be varied by changing the composition of the films and definitely depends upon the surface quality. In other words, interface property is the main factor in determining the effective metal work function as the metal contact with the dielectric. The interface property is affected by the texture, composition, orientation, and interfacial layers. The approaches to improve the interface property are divided into two categories: one is structure modification and another is interface treatment.

Changes to the structural morphology of the thin metal films can be further classified into phase, orientation, and texture. The structural modification is usually relied on the deposition technique and postdeposition anneal. On the other hand, chemical reactions are used to modify the film texture after film deposition, such as silicidation, metal alloying, nitridation, and oxidation. Silicidation and metal alloying can be formed by deposition of a two-layer stack followed by a thermal process [72, 73]. As the textures are changed, the intrinsic physical and chemical properties differ from the as-deposited films. This implies the work function should be changed after the texture modification. The phase transition often contains the changes in microstructure in the films as well as the texture. For example, the Ni_2Si transits to NiSi and $NiSi_2$ at 200 and 700 °C, respectively [74, 75]. The work function changes from 4.38 to 4.8 eV.

The interface treatment can be performed either with the surface treatment before metal deposition or with the ion implantation after metal deposition. The surface treatment can be plasma treatment or reactive gas annealing. The interface treatment locally change the texture, bonding, and even defects (dangling bonds) at the interface between metal gate and dielectric as well as incorporation of extra impurities. The changes in interface and incorporation of impurities will upset a balance within the primary interface. Chemical thermodynamic equilibrium occurred and led to a new static state. This self-reaction will regulate the charge distribution at interface and produce a new effective work function of gate electrode. For example, the work function of Mo film charges from 5.0 to 4.4 eV after N implantation and postannealing because as Mo film is implanted with N^+ ions, the texture structure of Mo film is destroyed [76]. The structure incorporating N impurities is rearranged after annealing. Mo_2N phase forms and N also piles up at interface between Mo film and gate dielectric. The local texture is thoroughly changed by implantation and annealing [77].

2.3.4 Metal Gate Structures

It is not easy that single metal layer can satisfy all the above criteria. The most promising gate structure is stack of two metal layers. The bottom layer serves as a threshold control layer. It is a thin barrier layer that needs to have proper work function, good adhesion, and chemical stability on gate dielectric. The top layer serves as a conduction layer. It is a thick layer that needs to have low resistivity and low stress, to protect the bottom layer from ion implantation, and to passivate the bottom layer against oxidation and chemical solution attack during oxide passivation and cleaning process, respectively. Tungsten (W) is one of the common metals used for this purpose [78]. It should be noted that the selected metal gate stack need to be compatible with subsequent processing.

2.3.5 Metal Gate/High-*k* Integration

Another important consideration in the selection of metal gate electrodes is the work function dependence in contact with high-*k* dielectrics. In general, the work function of a metal at a dielectric interface is different from its value in vacuum [79]. This variation needs to be taken into account when designing transistor gate stacks with alternative high-*k* gate dielectrics. The dependence of the metal work function on several gate dielectrics has been investigated using both experimental data and interface dipole theory [80]. To obtain suitable effective work function, metal gates on high-*k* materials for p-MOSFETs (n-MOSFETs) should have higher (lower) work function to compensate the effects of Fermi-level pinning. These complementary properties are likely to have a significant impact on future CMOS gate stack process while reactive metals are likely to react with the underlying gate dielectric.

2.3.6 Process Integration

Introducing new materials into the complex and well-established IC fabrication is not a straightforward work. For successful implementation, the proposed metal gate technology should be compatible with standard Si-based CMOS processing with respect to process integration and contamination, reproducible, as well as thermally and chemically stable. In general, one may use a combination of two or more metals on a single Si substrate to achieve the work function requirements discussed earlier. However, it must be noted that the integration of multiple metals on a single wafer poses significant process integration challenges. It is also important that these materials lend themselves easily to conventional thin-film deposition (physical or chemical vapor deposition) and reactive ion etching (RIE) techniques. These requirements will ensure that advanced gate CMOS devices can still be fabricated using conventional tools. The new gate materials also need to have thermal expansion coefficients that closely match those of the single crystalline Si substrate to ensure that no significant thermal stresses are introduced in the film during rapid temperature changes, and they must have good adhesion with the underlying dielectric. Since the gate electrode is deposited on top of the ultrathin gate dielectric, care has to be taken and preferably a one-step deposition process is desirable.

2.4
Integration of High-*k* Gate Dielectrics with Alternative Channel Materials

As pointed out in the above sections, the mobility degradation in metal gate/high-*k*/Si MOSFETs is a critical challenge to continue current CMOS scaling trends. Thus, it is desirable to replacing traditional Si or strained Si in the channel with higher carrier mobility materials for "future" high-speed devices. Fortunately, the use of alternative materials such as Ge and III–V as the channel semiconductor in MOS structures may overcome the limitations of Si-based CMOS downscaling [81–83]. The characteristics of several potential channel materials at 300 K are listed in Table 2.3, including their typical carrier mobility, bandgap, transition type, lattice constant, and dielectric constant [81, 84].

It can be seen that bulk hole and electron mobilities of Ge are 4.2 and 2.6 times higher than those of Si, respectively. In particular, the bulk hole mobility of Ge is the highest of all Group IV and III–V semiconductor materials. This enables more efficient source injection and shorter CMOS gate delay due to the low effective electron and hole masses. As the best candidate for replacement of the conventional Si p-channel, Ge has received considerable attention since it is relatively compatible with Si-technology [81, 85]. However, there are a lot of difficulties in fabricating germanium n-MOSFETs.

On the other hand, III–Vs promise to increase electron mobility by 6–24 times, which makes them great candidates for high-speed, low-power n-channel transistors [82, 86]. For example, the lowly doped $In_{0.53}Ga_{0.47}$ As material has bulk mobility nine times higher than silicon, although its hole mobility is very low. Therefore, the combination of Ge p-MOSFETs with III–V n-MOSFETs to produce high-speed CMOS circuits is a very promising idea that could revolutionize the microelectronics industry [87, 88], as shown in Figure 2.5. However, it is important to know that there are some massive hurdles to overcome before this idea can be put

Table 2.3 The basic properties of alternative channel materials at 300 K [81, 84].

	Si	Ge	GaAs	InAs	InSb	$In_{0.53}Ga_{0.47}As$
Bandgap, E_g (eV)	1.12	0.66	1.42	0.36	0.17	0.74
Electron affinity, χ (eV)	4.0	4.05	4.07	4.9	4.59	4.5
Electron mobility, μ_e (cm^2/V s)	1500	3900	8500	33 000	80 000	12 000
Hole mobility, μ_h (cm^2/V s)	450	1900	400	460	1250	300
Transition type	Indirect	Indirect	Direct	Direct	Direct	Direct
Lattice constant, a (nm)	0.543	0.565	0.565	0.606	0.648	0.587
Dielectric constant, k	11.9	16.0	13.1	15.1	17.7	13.9

into reality. In the following section, we begin the discussion of Ge or III–V channel with the associated high-k dielectrics development.

2.4.1 High-k/Ge Interface

Ge has been considered as a replacement channel material of Si for future high-speed CMOS technology since bulk Ge has higher hole mobilities than Si. In fact, the first transistor was fabricated in Bell Laboratories in 1947 using bulk Ge as the semiconducting material [89]. Unfortunately, Ge has suffered from the lack of a high-quality native oxide for device applications [90]. The native GeO_2 is thermodynamically unstable and water-soluble that hinders the processing and application of Ge CMOS devices in the semiconductor industry. Furthermore, it is more

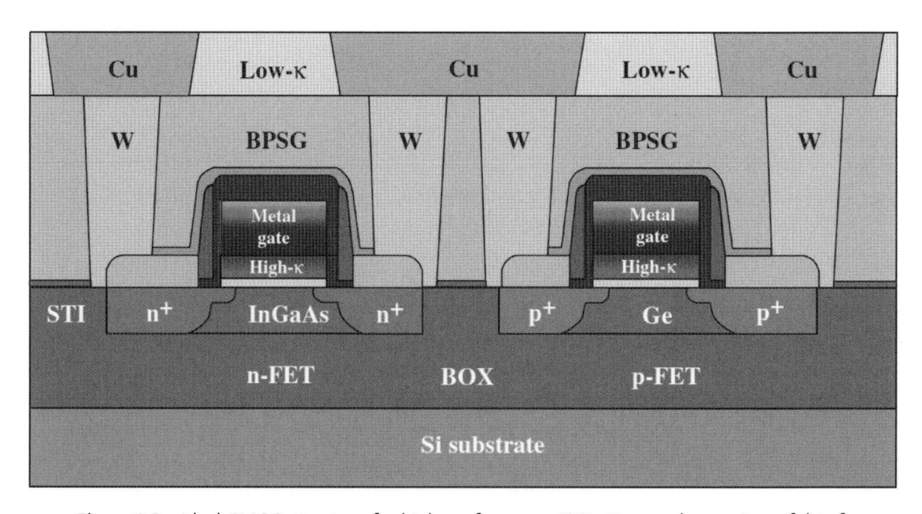

Figure 2.5 Ideal CMOS structure for high performance [87]. (For a color version of this figure, please see the color plate at the beginning of this book.)

difficult, compared to Si, to obtain stable oxides with low interface defect densities on Ge due to the chemical and electrical instabilities inherent to the GeO_2/Ge system. It is the reason that Si was used to replace Ge as a main channel material in CMOS technology in the past 40 years. As a result, the most challenging issue for potential future Ge MOSFETs is the development of effective Ge passivations and high-*k* gate dielectrics.

An effective passivation treatment should be chemically stable, protect the Ge channel surface from unwanted oxidation and contamination, and help minimize interface or surface-induced carrier recombination. In order to achieve these objectives, several methods have been proposed to passivate the clean Ge surfaces as discussed below. One of the possible ways to passivate Ge is to saturate dangling bonds on the surface with hydrogen. Due to the success of hydrogen passivation on conventional Si surfaces in the development of the semiconductor industry, hydrogen passivation on Ge substrates has been studied [91]. Clean oxygen-free H-terminated Ge surfaces can be obtained with HF solutions. Despite being successful for the fabrication of Si-based MOSFETs, this technique was found to be less effective on Ge. Bodlaki *et al.* has reported that the H-terminated Ge(111) surface is very unstable under ambient conditions [92]. In addition, rapid absorption of hydrocarbons has also been observed on H-terminated Ge(100) [93].

Another promising method is using sulfur to passivate a Ge surface before the deposition of high-*k* gate dielectrics. The prepared sulfur passivation layer can form a buffer layer to prevent the formation of interfacial GeO_x. Experimental [94] and theoretical [95] studies of low-pressure adsorption of sulfur on clean Ge(100) have shown that the sulfur is bridge bonded to the surface Ge atoms. Recently, Ge surface passivation with sulfur using aqueous ammonia sulfide $(NH_4)_2S$ was demonstrated [96]. The S-terminated Ge surface proves to be much more stable in air than H-terminated Ge surfaces [92].

Nitridation Ge surface has also been used to improve the reliability of Ge-based MOSFETs through the formation of strong Ge–N bonds [97, 98]. The nitrogen incorporation, with Ge to form a Ge oxynitride (GeON) or Ge nitride (Ge_3N_4) layer, provides higher thermal and chemical stability than native Ge oxides. High-quality thin GeON layer can be formed on germanium by nitridation of a thermally grown germanium oxide [99]. Ammonia (NH_3) is chosen as the nitriding agent due to its ability to incorporate more nitrogen into the oxynitride film than other nitriding species, such as nitrous oxide (N_2O) and nitric oxide (NO). With surface nitridation, the resulting film thickness can be scaled down to an EOT as thin as 1.9 nm with acceptable leakage [100]. It has also been found that the interface after thermal nitridation in ammonia is apparently better compared to that after a wet chemical treatment.

An alternative way to overcome the difficulty to passivate Ge/high-*k* interface is to shift the problem from a Ge surface to a Si surface. This requires, however, a careful deposition of Si on Ge in an epitaxial fashion, such that all Ge dangling bonds are tied up to a Si atom [101]. Using this method, the resulting Ge–Si interface is fully passivated with covalent bonds and should not introduce any states in the bandgap. Moreover, it is very important to control the Si layer thickness. A too thick Si layer will relax due to the lattice mismatch that will create numerous misfit dislocations at the interface and add to the EOT, while a too thin Si layer will be fully oxidized during the gate oxide deposition. In addition, annealing of Ge in $SiH_4 + N_2$ has been reported as the effective passivation method [102].

As high-*k* dielectrics are presently used in high-performance Si devices, another challenging issue for the Ge-based MOSFETs is the choice of high-*k* dielectrics. The above-mentioned basic requirements in high-*k*/Si gate stacks are also necessary in high-*k*/Ge gate stacks. Since the electron affinity of Ge is only 50 mV higher than that of Si (see Table 2.3), good high-*k* materials on a Si substrate would be expected to perform well on a Ge substrate from the band alignment point of view [103]. Furthermore, the volatility of Ge surface oxides or suboxides seems to make surface cleaning easier for high-*k* gate dielectric stack free of a performance limiting lower-*k* interfacial layer and thus overcome the EOT scaling barrier. However, high-*k* material selection on Ge is not straightforward due to interface reactions during processing.

In order to obtain a stable high-*k*/Ge interface, selection of an appropriate high-*k* material and interfacial layer is becoming very important. Various high-*k* candidates developed for Si or strained Si channels have been used in the fabrication of Ge-based MOSFETs. From the perspective of the thermodynamic stability in contact with Ge, binary metal oxides, such as Al_2O_3, ZrO_2, and HfO_2, have been the primary choices as high-*k* gate dielectrics [104]. In addition, several stable MGe_xO_y (where M stands for metal with high ion polarizability, such as Zr, Hf, and La) have also been proposed to potentially improve carrier mobility and interface stability [105, 106].

Unfortunately, one of the most promising high-*k* materials on a Si substrate, HfO_2, is unsuitable on a Ge substrate, since J_g is too large [105]. It is Ge interdiffusion into the HfO_2 layer that results in the poor electrical characteristics. First-principles calculations also suggest that Ge diffuses into interstitial and/or substitutional sites in the HfO_2 layer or forms a Ge–O complex, all of which generate electrically active levels in the HfO_2 layer [81]. As a result, a buffer layer, such as GeON [107], is necessary to stabilize the HfO_2/Ge interface. The optimum dielectric stack could be attained by rapid thermal nitridation (RTN) of Ge in NH_3 to form GeO_xN_y, followed by the capping dielectrics, such as HfO_2 [108]. Excellent electrical characteristics were obtained from MOS capacitors, with low leakage, good *C–V* characteristics, and reasonably low interface state densities, especially for p-MOS. High mobilities above $300\ cm^2/V\,s$ have been reported for Ge p-MOS.

On the other hand, experimental results show that sub-1.0-nm EOT ZrO_2 gate stacks on Ge without an intentionally prepared interfacial layer have promising *C–V* characteristics and lower leakage current densities than the HfO_2 counterparts [105, 109]. It has been reported that HfO_2 barely mixes with GeO_x, while ZrO_2 or Y_2O_3 can mix with GeO_2 [105, 110, 111]. These marked differences show that a high-quality interface of the high-*k*/Ge gate stack depends on the kind of high-*k* material. Four speculative models including intermixing model, GeO(g)-triggered metal germanide generation model, germanide oxidation model, and oxygen vacancy model have been proposed to explain the phenomenon [81].

In one word, high-*k* material selection is a critical issue for good electrical performance of Ge-based MOSFETs. Although much technological progress has been made on this subject, the properties of the high-*k*/Ge interface are still unsatisfactory. Greater understanding and much effort are needed for successful application of high-*k* dielectrics on Ge MOSFETs in the future.

2.4.2
High-k/III–V Interface

The experimental results for Ge-based n-MOSFETs so far have been far below the expected theoretical predictions. III–V compound semiconductors appear to be very attractive candidates as channel materials for high-performance n-MOSFETs due to their high electron mobilities [82, 112, 113]. Furthermore, bandgap engineering and direct bandgaps in III–V, not available in Si- and Ge-based systems, has provided great opportunities of novel device architectures. However, III–V materials have many significant and fundamental issues, which may prove to be severe bottlenecks to their implementation. The native oxides of III–V materials are present in a variety of bonding configurations and are thought to impact the interface state density significantly, resulting in poor device performance [114]. Similar to Ge channel material, a key restriction to the widespread use of III–V materials is the lack of an effective, stable, and natural insulator for III–V substrates, such as that available for the SiO_2/Si materials system. Extrinsic defects, such as surface antisite defects and high interface state density, are likely to cause inefficient Fermi-level response or Fermi-level pinning at the III–V semiconductor interface [115, 116].

As a result, finding an elegant surface passivation scheme and a high-quality gate insulator to form an interface with a sufficiently low density of interface states remains a major challenge for III–V compounds. In order to achieve high-performance III–V MOSFETs, several stringent criteria are required.

1) A low interfacial density of states (D_{it}) in the range of 10^{11} cm^{-2} eV^{-1};
2) Low electrical leakage current densities (J_g) of 10^{-8} A/cm^2 at a gate voltage of (V_{fb}) + 1 V;
3) Selected high-*k* dielectrics have a high dielectric constant and sufficient band offsets $\Delta E_C > 1$ eV to act as a barrier for both electrons and holes;
4) Thermodynamic stability at temperatures higher than 800 °C;
5) Scalability of high-*k* dielectrics down to 1 nm or less.

The Fermi-level pinning has been shown to occur at the III–V surface due to submonolayer oxygen coverage [117]. There have been many attempts in the past two decades to provide high-quality high-*k*/III–V interface and to prevent Fermi-level pinning issues [82, 118, 119]. The interfacial characteristics have been studied using high-resolution X-ray photoelectron spectroscopy, high-resolution transmission electron microscopy, capacitance–voltage, and leakage current density–electrical field measurements.

Numerous surface pretreatment methods have been used to passivate the III–V semiconductor surface before dielectric deposition. Early studies discovered that various sulfides could passivate the surface of III–V materials and improve the electronic properties [120]. Sulfur is less chemically reactive, although it has the same electron number in its outer shell as oxygen. A few monolayers of sulfur can passivate a pristine GaAs surface under certain conditions and prevent GaAs from oxidation to form arsenic oxides. It was reported that a class of sulfide including $(NH_4)_2S$, $Na_2S{\cdot}9H_2O$, and H_2S is able to passivate the GaAs surface and provide excellent device properties [120–122]. Another effective way is the use of an ultrathin amorphous or crystalline Si layer as an interfacial control layer (ICL) [123]. The Si ICL acts as a barrier layer preventing the reaction between oxygen and the III–V substrates and establishing more stable SiO_x bonds at the interface with high-*k* dielectrics. The Ga and As bonds with oxygen associated with the interface states are replaced by the Ga–Si and As–Si bonds [123, 124], thus drastically reducing interface state density from $\sim 5 \times 10^{12}$ to 2×10^{11} cm^{-2} eV^{-1} [125]. As a result, the Si ICL technique has been successfully implemented to enhancement mode GaAs and InGaAs MOSFETs, showing the most promising device performance in scaled devices [126, 127]. However, there still exist significant inherent issues with regard to implementation of Si ICL including (1) increase in EOT due to low-*k* fully or partially oxidized Si layer and (2) potential high-temperature Si interdiffusion sourced by a nonoxidized portion of Si interfacial control layer. Thus, several other surface pretreatment methods are at present under investigation on the III–V surface including exposure to NH_3 [128] and PH_3 [129] plasma.

In addition to the surface pretreatment, high-*k* dielectric materials have also been applied to III–V-based MOSFETs. Passlack and Hong discovered that amorphous $Ga_2O_3(Gd_2O_3)$ [GGO] dielectric film grown on a GaAs surface by electron-beam evaporation from single-crystal $Ga_5Gd_3O_{12}$ has the unpinned Fermi level in the interface between high-*k* and compound semiconductors [130, 131]. It should be noted that the sample was placed in the MBE chamber to remove the native oxide on the GaAs surface before the dielectric deposition. The produced MOSFETs with low leakage currents of 10^{-9}–10^{-8} A/cm^2 and low D_{it}'s in the range of $(4–9) \times 10^{11}$ cm^{-2} eV^{-1} for GGO on InGaAs were also demonstrated [132, 133].

On the other hand, atomic layer deposited high-*k* dielectrics including Al_2O_3, HfO_2, Y_2O_3, and HfAlO have been shown to improve the interface passivation of GaAs and InGaAs surfaces in recent years. Ye *et al.* demonstrated the first depletion-mode MOSFET work utilizing ALD-Al_2O_3 directly on GaAs(100) [134, 135], as shown in Figure 2.6. A sharp transition from crystalline III–V materials to amorphous Al_2O_3 was observed using HR-TEM, indicating the native oxides have been removed during ALD process. A subsequent more detailed study on inversion-mode surface-channel $In_xGa_{1-x}As$ MOSFETs with In concentrations of 20, 53, 65, and 75% integrated with ALD Al_2O_3, HfO_2, and HfAlO were done [82]. Moreover, the investigations of the interfacial characteristics using high-resolution synchrotron X-ray photoelectron spectroscopy have revealed that a self-cleaning mechanism exists to remove native oxides by ALD chemistry [136]. In the case of ALD–Al_2O_3 on $In_{0.53}Ga_{0.47}As$ substrate using trimethylaluminum TMA as a metal precursor[137], the reduction of native oxides was also found by HR-XPS. A more effective interfacial self-cleaning effect was subsequently observed from the ALD of HfO_2 on $In_{0.15}Ga_{0.75}As$ using tetrakis (ethylmethylamino) hafnium $[Hf(NCH_3C_2H_5)_4]$ as a metal precursor [138]. Recently, similar self-cleaning effects on $In_{0.2}Ga_{0.8}As$ [139], InAs [140], and InSb [141]

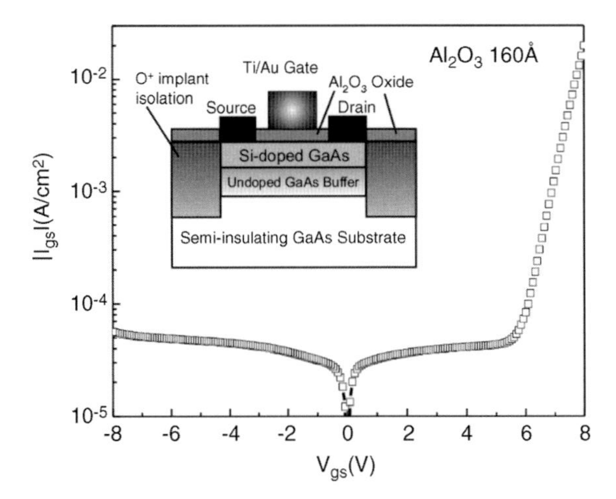

Figure 2.6 The first depletion-mode MOSFET work utilizing ALD-Al_2O_3 directly on GaAs (100) [134].

substrates were found using TMA or $Hf[N(CH_3)(C_2H_5)]_4$ as the metal precursor. The *in situ* "cleaning out" of native oxides on III–V compound semiconductors during ALD is beneficial in eliminating the interfacial native oxide layer between high-*k* dielectric and compound semiconductor. The abrupt high-*k*/semiconductor interface ensures the relaxation of the Fermi-level pinning phenomenon and improves the scalability of III–V compound semiconductor devices.

Successful integration of high-*k* dielectrics with III–V channel materials for future high-performance low-power logic applications calls for much improvement of the high-*k*/III–V interface quality. Although a lot of encouraging results have been obtained for III–V-based MOSFETs, the stringent demands in terms of electrical performance and oxide thickness scaling needed for highly scaled CMOS devices are not yet fully met. Further efforts are still necessary in understanding how to control the interfacial chemistry with high-*k* dielectrics and the subsequent effect on important electrical phenomenon such as Fermi-level pinning, frequency dispersion, and the nature of the associated defect states.

2.5 Summary

It is becoming more and more clear that CMOS scaling can be realized mainly by the introduction of new materials and/or innovative device architecture. Many promising high-*k* dielectrics have been proposed and investigated in the past decade. We have presented the basic requirements and several key challenges for high-*k* dielectrics. Metal gate electrodes have been shown to be more compatible with high-*k* gate dielectric materials than with poly-Si electrodes. The material requirements for metal gate CMOS technology and the challenges involved in the integration of metal gate with high-*k* dielectrics are also provided. Furthermore, we give an overview of the key challenges with respect to the integration of Ge and III–V semiconductors as high-mobility channels with high-*k* gate dielectrics. Although much progress has been made in fabricating novel gate dielectrics, metal gates, and new channel materials, there is room for development and many issues need to be investigated for better understanding.

References

1 Wilk, G.D., Wallace, R.M., and Anthony, J.M. (2001) High-κ gate dielectrics: current status and materials properties considerations. *J. Appl. Phys.*, **89**, 5243.

2 Locquet, J.-P., Marchiori, C., Sousa, M., Fompeyrine, J., and Seo, J.W. (2006) High-*K* dielectrics for the gate stack. *J. Appl. Phys.*, **100**, 051610.

3 Guha, S., Gusev, E., Copel, M., Ragnarsson, L., and Buchanan, D.A. (2002) Compatibility challenges for High-κ materials integration into CMOS technology. *MRS Bull.*, **27**, 226.

4 Robertson, J. (2004) High dielectric constant oxides. *Eur. Phys. J. Appl. Phys.*, **28**, 265.

5 Houssa, M. (2004) *High K Gate Dielectrics*, IOP, Berkshire.

6 Huff, H.R. and Gilmer, D.C. (2005) *High Dielectric Constant Materials: VLSI MOSFET Applications*, Springer.

7 Plummer, J.D. and Griffin, P.B. (2001) Material and process limits in silicon VLSI technology. *Proc. IEEE*, **89**, 240.

8 Robertson, J. (2000) Band offsets of wide-band-gap oxides and implications for future electronic devices. *J. Vac. Sci. Technol. B*, **18**, 1785.

9 Hubbard, K.J. and Schlom, D.G. (1996) Thermodynamic stability of binary oxides in contact with silicon. *J. Mater. Res.*, **11**, 2757.

10 Houssa, M., Pantisano, L., Ragnarsson, L.-Å., Degraeve, R., Schram, T., Pourtois, G., De Gendt, S., Groeseneken, G., and Heyns, M.M. (2006) Electrical properties of high-κ gate dielectrics: challenges, current issues, and possible solutions. *Mater. Sci. Eng. R Rep.*, **51**, 37.

11 de Almeida, R.M.C. and Baumvol, I.J.R. (2003) Reaction-diffusion in high-*k* dielectrics on Si. *Surf. Sci. Rep.*, **49**, 1.

12 Claflin, B. and Lucovsky, G. (1998) Interface formation and thermal stability of advanced metal gate and ultrathin gate dielectric layers. *J. Vac. Sci. Technol. B*, **16**, 2154.

13 Darmawan, P., Lee, P.S., Setiawan, Y., Lai, J.C., and Yang, P. (2007) Thermal stability of rare-earth based ultrathin Lu_2O_3 for high-*k* dielectric. *J. Vac. Sci. Technol. B*, **25**, 1203.

14 Zhao, X. and Vanderbilt, D. (2002) First-principles study of structural, vibrational, and lattice dielectric properties of hafnium oxide. *Phys. Rev. B*, **65**, 233106.

15 Rignanese, G.M., Gionze, X., Jun, G., Cho, K.J., and Pasquarello, A. (2004) First-principles investigation of high-κ dielectrics: comparison between the silicates and oxides of hafnium and zirconium. *Phys. Rev. B*, **69**, 184301.

16 Puurunen, R.L. (2005) Surface chemistry of atomic layer deposition: a case study for the trimethylaluminum/water process. *J. Appl. Phys.*, **97**, 121301.

17 Leskela, M. and Ritala, M. (2002) Atomic layer deposition (ALD): from precursors to thin film structures. *Thin Solid Films*, **409**, 138.

18 Niinisto, L., Paivasaari, J., Niinisto, J., Putkonen, M., and Nieminen, M. (2004) Advanced electronic and optoelectronic materials by atomic layer deposition: an overview with special emphasis on recent progress in processing of high-*k* dielectrics and other oxide materials. *Phys. Status Solidi A*, **201**, 1443.

19 Xiong, K., Robertson, J., and Clark, S.J. (2006) Defect energy states in high-*K* gate oxides. *Phys. Status Solidi B*, **243**, 2071.

20 Bersuker, G., Sim, J.H., Young, C.D., Choi, R., Zeitzoff, P.M., Brown, G.A., Lee, B.H., and Murto, R.W. (2004) Effects of pre-existing defects on reliability assessment of High-*K* gate dielectrics. *Microelectron. Reliab.*, **44**, 1509.

21 Foster, A.S., Sulimov, V.B., LopezGejo, F., Shluger, A.L., and Nieminen, R.N. (2001) Structure and electrical levels of point defects in monoclinic zirconia. *Phys. Rev. B*, **64**, 224108.

22 Gopireddy, D. and Takoudis, C.G. (2008) Diffusion-reaction modeling of silicon oxide interlayer growth during thermal annealing of high dielectric constant materials on silicon. *Phys. Rev. B*, **77**, 205304.

23 Foster, A.S., Lopez Gejo, F., Shluger, A.L., and Nieminen, R.M. (2002) Vacancy and interstitial defects in hafnia. *Phys. Rev. B*, **65**, 174117.

24 Cheng, Y.C. and Sullivan, E.A. (1974) Effect of coulomb scattering on silicon surface mobility. *J. Appl. Phys.*, **45**, 187.

25 Datta, S., Dewey, G., Doczy, M., Doyle, B.S., Jin, B., Kavalieros, J., Kotlyar, R., Metz, M., Zelick, N., and Chau, R. (2003) High mobility Si/SiGe strained channel MOS transistors with HfO_2/TiN gate stack. IEEE Int. Electron. Dev. Meeting Technical Digest, 653.

26 Zhu, J., Han, J.P., and Ma, T.P. (2004) Mobility measurement and degradation mechanisms of MOSFETs made with ultrathin high-*k* dielectrics. *IEEE Trans. Electron. Dev.*, **51**, 98.

27 Chowdhury, M.H., Mannan, M.A., and Mahmood, S.A. (2010) High-*k* dielectrics for submicron MOSFET. *Int. J. Emerging Technol. Sci. Eng.*, **2**, 1.

28 Wong, H. and Iwai, H. (2006) On the scaling issues and high-*k* replacement of ultrathin gate dielectrics for nanoscale MOS transistors. *Microelectron. Eng.*, **83**, 1867.

29 Fischetti, M., Cartier, E., and Neumayer, D. (2001) Effective electron mobility in Si inversion layers in metal-oxide-semiconductor systems with a high-κ insulator: the role of remote phonon scattering. *J. Appl. Phys.*, **90**, 4587.

30 Oshiyama, A. (1998) Hole-injection-induced structural transformation of oxygen vacancy in alpha-quartz. *Jpn J. Appl. Phys. 2*, **37**, L232.

31 Mutro, R.W., Gardner, M.I., Brown, G.A., Zeitzoff, P.M., and Huff, H.R. (2003) Challenges in gate stack engineering. *Solid State Technol.*, **46**, 43.

32 Hobbs, C., Fonseca, L., Dhandapani, V., Samavedam, S., Taylor, B., Grant, J., Dip, L., Triyoso, D., Hegde, R., Gilmer, D., Garcia, R., Roan, D., Lovejoy, L., Rai, R., Herbert, L., Tseng, H., White, B., and Tobin, P. (2003). Fermi level pinning at the polySi/metal oxide interface. Symposium on VLSI Technology Digest of Technical Papers, 9.

33 Ryu, B. and Chang, K.J. (2010) Defects responsible for the Fermi level pinning in n^+ poly-Si/HfO_2 gate stack. *Appl. Phys. Lett.*, **97**, 242910.

34 Soler, J.M., Artacho, E., Gale, J.D., Garcia, A., Junquera, J., Ordejo'n, P., and Sa'nchez-Portal, D. (2002) The SIESTA method for *ab initio* order-N materials simulation. *J. Phys. Condens. Matter*, **14**, 2745.

35 Tse, K. and Robertson, J. (2007) Defects and their passivation in high *K* gate oxides. *Microelectron. Eng.*, **84**, 663.

36 Quevedo-Lopez, M., El-Bouanani, M., Kim, M.J., Gnade, B.E., Wallace, R.M., Visokay, M.R., LiFatou, A., Chambers, J.J., and Colombo, L. (2003) Effect of N incorporation on boron penetration from p^+ polycrystalline-Si through $HfSi_xO_y$ films. *Appl. Phys. Lett.*, **82**, 4669.

37 Yeo, Y.-C., Ranade, P., King, T.-J., and Hu, C. (2002) Effects of high-κ gate dielectric materials on metal and silicon gate work functions. *IEEE Electron. Device Lett.*, **23**, 342.

38 Kim, Y.H., Onishi, K., Kang, C.S., Cho, H.-J., Nieh, R., Gopalan, S., Choi, R., Han, J., Krishnan, S., and Lee, J.C. (2002) Area dependence of

TDDB characteristics for HfO_2 gate dielectrics. *IEEE Electron. Device Lett.*, **23**, 594.

39 Liu, C.H., Lee, M.T., Chen, J.C.Y.L., Schruefer, K., Brighten, J., Rovedo, N., Hook, T., Khare, M., Shih, F.H., Wann, C., Chen, T.C., and Ning, T.H. (2001). Mechanism and process dependence of negative bias temperature instability (NBTI) for pMOSFETs with ultrathin gate dielectrics. IEEE Int. Electron. Dev. Meeting Technical Digest, 861.

40 Degraeve, R., Cartier, E., Kauerauf, T., Carter, R., Pantisano, L., Kerber, A., and Groeseneken, G. (2002) On the electrical characterization of high-κ dielectrics. *MRS Bull.*, **27**, 222.

41 Torii, K., Shirashi, K., Miyazaki, S., Yamabe, K., Boero, M., Chikyow, T., Yamada, K., Kitajima, H., and Arikado, T. (2004) Physical model of BTI, TDDB and SILC in HfO_2-based high-k gate dielectrics. IEEE Int. Electron. Dev. Meeting Technical Digest, 129.

42 Degraeve, R., Aoulaiche, M., Kaczer, B., Roussel, Ph., Kauerauf, T., Sahhaf, S., and Groeseneken, G. (2008) Review of reliability issues in high-*k*/metal gate stacks. The 15th International Symposium on the Physical and Failure Analysis of Integrated Circuits, 1.

43 Ribes, G., Mitard, J., Denais, M., Bruyere, S., Monsieur, F., Parthasarathy, C., Vincent, E., and Ghibaudo, G. (2005) Review on high-*k* dielectrics reliability issues. *IEEE T. Dev. Mater. Reliab.*, **5**, 5.

44 Watanabe, H. (2005) Statistics of grain boundaries in gate poly-Si. *IEEE T. Electron. Dev.*, **52**, 2265.

45 Josse, E. and Skotnichi, T. (1999) Polysilicon gate with depletion-or-metallic gate with buried channel: what evil worse? IEEE Int. Electron. Dev. Meeting Technical Digest, 661.

46 Sun, J.Y.-C., Wong, C., Taur, Y., and Hsu, C.-H. (1989) Study of boron penetration through thin oxide with p^+ polysilicon gate. Symposium on VLSI Technology Digest of Technical Papers, p. 17.

47 Ikenaga, E., Hirosawa, I., Kitano, A., Takata, Y., Muto, A., Maeda, T., Torii, K., Kitajima, H., Arikado, T., Takeuchi, A., Awaji, M., Tamasaku, K., Ishikawa, T., Komiya, S., and Kobayashi, K. (2005). Interface reaction of poly-Si/high-*k* insulator systems studied by hard X-ray photoemission spectroscopy. *J. Electron. Spectrosc.*, **144**, 491.

48 Yeo, Y.C. (2004) Metal gate technology for nanoscale transistors-material selection and process integration issues. *Thin Solid Films*, **462**, 34.

49 Maitra, K. and Misra, V. (2003) A simulation study to evaluate the feasibility of midgap workfunction metal gates in 25nm bulk CMOS. *IEEE Electron. Device Lett.*, **24**, 707.

50 Chang, L., Tang, S., King, T.-J., Bokor, J., and Hu, C. (2000) Gate length scaling and threshold voltage control double-gate MOSFETS. IEEE Int. Electron. Dev. Meeting Technical Digest, 719.

51 Michaelson, H.B. (1977) The work function of the elements and its periodicity. *J. Appl. Phys.*, **48**, 4729.

52 Lee, J., Zhong, H., Suh, Y.-S., Heuss, G., Gurganus, J., Chen, B., and Misra, V. (2002) Tunable work function dual metal gate technology for bulk and non-bulk CMOS. IEEE Int. Electron. Dev. Meeting Technical Digest, 359.

53 Westlinder, J., Schram, T., Pantisano, L., Cartier, E., Kerber, A., Lujan, G.S., Olsson, J., and Groeseneken, G. (2003) On the thermal stability of atomic layer deposited TiN as gate electrode in MOS devices. *IEEE Electron. Device Lett.*, **24**, 550.

54 Xuan, P. and Bokor, J. (2003) Investigation of NiSi and TiSi as CMOS gate materials. *IEEE Electron. Device Lett.*, **24**, 634.

55 Misra, V., Heuss, G.P., and Zhong, H. (2001) Use of metal-oxide-semiconductor capacitors to detect interactions of Hf and Zr gate electrodes with SiO_2 and ZrO_2. *Appl. Phys. Lett.*, **78**, 4166.

56 Murarka, S.P. (1993) *Metallization: Theory and Practice for VLSI and ULSI*, Butterworth-Heinemann, Boston.

57 Tsui, B.-Y. and Huang, C.-F. (2003) Wide range work function modulation of

binary alloys for MOSFET application. *IEEE Electron. Device Lett.*, **24**, 153.

58 Zhong, H., Hong, S.-N., Suh, Y.-S., Lazar, H., Heuss, G., and Misra, V. (2001) Properties of Ru-Ta alloys as gate electrodes for NMOS and PMOS silicon devices. IEEE Int. Electron. Dev. Meeting Technical Digest, 467.

59 Li, T.-L., Hu, C.-H., Ho, W.-L., Wang, H.C.-H., and Chang, C.-Y. (2005) Continuous and precise work function adjustment integratable dual metal gate CMOS technology using Hf-Mo binary alloys. *IEEE Trans. Electron. Dev.*, **52**, 1172.

60 Huang, T.S., Peng, J.G., and Lin, C.C. (1993) Thermal stability of Mo-Al Schottky metallizations on n-GaAs. *J. Vac. Sci. Technol. B*, **11**, 756.

61 Misra, V., Zhong, H., and Lazar, H. (2002) Electrical properties of Ru-based alloy gate electrodes for dual metal gate Si-CMOS. *IEEE Electron. Device Lett.*, **23**, 354.

62 Takeyama, M., Noya, A., Sase, T., and Ohta, A. (1996) Properties of TaN_x films as diffusion barriers in the thermally stable Cu/Si contact systems. *J. Vac. Sci. Technol.*, **14**, 674.

63 Pan, J., Woo, C., Yang, C.-Y., Bhandary, U., Gugilla, S., Krishna, N., Chung, H., Hui, A., Yu, B., Xiang, Q., and Lin, M.-R. (2003). Replacement metal-gate NMOSFETs with ALD TaN/EP-Cu, PVD Ta, and PVD TaN electrode. *IEEE Electron. Device Lett.*, **24**, 304.

64 Wakabayashi, H., Saito, Y., Takeuchi, K., Mogami, T., and Kunio, T. (2001) A dual-metal gate CMOS technology using nitrogen-concentration-controlled TiN_x film. *IEEE Trans. Electron. Dev.*, **48**, 2363.

65 Alshareef, H.N., Quevedo-Lopez, M., Wen, H.C., Huffman, C., El-Bouanani, M., and Gnade, B.E. (2008) Impact of carbon incorporation on the effective work function of WN and TaN metal gate electrodes. *Electrochem. Solid State Lett.*, **11**, H182.

66 Yu, H.Y., Li, M.F., and Kwong, D.L. (2004) Thermally robust HfN metal as a promising gate electrode for advanced MOS device application. *IEEE Trans. Electron. Dev.*, **51**, 609.

67 Sim, J.H., Wen, H.C., Lu, J.P., and Kwong, D.L. (2003) Dual work function metal gates using full nickel silicidation of doped poly-Si. *IEEE Electron. Device Lett.*, **24**, 631.

68 Liu, J. and Kwong, D.L. (2006) Investigation of work function adjustments by electric dipole formation at the gate/oxide interface in preimplanted NiSi fully silicided metal gates. *Appl. Phys. Lett.*, **88**, 192111.

69 Qin, M., Poon, V.M.C., and Ho, S.C.H. (2001) Investigation of polycrystalline nickel silicide films as a gate material. *J. Electrochem. Soc.*, **148**, G271.

70 Tavel, B., Skotnicki, T., Pares, G., Carrière, N., Rivoire, M., Leverd, F., Julien, C., Torres, J., and Pantel, R. (2001) Totally silicided ($CoSi_2$) polysilicon: a novel approach to very low-resistive gate without metal CMP nor etching. IEEE Int. Electron. Dev. Meeting Technical Digest, 825.

71 Joo, M.S., Park, C.S., Cho, B.J., Balasubramanian, N., and Kwong, D.-L. (2006) Interface configuration and Fermi-level pinning of fully silicided gate and high-K dielectric stack. *J. Vac. Sci. Technol. B*, **24**, 1341.

72 Larrieu, G., Dubois, E., Wallart, X., Baie, X., and Katcki, J. (2003) Formation of platinum-based silicide contacts: kinetics, stoichiometry, and current drive capabilities. *J. Appl. Phys.*, **94**, 7801.

73 Donaton, R.A., Jin, S., Bender, H., Zagrebnov, M., Baert, K., Maex, K., Vantomme, A., and Langouche, G. (1997) Formation of ultra-thin PtSi layers with a 2-step silicidation process. *Microelectron. Eng.*, **37–38**, 507.

74 Kittl, J.A., Lauwers, A., Hoffmann, T., Veloso, A., Kubicek, S., Niwa, M., van Dal, M.J.H., Pawlak, M.A., Demeurisse, C., Vrancken, C., Brijs, B., Absil, P., and Biesemans, S. (2006). Linewidth effect and phase control in Ni fully silicided gates. *IEEE Electron. Device Lett.*, **27**, 647.

75 Tinani, M., Mueller, A., Gao, Y., Irene, E.A., Hu, Y.Z., and Tay, S.P. (2001) *In situ* real-time studies of nickel silicide phase formation. *J. Vac. Sci. Technol. B*, **19**, 376.

76 Ranade, P., Choi, Y.-K., Ha, D., Agarwal, A., Ameen, M., and King, T.-J. (2002) Tunable work function molybdenum gate technology for FDSOI-CMOS. IEEE Int. Electron. Dev. Meeting Technical Digest, 363.

77 Juang, M.H., Lin, T.Y., and Jang, S.L. (2006) Formation of Mo gate electrode with adjustable work function on thin Ta_2O_5 high-*k* dielectric films. *Solid State Electron.*, **50**, 114.

78 Hu, J.C., Yang, H., Kraft, R., Rotondaro, A.L.P., Hattangady, S., Lee, W.W., Chapman, R.A., Chao, C.-P., Chatterjee, A., Hanratty, M., Rodder, M., and Chen, I.-C. (1997). Feasibility of using W/TiN as metal gate for conventional 0.13 μm CMOS technology and beyond. IEEE Int. Electron. Dev. Meeting Technical Digest, 825.

79 Yeo, Y.C., Ranade, P., King, T.-J., and Hu, C. (2002) Effects of high-*k* gate dielectric materials on metal and silicon gate workfunctions. *IEEE Electron. Device Lett.*, **23**, 342.

80 Yeo, Y.C., Ranade, P., Lu, Q., Lin, R., King, T.-J., and Hu, C. (2001) Effect of high-*K* dielectrics on the workfunctions of metal and silicon gate. Symposium on VLSI Technology Digest of Technical Papers, p. 49.

81 Kamata, Y. (2008) High-*k*/Ge MSOFETs for future nanoelectronics. *Mater. Today*, **11**, 30.

82 Ye, P.D. (2008) Main determinants for III–V metal-oxide-semiconductor field-effect transistors (invited). *J. Vac. Sci. Technol. A*, **26**, 697.

83 Heyns, M. and Tsai, W. (2009) Ultimate scaling of CMOS logic devices with Ge and III–V materials. *MRS Bull.*, **34**, 485.

84 Sze, S.M. (1981) *Physics of Semiconductor Devices*, 2nd edn, John Wiley & Sons, Inc., New York.

85 Shang, H., Frank, M.M., Gusev, E.P., Chu, J.O., Bedell, S.W., Guarini, K.W., and Ieong, M. (2006) Germanium channel MOSFETs: opportunities and challenges. *IBM J. Res. Dev.*, **50**, 377.

86 Hong, M., Kwo, J.R., Tsai, P., Chang, Y., Huang, M.-L., Chen, C., and Lin, T. (2007) III–V metal-oxide-semiconductor field-effect transistors with high kappa dielectrics. *Jpn J. Appl. Phys. 1*, **46**, 3167.

87 Sadana, D.K. (4 December 2005) IBM, III–V Substrate Engineering.

88 Takagi, S., Irisawa, T., Tezuka, T., Numata, T., Nakaharai, S., Hirashita, N., Moriyama, Y., Usuda, K., Toyoda, E., Dissanayake, S., Shichijo, M., Nakane, R., Sugahara, S., Takenaka, M., and Sugiyama, N. (2008). Carrier-transport-enhanced channel CMOS for improved power consumption and performance. *IEEE Trans. Electron. Dev.*, **55**, 21.

89 Bardeen, J. and Brattain, W.H. (1948) The transistor: a semiconductor triode. *Phys. Rev.*, **74**, 230.

90 Jackson, T.N., Ransom, C.M., and DeGelormo, J.F. (1991) Gate-self-aligned p-channel germanium MISFETs. *IEEE Electron. Dev. Lett.*, **12**, 605.

91 Deegan, T. and Hughes, G. (1998) An X-ray photoelectron spectroscopy study of the HF etching of native oxides on Ge(111) and Ge(100) surfaces. *Appl. Surf. Sci.*, **123**, 66.

92 Bodlaki, D., Yamamoto, H., Waldeck, D.H., and Borguet, E. (2003) Ambient stability of chemically passivated germanium interfaces. *Surf. Sci.*, **543**, 63.

93 Rivillon, S., Chabal, Y.J., Amy, F., and Kahn, A. (2005) Hydrogen passivation of germanium (100) surface using wet chemical preparation. *Appl. Phys. Lett.*, **87**, 253101.

94 Weser, T., Bogen, A., Konrad, B., Schnell, R.D., Schug, C.A., Moritz, W., and Steinmann, W. (1988) Chemisorption of sulfur on Ge(100). *Surf. Sci.*, **201**, 245.

95 Kruger, P. and Pollmann, J. (1990) 1st-principles theory of sulfur adsorption on semi-infinite Ge(001). *Phys. Rev. Lett.*, **64**, 1808.

96 Frank, M.M., Koester, S.J., Copel, M., Ott, J.A., Paruchuri, V.K., Shang, H., and

Loesing, R. (2006) Hafnium oxide gate dielectrics on sulfur-passivated germanium. *Appl. Phys. Lett.*, **89**, 112905.

97 Shang, H., Okorn-Schmidt, H., Chan, K.K., Copel, M., Ott, J.A., Kozlowski, P.M., Steen, S.E., Cordes, S.A., Wong, H.-S.P., Jones, E.C., and Haensch, W.E. (2002). High mobility p-channel germanium MOSFETs with a thin Ge oxynitride gate dielectric. *Int. Electron Dev. Meeting Technical Digest*, 441.

98 Kim, H., McIntyre, P.C., Chui, C.-O., Saraswat, K.C., and Cho, M.-H. (2004) Interfacial characteristics of HfO_2 grown on nitrided Ge(100) substrates by atomic-layer deposition. *Appl. Phys. Lett.*, **85**, 2902.

99 Hymes, D.J. and Rosenberg, J.J. (1988) Growth and materials characterization of native germanium oxynitride thin films on germanium. *J. Electrochem. Soc.*, **135**, 961.

100 Chui, C.O., Ito, F., and Saraswat, K.C. (2004) Scalability and electrical properties of germanium oxynitride MOS dielectrics. *IEEE Electron. Dev. Lett.*, **25**, 613.

101 Caymax, M., Houssa, M., Pourtois, G., Bellenger, F., Martens, K., Delabie, A., and VanElshocht, S. (2008) Interface control of high-*k* gate dielectrics on Ge. *Appl. Surf. Sci.*, **254**, 6094.

102 Wu, N., Zhang, Q., Zhu, C., Yeo, C., Whang, S.J., Chan, D.S.H., Li, M.F., Chin, A., Kwong, D.L., Du, A.Y., Tung, C.H., and Balasubramanian, N. (2004). Alternative surface passivation on germanium for metal-oxide-semiconductor applications with high-*k* gate dielectric. *Appl. Phys. Lett.*, **85**, 4127.

103 Robertson, J. and Falabretti, B. (2006) Band offsets of high *k* gate oxides on high mobility semiconductors. *Mater. Sci. Eng. B Solid*, **135**, 267.

104 Chui, C.O., Kim, H., McIntyre, P.C., and Saraswat, K.C. (2004) Atomic layer deposition of high-*k* dielectric for germanium MOS applications: substrate surface preparation. *IEEE Electron. Dev. Lett.*, **25**, 274.

105 Kamata, Y., Kamimuta, Y., Ino, T., and Nishiyama, A. (2005) Direct comparison of ZrO_2 and HfO_2 on Ge substrate in terms of the realization of ultrathin high-kappa gate stacks. *Jpn J. Appl. Phys. 1*, **44**, 2323.

106 Park, T.J., Kim, J.H., Jang, J.H., Seo, M., Hwang, C.S., and Won, J.Y. (2007) Improvements in the electrical properties of high-*k* HfO_2 dielectric films on $Si_{1-x}Ge_x$ substrates by postdeposition annealing. *Appl. Phys. Lett.*, **90**, 042915.

107 Sugawara, T., Oshima, Y., Sreenivasan, R., and McIntyre, P.C. (2007) Electrical properties of germanium/metal-oxide gate stacks with atomic layer deposition grown hafnium-dioxide and plasma-synthesized interface layers. *Appl. Phys. Lett.*, **90**, 112912.

108 Caymax, M., VanElshocht, S., Houssa, M., Delabie, A., Conard, T., Meuris, M., Heyns, M.M., Dimoulas, A., Spiga, S., Fanciulli, M., Seo, J.W., and Goncharova, L.V. (2006). HfO_2 as gate dielectric on Ge: interfaces and deposition techniques. *Mater. Sci. Eng. B Solid*, **135**, 256.

109 Chui, C.O., Ramanathan, S., Triplett, B.B., McIntyre, P.C., and Saraswat, K.C. (2002) Germanium MOS capacitors incorporating ultrathin high-kappa gate dielectric. *IEEE Electron. Dev. Lett.*, **23**, 473.

110 Nomura, H., Kita, K., Nishimura, T., and Toriumi, A. (2006) Interface layer control at Y_2O_3/Ge by N_2 and O_2 annealing on Ge(100) and Ge(111) surface. International Conference on Solid State Devices and Materials (SSDM) p. 406.

111 Kita, K., Nomura, H., Nishimura, T., and Toriumi, A. (2006) Impact of dielectric material selection on electrical properties of high-k/Ge devices. *ECS Trans.*, **3**, 71.

112 Frank, M.M., Wilk, G.D., Starodub, D., Gustafsson, T., Garfunkel, E., Chabal, Y.J., Grazul, J., and Muller, D.A. (2005) HfO_2 and Al_2O_3 gate dielectrics on GaAs grown by atomic layer deposition. *Appl. Phys. Lett.*, **86**, 152904.

113 Passlack, M., Droopad, R., Rajagopalan, K., Abrokwah, J., Gregory, R., and Nguyen, D. (2005) High mobility NMOSFET structure with high-*k* dielectric. *IEEE Electron. Dev. Lett.*, **26**, 713.

114 Robertson, J. (2009) Model of interface states at III–V oxide interfaces. *Appl. Phys. Lett.*, **94**, 152104.

115 Brammertz, G., Lin, H.-C., Martens, K., Mercier, D., Sioncke, S., Delabie, A., Wang, W.E., Caymax, M., Meuris, M., and Heyns, M. (2008) Capacitance–voltage characterization of GaAs-Al_2O_3 interfaces. *Appl. Phys. Lett.*, **93**, 183504.

116 Hasegawa, H., Forward, K.E., and Hartnagel, H.L. (1975) New anodic native oxide of GaAs with improved dielectric and interface properties. *Appl. Phys. Lett.*, **26**, 567.

117 Spicer, W.E., Newman, N., Spindt, C.J., Lilientalweber, Z., and Weber, E.R. (1990) Pinning and Fermi level movement at GaAs-surfaces and interfaces. *J. Vac. Sci. Technol. A*, **8**, 2084.

118 Dalapati, G.K., Sridhara, A., Wong, A.S.W., Chia, C.K., Lee, S.J., and Chi, D.Z. (2007) Interfacial characteristics and band alignments for ZrO_2 gate dielectric on Si passivated p-GaAs substrate. *Appl. Phys. Lett.*, **91**, 242101.

119 Mimura, T. and Fukuta, M. (1980) Status of the GaAs metal-oxide-semiconductor technology. *IEEE Trans. Electron. Dev.*, **27**, 1147.

120 Skromme, B.J., Sandroff, C.J., Yablonovitch, E., and Gmitter, T. (1987) Effects of passivating ionic films on the photoluminescence properties of GaAs. *Appl. Phys. Lett.*, **51**, 2022.

121 Wilmsen, C.W., Kirchner, P.D., Baker, J.M., McInturff, D.T., Pettit, G.D., and Woodall, J.M. (1988) Characterization of photochemically unpinned GaAs. *J. Vac. Sci. Technol. B*, **6**, 1180.

122 O'Connor, E., Long, R.D., Cherkaoui, K., Thomas, K.K., Chalvet, F., Povey, I.M., Pemble, M.E., Hurleya, P.K., Brennan, B., Hughes, G., and Newcomb, S.B. (2008) *In situ* H_2S passivation of $In_{0.53}Ga_{0.47}As$/InP metal-oxide-semiconductor capacitors with atomic-layer deposited HfO_2 gate dielectric. *Appl. Phys. Lett.*, **92**, 022902.

123 Hasegawa, H. and Akazawa, M. (2008) Surface passivation technology for III–V semiconductor nanoelectronics. *Appl. Surf. Sci.*, **255**, 628.

124 Ok, I., Kim, H.-S., Zhang, M., Kang, C.Y., Rhee, S.J., Choi, C., Krishnan, S.A., Lee, T., Zhu, F., Thareja, G., and Lee, J.C. (2006). Metal gate-HfO_2 MOS structures on GaAs substrate with and without Si interlayer. *IEEE Electron. Dev. Lett.*, **27**, 145.

125 Koveshnikov, S., Adamo, C., Tokranov, V., Yakimov, M., Kambhampati, R., Warusawithana, M., Schlom, D.G., Tsai, W., and Oktyabrsky, S. (2008) Thermal stability of electrical and structural properties of GaAs-based metal-oxide-semiconductor capacitors with an amorphous $LaAlO_3$ gate oxide. *Appl. Phys. Lett.*, **93**, 012903.

126 Ivanco, J., Kubota, T., and Kobayashi, H. (2005) Deoxidation of gallium arsenide surface via silicon overlayer: a study on the evolution of the interface state density. *J. Appl. Phys.*, **97**, 073711.

127 Sonnet, A.M., Hinkle, C.L., Jivani, M.N., Chapman, R.A., Pollack, G.P., Wallace, R.M., and Vogel, E.M. (2008) Performance enhancement of n-channel inversion type $In_xGa_{1-x}As$ metal-oxide-semiconductor field effect transistor using *ex situ* deposited thin amorphous silicon layer. *Appl. Phys. Lett.*, **93**, 122109.

128 Lu, H.L., Sun, L., Ding, S.J., Xu, M., Zhang, D.W., and Wang, L.K. (2006) Characterization of atomic-layer-deposited Al_2O_3/GaAs interface improved by NH3 plasma pretreatment. *Appl. Phys. Lett.*, **89**, 152910.

129 Lin, J., Lee, S., Oh, H.-J., Yang, W., Lo, G.Q., Kwong, D.L., and Chi, D.Z. (2008) Plasma PH_3-passivated high mobility inversion InGaAs MOSFET fabricated with self-aligned gate-first process and HfO_2/TaN gate

stack. IEEE Int. Electron. Dev. Meeting Technical Digest, 401.

130 Passlack, M., Hong, M., and Mannaerts, J.P. (1996) Quasistatic and high frequency capacitance–voltage characterization of Ga_2O_3-GaAs structures fabricated by *in situ* molecular beam epitaxy. *Appl. Phys. Lett.*, **68**, 1099.

131 Hong, M., Kwo, J., Kortan, A.R., Mannaerts, J.P., and Sergent, A.M. (1999) Epitaxial cubic gadolinium oxide as a dielectric for gallium arsenide passivation. *Science*, **283**, 1897.

132 Kwo, J., Murphy, D.W., Hong, M., Opila, R.L., Mannaerts, J.P., Masaitis, R.L., and Sergent, A.M. (1999) Passivation of GaAs using $(Ga_2O_3)_{(1-x)}(Gd_2O_3)_{(x)}$, $0<=x<=1.0$ films. *Appl. Phys. Lett.*, **75**, 1116.

133 Huang, Y.L., Chang, P., Yang, Z.K., Lee, Y.J., Lee, H.Y., Liu, H.J., Kwo, J., Mannaerts, J.P., and Hong, M. (2005) Themodynamic stability of $Ga_2O_3(Gd_2O_3)$/GaAs interface. *Appl. Phys. Lett.*, **86**, 191905.

134 Ye, P.D., Wilk, G.D., Kwo, J., Yang, B., Gossmann, H.-J.L., Frei, M., Chu, S.N.G., Mannaerts, J.P., Sergent, M., Hong, M., Ng, K.K., and Bude, J. (2003). GaAs MOSFET with oxide gate dielectric grown by atomic layer deposition. *IEEE Electron. Dev. Lett.*, **24**, 209.

135 Ye, P.D., Wilk, G.D., Yang, B., Kwo, J., Chu, S.N.G., Nakahara, S., Gossmann, H.J.L., Mannaerts, J.P., Hong, M., Ng, K.K., and Bude, J. (2003). GaAs metal-oxide-semiconductor field-effect transistor with nanometer thin dielectric grown by atomic layer deposition. *Appl. Phys. Lett.*, **83**, 180.

136 Wallace, R.M., McIntyre, P.C., Kim, J., and Nishi, Y. (2009) Atomic layer deposition of dielectrics on Ge and III–V materials for ultrahigh performance transistors. *MRS Bull.*, **34**, 493.

137 Huang, M.L., Chang, Y.C., Chang, C.H., Lee, Y.J., Chang, P., Kwo, J., Wu, T.B., and Hong, M. (2005) Surface passivation of III–V compound semiconductors using atomic-layer-deposition-grown Al_2O_3. *Appl. Phys. Lett.*, **87**, 252104.

138 Chang, C.H., Chiou, Y.K., Chang, Y.C., Lee, K.Y., Lin, T.-D., Wu, T.B., Hong, M., and Kwo, J. (2006) Interfacial self-cleaning in atomic layer deposition gate dielectric on $In_{0.15}Ga_{0.85}As$. *Appl. Phys. Lett.*, **89**, 242911.

139 Milojevic, M., Aguirre-Tostado, F.S., Hinkle, C.L., Kim, H.C., Vogel, E.M., Kim, J., and Wallace, R.M. (2008) Half-cycle atomic layer deposition reaction studies of Al_2O_3 on $In_{0.2}Ga_{0.8}As$ (100) surface. *Appl. Phys. Lett.*, **93**, 202902.

140 Lin, H.-Y., Wu, S.-L., Cheng, C.-C., Ko, C.-H., Wann, C.H., Lin, Y.-R., Chang, S.-J., and Wu, T.-B. (2011) Influences of surface reconstruction on the atomic-layer-deposited HfO_2/Al_2O_3/n-InAs metal-oxide-semiconductor capacitors. *Appl. Phys. Lett.*, **98**, 123509.

141 Hou, C.H., Chen, M.C., Chang, C.H., Wu, T.B., Chiang, C.D., and Luo, J.J. (2008) Effects of surface treatments on interfacial self-cleaning in atomic layer deposition of Al_2O_3 on InSb. *J. Electrochem. Soc.*, **155**, G180.

3
UV Engineering of High-*k* Thin Films

Ian W. Boyd

3.1
Introduction

For many decades, since the discovery of the laser, expensive laboratory-based UV systems have effectively established the agility of highly monochromatic radiation in the ultraviolet to initiate selective pyrolytic or photolytic pathways to induce heterogeneous substrate-based chemistry by dissociating liquids or gases. For application in industry, however, one must address the necessity for high production levels at minimal cost and laser-based systems are not necessarily ideal for such purposes. Alternative wavelength selective photon sources can offer more attractive processing conditions and a range of UV-based lamps has been developed for this purpose. Here, we review the use of the most efficient of these – dielectric barrier discharge (DBD) (excimer) lamp sources.

These sources tend to be significantly lower in capital cost and considerably cheaper and easier to operate. They can also be readily produced in a wide range of geometries. The energetic radiation emitted by these sources can be used to stimulate a range of physical, chemical and biological operations. From initial applications to the production of ozone for disinfection of water, their application has expanded to more general photochemical synthesis, polymerization, decomposition and deposition of thin films for micro- and nano-based thin-film technologies.

3.2
Gas Discharge Generation of UV (Excimer) Radiation

At pressures near or above atmospheric levels, dense rare gases have specific properties that enable electronic kinetic energy to be readily converted to excitation energy and stored initially in excited atomic and ionic states. At these pressures, the excitation is rapidly and efficiently channeled to a small number of low-level atomic and excimer energy states [1]. Excimer states (excited dimers and trimers) are unstable excited molecular complexes of molecules that do not possess a stable

High-k Gate Dielectrics for CMOS Technology, First Edition. Edited by Gang He and Zhaoqi Sun.

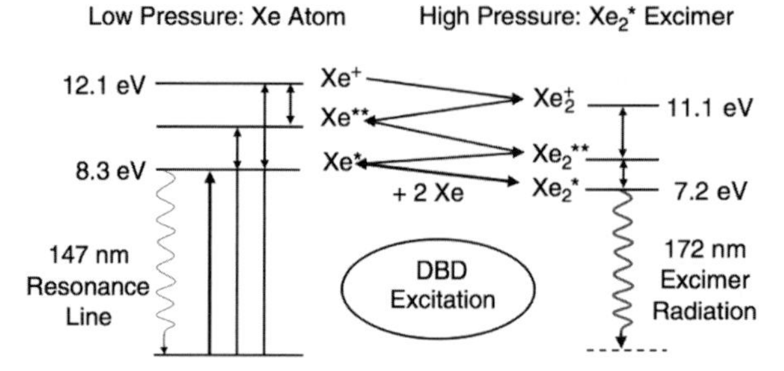

Figure 3.1 Schematic of both the resonance line generated at 147 nm at reduced pressure and the excimer light emitted at 172 nm around atmospheric pressure for a Xe DBD [2].

ground level under normal conditions. These can decompose through the effect of short-pulsed particle bombardment and give off their binding energies in the form of high-energy ultraviolet and vacuum-UV photons.

The Xe_2^* excimers are formed through the following principal routes:

$$\mathrm{Xe}^*(^3P_1, {}^3P_2) + 2\mathrm{Xe} \rightarrow \mathrm{Xe_2}^*\left(^3\sum\nolimits_u^{\,+}\right) + \mathrm{Xe}$$

and

$$\mathrm{Xe}^*(^1P_1, {}^3P_0) + 2\mathrm{Xe} \rightarrow \mathrm{Xe_2}^*\left(^1\sum\nolimits_u^{\,+}\right) + \mathrm{Xe}$$

Using Figure 3.1 to further elucidate this, it can be seen that the Xe_2^* excimer only forms at pressures above 0.1 bar. The excited Xe atom, Xe^*, is converted through a three-body interaction to the exciplex Xe_2^* – and this is quicker than the radiative decay of the Xe^* responsible for the 147 nm resonance emission.

By applying this concept, elementary but effective ultraviolet sources can be fabricated using a range of gas discharge mixtures to stimulate UV generation. Fluorescence radiation from these decaying complexes is usually confined to a narrow band of energies. The production of rare gas excimer radiation has an efficiency derived by theory from 45% up to 80% [3–7] with density of particles being formed up to about $10^{25}\,\mathrm{m}^{-3}$. The excited atoms and ionized species rapidly generate excimer complexes. The relaxation of the vibrational states is accelerated under these circumstances and photon emission from the lower 2 levels dominates. These levels only decay by light emission and not by interacting with the ground level. Rare gas atoms and ions create excimers very efficiently with typical emission half-widths of 10–15 nm. The positions of the peak emissions of the second excimer continuum from unmixed rare-gas excimers (in nm) are summarized in Table 3.1.

Table 3.1 Peak wavelength emissions (in nm) associated with various rare gas dimers.

He_2^*	Ne_2^*	Ar_2	Kr_2^*	Xe_2^*
74	83	126	146	172

Table 3.2 Peak emission wavelengths (in nm) for halogen and rare gas halides.

F_2^*	$ArCl^*$	ArF^*	$KrCl^*$	KrF^*	XeI^*	Cl_2^*	$XeBr^*$	Br_2^*	$XeCl^*$	I_2^*	XeF^*
157	175	193	222	248	253	259	282	289	308	342	354

UV Xe_2^* excimer UV lamps have been adopted in a number of industrial environments in recent years largely because of the accessibility of very pure silica from which they are manufactured. These glass tubes transmit sufficient VUV radiation at 172 nm with only modest absorption and reflection losses. For higher energy sources, special transparent materials, including CaF_2, LiF, or MgF_2, must be used and these are much more expensive and less commonly found. For wavelengths less than 130 nm such as the one obtained from the Ar_2^* excimer, there is a lack of transparent material from which windows can be fabricated. In such cases, Ar_2^* excimer radiation has been generated using through-flow windowless systems [8, 9].

Rare gas excimer emissions can occur over wavelength regions than those discussed above [10, 11] because of the generation of heteronuclear diatomic molecules. Combinations of rare gas halide exciplexes in the form RgX^* (Rg: Ar, Xe, Kr, Ne, He; X: Br, I, Cl, F) [3–7] can initiate bound-free B $\rightarrow$ X transitions resulting in spontaneous and highly efficient UV/VUV radiation and thinner (typically 2–4 nm) emission bandwidths [12, 13]. A selection of rare gas–halogen exciplexes with associated peak emission wavelengths is summarized in Table 3.2. The emission spectra for these excimers is also highlighted in Figure 3.2.

3.3 Excimer Lamp Sources Based on Silent Discharges

Excimer lamps, which can also be called DBD sources operating as nonequilibrium silent discharge systems, can be driven at operating pressures approaching several bar. Their design is distinguished by the fact that the metal electrodes are not directly exposed to the highly excited gas mixture. In practice, this is simpler to apply unlike using expensive high-energy electron beam sources or pulsed capacitor discharges offering at low operating rates.

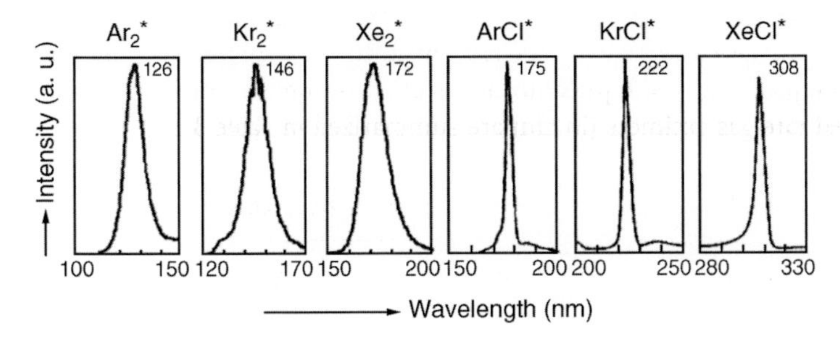

Figure 3.2 Spectral emission of several rare gases and rare gas halides.

Construction involves the use of Suprasil (high purity silica) that is transparent in the ultraviolet, which ideally could have protective coatings applied internally to avoid unwanted chemical etching of the surfaces or functional coatings to reduce the ignition voltage. Their glass-based construction acts as a protective dielectric for the electrodes but means that DC electromagnetic energy cannot pass, requiring alternating currents and voltages for their operation.

The time derivative of the applied voltage together with both the thickness and the dielectric constant determine the level of current displacement passing across the Suprasil. In addition to this, the applied electric field must be sufficiently large to initiate the gas breakdown. Switched mode power supplies providing frequencies from 20 kHz to 15 MHz and several hundred (peak-to-peak) volts to a few kV, stepped up using transformers or resonant circuits, provide a reliable and efficient driver of these gas discharge systems.

As depicted in Figure 3.3, though planar and other designs can be readily fabricated, highly efficient and powerful lamps are most often cylindrical in shape, enabling a fluid-cooling channel through the central electrode. The discharge gap width varies from 0.1 to a few mm with fill pressure between 0.1–50 bar. An additional buffer gas (Ne or He) may be included into the excimer-forming mix to encourage sparking and enable a degree of management of the energy distribution of the electrons.

For discharge geometries with several mm gap, a 1 bar excimer-forming gas pressure, and operational frequencies between 50 Hz and 14 MHz, ignition requires an applied voltage amplitude of several kV. For a much higher operating frequency range, the current limitation decreases and consequently dielectric barrier dielectrics are operated from about 10 MHz down to line frequency. For cylindrical geometric arrangements, an increase in radiation fluence can be achieved by utilizing a reflector, comprising polished 6061-grade aluminum, on the perimeter of the tube directly opposing the direction of the light emission to reflect radiation back to the

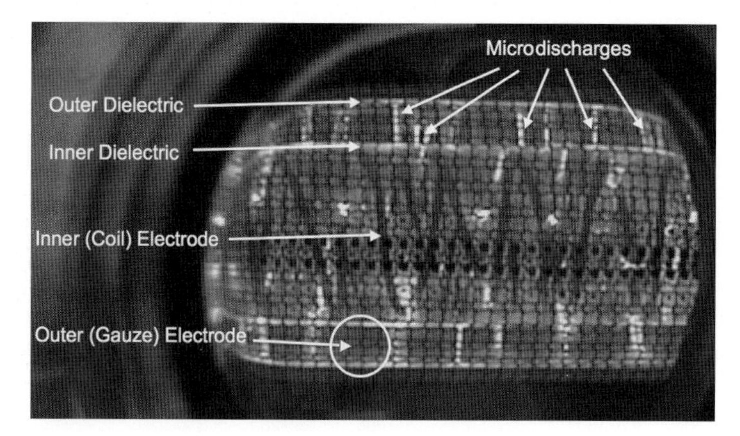

Figure 3.3 Micrograph of a Suprasil-based excimer lamp showing two concentric tubes with inner coiled electrode and all-enveloping outer gauze electrode sleeve. Light microdischarges form between the inner and the outer dielectrics.

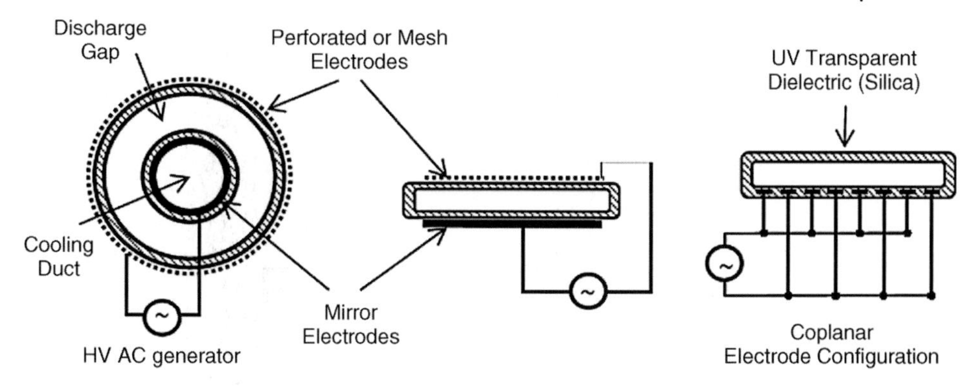

Figure 3.4 Sealed cylindrical (left) and planar dielectric-barrier discharge excimer lamp configurations [13].

center of the lamp. Other geometric designs are also possible (see Figure 3.4), and modern plasma screen displays utilize a flat plate arrangement.

Commercially available excimer sources operate at wavelengths of 308 nm ($XeCl^*$), 222 nm ($KrCl^*$), 172 nm ($Xe_2{}^*$), and 126 nm ($Ar_2{}^*$) with efficiencies ranging from 5% to 40%. $XeCl^*$ sources reaching lengths of 2 m have been constructed for application toward UV curing. $XeBr^*$ and XeI^* systems radiating at 282 [14, 15] and 253 nm [16], respectively, have also been successfully manufactured. Most excimer lamps concentrate their emission in narrow bands around their quoted wavelengths, though some also emit small amounts of light at other wavelengths, and this can develop with time as radiation-induced damage (color center formation) sets in with age and usage.

Operation at room temperature can be obtained and maintained by using a cooled water circuit to enable high electrical input power operation with little risk of thermal overload. At lower wavelengths, to overcome the limitation of radiation absorption loss to the glass cell, windowless or open configurations are used. An excellent illustration of this concept has been discussed by Elsner *et al.* [8] and is depicted in Figure 3.5. Here, the plasma discharge is ignited inside an enlarged cell housing both the excimer-forming elements and the work piece to be exposed.

3.4
Predeposition Surface Cleaning for High-*k* Layers

The cleaning of reticules for VUV lithography and in silicon manufacturing and glass substrates in the fabrication of LCDs are already very well established operations in industry. The expulsion of organics residing on surfaces as contamination, mainly aromatic and aliphatic hydrocarbons, promotes wettability and improves adhesion. Light from Xe lamps with photon energies at 7.25 eV is sufficiently strong to dissociate many organic molecular bonds, though not sufficiently high to decompose the CO_2 molecules formed as a result. Optimized cleaning occurs in mixtures of $O_{2/}$/Ar in a pressure range of 10–100 mbar.

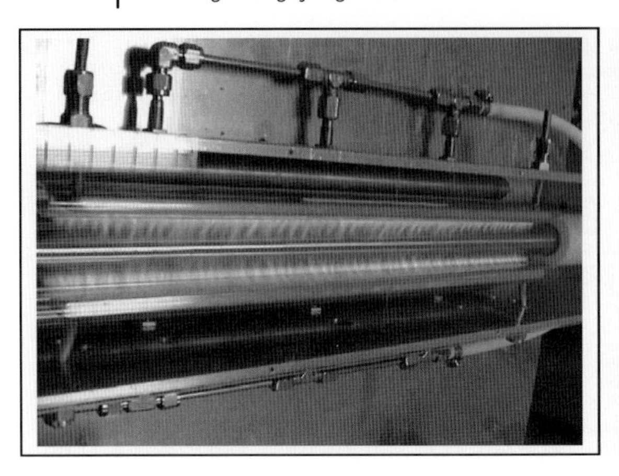

Discharge length: 30 cm
Number of
discharge zones: 2
RF-Voltage : Frequency 1 MHz
Amplitude 1.5 kV
RF-Power : 1-2 kW

Ar pressure: 1 bar
Ar gas flow: 5-10 1/h
Cooling water: 3 1/min

Maximum tolerable
residual oxygen content: 100 ppm

Figure 3.5 Windowless Ar_2^* excimer vacuum UV generator [8].

Argon gas is employed as a carrier and is transparent to the UV light, while the strongly absorbing oxygen molecule dissociates into the aggressively reactive components $O(^1D)$ and $O(^3P)$. The formation of ozone (O_3) is a consequence of the recombination of the O_2 molecules with O atoms and this in turn absorbs the UV light and photodecomposes yielding extremely reactive $O(^1D)$ species.

The cleaving of organics of the generalized form $C_xH_yO_z$ produces H_2O that also strongly absorbs this light and after further excitation homolyzes to form atomic hydrogen and hydroxyl radicals. The oxygen content within the overall gas mixture results in a most reactive zoology that contains O_3, and OH, $O(^1D)$, and $O(^3P)$ radicals, and when combined with the VUV irradiation, can be optimized in order to produce a powerful cleansing environment for the removal of surface contamination.

Stable end-products of CO_2, CO, H_2, O, and H_2 are formed from the intermediate molecular fragments and these are transported away as exhaust. This results in a cleaned surface with a wettability contact angle between 5 and 10° as opposed to 30–60° before VUV irradiation. Present technology utilizes arrays of parallel linear cylindrical Xe sources that generate power densities approaching 100 mW/cm^2. The chemical-free characteristic of this cleaning operation is most beneficial when the reduced scale of nanodevices and the additional challenges of the material and interfacial demands within the device structure are taken into consideration.

3.5 UV Photon Deposition of Ta_2O_5 Films

The relentless reduction of feature sizes in MOS technology as device geometries scale down has highlighted the fundamental limitation of the gate dielectric thickness. As the film thickness decreases, the tunneling current across the layer increases exponentially. For films as thin as 2 nm, leakage currents as high as 10 A/cm^2 are possible [17] and device performance is significantly altered, resulting in large amounts of power being dissipated. Additional difficulties in fabrication and reliability add to these fundamental quantum mechanical difficulties. One solution is to reduce this leakage while maintaining the same gate capacitance through the use of a thicker layer of material with a much larger dielectric constant, enabling an additional decrease in device area.

A number of candidate material systems have been advocated as replacements for the ubiquitous silicon dioxide or more recent SiN or SiO_xN_y mutations. A promising initial candidate was Ta_2O_5, and this was extensively studied because of its compatibility with submicron-scale fabrication and its thermal and chemical stability [18, 19].

Photo-CVD is commonly accomplished using lasers, which, although providing high irradiation levels, are exceedingly expensive while commercial systems offer only small beam areas (cm^2). VUV lamp sources are ideally placed to generate the necessary photon energy over larger areas more cheaply, and easily, to stimulate and initiate the required chemical processes.

Figure 3.6 outlines the excimer lamp setup utilized to grow thin high-k films. It is composed of two stainless steel vacuum cells (upper lamp section and lower reactor chamber) separated by a UV transmitting MgF_2 window. In this work, UV excimer sources of $KrCl^*$ ($\lambda = 222$ nm) or Xe_2^* ($\lambda = 172$ nm) constructed in a cylindrical geometry with output intensities up to 30 mW/cm^2 were employed [20].

Crystalline Si substrates (100 orientation, n-type, and 2–4 Ω cm resistivity) were positioned on a heated substrate holder (100–400 °C), and a low-pressure gaseous mix was passed above the surface. The gas mix was produced by vaporizing a tantalum metalorganic precursor and pushed into the cell using both an Argon gas carrier and N_2O as an oxidizing agent at a predetermined flow rate. A more complete description of this reactor is published elsewhere [21].

The photochemistry underpinning the Ta_2O_5 growth from the various precursors and N_2O follows essentially the same pathway. In the work described here, tantalum ethoxide ($Ta(OC_2H_5)_5$) and $Ta(OEt)_4(OCH_2CH_2NMe_2)$ were used. Each is indirectly dissociated following the photoreaction:

$$N_2O + h\nu \rightarrow O(^1D) + N_2\left(^1\sum\nolimits_g^+\right)$$

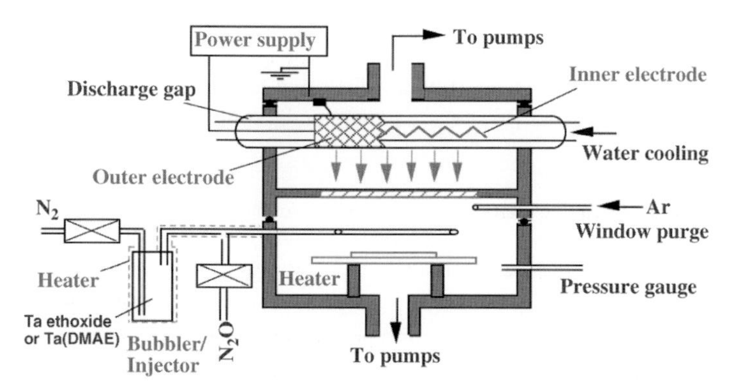

Figure 3.6 Outline of the photodeposition system incorporating UV excimer lamp chamber (top) and deposition chamber (bottom).

though there may also be some additional direct photodecompositional contributions. The resultant oxygen species O(^{1}D) actively reacts with the Ta precursor promoting its decomposition through a chain of reaction steps leading to Ta_2O_5 growth on the sample surface. It is found that while the deposition rate of a thermally grown film depends strongly on the temperature of the substrate, it is notably slow and almost insignificant at temperatures less than 400 °C with a sizeable activation energy of 1.97 eV, the UV photolytically deposited layers show little dependence on temperature and a much lower activation energy of only 0.078 eV. Substrate temperatures as low as 100 °C can be used when depositing with the UV light, and dielectric constants as high as 24 were measured. TEM studies revealed the presence of and SiO_2 layer at the Si–Ta_2O_5 interface, while current–voltage measurements indicated sizeable leakage currents.

Annealing of the deposited layers in an oxygen-rich environment was found to result in a dramatic decrease in leakage current in the Ta_2O_5 films. The UV annealing phenomenon is attributed not only to ozone formation but also to the presence of active species produced by the subsequent photolysis of O_2 to form O(^{1}D). This is adsorbed into the thin Ta_2O_5 layer, accepting electrons and filling voids, thereby reducing leakage currents. These active oxygen species, however, can also interact with Si at the Si/Ta_2O_5 interface or with Si species that might diffuse from the substrate into the interface region, leading to the creation of an undesirable native oxide layer. The possible contributors to the measured reduction in leakage current after the UV annealing operation include the following:

- Reduction in as-deposited particle size improvement in surface morphology.
- Densification of the as-deposited film and removal of impurities present.
- Active oxygen species reduces the density oxygen vacancies and of defects.
- A native oxide forms at the Si–Ta_2O_5 interface and on the Ta_2O_5 surface [22] by the reaction of Si and active oxygen species leading to improved interface.
- Active oxygen species react with, and reduce or remove suboxides present, culminating in improved stoichiometry.

Further electrical studies showed that following an optimized anneal step, the leakage current could be decreased (by three orders of magnitude) as summarized in Table 3.3.

The density of fixed oxide charge is additionally found to decrease with further annealing. The flatband voltage (V_{FB}) was negative suggesting that positive fixed

Table 3.3 Electrical characteristics of Ta_2O_5 layers UV annealed for different times.

Annealing time (h)	Leakage current density at 1 MV/cm (A/cm^2)	V_{FB} (V)	Fixed charge density (cm^{-2})
0	2.19×10^{-5}	−1.8	2.9×10^{11}
0.5	2.55×10^{-6}	−0.8	2.3×10^{11}
1	1.63×10^{-8}	−0.1	3.7×10^{10}

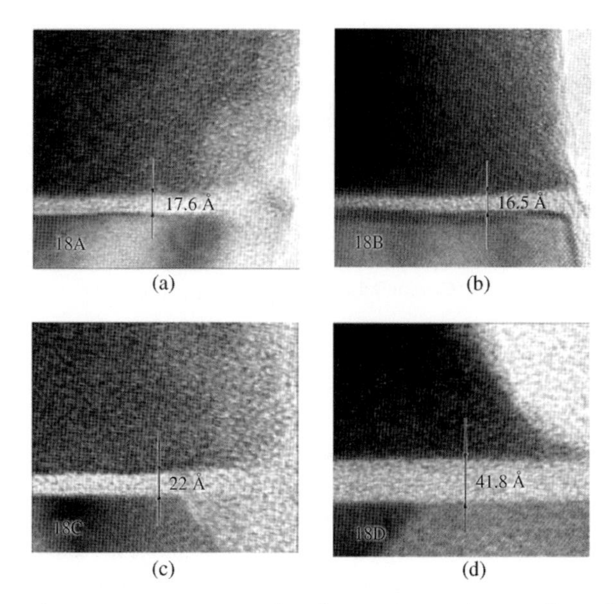

Figure 3.7 TEM micrographs of (a) as-deposited Ta_2O_5 films, (b) UV annealed for 10 min, (c) UV annealed for 15 min, and (d) UV annealed for 40 min, clearly showing the presence and growth of the interface SiO_2 layer.

charges were present close to the Si/Ta_2O_5 interface. With no anneal, V_{FB} was -1.8 V and decreased to -0.8 V and -0.1 V after annealing for 30 min and 60 min, respectively, demonstrating a decrease in positive charge at the surface.

As indicated above, the annealing step not only reduced leakage but also induced the formation of an interface SiO_2 layer. This is clearly shown in Figure 3.7. Thus, one has to balance the reduction of leakage current with the added presence of the silicon oxide, which serves to reduce the overall high-*k* contribution of the Ta_2O_5 layer.

Another UV-based processing step can also be used to help suppress an extensive silicon oxide interfacial layer from forming between the silicon and the high-*k* layer. Prior to photodeposition, the silicon substrates can be treated to a UV anneal with N_2O, using background temperatures around 350 °C. Such a step using N_2O has previously been reported, but typically without the photodissociating UV radiation and only at temperatures higher than 800 °C. Figure 3.8 shows how such a low-temperature UV–N_2O immersion process for only 10 or 20 min can lead to only a stunted SiO_2 interfacial layer formation. The growth of interface silicon oxide films as thick as 39 Å can be avoided, with only 14 Å forming under the Ta_2O_5 after a 20 min treatment of the silicon.

UV photochemical operations effected at reduced temperatures minimize complications such as distribution of dopant, atomic diffusion, and high-temperature-induced defect generation. Furthermore, the sample is not exposed to the damaging effect of ion bombardment that may prevail in alternative physical deposition systems. Optical radiation can initiate distinct chemical steps in the vapor

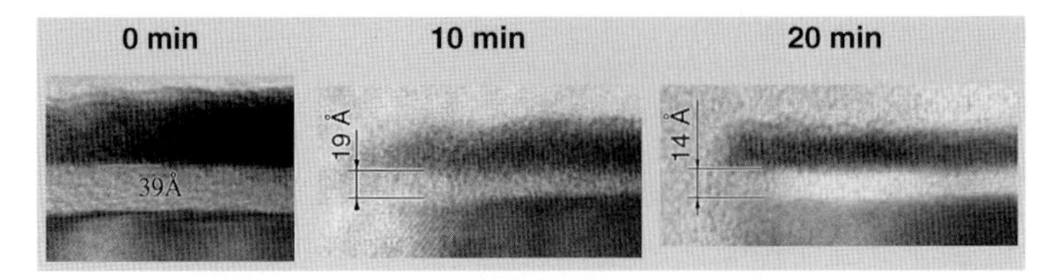

Figure 3.8 The growth of a 39 Å interface layer of SiO_2 under deposited Ta_2O_5 films without any predeposition treatment (left) has been progressively suppressed following 10 min (center) and 20 min UV treatment (right) with N_2O at 350 °C, to only 14 Å.

(e.g., active oxygen generation) adjacent to or on the surface, significantly decreasing active temperature dependency and enabling growth to occur at considerably reduced temperature.

Low dielectric constant polymer layers [23, 24] also hold huge potential for UV processing due to their relatively simple patterning process and large thermal stability. UV radiation plays an important role in polyamic curing where studies indicate [25, 26] the total conversion of the polyamic acid film at only 150 °C to a polyimide with current leakages decreased by an order of magnitude comparable to conventional methods.

3.6
Photoinduced Deposition of $Hf_{1-x}Si_xO_y$ Layers

Another attractive candidate high-*k* dielectric material is hafnia (HfO_2) that satisfies many of the prerequisites for the replacement of SiO_2 since it has a wide (around 6 eV) bandgap, high value ($k = 25$) dielectric constant, high level of stability with silicon surfaces based on equilibrium condition Gibbs free energy analysis, and band offsets with Si of 1.5 eV [27]. There are many reports, however [28–32], of the appearance of low-*k* intermediate layers between the Si and the HfO_2 interface during HfO_2 film growth or as a result of annealing at elevated temperatures.

The low-*k* interface layers (hafnium silicate or SiO_2) limit the maximal capacitance of the gate stack that can be achieved [33]. In this case, it is presumed that Hf ions and atomic oxygen or oxygen ions interdiffuse either in a grain boundary or in a vacancy sublattice such that these less desirable intermediates form [34].

The use of silicates themselves as gate dielectrics has actually been proposed [35, 36]. This work, by Wilk *et al.*, focused predominantly on dielectrics produced by Hf and Zr reactive sputtering. For a Hf silicate layer containing 6% atomic Hf, a refractive index value of $k = 11$ and an equivalent oxide thickness (EOT) around 16 Å has been obtained. The total permittivity of the Hf silicate layer is, of course, less than the value for the unblended metal oxide.

The continued growth of the low-*k* interface material during silicate growth can be largely eliminated because of its high blocking level to the diffusion of oxygen [33–35]. Furthermore, the inclusion of silicon dioxide into amorphous ZrO_2 or HfO_2 layers aids in their stability when undergoing annealing steps at high temperatures and enhances the carrier mobility at the interface [35–37]. Because of this, of the many possible dielectrics to be directly contacted to Si, Zr and Hf silicates are among the most attractive candidates [38].

A number of techniques have historically been employed to grow hafnium silicate layers, including DC and RF sputtering [33, 35, 39], atomic layer deposition (ALD) [40–42], oxidation of metal silicide by an ozone/UV radiation approach [43], and a variety of chemical vapor deposition methods, including MOCVD [44–46], thermal deposition [47], and PECVD [47, 48]. The growth, at temperatures up to 450 °C, of $Hf_{1-x}Si_xO_y$ layers directly on to single-crystal Si by photo-CVD using excimer UV lamps is summarized here. Over the years, UV excimer lamp-induced photo-CVD has also been extensively applied to the deposition of a wide variety of oxides, including TiO_2, ZrO_2 HfO_2, CeO_2, and TiO_2–HfO_2 [49–52] thin films.

The photo-CVD system shown in Figure 3.6 was used with a 222 nm source with flowing alkoxide complexes of Hf(IV)bis-*t*-butoxide-(bis)1-methoxy-2-methyl-2-propoxide, $Hf(OBut)_2(mmp)_2$, and tetraethoxysilane $Si(OC_2H_5)_4$ [53]. The Si/(Si + Hf) atomic ratio was varied from 10 to 60% in 0.05 M mixtures in a solution of octane, and Ar gas at 60 sccm and oxidizing N_2O at 20 sccm flow rate were employed to transport the precursor to the showerhead, held at a temperature of 110 °C, to the RCA-cleaned 450 °C c-Si substrates [51, 52].

Figure 3.9 shows the growth rate at 450 °C and 0.1 mbar of about 6 nm/min for 300 s, where the refractive index decreases with an increase in the Si/(Si + Hf) ratio from 1.870 to 1.780. The lowest growth rate of 5.25 nm/min is found at the largest ratio of 60 atomic % while the maximum deposition rate (6.4 nm/min) is found for the smallest Si/(Si + Hf) ratio (10 atomic %). X-ray diffraction confirmed that the hafnium silicate layers were amorphous up to 30 nm in thickness [53].

Figure 3.10a compares the XPS spectra of UV-deposited HfO_2 and HfSiO films. The $4f_{5/2}$ and $4f_{7/2}$ components at 19.50 and 18.0 eV, respectively, are at higher binding energies relative to the pure oxide, as expected [31, 32]. Figure 3.10b exhibits

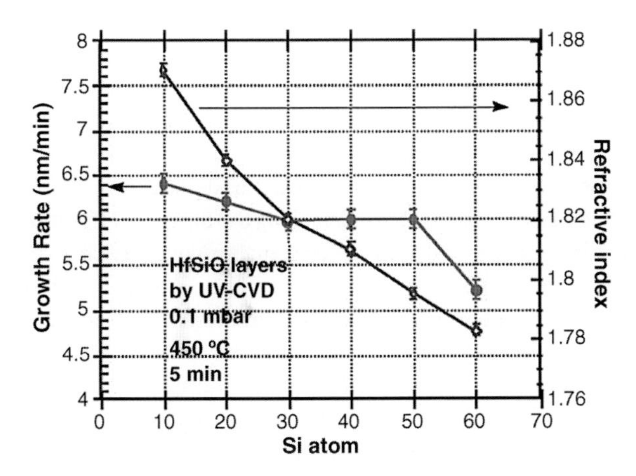

Figure 3.9 Refractive index and deposition rate of HfSiO films UV deposited on c-Si as a function of Si/(Si + Hf) ratio.

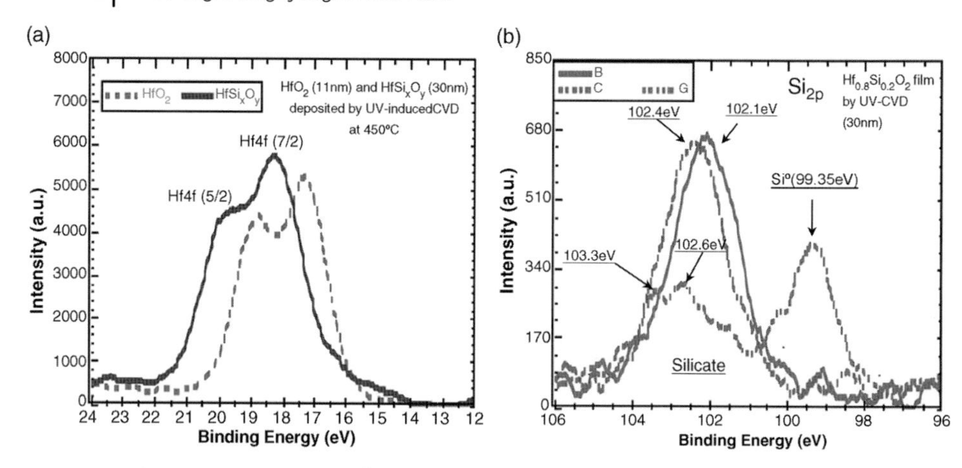

Figure 3.10 Comparison of XPS spectra of UV-deposited HfO_2 and HfSiO layers (a) and binding energies of Si (2p) in HfSiO before (B) and after etching (C = 15 min, G = 45 min) (b).

the peaks of the Si 2p binding energy at 101–105 eV [33] from the hafnium silicate after etching with Ar^+ for different times.

The 102.1 eV surface feature shifting to 102.4 and 102.6 eV after progressive etching clearly indicates the presence of Hf–O–Si bonding [34]. The strong 99.35 eV feature relates to Si–Si bonding that appears through the film from the silicon after strong (45 min) etching and the weaker shoulder around 103.3 eV associated with the 102.6 eV peak is a result of the presence of Si–O–Hf and Si–O_x bonds at the interface.

UV annealing has been shown to increase the Si–O–Hf bond concentration indicated by an increase in FTIR absorption around 1000 cm^{-1} and a shift of the absorption peak position after 15 min annealing from 970 cm^{-1} (as deposited) to 1014 cm^{-1} [53]. This indicates that annealing with UV radiation can also possibly convert any interface suboxides into stoichiometric $HfSiO_4$.

The ratio of Si/(Si + Hf) in the films is strongly influenced by the ratio in the precursor solution, as expected, though unexpectedly at lower ratios it has a stronger influence than at higher ratios [53]. The oxygen content in the films also deviates from stoichiometric values, as shown by the XPS data in Figure 3.11a. Oxygen is deficient in the lower (from 10–30% atomic) ratio range, while stoichiometry is obtained, with even a light excess at the higher (40–60% atomic) ratio range, from Fourier transform infrared (FTIR) studies of these layers indicate that increasing the Si/(Si + Hf) ratio in the deposited hafnium silicate layers tends to create a higher concentration of Si–O–Si bonds in the layers [53].

Figure 3.11b shows the carbon content, analyzed by XPS, in the UV as-deposited films. It can be seen to be quite low, typically around 4 at.% and to decrease with increasing Si/(Si + Hf) ratio. At a Si/(Si + Hf) ratio of 60%, for example, the carbon concentration was as low as 1.4 at.%. This shows that the carbon residues within the as-grown layers must originate primarily from the $Hf(OBut)_2(mmp)_2$ precursor used. These experimental observations indicate that the UV excimer lamp approach to high-*k* thin film deposition is an attractive prospect.

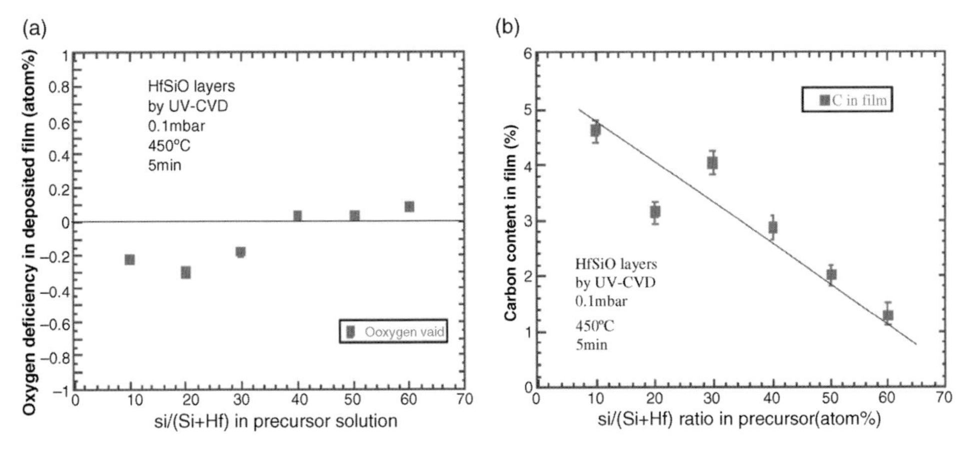

Figure 3.11 (a, b) Oxygen and carbon content, respectively, in excimer UV-deposited HfSiO layers as a function of Si/(Si + Hf) ratio in precursor.

3.7 Summary

The efficiency, simplicity, and utility factor of excimer UV lamps as sources to produce powerful narrow band incoherent UV and VUV radiation have been demonstrated. Their use and potential to be further exploited to large-area processing has been discussed. In this chapter, the application toward the growth and annealing of high-*k* thin films indicates not only their suitability but also their significant versatility.

Growth rates of between 5 and 10 nm/min are considered acceptable for application purposes. Control of the photochemistry also brings an additional dimension to the film growth. Thus, engineering at the substrate–film interface is possible, for the first time at low temperatures, through control of the oxygen species generated by the UV photons. Such excimer lamp annealing processes can be used not only for UV deposited thin films but also for those prepared by alternative methods.

References

1 Lorents, D.C. (1976) Physics of electron-beam excited rare-gases at high densities. *Physica B&C*, **82**, 19.
2 Kogelschatz, U. (2002) Industrial innovation based on fundamental physics, *Plasma Source Sci. Technol.*, **11**, A1.
3 Rhodes, Ch.K. (ed.) (1979) *Excimer Lasers*. Springer, New York.
4 Eliasson, B. and Kogelschatz, U. (1988) UV excimer radiation from dielectric-barrier discharges. *Appl. Phys. B*, **46**, 229.
5 Ivanov, V.V., Klopovskii, K.S., Mankelevich, Yu.A., Rakhimov, A.T., Rakhimova, T.V., Rulev, G.B., and Saenko, V.B. (1996) Experimental and theoretical study of the efficiency of an excimer lamp pumped by a pulse distributed discharge in xenon. *Laser Phys.*, **6**, 564.
6 Oda, A., Sugawara, H., Sakai, Y., and Akashi, H. (2000) Estimation of the light output power and efficiency of Xe barrier discharge excimer lamps using a

one-dimensional fluid model for various voltage waveforms. *J. Phys. D Appl. Phys.*, **33**, 1507.

7 Mildren, R.P. and Carman, R.J. (2001) Enhanced performance of a dielectric barrier discharge lamp using short-pulsed excitation. *J. Phys. D Appl. Phys.*, **34**, L1.

8 Elsner, C., Lenk, M., Prager, L., and Mehnert, R. (2006) Windowless argon excimer source for surface modification. *Appl. Surf. Sci.*, **252**, 3616, and references therein.

9 Lomaev, M.I., Skakun, V.S., Tarasenko, V.F., Shitts, D.V., and Lisenko, A.A. (2006) A windowless VUV excilamp. *Tech. Phys. Lett.*, **32**, 590.

10 Kubodera, S., Honda, M., Kitahara, M., Kawanaka, J., Sasaki, W., and Kurosawa, K. (1995) Extended broad-band emission in vacuum-ultraviolet by multi-rare-gas silent discharges. *Jpn. J. Appl. Phys. 2*, **34**, L618.

11 Gerasimov, G.N., Volkova, G.A., and Zvereva, G.N. (1998) *Proceedings of the 8th International Symposium on Science Technology Light Sources* (ed. G. Babucke), Griefswald, Germany, **LS-8**, p. 248.

12 Shuaibov A.K., Shimon, L.L., and Shervera, I.V. (1998) A multiwave electric-discharge lamp on halides of inert gases. *Instrum. Exp. Tech.*, **41**, 427.

13 Shuaibov, A.K., Shimon, L.L., Dashchenko, A.I., and Shervera, I.V. (2001) An electric discharge emitter operating simultaneously in the 308 [XeCl (B-X)], 258 [Cl_2 (D′-A′)], 236 [XeCl (D-X)], 222 [KrCl (B-X)], 175 [ArCl (B-X)], and 160 [H_2 (B-X)] nm bands. *Tech. Phys.*, **46**, 207.

14 Zhang, J.Y. and Boyd, I.W. (1996) Efficient XeI excimer ultraviolet sources from a dielectric barrier discharge. *J. Appl. Phys.*, **84**, 1174.

15 Zhang, J.Y. and Boyd, I.W. (1996) Investigation of high efficient excimer UV sources from a dielectric barrier discharge in rare gas/halogen sources. *J. Appl. Phys.*, **80**, 663.

16 Kogelschatz, U., Esrom, H., Zhang, J.-Y., and Boyd, I.W. (2000) High-intensity sources of incoherent UV and VUV excimer radiation for low-temperature materials processing. *Appl. Surf. Sci.*, **168**, 29.

17 Mueller, D.A., Sorsch, T., Moccio, S., Baumann, F.H., Evans-Lutterodt, K., and Timp, G. (1999) *Nature*, **399**, 758.

18 Zhang, J.Y., Lim, B., Boyd, I.W., and Dusastre, V. (1998) Characteristics of high quality tantalum oxide films deposited by photo induced chemical vapor deposition. *Appl. Phys. Lett.*, **73**, 2299.

19 Chaneliere, C., Autran, J.L., Balland, B., and Devine, R.A.B. (1998) Tantalum pentoxide thin films for advanced dielectric applications. *Mater. Sci. Eng. Rep.*, **22**, 269.

20 Zhang, J.Y., Lim, B., and Boyd, I.W. (1998) Thin tantalum pentoxide films deposited by photo-induced CVD. *Thin Solid Films*, **336**, 340.

21 Zhang, J.Y., Boyd, I.W., Mooney, M.B., Hurley, P.K., O'Sullivan, B., Beechinor, T., Kelly, P.V., Crean, G.M., and Senateur, J.P. (1999) Photo-induced CVD of tantalum pentoxide dielectric films using an injection liquid source. *Ultrathin SiO_2 and High-k Materials for ULSI Gate Dielectrics (Materials Research Society Proceedings)*, vol. **567**, Materials Research Society, p. 397.

22 Gellert, B. and Kogelschatz, U. (1991) Generation of excimer emission in dielectric barrier discharges. *Appl. Phys. B*, **52**, 14.

23 Singer, P. (1994) New interconnect materials: chasing the promise of faster chips. *Semiconduct. Int.*, **34**, 52–56.

24 Murarka, S.P. (1996) Low dielectric constant materials for interlayer dielectric applications. *Solid State Tech.*, **39**, 83.

25 Zhang, J.Y. and Boyd, I.W. (1998) UV light-induced deposition of low dielectric constant organic polymer for interlayer dielectrics. *Opt. Mater.*, **9**, 251.

26 Zhang, J.Y. and Boyd, I.W. (2000) Low dielectric constant porous silica films formed by photo-induced sol–gel processing. *Mater. Sci. Semiconduct. Process.*, **3**, 345.

27 Lee, P.F., Dai, J.Y., Wong, K.H., Chan, H.L.W., and Choy, C.L. (2003) *Appl. Phys. Lett.*, **82**, 2419.

28 Zhang, J.Y., Fang, Q., and Boyd, I.W. (1999) Growth of tantalum pentoxide film by pulsed laser deposition. *Appl. Surf. Sci.*, **138**, 320.

29 Kaliwoh, N., Zhang, J.Y., and Boyd, I.W. (2000) Titanium dioxide films prepared by photo-induced sol–gel processing using 172nm excimer lamps. *Surf. Coat. Technol.*, **125**, 424.

30 Zhang, J.Y., Lim, B., and Boyd, I.W. (1998) Thin tantalum pentoxide films deposited by photo-induced CVD. *Thin Solid Films*, **336**, 340.

31 Zhang, J.Y., Bie, L.J., and Boyd, I.W. (1998) Formation of high quality tantalum oxide thin films at 400 °C by 172nm radiation. *Jpn. J. Appl. Phys. 2*, **37**, L27.

32 Fang, Q., Zhang, J.Y., Wang, Z.M., Wu, J.X., O'Sullivan, B.J., Hurley, P.K., Leedham, T.L., Audier, M.A., Senateur, J.P., and Boyd, I.W. (2004) Interface of ultrathin HfO_2 films deposited by UV-photo-CVD. *Thin Solid Films*, **453**, 203.

33 Wilk, G.D., Wallace, R.M., and Anthony, J.M. (2001) High-*k* gate dielectrics: current status and materials properties considerations. *J. Appl. Phys.*, **89**, 5243.

34 Kirsch, P.D., Kang, C.S., Lozano, J., Lee, J.C., and Ekerdt, J.G. (2002) Electrical and spectroscopic comparison of HfO_2/Si interfaces on nitride and un-nitrided Si (100). *J. Appl. Phys.*, **91**, 4353.

35 Wilk, G.D. and Wallace, R.M. (1999) Electrical properties of hafnium silicate gate dielectrics deposited directly on silicon. *Appl. Phys. Lett.*, **74**, 2854.

36 Wilk, G.D., Wallace, R.M., and Anthony, J.M. (2000) Hafnium and zirconium silicates for advanced gate dielectrics. *J. Appl. Phys.*, **87**, 484.

37 Ritala, M., Kukli, K., Rahtu, A., Räisänen, P.I., Leskelä, M., Sajavaara, T., and Keionen, J. (2000) Atomic layer deposition of oxide thin films with metal alkoxides as oxygen sources. *Science*, **288**, 319.

38 Vainonen-Ahlgren, E., Tois, E., Ahgren, T., Khriachtchev, L., Marles, J., Haukka, S., and Tuominen, M. (2003) Atomic layer deposition of hafnium and zirconium silicate thin films. *Comp. Mater. Sci.*, **27**, 65.

39 Callegari, A., Cartier, E., Gribelyuk, M., Okorn-Schmidt, H.F., and Zabel, T. (2001) Physical and electrical characterization of hafnium oxide and hafnium silicate sputtered films. *J. Appl. Phys.*, **90**, 6466.

40 Cho, M.H., Roh, Y.S., and Whang, C.N. (2002) Thermal stability and structural characteristics of HfO_2 films on Si (100) grown by atomic-layer deposition. *Appl. Phys. Lett.*, **81**, 472.

41 Senzaki, Y., Park, S., and Chatham, H. (2004) Atomic layer deposition of hafnium oxide and hafnium silicate thin films using liquid precursors and ozone. *J. Vac. Sci. Technol. A*, **22**, 1175.

42 Kim, W.K., Rhee, S.W., and Lee, N.I. (2004) Atomic layer deposition of hafnium silicate films using hafnium tetrachloride and tetra-*n*-butyl orthosilicate. *J. Vac. Sci. Technol. A*, **22**, 1285.

43 Addepalli, S., Sivasubramani, P., El-Bouanani, M., Kim, M.J., Gnade, B.E., and Wallace, R.M. (2004) Deposition of Hf-silicate gate dielectric on Si_xGe_{1-x} (100): detection of interfacial layer growth. *J. Vac. Sci. Technol. A*, **22**, 616.

44 Bastos, K.P., Morais, J., and Miotti, L. (2002) Oxygen reaction-diffusion in metalorganic chemical vapor deposition HfO_2 films annealed in O_2. *Appl. Phys. Lett.*, **81**, 1669.

45 Kim, J. and Yong, K. (2004) Characterization of hafnium silicate thin films grown by MOCVD using a new combination of precursors. *J. Cryst. Growth*, **263**, 442.

46 Roberts, J.L., Marshall, P.A., and Jones, A.C. (2004) Deposition of hafnium silicate films by liquid injection MOCVD using a single source or dual source approach. *J. Mater. Chem.*, **14**, 391.

47 Rangarajan, V., Bhandari, H., and Klein, T.M. (2002) Comparison of hafnium silicate thin films on silicon (100) deposited using thermal and plasma enhanced metal organic chemical vapor deposition. *Thin Solid Films*, **419**, 1.

48 Kato, H., Nango, T., Miyagawa, T., Katagiri, T., Seol, K.S., and Ohki, Y. (2002) Plasma-enhanced chemical vapor deposition and characterization of high-permittivity hafnium and zirconium silicate films. *J. Appl. Phys.*, **92**, 1106.

49 Wang, Z.M., Fang, Q., Zhang, J.Y., Wu, J.X., Di, Y., Chen, W., and Boyd, I.W. (2004) Growth of titanium silicate thin

films by photo-induced chemical vapor deposition. *Thin Solid Films*, **453**, 167.

50 Fang, Q., Zhang, J.Y., Wang, Z.M., Wu, J.X., O'Sullivan, B.J., Hurley, P.K., Leedham, T.L., Audier, M.A., Senateur, J.P., and Boyd, I.W. (2003) Characterisation of HfO_2 deposited by photo-induced chemical vapour deposition. *Thin Solid Films*, **427**, 391.

51 Fang, Q., Zhang, J.Y., Wang, Z.M., Wu, J.X., O'Sullivan, B.J., Hurley, P.K., Leedham, T.L., Davies, H., Audier, M.A., Jiminez, C., Senateur, J.P., and Boyd, I.W. (2003). Characterisation of HfO_2 deposited by photo-induced chemical vapour deposition. *Thin Solid Films*, **428**, 248.

52 Fang, Q., Zhang, J.Y., Wang, Z.M., He, G., and Boyd, I.W. (2003) High-*k* dielectrics by UV photo-assisted chemical vapour deposition. *Microelectron. Eng.*, **66**, 621.

53 Liu, M., Zhu, L.Q., He, G., Wang, Z.M., Wu, J.X., Zhang, J.Y., Liaw, I., Fang, Q., and Boyd, I.W. (2007) $Hf_{1-x}Si_xO_y$ dielectric films deposited by photo-induced chemical vapour deposition (UV-CVD). *Appl. Surf. Sci.*, **253**, 7869.

4
Atomic Layer Deposition Process of Hf-Based High-*k* Gate Dielectric Film on Si Substrate

Tae Joo Park, Moonju Cho, Hyung-Suk Jung, and Cheol Seong Hwang

4.1
Introduction

Hf-based gate dielectrics with a metal gate have been implemented in the mass production of state-of-the-art complementary metal oxide semiconductor field effect transistors (CMOSFETs) because SiO_2 gate dielectrics have already reached their physical limit due to the high leakage current and reliability concerns [1]. With this radical change in the gate dielectric material, other related materials, such as gate metal and spacers, as well as integration processes have also undergone unprecedented innovation over the past decade. There are several other notable changes in characterizing the device properties, which are still highly important research areas in microelectronics and are discussed in other chapters of this book.

Some of the critical issues and related material processing are closely related to the material properties of high-*k* dielectric films, which are determined largely by the processing conditions of thin films. Atomic layer deposition (ALD) has been the industry standard process of the high-*k* gate dielectrics as well as several related materials. The merits of ALD include accurate thickness control, low thermal budget, good uniformity, and conformality. The conformality has not been the critical requirement for conventional planar MOSFETs but its importance is increasing as the three-dimensional structures, e.g., FinFET, are gaining more focus. The low thermal budget is another merit of ALD compared to chemical vapor deposition (CVD), particularly the gate-last integration scheme.

ALD of high-*k* oxide films proceeds by alternately exposing the substrates to precursor and oxygen source that undergo self-terminating reactions with each other on the substrate surface. The metal half-cycle of ALD involves exposure of the growth surface to a gas-phase Hf precursor. After a purge of the reactor, the substrate is exposed to the oxygen source, which is followed by a subsequent purge, resulting in the formation of HfO_2 films. The thickness of the HfO_2 films can be controlled by repeating this reaction cycle. The selection of Hf precursor and oxygen source strongly affects the physical and electrical properties of the ALD HfO_2 films. As the

High-k Gate Dielectrics for CMOS Technology, First Edition. Edited by Gang He and Zhaoqi Sun.

ALD depends more on the specific chemistry and reaction route, compared to CVD. As ALD has evolved rapidly over the past several decades (ALD was first suggested by Professor Suntola in the early 1980s. [2]), a wide range of chemistry has become available, including inorganic and organic precursors.

This chapter examines the issues related to the ALD processes of HfO_2-based dielectric materials. The topics include the film properties and interactions with the Si substrate depending on the types of Hf precursor and oxygen sources. Although some chemistry aspects of ALD process itself are discussed but not mainly focused in this chapter. This chapter shows that various structural, physicochemical, and electrical characteristics of dielectric films are closely related to the details of the ALD process and materials. The influence of adding different elements, such as Si, Al and Zr, during the ALD HfO_2 deposition to improve the thermal stability and electrical characteristics of ALD HfO_2 will be discussed too.

4.2
Precursor Effect on the HfO_2 Characteristics

4.2.1
Hafnium Precursor Effect on the HfO_2 Dielectric Characteristics

Selection of the Hf precursor is the most critical factor for determining the basic performance of ALD HfO_2 films. This is related to the saturation growth rate, impurity concentration, and O/Hf ratio, which are the most important parameters, are determined primarily by the types of Hf- precursor. The growth rate is not just the factor that determines the film thickness for a given time, but it is a crucial factor that determines the interaction between the film and the substrate. The interaction between the film/atmosphere and the Si substrate governs the formation of the interfacial layer, which has crucial importance in determining the device performance. Of course, other factors, such as the growth temperature, types of oxygen sources, and so on, also affect the formation of the interfacial layer. In this section, the experimental results from the five different types of Hf precursors, one inorganic and four metal-organic, are described.

4.2.1.1 Hafnium Chloride ($HfCl_4$)

$HfCl_4$ is the most extensively studied precursor for HfO_2 applications in MOSFETs, combined with ALD [3, 4]. In contrast to other metal-organic (MO) Hf precursors, the superior thermal stability of the $HfCl_4$ makes the ALD temperature range very wide. Commonly, ALD process using a $HfCl_4$ precursor is performed at 300 °C to obtain a uniform layer at a moderate speed. The saturation growth rate of HfO_2 is ~0.05 nm/cycle at 300 °C [4–6]. The film growth rate decreases with increasing deposition temperature at temperatures between 200 and 300 °C [5], but processing in lower temperature than 275 °C leaves behind a significant amount of chlorine residue in the HfO_2 layer [7]. Although the leakage current density as a function of capacitance equivalent oxide thickness (CET) shows a similar profile to SiO_2 [4, 8], adopting alternative precursors and additional reaction barrier layers can improve this [6, 7, 9–11].

Here, the physical characteristics of the HfO_2 layer deposited on a Si substrate by ALD with $HfCl_4$ and H_2O at temperatures ranging from 200 to 400 °C [5] are described first. At 200 °C, an interfacial layer between the HfO_2 bulk and Si substrate grows with increasing number of deposition cycles. On the other hand, over 300 °C, the interfacial layer appears to dissolve into the HfO_2 bulk layer resulting in a decreased thickness over 100 cycles. Postannealing at 800 °C in a N_2 atmosphere increases the interfacial layer, which reduces the dielectric constant of the entire dielectric stack and increases the CET. Therefore, adopting an interfacial barrier layer to suppress SiO_x diffusion is essential for obtaining good high-*k* gate oxide properties after postannealing. It is extremely important to reduce interfacial oxide growth, which is critical in decreasing the CET.

First, the SiN_x gate dielectric was introduced as a reaction barrier layer (RBL) between SiO_2 and HfO_2 dielectrics using an ALD technique [12]. Here, the SiN_x RBL was deposited by either CVD or ALD. The SiN_x RBL suppresses Si and O (inter-) diffusion during HfO_2 film deposition and postdeposition annealing (PDA), resulting in a small increase in the CET after postannealing up to 1000 °C, as shown in Figure 4.1a. Increasing the dielectric constant of the interfacial layer is another method for decreasing the CET. The increase in CET is reduced further by adding a small amount of oxygen ($\sim$1%) to the annealing atmosphere. This appears to be a consequence of enhanced HfO_2 diffusion into the SiN_x RBL, resulting in a higher *k* interfacial layer (Figure 4.1b). Nevertheless, high hysteresis and fixed charge density problems remain even with the improvement in CET. Therefore, an investigation of Al_2O_3 RBL has been followed.

The adoption of an Al_2O_3 RBL also suppresses Si diffusion to a certain degree during HfO_2 film deposition and postannealing [13]. These characteristics result in a smaller increase in equivalent oxide thickness (EOT) after postannealing at temperatures > 650 °C, compared to a single HfO_2 layer film. Below 650 °C, the EOT decreases with increasing PDA temperature as a result of densification. The increase

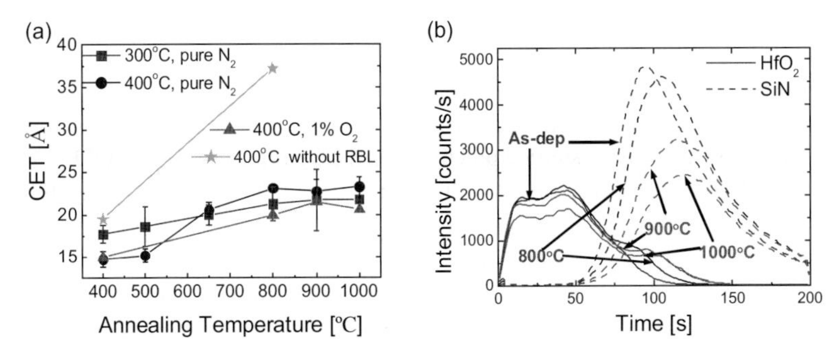

Figure 4.1 (a) CET of the HfO_2/SiN_x dielectric deposited at 300 or 400 °C as a function of the postannealing temperature. (b) SIMS depth profiles of HfO_2 and SiN_x signals as a function of the postannealing temperature [12].

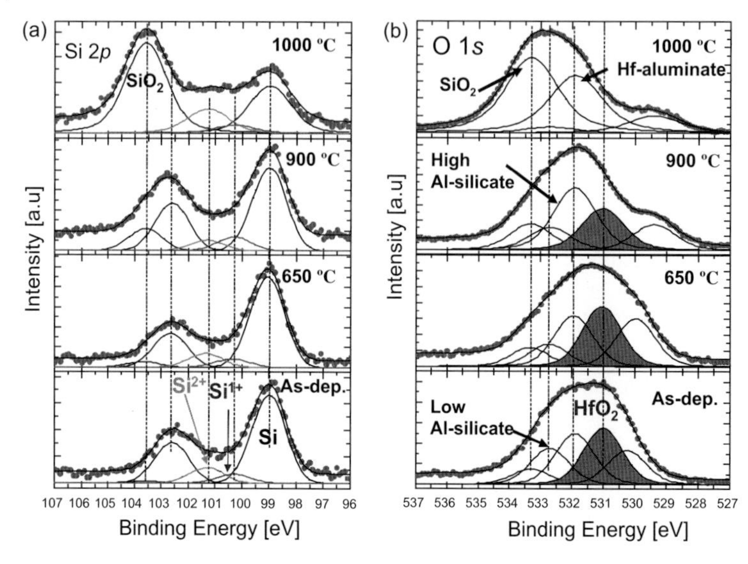

Figure 4.2 XPS results of (a) Si 2p, (b) O 1s in HfO_2/Al_2O_3 stacks before and after postannealing [13].

in EOT over 650 °C is due mainly to SiO_2 interfacial layer formation by residual oxygen, as shown in Figure 4.2a. The RBL is a mixture of Al_2O_3, Al-silicate, and SiO_2. A detailed X-ray photoelectron spectroscopy (XPS) study revealed critical differences in phase stability after PDA at 1000 °C in Al_2O_3 or HfO_2 single layers, or a HfO_2/Al_2O_3 stack. The Al_2O_3 single layer (or Al-silicate with a high Al concentration) reacts with Si/O to form an Al-silicate with a medium Al concentration after annealing. The formation of interfacial SiO_2 is serious. On the other hand, the HfO_2 layer on top of the Al_2O_3 RBL absorbs Al during high-temperature annealing, resulting in a crystallized Hf-aluminate with a small amount of Al-silicate remaining (refer to Figure 4.2b). The formation of interfacial SiO_2 is as serious as that in the Al_2O_3 single layer.

Very low leakage current density of 8×10^{-6} A/cm^2 at -1 V is obtained at an EOT of 1.3 nm from a HfO_2/Al_2O_3 capacitor with a poly-Si gate even after postannealing at 1000 °C. On the other hand, a rather high fixed charge density at the Al_2O_3/SiO_2 interface induces a large flatband voltage (V_{FB}) shift that cannot be removed by PDA. The crystalline Hf-aluminate/SiO_2 interface increases the V_{FB} shift further after postannealing at 1000 °C. To improve the electrical properties of HfO_2/Al_2O_3 stacks, a range of oxygen sources were applied in the Al_2O_3 RBL deposition procedure [14]. Among the oxygen sources, including remote O_2 plasma, H_2O, and O_3, O_3 shows superior quality due to its high oxidation power. Although the interfacial oxidation of Si is observed from the as-deposited state, the stable O_3–Al_2O_3 layer reduces the increase in CET after postannealing, and decreases the hysteresis and leakage current. The relatively high interfacial fixed charge density at the Al_2O_3/SiO_2 interface is still a problem. Nevertheless, the fixed charges at the HfO_2/Al_2O_3, which have the opposite sign to that of the Al_2O_3/SiO_2 interface charges, can reduce the total

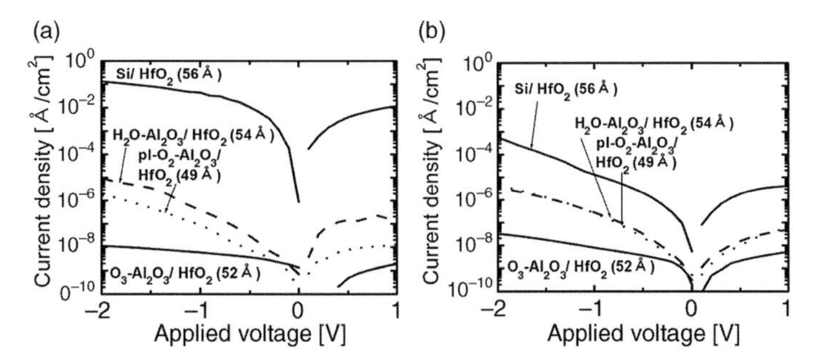

Figure 4.3 Leakage current densities of various films (a) at the as-deposited state and (b) after postannealing [14].

V_{FB} shift by a proper combination of the thicknesses of the Al_2O_3 and HfO_2 layers. Primarily, the O_3–HfO_2/Al_2O_3 stack shows an extremely low leakage current density (Figure 4.3), suggesting its significance as a gate dielectric in future low-power MOSFET devices.

Another experiment also revealed superior characteristics of the O_3 oxygen source combined with the $HfCl_4$ precursor [15]. Owing to the higher oxygen concentration in the films grown by O_3, rapid thermal annealing (RTA) at 750 °C in a N_2 atmosphere produces a thicker interfacial SiO_2 and Hf-silicate, which increases the CET. Moreover, all the other electrical properties, including the fixed charge density, interface trap density, leakage current density, and hysteresis in the capacitance–voltage (C–V) curves, are superior to those of the films grown with H_2O. In particular, O_3–HfO_2 films show a low D_{it} level, which is comparable to that of SiO_2/Si (see Figure 4.4), due to the presence of a high-quality interfacial SiO_2 from the O_3 oxygen source.

Another study on the chemical states at the HfO_2/Si interface using angle-resolved high-resolution XPS (HRXPS) was also performed using nondestructive depth profiling [16]. The HfO_2 is deposited with $HfCl_4$ and H_2O in this case. The Hf 4f

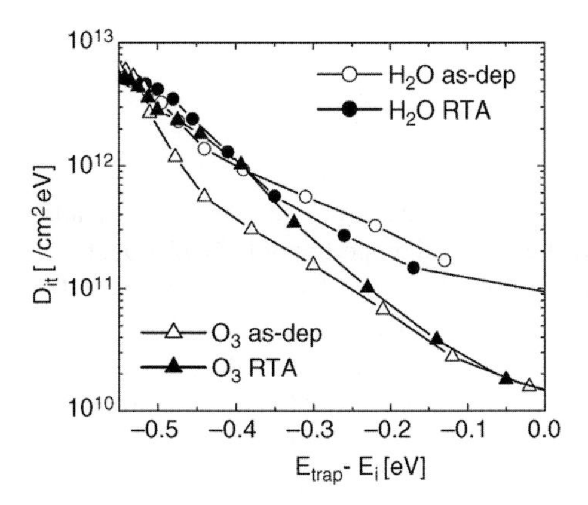

Figure 4.4 D_{it} as a function of the trap energy (E_T)–intrinsic Fermi energy (E_i) from various HfO_2 stacks [15].

chemical state does not change with depth, suggesting that no Hf compounds other than HfO_2 are formed at the interface. This is in direct contradiction to the model proposed previously, where the chemical state of the HfO_2/Si interface consists of Hf-silicate. A new model is proposed from the HRXPS of Hf 4 f, O1s, and Si 2p core levels and their depth profile. The model states that the chemical states of the HfO_2/Si interface consist mainly of Si_2O_3 and strained SiO_2. This model is also consistent with the dielectric constant of 4.8 for the interfacial layer, which is intermediate between that of SiO_2 (3.9) and SiO (6). Therefore, further study will be needed to identify more clearly the interface between the film and the substrate.

The reliability characteristics of the $HfCl_4$-based HfO_2 layer was also examined using the time-dependent dielectric breakdown (TDDB) characteristics [17]. This study focused on the effect of chlorine (Cl) residue in the HfO_2 layer. HfO_2 was deposited at various H_2O pulse times from 0.3 to 10 s. Increasing H_2O pulse time caused a more than one order of magnitude decrease in the residual chlorine content. On the other hand, the TDDB lifetime was not affected, and charge pumping (CP) analysis revealed a negligible trap charge difference between high Cl content and low Cl content films, as shown in Figure 4.5a. Mobility is also not degraded in high Cl content films. Figure 4.5b further confirms this observation by first-principles calculations, which demonstrates that the residual chlorine does not form additional trap energy levels inside the HfO_2 bandgap.

4.2.1.2 Tetrakis Dimethylamido Hafnium [$HfN(CH_3)_2]_4$

There is a plenty of MO chemistry of the Hf precursor, such as alkoxides, alkyl-amido compounds, and amidinates. To understand the effects of using MO Hf precursors on the physical/electrical characteristics in the HfO_2 layer, nitrogen-incorporated Hf $[N(CH_3)_2]_4$ precursor and H_2O oxygen source were used [11]. The growth rate of the HfO_2 film was ~0.08 nm/cycle at 300 °C. A thin $Si(O)N_x$ interfacial layer was formed simultaneously at the HfO_2/Si interface from the early stage of HfO_2 film deposition.

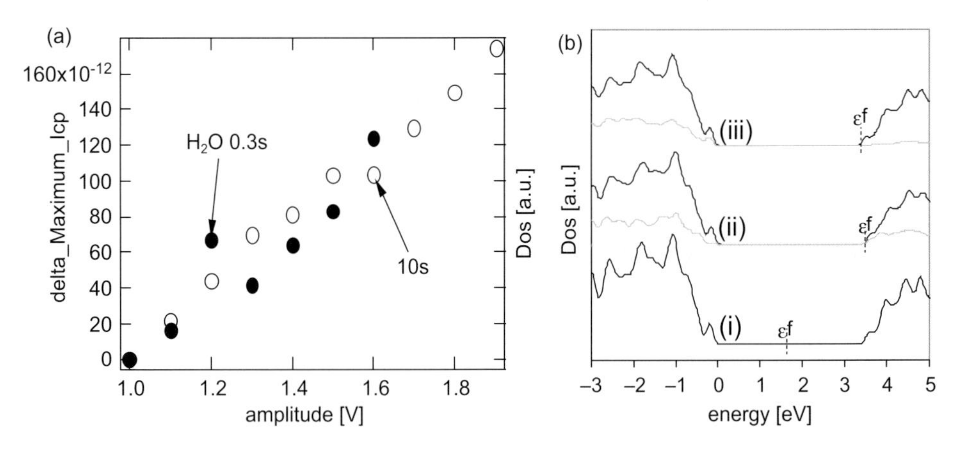

Figure 4.5 (a) Maximum charge pumping current (I_{cp}) versus gate voltage pulse amplitude. (b) Total and partial density of states (DOS and PDOS) for (i) an ideal monoclinic hafnium, (ii) a tetravalent, and (iii) a trivalent chlorine doped [17].

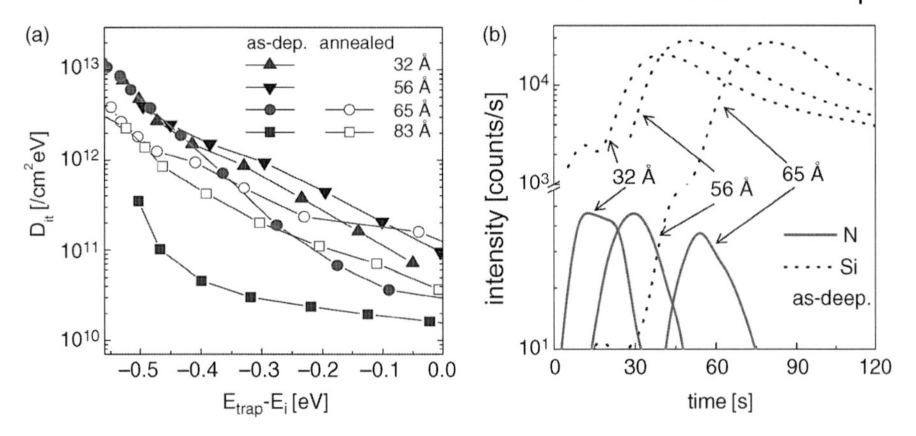

Figure 4.6 (a) D_{it} of the HfO_2 films deposited using the $Hf[N(CH_3)_2]_4$ with various thicknesses as a function of $E_{trap}–E_i$ and (b) low-energy SIMS depth profile results of the 32, 52, and 65 Å thick films [11].

The CET and thermal stability of these HfO_2 thin films were improved considerably due to this thin *in situ* $Si(O)N_x$ RBL, which prevents substrate Si diffusion into the HfO_2 film. The lower CET and leakage current density shows a high-*k* property without adopting intentional barrier layers. In addition, D_{it} shows a low $10^{10}\,cm^{-2}$ eV^{-1} value near the mid-gap energy state due to the presence of a $Si(O)N_x$ layer, even in a reduced thickness, as shown in Figure 4.6. The effect of the O_3 oxygen source was also examined with the $Hf[N(CH_3)_2]_4$ precursor at a deposition temperature of 300 °C [8]. The O_3–HfO_2 film contains a much smaller carbon impurity concentration and shows a more amorphous structure from HRTEM analysis than the films grown with H_2O. Because of these structural and chemical improvements in the O_3–HfO_2 film, a three orders of magnitude smaller leakage current density was obtained, as shown in Figure 4.7a, at both the as-deposited and the postannealed states. This was attributed to the higher interfacial potential barrier for the Fowler–Nordheim (FN) tunneling in the O_3–HfO_2 film. The H_2O–HfO_2 films show a negative V_{FB}, as shown in Figure 4.7b, which is induced from the positive fixed charges associated with the $Si(O)N_x$ interfacial layer. These pre-existing defects in the $Si(O)N_x$ interfacial layer results in high hysteresis. The dielectric constant of the as-deposited O_3–HfO_2 film is ∼24 and the leakage current density is $1.6 \times 10^{-7}\,A/cm^2$ at a CET of 1.49 nm.

The effects of the carbon residue from the $Hf[N(CH_3)_2]_4$ precursor on the leakage current density and the TDDB reliability characteristics were further investigated [18]. The use of a higher O_3 density generates HfO_2 stacks with less carbon residue in the film. Although low-O_3(160 g/Nm^3) HfO_2 exhibits better thermal stability in the CET with increasing PDA temperature, the overall gate leakage current density (J_g) versus CET is inferior to that of the high-O_3(390 g/Nm^3) HfO_2. This is due to the higher leakage current of the low-O_3 (160g/Nm^3) HfO_2 film. Here, the test sample has a simple metal (Pt)–insulator–semiconductor (MIS) capacitor geometry. First-principles calculations show that the interstitial carbon atoms in the HfO_2 produces

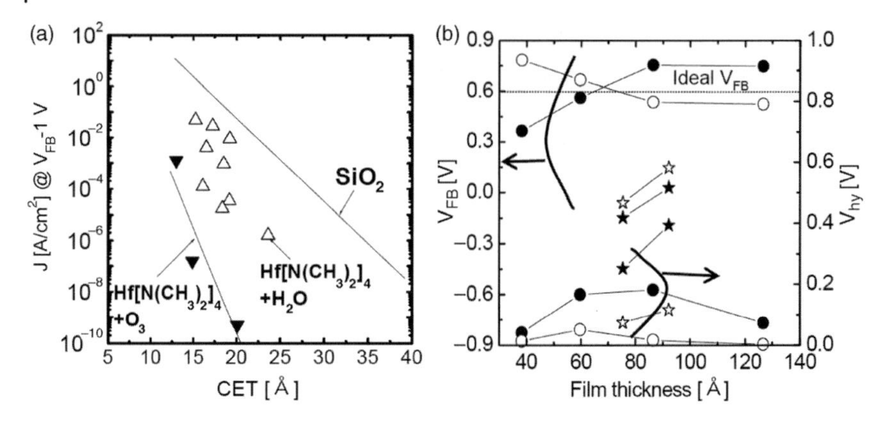

Figure 4.7 (a) J_g at $V_{FB} - 1$ V versus the CET of the Hf[N$(CH_3)_2]_4$-HfO_2 films grown using H_2O or O_3 and (b) the variations in V_{FB} and hysteresis (V_{hy}) of the H_2O-HfO_2 at the as-deposited state (closed star) and after postannealing (opened star), and O_3-HfO_2 at the as-deposited (closed circle), and after postannealing (opened circle) [8].

deep acceptor like trap states in the bandgap, which might enhance the electrical conduction by a trap-mediated conduction mechanism [18, 19]. Accordingly, the 10-year lifetime in TDDB reliability of high-O_3 HfO_2 is guaranteed at -1 V, whereas that of low-O_3 HfO_2 is obtained only at voltages < -0.8 V. Furthermore, the SiO_2 interfacial layer thickness effect on the Hf[N$(CH_3)_2]_4$–HfO_2 stacks was examined by adopting 0.6 and 1.4 nm-thick SiO_2 layers grown with an *in situ* O_3 oxidation process at 300 °C [20]. The interface oxide is grown further during postannealing by the excessive oxygen within the HfO_2 layer. A thicker SiO_2 layer is a better O and Si diffusion barrier. The HfO_2/thick-SiO_2 stack is thermally unstable even after PDA at 1000 °C because Hf-silicate formation is accompanied by the consumption of the interfacial SiO_2 layer. In the case of a HfO_2/thin-SiO_2 stack, Hf-silicate is formed by enhanced Si diffusion from the Si substrate during the same PDA, and inferior thermal stability appears.

The field-effect transistor characteristics with HfO_2 films deposited with the Hf[N$(CH_3)_2]_4$ precursor were followed for a device study [21]. Voltage-induced degradation (VID) near the region of the threshold voltage (V_{th}) was a major focus. The devices with a thin 0.7–0.8 nm SiO_2 layer oxidized *in situ* at a low temperature of 300 °C showed no VID, whereas other samples without the intervening SiO_2 layer showed serious VID (Figure 4.8a; "O_3- SiO_2" stands for the *in situ* O_3-oxidized SiO_2). This was attributed to the smaller interfacial trap density (D_{it}) near the bandgap edge in the device with *in situ* O_3 oxidation process as shown in Figure 4.8b. The gate leakage current is not the major reason for the VID behavior. XPS showed that the thin O_3-oxidized SiO_2 works well as a reaction barrier between the HfO_2 and the Si substrate, which results in a smaller D_{it}.

The impact of the oxygen concentration during HfO_2 deposition on the transistor performance was also reported [22]. In the whole PDA temperature range up to 1000 °C, the film with the lower oxygen concentration (O_3 concentration ~150 g/Nm^3) showed better dielectric performance in terms of the J_g versus CET. This

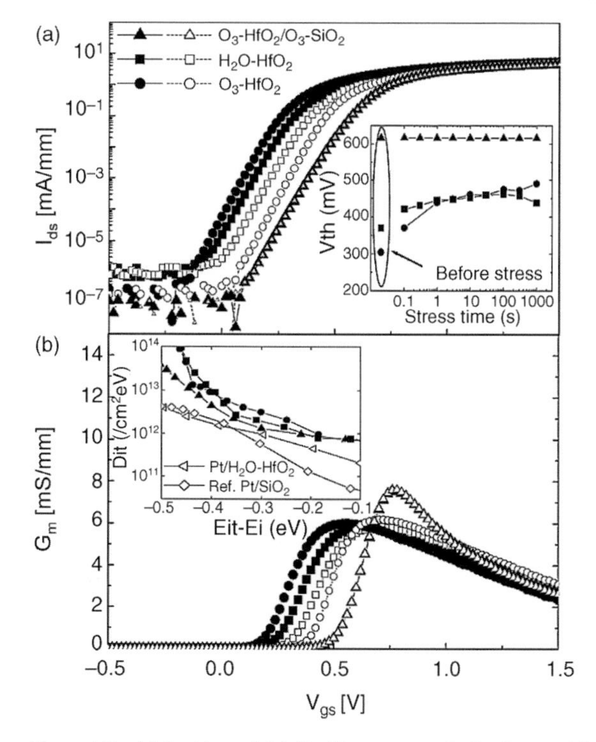

Figure 4.8 (a) I_{ds}–V_{gs} and (b) G_m–V_{gs} curves at the fresh state (closed symbols) and after voltage stress for 1000s (open symbols); insets in (a) and (b) show the changes in V_{th} as a function of the stress time and D_{it} as a function of E_{it}–E_i, respectively [21].

is in contradiction to the MIS capacitor case discussed above, which is probably due to the different gate material (poly-Si in MOSFET versus Pt in MISCAP) and process conditions. The low D_{it} level (~5×10^{10} cm^{-2} eV^{-1}) was maintained only up to a PDA temperature of 700 °C in the case of a low oxygen concentration. A higher PDA temperature or higher oxygen concentration seriously degrades D_{it} because of the interfacial oxidation of the Si(O)N_x RBL introduced from the Hf[N$(CH_3)_2]_4$ precursor.

The use of a low O_3 concentration produces a stoichiometric HfO_2 film that minimizes D_{it} degradation during MOSFET fabrication with the conventional poly-Si electrode process. D_{it} of a HfO_2 film can be as small as that of a SiO_2 film. On the other hand, even with such an optimized HfO_2 layer, Figure 4.9a and b shows that the electron mobility of a MOSFET is still ~65% of that of SiO_2. This is due to carrier scattering by nonnegligible fixed charges and long-range optical phonons.

4.2.1.3 **Tetrakis Ethylmethylamino Hafnium (Hf[N(C_2H_5)(CH_3)]$_4$)**

The adoption of nitrogen-containing Hf[N$(CH_3)_2]_4$ precursor in the ALD process formed an *in situ* Si(O)N_x RBL that suppressed the interfacial reaction between the HfO_2 dielectric and the Si substrate. This enhanced the thermal stability of the capacitance density and reduced D_{it}. Although D_{it} of the capacitor is excellent, the electron mobility of the MOSFET with a conventional poly-Si electrode was almost

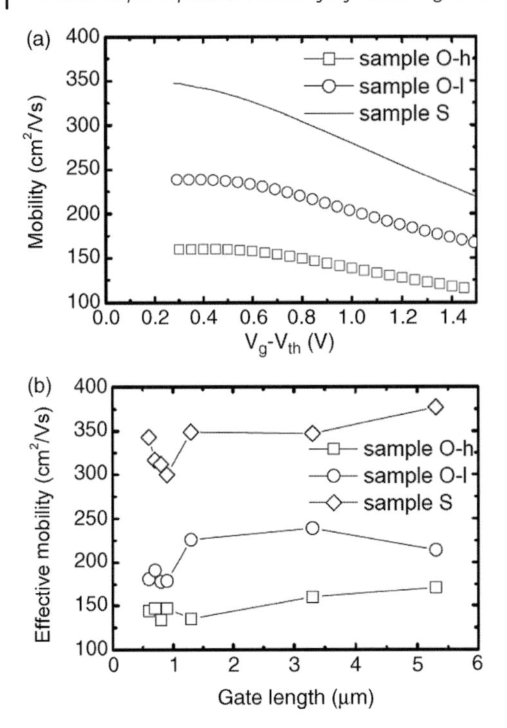

Figure 4.9 (a) Electron mobility versus V_g overdrive of the devices with thermally grown SiO_2 (S), low-O_3 HfO_2 (O-l), and high-O_3 HfO_2 (O-h) at a gate length of 0.5 um; (b) effective electron mobility variation as a function of gate length [22].

half of that from the SiO_2 dielectric due largely to the degraded D_{it} ($\sim 10^{12}\,cm^{-2}$ eV^{-1}). Another N-containing Hf precursor, $Hf[N(C_2H_5)(CH_3)]_4$), was used to examine the possible improvement keeping the *in situ* $Si(O)N_x$ RBL.

Capacitors using a poly-Si gate electrode were fabricated using the $Hf[N(C_2H_5)(CH_3)]_4$ precursor and either an O_2 or N_2O oxygen source [23]. The N_2O oxygen source increases the interfacial $Si(O)N_x$ RBL formation during deposition, which further suppresses the interaction between the HfO_2 layer and Si substrate. In addition, the D_{it} by the poly-Si gate process is minimized in the N_2O–HfO_2 case. The N_2O–HfO_2 stack showed an approximately one order of magnitude lower D_{it} than the O_2–HfO_2 stack near the mid-gap. The $Si(O)N_x$ RBL enhanced by the N_2O oxygen source also minimizes EOT degradation so that an EOT of 1.52 nm was obtained with a leakage current density of $1.5 \times 10^{-6}\,A/cm^2$ at -1 V, even after a dopant activation annealing at 1000 °C.

4.2.1.4 *tert*-Butoxytris[Ethylmethylamido] Hafnium ($HfO^tBu[NEtMe]_3$)

This section discusses the usefulness and limitations of heteroleptic-Hf precursor (*tert*-butoxytris[ethylmethylamido] hafnium ($HfO^tBu[NEtMe]_3$). The structure of $HfO^tBu[NEtMe]_3$ is similar to that of tetrakis(ethylmethylamido)hafnium (Hf(NEtMe)$_4$) except that one of its four amido ligands is replaced by a *tert*-butoxy ligand.

This heteroleptic structure largely improves the ALD growth rate (0.16 nm/cycle) and Hf density [24]. The *in situ* formed interfacial $Si(O)N_x$ layer was also found from XPS and secondary ion mass spectroscopy (SIMS). The Hf mass per unit volume in the HfO_2 film is 7.6 g/cm^3, which is improved by ~20% compared to the $[N(CH_3)_2]_4$-HfO_2.

The HfO_2 film deposited with O_3 showed a self-regulated ALD growth behavior at a growth temperature of 300 °C with highest growth rate as 0.16 nm/cycle among those reported for HfO_2 ALD processes. The higher Hf density induces anticrystallization properties in the as-grown film. Consequently, the amorphous phase of the HfO_2 film is retained up to ~15 nm. The amorphous nature and higher Hf density are also maintained after PDA, which strongly enhances the thermal stability of the electrical performance. The CET of the films with thicknesses ranging from 4 to 13 nm is relatively constant up to a PDA temperature of 1000 °C (Figure 4.10). However, the CET saturated at ~ 2 nm even with the decrease in physical thickness of the dielectric layer. This is related to the serious oxidation of the Si substrate induced by using this precursor. Therefore, further optimization, in terms of the O_3 concentration, precursors and O_3 feeding time, and process sequence, will be needed to fully utilize the positive aspect of this precursor.

4.2.1.5 ***tert*-Butoxide Hafnium ($Hf[OC_4H_9]_4$)**

A N-free metal-organic precursor was investigated to observe the nitrogen effect. *tert*-Butoxide hafnium was used to grow a 2 nm-thick ALD HfO_2 on a 1 nm O_3–SiO_2 layer,

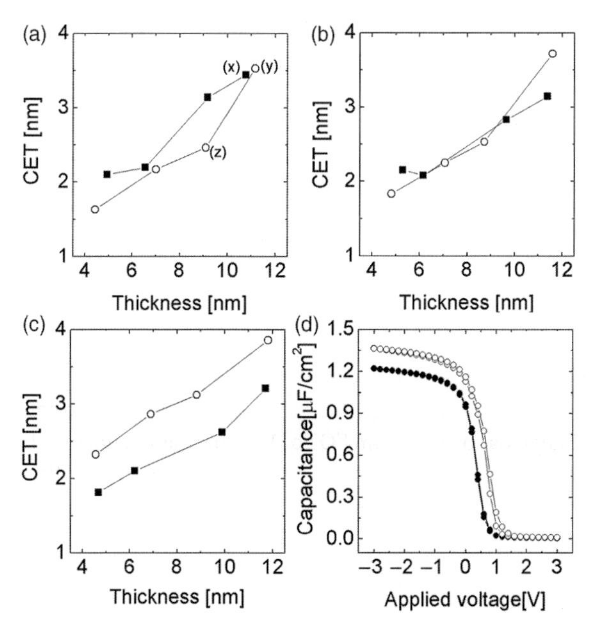

Figure 4.10 Changes in the CET of (a) as-grown, (b) 700 °C annealed, and (c) 1000 °C annealed HfO_2 films using $HfO^tBu[NEtMe]_3$ (closed circles) and $[HfN(CH_3)_2]_4$ (open circles), respectively, as a function of the film thickness. (d) *C–V* curves of the HfO_2 films annealed at 700 °C [24].

which was formed immediately before the ALD by *in situ* O_3 oxidation of the Si substrate. The stack shows less carbon residue remaining in the HfO_2 film, and no V_{FB} shift with thermal treatment, a lower leakage current [25].

The lack of N in the interfacial layer results in a stable V_{FB}. A similar CET from with/without the *in situ* O_3–SiO_2 interfacial layer was obtained before and after PDA. Furthermore, the O_3–SiO_2 layer enhances the layer-by-layer growth behavior of the HfO_2 film resulting in an atomically uniform amorphous HfO_2 microstructure, even after RTA. This improvement contributes to the better reliability in the V_{FB} of the film, as shown in Figure 4.11. A very small D_{it} ($3.3 \times 10^{10}\,cm^{-2}\,eV^{-1}$) near the mid-gap was obtained. Therefore, interposing a thin *in situ* O_3–SiO_2 layer between the HfO_2 and the Si substrate is an effective way of improving the interface properties of the HfO_2/Si gate dielectric stack. The lack of a N-containing interfacial layer may have induced the severe interfacial reaction during the PDA, but the much thinner layer (2 nm) compared to the previous cases (~ 4–10 nm) using other precursors reduced the adverse reaction. This was attributed to the lower total excess of oxygen in the HfO_2 film, or the more stoichiometric composition of this film.

4.2.2 Oxygen Sources and Reactants

4.2.2.1 H_2O versus O_3

The oxygen source greatly affects the properties of ALD high-*k* films because the oxygen source in the ALD process is not only the source of oxygen in the film but also the reaction agent that removes or exchanges the ligand molecules bonded to the metal ion in the metal precursor during the ALD reaction [26, 27]. In addition, the interfacial reactions with the Si substrate during ALD growth, such as intermixing of the related elements and Si surface oxidation, depend heavily on the type of oxygen

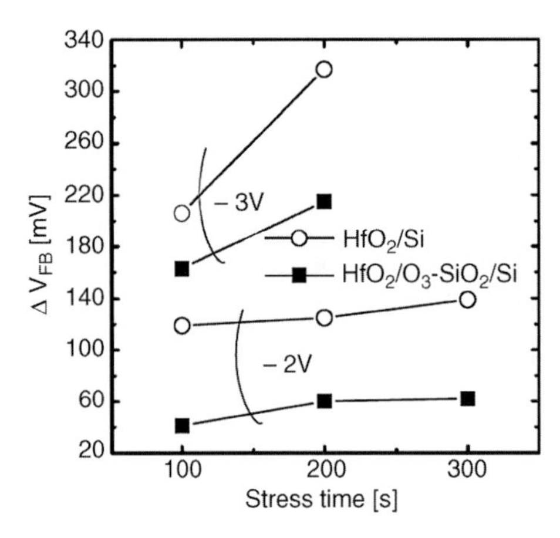

Figure 4.11 V_{FB} variations as a function of the constant voltage stress (at −2 and −3 V) on the HfO_2 films grown with/without O_3–SiO_2 interfacial layer [25].

source. Therefore, the effects of various oxygen sources in the ALD process on the high-*k* film properties should be understood.

H_2O has been used as a popular oxygen source from the early stages of the ALD process development because it provides a very facile ligand exchange reaction with one of the best ALD precursors, trimethyl aluminum (TMA), which has been used for the typical Al_2O_3 ALD process [28]. On the other hand, as the chemical structure of the metal precursors becomes much more complicated for various purposes, such as thermal stability and efficiency of the chemical reaction, the reactivity and oxidation power of H_2O is not sufficient to remove or exchange ligand molecules completely during ALD reaction, leaving residual impurities in the film. Therefore, O_3, which is an oxidizing agent with stronger oxidation power and high reactivity than H_2O, has been used as an alternative oxygen source for the ALD of several high-*k* films. O_3 and H_2O, as an oxygen source in the ALD of high-*k* films, are compared below on the basis of several related studies.

The growth behaviors of the ALD high-*k* film on a Si substrate, such as film growth rate, Si diffusion, and initial growth behavior, are influenced significantly by the oxygen source. This is because the Si surface status and ALD reaction mechanism are affected by the oxygen source. The incubation cycle was observed in the ALD HfO_2 film grown using H_2O on the HF-last Si substrate [29], which originates from the lack of reactive sites. This unfavorable nucleation behavior induced island-type film growth.In contrast, O_3 oxidizes the Si surface from the initial stages of the film growth, which effectively eliminates the incubation cycle, as shown in Figure 4.12a, where the thicknesses of the HfO_2 film grown using O_3 and H_2O on a Si wafer as a function of the deposition cycle number are displayed [30]. The very thin surface SiO_2 layer formed by O_3 during ALD growth provides the films with a sufficiently high density of reactive sites that induces ideal layer-by-layer growth. Figure 4.12b shows the interfacial layer thickness of HfO_2 films with the thickness of 2–4 nm grown on the Si with HF-last, chemical oxide, or

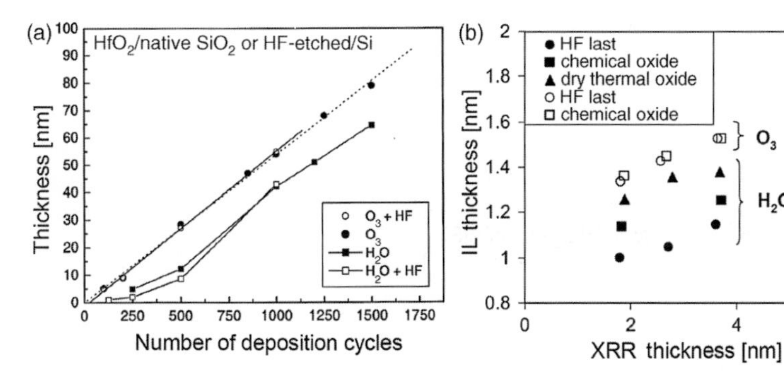

Figure 4.12 (a) The dependence of HfO_2 film thickness on the number of deposition cycles. The inset labels denote the oxygen source. (b) Interfacial layer thickness determined by subtracting the X-ray reflectometry (XRR) HfO_2 thickness from the spectroscopic ellipsometry (SE) thickness for 2–4 nm HfO_2 films deposited with Hf last, chemical oxide, and dry thermal oxide [30, 31].

dry thermal oxide. The O_3 oxidized HF-last Si surface to a similar thickness with the chemical oxide by IMEC-clean during the initial film growth stage. But in the H_2O process, the interfacial layer was clearly thinner than the intentionally grown interfacial SiO_2, such as chemical and thermal oxide [31].

The growth rate of the ALD high-*k* film is closely related to the oxygen source. The growth rate of the HfO_2 film grown using O_3 was slightly higher than that of the film grown using H_2O regardless of the Hf precursor, such as alkylamino ligand based on Hf precursors, cyclopentadienyl-series of Hf precursors, and $HfCl_4$ [30–32]. Stronger oxidation power and higher reactivity of O_3 compared to H_2O resulted in a higher growth rate of film. The stronger oxidation potential of the oxygen source removes the ligands more efficiently during the precursor pulse/adsorption step, which reduces the steric hindrance. A similar influence of the oxygen source during the precursor pulse step was reported in the case of a plasma activated N_2O oxygen source by Won *et al.* [33]. This was also observed for the ALD of another important high-*k* film for capacitor application, TiO_2, by Choi *et al.* [34] and Won *et al.* [33]. The N ions temporarily present on the surface appear to play a key role in enhancing the removal of ligands, which eventually increases the growth rate [35].

The Si diffusion behavior during ALD affects the Si concentration in the high-*k* film, the interfacial layer formation, the effective *k*-value, reliability, and interface properties of the film. This issue is also closely related to the type of oxygen source because the physical density, crystallinity, and film morphology, which affect the Si diffusion behavior, should be influenced by the type of oxygen source. One of the important factors for determining the morphology of the high-*k* film is the initial film growth behavior. The initial island-like film growth in the HfO_2 grown using H_2O results in a smaller grain size (larger number of grain boundary through the film) in the film compared to the film grown using O_3 [30], which enhances Si diffusion into the high-*k* film during ALD and PDA [15]. For the same reason, the high-*k* film grown with H_2O has a higher surface roughness than the film grown with O_3 [15, 30]. The lower physical density of the high-*k* film due to the higher impurity concentration could enhance Si diffusion. The interfacial layer growth during ALD, which results in the high EOT, was enhanced in the O_3 process due to severe surface oxidation of the Si substrate. Park *et al.* recently reported that the O_3 process suppressed a silicate phase formation by retarding Si diffusion into the high-*k* film due to the higher physical density of the film, but it was also found to increase the interfacial layer growth during ALD compared to the H_2O process [36].

The residual impurities from the precursor molecules are frequently observed in ALD high-*k* films. Understanding the effects of the oxygen source on the incorporation of these impurities is essential because impurities in the gate dielectric high-*k* film, even in minute quantities, affect the performance and reliability of the devices significantly. It was reported that the strong oxidation power and high reactivity of O_3 reduced the impurity concentration in the ALD high-*k* film, which originates from the precursor molecules, such as Cl, C, H, N, and so on [8, 15, 36]. SIMS depth profiling showed that the Cl concentration in the ALD HfO_2 film grown with $HfCl_4$ was far lower in the case of the films grown using the O_3 oxygen source compared to that of the film grown using H_2O [15]. Niinistö *et al.* also reported that the concentrations of Cl, C and H impurities in the HfO_2 films grown at 300–400 °C using various cyclopentadienyl-type precursors and O_3 were lower than those in the film grown with H_2O [30]. Cho and co-workers reported that the O_3 process reduced the C impurity concentration in the HfO_2 grown with tetrakis(dimethylamino) hafnium compared to the H_2O process [15]. Therefore, O_3 as an oxygen source is essential for decreasing the residual impurity contamination level in the ALD film.

Swerts *et al.* reported that the C impurity concentration had a different temperature dependence according to the O_3 and H_2O oxygen source. Although the C impurity concentration in the HfO_2 grown using H_2O increased with increasing growth temperature, that in the HfO_2 grown using O_3 decreased [31]. As a result, the O_3 process results in a lower C concentration in the HfO_2 film grown at 345 °C but a higher C concentration in the film grown at 285 °C. On the other hand, Liu *et al.* reported that H_2O as an oxygen source decreased the C and H impurity concentrations in the HfO_2 film grown with tetrakis(ethylmethylamido)hafnium compared to O_3 at growth temperatures of 200–320 °C [32]. While the effects of the oxygen source on the impurity concentration in the ALD high-*k* film are controversial, Park *et al.* recently reported the detailed XPS results on this issue, where they observed the impurity behaviors in ALD La_2O_3 films using high-resolution *in situ* XPS.

Figure 4.13 shows the C 1s and N 1s core-level spectra of the La_2O_3 films grown using H_2O and O_3 [36]. The C- and N-related residual impurities, which originated from the incomplete reactions of precursor molecules during ALD growth, were suppressed in the films grown using O_3. However, the use of O_3 resulted in an undesirable La-carbonate phase in the film [36]. It must be noted that this could be specific problem of the La-based material; Hf- or Zr-based materials may have different behaviors. Most previous results on the impurity concentration were evaluated by SIMS, which barely distinguishes the impurity binding status. XPS, especially *in situ* XPS, is a valuable tool in this regard.

The different chemical and physical properties of the ALD high-*k* films grown using H_2O and O_3 induce different electrical properties of the films. The permittivity

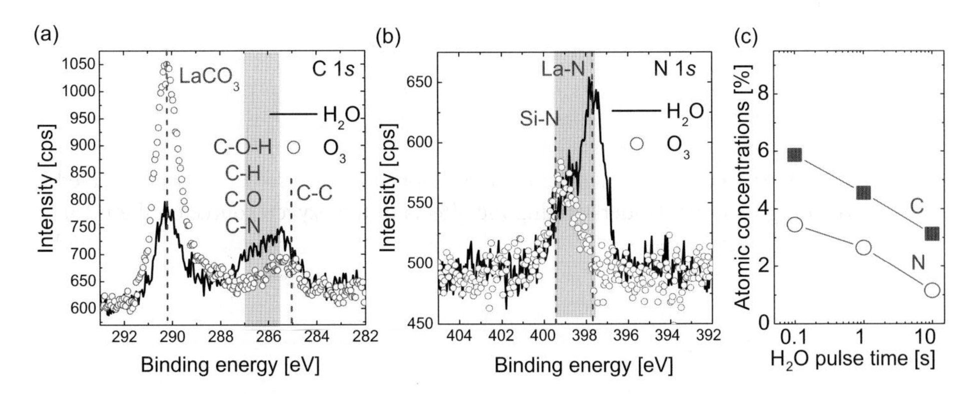

Figure 4.13 (a) C 1s and (b) N 1s core-level spectra of the La_2O_3 films grown using H_2O and O_3 with 30 ALD cycles; (c) C and N concentration in the film estimated by XPS as a function of the H_2O feeding time [36].

of amorphous/monoclinic, cubic, and tetragonal HfO_2 (bulk) is known to be 16–24, 29, and 40, respectively [37–40]. Park *et al.* reported that the effective permittivity, including the interfacial layer of the HfO_2 film grown using $HfCl_4$, is 14.7 and 15.6 for the H_2O and O_3 process, respectively [15]. The permittivity of the film grown using O_3 is slightly higher than that of the film grown using H_2O. Cho *et al.* and Kukil *et al.* also reported that the effective permittivity of a HfO_2 film grown using an amino-type precursor ($Hf[N(CH_3)_2]_4$ and $Hf[N(C_2H_5)(CH_3)]_4$) was 14.5–14.9 and 15.6 for the H_2O and O_3 process, respectively [8, 41]. Figure 4.14 shows the typical J_g–V and J_g–E curves for the HfO_2 film grown using H_2O and O_3 before and after PDA. The breakdown voltage (field) of the HfO_2 film was increased significantly by the O_3 process compared to the H_2O process due to the lower impurity concentration in the film [8, 15, 30]. On the other hand, Niinistö *et al.* reported a slightly higher effective permittivity of HfO_2 grown with H_2O (~13) than that of the HfO_2 film grown with O_3 (~12) [30]. The slightly better dielectric properties (J_g versus EOT) of the HfO_2 film with the grown H_2O due to the thinner interfacial layer were also observed by Swerts *et al.* [31]. Therefore, the merits of O_3 over H_2O are still somewhat controversial.

The fixed charge density and charge trap (border trap) density of the high-*k* film can be estimated from the V_{FB} shift and hysteresis characteristics of the MOS capacitors, respectively. Fewer (positive) fixed charges and charge traps were observed in the HfO_2 film grown using O_3 compared to that grown using H_2O [8, 15]. Niinistö *et al.* reported fewer charge traps (border traps) in the HfO_2 film grown using O_3, but a higher (positive) fixed charge density [30]. The D_{it} of the high-*k* film is related directly to the carrier mobility in the MOSFETs. Although it was reported that the interfacial layer grown at the interface between the high-*k* film and the Si substrate during ALD growth using O_3 improved the D_{it} ($<\sim 10^{11}\ cm^{-2}\ eV^{-1}$ at the mid-gap energy) [42], Baldovino *et al.* reported that D_{it} of the high-*k* film was improved by partial passivation of the interface state as a result of the H from the H_2O molecule during ALD growth [43]. Swerts *et al.* reported that the breakdown field (E_{BD}) of the film is 5–6 and

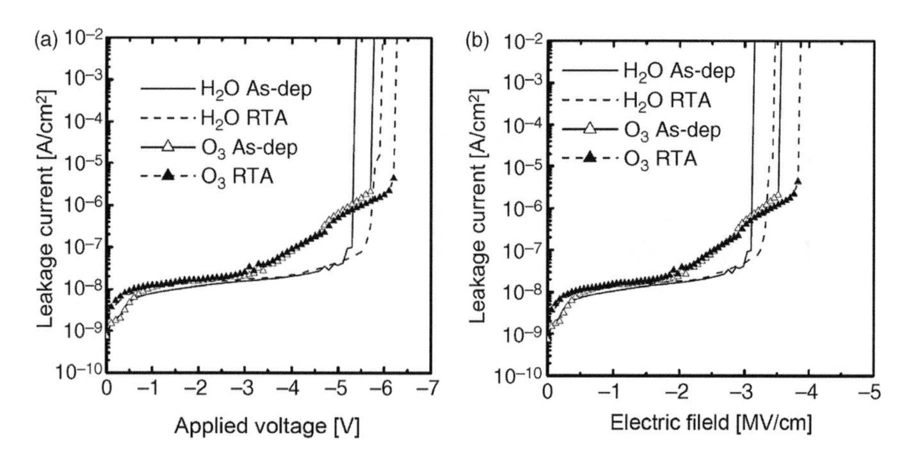

Figure 4.14 (a) J_g–V and (b) J_g–E curves of the MIS capacitor at the as-deposited state and after RTA at 750 °C of the HfO_2 film grown with H_2O and O_3 as oxygen sources [15].

5.5–7 MV/cm for the H_2O and O_3 process [31], respectively. The lower impurity concentration in the film grown using O_3 might be the reason for the higher E_{BD}. This trend is in agreement with Figure 4.14.

4.2.2.2 O_3 Concentration

The oxidation power or reactivity of the oxygen source affects the film bulk properties, such as physical density, impurities level, oxygen concentration, and the interface reactions with the Si substrate. Therefore, the O_3 concentration in the high-k metal–oxide ALD process is essential to the electrical, physical, and chemical properties of the film. As stated previously, O_3 oxidizes the Si substrate forming a more severe interfacial layer than H_2O during ALD growth due to the stronger oxidation power [8]. Hence, a high O_3 concentration induces a thicker interfacial layer. Therefore, the effective permittivity of the HfO_2 film including interfacial layer grown using O_3 with the lower concentration is higher than that with the higher concentration [44]. The HfO_2 film grown using a low O_3 concentration showed better dielectric performance in the J_g versus EOT plot [45]. On the other hand, the HfO_2 film grown using a high concentration of O_3 showed better thermal stability against performance degradation after a high-temperature PDA because the thicker interfacial layer suppressed the interfacial reactions during PDA and integration process steps [45]. On the other hand, Cho *et al.* and Park *et al.* reported that an excess oxygen concentration in the film grown using a high O_3 concentration increases the EOT further after PDA due to the more serious interfacial oxidation [18, 22].

The D_{it} also depends on the O_3 concentration during ALD growth. Figure 4.15 shows the changes in D_{it} as a function of the energy level in the Si bandgap of the HfO_2 films grown using O_3 at a concentration of 160 and 370 g/Nm^3 after PDA at temperatures ranging from 600 to 1000 °C. The D_{it} of the HfO_2 film grown using lower O_3 concentration was low over the entire energy range until the PDA temperature was 700 °C [42], whereas it increases at higher PDA temperature. The sample grown with the higher O_3 concentration always showed a high D_{it} level.

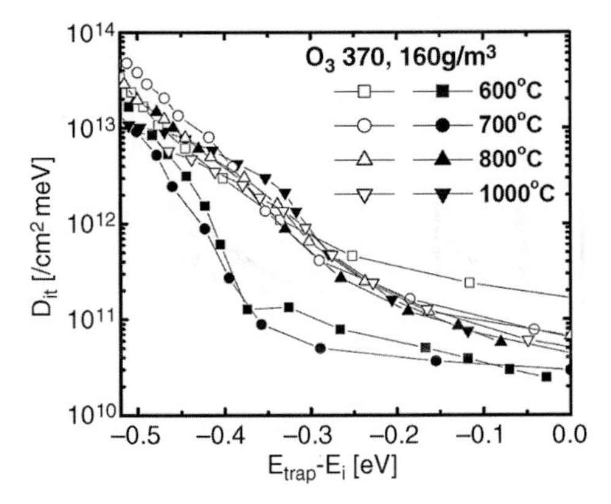

Figure 4.15 D_{it} as a function of the energy level in the Si bandgap of the HfO_2 films grown using O_3 at a concentration of 160 and 370 g/Nm^3 after PDA at 600, 700, 800, and 1000 °C [42].

The D_{it} degradation with increasing PDA temperature was attributed to the serious interfacial reaction. The lower D_{it} of the film grown using lower O_3 concentration is supported by Kamiyama *et al.* and Chung *et al.*, where the humps observed in the *C–V* curve reflecting the D_{it} of the film increased with increasing O_3 concentration [44, 45]. The excess oxygen in the film grown in a higher O_3 concentration induced excess bonding of interfacial Si with oxygen, resulting in serious D_{it} degradation [42]. The electrically unstable interfacial layer may have been forced to grow by O_3 at a low temperature (typical ALD temperature window of 250–300 °C) [44].

As mentioned above, O_3 usually decreased the impurity level in the high-*k* film due to the high reactivity and strong oxidation power. Hence, a higher concentration O_3 reduces the impurity level further. Figure 4.16 shows the SIMS depth profiles of the HfO_2 films grown using O_3 at concentrations of 25, 50, and 200 g/Nm^3. The C and N impurity concentrations decreased with increasing O_3 concentration [18, 44], which affect the reliability of the high-*k* films, such as breakdown voltage (time-zero dielectric breakdown, TZDB) and TDDB. The E_{BD} of the films by TZDB increased with increasing O_3 concentration before and after PDA [45]. TDDB analysis with various constant voltage stresses (CVS) and the lifetime extrapolation shows that the HfO_2 films grown in a high O_3 concentration have a better long-term reliability and longer device lifetime [18]. Up to the hard breakdown in both TDDB and TZDB, the localized current path, that is, percolation path, formation is expected to be enhanced by the carbon impurities in the film. Therefore, the lower carbon impurity concentration in the film results in better reliability [46, 47].

There is another important point regarding O_3 production. O_3 gas is generated mostly with the assistance of a small quantity of N_2 gas to improve the generation efficiency. The corrosive nitric acid vapor [48], which is generated as a by-product during the O_3 generation process, etches the stainless steel gas delivery tubes resulting in metal impurities in the film. These kinds of unexpected metal impurities can act as high leakage current paths and charge trapping sites in the gate dielectric film, which may cause high power consumption, poor long-term reliability and

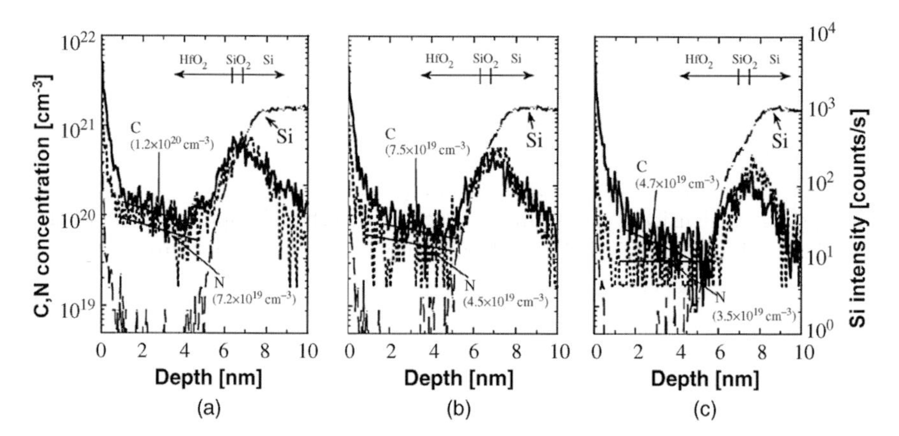

Figure 4.16 SIMS depth profiles of the ALD HfO_2/Si structures. The O_3 concentrations used were (a) 25, (b) 50, and (c) 200g/Nm^3, respectively.

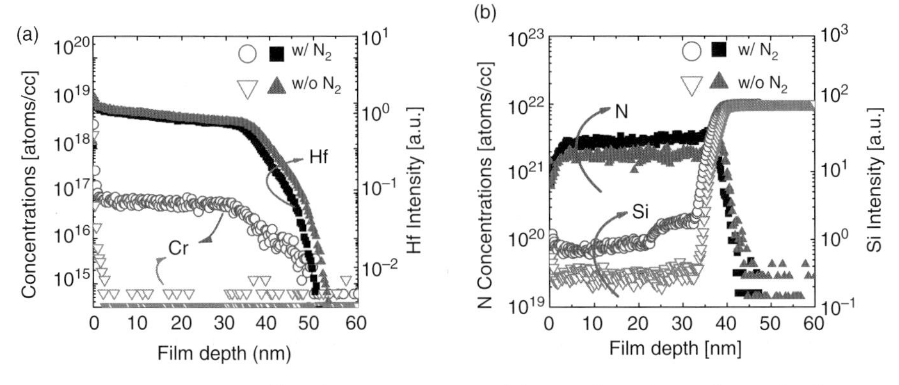

Figure 4.17 SIMS depth profiles of (a) Cr impurities and (b) N and Si in the as-deposited HfO_2 films grown using O_3 generated with and without N_2 assistance [52].

threshold voltage instability of MOSFETs [49–51]. Recently, Park *et al.* reported the characteristics of a HfO_2 film grown using O_3 generated with the ultrashort gap discharge technique without N_2 assistance [52]. The SIMS depth profiles of the HfO_2 film grown using O_3 generated with and without N_2 assistance in Figure 4.17 showed fewer Cr, N and Si impurities in the HfO_2 film grown using O_3 generated without N_2 assistance compared to the case with N_2 assistance [52]. Cr was originated from the etching of stainless steel tubing. The suppressed Si diffusion into the film originated from the higher physical density of the film grown using O_3 generated without N_2 assistance due to lower impurity concentration. The V_{FB} and hysteresis after CVS before and after PDA were suppressed in the film grown using O_3 generated without N_2 assistance [52]. The TZDB characteristic was also improved by O_3 generated without N_2 assistance. Therefore, the ALD process using O_3 generated without N_2 assistance is promising for the low impurity concentration in the film.

4.2.2.3 **Reactants for *In Situ* N Incorporation**

Nitrogen incorporation into the high-*k* film has been studied extensively for several years [53, 54]. The appropriate concentration of nitrogen at the interface between the high-*k* film and the Si substrate improves the carrier mobility [53]. This is in contradiction with the SiO_2 gate dielectric, suggesting that a high-*k*/interfacial layer/Si interface generally has a more defective nature, even in the presence of a SiO_2-like interfacial layer. In addition, the nitrogen in the high-*k* film suppresses the leakage current density by passivating the electrical defects and improves the thermal stability, such as Si out-diffusion and crystallization temperature. Nitrogen incorporation is usually performed with thermal or plasma annealing using NH_3 or N_2O gas [55, 56]. On the other hand, the high thermal budget, serious plasma damage and excessive nitrogen incorporation at the interface by these postdeposition nitridation technique result in mobility degradation and impaired reliability [53, 54]. The effectiveness of using a N-containing precursor (alkyl-amino precursors) in incorporating the interfacial $Si(O)N_x$ layer was discussed in detail above. The N-concentration incorporated using only a N-containing precursor in combination

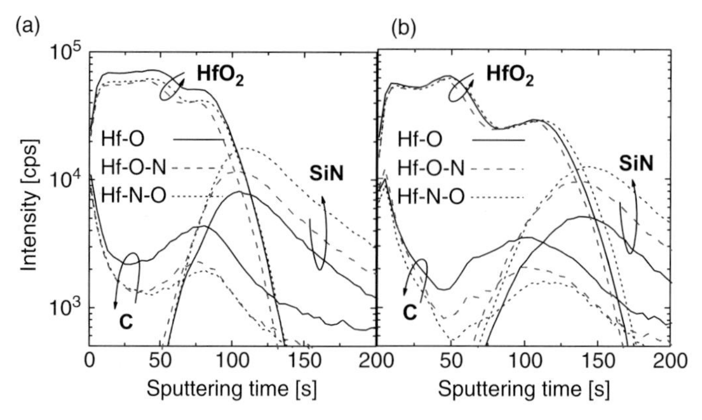

Figure 4.18 SIMS depth profiles of HfO_2 films grown with *in situ* NH_3 pulse before (Hf–N–O) or after oxygen source (O_3) pulse (Hf–O–N), (a) before and (b) after PDA at 1000 °C [57].

with the highly oxidizing O_3 was not sufficiently high. The use of a reactant containing nitrogen, such as NH_3, NH_4OH, and low-temperature N_2/O_2 mixture plasma enables to control the nitrogen profile in a high-k film without a high thermal budget and plasma damage during ALD growth [57–60]. Figure 4.18 presents the SIMS depth profiles of HfO_2 films grown with *in situ* NH_3 pulse before (Hf–N–O) or after oxygen source (O_3) pulse (Hf–O–N), (a) before and (b) after PDA at 1000 °C [57]. The Si(O)N_x layer was formed successfully by *in situ* nitridation using a NH_3 pulse, where the N concentration was certainly higher than the case without the intermediate NH_3 pulse steps. The *in situ* Si(O)N_x layer has several benefits, such as suppressed Si out-diffusion into the film, reduced leakage current density, and improved D_{it}, as mentioned above [11, 57]. The *in situ* nitridation of the ALD Ta_2O_5 film using NH_4OH/H_2O vapor also suppressed the stretch-out of the *C–V* curve, which suggests improved D_{it}. The appropriate *in situ* nitridation using low-temperature N_2/O_2 mixture plasma during ALD growth was reported to improve D_{it} of the film [60]. The nitrogen atom in the high-*k* film was reported to passivate the electrical defects by oxygen vacancies [61].

Achieving the direct tunneling current behavior in the Ln J_g–$E^{1/2}$ plot of the HfO_2 film grown with *in situ* NH_3 pulse in Figure 4.19a suggests that the electrical defects possibly related to oxygen vacancies were almost eliminated by the *in situ* NH_3 pulse. The HfO_2 films grown without the *in situ* NH_3 pulse showed defect-mediated leakage behavior, such as Poole–Frenkel or trap-assisted tunneling behavior. The schematic band diagrams of HfO_2 films in Figure 4.19c show the direct tunneling mechanism without electrical defects in the HfO_2 film grown using the *in situ* NH_3 pulse while Figure 4.19b shows the adverse influence of the defects on the current conduction. The dielectric breakdown characteristics were also improved by *in situ* nitridation during ALD. The improvement in the dielectric breakdown by the *in situ* NH_3 pulse was attributed mainly to the low C impurity concentration in the film [18, 57]. The SIMS depth profiles in Figure 4.18 show that the C impurity concentration was reduced by the *in situ* NH_3 pulse. Since the interstitial C atom induces a deep defect state from the Si conduction band edge, the reduced C impurity concentration by

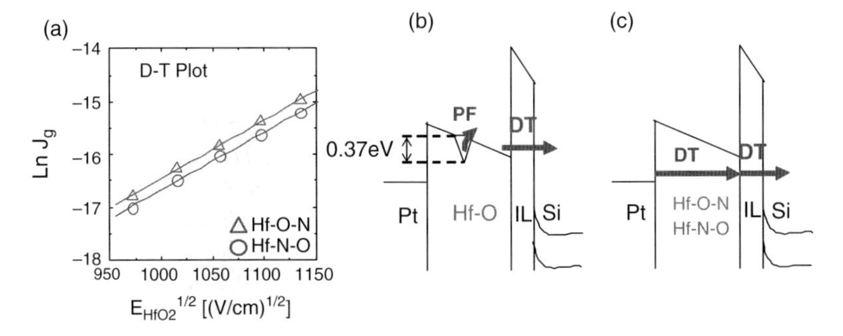

Figure 4.19 (a) Direct tunneling plot of HfO_2 films grown with *in situ* NH_3 pulse before (Hf–N–O) or after oxygen source (O_3) pulse (Hf–O–N) at 60 °C. (b) Band diagrams of the HfO_2 films without *in situ* NH_3 pulse (Hf–O), showing the P–F conduction mechanism in the upper layer and direct tunneling mechanism in the interfacial layer. (c) Band diagrams of Hf–N–O and Hf–O–N samples, showing the direct tunneling conduction mechanism across all dielectrics [57].

in situ NH_3 pulse contributes to the better breakdown characteristics. Improved breakdown characteristics were also observed in the high-*k* film grown using low-temperature N_2/O_2 mixture plasma and NH_4OH/H_2O vapor [58, 59].

4.3
Doped and Mixed High-*k*

Further scaled MOSFETs (gate length < 20 nm) require post-HfO_2 gate dielectric materials with superior properties to those of conventional high-*k* materials [62, 63]. One of the popular methods for improving the properties of high-*k* films is based on the doping or mixing of another material into the conventional binary oxide high-*k* films. The term "doping" has been used for the much smaller concentration of impurities in single-crystalline semiconductors, and, thus, it may not be an appropriate term. However, this term is generally used in this field. The permittivity of HfO_2 was reported to be increased by doping with ZrO_2, Al_2O_3, SiO_2 and so on, where the crystalline structures of HfO_2 were modified by doping from monoclinic (~15–17) to tetragonal or cubic phases with a higher permittivity (~30–40) [37–40]. Another way of enhancing the permittivity is to mix high-*k* materials (e.g., HfO_2 and ZrO_2) with higher *k* materials (e.g., TiO_2). Superior properties to the binary transition metal oxide can be acquired if the permittivity and band gap offsets can be optimized. On the other hand, the reduced gate leakage current can be obtained by suppressing the crystallization of the high-*k* films using La_2O_3 as a dopant [64] because these can suppress the formation of grain boundaries that act as a path for the high leakage current. Among the several methods, this section covers the effects of doping on the properties of a HfO_2 film. Each dopant (or alloying element), such as Al_2O_3, SiO_2, and ZrO_2, has a unique feature as well as common effects, which are discussed and summarized in the following sections, with ZrO_2 being discussed first.

4.3.1 Zr-Doped HfO_2

Although ZrO_2 has almost identical structural and chemical properties to those of HfO_2 [65], it has not received as much attention as HfO_2 because of its relatively insufficient thermal stability with Si, which may induce chemical reactions with the poly-Si gate electrode as well as the channel region [66]. ZrO_2 also has a slightly smaller bandgap compared to HfO_2, which can cause a leakage current issue [63]. However, as the poly-Si may not be used as the gate material in future MOSFETs, ZrO_2 certainly deserves renewed interests.

The band structures of the as-deposited HfO_2 and ZrO_2 were examined experimentally by combining XPS and Auger electron spectroscopy–reflective electron energy-loss spectroscopy (AES-REELS). Here, HfO_2, ZrO_2, and $Hf_{1-x}Zr_xO_y$ films were grown by ALD at a wafer temperature of 280 °C, of which details can be found in Ref. [67]. ZrO_2 showed a lower conduction band offset (CBO) than HfO_2, whereas the valence band offset (VBO) values remained same [68]. Evolution of the CBO in the $Hf_{1-x}Zr_xO_y$ composite films was investigated by synchrotron X-ray absorption spectroscopy (XAS). XAS with various incident photon energies near the O *K*-edge absorption (~530 eV) reflects the electronic transition from the O 1s core state to the O 2p unoccupied states. Because the O 2p states are well hybridized with the nearest orbital states, O *K*-edge XAS can be used to identify the unoccupied electronic structure of the nearest Hf and Zr ions. Figure 4.20 shows the O *K*-edge XAS spectra of the $Hf_{1-x}Zr_xO_y$ films. The features can be attributed to the composition of Hf 5d and Zr 4d states. As the Zr concentration increases, the feature near the photon energy of 532 eV grows, reducing the CB maximum (CBM) energy. The relative CBM [$-E_{CBM} = E_{CBM}(HfO_2) - E_{CBM}$] from that of HfO_2 can be deduced from an extrapolation of the main features at the steepest slopes. The results are plotted in the inset of Figure 4.20. The values are consistent with the evolution in the bandgap energy, suggesting that the decrease in bandgap with the introduction of Zr atoms is mainly due to the lower CBM energy but the VBM energy is nearly constant.

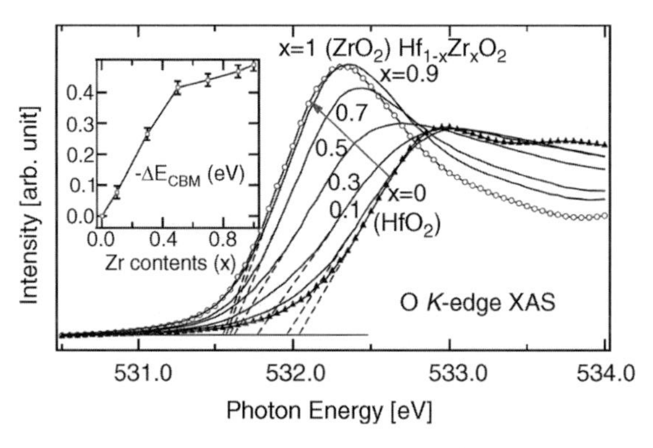

Figure 4.20 O *K*-edge XAS spectra of the $Hf_{1-x}Zr_xO_y$ films. The relative CBM [$-E_{CBM} = E_{CBM}(HfO_2) - E_{CBM}$] from that of HfO_2 can be deduced by extrapolations of the main features at the steepest slopes. This is plotted in the inset of the figure [74].

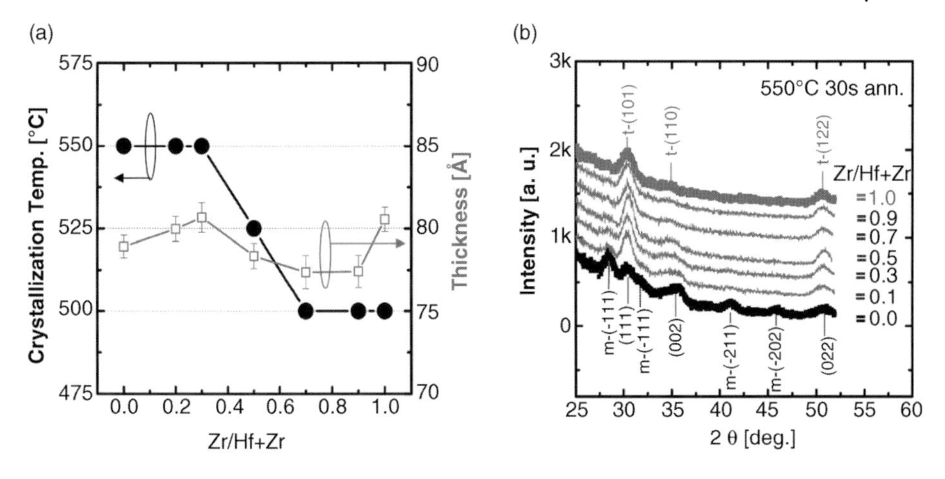

Figure 4.21 The crystallization temperature as a function of Zr compositions in $Hf_{1-x}Zr_xO_y$. (b) GAXRD spectra of $Hf_{1-x}Zr_xO_y$ with various Zr compositions after annealing at 550 °C.

ZrO_2 has a monoclinic, tetragonal and cubic crystal structure depending on the temperature and pressure [69]. In particular, the tetragonal phase shows a higher dielectric constant and smaller grain size in a crystallized film than the monoclinic phase [70, 71]. GAXRD (incident angle of 1°) was performed on annealed $Hf_{1-x}Zr_xO_y$ films with various Zr compositions to understand the effect of the Zr composition on the crystallization temperature and crystalline phase of $Hf_{1-x}Zr_xO_y$ films, as shown in Figure 4.21. ZrO_2 crystallized at lower annealing temperatures than HfO_2. In addition, the crystallization temperature decreased with increasing Zr composition in the $Hf_{1-x}Zr_xO_y$ films. The crystallized HfO_2 films show a strong peak near $2\theta = 28.5°$, indicating that the crystallized HfO_2 films have a monoclinic phase [71–73], whereas ZrO_2 does not show a peak near $2\theta = 28.5°$, and a relatively stronger peak near $2\theta = 30.5°$, which corresponds to the (101) peak for the tetragonal phase [71–73]. The peaks corresponding to the monoclinic phase of $2\theta = 28.5°$ decreased with increasing Zr content, suggesting that the incorporation of Zr in HfO_2 causes a monoclinic to tetragonal transformation of the microstructure. This indicates that the Zr composition in annealed $Hf_{1-x}Zr_xO_y$ films plays an important role in the crystalline structure transformation. A change in the local structure near the Hf and Zr atoms with increasing Zr concentration was observed by X-ray absorption fine structure analysis, showing a monoclinic to tetragonal martensitic transition [74].

The degradation mechanism of Hf-based gate dielectrics under the positive bias temperature instability (PBTI) stress in n-MOSFET was reported to be electron trapping at the preexisting trap sites, which are relatively shallow traps originating from oxygen vacancies [75]. The grain morphology of oxide is also considered to be an important factor that controls the electrical properties and reliability characteristics because oxygen easily diffuses along the grain boundaries to passivate the oxygen vacancies at the grain boundaries or inside the grains. In addition, the crystalline phase is related to the grain size of the crystallized films. As shown in Figure 4.21b, the crystallized ZrO_2 and HfO_2 have a tetragonal and monoclinic phase, respectively. The stabilization of the tetragonal phase in ZrO_2 thin films can be explained by surface energy effects [76]. At room temperature, the monoclinic phase is energetically preferred in bulk materials. On the other hand, the surface energy of the tetragonal phase is generally lower than that of the monoclinic phase. Therefore, the tetragonal phase can be stabilized when the grain size of the oxide material (or thin film) is small, where the gain in the surface energy term may compensate for the loss of bulk energy by the monoclinic to tetragonal phase transformation. As a result, the grain size of the tetragonal phase film is smaller than that of the monoclinic phase film in both HfO_2 and ZrO_2 [72]. By measuring the surface morphology of annealed ZrO_2 and HfO_2 by atomic force microscopy (AFM), ZrO_2 was found to have more uniform and smaller grain size than the annealed HfO_2 [72]. In terms of the V_{th} shift under PBTI stress in n-MOSFET, ZrO_2 shows a much smaller V_{th} shift than that of HfO_2 [67]. This suggests that the origin of the V_{th} shift reduction in ZrO_2 is related to the reduced electron trap density in the bulk ZrO_2. The reduced electron trapping under PBTI stress suggests that the oxygen vacancy concentration is lower in the ZrO_2 film ($Hf_{1-x}Zr_xO_y$) compared to the HfO_2 film. This difference can be attributed to the different passivation processes of the oxygen vacancies in the two films. ZrO_2 having smaller and more uniform grain size makes it easier for oxygen to diffuse into the grains and near to grain boundaries. These diffused oxygen atoms may reduce the oxygen vacancy concentration. The oxygen can be supplied by the residual oxygen in the gas phase or from the oxide film itself.

4.3.2 Si-Doped HfO_2

The existence of a Si-related phase in the high-*k* film is considered undesirable because it normally originates from Si diffusion into the film during ALD growth or PDA resulting in an increased EOT due to the low permittivity of SiO_2 (~3.9). On the other hand, Böscke *et al.* and Tomida *et al.* recently reported that the intentional incorporation of an appropriate amount of Si (doping) increased the permittivity of the high-*k* film by modifying the crystalline structure of the film [77, 78]. This is a different mechanism from that observed in the case of ZrO_2 doping discussed above. There are several doping elements that would or would not induce a structural transition, which has been recently dealt with in detail by Lee *et al.* [37] using a first-principles study. Si with a concentration $<10\%$ could modify the crystalline structure of the HfO_2 film [78]. The permittivity of the Si-doped HfO_2 increased abruptly at the PDA temperature of 700 °C, whereas those of the two films were similar until 600 °C [77]. This suggests that the pure HfO_2 film was crystallized to a monoclinic phase with similar permittivity to the amorphous phase ($<\sim 20$), but the Si-doped HfO_2 crystallized to the tetragonal phase with higher permittivity.

Crystallization of an amorphous high-*k* film generally induces a defective structure related to the grain boundaries, which would act as a detrimental J_g path through the film, resulting in a significant increase in J_g. On the other hand, Si doping reduced this type of defect formation by stabilizing the tetragonal phase, which suppressed the increase in J_g significantly after high-temperature PDA [77]. The J_g for the crystalline (monoclinic) HfO_2 film shows a stronger area dependence than the amorphous HfO_2 film, which suggests that local defects, as mentioned above, were induced by crystallization (monoclinic). On the other hand, the area statistics of J_g for amorphous and tetragonal Si-doped HfO_2 film were similar because the local defect formation during crystallization was suppressed by tetragonal phase stabilization [77]. Films with a highly localized current path show a less area-dependent J_g, whereas the films with more uniformly distributed current paths should have an area-dependent J_g when J_g was calculated simply by dividing the measured current by nominal electrode area.

Optimization of the concentration of Si dopant in the high-*k* film is important. The Si-doped HfO_2 film with Si concentrations ranging from 4 to 10% crystallized into the cubic (or tetragonal) phase, but the pure HfO_2 film crystallized into the monoclinic phase. This phase transformation should affect the permittivity of the films [78]. Figure 4.22 shows the permittivity of the Si-doped HfO_2 films as a function of the Si concentration. The permittivity of the amorphous Si-doped HfO_2 film decreased with increasing Si concentration because of the lower permittivity of the SiO_x phase. On the other hand, the crystallized Si-doped HfO_2 films with the proper Si concentrations have higher permittivity ($\sim$27) than the amorphous films. The high Si concentration suppressed the crystallization of the film so there was no observable change in the permittivity of the film. Nevertheless, such Si composition-dependent crystallization behaviors have been observed from rather thick (>10 nm) films, which have little relevance to actual MOSFET applications. The more relevant thinner films could exhibit a different crystallization behavior, as discussed in the case of Al-doping shown below.

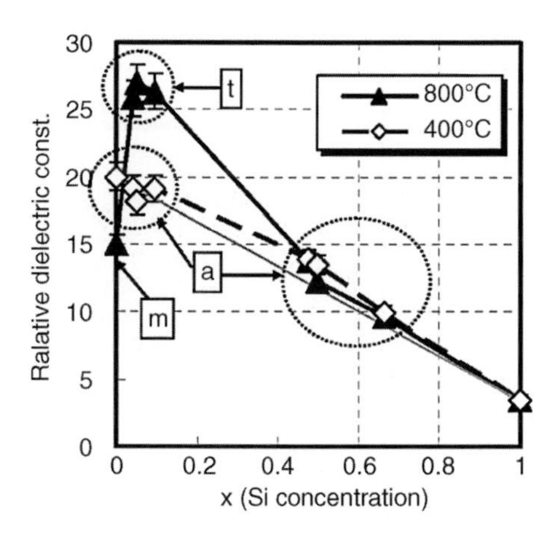

Figure 4.22 *k*-value as a function of the Si concentration (*x*) and annealing temperature. The characters *a*, *m* and *t* denote amorphous, monoclinic and tetragonal structure, respectively [78].

4.3.3 Al-Doped HfO_2

Al_2O_3 is frequently used to improve the thermal stability of high-k HfO_2 films [79]. Al_2O_3 suppresses the crystallization of high-k films and the Si out-diffusion into the high-k film after PDA at a high temperature. Si diffusion into the high-k films was enhanced by the increased annealing temperature, which is suppressed by increasing Al_2O_3 concentration in the film [79]. The detrimental Si diffusion from the substrate can be suppressed in principle by doping the HfO_2 film with other cations, which can be understood from basic thermodynamic considerations; the mixing enthalpy and entropy term of the doped HfO_2 stabilizes the high-k layer so that the driving force for Si diffusion decreases.

The enhanced thermal stability of the film due to Al_2O_3 was also confirmed by XRD analysis of the $HfAlO_x$ films with various Al_2O_3 concentrations [79]. The $HfAlO_x$ film with an Al_2O_3 concentration of ~33% remained amorphous even after PDA at 900 °C. Despite these advantages, the lower permittivity of Al_2O_3 (~9) could be a critical drawback increasing the EOT of high-k films. On the other hand, Park and Kang reported the permittivity enhancement of the HfO_2 film by Al_2O_3 doping [80]. The Al_2O_3 was incorporated by the insertion of a few Al-O ALD cycles during the ALD of HfO_2 film, which stabilized the tetragonal phase of HfO_2 after high-temperature annealing. This is in coincidence with the theoretical expectations by Lee *et al.* [37]. Figure 4.23 shows the XRD patterns of $HfAlO_x$ films with various cycle ratios in the Al-doped HfO_2 ALD (1: number of HfO_2 cycles) after PDA [80]. The XRD peaks corresponding to the tetragonal phases appeared in the $HfAlO_x$ films as the Al-concentration increases, even though only the monoclinic phase was observed in the pure HfO_2 film. The permittivity of the film depends on the crystalline structure, as

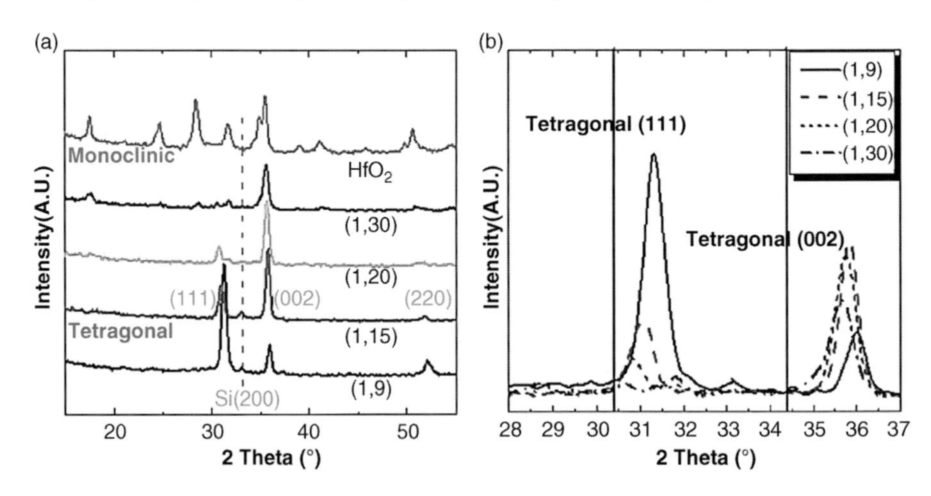

Figure 4.23 (a) XRD patterns of Hf-aluminate films with respect to the number of unit cycles in a HfO_2 subcycle after the annealing process at 700 °C recorded by θ–2θ scanning. (b) A superimposed view of the XRD patterns of the Hf-aluminate films showing the peak position shift toward larger 2θ values [80].

mentioned above. The permittivity of the $HfAlO_x$ films with the appropriate Al_2O_3 concentration increased significantly to ~47 after PDA [80]. The enhancement in the permittivity by the modified crystalline structure of the film overwhelmed the negative effect from the low permittivity of Al_2O_3. Nevertheless, these are the situations where the films are thick enough to be crystallized, which is of little relevance to actual applications. The film properties are also affected by several other factors, such as the diffusion of the substrate elements, impurity level in the film, and bandgap energy of extremely thin film, as well as the doping-induced crystallinity of the high-*k* films. Therefore, a study of these other parameters should lead to a better understanding of the properties of doped high-*k* film. In addition, in the ultrathin film regime where the high-*k* film is difficult to crystallize, the improved properties of the doped high-*k* film are difficult to explain using only the crystalline structure modification model. Therefore, it is essential to examine the properties of the doped high-*k* films from several other viewpoints, such as the chemical structure, the state of electrical defects, electronic structures, crystalline structures, and interfacial reactions.

Figure 4.24 shows the SIMS depth profiles of HfO_2 and Al-doped HfO_2 films with various Al concentrations. The Al/(Hf + Al) ratio varies from 10.8 (HAO1) to 18% (HAO3). A lower C concentration in the Al-doped HfO_2 film was observed compared to the pure HfO_2 film, which is closely related to the oxygen vacancies (V_O) in the film.

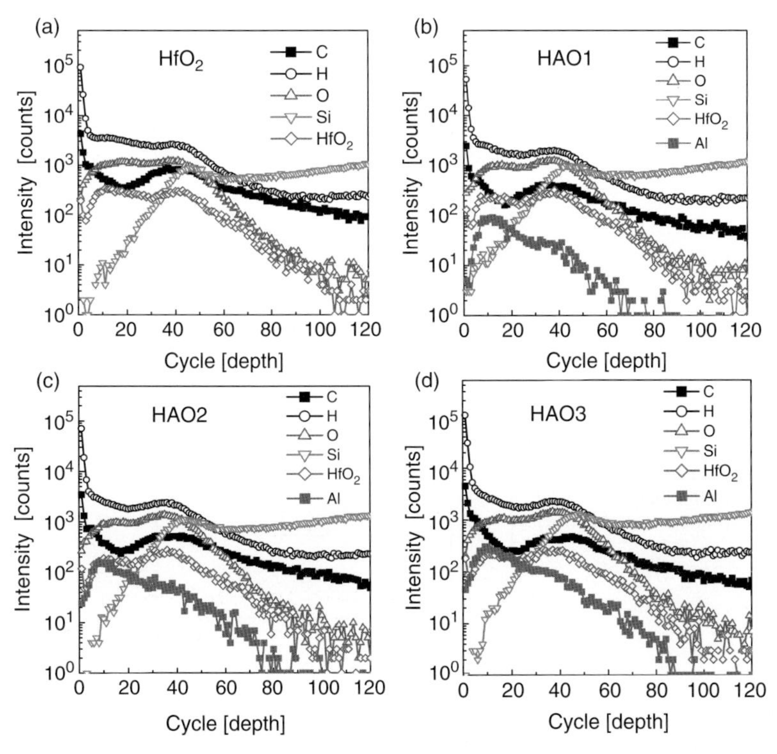

Figure 4.24 SIMS depth profiles for the (a) control HfO_2 and (b–d) Al-doped HfO_2 films with various Al concentrations. The Al/(Hf + Al) ratio varies from 10.8 (HAO1) to 18% (HAO3) [40].

The V_O in the HfO_2 film decreased from 12.5 to <8.3% by Al-doping of HfO_2 [40]. The reduction of V_O is supported by the first-principles calculations by Wang *et al.* who reported that the Al in HfO_2 led to the passivation of V_O [81]. In addition, doping materials, such as Ba and La, decreased the V_O concentrations in HfO_2 films considerably [82, 83]. The changes in the V_O concentration in the film should affect the electrical defects in the film.

Figure 4.25 shows the reconstructed band structure of (a) pure HfO_2 and (b) Al-doped HfO_2 with the electrical defect level using the temperature-dependent J_g–V characteristics, valence band and O 1s loss spectra. The apparent decrease in the V_O density, which induced the relatively shallow level traps, led to the appearance of deep-level defects in the electrical measurements. This means that the shallow defect states were removed primarily by Al-doping. The bandgap energy increased slightly by Al-doping due to the low V_O concentration [84]. The Al-O material itself in the HfO_2 lattice is not expected to increase the energy gap due to the lower chance of hybridizing the Hf 5d and Al 2p orbitals. As a result, Al-doping into the HfO_2 film

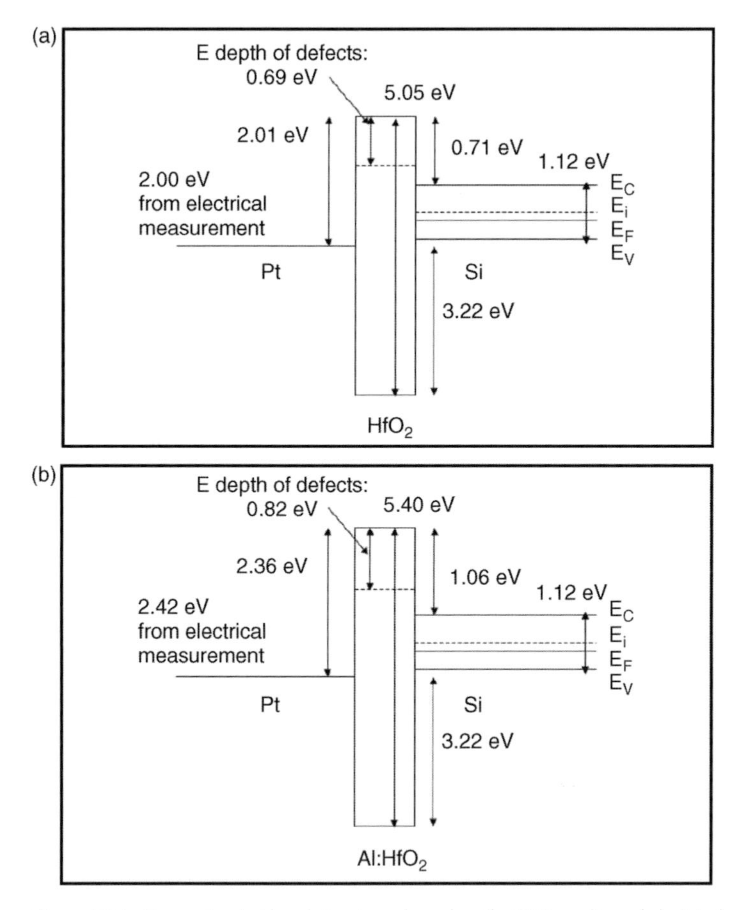

Figure 4.25 Reconstructed band structures based on the XPS results and electrical measurements of (a) the control HfO_2 and (b) Al-doped HfO_2 films (Al:HfO_2) on Si [40].

decreased the J_g level significantly. A plot of J_g versus EOT for the as-deposited films shows the improved electrical properties of the Al-doped HfO_2 films with a two orders of magnitude decrease in J_g compared to the nondoped HfO_2 film [40]. There was no change in the permittivity (~24) of the Al-doped HfO_2 films because the films were in the as-deposited state and were extremely thin. In addition, PDA induced crystallization of the ultrathin (~ 3 nm) Al-doped film into the monoclinic phase, not the tetragonal or cubic phase [40]. This is a certainly deviated behavior from that of the thicker films. Hence, there are several other factors which help to improve the properties of Al-doped HfO_2 than modification of the crystalline structure.

4.4 Summary

This chapter discussed the high-*k* properties of ALD grown HfO_2 gate dielectric films on a Si substrate for high-performance MOSFET applications. For that purpose, the growth behavior of the ALD films depending on the types of Hf precursor and oxygen source was reviewed. The accompanying changes in the structural, chemical, and electrical properties of the films were discussed with a focus on the interaction with the substrate (interfacial layer formation and Si diffusion). Although considerable effort has been made to make the HfO_2 films as defect free and electrically reliable as thermally grown SiO_2, unfortunately, this goal has not been achieved. This may somehow be a natural consequence of the deposition process since it uses an organic or inorganic Hf precursor on the chemically very active Si material. Nevertheless, due to the immense improvement made in various deposition techniques, materials and the understanding of the process and material properties, the materials are already in use for high-performance Si chips.

The further scaling of the MOSFETs requires improvement in the performance of the gate stack technology, including the implementation of an even higher *k*-value material with appropriate gate metal technology. A short-term solution for this requirement may come from the doping technique, where the pre-existing HfO_2 dielectrics are literally doped to enhance dielectric properties, sometimes by a phase transition to the higher *k* phase. Although the doping-related phase transitions have been expected theoretically and confirmed experimentally, this may not necessarily be the case depending on the high-*k* film thickness.

Eventually, new materials with far higher *k* values, combined with a scavenging technique to reduce the interfacial layer, should be developed. In addition, studies on three-dimensional structures and high-mobility substrates, as well as high-*k* or higher k dielectrics will be indispensable for future devices.

References

1 http://www.intel.com/technology/index.htm.

2 Suntola, T. (1984) Atomic layer epitaxy. Proceedings of the 16th International Conference on Solid State Devices and Materials, p. 647.

3 Delabie, A., Caymax, M., Brijs, B., Brunco, D.P., Conard, T., Sleeckx, E.,

Elshocht, S.V., Ragnarsson, L., Gendt, S.D., and Heyns, M.M. (2006) Scaling to Sub-1nm equivalent oxide thickness with hafnium oxide deposited by atomic layer deposition. *J. Electrochem. Soc.*, **153** (8), F180.
4 Green, M.L., Ho, M.-Y., Busch, B., Wilk, G.D., Sorsch, T., Conard, T., Brijs, B., Vandervorst, W., Räisänen, P.I., Muller, D., Bude, M., and Grazul, J. (2002). Nucleation and growth of atomic layer deposited HfO_2 gate dielectric layers on chemical oxide Si–O–H and thermal oxide SiO_2 or Si–O–N underlayers. *J. Appl. Phys.*, **92**, 7168.
5 Cho, M., Park, J., Park, H.B., Hwang, C.S., Jeong, J., and Hyun, K.S. (2002) Chemical interaction between atomic-layer-deposited HfO_2 thin films and the Si substrate. *Appl. Phys. Lett.*, **81**, 334.
6 Triyoso, D.H., Hegde, R.I., Zollner, S., Ramon, M.E., Kalpat, S., Gregory, R., Wang, X.-D., Jiang, J., Raymond, M., Rai, R., Werho, D., Roan, D., White, Jr., B.E., and Tobin, P.J. (2005). Impact of titanium addition on film characteristics of HfO_2 gate dielectrics deposited by atomic layer deposition. *J. Appl. Phys.*, **98**, 054104.
7 Kukli, K., Aaltonen, T., Aarik, J., Lu, J., Ritala, M., Ferrari, S., Hårsta, A., and Leskelä, M. (2005) Atomic layer deposition and characterization of HfO_2 films on noble metal film substrates. *J. Electrochem. Soc.*, **152**, F75.
8 Cho, M., Jeong, D.S., Park, H.B., Lee, S.W., Park, T.J., Hwang, C.S., Jang, G.H., and Jeong, J. (2004) Comparison between atomic-layer-deposited HfO_2 films using O_3 or H_2O oxidant and $Hf[N(CH_3)_2]_4$ precursor. *Appl. Phys. Lett.*, **85**, 5953.
9 Schaeffer, J.K., Samavedam, S.B., Gilmer, D.C., Dhandapani, V., Tobin, P.J., Mogab, J., Nguyen, B.-Y., White, B.E., Jr., Dakshina-Murthy, S., Rai, R.S., Jiang, Z.-X., Martin, R., Raymond, M.V., Zavala, M., La, L.B., Smith, J.A., Garcia, R., Roan, D., Kottke, M., and Gregory, R.B. (2003). Physical and electrical properties of metal gate electrodes on HfO_2 gate dielectrics. *J. Vac. Sci. Technol. B*, **21**, 11.
10 Conley, J.F., Jr., Ono, Y., Solanki, R., Stecker, G., and Zhuang, W. (2003) Electrical properties of HfO_2 deposited via atomic layer deposition using $Hf(NO_3)_4$ and H_2O. *Appl. Phys. Lett.*, **82**, 3508.
11 Cho, M., Park, H.B., Park, J., Lee, S.W., Hwang, C.S., Jang, G.H., and Jeong, J. (2003) High-*k* properties of atomic-layer-deposited HfO_2 films using a nitrogen containing $Hf[N(CH_3)_2]_4$ precursor and H_2O oxidant. *Appl. Phys. Lett.*, **83**, 5503.
12 Cho, M., Park, J., Park, H.B., Hwang, C.S., Jeong, J., Hyun, K.S., Kim, Y.-W., Oh, C.-B., and Kang, H.-S. (2002) Thermal stability of atomic-layer-deposited HfO_2 thin films on the SiN_x-passivated Si substrate. *Appl. Phys. Lett.*, **81**, 3630.
13 Cho, M., Park, H.B., Park, J., Hwang, C.S., Lee, J.-C., Oh, S.-J., Jeong, J., Hyun, K.S., Kang, H.-S., Kim, Y.-W., and Lee, J.-H. (2003). Thermal annealing effects on the structural and electrical properties of HfO_2/Al_2O_3 gate dielectric stacks grown by atomic layer deposition on Si substrates. *J. Appl. Phys.*, **94**, 2563.
14 Cho, M., Park, H.B., Park, J., Lee, S.W., Hwang, C.S., Jeong, J., Kang, H.S., and Kim, Y.W. (2005) Comparison of properties of an Al_2O_3 thin layer grown with remote O_2 plasma, H_2O, or O_3 as oxidants in an ALD process for HfO_2 gate dielectrics. *J. Electrochem. Soc.*, **152**, F49.
15 Park, H.B., Cho, M., Park, J., Lee, S.W., Hwang, C.S., Kim, J.-P., Lee, J.-H., Lee, N.-I., Lee, J.-C., and Oh, S.-J. (2003) Comparison of HfO_2 films grown by atomic layer deposition using $HfCl_4$ and H_2O or O_3 as the oxidant. *J. Appl. Phys.*, **94**, 3641.
16 Lee, J.-C., Oh, S.-J., Cho, M., Hwang, C.S., and Jung, R. (2004) Chemical structure of the interface in ultrathin HfO_2/Si films. *Appl. Phys. Lett.*, **84**, 1305.
17 Cho, M., Degraeve, R., Pourtois, G., Delabie, A., Ragnarsson, L., Kauerauf, T., Groeseneken, G., Gendt, S.D., Heyns, M., and Hwang, C.S. (2007) Study of the reliability impact of chlorine precursor residues in thin atomic-layer-deposited HfO_2 layers. *IEEE T. Electron. Dev.*, **54**, 752.
18 Cho, M., Kim, J.H., Hwang, C.S., Ahn, H.-S., Han, S., and Won, J.Y. (2007)

Effects of carbon residue in atomic layer deposited HfO_2 films on their time-dependent dielectric breakdown reliability. *Appl. Phys. Lett.*, **90**, 182907.

19 Yang, C.W., Fang, Y.K., Chen, C.H., Chen, S.F., Lin, C.Y., Lin, C.S., Wang, M.F., Lin, Y.M., Hou, T.H., Chen, C.H., Yao, L.G., Chen, S.C., and Liang, M.S. (2003). Effect of polycrystalline-silicon gate types on the opposite flatband voltage shift in n-type and p-type metal–oxide–semiconductor field-effect transistors for high-*k*-HfO_2 dielectric. *Appl. Phys. Lett.*, **83**, 308.

20 Park, T.J., Kim, J.H., Seo, M.H., Jang, J.H., and Hwang, C.S. (2007) Improvement of thermal stability and composition changes of atomic layer deposited HfO_2 on Si by *in situ* O_3 pretreatment. *Appl. Phys. Lett.*, **90**, 152906.

21 Park, J., Cho, M., Park, H.B., Park, T.J., Lee, S.W., Hong, S.H., Jeong, D.S., Lee, C., Choi, J., and Hwang, C.S. (2004) Voltage-induced degradation in self-aligned polycrystalline silicon gate n-type field-effect transistors with HfO_2 gate dielectrics. *Appl. Phys. Lett.*, **85**, 5965.

22 Park, J., Park, T.J., Cho, M., Kim, S.K., Hong, S.H., Kim, J.H., Seo, M., Hwang, C.S., Won, J.Y., Jeong, R., and Choi, J.-H. (2006) Influence of the oxygen concentration of atomic-layer-deposited HfO_2 gate dielectric films on the electron mobility of polycrystalline-Si gate transistors. *J. Appl. Phys.*, **99**, 094501.

23 Lee, S.W., Hong, S.H., Park, J., Cho, M., Park, T.J., Hwang, C.S., Kim, Y.-S., Lim, H.J., Lee, J.-H., and Won, J.Y. (2005) Fabrication of HfO_2 thin-film capacitors with a polycrystalline Si gate electrode and a low interface trap density. *Electrochem. Solid State Lett.*, **8**, F32.

24 Seo, M., Min, Y.-S., Kim, S.K., Park, T.J., Kim, J.H., Na, K.D., and Hwang, C.S. (2008) Atomic layer deposition of hafnium oxide from tertbutoxytris (ethylmethylamido)hafnium and ozone: rapid growth, high density and thermal stability. *J. Mater. Chem.*, **18**, 4324.

25 Park, H.B., Cho, M., Park, J., Lee, S.W., Park, T.J., and Hwang, C.S. (2004) Improvements in reliability and leakage current properties of HfO_2 gate dielectric films by *in situ* O_3 Oxidation of Si substrate. *Electrochem. Solid State Lett.*, **7**, G254.

26 Shevjakov, A.M., Kuznetsova, G.N., and Aleskovskii, V.B. (1965) Chemistry of high temperature materials. Proceedings of the Second USSR Conference on High-Temperature Chemistry of Oxides, Leningrad, USSR, p. 26.

27 Suntola, T. and Antson, J. (1977) Method for Producing Compound Thin Films, U.S. Patent No. 4,058,430.

28 Puurunen, R.L. (2005) Surface chemistry of atomic layer deposition: A case study for the trimethylaluminum/water process. *J. Appl. Phys.*, **97**, 121301.

29 Gusev, E.P., Cabral, C., Jr., Copel, M., D'Emic, C., and Gribelyuk, M. (2003) Ultrathin HfO_2 films grown on silicon by atomic layer deposition for advanced gate dielectrics applications. *Microelectron. Eng.*, **69**, 145.

30 Niinistö, J., Putkonen, M., Niinistö, L., Arstila, K., Sajavaara, T., Lu, J., Kukli, K., Ritala, M., and Leskelä, M. (2006) HfO_2 films grown by ALD using cyclopentadienyl-type precursors and H_2O or O_3 as oxygen source. *J. Electrochem. Soc.*, **153**, F39.

31 Swerts, J., Peys, N., Nyns, L., Delabie, A., Franquet, A., Maes, J.W., Elshocht, S.V., and De Gendt, S. (2010) Impact of precursor chemistry and process conditions on the scalability of ALD HfO_2 gate dielectrics. *J. Electrochem. Soc.*, **157**, G26.

32 Liu, X., Ramanathan, S., Longdergan, A., Srivastava, A., Lee, E., Seidel, T.E., Barton, J.T., Pang, D., and Gordon, R.G. (2005) ALD of hafnium oxide thin films from tetrakis (ethylmethylamino) hafnium and ozone. *J. Electrochem. Soc.*, **152**, G213.

33 Won, S.-J., Suh, S., Lee, S.W., Choi, G.-J., Hwang, C.S., and Kim, H.J. (2010) Substrate dependent growth rate of plasma-enhanced atomic layer deposition of titanium oxide using N_2O gas. *Electrochem. Solid State Lett.*, **13**, G13.

34 Choi, G.-J., Kim, S.K., Won, S.-J., Kim, H.J., and Hwang, C.S. (2009) Plasma-enhanced atomic layer deposition of TiO_2 and Al-doped TiO_2 films using

N_2O and O_2 reactants. *J. Electrochem. Soc.*, **156**, G138.

35 Won, S.-J., Kim, J.-Y., Choi, G.-J., Heo, J., Hwang, C.S., and Kim, H.J. (2009) The formation of an almost full atomic monolayer via surface modification by N_2O-plasma in atomic layer deposition of ZrO_2 thin films. *Chem. Mater.*, **21**, 4374.

36 Park, T.J., Sivasubramani, P., Coss, B.E., Kim, H.-C., Lee, B., Wallace, R.M., Kim, J., Rousseau, M., Liu, X., Li, H., Lehn, J.-S., Hong, D., and Shenai, D. (2010). Effects of O_3 and H_2O oxidants on C and N-related impurities in atomic-layer-deposited La_2O_3 films observed by *in situ* X-ray photoelectron spectroscopy. *Appl. Phys. Lett.*, **97**, 092904.

37 Lee, C.-K., Cho, E., Lee, H.-S., Hwang, C.S., and Han, S. (2008) First-principles study on doping and phase stability of HfO_2. *Phys. Rev. B*, **78**, 012102.

38 Fujimori, H., Yashima, M., Sasaki, S., Kakihana, M., Mori, T., Tanaka, M., and Yoshimura, M. (2001) Cubic-tetragonal phase change of yttria-doped hafnia solid solution: high-resolution X-ray diffraction and Raman scattering. *Chem. Phys. Lett.*, **346**, 217.

39 Bernay., C., Ringuedé, A., Colomban, P., Lincot, D., and Cassir, M.J. (2003) Yttria-doped zirconia thin films deposited by atomic layer deposition ALD: a structural, morphological and electrical characterisation. *J. Phys. Chem. Solids*, **64**, 1761.

40 Park, T.J., Kim, J.H., Jang, J.H., Lee, C.-K., Na, K.D., Lee, S.Y., Jung, H.S., Kim, M., Han, S., and Hwang, C.S. (2010) Reduction of electrical defects in atomic layer deposited HfO_2 films by Al doping. *Chem. Mater.*, **22**, 4175.

41 Kukli, K., Ritala, M., Sajavaara, T., Keinonen, J., and Leskelä, M. (2002) Atomic layer deposition of hafnium dioxide films from hafnium tetrakis (ethylmethylamide) and water. *Chem. Vapor Depos.*, **8**, 199.

42 Park, J., Cho, M., Kim, S.K., Park, T.J., Lee, S.W., Hong, S.H., and Hwang, C.S. (2005) Influence of the oxygen concentration of atomic-layer-deposited HfO_2 films on the dielectric property and interface trap density. *Appl. Phys. Lett.*, **86**, 112907.

43 Baldovino, S., Spiga, S., Scarel, G., and Fanciulli, M. (2007) Effects of the oxygen precursor on the interface between (100) Si and HfO_2 films grown by atomic layer deposition. *Appl. Phys. Lett.*, **91**, 172905.

44 Kamiyama, S., Miura, T., and Nara, Y. (2006) Impact of O_3 concentration on ultrathin HfO_2 films deposited on HF-cleaned silicon using atomic layer deposition with $Hf[N(CH_3)(C_2H_5)]_4$. *Electrochem. Solid State Lett.*, **9**, G285.

45 Chung, K.J., Park, T.J., Sivasubramani, P., Kim, J., and Ahn, J. (2010) Effect of ozone concentration on atomic layer deposited HfO_2 on Si. *ECS Trans.*, **28**, 221.

46 Degraeve, R., Kerber, A., Roussel, P., Cartier, E., Kauerauf, T., Pantisano, L., and Groeseneken, G. (2003). Effect of bulk trap density on HfO_2 reliability and yield. IEEE Technical Digest of International Electron Devices Meeting, p. 935.

47 Kyuno, K., Kita, K., and Toriumi, A. (2005) Evolution of leakage paths in HfO_2/SiO_2 stacked gate dielectrics: a stable direct observation by ultrahigh vacuum conducting atomic force microscopy. *Appl. Phys. Lett.*, **86**, 063510.

48 Tabata, Y., Okihara, Y., Ishikawa, M., and Saitsu, T., and Yotsumoto, H. (2008) Ozone generator system and ozone generating method. U.S. Patent No. 7,382,087.

49 Liao, C.C., Cheng, C.F., Yu, D.S., and Chin, A. (2004) The copper contamination effect of Al_2O_3 gate dielectric on Si. *J. Electrochem. Soc.*, **151**, G693.

50 Pan, T.-M., Ko, F.-H., Chao, T.-S., Chen, C.-C., and Chang-Liao, K.-S. (2005) Effects of metallic contaminants on the electrical characteristics of ultrathin gate oxides. *Electrochem. Solid State Lett.*, **8**, G201.

51 Choi, B.D. and Schroder, D.K. (2001) Degradation of ultrathin oxides by iron contamination. *Appl. Phys. Lett.*, **79**, 2645.

52 Park, T.J., Chung, K.J., Kim, H.-C., Ahn, J., Wallace, R.M., and Kim, J. (2010) Reduced metal contamination in atomic-layer-deposited HfO_2 films grown on Si using O_3 oxidant generated without N_2

assistance. *Electrochem. Solid State Lett.*, **13**, G65.

53 Nieh, R.E., Kang, C.S., Cho, H.-J., Onishi, K., Choi, R., Krishnan, S., Han, J.H., Kim, Y.-H., Akbar, M.S., and Lee, J.C. (2003) Electrical characterization and material evaluation of zirconium oxynitride gate dielectric in TaN-gated NMOSFETs with high-temperature forming gas annealing. *IEEE T. Electron. Dev.*, **50**, 333.

54 Cho, H.J., Park, D.-G., Yeo, I.S., Roh, J.-S., and Park, J.W. (2001) Characteristics of TaO_xN_y gate dielectric with improved thermal stability. *Jpn. J. Appl. Phys.*, **40**, 2814.

55 Kang, M.S., Chung, T.H., and Kim, Y. (2006) Plasma enhanced chemical vapor deposition of nitrogen-incorporated silicon oxide films using TMOS/N_2O gas. *Thin Solid Films*, **506**, 45.

56 Maikap, S., Lee, J.-H., Mahapatra, R., Pal, S., No, Y.S., Choi, W.-K., Ray, S.K., and Kim, D.-Y. (2005) Effects of interfacial NH_3/N_2O-plasma treatment on the structural and electrical properties of ultra-thin HfO_2 gate dielectrics on p-Si substrates. *Solid State Electron.*, **49**, 524.

57 Kim, J.H., Park, T.J., Cho, M., Jang, J.H., Seo, M., Na, K.D., Hwang, C.S., and Won, J.Y. (2009) Reduced electrical defects and improved reliability of atomic-layer-deposited HfO_2 dielectric films by *in situ* NH_3 injection. *J. Electrochem. Soc.*, **156**, G48.

58 Lim, J.W. and Yun, S.J. (2004) Electrical properties of alumina films by plasma-enhanced atomic layer deposition. *Electrochem. Solid State Lett.*, **7**, F45.

59 Maeng, W.J., Lim, S.J., Kwon, S.-J., and Kim, H. (2007) Electrical property improvements of high-*k* gate oxide by *in situ* nitrogen incorporation during atomic layer deposition. *Appl. Phys. Lett.*, **90**, 062909.

60 Maeng, W.J. and Kim, H. (2007) Atomic scale nitrogen depth profile control during plasma enhanced atomic layer deposition of high *k* dielectrics. *Appl. Phys. Lett.*, **91**, 092901.

61 Umezawa, N., Shiraishi, K., Ohno, T., Watanabe, H., Chikyow, T., Torii, K., Yamabe, K., Yamada, K., Kitajima, H., and Arikado, T. (2005) First-principles studies of the intrinsic effect of nitrogen atoms on reduction in gate leakage current through Hf-based high-*k* dielectrics. *Appl. Phys. Lett.*, **86**, 143507.

62 Wilk, G.D., Wallace;, R.M., and Anthony, J.M. (2000) Hafnium and zirconium silicates for advanced gate dielectrics. *J. Appl. Phys.*, **87**, 484.

63 Robertson, J. (2006) High dielectric constant gate oxides for metal oxide Si transistors. *Rep. Prog. Phys.*, **69**, 327.

64 Yamamoto, Y., Kita, K., Kyuno, K., and Toriumi, A. (2006) Structural and electrical properties of $HfLaO_x$ films for an amorphous high-*k* gate insulator. *Appl. Phys. Lett.*, **89**, 032903.

65 Zheng, W., Bowen, K.H., Li, J., Dabkowska, I., and Gutowski, M. (2005) Electronic structure differences in ZrO_2 vs HfO_2. *J. Phys. Chem.*, **109**, 11521.

66 Gutowski, M., Jaffe, J.E., Liu, C.-L., Stoker, M., Hegde, R.I., Rai, R.S., and Tobin, P.J. (2002) Thermodynamic stability of high-*K* dielectric metal oxides ZrO_2 and HfO_2 in contact with Si and SiO_2. *Appl. Phys. Lett.*, **80**, 1897.

67 Jung, H.-S., Park, T.J., Kim, J.H., Lee, S.Y., Lee, J., Oh, H.C., Na, K.D., Park, J.-M., Kim, W.-H., Song, M.-W., Lee, N.-I., and Hwang, C.S. (2009). Systematic study on bias temperature instability of various high-*k* gate dielectrics; HfO_2, $HfZr_xO_y$ and ZrO_2. Proceedings of International Reliability Physics Symposium, p. 971.

68 Jung, H.-S., Jang, J.H., Cho, D.-Y., Jeon, S.-H., Kim, H.K., Lee, S.Y., and Hwang, C.S. (2011) The effects of postdeposition annealing on the crystallization and electrical characteristics of HfO_2 and ZrO_2 gate dielectrics. *Electrochem. Solid State Lett.*, **14**, G17.

69 Green, D.F., Hannink, R.H.J., and Swain, M.V. (1989) *Transformation Toughening of Ceramics*, vol. 1, CRC Press, Boca Raton, FL.

70 Kim, S.K. and Hwang, C.S. (2008) Atomic layer deposition of ZrO_2 thin films with high dielectric constant on TiN substrates. *Electrochem. Solid State Lett.*, **11**, G9.

71 Müller, J., Böscke, T.S., Schröder, U., Reinicke, M., Oberbeck, L., Zhou, D.,

Weinreich, W., Kücher, P., Lemberger, M., and Frey, L. (2009) Improved manufacturability of ZrO_2 MIM capacitors by process stabilizing HfO_2 addition. *Microelectron. Eng.*, **86**, 1818.

72 Hegde, R.I., Triyoso, D.H., Samavedam, S.B., and White, B.E. (2007) Hafnium zirconate gate dielectric for advanced gate stack applications. *J. Appl. Phys.*, **101**, 074113.

73 Triyoso, D.H., Hegde, R.I., Schaeffer, J.K., Roan, D., Tobin, P.J., Samavedam, S.B., White, B.E., Gregory, R., and Wang, X.-D. (2006) Impact of Zr addition on properties of atomic layer deposited HfO_2. *Appl. Phys. Lett.*, **88**, 222901.

74 Cho, D.-Y., Jung, H.-S., and Hwang, C.S. (2010) Structural properties and electronic structure of HfO_2-ZrO_2 composite films. *Phys. Rev. B*, **82**, 094104.

75 Bersuker, G., Sim, J.H., Park, C.S., Young, C.D., Nadkarni, S.V., Choi, R., and Lee, B.H. (2007) Mechanism of electron trapping and characteristics of traps in HfO_2 gate stacks. *IEEE Trans. Dev. Mater. Reliab.*, **7**, 138.

76 Cho, D.-Y., Jung, H.-S., Kim, J.H., and Hwang, C.S. (2010) Monocliniclike local atomic structure in amorphous ZrO_2 thin film. *Appl. Phys. Lett.*, **97**, 141905.

77 Böscke, T.S., Govindarajan, S., Kirsch, P.D., Hung, P.Y., Krug, C., Lee, B.H., Heitmann, J., Schröder, U., Pant, G., Gnade, B.E., and Krautschneider, W.H. (2007) Stabilization of higher-tetragonal HfO_2 by SiO_2 admixture enabling thermally stable metal-insulator-metal capacitors. *Appl. Phys. Lett.*, **91**, 072902.

78 Tomida, K., Kita, K., and Toriumi, A. (2006) Dielectric constant enhancement due to Si incorporation into HfO_2. *Appl. Phys. Lett.*, **89**, 142902.

79 Yu, H.Y., Wu, N., Li, M.F., Zhu, C., Cho, B.J., Kwong, D.-L., Tung, C.H., Pan, J.S., Chai, J.W., Wang, W.D., Chi, D.Z., Ang, C.H., Zheng, J.Z., and Ramanathan, S. (2002). Thermal stability of $(HfO_2)_x(Al_2O_3)_{1-x}$ on Si. *Appl. Phys. Lett.*, **81**, 3618.

80 Park, P.K. and Kang, S.W. (2006) Enhancement of dielectric constant in HfO_2 thin films by the addition of Al_2O_3. *Appl. Phys. Lett.*, **89**, 192905.

81 Wang, X.F., Li, Q., Egerton, R.F., Lee, P.F., Dai, J.Y., Hou, Z.F., and Gong, X.G. (2007) Effect of Al addition on the microstructure and electronic structure of HfO_2 film. *J. Appl. Phys.*, **101**, 013514.

82 Umezawa, N. (2009) Effects of barium incorporation into HfO_2 gate dielectrics on reduction in charged defects: first-principles study. *Appl. Phys. Lett.*, **94**, 022903.

83 Umezawa, N., Shiraishi, K., Sugino, S., Tachibana, A., Ohmori, K., Kakushima, K., Iwai, H., Chikyow, T., Ohno, T., Nara, Y., and Yamada, K. (2007) Suppression of oxygen vacancy formation in Hf-based high-*k* dielectrics by lanthanum incorporation. *Appl. Phys. Lett.*, **91**, 132904.

84 Suzuki, K., Ito, Y., and Miura, H. (2006) Quantum chemical molecular dynamics analysis of the effect of oxygen vacancies and strain on dielectric characteristic of HfO_2 films. Proceedings of the 2006 International Conference on Simulation of Semiconductor Processes and Devices, p. 165.

5
Structural and Electrical Characteristics of Alternative High-κ Dielectrics for CMOS Applications

Fu-Chien Chiu, Somnath Mondal, and Tung-Ming Pan

5.1
Introduction

The complementary metal oxide semiconductor (CMOS) field-effect transistors (FET) are the most important element for all electronic circuits. High-performance and low power consumption are the major advantages of CMOS technology. Over the past four decades, continuous improvements of its performance have been achieved by means of scaling device dimension to the smallest feature size. A smaller feature size of the device requires less material and consumes less power. The costs per transistor and per function are greatly reduced in any integrated circuits. Since the size becomes smaller, the carrier transportation time across the channel of transistor will be shortened accordingly. Higher operation speed will then be realized by this downsizing effect. In addition, more components in a chip are possible to follow the famous Moore's law [1]. There are many limiting factors for scaling down the device dimensions. Lithography, the need of very short wavelengths of light for device patterning, is the major constraint for ultrascaled devices. Hence, the gate dielectric material is a critical modification to continue this scaling trend. In particular, the gate dielectric thickness in modern CMOSFET circuits has already reached its atomic dimension to cause a high gate leakage current, which is a main constraint to improving device performances in future.

The traditional gate dielectric material is SiO_2 that has several significant properties employed in semiconductor technology. In particular, SiO_2 is used as the gate dielectric in CMOS transistor. The major advantages of SiO_2 are listed as follows:

i) It can be grown by simple reaction of silicon with oxygen and can be controlled in a very precise manner with a minimum number of process parameters. Excellent control over thickness is possible due to its consistent deposition technique.
ii) High thermal conductivity of SiO_2 facilitates its use in power electronic devices. No thermal well or thermal via holes (metal vias) are needed to handle the large thermal budget.

High-k Gate Dielectrics for CMOS Technology, First Edition. Edited by Gang He and Zhaoqi Sun.

iii) Attractive mechanical properties, such as outstanding thermal stability, high strength, and stiffness with low expansion coefficient, lead to a high yield of production.
iv) Chemically inert SiO_2 forms excellent Si/SiO_2 interface with very low intrinsic interface defects that can be passivated very efficiently by simple postmetallization annealed in hydrogen ambient.
v) Very low moisture absorption and low permeability to ionic contamination result in excellent reliability of the devices.
vi) Large optical bandgap of SiO_2 (about 9 eV) confers a high breakdown field and an excellent electrical isolation. The conduction band and valence band offsets with respect to silicon are rather high, which makes the SiO_2 an excellent gate oxide.
vii) Thermal stability and compatibility between SiO_2 dielectrics and polysilicon gate electrode are also promising features in its successful integration with almost all devices.

All these brilliant properties of SiO_2 work well over the thickness of dielectric layer of 1.5 nm [2]. However, to accommodate more devices in a single chip, the scaling of SiO_2 dielectric thickness is going to atomic dimension and it becomes increasingly difficult to control devices and is a major technical barrier. There are many concerns about the scaling of gate dielectric thickness. First is the gate leakage current due to direct tunneling of charge carrier through the thin dielectric layer. The tunneling current increases exponentially with decreasing gate oxide thickness [3]. For example, Brar *et al.* shows the exponential dependence of leakage current with gate dielectric thickness [4]. Figure 5.1 shows the gate leakage current depends on the dielectric thickness of a MOS transistor. Hence, to reduce the gate leakage current and maintain the high gate capacitance, silicon dioxide has to be substituted by a high dielectric constant (high-κ) material, which provides a physically thicker film for the

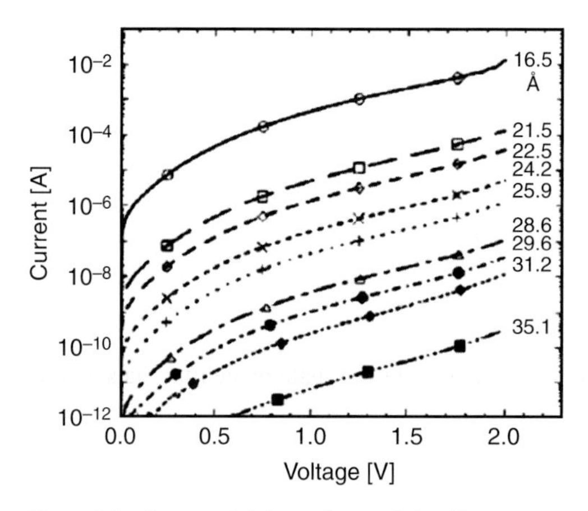

Figure 5.1 Exponential dependence of the direct quantum tunneling current on the dielectric oxide thickness (t_{ox}) of $Al/SiO_2/n\text{-}Si$ MOS structures.

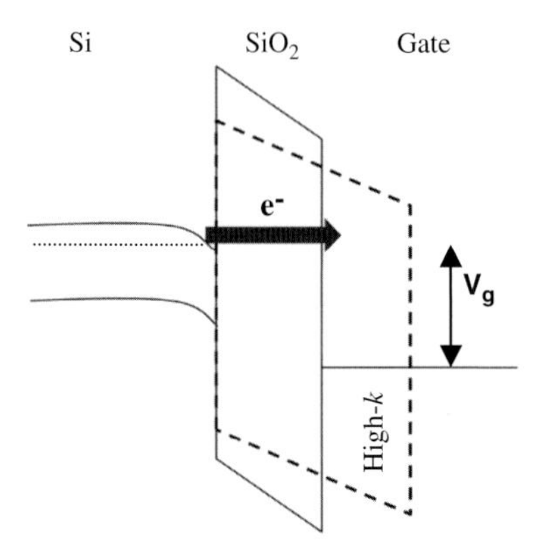

Figure 5.2 Schematic band diagram of direct tunneling through n-Si/SiO_2/metal gate structure. Electron tunnels from silicon substrate to the gate easily for thin SiO_2 layer, but thick high-κ layer (dashed line) hinders this effect at the same applied bias.

same electrically equivalent oxide thickness (EOT, t_{eq}). A schematic representation of implementation of high-κ oxide to improve leakage current is shown in Figure 5.2. The total capacitance of FET channel will remain constant to decrease the leakage current.

For an ultrathin SiO_2 layer, the CMOS device reliability is another concern for engineer. During normal operation, as charge carrier flows through the channel to the dielectric, both bulk and interfacial defects are generated causing permanent breakdown of dielectric film and hence failure of devices [5–10]. Dielectric breakdown or quasi-breakdown also limits the maximum gate voltage ($V_{g,max}$) that can be applied to a MOSFET. $V_{g,max}$ decreases as the thickness of the film decreases. Below a certain value of SiO_2, thickness in reliability is more critical than the leakage issue. A traditional SiO_2 dielectric can no longer be used in this limit. Hence, we are in great demand in optimizing the alternative high-κ dielectric materials to replace SiO_2 for future CMOS applications. Many alternative high-κ dielectrics have been explored over the years, but choice of high-κ dielectric and its processing of adequate quality for a gate dielectric is still unknown.

The high leakage current due to direct tunneling of electron through ultrathin SiO_2 raises another issue for a power handling capability of devices. Especially, the power efficiency of the devices decreases to unacceptable values [10–19]. In addition, time-dependent dielectric breakdown and bias temperature instabilities of the ultrathin dielectric layer hinder the use of a traditional gate dielectric material for future integrated circuit (IC) technology.

To challenge the above critical issues, the use of other gate oxide materials with a higher κ value could permit similar CMOS transistor performances with drastically reduced leakage currents and also improve the reliability of the devices because of the greater physical thickness of the gate dielectric. Besides having a higher dielectric constant of material than that of SiO_2 ($\kappa_{SiO2} = 3.9$), the high-κ material as alternative gate dielectrics must fulfill other critical criteria. These include (1) a large bandgap and large band offsets with silicon, (2) a low density of defects at the dielectric/silicon interface, (3) stability in thermodynamic respect in contact with silicon, (4) a good film morphology, and (5) compatibility with the gate electrode.

In this chapter, the structural and electrical properties of high-κrare-earth (RE) metal oxides are discussed to replace a traditional SiO_2 dielectric for CMOS applications. This chapter has three goals. First, the process compatibility is described in view of different deposition methods to obtain adequate quality of the gate oxide. Then, structural and electrical properties are explored toward the successful application of high-κ RE oxide dielectrics as an alternative gate oxide in CMOS applications.

5.2
Requirement of High-*k* Oxide Materials

As discussed before, the SiO_2 is a traditional gate dielectric in CMOS applications due to its inherent characteristics over silicon IC technology. Technical challenges are coming in the way of scaling down the dielectric thickness beyond a certain value. Hence, the replacement of a SiO_2 dielectric with a new, high dielectric constant material is in great demand. The requirements for the successful implementation of new high-κ gate oxide materials in CMOS applications are described next.

i) **High-κ value.**
   The International Technology Roadmap for Semiconductor (ITRS) suggests the process parameters will scale in years to come taking the device performance to higher efficiency level [20]. According to ITRS 2010, the EOT for next generation of a low standby power and a high-performance logic technology will be below 10 Å. Table 5.1 represents the gate length of transistor, EOT of gate oxide, and allowable gate leakage current for next generation of a high performance logic technology. The use of a high-κ gate dielectric will facilitate to improve the device performance taking the EOT of the gate oxide to minimal value. The κ-value of the gate dielectric must be large enough such that it can be used for a long time in this scaling era. However, the extremely large value of dielectric constant is problematic in CMOS applications due to a low band offset between conduction and valence band of silicon and a large fringing field at the source and drain. There are various oxide materials with high dielectric constant, but lack of proper technological development and the understanding of properties of an ultrathin gate dielectric material obstruct the successful application of CMOS devices. Few gate oxide dielectrics with their corresponding κ-values are listed in Table 5.2.
ii) **Large bandgap.**
   The gate leakage current of a MOS transistor is the focus of another major attention with respect to the scaling down of device size. Figure 5.3 shows the acceptable gate leakage density of a high performance logic circuit in future. The data are extracted from the ITRS 2010. For this aggressive high-performance

Table 5.1 ITRS 2010 specified EOTs for next generation of a low standby power and a high-performance logic circuit along with their corresponding technology node and production year.

Year	Node (nm)	L_g (nm)	EOT (nm)		$J_{g,leak}$ (kA/cm^2)
			Low standby	High performance	
2009	52	29	1.2[a]	1[a]	0.65
2010	45	27	1[a]	0.95[a]	0.83
2011	40	24	1.2[a]	0.88[a]	0.9
2012	36	22	1[a]	0.75[a]	1
2014	28	18	0.95[b]	0.55[a]–0.68[b]	1.2
2016	23	15.3	0.85[b]–1.1[c]	0.57[b]–0.7[c]	1.4
2018	18	12.8	1[c]	0.54[b]–0.64[c]	1.7
2020	14	10.7	0.9[c]	0.59[c]	2.1
2022	11	8.9	0.8[c]	0.55[c]	2.5
2024	9	7.4	0.7[c]	0.5[c]	2.9

a) Extended planar bulk
b) UTF FD
c) Metal gate
Allowable gate leakage is also specified (L_g: physical gate length; $J_{g,leak}$: maximum gate leakage current density).

technology, excellent leakage performances are needed. This requires a high optical bandgap (E_G) material with acceptable band offset with silicon. To inhibit the Schottky emission of electron or holes through the silicon–oxide interface is required a larger band offset than 1 eV [21–23]. To maintain this minimum leakage current and good band offset on each side, an oxide material with a higher bandgap value than 5 eV is required. In practice, metal oxides with a higher bandgap have a smaller κ-value (Table 5.2). Hence, it is in demand to choose a gate

Table 5.2 The dielectric constant (κ) and bandgap (E_G) values of some alternative gate oxides to replace a traditional SiO_2 gate dielectric for advanced CMOS applications.

Oxide	*k*-value	E_G (eV)	Oxide	*k*-value	E_G (eV)
Al_2O_3	9	8.8	Sm_2O_3	10–43	5.0
Ta_2O_5	22	4.4	Eu_2O_3	15	4.3
HfO_2	25	5.8	Gd_2O_3	14–23	5.4
ZrO_2	25	5.8	Tb_2O_3	12	3.8
TiO_2	80	3.5	Dy_2O_3	14	4.9
Y_2O_3	15	6	Ho_2O_3	12	5.3
La_2O_3	8–23	5.5	Er_2O_3	7–14	5.3
CeO_2	52	3.78	Tm_2O_3	7–22	5.4
Pr_2O_3	30	3.8	Yb_2O_3	13	4.9
Nd_2O_3	27	4.6	Lu_2O_3	14	5.5

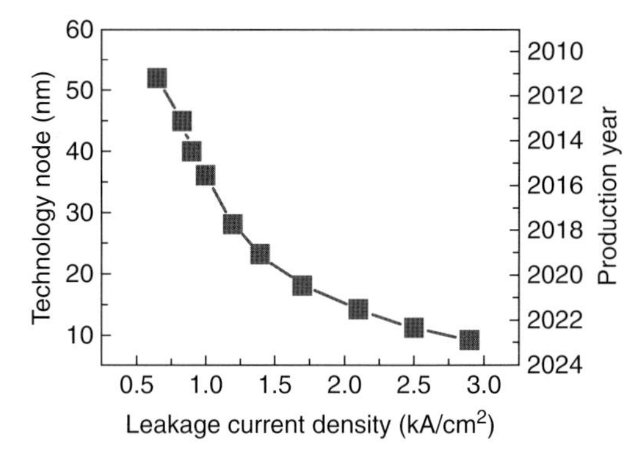

Figure 5.3 The leakage current density specification of high-performance circuits versus technology node and production year. This data is drawn from the ITRS 2010.

oxide with acceptable bandgap (>5 eV) and high κ-value. Various lanthanide oxides and their silicates satisfy this criterion for CMOS applications.

iii) **High thermal stability.**

The interface between high-κ oxide and silicon is usually not stable because of the formation of a silicide or silicate layer at high temperature. During device fabrication process, several thermal treatments are exposed on gate oxide. The effects of thermal treatment on gate dielectrics have four issues. First, the formation of a thin SiO_2 layer occurs at high-κ/silicon interface. The low κ-value of an interfacial SiO_2 layer pulls back the advantages of the high-κ oxide. Second, to remove deep trap centers from the gate oxide, high-temperature annealing is crucial. At this high temperature, these high-κ materials may decompose to form more stable silicide at interface or oxygen vacancies (defects) may appear in the bulk oxide layer when annealed in nitrogen environment [24, 25]. Finally, crystallization may occur in as-deposited amorphous oxide layer. Crystallization of a thin-film layer increases the leakage current by extra leakage path through the grain boundary. Hence, a high-κ oxide material with high thermal stability is strongly required for future CMOS application.

iv) **Excellent interface quality and surface topography**. Epitaxial growth of the high-κ oxide film on silicon is challenging; rather it has similar lattice constant [26]. Since the high-κ metal oxides are more ionic and have large coordination numbers, the interface bonding with silicon will be more unstable and complicated compared to the SiO_2, leading to a higher density of interface trap. Also, the silicon dangling bond with metal oxide will affect the interfacial defect states. For aggressive device performances of CMOS device, the oxide/silicon interface must be free from interfacial defects since most carriers in the channel flow within the angstrom of this oxide layer. Furthermore, the surface topography of gate dielectric layer also influences device performance. For example, in polycrystalline dielectrics (high roughness of film), the grain boundaries of the nanocrystal may introduce extra defects in the interface. The amorphous film of the dielectric layer could improve the electrical performances of the device by configuring its interface bonding to minimize the number of defect states.

v) **Defects.** At low-temperature processing, most high-κ materials arise with large oxygen vacancies, which produces extra energy state in the bandgap and behaves as charge trap center due to incomplete reaction with oxygen. Charge trapped in this trapping center causes shift in gate threshold voltage (V_{TH}) of a CMOS device. In addition, these trapped charges lead to a lower channel carrier mobility by scattering effect. Moreover, the electrical stress across these high-κ thin dielectric layers gives rise to extra electronic states in the bandgap. The stress inducing charge trap strongly varies with time and hence the device performance changes with timescale. These defects are undesirable for future high-density and high-performance CMOS devices. Many efforts are carried out to reduce these defect densities toward the application of a high-κ oxide as a gate insulator in CMOS devices.

5.3 Rare-Earth Oxide as Alternative Gate Dielectrics

In view of all the above requirements for new alternative gate dielectrics, the RE metal oxide materials will be the promising candidate in future CMOS applications. The permittivity values of RE oxide materials are comparable or exceed that of other high-κ materials. Binary RE oxide materials with their corresponding κ-values and bandgaps are listed in Table 5.2. Most of these oxides have a higher conduction band offset with silicon and result in a lower leakage current although the higher permittivity of material leads to a higher tunneling efficiency. However, a higher tunneling current can be suppressed by a higher electron effective mass or barrier. The high electron effective mass in some RE oxide materials due to the narrow f-type conduction band facilitates the application of such material. The RE oxide films, such as La_2O_3, Pr_2O_3, Gd_2O_3, and Nd_2O_3, have been investigated due to their large bandgaps, high dielectric constants, and good thermodynamic stability on silicon even at high temperatures, that is, when heated in contact with silicon it does not directly react to form silicide, metal, or silicon oxide [27]. Another attractive feature of Pr_2O_3, Gd_2O_3, and Nd_2O_3 is their relatively close lattices match those of silicon ($2a_{Si} = 1.090$ nm), which offers the possibility of epitaxial growth and thus eliminating problems associated with grain boundaries in polycrystalline films.

The thermodynamics of the RE oxide materials predicts their chemical stability on silicon. The closer lattice match on silicon is most advantageous for successful implication of RE oxide materials as a high-κ gate oxide for future CMOS applications. For instance, epitaxial Sc_2O_3, CeO_2, Pr_2O_3, and Gd_2O_3 films have been grown on Si surface [28–31], and a small EOT value of 0.38 nm for a 5 nm thick CeO_2 film has been reported [29]. The process compatibility with structural and electrical properties of such RE oxide material is discussed in detail in the following section.

5.4
Structural Characteristics of High-κ RE Oxide Films

5.4.1
Process Compatibility

Over the past four decades, the scaling down of Si CMOS technology continues to follow the prediction of Moore's law. According to previously discussed issues of ultrathin dielectrics, the SiO_2 gate oxide has to be replaced by proper high-κ dielectrics, enabling larger tunnel barriers at approximately the same drive current. In industrial CMOS, such high-κ dielectrics were introduced beyond the 45 nm technology node, often in combination with metal gates [32]. However, the progress toward attaining high-performance high-κ CMOSFET has been hampered because of the lack of good thermodynamic stability of the high κ/Si dielectric interface. Most advanced atomic layer deposition (ALD) and traditionally used sputtering techniques for semiconductor manufacturing inevitably cause the formation of a thicker interfacial layer of ~1.0 nm due to chemical intermixing. The material requirements for the alternative gate dielectric are very challenging to achieve performance comparable to SiO_2. This includes dielectric constant, bandgap, conduction band offset, reliability, channel mobility, recrystallization temperature (for amorphous dielectrics), low oxygen diffusivity, thermodynamic stability in contact with Si at temperature exceeding 800 °C, high-quality interface with Si with a smaller interfacial state density (D_{it}), and a lower leakage conduction than SiO_2 at an EOT less than 1 nm [16]. Moreover, there are demanding issues for process integration compatibility, such as morphology, interfacial structure and reaction, thermal stability, and gate compatibility. There also exist fundamental limitations, such as interface state, fixed charge, dopant depletion in poly-Si gate, dopant diffusion, and increasing field in the channel region.

According to ITRS, more aggressive scaling is required in gate oxide to achieve minimum EOT for 22 nm technology node application. This requires stringent control of interfacial oxide layer, either by decreasing its thickness or by increasing its κ-value. A good interface necessitates an amorphous oxide as gate dielectrics to the silicon substrate. Amorphous oxides designate a low-cost solution; nevertheless, the challenge is to maintain these materials amorphous in nature even after postdeposition high-temperature process in order to evade the surface roughness and additional leakage due to the formation of grain boundaries. Another technique is based on the development of epitaxial high-κ oxides directly on silicon surfaces.

RE thin oxides offer interesting properties to execute as alternative gate dielectrics because they possess a high dielectric constant, a high bandgap, and suitable conduction and valence band offsets with respect to Si, sufficiently high breakdown strength, extremely low leakage current, and well-behaved interface properties at the high κ/Si interface [33]. In addition, their foremost advantages for advanced CMOS technology include thermal stability, a feature to shift the work function of a metal gate toward n-type and thus engineer a transistor threshold voltage, and a thinner low-k interfacial layer [34].

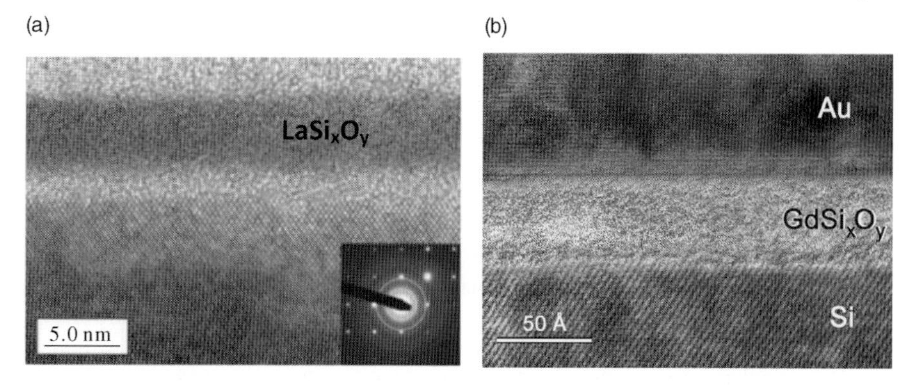

Figure 5.4 High-resolution transmission electron micrograph of a thin (a) La–silicate and (b) Gd–silicate film.

An important consideration for choosing an alternative high-*k* dielectric is its material compatibility with Si, and metal silicates have attracted much attention in recent years [21, 35, 36]. The presence of a silicate layer causes improved high-κ oxide/silicon interface stability and reduced leakage currents. This in turn has generated interest in the deposition of RE silicates (Figure 5.4), such as La-silicate, Pr-silicate, Nd-silicate, Gd-silicate, and Er-silicate [37–41]. A silicate formation is known to occur when an RE oxide is in contact with a Si-containing dielectric or a silicon substrate in the presence of oxygen [36]. Ono and Katsumata [42] reported that the peak height of Si–O–RE bonds postannealing in O_2 and N_2 ambient decreased as the atomic number increased, as shown in Figure 5.5. The light RE metal oxides with larger ionic radii, Si atoms from the substrate diffuse easily into the space in the RE_2O_3 film and form Si–O–RE bonds inside, causing the formation of an interfacial RE silicate layer between the RE_2O_3 and the Si substrate. Therefore, it can be used to

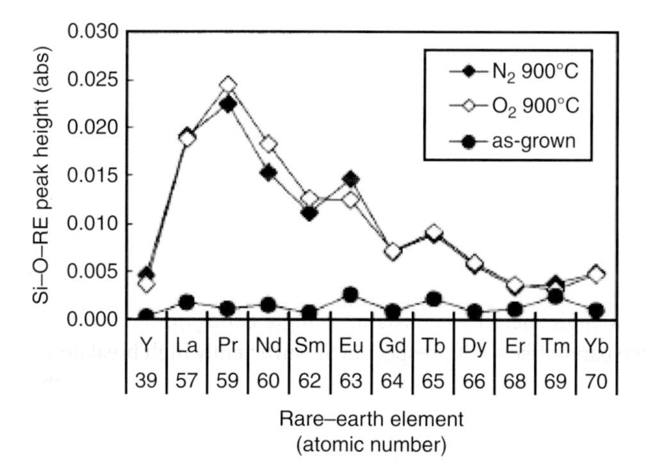

Figure 5.5 The peak height of Si–O–RE bonds for thin RE_2O_3 films as-grown and annealed at 900 °C for 30 min in O_2 and N_2 ambient.

consume the interfacial layer between high-κ and silicon substrate or to increase its κ-value.

5.4.2
X-Ray Diffraction Analysis

The crystal structure and growth orientation of high-κ RE oxide materials at different process conditions are described in this section. X-ray diffraction (XRD) is a technique used to determine the crystallographic structure of natural and synthetic materials. The crystallographic structure and orientation of RE oxide films are related to the deposition process parameters and annealing temperatures. Deposition of some RE oxide (RE_2O_3) films on Si(100) substrate by ALD at ~300 °C is summarized in Figure 5.6. The presence of multiple peaks in the XRD data expresses the polycrystalline grain growth with cubic structure [43]. All films have preferred orientation at cubic (400) direction, while exceptionally the Nd_2O_3 film has hexagonal (101) orientation. However, other orientations at (222), (411), (440), and (622) directions also present with less probability.

It has been reported that the change in preferred orientation of RE_2O_3 film is associated with the deposition temperature. For example, Er_2O_3 film orientation changes from (400) to (222) over switching the deposition temperature from 300 to 350 °C, whereas below 250 °C it is amorphous [44]. In bulk form of these RE oxides, the preferred crystal structure depends on the process temperature and pressure. For example, the cubic-phase structure of Sm_2O_3 in bulk is most stable at low temperature.

The crystalline structure of RE_2O_3 film is also related to the oxygen content and postdeposition annealing (PDA) temperature. Sm_2O_3 films annealed at 600 °C

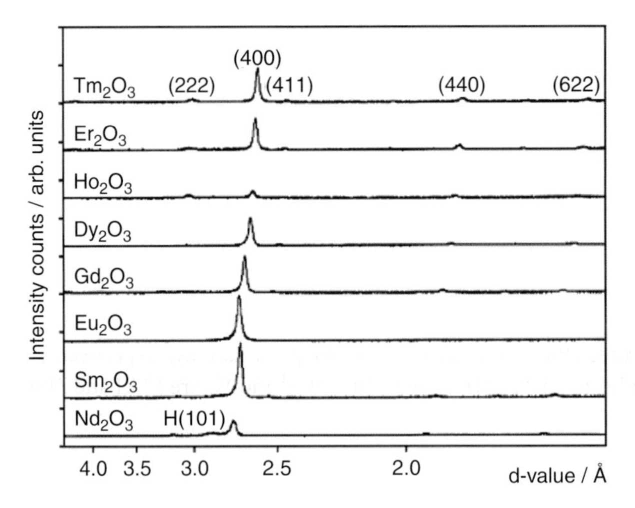

Figure 5.6 XRD patterns of RE_2O_3 films deposited at 300 °C showing the films to have polycrystalline and cubic structure.

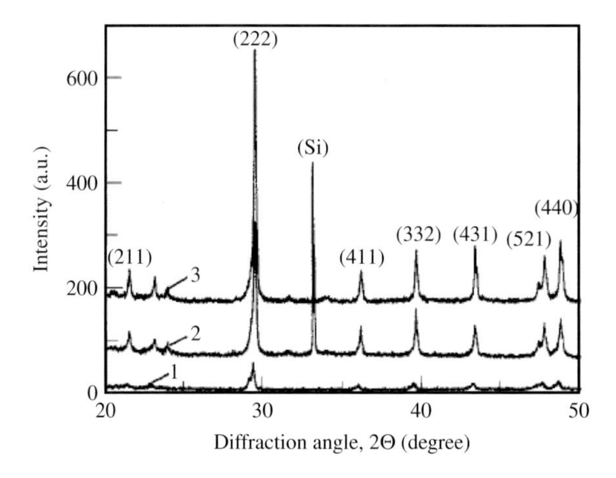

Figure 5.7 XRD measurement of as-deposited Er_2O_3 films: (1) 4.5 nm film, (2) 50 nm film, and (3) 100 nm film.

exhibit a weak reflection from (400) plane, while those annealed at 800 °C show a strong reflection from (444) plane. In addition, the same Sm_2O_3 film deposited under a 15/10 (Ar/O_2) flow condition reveals the strongest reflection from (400) plane [45]. It is reported that for the lighter RE oxides, such as Nd_2O_3, the hexagonal structure is the dominant phase at higher temperatures, while the cubic structure exists at low temperatures (<600 °C) [46]. In some RE oxides, for example, Gd_2O_3 film, the crystal orientation strongly depends on the film thickness [47]. In addition, the significant differences in sharpness, intensity, and orientation of the diffraction peak for Er_2O_3 films deposited on Si(100) substrates by electron beam evaporation were related to the film thickness, as shown in Figure 5.7. These films are found to be cubic structure ($a_0 = 10.5$ Å) and microcrystalline with grain sizes of 20–50 nm [48].

Zhao *et al.* [49] demonstrated that both the crystal structure and the κ-value of some RE oxide films change with the time exposed to the air or amount of moisture absorption. For example, Figure 5.8a shows the amount of hexagonal $La(OH)_3$ in the La_2O_3 film increases with time. According to the Clausius–Mossotti relation, the κ is determined by

$$\kappa = \frac{1 + (2/3)(4\pi\alpha^{T}/V_{m})}{1 - (1/3)(4\pi\alpha^{T}/V_{m})} \quad (5.1)$$

where V_m and α^T represent the molar volume and total polarizability, respectively, in the dielectric film. For hexagonal $La(OH)_3$, the values of α^T and V_m are 12.81 and 71 Å, respectively. From Equation 5.1, the κ-value of hexagonal $La(OH)_3$ is about 10, which is lower compared to that of La_2O_3 (35) film. Therefore, the effective permittivity of La_2O_3 film after exposure to an air environment will be degraded. To overcome the problem of moisture absorption, the incorporation of Ti or Y into the RE dielectric films can result in improved physical and electrical properties (a thinner interfacial

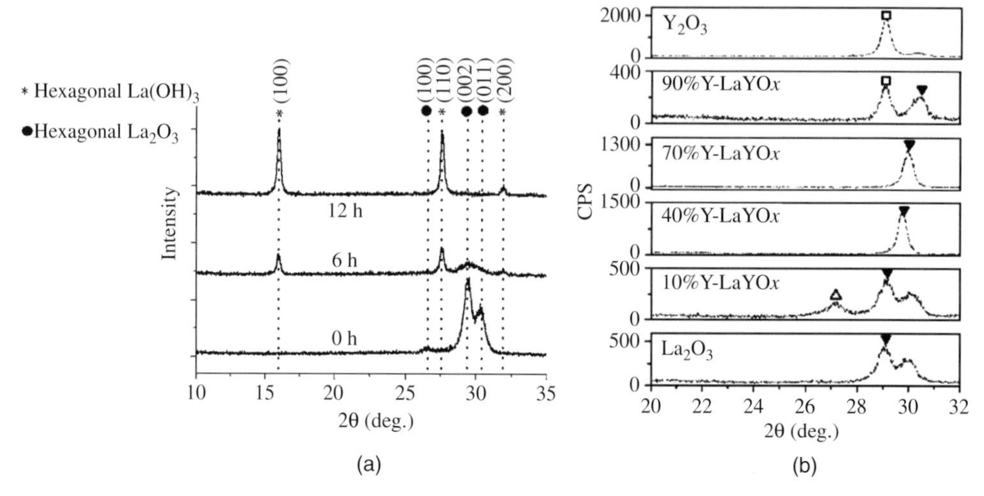

Figure 5.8 XRD patterns of (a) La_2O_3 films on silicon after exposure to the air for different times and (b) $LaYO_x$ films with different Y concentrations after annealing at 600 °C (Solid inverted triangle: hexagonal $LaYO_x$ (002); open square: cubic Y_2O_3 (222); and open triangle: cubic La_2O_3 (222)).

layer, lower solubility in water, higher capacitance, and lower leakage current) because it decreases the extent of the reaction of the dielectric film with water [50, 51]. Zhao *et al.* [51] indicated that the $LaYO_x$ films containing 40–70% Y concentrations after annealing at 600 °C are well crystallized in the hexagonal phase to resist the moisture absorption, as shown in Figure 5.8b. Moreover, these films have a higher κ-value than La_2O_3 film, suggesting the better crystallinity. Another factor is that these films were crystallized in the hexagonal phase rather than the cubic phase. It has been reported that the V_m values of hexagonal phase of RE oxide films are smaller compared to those of the cubic phase oxides [52]. For La_2O_3 film, there will be about 10% shrink in the V_m value after the transformation from cubic phase to hexagonal [53]. Consequently, based on the Clausius–Mossotti relation, the hexagonal RE oxides show a higher κ-value than cubic RE oxides. If α^T is assumed to be a constant but the V_m changes, the smaller V_m will increase the κ-value. For hexagonal La_2O_3 film, the permittivity of hexagonal La_2O_3 is about 35 because α^T and V_m are 18.17 and 82.7 [51], respectively. The same method is applied to hexagonal Y_2O_3 (222) to estimate the permittivity, whereas the permittivity of the cubic-phase Y_2O_3 is about 11 [33, 51]. In addition, the peak of hexagonal (002) $LaYO_x$ films gradually shifts to a larger 2θ with the increasing Y concentration. This shift is attributed to the decrease in the lattice parameter due to a smaller ionic radius of Y^{3+} compared to that of La^{3+}.

5.4.3
Atomic Force Microscope Investigation

The surface roughness of a gate dielectric film plays a major role in the electrical performances of CMOS transistor. To investigate the surface morphology, specifically

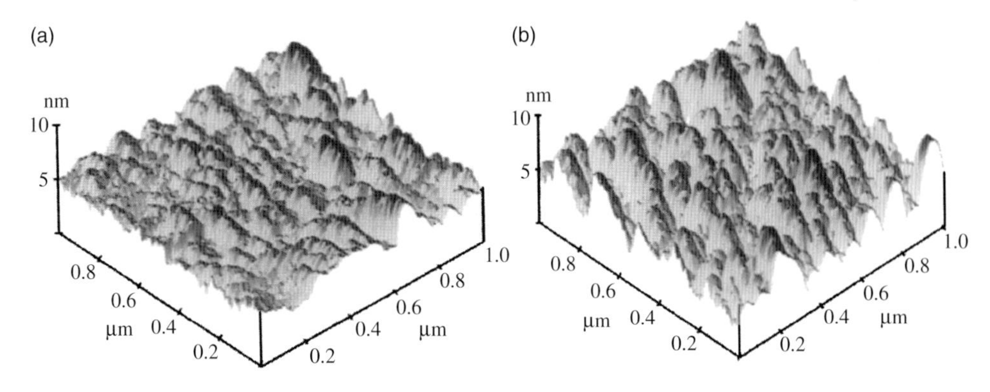

Figure 5.9 AFM images of 50 nm Er_2O_3 films for (a) as-deposited and (b) annealed at 750 °C.

the root mean square roughness (R_{rms}) and the difference between the highest and the lowest points in the surface (R_{max}), atomic force microscope was used in tapping mode for the sample scan. The surface roughness of RE oxide films strongly depends on different film thicknesses, annealing temperatures, and process conditions. Figure 5.9 shows an AFM surface image of an as-deposited 50 nm-thick Er_2O_3 oxide film, revealing R_{rms} and R_{max} values of 0.82 and 5.96 nm, respectively. We observed a small increase of the roughness to $R_{rms} = 1$ and $R_{max} = 7$ nm with annealing temperature [54]. A 100 nm-thick Er_2O_3 film has higher R_{rms} and R_{max} values of 1.1 and 7.3 nm, respectively, than that of a 50 nm-thick film. For E-beam deposition, the surface roughness of the Lu_2O_3 film annealed at 600 °C ($R_{rms} = 1.1$ nm) is larger compared to that of the as-deposited film ($R_{rms} = 0.45$ nm) [55]. In contrast, the Lu_2O_3 film by ALD after annealing at 950 °C ($R_{rms} = 0.24$ nm) exhibited a smoother surface than that of the as-grown film ($R_{rms} = 0.87$ nm) [56]. In this ALD method, the surface roughness of Lu_2O_3 film decreased with increasing the annealing temperature. This different result may be attributed to different film thicknesses and deposition methods. Furthermore, Ohmi *et al.* [57] reported that the R_{rms} values of La_2O_3 films deposited on Si(100) substrates at room temperature, 250, and 400 °C exhibited 0.442, 0.195, and 0.339 nm, while those for the Yb_2O_3 films at room temperature and 250 °C were 0.117 and 0.189 nm, respectively. The surface roughness may depend on the difference of lattice energy of the materials. Thin Yb_2O_3 film has a higher lattice energy than La_2O_3 film [57], resulting in a smooth surface. In addition, the surface roughness of RE dielectric films is related to the method of film deposition. At the same thickness, the Er_2O_3 film grown by ALD at 300 °C exhibits a smoother surface of 0.16 nm than that of E-gun deposition [43].

The surface roughness of the Sm_2O_3, Dy_2O_3, and Tm_2O_3 films after rapid thermal annealing (RTA) at various temperatures are shown in Figure 5.10a. The surface roughness of the Dy_2O_3 and Tm_2O_3 films clearly increased upon increasing the PDA temperature. The increase in surface roughness after RTA treatment is plausibly due to grain growth and/or crystallization during annealing. On the other hand, the surface roughness of the Sm_2O_3 film clearly decreased upon

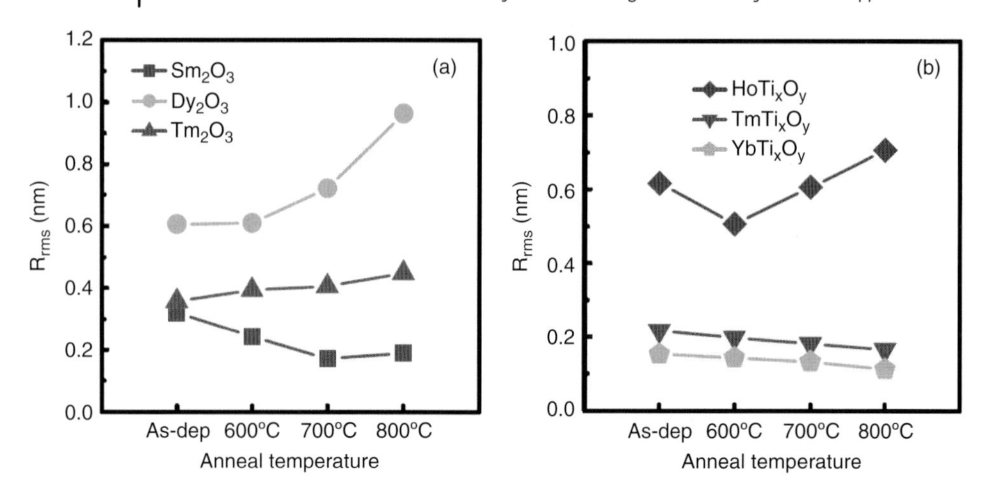

Figure 5.10 Surface roughness as a function of annealing temperature for (a) Sm_2O_3, Dy_2O_3, and Tm_2O_3 and (b) $HoTi_xO_y$, $TmTi_xO_y$, and $YbTi_xO_y$ films.

increasing the RTA temperature but increased at 800 °C [45]. During high-temperature annealing, the oxygen atom removed from the film mostly migrated to the interface and increased the formation of an amorphous silica layer, leading to a higher surface roughness [14]. It has been reported that the roughness at the dielectric/Si and dielectric/gate interfaces can significantly affect the leakage current through the gate dielectric [16]. The moisture absorption should be the intrinsic feature of the RE oxide film due to the oxygen vacancies in the film [51]. To solve the issue of moisture absorption and poor crystallinity, the addition of Ti or TiO_2 into RE oxide materials exhibited excellent physical properties and electrical characteristics [50, 58]. Figure 5.10b shows the R_{rms} values of the $HoTi_xO_y$, $TmTi_xO_y$, and $YbTi_xO_y$ films before and after annealing at different temperatures [59]. The surface roughness of the $HoTi_xO_y$ films clearly increased upon increasing the RTA temperature, except for as-deposited film. In contrast, the surface roughness of the $TmTi_xO_y$ and $YbTi_xO_y$ films decreased upon increasing the RTA temperature. This is condensed during annealing as density increases. The availability of defect sites due to lower density can reduce self-diffusion of thulium and ytterbium resulting in lower grain boundary velocity during annealing and therefore small and more uniform grains.

Surface pretreatment with N_2O or NH_3 before deposition of these metal oxides on Si substrates is required to achieve high-quality gate dielectrics. The incorporation of N atoms helps the interfacial layer to (i) function as a barrier hindering the diffusion of both Si and O atoms, (ii) effectively reduce the equivalent oxide thickness and leakage current, (iii) passivate O atom vacancy states, and (iv) enhance structural stability. Moreover, the N atom content in the interfacial layer prevents degradation of the mobility with Hf-aluminate gate dielectrics [14]. Generally, N atom incorporation into gate oxides is performed through the high-temperature or plasma annealing of oxide films in a nitrogen environment. Figure 5.11 depicts the surface roughness of

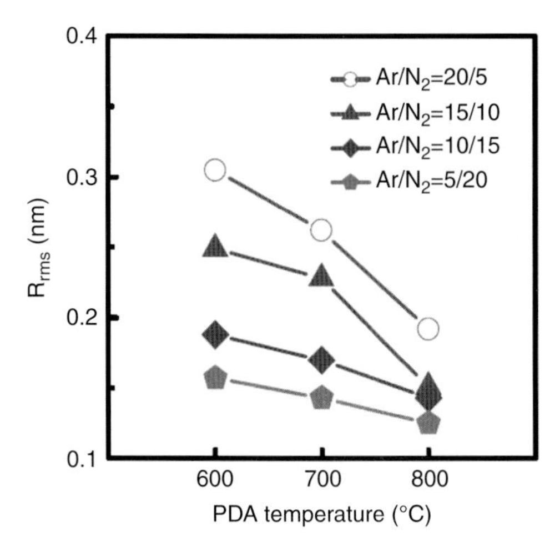

Figure 5.11 Surface roughnesses of the NdO_xN_y films plotted as a function of the PDA temperature for each of the flow ratios.

the NdO_xN_y films as a function of the PDA temperature for different flow conditions [60]. The surface roughness clearly decreased upon increasing the N atom content. This behavior is due to an increased number of N atoms in the film reducing the clustering of grains, thus decreasing the surface roughness of the NdO_xN_y film. The surface roughness of the NdO_xN_y films decreased upon increasing the PDA temperature. Green *et al.* [61] reported that the content of incorporated N atoms increases with increasing oxidation temperature. Therefore, the Si−O−N layer influences the local structure of the interface, causing a smoother surface.

5.4.4
Transmission Electron Microscopy Technique

In order to understand the device physics, precise measurement of the RE_2O_3 film and interfacial layer thicknesses is a very critical parameter. High-resolution transmission electron microscopy (HRTEM) is a reliable tool for determination of these thicknesses. Intermediate silicate layers at the interface between RE oxide films and Si were produced unintentionally when deposited directly on Si and present one of the greatest challenges for these RE dielectrics. The silicate formation is likely due to excess oxygen incorporated into the oxide layer during the film growth or the oxygen in the annealing atmosphere. The reaction of RE oxide with silicon can be presented as

$$RE_2O_3 + Si + O_2 \rightarrow RE_2SiO_5 \qquad (5.2)$$

where O_2 comes from the environment and goes through the oxide layer to react with the underlying Si. This is attributed to superstoichiometric oxygen in the as-grown oxide layer at equilibrium conditions. The TEM studies on the RE-based oxide film indicate an amorphous or very poorly crystalline microstructure even when grown at

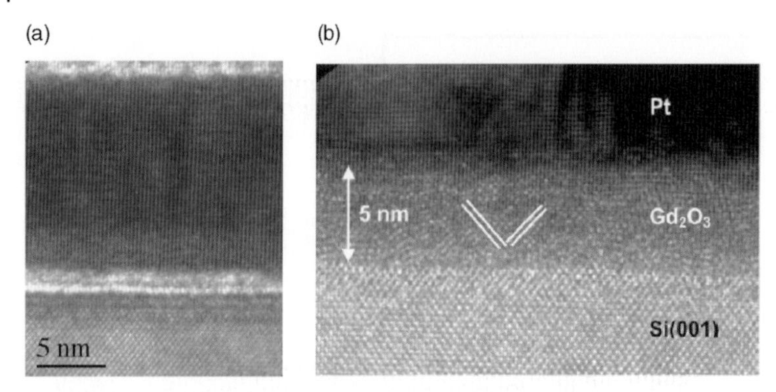

Figure 5.12 HRTEM image of (a) Gd_2O_3 film on Si(001) after 10 min O_2 anneal at 780 °C and (b) Gd_2O_3 film grown on Si(001) by MBE at 600 °C under oxygen partial pressure of 5×10^{-7} mbar.

high temperature. Few examples of such critical HRTEM investigation of RE oxides for different process conditions are explained in this section. Figure 5.12a shows that RTA in oxygen caused Gd_2O_3 films deposited directly on Si(001) to form a three-layer structure: polycrystalline Gd_2O_3 layer (top), mixed $(SiO_2)_x(Gd_2O_3)_{1-x}$ layer (intermediate), and SiO_y layer (bottom) [62]. An amorphous structure can be seen for intermediate and bottom layers. The partial oxygen pressure during the interface formation and during growth is a very crucial parameter. Too low oxygen content can result in the formation of a silicide-like layer. On the other hand, too high oxygen content might oxidize the Si surface, leading to a lower κ interfacial SiO_x layer. Czernohorsky *et al.* [63] demonstrated that thin Gd_2O_3 film grown on Si(001) by molecular beam epitaxy (MBE) under specific conditions exhibited no interfacial layer at the Gd_2O_3 and Si substrate interface, as shown in Figure 5.12b. The optimal control of oxygen partial pressure during the interface formation and/or during the subsequent MBE growth can prevent any kind of silicide inclusions, while avoiding the formation of interfacial SiO_x.In order to reduce the thickness of a SiO_y layer, Choi *et al.* [64] proposed that the Er-silicate film is formed by the interfacial reaction between Er and SiO_2 films. With increasing annealing temperature, a large amount of Er atoms from Er metal diffuse toward Er-silicate film, as shown in Figure 5.13. This result leads to the change in Er-silicate bonding state from Si rich to Er rich. In addition, the continual Er diffusion through Er-silicate film, results in further reaction with the remaining SiO_2 film and thus leading to the increase in the thickness of Er silicate and decrease in the thickness of SiO_2 film. It is found that the κ-value of a W/Er/SiO_2 gate stack for as-deposited is 7.4, whereas annealed samples at 200, 250, 300, and 350 °C become 8.8, 9.1, 9.4, and 11.4, respectively [64]. This behavior is due to the chemical bonding change of Er silicate from Si-rich to Er-rich silicate induced by the Er diffusion.

Recently, many efforts have been focused on RE oxides that can be grown epitaxially on (001)-oriented Si substrates. Various RE oxide dielectrics, such as Er_2O_3 and Pr_2O_3, have been epitaxially grown on Si(001) substrates [65, 66], but the unfavorable orientation may cause the twinning and other defects. Dimoulas *et al.* [67] reported

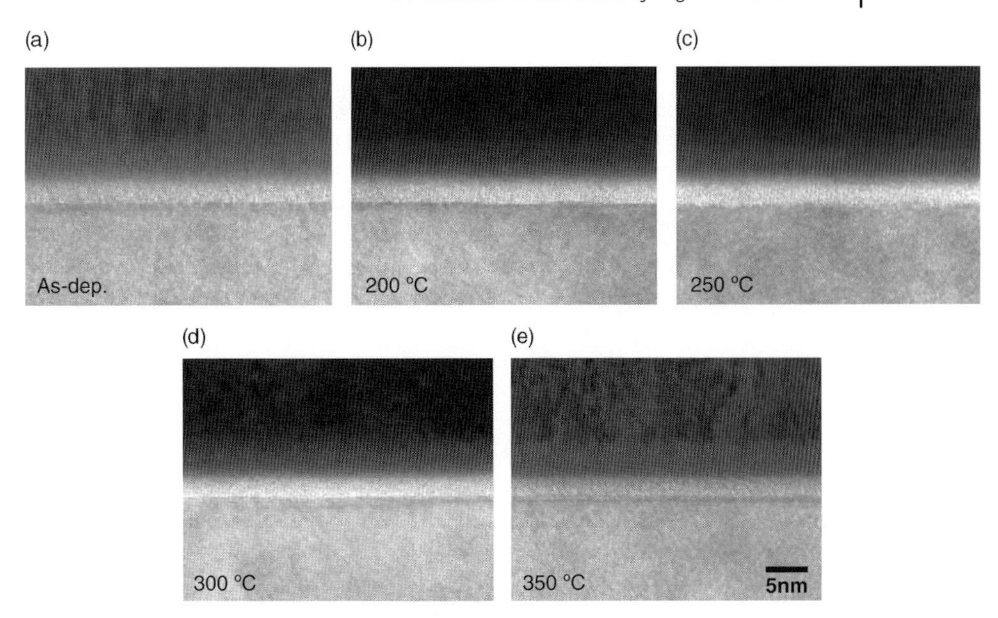

Figure 5.13 HREM images of the $W/Er/SiO_2$ gate stacks before and after annealing at the temperatures in the range of 200–350 °C.

that a high-quality $La_2Hf_2O_7$ (LHO) film can be epitaxially deposited with a cube-on-cube pattern on Si(001) by MBE. The spacing of the LHO film has a lattice mismatch of 0.74% with respect to spacing twice that of Si(001) at room temperature and effectively zero at 800 °C. Wei *et al.* [68] investigated that the interface and crystallinity of the LHO film were deposited on Si(001) by pulsed laser deposition (PLD) at 780 °C, as shown in Figure 5.14a. The orientation relationship between LHO and Si was

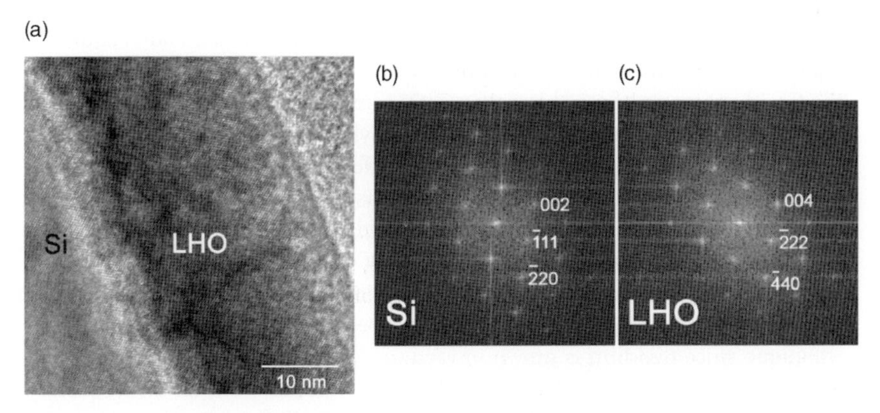

Figure 5.14 (a) Cross-sectional HRTEM image, the digital FFT patterns of the (b) Si substrate, and (c) LHO layer showing LHO dielectric grown on Si substrate at 780 °C.

determined as $[110]_{LHO}||[110]_{Si}$ and $[001]_{LHO}||[001]_{Si}$. The in-plane epitaxy of [001] LHO, parallel to [001] Si, suggests an excellent lattice match, that is, the spacing of LHO [001] lattice (10.78 Å) matches twice that of Si [001] (10.86 Å). Figure 5.14b and c showed that the digital fast Fourier transform (FFT) patterns indicate a pyrochlore structure in the LHO film. The FFT pattern of the LHO layer is almost the same as that of the Si substrate, and some diffraction patterns in LHO layer are absent. This finding is probably attributed to the indistinguishable contrast of the atom patterns between La and Hf. Moreover, no interfacial layer between LHO and Si substrate was observed in HRTEM image. However, an ultrathin thickness of an amorphous interfacial layer was found in few areas. This layer is probably caused by the oxygen diffusion during the thermal treatment in oxygen ambient after films' growth.

5.4.5 X-Ray Photoelectron Spectroscopy Analysis

The description and understanding of the local atomic environment of ultrathin and thin RE oxide films are of paramount importance because it determines the electronic properties by its first and second atomic shell interactions. The main advantages of X-ray photoelectron spectroscopy (XPS) are a rather direct structural interpretation, atomic selectivity, and a high precision in the measurement of interatomic distances. Cerium oxide (CeO_2) is possibly one of the most studied materials among various RE oxides because of its technological importance and the interesting physical properties, inherent to the Ce f-electron system [69]. CeO_2 film is a hard material with a relatively high dielectric constant and wide electronic bandgap [33]. CeO_2 film is a cubic structure [calcium fluorite (CaF_2)], indicating that Ce atom is surrounded by eight oxygen atoms in an eightfold coordination. The electronic properties of this film are related to a largely localized or delocalized 4f electronic state of Ce [70]. The electronic structure of CeO_2 dielectric is associated with unoccupied 4f states of Ce^{4+} ($4f^0$) [71]. Different 4f configurations for Ce^{4+} and Ce^{3+} lead to different core level and valence band structures. In order to realize the chemical state of CeO_2 film by PLD method [72], the Ce 3d XPS measurement was performed and is shown in Figure 5.15. Cerium has two common oxidation states (3^+ and 4^+). In general, the Ce 3d spectrum of stoichiometric CeO_2 film is composed of three spin orbit doublets (six peaks), which result from Ce 4f hybridization in both the initial and the final states. These peaks are ascribed to different 4f configurations. However, some additional features were observed in measured spectrum, apart from peaks corresponding to stoichiometric CeO_2. Two additional spin orbit doublets in the spectra are an evidence of a partial formation of Ce^{3+} ions in CeO_2. The presence of Ce^{3+} can be understood by the transformation of Ce^{4+} to Ce^{3+}. On the other hand, during the film growth, CeO_2 film is expected to lose of oxygen in the PLD technique since this film is grown in vacuum. When oxygen atom is released in the form of an oxygen molecule, two electrons are left behind. Consequently, these electrons are trapped and localized at the 4f level of two Ce atoms changing its valency from 4^+ to 3^+.

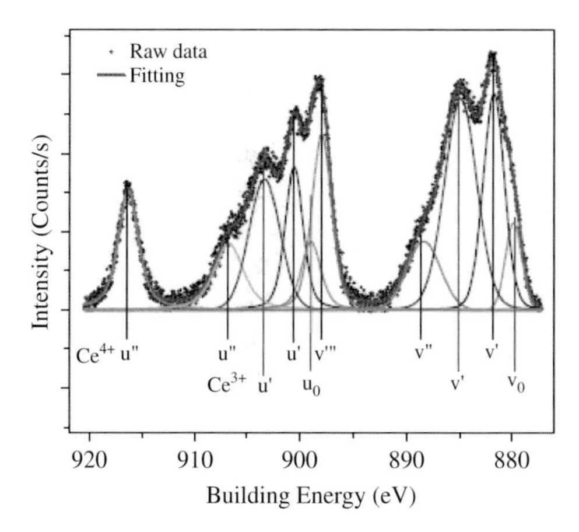

Figure 5.15 Ce 3d XPS spectra of CeO_2 film; the solid curves represent Ce^{4+} contributions and the dotted curves show the Ce^{3+} related doublets ν', u' and ν_0, u_0.

Thin RE oxide films on Si form interfacial silicate layers after annealing at several hundred degrees. The ion radius of RE oxides decreases with increasing atomic number. This effect is also known as RE oxide contraction. It has been reported that the thickness of the interfacial layer decreases significantly with higher atomic numbers and attribute this to the decrease in the ion radius [42]. Consequently, the interfacial layer formed during the high-temperature annealing is most pronounced for Pr_2O_3 film due to its large ion radius compared to the radii of other RE oxides. If hexagonal Pr_2O_3 films are annealed in 1 bar nitrogen or 10^{-5} mbar oxygen atmosphere [73, 74], an interfacial layer is formed and the films are transformed to cubic Pr_2O_3 [74], which is stable under normal temperature and pressure. The lattice constant of bulk cubic Pr_2O_3 is about two times larger than that of Si with a lattice mismatch of 2.7%. Therefore, cubic Pr_2O_3 films form antiphase domains on Si (111) substrates because cubic Pr_2O_3 can nucleate on different sublattices on Si(111). PrO_2 (fluorite structure) has the big advantage that its lattice constant matches almost exactly the lattice constant of Si. Weisemoeller [75] *et al.* demonstrated that a hexagonal Pr_2O_3 film was deposited on Si(111) by MBE after annealing in 1 atm oxygen at different temperatures, ranging from 100 to 700 °C. For different stoichiometric Pr_2O_3 films, $3d_{5/2}$ and $3d_{3/2}$ peaks can be expected at 935 and 955 eV (Figure 5.16a), respectively. The most prominent one can be seen at 931 eV and a sharp satellite peak should build up for single crystalline and stoichiometric PrO_2. However, the intensity of the main Pr 3d peaks does not disappear as fast as the intensity of the PrO_2 (specific Pr $3d_{3/2}$ peak at 967 eV). The O 1s XPS signal of the Pr_2O_3 sample is shown in Figure 5.16b. At the beginning, the double-peak structure visible is due to a stoichiometric PrO_2 structure. For 18 min of sputtering, the binding energy shifts from the PrO_2 state toward higher bindings during the sputter process. This confirms the oxygen loss due to preferential sputtering of oxygen. After 50 min of sputtering, a single O 1s peak appears at 531.4 eV, which can be attributed to the

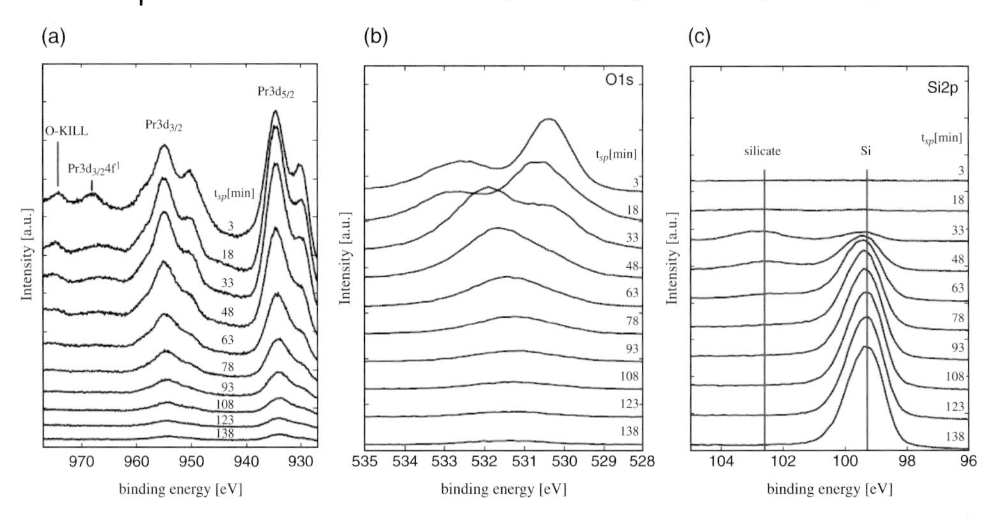

Figure 5.16 Sputter XPS measurement of the (a) Pr 3d, (b) O 1s, and (c) Si 2p region of a 5 nm Pr_2O_3 on Si(111) sample after PDA at 450 °C in 1 bar oxygen.

formation of a Pr-silicate layer at the Pr_2O_3 film and Si substrate interface. Figure 5.16c presents complementary results for Si 2p photoelectrons. No Si signal is visible after 3 and 18 min of sputtering. After 33 min of sputtering, Si 2p peak is detected at a binding energy of 99.3 eV. Another peak appears at a binding energy of 102.6 eV. Therefore, we attribute this peak to the formation of SiO_2-rich silicate at the oxide film and Si substrate interface. In the following sputter cycles, the silicate peak disappears for 60 min and only the bulk Si remains. A very thin SiO_2-rich silicate is formed at the oxide film and substrate interface because the PrO_2 film loses oxygen during the sputter process.

Recently, many RE_2O_3 films have been investigated as a promising candidate for a gate dielectric material. Moreover, the incorporation of titanium or TiO_2 into RE dielectric materials has attracted considerable attention as a method to obtain suitable structures for CMOS device applications [50, 58]. It has already been reported that MBE deposited Er_2O_3 thin film is a promising gate dielectric material due to its high dielectric constant and excellent electrical properties [33]. The XPS analysis of such TiO_2-incorporated RE oxide is described in this section as an example. Figure 5.17 shows the Er 4d, Ti 2p, O 1s, and Si 2p XPS spectra for the $ErTi_xO_y$/Si interface before and after RTA in N_2 ambient. Oxygen bonding to titanium and erbium is expected to shift the Er 4d peak toward higher binding energies (169.8 eV corresponding to $ErTi_xO_y$) as compared to Er-O (168.5 eV corresponding to Er_2O_3) in Er_2O_3 by considering the coordination number of titanium. After annealing is performed at 700 °C, the Er 4d peak at 169.8 eV corresponding to $ErTi_xO_y$ bonds presents at the interface as shown in Figure 5.17a. The Er 4d peak position of the film after PDA at 800 °C shifts to a higher binding energy by only 0.1 eV, indicating a relatively less amount of Er reaction with Si resulting in a thinner erbium silicate layer at the interface. The Ti 2p doublet (Ti $2p_{1/2}$ and Ti $2p_{3/2}$ at 465.6 and 459.7 eV, respectively)

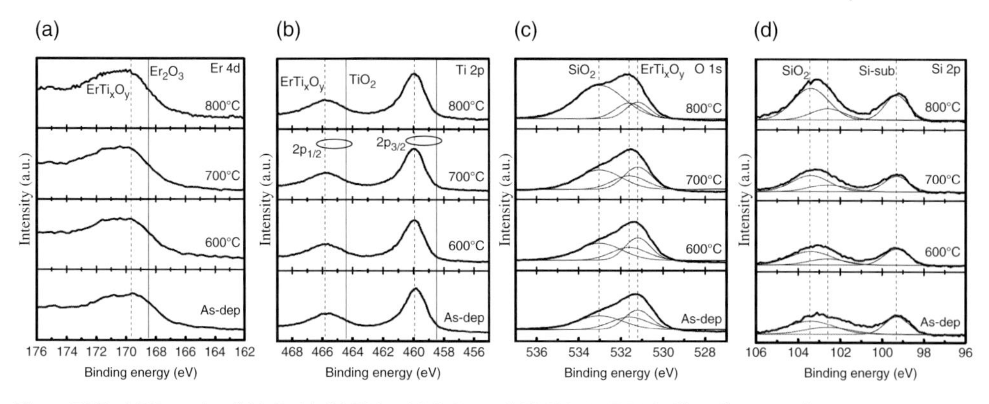

Figure 5.17 XPS results of (a) Er 4d, (b) Ti 2p, (c) O 1s, and (d) Si 2p in $ErTi_xO_y$ film after annealing at various temperatures.

is shifted to higher binding energy compared to the TiO_2 reference position (Ti $2p_{1/2}$ and Ti $2p_{3/2}$ at 464.3 and 458.7 eV, respectively), as shown in Figure 5.17b. This shift was attributed to Ti in $ErTi_xO_y$ compound. There is not an obvious shift in Ti 2p doublet peak position with increasing PDA temperature, suggesting the Er–O–Ti type of bond. Furthermore, no evidence for Ti–Si bonds from silicides was observed in the Ti 2p spectra. The O 1s spectra in Figure 5.17c can be deconvoluted to three chemical states. The low binding energy state at 531.2 eV can be related to O in $ErTi_xO_y$. The median binding energy state at 531.6 eV can be attributed to interfacial O atoms in nonstoichiometric $ErSi_xO_y$. The high binding energy state at 533 eV can be related to O in SiO_2. The feature at 531.6 eV is distinctly different from both O in SiO_2 (high-energy feature at 533 eV) and O in $ErTi_xO_y$ (low-energy feature at 531.2 eV). It is reasonable to attribute this chemical state to a mixture of $ErTi_xO_y$ and SiO_2. The as-grown film seems to be composed mainly of $ErTi_xO_y$, Er-silicate, and SiO_2. The O 1s peak intensity corresponding to $ErTi_xO_y$ is rather constant up to 700 °C but suddenly decreases at 800 °C, whereas the O 1s peak intensity corresponding to SiO_2 suddenly increases at 800 °C. This suggests that the oxygen moving from the $ErTi_xO_y$ film was mostly consumed by the formation of SiO_x. Furthermore, the intensity of the O 1s peak at 531.2 eV corresponding to $ErTi_xO_y$ remain almost unaltered with increasing annealing temperature, suggesting that a well-crystallized $ErTi_xO_y$ structure resulting in a higher thermal stability and a lower diffusivity of oxygen. The Si 2p XPS spectra of the film after different annealing temperatures were composed of three different component peaks at binding energies of 99.3, 102.6, and 103.4 eV, as indicated in Figure 5.17d. The Si 2p peak position at 99.3 eV was assigned to the Si substrate, whereas the Si 2p peak located at 103.4 eV mainly corresponds to the SiO_2. The Si 2p peak at 102.6 eV can be attributed to silicate-type bonding (Si–O–Er). The Si 2p peak at 102.6 and 103.4 eV of the film after annealing at 700°C remains a smaller value, indicating that there is less silicate and silicon oxide formed at the interface with silicon substrate. In addition, the single strong peak at 103.4 eV of the sample after PDA at 800 °C was observed, indicating the formation of an amorphous silica layer.

5.5 Electrical Characteristics of High-κ RE Oxide Films

5.5.1 The Threshold Voltage, Flatband Voltage, Interface Trap, and Fixed Charge

The threshold voltage of a MOSFET is usually defined as the gate voltage where an inversion layer forms at the interface between the gate oxide and the substrate of the transistor. The threshold voltage for large geometry is given by

$$V_{\mathrm{TH}} = V_{\mathrm{FB}} + 2\phi_{\mathrm{F}} + \frac{\sqrt{2q\kappa_{\mathrm{s}}\varepsilon_0 N_{\mathrm{A}}(2\phi_{\mathrm{F}} - V_{\mathrm{BS}})}}{C_{\mathrm{ox}}} \tag{5.3}$$

where V_{FB} is the flatband voltage, ϕ_{F} is the potential difference between the Fermi level and the intrinsic Fermi level in silicon, C_{ox} is gate oxide density, κ_{s} is the semiconductor permittivity, ε_0 is the permittivity of vacuum, N_{A} is the doping concentration, and V_{BS} is the substrate–source voltage. The flatband voltage is determined by the metal–semiconductor work function difference (Φ_{MS}) and the various oxide charges through the relation

$$V_{\mathrm{FB}} = \Phi_{\mathrm{MS}} - \frac{Q_{\mathrm{it}}(\phi_{\mathrm{s}})}{C_{\mathrm{ox}}} - \frac{Q_{\mathrm{f}}}{C_{\mathrm{ox}}} - \gamma\frac{Q_{\mathrm{ot}}}{C_{\mathrm{ox}}} - \gamma\frac{Q_{\mathrm{m}}}{C_{\mathrm{ox}}} \tag{5.4}$$

where Q_{it} is the interface trapped charge, Q_{f} is the fixed oxide charge, Q_{ot} is the oxide trapped charge, Q_{m} is the mobile ionic charge, ϕ_{s} is the surface potential, and γ is the charge distribution factor. It is defined by

$$\gamma = \frac{\int_0^{t_{\mathrm{ox}}} x/t_{\mathrm{ox}}\varrho(x)\mathrm{d}x}{\int_0^{t_{\mathrm{ox}}} \varrho(x)\mathrm{d}x} \tag{5.5}$$

where t_{ox} is the gate oxide thickness and $\varrho(x)$ is the oxide trapped or mobile ionic oxide charge per unit volume [76]. The fixed oxide charge is located very close to the oxide and substrate interface. Mobile ionic and oxide-trapped charges, however, may be distributed through the oxide film. The effect on flatband voltage is greatest when the charge (Q_{it} and Q_{f}) is located at interface between the oxide and the substrate. Most of the alternative high-κ dielectric candidates examined to have a substantial amount of interface-trapped and fixed oxide charges, which could present significant issues for CMOS applications. In addition, a reproducible V_{FB} (correspondingly V_{TH} for transistors) value is needed for a stable and reliable transistor. Thus, a hysteretic change in V_{FB} of less than 20 mV is often required [12]. This hysteresis phenomenon is caused mainly by the stress-induced defect formation, chemical contaminations, mobile ions, inner interface oxide traps, and border traps [77]. Scandate thin films, such as $LaScO_3$, $DyScO_3$, and $GdScO_3$ [78, 79], grown by different deposition techniques have high κ-values (>20) and large optical bandgaps (>5 eV) and band offsets (2–2.5 eV) compared to silicon. Lopes *et al.* [80] reported MBE grown $LaScO_3$

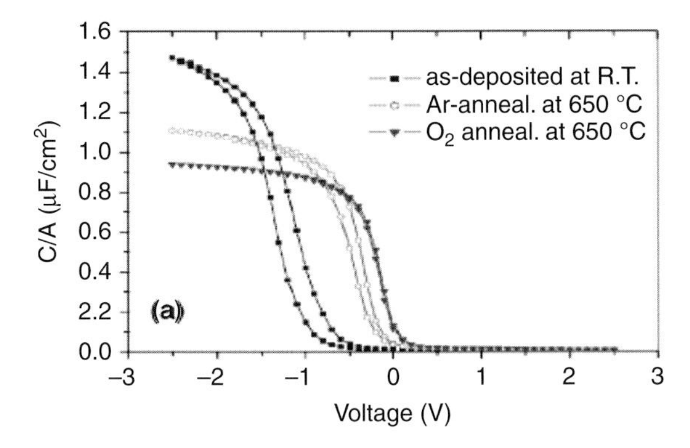

Figure 5.18 *C–V* curves at 100 kHz for $Pt/LaScO_3/p$-Si capacitor stacks fabricated with as-deposited and annealed at 650 °C in Ar or O_2 ambient.

thin films on Si(100) after PDA in O_2 or inert Ar atmosphere at 650 °C. The *C–V* curves were measured under forward and reverse bias sweeps at a frequency of 100 kHz, as shown in Figure 5.18. For the as-deposited film, a large hysteresis of ~235 mV, large negative V_{FB} of −0.75 V, and without saturation in the accumulation region was observed in the *C–V* curves. In contrast, the $LaScO_3$ film after PDA at 650 °C in O_2 atmosphere exhibited a small hysteresis of ~25 mV and a low V_{FB} of −0.1 V. The negative flatband voltage and the hysteresis are due to interface-trapped and fixed oxide charges, probably related to the existence of oxygen vacancies and broken bonds in the oxide film and at the interface between the oxide and the substrate. Consequently, the O_2 annealing is more effective in the suppression of such oxygen-related defects compared to the annealing in an Ar gas. On the other hand, a reduction in the capacitance in the accumulation region is also found, which is related to the growth of a lower κ interfacial layer at the oxide/Si interface after annealing in Ar or O_2.

The MBE growth of high-κ dielectrics on Si substrate is an alternative method to solve polycrystalline problem because it provides high-quality single crystalline oxide. Consequently, the excellent controllability of interface property between single crystalline oxide and silicon ensures almost no interfacial layer between oxide and silicon. Sun *et al.* [81] demonstrated the optimization of forming gas annealing treatment for both $Pt/Gd_2O_3/Si(100)$ and $W/Gd_2O_3/Si(100)$ gate stacks to passivate these dangling bonds and improve the flatband voltage. Figure 5.19a shows the *C–V* plots of the Nd_2O_3 film measured under different frequency at 100 °C [82]. From the figure, it is found that *C–V* plot taken at 1 MHz exhibited a typical high-frequency *C–V*. Moreover, capacitance density at inversion region decreases with increasing frequency. Finally, a typical quasi-static *C–V* plot was observed at the frequency of 500 Hz. An anomalous hump was found in the inversion region of quasi-static *C–V* curves. This hump is attributed to be induced by the interface-trapped charges. The difference in the *C–V* curves between quasi-static and high frequency indicates the existence of a certain number of interface traps. The density of interface traps in terms of the measured maximum conductance is as follows:

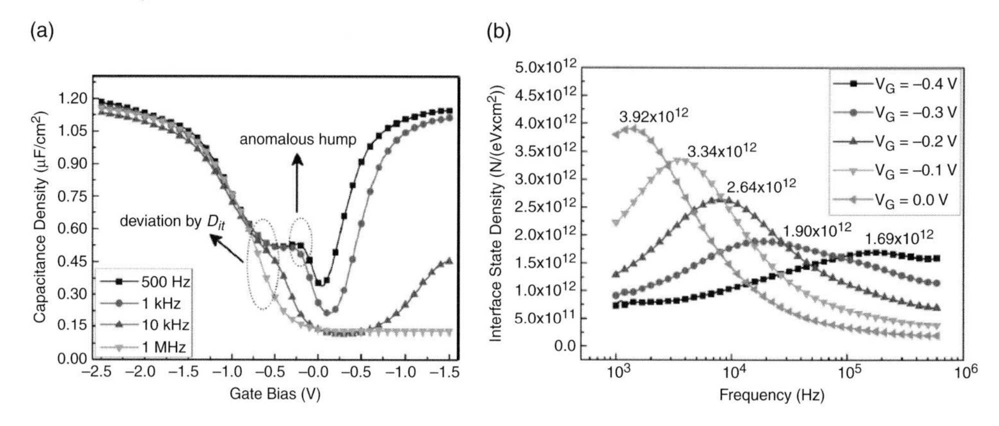

Figure 5.19 (a) Frequency-dependent quasistatic *C–V* curves and (b) interface state density as a function of frequency at V_{FB} of single crystalline Nd_2O_3 on Si(111) substrate.

$$D_{it} = \frac{2.5}{q}\left(\frac{G_p}{\omega}\right)_{max} \tag{5.6}$$

where G_p is the equivalent parallel conductance of MOS capacitor. The G_p/ω represents the maximum value of G_p/ω versus frequency plot at different gate biases [76]. The D_{it} value illustrates a maximum at a particular frequency shown in Figure 5.19b. It is clear that the D_{it} value decreases with decreasing gate voltage. The Nd_2O_3 film measured under a gate voltage of 0 V showed a large D_{it} value (3.92×10^{12} eV^{-1} cm^{-2}), while the one measured under a gate voltage of −4 V had a low D_{it} value (1.69×10^{12} eV^{-1} cm^{-2}). These D_{it} values of Nd_2O_3 film are higher compared to a traditional SiO_2 film (~10^{10} eV^{-1} cm^{-2}).

5.5.2 Leakage Mechanism

Dielectric is a material in which the electrons are very tightly bonded. The electric charges in dielectrics will respond to an applied electric field through the change in dielectric polarization. Dielectric materials are nearly insulators in which the electrical conductivity is very low and the energy bandgap is large. In general, the value of energy bandgap of insulators is set to be larger than 3 or 5 eV. Although not all dielectrics are insulators, all insulators are typical dielectrics. At 0 K, the valence band is completely filled and the conduction band is completely empty. Thus, there is no carrier for electrical conduction. When the temperature is larger than 0 K, there will be some electrons thermally excited from the valence band and also from the donor impurity level to the conduction band. These electrons will contribute to the current transport of the dielectric material. Similarly, holes will be generated by acceptor impurities and vacancies will be left by excited electrons in the valence band. The conduction current of insulators at normal applied electric field will be very small because their conductivities are inherently low, on the order of 10^{-20}–10^{-8} Ω^{-1} cm^{-1}. However, the conduction current through the dielectric film is noticeable when a relatively large electric field is applied. These noticeable conduction currents are owing to many different conduction mechanisms, which is critical to the applications of the dielectric films. For example, the gate dielectric of MOSFETs, the capacitor dielectric of DRAMs, and the tunneling dielectric of Flash memory devices are of top importance to the applications of integrated circuits. In these cases, the conduction current must be lower than a certain level to meet the specific reliability criteria under normal operation of the devices. Accordingly, the study of the various conduction mechanisms through dielectric films is of great importance to the success of the integrated circuits.

Among the conduction mechanisms investigated, some electrical properties depend on the electrode–dielectric contact. These conduction mechanisms are called electrode-limited conduction mechanisms or injection-limited conduction mechanisms. There are other conduction mechanisms that depend only on the properties of the dielectric itself. These conduction mechanisms are called bulk-limited conduction mechanisms or transport-limited conduction mechanisms [83–92]. The methods to distinguish these conduction mechanisms are essential because there are a number of conduction mechanisms that may all contribute to the conduction current through the dielectric film at the same time. Since several conduction mechanisms depend on the temperature in different ways, measuring the temperature-dependent conduction currents may afford us a helpful way to know the constitution of the conduction currents. The electrode-limited conduction mechanisms include (1) Schottky or thermionic emission, (2) Fowler–Nordheim tunneling, (3) direct tunneling, and (4) thermionic field emission. The bulk-limited conduction mechanisms include (1) Poole–Frenkel emission, (2) hopping conduction, (3) ohmic conduction, (4) space charge-limited conduction, (5) ionic conduction, and (6) grain boundary-limited conduction.

5.5.2.1 **Schottky or Thermionic Emission**

Figure 5.20 shows the MOS energy band diagram when the metal electrode is under negative bias with respect to the dielectric and the semiconductor substrate. If the electrons can gain enough energy provided by thermal activation, the electrons in the

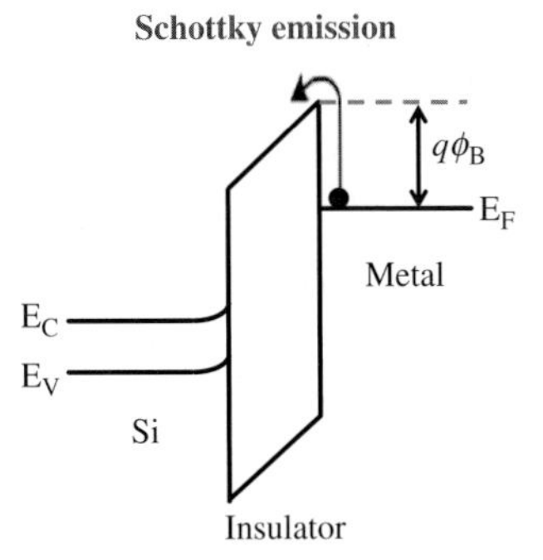

Figure 5.20 Schematic energy band diagram of thermionic emission or Schottky emission.

metal may overcome the energy barrier ($q\phi_B$) to go to the dielectric. The energy barrier height at the metal–dielectric interface may be lowered by the image force, which is called the Schottky effect. This conduction mechanism due to electron emission from the metal to the dielectric layer is called thermionic emission or Schottky emission. Thermionic emission is one of the most often-observed conduction mechanisms in dielectric films, especially at relatively high temperatures. The expressions of Schottky emission is

$$J = A^* T^2 \exp\left[\frac{-q\left(\phi_B - \sqrt{qE/4\pi\varepsilon_r\varepsilon_0}\right)}{k_B T}\right], \quad A^* = \frac{4\pi q k_B^2 m^*}{h^3} = \frac{120 m^*}{m_0} \tag{5.7}$$

where J is the current density, A^* is the effective Richardson constant, m_0 is the free electron mass, m^* is the effective electron mass in dielectric, T is the absolute temperature, q is electronic charge, $q\phi_B$ (Φ_B) is the Schottky barrier height (i.e., conduction band offset), E is the electric field across the dielectric, k_B is the Boltzmann's constant, h is the Planck's constant, ε_0 is the permittivity in vacuum, and ε_r is the optical dielectric constant (i.e., the dynamic dielectric constant). In view of the classical relationship between dielectric and optical coefficients, the dynamic dielectric constant should be close to the square of the refractive index (i.e., $\varepsilon_r = n^2$) [93].

Figure 5.21 indicates the current density–electric field (J–E) characteristics of Al/CeO_2/p-Si MOS capacitors biased in accumulation mode at temperatures ranging from 300 to 500 K. Based on the optical characterization of CeO_2 films, the refractive index (n) at 632.8 nm is about 2.33. Therefore, the dynamic dielectric constant (ε_r) is about 5.43. Figure 5.21 shows the simulations of Schottky emission and the

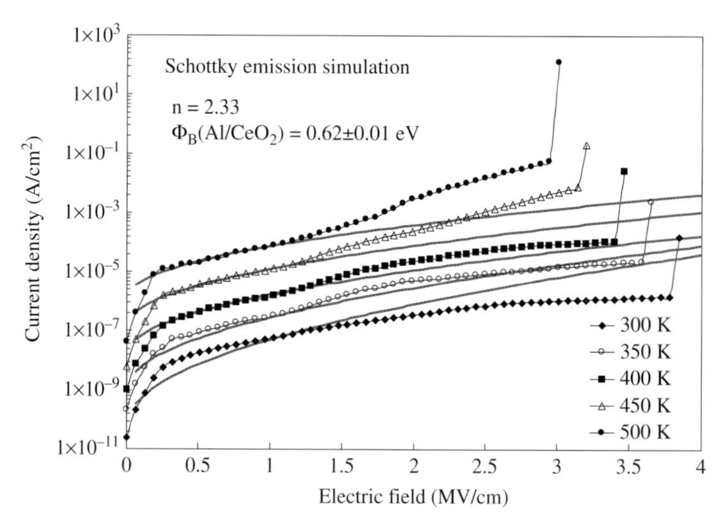

Figure 5.21 Characteristics of J–E plot (symbols) and simulation of Schottky emission (lines) for the Al/CeO_2/p-Si MOS capacitors.

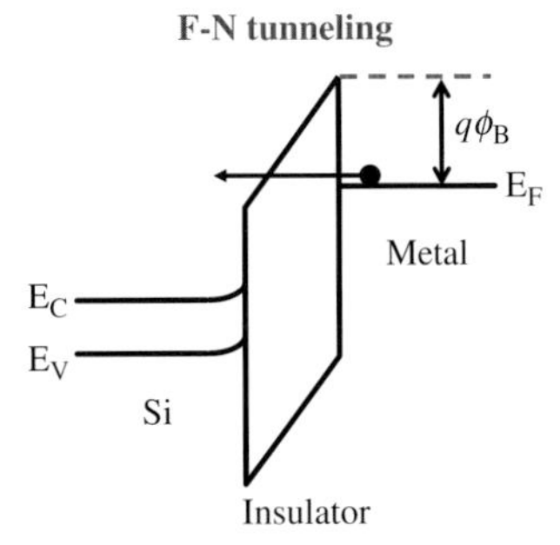

Figure 5.22 Schematic energy band diagram of Fowler–Nordheim tunneling.

measured data in [93]. At high temperatures (≥400 K) and in medium electric fields (0.3–1.3 MV/cm), the experimental results match the Schottky emission theory very well. The corresponding conduction band offset between Al and CeO_2 is then determined to be 0.62 ± 0.01 eV.

5.5.2.2 **Fowler–Nordheim Tunneling**

Based on classical physics, when the energy of the incident electrons is smaller than the potential barrier, the electrons will be reflected. However, quantum mechanism predicts that the electron wave function will penetrate through the potential barrier when the barrier is thin enough (<100 Å). Hence, the probability of electrons existing at the other side of the potential barrier is not zero because of the tunneling effect. Figure 5.22 shows the schematic energy band diagram of Fowler–Nordheim (FN) tunneling. FN tunneling occurs when the applied electric field is large enough so that the electron wave function may penetrate through the triangular potential barrier into the conduction band of the dielectric. Then, the expression of the FN tunneling current is

$$J = \frac{q^3 E^2}{8\pi h q \phi_B} \exp\left[\frac{-8\pi\sqrt{2m^*}(q\phi_B)^{3/2}}{3qhE}\right] \tag{5.8}$$

For FN tunneling mechanism, it is mandatory to include some resistance effect, *R*, which may come from device layout, substrate doping, and contact coefficient. The expression of the FN tunneling current is [94]

$$I_{FN} = A(V_{OX} - I_{FN}R)^2 \exp\left(-\frac{B}{V_{OX} - I_{FN}R}\right) \tag{5.9}$$

where *A* is a constant proportional to the injecting area and $V_{ox} = V - \Delta V$, with ΔV a correction due to the flatband voltage and due to the local band bending at the silicon electrode. $B = 8\pi(2m^*)^{1/2}\phi_B{}^{3/2}t_{eff}/3hq$, where t_{eff} is the effective thickness.

Figure 5.23 shows the *I–V* data measured on devices with different and identical gate areas (*S*) of a La_2O_3 MOS capacitor [94]. The inset graph indicates a schematic representation of the localized leakage path through the thin La_2O_3 oxide layer. The t_{ox} is the nominal oxide thickness. Note that the conduction current mostly occurs at the

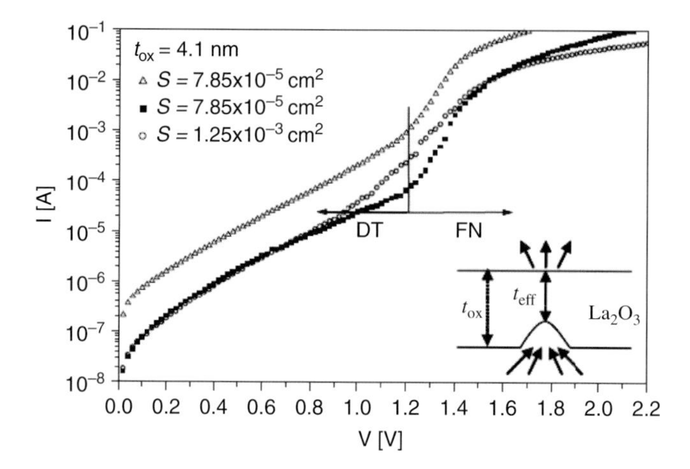

Figure 5.23 Experimental I–V characteristics measured on devices with different identical gate areas. The arrows indicate the transition between direct (DT) and FN tunneling.

thinnest part of the oxide layer with effective thickness t_{eff} because of the higher transmission probability.

A typical I–V characteristic of FN tunneling that is, $\ln(I/V^2)$ versus $1/V$, is shown in Figure 5.24. Miranda *et al.* showed that the curvature exhibited cannot be accounted for by solely invoking the standard FN model ($R = 0\,\Omega$). By minimizing the linear correlation coefficient of the data distribution, R can be evaluated. In an alternative way, R can be obtained by graphical analysis of the voltage deviation from the

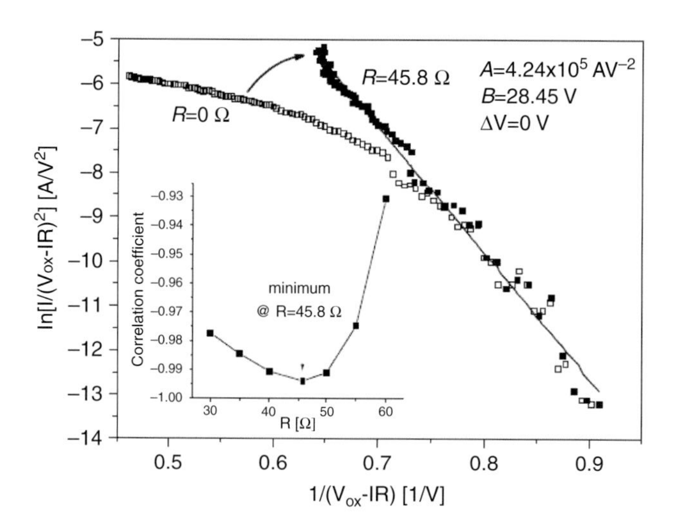

Figure 5.24 FN plots for a typical sample ($t_{ox} = 3.3$ nm). Blank squares correspond to the standard FN approach ($R = 0$). Filled squares are found varying the series resistance R and minimizing the linear correlation coefficient. The inset shows the correlation coefficient as a function of R.

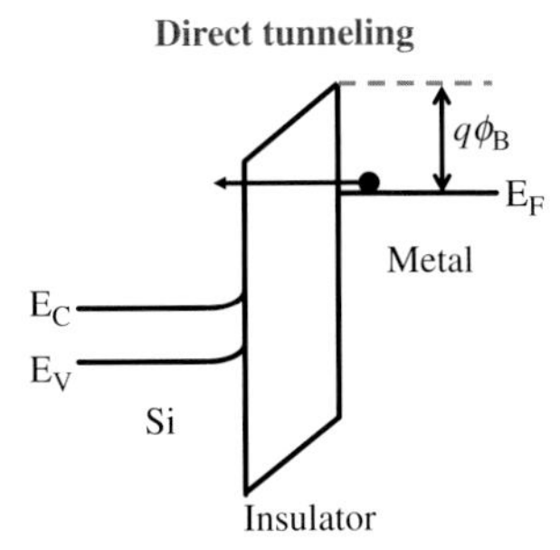

Figure 5.25 Schematic energy band diagram of direct tunneling.

theoretical FN characteristic, but it needs some assumption of parameters A and B. $R = 45.8\,\Omega$ is extracted for this device in the case of $\Delta V = 0$ V. Once R has been fixed, A and B can be evaluated by the standard approach. The theoretical FN characteristic is represented by the solid line in Figure 5.24.

5.5.2.3 **Direct Tunneling**

Figure 5.25 shows the schematic energy band diagram of direct tunneling. The expression of the direct tunneling current density is [95] as follows:

$$J = \frac{q^2}{8\pi h\varepsilon\phi_B} C(V_G, V, t, \phi_B)\exp\left\{-\frac{8\pi\sqrt{2m^*}(q\phi_B)^{3/2}}{3hq|E|}\left[1-\left(1-\frac{|V|}{\phi_B}\right)^{3/2}\right]\right\} \tag{5.10}$$

where t is the thickness of the dielectric, V is the voltage across the dielectric, and the other notations are the same as defined before. The correction function C can be expressed as

$$C(V_G, V, t, \phi_B) = \exp\left[\frac{20}{\phi_B}\left(\frac{|V|-\phi_B}{\phi_0}+1\right)^{\alpha}\left(1-\frac{|V|}{\phi_B}\right)\right]\frac{V_G}{t}N \tag{5.11}$$

where α is a fitting parameter and $q\phi_0$ is the Si/dielectric band offset. The tunneling current components include electron tunneling from the conduction band (E_{CB}), electron tunneling from the valence band (E_{VB}), and hole tunneling from the valence band (H_{VB}). N is an auxiliary function that is used as an indicator of carrier population for E_{CB} and H_{VB} cases or transmission probability for the E_{VB} case. For E_{CB} and H_{VB}, tunneling processes in both the inversion and the accumulation regimes, N is given by

$$N = \frac{\varepsilon}{t}\left\{n_{inv}\,\nu_T \ln\left[1+\exp\left(\frac{V_{G,eff}-V_{TH}}{n_{inv}\nu_T}\right)\right]+\ln\left[1+\exp\left(\frac{V_G-V_{FB}}{\nu_T}\right)\right]\right\} \tag{5.12}$$

where ν_T (= kT/q) is the thermal voltage, V_{TH} is threshold voltage, V_{FB} is flatband voltage, and $V_{G,eff} = V_G - V_{poly}$ is the effective gate voltage after accounting for the voltage drop across the polysilicon depletion region. The rate of increase in the subthreshold carrier density with V_G is indicated by the swing parameter n_{inv}, where $n_{inv} = S/v_T$ and S is the subthreshold swing that is positive for NMOS and negative for PMOS. For E_{VB} tunneling process, N can be written as

$$N = \frac{\varepsilon}{t}\left\{3v_T \ln\left[1+\exp\left(\frac{q|V|-E_g}{3k_B T}\right)\right]\right\} \tag{5.13}$$

Equation 5.10 can be simplified using a binomial expansion and neglecting higher order terms, which leads to

$$J = \exp\left[-\frac{8\pi\sqrt{2m_{eff}}(q\phi_B)^{3/2}}{3hq|E|}\frac{3|V|}{2\phi_B}\right] = \exp\left[-\frac{8\pi\sqrt{2q}}{3h}(m_{eff}\phi_B)^{1/2}\kappa t_{ox,eq}\right] \tag{5.14}$$

Yeo *et al.* [96] indicated the scaling limits of alternative gate dielectrics based on their direct tunneling characteristics and gate leakage requirements for future CMOS technology. The tunneling leakage current for a given EOT ($t_{ox,eq}$) is dependent not only on the κ-value of a gate dielectric but also tunneling barrier height (Φ_B) and tunneling effective mass (m_{eff}). They introduced a figure of merit to compare the relative advantages of gate dielectric candidates. The figure of merit is given by $f = (m_{eff}\Phi_B)^{1/2}\kappa$. Figure 5.26 indicates the scaling limit for several gate dielectrics when V_G or V_{dd} is specified to be 1.0 V and the maximum tolerable gate current density $J_{G,limit}$ is 1 A/cm^2. A dielectric with a larger figure of merit possesses a lower scaling limit EOT, as shown in the inset of Figure 5.26. The EOT scaling limit of rare-earth oxide La_2O_3 is very low (about 4 Å) compared to other high-k dielectric material.

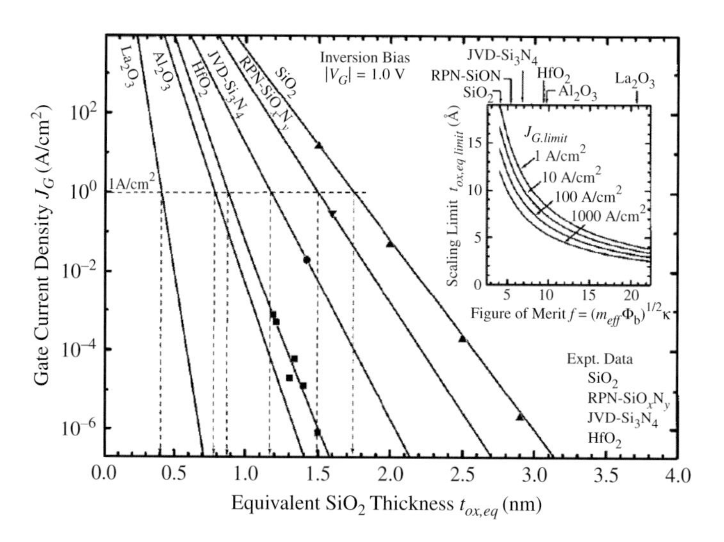

Figure 5.26 Scaling limit for several gate dielectrics when $V_{dd} = 1.0$ V and $J_{G,limit} = 1$ A/cm^2.

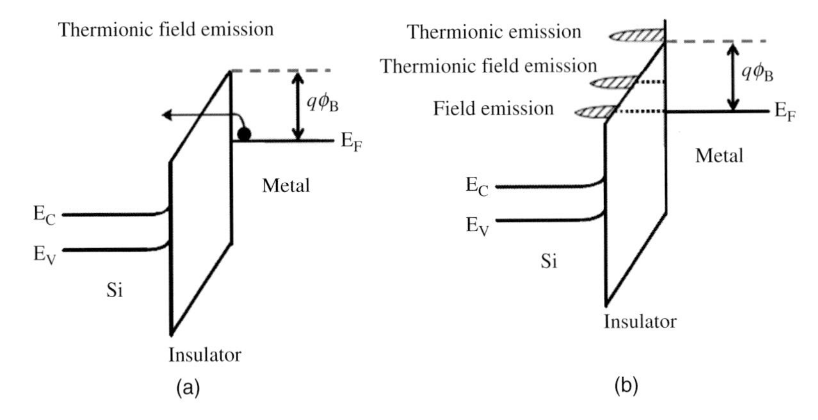

Figure 5.27 (a) Schematic energy band diagram of thermionic field emission. (b) Comparison of thermionic field emission, thermionic emission, and field emission.

5.5.2.4 **Thermionic Field Emission**

Figure 5.27a shows the schematic energy band diagram of thermionic field emission. The comparison of thermionic field emission, thermionic emission, and field emission is illustrated in Figure 5.27b. The current density owing to thermionic field emission can be approximated as [90]

$$J = \frac{q^2\sqrt{m^*}(k_B T)^{1/2} E}{8\hbar^2 \pi^{5/2}} \exp\left(-\frac{q\phi_B}{k_B T}\right) \exp\left[\frac{\hbar^2 q^2 E^2}{24m^*(k_B T)^3}\right] \tag{5.15}$$

5.5.2.5 **Poole–Frenkel Emission**

Poole–Frenkel emission involves a mechanism that is very similar to Schottky emission, namely, the thermal excitation of electrons may emit from traps into the conduction band of the dielectric. Therefore, Poole–Frenkel emission is sometimes called the internal Schottky emission. Considering an electron in a trapping center, the Coulomb potential energy of the electron can be reduced by an applied electric field across the dielectric film. The reduction in potential energy may increase the probability of an electron being thermally excited out of the trap into the conduction band of the dielectric. The schematic energy band diagram of Poole–Frenkel emission is shown in Figure 5.28.

The general expression of Poole–Frenkel emission is

$$J = q\mu N_C E \exp\left[-\frac{q\left(\phi_T - \sqrt{qE/\pi\varepsilon_i\varepsilon_0}\right)}{k_B T}\right] \tag{5.16}$$

where μ is the electronic drift mobility, N_C is the density of states in the conduction band, $q\phi_T(=\Phi_T)$ is the trap energy level, and the other notations are the same as defined before. Since Poole–Frenkel emission is owing to the thermal activation under an electric field, this conduction mechanism is often observed at high

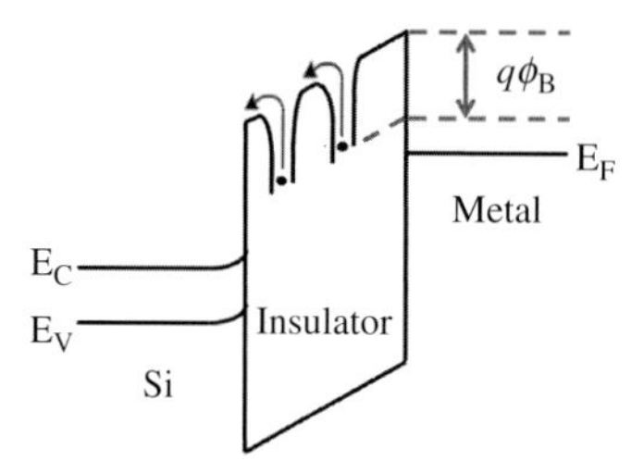

Figure 5.28 Schematic energy band diagram of Poole–Frenkel emission.

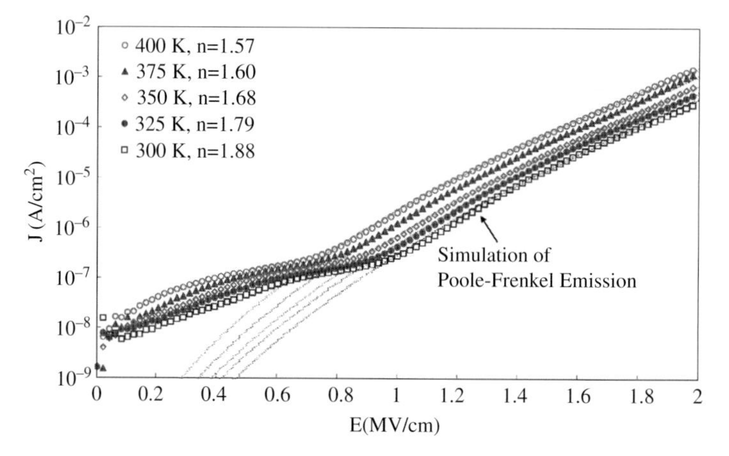

Figure 5.29 Characteristics of *J*–*E* plot and simulation of PF emission for the laminated Pr_2O_3/SiON MOS capacitors at high fields.

temperature and high electric field. Chiu *et al.* [97] reported that the dominant conduction mechanism through Pr_2O_3 is the Poole–Frenkel emission at high electric field (>2 MV/cm) and temperature (300–400 K) using the metal oxide semiconductor (MOS) structure, as shown in Figure 5.29. The trap energy level in Pr_2O_3 determined from the Arrhenius plot is about 0.56 ± 0.01 eV, as shown in Figure 5.30.

5.5.2.6 Hopping Conduction

Hopping conduction is due to the tunneling effect of trapped electrons "hopping" from one trap site to another in the film. Figure 5.31 shows the schematic energy band diagram for hopping conduction. The expression of hopping conduction is [87, 92, 97]

$$J = qanv \exp\left(\frac{qaE}{k_B T} - \frac{E_a}{k_B T}\right) \tag{5.17}$$

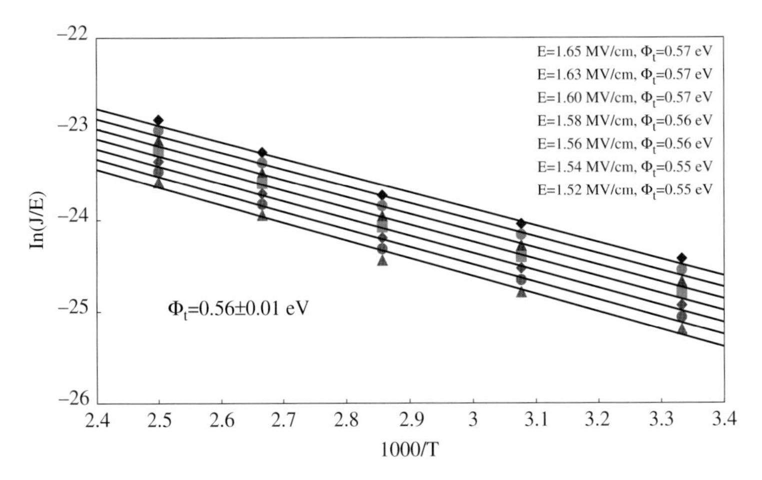

Figure 5.30 Arrhenius plot of the PF emission for the laminated Pr_2O_3/SiON MOS capacitors at high fields.

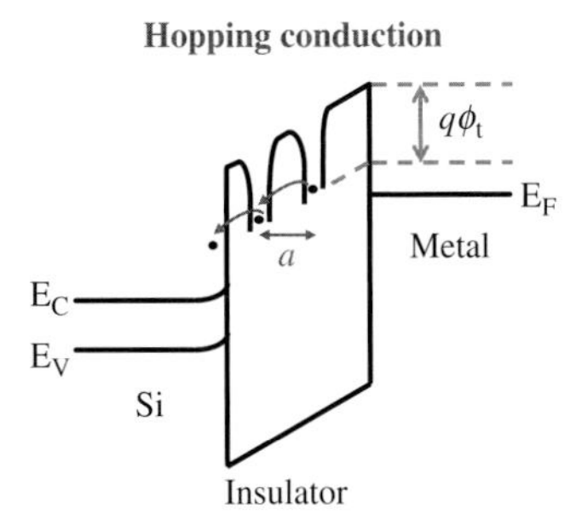

Figure 5.31 Schematic energy band diagram of hopping conduction.

where a is the mean hopping distance, n is the electron concentration in the conduction band of the dielectric, ν is the frequency of thermal vibration of electrons at trap sites, and E_a is the activation energy (i.e., the energy from the trap states to the dielectric conduction band); the other terms are as defined above.

The PF emission corresponds to the thermionic effect and the hopping conduction corresponds to the tunnel effect. In PF emission, the carriers can overcome the trap barrier through the thermionic mechanism. However, in hopping conduction, the carrier energy is lower than the maximum energy of the potential barrier between two trapping sites. In such case, the carriers can still transit using the tunnel mechanism.

Chiu *et al.* [97] reported that the experimental J–E data match the simulated hopping conduction curves very well from 300 to 400 K at low electric fields (<0.6 MV/cm) in a Pr_2O_3 MOS structure, as shown in Figure 5.32. This implies that the hopping conduction dominates the current transportation in Pr_2O_3 at low fields. From the simulations of hopping conduction, the mean hopping distance in the films

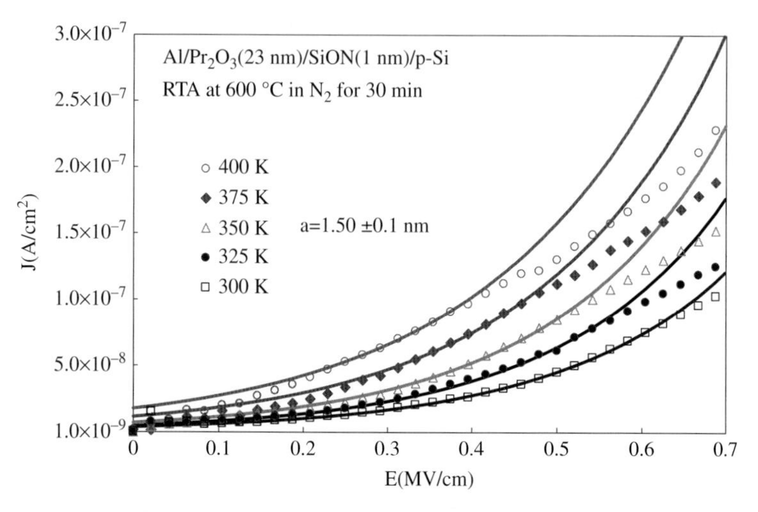

Figure 5.32 Characteristics of *J*–*E* plot and simulation of hopping conduction for the laminated Pr_2O_3/SiON MOS capacitors at low fields.

was determined to be about 1.5 ± 0.1 nm. Using the slopes of Arrhenius plot at low fields, the activation energy was determined to be about 50 ± 1 meV at temperatures ranging from 300 to 400 K, as shown in Figure 5.33. This suggests that a quite shallow trapping level in Pr_2O_3 film dominates the hopping conduction at low fields.

5.5.2.7 **Ohmic Conduction**

Ohmic conduction is caused by the movement of mobile electrons in the conduction band and holes in the valence band. In this conduction mechanism,

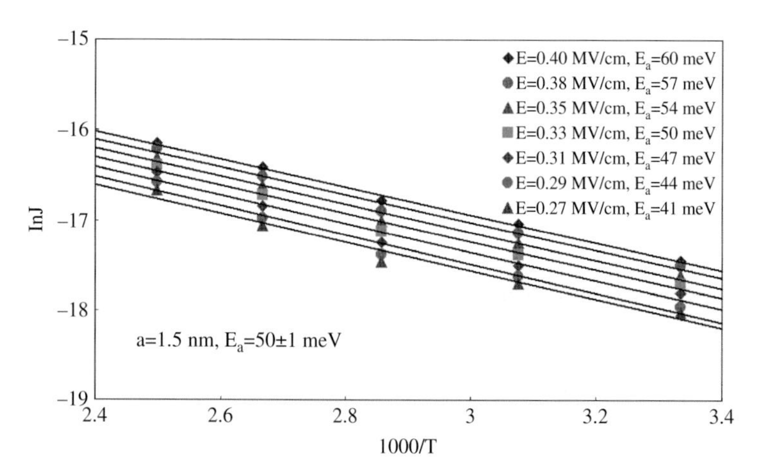

Figure 5.33 Arrhenius plot of the hopping conduction at low fields for the laminated Pr_2O_3/SiON MOS capacitors.

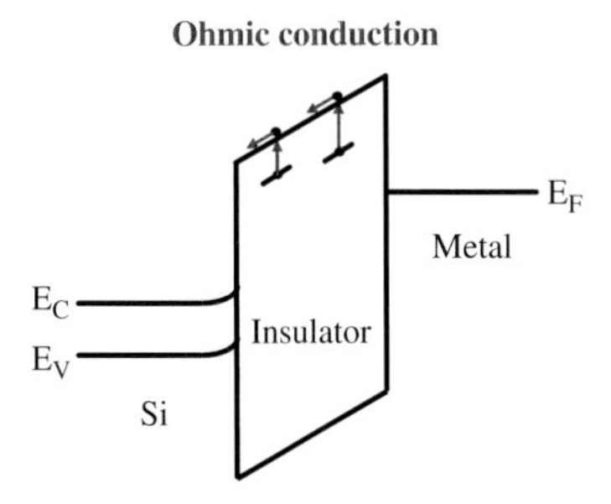

Figure 5.34 Schematic energy band diagram of Ohmic conduction.

a linear relationship exists between the current and the electric field. Figure 5.34 shows a schematic energy band diagram of the ohmic conduction due to electrons. Although the energy bandgap of dielectrics is by definition large, there will still be a small number of carriers that may be generated owing to the thermal excitation. For example, the electrons may be excited to the conduction band, either from the valence band or from the impurity level. The carrier numbers will be very small but they are not zero. The current density of Ohmic conduction can be expressed as

$$J = \sigma E = nq\mu E, \quad n = N_C \exp\left[\frac{-(E_C - E_F)}{k_B T}\right] \tag{5.18}$$

where σ is electrical conductivity, n is the number of electrons in the conduction band, μ is electron mobility, and N_C is the effective density of states of the conduction band.

Because the energy bandgap of a dielectric is very large, we can assume that the Fermi level (E_F) is close to the middle of the energy bandgap, that is, $E_C - E_F = E_g/2$. In this case, the ohmic conduction current is $J = q\mu E N_C \exp(-E_g/2k_B T)$. The magnitude of this current is very small. This current mechanism may be observed if there is no significant contribution from other conduction mechanisms of current transport in dielectrics [92]. The Ohmic conduction current due to mobile electrons in the conduction band or similarly holes in the valence band is linearly dependent on the electric field. Usually, this current may be observed at very low voltage in the current–voltage (I–V) characteristics of the dielectric films.

5.5.2.8 **Space Charge-Limited Conduction**

The mechanism of space charge-limited (SCL) conduction is similar to the transport conduction of electrons in a vacuum diode. The thermionic cathode of a vacuum diode can emit electrons with a Maxwellian distribution of initial velocities (ν). The corresponding charge distribution can be written by the Poisson equation:

$$\frac{\partial^2 V}{\partial x^2} = -\frac{\varrho(x)}{\varepsilon_0} \tag{5.19}$$

Moreover, in the steady state, with the condition $v(x) = [2qV(x)/m]^{1/2}$, the continuity equation is

$$j_x(x) = qn(x)v(x) \tag{5.20}$$

The current density–voltage (*J*–*V*) characteristic of a vacuum diode is governed by the Child's law:

$$J = \frac{4\varepsilon_0}{9}\left(\frac{2e}{m}\right)^{1/2}\frac{V^{3/2}}{d^2} \tag{5.21}$$

In a solid material, the space charge-limited current is caused by the injection of electrons at an ohmic contact. The continuity equation should include the diffusion component and can be written as

$$j_x = qn(x)v(x) + qD\frac{dn}{dx} \tag{5.22}$$

A typical *J*–*V* relation plotted in a log–log curve for space charge-limited current is shown in Figure 5.35. The *J*–*V* characteristics in the log *J*–log *V* plane are bound by three limited curves, namely, Ohm's law ($J_{Ohm} \propto V$), trap-filled limit (TFL) current ($J_{TFL} \propto V$), and Child's law ($J_{Child} \propto V^2$).

$$J_{Ohm} = qn_0\mu\frac{V}{d} \tag{5.23}$$

$$J_{TFL} = \frac{9}{8}\mu\varepsilon\theta\frac{V^2}{d^3} \tag{5.24}$$

$$J_{Child} = \frac{9}{8}\mu\varepsilon\frac{V^2}{d^3} \tag{5.25}$$

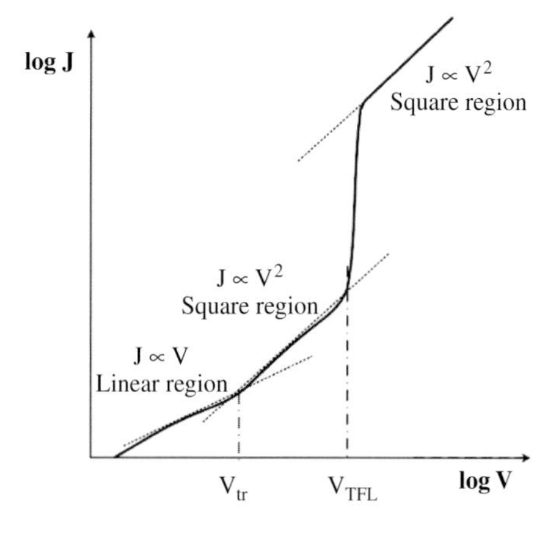

Figure 5.35 A typical current density–voltage (*I*–*V*) characteristic of space charge-limited conduction current. V_{tr} is transition voltage. V_{TFL} is trap-filled limit voltage.

$$J_{tr} = \frac{9}{8}\frac{qn_0 d^2}{\varepsilon\theta} \tag{5.26}$$

$$V_{TFL} = \frac{qN_t d^2}{2\varepsilon} \tag{5.27}$$

$$\tau_c = \frac{d^2}{\mu\theta V_{tr}} \tag{5.28}$$

$$\tau_d = \frac{\varepsilon}{qn\mu\theta} \tag{5.29}$$

where n_o is the concentration of the free charge carriers in thermal equilibrium, θ is the ratio of the free carrier density to total carrier (free and trapped) density, V is the applied voltage, d is the thickness of thin films, ε is the dielectric constant, N_t is the trap density, n is the concentration of the free carrier in the insulator, the other terms are as defined above. It is noticed that Equation 5.25 is the Mott–Gurney relation that indicates the space-charge-limited current under the condition of single type of carriers and no traps. If there are traps in the dielectric, the SCL conduction can be described by Equation 5.24 based on the assumption of monoenergetical trapping levels in the dielectric [87–89, 92].

At low applied voltages ($V < V_{tr}$), J–V characteristics followed the Ohm's law, which implies that the density of thermally generated free carriers (n_0) inside the films is larger than the injected carriers [98]. This ohmic mode takes place in the electrically quasi-neutral state corresponding to the situation when partial trap centers are filled at weak injection. When the transition from the ohmic to the space-charge-limited region, the carrier transit time (τ_c) at V_{tr} (the minimum voltage required for the transition) becomes equal to the dielectric relaxation time (τ_d) [84]. The onset of the departure from Ohm's law or the onset of the SCL conduction takes place when the applied voltage reaches the value of V_{tr}. Accordingly, $\tau_c \cong \tau_d$ can be extracted at the transition point V_{tr}. If the applied voltage V is smaller than V_{tr}, then the carrier transit time τ_c is larger than the dielectric relaxation time τ_d. This implies that the injected carrier density n is small in comparison with n_0 and that the injected carriers will redistribute themselves with a tendency to maintain electric charge neutrality internally in a time comparable to τ_d. Consequently, the injected carriers have no chance to travel across the insulator. The redistribution of the charges is known as dielectric relaxation. The ohmic behavior can be observed only after these space charge carriers become trapped.

In the case of strong injection, the traps are filled up and a space charge appears. When $V > V_{tr}$ and $\tau_c \sim \tau_d$ or $\tau_c < \tau_d$, the injected excess carriers dominate the thermally generated carriers since the injected carrier transit time is too short for their charges to be relaxed by the thermally generated carriers. Note that for $V < V_{tr}$, τ_c increases with decreasing V but τ_d remains almost constant; while for $V > V_{tr}$, τ_c decreases with increasing V and τ_d also decreases with increasing V since the increase in V causes an increase in free carrier density in the dielectric. The increase of applied voltage may increase the density of free carriers resulting from injection to such a value that the Fermi level (E_{Fn}) moves up above the electron trapping level (E_t). The trap-filled limit (TFL) is the condition for the transition from the trapped J–V characteristics to the trap-free J–V characteristics. It can be imagined that after all traps are filled up, the subsequently injected carriers will be free to move in the thin films, so that at the subthreshold voltage (V_{TFL}) to set on this transition, the current will rapidly jump from its low trap-limited value to a high trap-free SCL current. V_{TFL} is defined as the voltage required to fill the traps or, in other words, as the voltage at which Fermi level (E_{Fn}) passes through E_t.

In the case of very strong injection, all traps are filled and the conduction becomes the space-charge-limited (Child's law). Thus a space charge layer in the insulator builds up and the electric field cannot be regarded as constant any longer. While the bias voltage reaches V_{TFL} in the strong injection mode, the traps get gradually saturated, which means the Fermi-level gets closer to the bottom of the conduction band. This results in a strong increase of the number of free electrons thus explaining the increase of the current for $V = V_{TFL}$. For the voltage $V > V_{TFL}$ the current is fully controlled by the space charge, which limits the further injection of free carriers in the dielectric. Square law dependence of the current ($J \sim V^2$, Child's law) is the consequence of the space charge controlled current. A report showed that the dominant conduction mechanism through the polycrystalline La_2O_3 thin films is space-charge-limited current [98], as shown in Figure 5.36. Three different regions, Ohm's law region, trap filled limited region, and Child's law region are observed in the current density-voltage (J–V) characteristics at room temperature. Based on the study of space charge limited conduction, the parameters of V_{tr}, V_{TFL}, electronic mobility in polycrystalline La_2O_3, trap density (N_t), ratio of the free carrier density to total carriers (θ) and density of states in the conduction band (N_C) were determined.

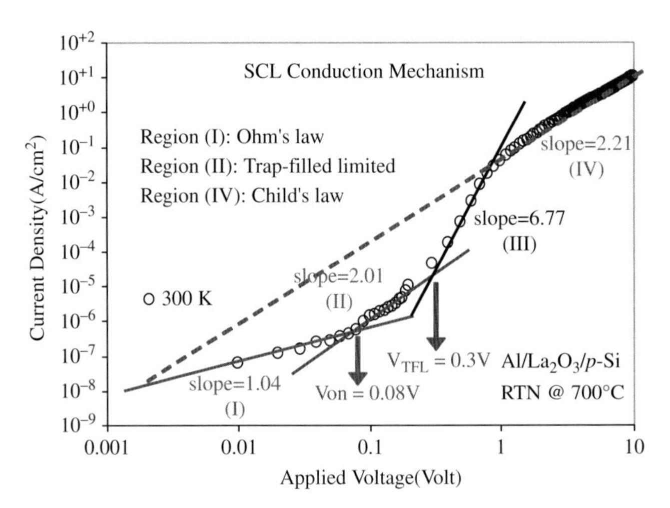

Figure 5.36 Logarithm of the current density for Al/La_2O_3/p-Si structure plotted as a function of the logarithm of the applied voltage under negative bias at room temperature.

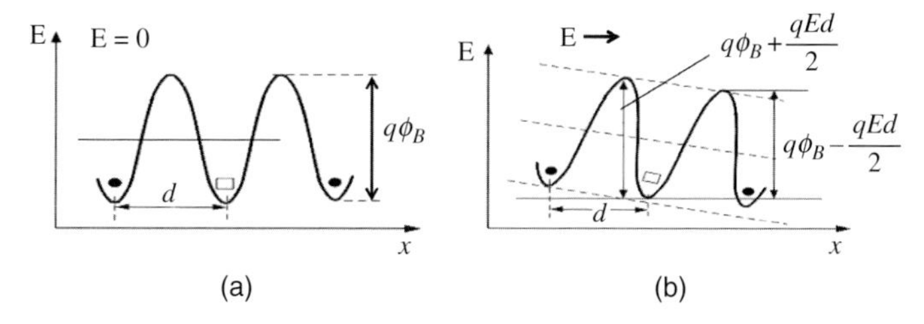

Figure 5.37 Energy band diagram of ionic conduction (a) without the applied electric field and (b) with the applied electric field. *E* is the applied electric field, *d* is the spacing between ionic sites, and $q\phi_B$ is the potential energy barrier.

5.5.2.9 Ionic Conduction

Ionic conduction results from the movement of ions under an applied electric field. The movement of the ions may come from the existence of lattice defects in the dielectrics. Under the influence of an external electric field, the ions may jump over a potential barrier from one defect site to another. Figure 5.37a shows a schematic energy band diagram of ionic conduction without the applied field. Figure 5.37b shows the condition with applied field. The ionic conduction current can be expressed as:

$$J = J_0 \exp\left[-\left(\frac{q\phi_B}{k_B T} - \frac{Eqd}{2k_B T}\right)\right] \tag{5.30}$$

where J_0 is the proportional constant, $q\phi_B$ is the potential barrier height, E is the applied electric field, and d is the spacing of two nearby jumping sites [92]. Because of the large mass of ions, this mechanism is usually not important for the applications of dielectric films in CMOS technology.

5.5.2.10 Grain Boundary-Limited Conduction

In a polycrystalline dielectric material, the resistivity of the grain boundaries may be much higher than that of the grains. Therefore, the conduction current could be limited by the electrical properties of the grain boundaries. In this case, the conduction mechanism is called the grain boundary-limited conduction [92, 99]. The grain boundary will build a grain boundary potential energy barrier (Φ_B) that is inversely proportional to the relative dielectric constant of the dielectric material. The potential energy barrier can be written as

$$\Phi_B = q\phi_B = \frac{q^2 n_b^2}{2\varepsilon N} \tag{5.31}$$

where n_b is the grain boundary trap density, ε is the relative dielectric constant of the dielectric material, and N is the dopant concentration. From Equation 5.31, it can be seen that the dielectric constant can significantly affect the potential energy barrier at the grain boundaries.

Figure 5.38a shows the charge distribution across an electron-trapped grain boundary and the existence of depletion regions next to the grain boundary.

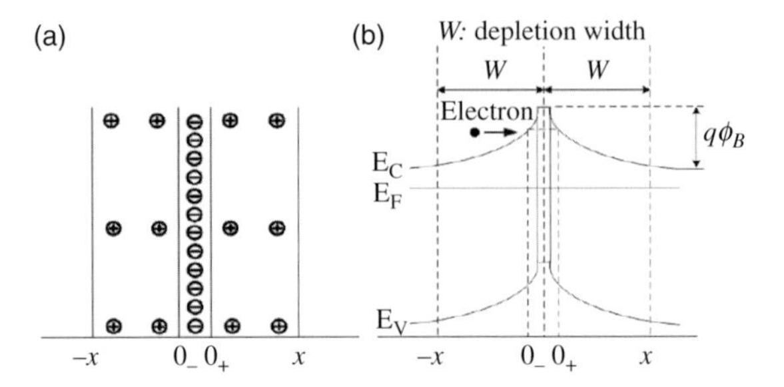

Figure 5.38 (a) The charge distribution of an electron-trapping grain boundary and (b) the resulting potential energy barrier at the grain boundary. $q\phi_B$ is the potential barrier and b is the depletion width.

The potential energy barrier at the grain boundary is shown in Figure 5.38b due to the charge distribution close to the grain boundary. Figure 5.39 indicates an energy band diagram of the grain boundary-limited conduction in a metal–insulator–metal, MIM, structure.

5.5.3 High-κ Silicon Interface

For all thin gate dielectrics, the interface with silicon plays a key role in determining the overall electrical properties. Reactions between the dielectric and the Si will almost lead to a failure of the dielectric in the MOSFETs. For instance, if a reaction product is a conductor, its presence will screen the applied electric field from reaching the silicon channel, destroying MOSFET's performance. If a reaction product is an insulator with lower κ, its low κ in series with the desired high-κ dielectric will rapidly nullify the benefit of the high-κ dielectric. Furthermore, the increase in dangling bonds that can accompany a reaction will increase the interface-trapped charge density.

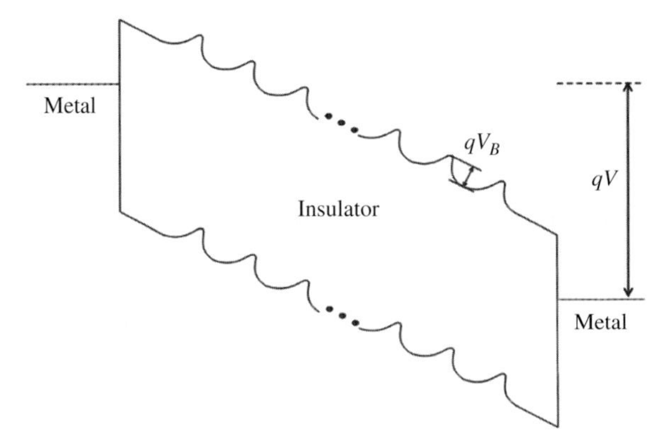

Figure 5.39 Schematic energy band diagram for grain boundary-limited conduction of a metal–insulator–metal structure.

Alternative high-κ gate dielectrics need to be thermodynamically stable in contact with Si to withstand the 900–1000 °C dopant drive-in anneal [12]. If high-κ gate dielectrics could be combined with the existing CMOS technology, the material properties available for use in microelectronics would be drastically enhanced, allowing improved functionality and performance to be realized. Unfortunately, direct deposition of high-κ dielectrics on Si is frequently accompanied by extensive interdiffusion or chemical reactions, which degrades the properties of the dielectric, the underlying Si, or both. A useful method to avoid reactions between high-κ dielectric and Si is to select a dielectric that is thermodynamically stable in contact with Si. To comprehensively assess the thermodynamic stability of binary oxides in contact with silicon at 1000 K, a thermodynamic approach was used by Hubbard and Schlom [27]. They indicated that rare-earth oxide (except for the radioactive Pm_2O_3) with some other high-k oxide, for example, ZrO_2, HfO_2, Al_2O_3, and Y_2O_3, are thermodynamically stable in contact with Si, whereas Ta_2O_5 and TiO_2 are not.

Most of the high-κ metal oxide systems have been shown to have an unstable interface with silicon. They react with Si under equilibrium conditions to form an undesirable interfacial layer. Studies showed that the lowering of Gibbs free energy ($\Delta G^o < 0$) for Ta_2O_5 and TiO_2 will lead to the reactions between Si and binary oxides at 1000 K [27, 100]. Therefore, these materials require an interfacial reaction barrier. A great deal of material research has been devoted to overcoming these obstacles through the identification of compatible buffer layers to be used between Si and desired oxide layers. When selecting a material for use as a buffer layer between Si and a particular oxide, the effects of interdiffusion, chemical reactions, thermal expansion match, crystal structure, lattice match, and so on must be considered. Although ZrO_2 was reported to be stable in contact with Si [27], the formation of Zr-silicide at the poly-Si/ZrO_2 interface was found to take place [101–103]. Meanwhile, the HfO_2/Si interface is found to be stable with respect to formation of silicides, in contradiction of ZrO_2/Si interface [101]. Chiu *et al.* showed that an interfacial layer was formed between the 500 °C annealed ZrO_2 and the Si substrate [104]. Also, an interfacial layer was found at the interfaces of La_2O_3/Si [105] and HfO_2/Si [106] after 700 °C annealing. Gutowski *et al.* indicated that a silicate-like interfacial layer was formed between ZrO_2 (HfO_2) and SiO_2 that is frequently used as the buffer layer [101].

Several research groups have recently focused on the improvement of the thermal stability of high-k gate dielectrics to overcome the problems of insufficient immunity to oxygen or impurity diffusion during the subsequent thermal processes [107]. The proposed schemes include NH_3 nitridation of Si surface [108–110], addition of Al into HfO_2 ($HfAl_xO_y$) [111] or ZrO_2 ($ZrAl_xO_y$) [112], incorporation of nitrogen into HfO_2 (HfO_xN_y) [113], ZrO_2 (ZrO_xN_y) [114], or HfSiON [115], capping an HfO_2 layer with a nitrogen-incorporated layer [116], and SiN/HfO_2/SiON gate stack. These proposals need additional process step or involve hydrogen impurity from NH_3. Although addition of Al may increase the crystallization temperature of HfO_2, the dielectric constant is degraded [117]. The crystallization of dielectric films can be suppressed by nitrogen incorporation into HfO_xN_y without any degradation in EOT [107]. The EOT enlargement of the HfO_xN_y can be suppressed to smaller than 0.5 Å even after 950 °C postmetal annealing. This is a great reduction compared to the EOT increase in the cases of the pure HfO_2 (4.5 Å) [107], HfO_2 capped with a nitrogen incorporated layer (3 Å) [116], and Si–substrate surface nitride layer (1.5 Å) [108]. The nitrogen incorporation into the HfSiON film is able to improve the microscopic homogeneity of the film, enhances the dielectric constant, suppresses the boron penetration, and remains amorphous even after 1100 °C annealing [115]. However, the dielectric constant of HfSiON is smaller than 13 [115]. NH_3 nitridation (i.e., bottom nitridation, BN) of Si surface prior to high-*k* dielectric deposition was shown to be effective in achieving low EOT and preventing boron penetration. However, this technique may result in higher interface charge, higher hysteresis, and reduced channel mobility [108, 109, 113]. Top nitridation (TN) on the HfO_2 was demonstrated by the deposition of HfO_xN_y on an HfO_2 layer to improve the thermal stability, diffusion immunity, and MOSFET characteristics without degrading interface quality [107, 110, 113]. Except interface quality that is significantly degraded by the presence of nitrogen at the dielectric/Si interface [107], fully nitrided HfO_2 (i.e., HfO_xN_y) films are expected to have advantages over TN, BN, and pure HfO_2 [113]. Morisaki *et al.* demonstrated the high performance and high reliability of ultrathin poly-Si-gated SiN/HfO_2/SiON high-κ stack dielectrics with EOT smaller than 1 nm [118]. The top SiN layer was found to be important for suppression of the dopant penetration (i.e., boron penetration for PMOSFETs and phosphorus penetration for NMOSFETs) and the reaction between the HfO_2 and the poly-Si at 1050 °C. The bottom SiON layer was needed to achieve high reliability in spite of the mobility degradation. A low leakage current of 5–6 orders of magnitude reduction was achieved. The suppression of dopant penetration, prevention of interfacial reaction, and excellent reliability of time-dependent dielectric breakdown (TDDB) and hot electron integrity (HCI) were obtained [118]. In addition, Cheng *et al.* indicated that the gate stack architecture plays an important role in the device performance [119]. The short-channel performance of a double-layer structure with an underlying low-*k* dielectric (e.g., SiO_2) is much improved over the single layer high-κ structure at the same EOT because of enhanced controllability of the channel charge [119]. This asserts that the use of multiple layer gate stack structures is a good approach to the high-κ gate dielectric used in future sub-100 nm technology nodes.

Most of the reported high-κ dielectrics have high interface charge density D_{it} ($\sim 10^{11}$–$10^{12}\,eV^{-1}\,cm^{-2}$) and exhibit a substantial flatband voltage shift >300 mV [12]. Recently, Chiu *et al.* [120] reported that the interface-trapped charge density D_{it} between cerium dioxide (CeO_2) and silicon can be lower than $10^{11}\,eV^{-1}\,cm^{-2}$ because the lattice constant of CeO_2 matches with Si ($\Delta a < 1\%$) [121, 122]. In order to maintain high-quality interface and channel mobility, it is expected to have low defects or impurities present at or close to the high-κ/Si interface because of the dopant penetration, oxygen diffusion, metal outdiffusion, or reactions between high-*k* material and Si. It has been realized that hydrogen atom (H) plays an important role in MOS device performance [123]. The relatively high background level of H-containing molecules in silicon microelectronics processing ambiences (e.g., NH_3 and SiH_4 in CVD reactors), and the high diffusivity of H in SiO_2, supports the belief that H is ubiquitous in the Si/SiO_2 systems. In fact, nuclear reaction analysis experiments have revealed that hydrogen or deuterium (D_2) concentrations in SiO_2 films can be as high as the 10^{21} cm^{-3} ranges [123]. Hydrogen annealing below ~600 °C has been proven to passivate the dangling bonds at the interface, which in turn further reduces the concentration of surface states [123]. High-temperature (400–600 °C) forming gas (FG) annealing improved channel carrier mobility and subthreshold swing in both n- and p-MOSFET with HfO_2 gate dielectrics [124]. The improvement has been correlated with the reduction in interface state density. Choi *et al.* showed that both FG [124, 125] and D_2 [125] anneal exhibited the improved interface qualities and resulted in better MOSFET characteristics such as higher drive current, lower subthreshold swing, and higher mobility in comparison with control samples. Despite the improved performance, FG annealing potentially results in adverse effects on HfO_2 MOSFET properties such as lower breakdown voltage and degraded negative bias temperature instability (NBTI) characteristics. D_2 anneal has been found to be effective in reducing NBTI [126] and improving hot carrier reliability [127]. D_2 annealed samples showed slightly better I_{ds} and G_m properties than FG annealed ones. Unlike FG anneal, D_2 anneal showed negligible degradation of reliability on breakdown voltage (V_{BD}), HCI, NBTI, and trapped charge [125].

5.5.4 Band Alignment

Most of the current research in high-κ gate dielectrics is concentrated on binary metal oxide and silicates. The candidate materials provide desirable properties that make them potential replacements for current conventional oxynitrides. The major motivation for exploring these materials is their wide bandgap and large conduction band offset (ΔE_C, i.e., high-*k*/Si barrier height) for reduced gate leakage current of MOSFETs. In order to obtain low leakage currents, it is desirable to find a gate dielectric that has a sufficient ΔE_C value to Si. Figure 5.40 shows the dielectric energy

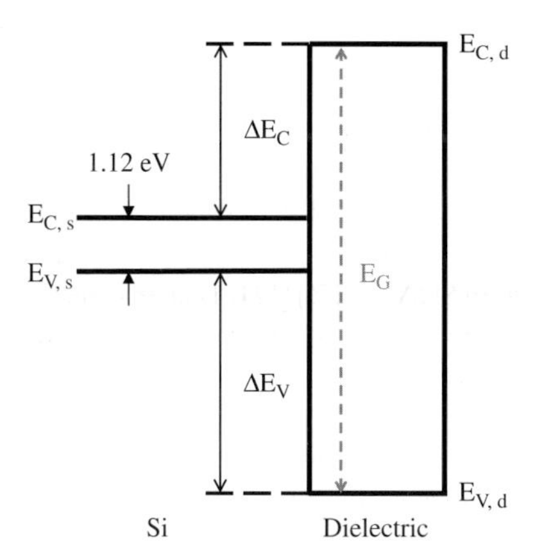

Figure 5.40 Dielectric energy bandgap (E_G) and associated conduction band offset (ΔE_C) between dielectric and silicon.

bandgap (E_G) and associated conduction band offset (ΔE_C) between dielectric and silicon. If the ΔE_C is smaller than 1.0 eV, it will likely preclude using these oxides in gate dielectric application since thermal emission or tunneling of carriers would lead to unacceptable high leakage currents [12]. Robertson reported that $\Delta E_C = 2.3$ eV for La_2O_3 and Y_2O_3, and $\Delta E_C \sim 1.5$ eV for ZrO_2, HfO_2, $ZrSiO_4$, and $HfSiO_4$ [21, 128]. For high-k materials, the dielectric constant and bandgap of a given material generally exhibit an inverse relationship [21]. Studies showed that the gate leakage current of HfO_2 and ZrO_2 gate stacks was three–six orders of magnitude lower than that of SiO_2 with the same EOT [12]. Compared to ZrO_2 and HfO_2, the leakage current density of La_2O_3 can be even lower to around two–four orders of magnitude [129].

For CMOS scaling in the long term, the ITRS roadmap predicted that poly-Si gate technology will likely be phased out beyond the 45 nm node, after which a metal gate substitute is required [20]. Replacing the poly-Si gate with a metal gate may dispose of sheet resistance constraints and dopant depletion that reduces the capacitance equivalent thickness by several angstroms. To overcome the problems of high gate resistance and poly gate depletion, the active dopant density in the poly-Si gate material must be increased. The 2001 ITRS roadmap pointed out that the active dopant density must be greater than 1.87×10^{20} cm^{-3} at $L_G = 25$ nm CMOS technology generation for the depletion layer of poly-Si gate to be less than 25% of the EOT. This presents a considerable difficulty since the active poly-Si dopant density at the gate–dielectric interface saturates at 6×10^{19} cm^{-3} and 1×10^{20} cm^{-3} for p$^+$- and n$^+$- doped poly-Si, respectively. Insufficient active dopant density in the gate leads to a significant voltage drop across the gate depletion layer and increases the EOT. Consequently, there is immense interest in metal gate technology. The metal gate material not only eliminates the problems of gate depletion and boron penetration but also greatly reduces the gate sheet resistance. However, metal gate candidates must possess compatible work function and thermal/chemical interface stability with the underlying dielectric. The work function of a metal is a measure of the minimum energy required to extract an electron from the surface of the metal. It is commonly measured in vacuum by making use of phenomenon such as the thermionic emission or photoelectric effect. For the measurement of a metal work function in a metal–dielectric system, the most common method is to extract the flatband voltage from the C–V characteristics of a MOS capacitor. Internal photoemission may also be used to extract the barrier height at the interface of a metal–dielectric system. The metal work function may be deduced from the barrier height and the electron affinity of the dielectric. A key issue for gate electrode materials will be the control of the work function of the gate electrode after further CMOS processing. Metal gates with work functions near the conduction and valence band edges of Si ($\Phi_m \sim 4$ eV and $\Phi_m \sim 5$ eV) will likely be required for future bulk CMOS devices [130]. At present, there are two basic approaches to achieve successful metal electrodes, namely, a single mid-gap metal or two separate metals. The energy band diagrams relative to these two approaches are shown in Figure 5.41.

Metal interdiffusion gate (MIG) technology [131] was recently adopted to achieve low threshold voltages for both nMOSFETs and pMOSFETs with the Ti–Ni metal

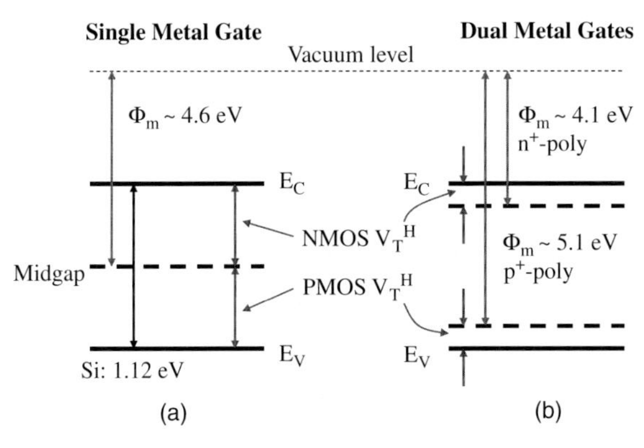

Figure 5.41 Energy band diagrams of threshold voltages for CMOS devices using (a) mid-gap metal gates and (b) dual metal gates. E_C is the conduction band edge, E_V is the valence band edge, V_{TH} is the threshold voltage, and Φ_m is the work function of the metal indicated.

combination, which produces gate electrodes with work functions of 3.9 eV and 5.3 eV for nMOSFETs and pMOSFETs, respectively. The work functions of molybdenum (Mo) [132] and titanium nitride (TiN_x) [133] can be reduced by nitrogen implantation. Hence, Mo or TiN_x may be a potential candidate for a single-metal dual-work function technology for future CMOS technology. (110)-Mo exhibited a high work function value (4.95 eV) appropriate for pMOSFETs; and with nitrogen implantation, a gate work function of 4.53 eV (0.42 eV reduction) was achieved for the nMOSFETs [132]. Wakabayashi *et al.* indicated that the work function shift of TiN is about 0.11 eV by low-energy nitrogen ion implantation into the TiN film when the N/Ti signal ratio is 0.87 [133]. Elemental metals with lower work functions (e.g., Al, Ta, Ti, Zr, and Hf) have a problem of the reaction between the meal and the underlying dielectric [134, 135]. One way to achieve good thermal stability is to introduce N and/or Si into the low work function metals [134]. TaN with a work function of 4.15 eV is thermally stable up to 1000 °C, which is appropriate for nMOSFETs [136]. The work functions of TaSiN and TiN on HfO_2 were extracted to be 4.4 eV [137] and 4.8 eV [133], respectively, which makes them appropriate for nMOSFETs and pMOSFETs. In addition, the thin films of transition metal oxides, such as ruthenium oxide (RuO_2) and iridium oxide (IrO_2), are also attractive for pMOSFETs due to their large work function (~5.1 eV), low resistivity (~65 μΩ cm), and excellent thermal stability [138, 139].

Fully silicided (FUSI) gate is an attractive metal gate alternative because of the relative simplicity of process integration [140, 141]. The FUSI/high-κ gate stacks may significantly reduce gate leakage (six–seven orders of magnitude) and improve driving current owing to elimination of polydepletion effect [140]. There are several silicides that have been studied and implemented for junction contacts with salicide processes, such as $TiSi_2$, $CoSi_2$, WSi_2, $MoSi_2$, $TaSi_2$, PtSi, and NiSi [141, 142]. The work function of silicides can be engineered by either adjusting the bulk silicide work function or changing the electrical properties of the silicide gate dielectric interface [143]. In the first, silicon substitutional dopants (e.g., P, As, Sb, and B) can be implanted into the poly-Si gate prior to silicidation. In the second, metal substitutional impurities can be used to change the chemical composition of silicides. Fully $CoSi_2$ silicided gate with work function of 4.6 eV was successfully integrated with MOS technology for the first time [144]. Kedzierski *et al.* showed that the work function of CoSi and $CoSi_2$ is 4.47 and 4.54 eV corresponding to the silicide reaction temperature at 500 °C and 750 °C, respectively [143]. Cabral and coworkers showed that the $CoSi_2$ FUSI process consisting of a two-step anneal (550/700 °C) may result in voiding at the dielectric interface and dielectric reliability issues [141]. The work function of $CoSi_2$ with intrinsic poly-Si may change from 4.5 eV to 4.26 eV with phosphorus predoped poly-Si, which is appropriate for nMOSFET fabrication [141]. Nickel silicide (NiSi) has been applied to VLSI process because of its low resistivity, low process temperature, and immunity from narrow line effect [145]. The temperature of silicide formation is important to its work function, but the work function is stable with the phase of silicide [142]. Takahashi *et al.* showed that the key to work function tuning is the control of FUSI Ni-silicide phase that can be made by changing Ni film thickness prior to silicidation anneal [146]. Iijima *et al.* showed that the temperature of Ni-silicide formation is 280–350°C and 350–750°C for the phase of Ni_2Si (~34 μΩ cm) and NiSi (~14 μΩ cm), respectively [147]. When the sintering temperature of Ni-silicide is lower than 350 °C, the work function of Ni_2Si is about 4.72 eV [142]. The work function of NiSi on n^+, p^+, and undoped poly-Si is 4.6, 5.0, and 4.9 eV, respectively [142]. The work function of n^+-doped NiSi is closer to the mid-gap of Si (4.61 eV) and is stable from 400 to 800 °C [142]. The silicidation-induced impurity segregation provides a good way of adjusting the work function of NiSi gates. A submonolayer amount of the implanted impurity will segregate into the boundary between the silicide and the remaining poly-Si while the silicide is formed. When the silicide front reaches the gate dielectric interface, the impurities are fixed at that interface, which changes the work function [143, 148]. However, incompleteness of gate silicidation and variation in gate silicide phase are two serious challenges of NiSi-gate integration [149]. Cabral *et al.* also indicated that the undoped NiSi gate with a mid-gap work function of 4.65 ± 0.05 eV is suitable to tailor the process to achieve dual work functions [141]. Studies showed that predoping poly-Si gate with P, As, and Sb [143, 150] allows for movement of the work function toward the nMOSFET direction ($\Delta\Phi_m < 0$). Among these, Sb is the most effective ($\Delta\Phi_m \sim -0.4$ eV) [141]. Shift in the pMOSFET direction ($\Delta\Phi_m > 0$) can be achieved either by B [43, 50] and Al [40, 41] predoping of the poly-Si or by using a Ni(Pt)Si alloy [140, 141] silicide. However, B is not favorable due to the problematic penetration into the dielectric and the minimal segregation [140, 150]. The shift of NiSi gate work function increases with increasing predoping concentration with P or B [143, 150]. On the other hand, the increased doping of As is not effective in the work function modulation [150]. A NiSi gate predoped with As provides the work function of 4.39 eV, which comes to 4.29 eV when coimplanted with P [150]. Either the Al predoping or Ni(Pt)Si alloy silicide may result in work function shift of about 0.15 eV ($\Delta\Phi_m \sim +0.15$ eV) [140, 141]. The combination of Al predoped poly-Si and Ni(Pt)Si can be used to achieve a work function shift of about 0.30 eV, which leads to

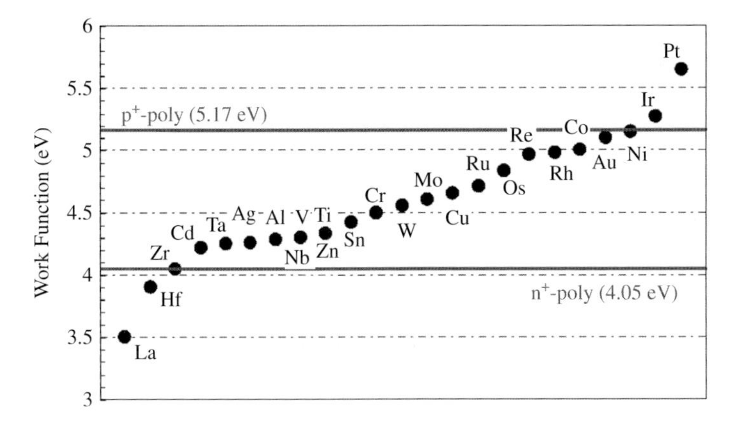

Figure 5.42 Work functions of metal elements.

a pMOSFET work function within 0.2 eV of the valence band edge [141]. The control of work function of the Co–Ni alloy silicide can be carried out by varying x in a $Co_{(1-x)}Ni_{(x)}Si_2$ alloy [143]. When x is equal to 0.5, the $Co_{0.5}Ni_{0.5}Si_2$ work function is 4.59 eV.

For HfSiON high-κ transistors, Takahashi *et al.* indicated that the work function can be adjusted by the phase-controlled FUSI technique [146]. The phase of Ni-silicide associated with its composition is controlled by the changes in Ni film thickness and silicidation sintering temperature. The effective work functions of Ni_3Si and $NiSi_2$ on HfSiON are 4.8 eV and 4.4 eV, respectively. Meanwhile, the dopant segregation method did not work on HfSiON (remains at 4.5 eV), even though it was successful in V_{TH} control of NiSi on SiO_2, [146]. Lin *et al.* indicated that the extracted work function of the FUSI NiSi gate on La_2O_3 MOSFETs was 4.42 eV [151]. Takahashi *et al.* indicated that the same NiSi gate work function of 4.6 eV is achieved for the gate dielectrics of SiO_2, SiON, and HfO_2 [146]. This implies that the NiSi gate may not undergo the effect of Fermi-level pinning. Based on their claims, the Si—Hf bonds built at the poly/HfO_2 interface are broken and replaced by Ni—Hf or Ni—O bonds, hence suppressing any pinning. Although the work function of TaC is about 3.7 eV in the vacuum [152], the effective work function of TaC on HfO_2 and HfSiON is 4.18–4.29 eV [152–154] and 4.45–4.48 eV [154], respectively. Figure 5.42 shows the work functions of several metal elements [155]. Figure 5.43 shows the work functions of metal elements, metal alloys, metal nitrides, metal carbides, and transition metal oxides. Figure 5.44 shows the work functions of metal alloy silicides and metal silicides with predoped poly-Si.

Metal work functions on high-k dielectrics are usually observed to differ from their values on SiO_2 or in vacuum because of Fermi-level pinning at the metal/high-k interface [156, 157]. For bulk CMOS devices, the metal work functions are required to be within 0.2 eV of the conduction band edge (E_C) and valence band edge (E_V) of Si for nMOSFETs and pMOSFETs, respectively. Achieving such work functions on high-k dielectric is still a challenge due to Fermi-level pinning and

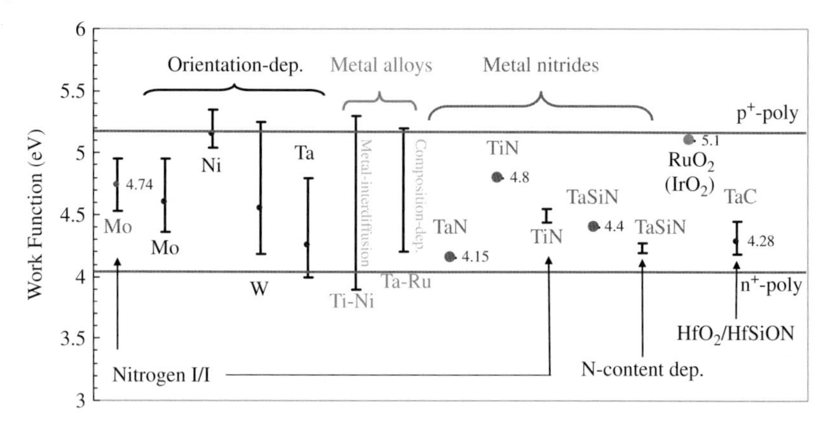

Figure 5.43 Work functions of metal elements, metal alloys, metal nitrides, metal carbides, and transition metal oxides.

thermal instability at the gate/dielectric interface [158–161]. Fermi-level pinning is a mechanism to cause high threshold voltages for metal gates on high-*k* gate dielectrics. Metal Fermi levels tend to be pinned to charge neutrality level (CNL) of the dielectric. This tendency increases with the electronic component of dielectric constant (ε_∞), which is equal to the square of the refractive index (n) of the

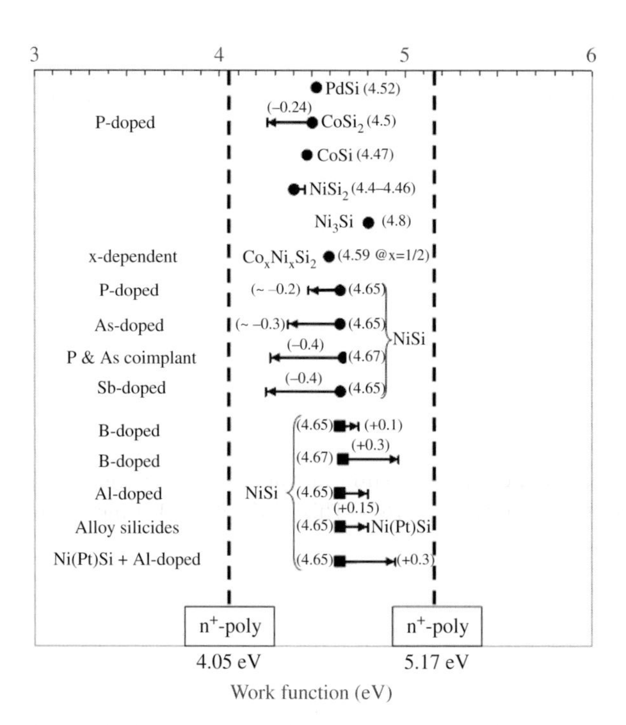

Figure 5.44 Work functions of metal alloy silicides and metal silicides with predoped poly-Si.

dielectrics (i.e., $\varepsilon_\infty = n^2$). By definition, CNL is the highest occupied surface state of the dielectric [162]. In general, the CNL can be thought of as a local Fermi level. In the case of a neutral surface, the surface states with energy smaller than CNL are occupied, while those with energy larger than CNL are empty. Those surface states are due to the dangling bonds and defects resulting from the surface creation and are energetically located in the dielectric bandgap.

When an interface is formed between a metal and a semiconductor or a dielectric, Schottky model indicates that there is no charge transfer across the interface and the barrier height for the electrons is given by the difference between the work function of metal in vacuum ($\Phi_{m,vac}$) and the electron affinity of semiconductor (χ) [163]. However, it was observed experimentally that the Schottky model is not generally obeyed owing to the presence of intrinsic interface states. To explain this observation, Bardeen proposed the surface state model [164] in which he postulates a high density of surface states on the order of 1 per surface atom at a well-defined energy relative to the conduction band edge of the semiconductor. These surface states act to pin the metal Fermi level. Heine pointed out that the wave functions of electrons in the metal decay into the semiconductor in the energy range where the conduction band of the metal overlaps the semiconductor bandgap [165]. These resulting states in the forbidden gap are known as metal-induced gap states (MIGS) or simply intrinsic states established from self-consistent pseudo-potential calculations of the electronic structure of materials' interfaces [166]. These interface states are predominantly donor-like close to E_V, and mostly acceptor-like close to E_C. The energy level in the bandgap at which the dominant character of the interface states changes from donor-like to acceptor-like is called the charge neutrality level, E_{CNL} [167]. Charging of these interface states creates a dipole that tends to drive the band line-up toward a position that would give zero dipole charge. This interface dipole drives the band alignment so that metal Fermi level $E_{F,m}$ goes toward $E_{CNL,d}$ (E_{CNL} in the dielectric). Therefore, the effective metal work function ($\Phi_{m,eff}$) differs from the metal work function in vacuum ($\Phi_{m,vac}$). Yeo *et al.* applied the interface dipole theory to explain the dependence of work function of metal gates on the underlying gate dielectric [156, 157]. A nice agreement with experimental data was made. Figure 5.45 shows that the electron Schottky barrier (Φ_n) is reduced to $\Phi_n = S(\Phi_{m,vac} - \Phi_{CNL,d}) + (\Phi_{CNL,d} - \chi_d)$owing to the charge transfer between the metal and interface states lying in the dielectric bandgap [21], where $\Phi_{CNL,d}$ is the energy difference between vacuum level E_{vac} and $E_{CNL,d}$, χ_d is the dielectric electron affinity, and S is the dielectric pinning strength (i.e., Schottky barrier slope) that is a slope parameter accounting for dielectric screening and depends on the electronic component of the dielectric constant ε_∞ [168]. This shift of work function is proportional to the difference between $E_{CNL,d}$ and $E_{F,m}$, or, equivalently, the difference between $\Phi_{m,vac}$ and $\Phi_{m,eff}$. Hence, $\Phi_{m,eff}$ is given by $\Phi_{m,eff} = \Phi_{CNL,d} + S(\Phi_{m,vac} - \Phi_{CNL,d}) = \Phi_n + \chi_d$ [156, 157]. It is noted that S is a dimensionless slope of barrier height to metal work function, $S = \partial\Phi_n/\partial\Phi_M$ [169]. Monch revealed that the slope parameter S obeys an empirical relationship given by $S = 1/[1 + 0.1(\varepsilon_\infty - 1)^2]$ and $\varepsilon_\infty = n^2$, where n is the refractive index [21]. Materials with a lower S tend to pin the metal Fermi level more effectively to $E_{CNL,d}$.

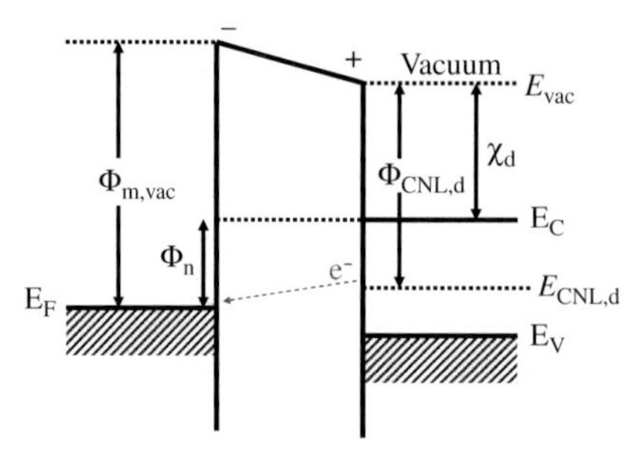

Figure 5.45 Band diagram of the charge transfer effect between the metal and the dielectric interface on the Schottky barrier height. Φ_n is electron Schottky barrier, $\Phi_{CNL,d}$ is the energy difference between vacuum level E_{vac} and $E_{CNL,d}$, S is the Schottky barrier slope, and χ_d is the dielectric electron affinity.

The maximum value for S is unity, which corresponds to no pinning of the metal Fermi level.

Studies showed that Fermi-level pinning at the poly-Si/metal oxide interface causes high threshold voltages in MOSFET devices [158, 162, 170]. The pinning locations are just below E_C and above E_V of poly-Si for HfO_2 and Al_2O_3 owing to the interfacial bonds of Si––Hf and Si–O–Al, respectively. Oxygen vacancies at poly-Si/HfO_2 interface may also result in Fermi-level pinning. The interfacial Hf (IV-B element) atoms act metal-like, whereas the interfacial Al (III-A element) atoms act more like p-type dopants for Si. Therefore, the interfacial Hf and Al atoms may increase the polysilicon depletion for p-doped and n-doped poly-Si gates, respectively. Schaeffer *et al.* showed that when a low (high) vacuum work function metal is deposited on HfO_2, it shifts to a higher (lower) work function value [152]. This implies that when various metals are deposited on HfO_2, $\Phi_{m,eff}$ may shift toward about 4.4 eV, which is close to the average of 4.1 eV (n^+-poly work function) and 5.1 eV (p^+-poly work function). They indicated that these work function shifts are especially pronounced for TaC ($\Phi_{m,vac} = 3.7$ eV, $\Phi_{m,eff} = 4.18$ eV), LaB_6 ($\Phi_{m,vac} = 2.7$ eV, $\Phi_{m,eff} = 4.3$ eV), and Pt ($\Phi_{m,vac} = 5.6$ eV, $\Phi_{m,eff} = 4.6$ eV) electrodes deposited on HfO_2 [152]. The possibility of the shift of work function may come from intrinsic and/or extrinsic interface states. In the intrinsic models, charges are exchanged between the metal Fermi level and the dielectric gap states, which forms interface dipoles and shifts $\Phi_{m,eff}$ followed by $\Phi_{m,eff} = \Phi_{CNL,d} + S(\Phi_{m,vac} - \Phi_{CNL,d})$. $\Phi_{CNL,d}$ is about 4.4–4.5 eV and S is about 0.1–0.2 for HfO_2 with the metal gates of LaB_6 and Pt [152]. The pinning strengths of HfO_2 were determined to be 0.53 by the empirical model [21] and 0.52 by the experimental data [171]. Since the pinning strength of HfO_2 for TaC, LaB_6, and Pt is much greater than that predicted from the electronic contribution to the dielectric constant ($S = 0.53$), extrinsic states are believed to exist at the metal–HfO_2 interface [171]. In addition, in the case of polysilicon gates, ($\Phi_{CNL,d} = 4.4$ eV, $S = 0.2$) for HfO_2 and ($\Phi_{CNL,d} = 4.7$ eV, $S = 0.4$) for Al_2O_3 were obtained by Hobbs *et al.* [170]. Yu *et al.* indicated that Fermi-level pinning of the metal work function increases with increasing annealing temperature [160]. They proposed a metal–dielectric interface model to explain the work function thermal instability by taking the role of extrinsic states into account. Fermi-level pinning for metal gates of TiN, Ta, TaTi, TaN, TaPt, and TaTiN may enhance at higher annealing temperature [160], and the work function of these metals will converge as the annealing temperature increases. The Fermi levels of these metal gates are pinned at about 4.7–4.8 eV below the vacuum level ($\Phi_{m,eff} = 4.7–4.8$ eV) [160]. Extrinsic states, usually relative to bonding defects, drive the Fermi-level pinning and the convergence of work function. The produced interface dipoles may drive the Fermi level toward the pinning location. However, the work functions of HfN, Ru, and TaSiN change little with annealing temperature [160]. This indicates the absence of extrinsic states at the metal/oxide interfaces or the close alignment between metal work function and Fermi pinning level. A study showed that the pinning effect is more significant for SiO_2 than for HfO_2 gate dielectric [160]. Therefore, the generation of extrinsic states or interfacial bonding defects upon annealing is less crucial to metal gates on HfO_2 compared to metal gates on SiO_2.

Studies showed that a monolayer of SiO_2 at the metal/dielectric interface could relax the effect of Fermi-level pinning [156, 157]. The effective work functions of poly-Si or poly-SiGe gates exhibit less dependence on the gate dielectric material, which implies that the poly-Si or poly-SiGe gates are less vulnerable to Fermi-level pinning compared to the metal gates. In general, high work function metals (e.g., Pt and Au) are resistant to chemical or plasma etching. Hence, the gate patterning of high work function metals would be particularly challenging. Meanwhile, Low work function metals (e.g., Mg and Hf) are extremely reactive, and this might introduce extrinsic interface states due to defects arising from an interfacial reaction. Defect-related extrinsic interface states should be distinguished from intrinsic interface states. To avoid the need for either an inert or a reactive metal gate, a possible solution is to introduce at least a monolayer of SiO_2 at the metal–dielectric interface to achieve a large *S*. However, this approach results in a lower effective dielectric constant for the gate dielectric stack and suffers from gate leakage current and power consumption. FUSI metal gates have recently drawn attention because of the process compatibility with the conventional poly-Si technology [140, 171]. Due to the Fermi-level pinning, the effective work functions ($\Phi_{m,eff}$) of FUSI NiSi and PtSi on $HfO_x(N)$ are rigidly fixed at 4.43 eV and 4.69 eV, respectively [172]. Accordingly, the impurity doping does not help changing $\Phi_{m,eff}$ at all. The $HfO_x(N)$ pinning strength is 0.51 even with doped FUSI gate electrodes. Flatband voltage shift of FUSI PtSi MOSFETs is not affected by Si deposition processes or growth ambient. Fermi-level pinning does not take place at the Pt/HfO_2 interface because of the slight difference of $\Phi_{m,eff}$ values between Pt gates on HfO_2 and SiO_2. In addition, FUSI Ni-silicided gates were demonstrated to eliminate Fermi-level pinning at the poly-Si/HfO_2 dielectric interface in pMOSFETs [173]. The control of Si atom content at the $PtSi_x$/HfO_2 interface is a key factor to relax the pinning effect by reducing the number of Hf–Si bonds [172]. Depinning can be realized by decreasing Si content in $PtSi_x$ and by inserting a SiN cap layer between PtSi and HfO_2. Experimental data showed that increasing Pt/Si ratio may relax the pinning effect, which can be realized by increasing the thickness ratio between Pt and undoped amorphous Si on HfO_2. Experimental results showed that

$\Phi_{m,eff}$ for $PtSi_x$ gates recovered from 4.61 to 4.86 eV as the Pt/Si ratio increased from 1/1 to 10/1. Because of the reduction in Si–Hf interaction, the pinning effect can also be relaxed by inserting a SiN cap layer between PtSi and HfO_2. However, the SiN film thicker than 2 nm is needed to completely cut off the Hf–Si interaction. Therefore, partial silicidation technology may provide a feasible method for advanced metal gate CMOS.

Substituted aluminum (SA) on high-k dielectrics was demonstrated as an alternative of forming a metal gate [174]. Full substitution of poly-Si with Al can be achieved in Ti/Al/poly-Si/HfAlON gate structure by a low-temperature annealing at 450 °C [175]. Owing to the strong driving force of silicon diffusion toward Ti, poly-Si is replaced with Al under the presence of Ti on top of Al during low-temperature annealing. The substituted Al gate on HfAlON dielectric exhibits a low work function of 4.25 eV, which is quite suitable for bulk nMOSFETs. Substituted aluminum process is fully free from the problems of Fermi-level pinning and thermal instability. In addition, compared to FUSI metal gate, substituted aluminum process exhibits improved uniformity in leakage current distribution. Because substituted aluminum process can be lowered to 450 °C, it is also suitable for FUSI NiSi process. Therefore, a possible integration scheme for dual metal gates on high-k gate dielectric can be implemented by using FUSI NiSi for pMOSFETs and SA for nMOSFETs. A study showed that high work function (4.9 eV) on high-k gate dielectric has been achieved by using FUSI Pt_xSi gate without boron predoping of poly-Si, which is suitable for bulk pMOSFETs [176]. For Pt_xSi formation, Pt was deposited on top of poly-Si by physical vapor deposition (PVD) right after diluted HF (DHF) cleaning. A Ti capping layer was sequentially deposited on top of Pt. It was found that Pt can have similar effects of substitution to those observed in substituted aluminum gate if the Pt concentration is high enough in Pt_xSi. To achieve Pt_xSi with such high x, Si movement in Pt should be as easy as possible, and Pt should be thick enough compared to poly-Si. Preexisting Hf–Si bonds may be broken and replaced by Pt–Si bonds owing to the lower heat of formation of Hf-Si versus Pt-Si. The lower concentration of Pt in Pt_xSi formed without Ti capping layer could be due to oxygen diffused near the interface. Oxygen can diffuse preferably along grain boundaries of polycrystalline Pt film and then suppress the movement of Si along the grain boundaries of Pt, resulting in a high concentration of Si in Pt_xSi. Using Ti capping layer on Pt in the FUSI process, high concentration of Pt in FUSI Pt_xSi can be achieved. This is quite important to achieve high work function and reduced Fermi-level pinning on high-k dielectric. Experimental results showed that the extracted work functions of Pt_xSi on SiO_2 and HfAlON are 4.98 and 4.9 eV, respectively, with Ti capping layer, and 4.92 and 4.75 eV without Ti capping layer [176]. Both higher work function and reduced Fermi-level pinning result from higher Pt and lower Si concentration in Pt_xSi by Ti capping. Studies showed that Fermi-level pinning will be worsened by the presence of Si at metal–dielectric interface [158, 160]. Therefore, Fermi-level pinning can be reduced by lowering Si in Pt_xSi using Ti capping. In addition, the thermal budget in Pt_xSi process is reduced, which is of benefit to reduce Fermi-level pinning. A combination of SA and Pt_xSi gates is favorable for dual metal gates in bulk Si CMOS technology because both work functions are within or very close to 0.2 eV of E_C and E_V of Si.

5.5.5 Channel Mobility

The ideal scaling trend maintains a constant electric field across the gate dielectric for a given technology node, which reduces the operating voltage and the transistor dimension by the same factor. However, the feature dimensions have been reduced more rapidly than the operating voltage beyond deep submicron technologies. Accordingly, the aggressive scaling trend increases the electric field across the gate dielectric and the effective electric field in the channel region. These increased electric fields lead to enhancement of phonon scattering and interface roughness scattering, thereby decrease the channel carrier mobility [12]. In advanced high-performance CMOS transistors, it is likely to incorporate highly strained Si and SiGe channel for enhanced carrier transport and high-κ/metal gate stack for low gate leakage [130]. To strain the silicon channel, the following techniques can be used: (a) biaxial in-plane tensile strain using epitaxial SiGe [177], (b) strain by a tensile film [178], and (c) strain by mechanical force [179]. MOSFET devices with HfO_2/poly-Si gate stack exhibit significant degradation in electron mobility compared to SiO_2/poly-Si devices. HfO_2/TiN metal gate stack may recover a major degradation of channel mobility, which is close to the SiO_2/poly-Si level. Using the technique of biaxial tensile strain in Si (10% and 15% Se virtual substrate), the channel mobility can be further enhanced. Because of the biaxial compressive strain in the SiGe layer, the enhancement of PMOS hole mobility can be increased with increasing Ge percentage in the SiGe layer. The improvement of hole mobility does not reduce at high vertical electric field in the channel as in the case of biaxial tensile strain in Si. Furthermore, more than three orders of magnitude reduction in gate leakage current was achieved in HfO_2/TiN gate on both strained and unstrained Si compared to the SiO_2/poly-Si control.

Charge trapping is a main issue in high-κ CMOS transistors. Several negative aspects of charge trapping in high-κ device performance were revealed, for example, DC mobility degradation, threshold voltage instability, C–V and I–V hysteresis, and poor reliability [180]. Charge trapping in high-κ MOSFETs was recently investigated using short single pulse I_d–V_g measurements [181], as shown in Figure 5.46a. Due to the charge trapping, drain current (I_d) cannot follow the same line when the gate voltage sweeps back during pulse time as shown in Figure 5.46b. The enhanced threshold voltage shift can be observed from the pulsed I–V technique. This implies that a significant fraction of the trapped charges in high-κ MOSFETs may be detrapped during the down trace of a traditional DC measurement [182]. Charge trapping is bias-dependent, and the threshold voltage shift between the I_d–V_g curves increases as the V_g pulse height increases [183]. The threshold voltage shift, $\Delta V_{TH} = Q/C_{eff}$, where C_{eff} is the effective capacitance of the trapped charge Q, can be estimated according to the current decrease ΔI_d at the given voltage V_g, $\Delta V_{TH} = (\Delta I_d/I_d)(V_g - V_{TH})$ [183]. Charge trapping may be completely reversible with a negative

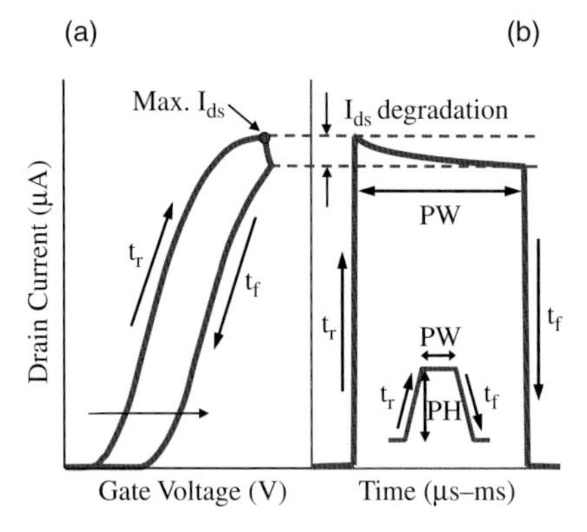

Figure 5.46 (a) "Single-pulse" I–V methodology. (b) I_{ds} degradation of high-κ devices because of the impact of charge trapping during the pulse width period. PW is the pulse width, PH is the pulse height, t_r is the rise time, and t_f is the fall time.

bias. Figure 5.47 shows rapid drain current decrease during the short voltage pulse applied to the gate [181]. The drain current increases with shorter rise times compared to the slower DC I_{ds}–V_{gs}. A report indicated that fast charge trapping may take place in a timescale smaller than a few microseconds [181]. In addition, the fast electron trapping is a main source of degradation in the DC characteristics of the high-κ transistors, which strongly depends on the thickness of interfacial layer (SiO_2) in $HfSi_xO_y$ gate stacks. As the SiO_2 thickness increases from 0 to 1.9 nm, the characteristic electron trapping time (τ_c) increases from 1 μs to 1 ms. This implies that the decrease in interfacial layer thickness may increase the efficiency of electron tunneling, thereby decreasing τ_c [181, 184]. Increasing the gate bias will lead to a larger ΔV_{TH} for the high-k gate stacks except for the thick SiO_2 interfacial layer (>1.9 nm), for which the trapping can be negligible. Furthermore, significant enhancement of drain current in linear region ($I_{d,lin}$) can be observed at ultrashort pulsed I–V measurements [181, 185, 186], which indicated that the low DC mobility of high-k device results from the fast transient charging effects (FTCE), as shown in Figure 5.47. Generally, $I_{d,lin}$ is defined as the drain current measured at $V_{gs} = V_{dd}$ and $V_{ds} = 0.1$ V. Compared to the DC I_d–V_g, $I_{d,lin}$ increases with shorter rise time, as

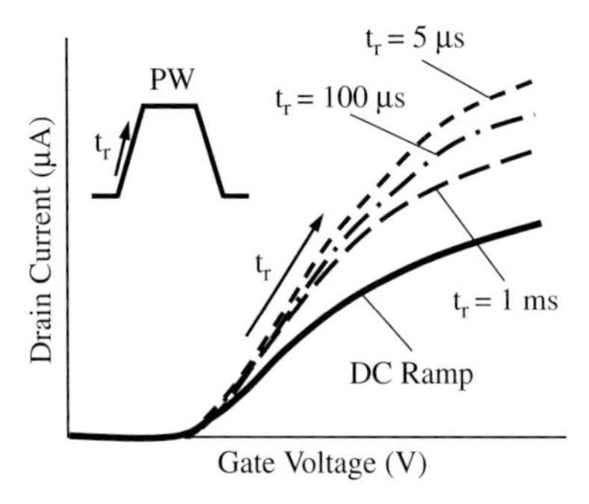

Figure 5.47 "Single-pulse" I–V methodology on a high-κ MOS device showing the impact of charge trapping on the rise time. I_{ds} increases with shorter rise times.

shown in Figure 5.47. This implies that the carrier mobility may be underestimated using the DC I–V measurement for high-k devices. The FTCE model predicts that thicker high-k layer leads to worse DC mobility degradation because of the larger amount of available electron traps [186]. This implies that the scattering induced from phonons and/or fixed charges may not be the dominant effects of channel mobility degradation in high-κ devices. In comparison with poly gate devices, the DC mobility of metal gate devices can be improved since FTCE is lower for metal gate devices [185, 186]. During conventional DC measurements, charge trapping may prevent evaluation of the intrinsic properties of high-k dielectrics. The capability of nanosecond regime (~35 ns) measurements is, therefore, required to achieve trap-free pulse I–V characteristics [181]. For $t_p << \tau_c$, the intrinsic I_d with negligible trapping can be achieved; and for $t_p > \tau_c$, the reduced I_d owing to trapping can be observed, where t_p is the sum of the gate input pulse rise time and pulse width, and τ_c is the characteristic electron trapping time [181].

The chemical bonding involving d-electrons results in highly vulnerable high-κ dielectrics with the formation of structural defects during the deposition process. Electrically active defects may also be introduced into the high-κ dielectrics during the processes of gate electrode deposition and/or thermal annealing owing to high diffusivity of various species in high-k materials. These defects may be a source of fixed charges and electron traps. Transistor characteristics under the electrical stress may be deteriorated by the defect generation and/or by the electron trapping/detrapping at the preexisting defects. The latter reflects initial quality of the high-k film rather than its electrical stability [183]. A report showed that time-dependent threshold voltage of high-k transistors during the electrical stress came from the electron trapping at preexisting defects in the high-k dielectric rather than stress-induced trap generation [183]. For positive gate bias, the preexisting defects in the high-κ dielectric (e.g., HfO_2) are charged by carrier tunneling through the interfacial SiO_2 layer. When the bias is reversed, the trapped charges tunnel back to the Si substrate. The rapid shift with gate bias corresponds to the efficient charging and discharging of defects in the high-κ layer [187, 188]. A defect band model is proposed to explain the reasons of the charge trapping and detrapping for the V_{TH} instability. The physical origin of the defect band may come from the oxygen vacancies, chlorine impurities, or water related defects generated by the precursor chemistry [188]. According to the high-frequency C–V characteristics, negative charges can be created within the high-k dielectric when a positive bias is applied to the gate. When the starting voltage is -1 V or 0 V, either "full" or "partial" V_{TH} recovery is achieved. Therefore, the V_{TH} instability is strongly affected by the measurement procedure [187]. Parallel shift of I_d–V_g curve indicates that the interface quality is not degraded significantly during the stressing despite of significant electron charging [186, 189, 190]. A study shows that charge trapping rate ($\Delta V_{TH}/\Delta V_g$) is not noticeably changed by the repeated stresses (e.g., 2 V for 500 s) [183]. This implies that no significant generation of the electron traps occurred during the pulse time. A rapid turn around of V_{TH} shift is observed after turning off the FN stress. The relaxation behavior follows a typical exponential function, and there is a residual shift even after a long wait time (e.g., 10^4 s).

This indicates that the V_{TH} shift cannot be 100% restored. There are two factors that shift V_{TH} during the stress. One is the electrons filling the existing traps and the other is the charged defects created during the stress. A part of the electrons is detrapped spontaneously to reduce the instant built-in potential after the stress is removed. Meanwhile, the residual V_{TH} shift indicates that there remains some amount of trapped electrons and charged damage. The relaxation rate is primarily determined by both the stress bias and the duration of stress [180]. At higher stress voltage, the device shows faster stress relaxation. At the same bias, the shorter stress has faster relaxation. A report showed that dielectric relaxation currents may come from the electron trapping and detrapping in high-κ dielectric stacks [191]. According to the substrate injection experiments, the relaxation current is mainly due to the traps near the SiO_2/high-κ interface [184, 191]. The dielectric relaxation effect may affect the device performances, for example, drive current variation and high-frequency properties. The relaxation current is proportional to t^{-1} using the tunnel-front model [191] if the distribution of the traps near the SiO_2/high-κ interface is uniform. The relaxation current in high-κ dielectric stacks may be affected by the thickness, temperature, and gate voltage polarity.

5.5.6
Dielectric Breakdown

Dielectric breakdown is the irreversible local change of the dielectric insulation property. In most cases, the local change in properties results in the product failure, namely, the product is to stop intended functioning. In the case of dielectric breakdown, the conductivity of a tiny spot developed within the dielectric area is strikingly larger than that of the rest of the dielectric area in which the conductivity remains nearly the same. This conducting spot dominates the current flow and then changes the electrical characteristics of the dielectric. Generally, a breakdown happens after a certain period of time as the dielectric is subjected to an electrical stress at product operation or elevated conditions. There are two kinds of dielectric breakdown, that is, intrinsic breakdown and extrinsic breakdown. Intrinsic breakdown corresponds to the inherent material property of flawless dielectrics that nevertheless wear out over time and eventually fail at the moment of breakdown. It is generally assumed that intrinsic dielectrics are defect free. For a wear-out process, intrinsic breakdown exhibits an increasing failure rate in the reliability bathtub curve. This implies that within a relative short period of time after the intrinsic breakdown of the first of the samples, the entire population stressed at the same time will fail. This results in a steep slope of the statistical failure distributions. Note that this bathtub curve can be divided into three regions, namely, a serious early failure with decreasing failure rate is initially observed, followed by a constant random low failure rate, and ultimately a wear-out region with increasing failure rate. Compared to intrinsic breakdown, extrinsic breakdown takes place much earlier, which distributes over a wide time range before the intrinsic breakdown and causes a wide or shallow distribution before the intrinsic distribution. The extrinsic distribution typically exhibits a decreasing failure rate. The early failure does not come from wear out but from flaws in the dielectric, which can be improved by using the method of burn-in. The flaws in dielectrics are most often due to manufacturing process defects. Because intrinsic failure reflects wear out and material properties, all the discussions of dielectric breakdown are related to intrinsic dielectrics in the following.

McPherson *et al.* reported the reliability trade-offs associated with the high-κ material selection [192]. They found that the fundamental relationship between the dielectric breakdown (E_{bd}) and the dielectric constant is approximately $E_{bd} \sim (\kappa)^{-1/2}$. Experimental data suggested that the local electric field (E_{loc}) in the high-κ materials plays a very important role in the observed $E_{bd} \sim (\kappa)^{-1/2}$ behavior. The local electric field is a superposition of the externally applied field E_{ext} and the dipolar field leading to the Lorentz relation or Mossotti field: $E_{loc} = [(2 + \kappa)/3]E_{ext}$ [193]. In high-k materials, the local electric field can be very large even though the applied field and current injection may be rather low. The local electric field may distort and weaken polar molecular bonds such that the bond strength becomes very susceptible to breakage by standard Boltzmann processes and/or hole capture [194]. As the molecular bonds break, a conductive path develops, resulting in electrical/thermal breakdown of the dielectric. The field acceleration parameter (γ) of TDDB is given by the physical/molecular model [193, 194]: $\gamma = p_o(2 + \kappa)/3k_BT$, where p_o is the molecular dipole moment component opposite to the local field, k_B is Boltzmann's constant, and T is the temperature. This model predicts that the field acceleration parameter increases as the dielectric constant increases. Therefore, the trends of "lower E_{bd}" and "higher field acceleration parameter γ" with higher dielectric constant are clearly predicted in the TDDB data. A report showed that the TDDB field acceleration parameters (γ) are 3.5, 6.5, 13.6, and 60 for SiO_2 ($\kappa = 3.9$), HfSiON ($\kappa = 7$), Ta_2O_5 ($\kappa = 26$), and PZT ($\kappa = 250$), respectively [195].

In general, the degree of ionic bonding of high-κ dielectrics is larger than that of SiO_2 [128]. In SiO_2, silicon forms four directional covalent bonds to oxygen atoms. In ZrO_2 and HfO_2, they may form cubic crystals with the fluorite structure in which Zr or Hf is eightfold coordinated [128]. Also, they may form tetragonal and orthorhombic crystals in which Zr or Hf is sevenfold coordinated. For the structure of crystalline La_2O_3, La is sevenfold coordinated with four bonds and three longer bonds [128]. A report indicated that the coordination number (CN) plays a major role in E_{bd} [192]. For example, if Hf takes on cubic symmetry (CN = 8) in HfO_2, then the dielectric breakdown strength (E_{bd}) is expected to be about 6.7 MV/cm. If, however, Hf takes on a dominant tetragonal symmetry (CN = 4), the expected E_{bd} is only about 3.9 MV/cm. In an amorphous network, one might expect to find a combination of eightfold and fourfold coordinated metal ions, leading to a significant impact on the dielectric breakdown strength. In general, higher symmetry (higher CN) tends to result in higher dielectric breakdown strength. Higher CN indicates that more bonds have to be broken (higher activation energy) for a permanent metal ion displacement [192].

Gate dielectrics reliability has always been a major concern throughout all CMOS generations. Stathis *et al.* showed that the SiO_2 gate oxide smaller than 2.2–2.6 nm would encounter the stringent 10 year reliability criteria [196]. For assessment of the dielectric reliability, it is clearly not feasible to test individual devices for 10 years prior to product incorporation. Consequently, testing must be accelerated at higher voltages and/or temperatures than actually experienced by typical devices. Making reliability projection from accelerated to actual condition requires proper scaling for area, voltage, temperature, and the failed fraction of devices [196, 197]. A fundamental mechanism for oxide breakdown in the ultrathin SiO_2 regime is described by the percolation model [198, 199]. This model indicates that ultrathin oxide breakdown is related to the buildup of many traps/defects within the SiO_2 layer, where after a certain amount of stress, a complete path of traps/defects is constructed across the oxide thickness [198]. Percolation models and a cell-based analytic model [200] hypothesize that defective cells are generated at random in the dielectric bulk during the electrical stress. Each trap has an effective sphere of influence. When two spheres touch or overlap, the conduction between the two traps is possible. The dielectric breakdown is triggered by the formation of defective cells in a conducting path (i.e., breakdown path) between the cathode and the anode. The breakdown path for thicker oxides consists of a larger number of defects and the spread on the defect density necessary to generate such a large path is smaller. As a result, the percolation model predicts that the Weibull slope (β) of the charge-to-breakdown (Q_{bd}) distribution decreases as oxide thickness decreases for intrinsic oxides. The charge to breakdown is a statistically distributed value that can be obtained from the electrical tests. In general, the statistics of dielectric breakdown is described by the Weibull distribution [198],

$$F(Q) = 1 - \exp[-(Q/\alpha)^{\beta}] \tag{5.32}$$

$$W(Q) = Ln[1 - Ln(1 - F)] = \beta Ln(Q/\alpha) \tag{5.33}$$

where $F(Q)$ is the cumulative distribution function of failure, Q is the charge to breakdown, α is the charge to breakdown at approximately 63rd percentile, and β is the Weibull shape factor, often called Weibull slope. If this percolation concept is still valid in high-k dielectrics, the Weibull slope should be proportional to the number of traps in that path, and its value will increase as the physical oxide thickness increases. Since the Weibull slope is applied to estimate the lifetime of the chip's total gate oxide area and the percentage of failures, a reduced Weibull slope corresponds directly to a decrease in reliability [195].

During the electrical stress, it is reasonable to assume that there are many kinds of defects generated. Each kind of defect has its own defect properties, for example, defect size, capture cross section, characteristic time constant, and so on. One major kind of defect may dominate the trapping behaviors and possesses the largest defect size and the shortest time constant in the duration of electrical stressing. Therefore, Chiu proposed a modified percolation model [201] of dielectric breakdown, as shown in Figure 5.48 in which the schematic pictures of defect generation and breakdown condition are indicated. This model claims that the major kind of defect (with size a_1) and many other kinds of defects (with a number of smaller sizes $a_2, a_3 \ldots$) are formed during the electrical stress. In the modified percolation model, the event of

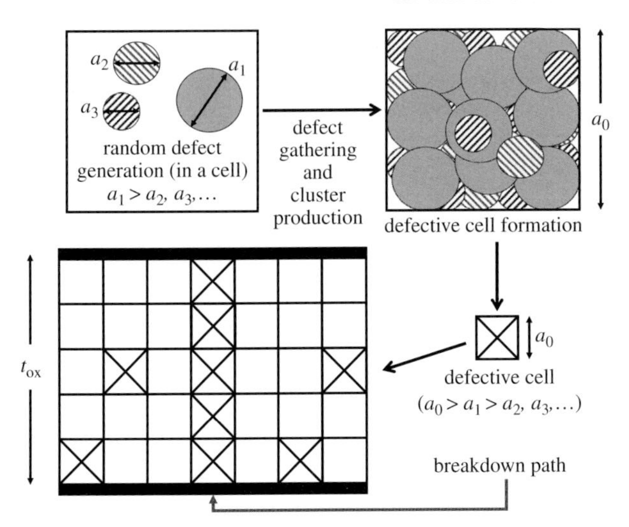

Figure 5.48 Schematic pictures of defects generation in the events of "stressing" and "breakdown."

"stressing" is defined as the defects are produced during the electrical stress. During the stressing event, all kinds of defects are randomly generated in the dielectric bulk and are potentially gathered into a "cluster." The cluster is a constructed larger defect that eventually forms the single defective cell (with a large size a_0) at the onset of dielectric breakdown. At the onset of dielectric breakdown, the mass production of the clusters takes place, which is defined as the event of "breakdown." In this situation, the defective cells are formed simultaneously and connect each other into a column between cathode and anode. Hence, the dielectric breakdown occurs. In other words, the breakdown path is instantly constructed by the defective cell in the event of "breakdown." So, "$a_0 > a_1 > a_2, a_3 \ldots$." is inferred from the above argument.

Reports showed that the measurements of charge trapping and stress-induced leakage current (SILC) may provide information of the oxide defects generated during the electrical stress [202–204]. The trap capture cross sections (σ_S) in CeO_2 [202], HfO_2 [203], and SiO_2 [202–204] in the stressing event were determined to be about 4.3×10^{-18}, 1.5×10^{-17}, and 1.5×10^{-16} cm^2, respectively. These values should correspond to the major kind of defects generated in the "stressing" event. To determine the defect size in the "breakdown" event, the cell-based analytical model can be used. The mean size (a_0) of stress-induced defects in the breakdown paths can be determined by the dependence of β on t_{ox}: $a_0 = n(\Delta\beta/\Delta t_{ox})^{-1}$, where $\Delta\beta$ denotes the change in Weibull slope, Δt_{ox} denotes the change in oxide thickness, and n is associated with a power law relationship between the fraction of defective cells (λ) of dielectric breakdown and charge to breakdown per unit area (Q), namely, $\lambda \propto Q^n$. The exponent n was determined to be 1 and 0.6 by Stathis [10] and Degraeve [199], respectively. Figure 5.49 shows the thickness dependence of Weibull slopes for CeO_2, HfO_2, and SiO_2. The Weibull slope β decreases linearly with decreasing dielectric

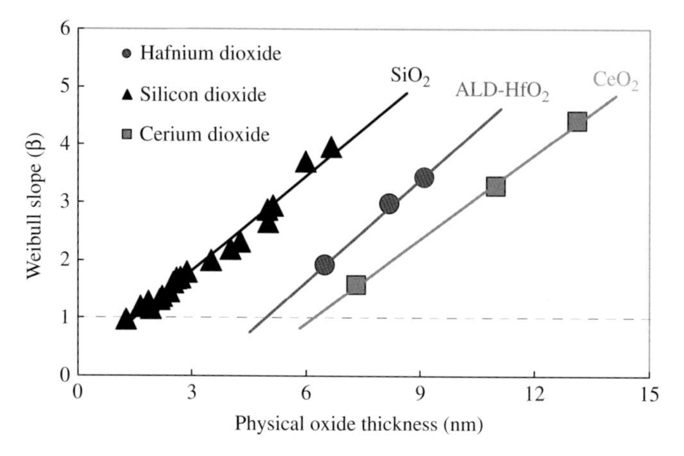

Figure 5.49 Thickness dependence of Weibull slopes for CeO_2, HfO_2, and SiO_2.

thickness. In the calculation of a_0, $n=1$ and 0.6 were used. These determined n values are based on SiO_2. Assume it is also valid for CeO_2 and HfO_2. Therefore, the defect sizes (a_0) in CeO_2 and HfO_2 in the breakdown event can be obtained. If $n=1$, the defect sizes of CeO_2, SiO_2, and HfO_2 were estimated to be about 2.12, 1.83, and 1.71 nm, respectively. If $n=0.6$, then $a_0=1.27$, 1.10, and 1.03 nm for CeO_2, SiO_2, and HfO_2, respectively. Note that it is worth studying whether the "n" value depends on the dielectric material even though $n=1$ and 0.6 is used in this work.

Chiu reported that choosing $n=1$, the defect size in CeO_2 at the onset of dielectric breakdown was estimated to be about 2.12 nm [201]. This implies that the capture cross section of the defects generated in CeO_2 is on the order of $3.5\times10^{-14}\,cm^2$ in the breakdown event. Studies [205, 206] showed that the defect sizes in SiO_2 and HfO_2 are about 1.83 and 1.71 nm, respectively. Therefore, the capture cross sections (σ_B) of the defects generated in SiO_2 and HfO_2 in the breakdown event are about 2.6×10^{-14} and $2.3\times10^{-14}\,cm^2$, respectively. The results of the defect sizes in CeO_2, SiO_2, and HfO_2 in the breakdown event are close to each other. This implies that the same dielectric breakdown mechanism may exist in these oxides. Table 5.3 lists the capture cross sections (σ_S and σ_B) and defect sizes in the events of "stressing"

Table 5.3 Capture cross sections and defect sizes in the events of "stressing" and "breakdown" in CeO_2, HfO_2, and SiO_2.

Material	σ_S (cm^2)	σ_B (cm^2)	a_0 (nm)
CeO_2	4.3×10^{-18}	3.5×10^{-14}	2.12
HfO_2	1.5×10^{-17}	2.3×10^{-14}	1.71
SiO_2	1.5×10^{-16}	2.6×10^{-14}	1.83

σ_S: capture cross section of the generated defects in the event of "stressing."
σ_B: capture cross section of the generated defects in the event of "breakdown."
a_0: size of the generated defects in the event of "breakdown" when $n=1$.

and “breakdown” in CeO_2, HfO_2, and SiO_2. In the stressing events, the measurements of SILC and charge trapping give information on the generated defects that are classified into the major kind of defects with size a_1 during the electrical stress. It is obvious that the capture cross sections of the traps/defects generated in the “stressing” event is strikingly smaller than those in the breakdown event. A report shows that the capture cross section of neutral electron traps ranges between 10^{-18} and 10^{-14} cm^2 [207]. In the events of stressing and breakdown, the determined σ_S and σ_B are about on the orders between 10^{-18} and 10^{-14} cm^2. Therefore, the defects generated in both the events of stressing and breakdown belong to the neutral centers. The neutral centers may result in the dielectric degradation during the electrical stress and lead to the dielectric breakdown consequently. Every kind of defect generated during the stressing is gathered into a cluster. This cluster is a constructed larger defect that grows with time during the stressing. The dielectric breakdown is triggered when the cluster enlarges close to the critical size, which induces the positive feedback in the gathering of clusters. At the moment of dielectric breakdown, the mass production of the clusters takes place. In this situation, the defect cells are formed simultaneously and connect each other into a breakdown path between cathode and anode immediately. Consequently, the dielectric breakdown occurs when the capture cross section of defect cell reaches the critical size. Experimental results showed that the capture cross sections of defect cell in CeO_2, SiO_2, and HfO_2 in the breakdown event were determined to be on the order of 10^{-14}–10^{-13} cm^2 in the literature [201, 205, 206]. This finding supports the argument of the kinetics of dielectric breakdown in the modified percolation model [201].

5.6 Conclusions and Perspectives

In this chapter, recent works on rare-earth-based oxides and silicates have been reviewed with an emphasis on material suitability for integration in CMOS technology. According to ITRS, the silicon technology will shrink more with the advancing of Moore’s law for the next few couple of decades. With the trend of device size downscaling, it has been suggested that a new type of gate dielectric with rare-earth oxide will take position of conventional gate oxide with the same or less EOT value. Structural and electrical behaviors associated with several serious issues, such as process compatibility and leakage current, have been discussed in this report. To be the alternative gate dielectrics, the new selected oxides must fulfill the criteria of satisfactory κ-value and bandgap, high thermal and chemical stability, excellent interface quality and surface topology, and low defect density. It has been shown that the RE oxides and their silicates have potential to be implemented in the working devices and the necessary processing techniques to obtain the adequate quality for the use of gate dielectric. In Section 5.4.1, the process compatibility of several rare-earth oxides for existing CMOS technology was demonstrated. Successful implementation of rare-earth oxide as the gate dielectric was realized and optimized by the analytical tools of XRD and AFM. In order to achieve high performances of devices, the gate dielectrics need to be optimized further. Interfacial and atomic environment of ultrathin and thin RE oxide dielectrics was explored for better understanding of the device physics, which gives some useful information of other RE_2O_3 films. Relationship among flatband voltage, threshold voltage, and trap charges was explored in Nd_2O_3 and lanthanum scandate. It is believed that main oxide defects are due to the large oxygen vacancies and interfacial trap sites. So, the improvement of flatband voltage shift and low defect density is required for rare-earth oxides. PDA in Ar and O_2 ambient shows satisfactory results for defect density.

Detailed analyses of the leakage conduction mechanisms in some selected rare-earth oxides were discussed. Different conduction mechanisms through high-κ RE oxides were found, which may depend on the deposition method, electrode material, applied field range, and so on. Since high-κ/silicon interfacial property plays a key role in the device performance, different barrier layers were investigated to improve and optimize the interface quality for the high-κ oxides. It has been shown that the interface trap density could be minimized for CeO_2 MOSFETs because of lattice matching between CeO_2 and Si. Optimized work function engineering can maintain large enough band offset to control the leakage current through the dielectrics. In this chapter, the mobility of charge carriers was also studied. The reduced mobility in high-κ devices with respect to standard SiO_2 ones is attributed to the effects of phonon scattering, interface surface roughness scattering, and charge trapping. An oxynitride film replacement may improve the carrier mobility in spite of the expected scarification of κ-value. Dielectric degradation mechanisms are also elucidated by the time-dependent dielectric breakdown of rare earth oxide dielectrics. Nevertheless, the use of high-κ rare-earth oxides may encounter some crucial issues. Hence, further rigorous and detailed investigations are needed to completely realize the material characteristics for process integration in practical devices.

References

1 Moore, G.E. (1965) Cramming more components onto integrated circuits. *Electronics*, **38**, 114.

2 Chau, R., Kavalieros, J., Roberds, B., Schenker, R., Lionberger, D., Barlage, D., Doyle, B., Arghavani, R., Murthy, A., and Dewey, G. (2000) 30 nm physical gate length CMOS transistors with 1.0ps n-MOS and 1.7ps p-MOS gate delays. IEEE Technical Digest of International Electron Devices Meeting, 45.

3 Lo, S.H., Buchanan, D.A., Taur, Y., and Wang, W. (1997) Quantum-mechanical modeling of electron tunneling current from the inversion layer of ultra-thin-oxide nMOSFET's. *IEEE Electron. Device Lett.*, **18**, 209.

4 Brar, B., Wilk, G.D., and Seabaugh, A.C. (1996) Direct extraction of the electron tunneling effective mass in ultrathin SiO_2. *Appl. Phys. Lett.*, **69**, 2728.

5 Harari, E. (1978) Dielectric breakdown in electrically stressed thin films of thermal SiO_2. *J. Appl. Phys.*, **49**, 2478.

6 DiMaria, D.J., Cartier, E., and Arnold, D. (1993) Impact ionization, trap creation, degradation, and breakdown in silicon dioxide films on silicon. *J. Appl. Phys.*, **73**, 3367.

7 DiMaria, D.J. and Cartier, E. (1995) Mechanism for stress-induced leakage currents in thin silicon dioxide films. *J. Appl. Phys.*, **78**, 3883.

8 Degraeve, R., Groeseneken, G., Bellens, R., Ogier, J.L., Depas, M., Roussel, P.J., and Maes, H. (1998) New insights in the relation between electron trap generation and the statistical properties of oxide breakdown. *IEEE Trans. Electron. Dev.*, **45**, 904.

9 Houssa, M., Nigam, T., Mertens, P.W., and Heyns, M.M. (1998) Model for the current–voltage characteristics of ultrathin gate oxides after soft breakdown. *J. Appl. Phys.*, **84**, 4351.

10 Stathis, J.H. (1999) Percolation models for gate oxide breakdown. *J. Appl. Phys.*, **86**, 5757.

11 Halter, J.P. and Najm, F.N. (1997) A gate-level leakage power reduction method for ultra-low-power CMOS circuits. Custom Integrated Circuits Conference Proceedings of the IEEE, 475.

12 Wilk, G.D., Wallace, R.M., and Anthony, J.M. (2001) High-κ gate dielectrics: current status and materials properties considerations. *J. Appl. Phys.*, **89**, 5243.

13 Yeo, Y.C., King, T.J., and Hu, C. (2002) Direct tunneling leakage current and scalability of alternative gate dielectrics. *Appl. Phys. Lett.*, **81**, 2091.

14 Wallace, R.M., and Wilk, G.D. (2003) High-κ dielectric materials for microelectronics. *Crit. Rev. Solid State,* **28**, 231.

15 Robertson, J. (2004) High dielectric constant oxides. *Eur. Phys. J. Appl. Phys.*, **28**, 265.

16 Huff, H. and Gilmer, D. (2005) *High Dielectric Constant Materials, Series in Microelectronics*, Springer, vol. 16.

17 Robertson, J. (2006) High dielectric constant gate oxides for metal oxide Si transistors. *Rep. Prog. Phys.*, **69**, 327.

18 Locquet, J.P., Marchiori, C., Sousa, M., Fompeyrine, J., and Seo, J.W. (2006) High-κ dielectrics for the gate stack. *J. Appl. Phys.*, **100**, 051610.

19 Kim, Y.B. (2010) Challenges for nanoscale MOSFETs and emerging nanoelectronics. *Trans. Electr. Electron. Mater.*, **11**, 93.

20 International Technology Roadmap for Semiconductors (ITRS) (2010) 2010 Edition access date 10.03.2011 (http://www.itrs.net/Links/2010ITRS/Home2010.htm).

21 Robertson, J. (2000) Band offsets of wide-band-gap oxides and implications for future electronic devices. *J. Vac. Sci. Technol. B*, **18**, 1785.

22 Robertson, J. and Chen, C.W. (1999) Schottky barrier heights of tantalum oxide, barium strontium titanate, lead titanate, and strontium bismuth tantalate. *Appl. Phys. Lett.*, **74**, 1168.

23 Puthenkovilakam, R. and Chang, J.P. (2004) Valence band structure and band alignment at the ZrO_2/Si interface. *Appl. Phys. Lett.*, **84**, 1353.

24 Wong, H., Ng, K.L., Zhan, N., Poon, M.C., and Kok, C.W. (2004) Interface bonding structure of hafnium oxide prepared by direct sputtering of hafnium in oxygen. *J. Vac. Sci. Technol. B*, **22**, 1094.

25 Zhan, N., Poon, M.C., Kok, C.W., Ng, K.L., and Wong, H. (2003) XPS study of the thermal instability of HfO_2 prepared by Hf sputtering in oxygen with RTA. *J. Electrochem. Soc.*, **150**, F200.

26 Peacock, P.W. and Robertson, J. (2004) Bonding, energies, and band offsets of Si-ZrO_2 and HfO_2 gate oxide interfaces. *Phys. Rev. Lett.*, **92**, 057601.

27 Hubbard, K.J. and Schlom, D.G. (1996) Thermodynamic stability of binary oxides in contact with silicon. *J. Mater. Res.*, **11**, 2757.

28 Chen, C.P., Hong, M., Kwo, J., Cheng, H.M., Huang, Y.L., Lin, S.Y., Chi, J., Lee, H.Y., Hsieh, Y.F., and Mannaerts, J.P. (2005) Thin single-crystal Sc_2O_3 films epitaxially grown on Si (111)-structure and electrical properties. *J. Cryst. Growth*, **278**, 638.

29 Nishikawa, Y., Matsushita, D., Satou, N., Yoshiki, M., Schimizu, T., Yamaguchi, T., Satake, H., and Fukushima, N. (2004) Tensile strain in Si due to expansion of lattice spacings in CeO_2 epitaxially gown on Si(111). *J. Electrochem. Soc.*, **151**, F202.

30 Osten, H.J., Bugiel, E., and Fissel, A. (2002) Epitaxial praseodymium oxide: a new high-κ dielectric. *Material Research Society Proceedings*, **744**, M1.5.

31 Kwo, J., Hong, M., Busch, B., Muller, D.A., Chabal, Y.J., Kortan, A.R., Mannaerts, J.P., Yang, B., Ye, P., Gossmann, H., Sergent, A.M., Ng, K.K., Bude, J., Schulte, W.H., Garfunkel, E., and Gustafsson, T. (2003) Advances in high-κ gate dielectrics for Si and III–V semiconductors. *J. Cryst. Growth*, **251**, 645.

32 Gusev, E.P., Narayanan, V., and Frank, M.M. (2006) Advanced high-κ dielectric stacks with polySi and metal gates: recent progress and current challenges. *IBM J. Res. Dev.*, **50**, 387.

33 Fanciulli, M. and Scarel, G. (2007) *Rare Earth Oxide Thin Film: Growth, Characterization, and Applications, Series in Applied Physics*, Springer, vol. 106.

34 Pantisano, L., Schram, T., O'Sullivan, B., Conard, T., De Gendt, S., Groeseneken, G., Zimmerman, P., Akheyar, A., Heyns, M.M., Shamuilla, S., Afanas'ev, V.V., and Stesmans, A. (2006) Effective work function modulation by controlled dielectric monolayer deposition. *Appl. Phys. Lett.*, **89**, 113505.

35 Mitrovic, I.Z., Buiu, O., Hall, S., Bungey, C., Wagner, T., Davey, W., and Lu, Y. (2007) Electrical and structural properties of hafnium silicate thin films. *Microelectron. Reliab.*, **47**, 645.

36 Van Elshocht, S., Adelmann, C., Conard, T., Delabie, A., Franquet, A., Nyns, L., Richard, O., Lehnen, P., Swerts, J., and De Gendt, S. (2008) Silicate formation and thermal stability of ternary rare earth oxides as high-κ dielectrics. *J. Vac. Sci. Technol. A*, **26**, 724.

37 Copel, M., Cartier, E., and Ross, F.M. (2001) Formation of a stratified lanthanum silicate dielectric by reaction with Si(001). *Appl. Phys. Lett.*, **78**, 1607.

38 Lupina, G., Schroeder, T., Wenger, C., Dabrowski, J., and Mussig, H.-J. (2006) Thermal stability of Pr silicate high-κ layers on Si(001). *Appl. Phys. Lett.*, **89**, 222909.

39 Pan, T.M., Lee, J.D., Shu, W.H., and Chen, T.T. (2006) Structural and electrical properties of neodymium oxide high-κ gate dielectrics. *Appl. Phys. Lett.*, **89**, 232908.

40 Gupta, J.A., Landheer, D., McCaffrey, J.P., and Sproule, G.I. (2001) Gadolinium silicate gate dielectric films with sub-1.5nm equivalent oxide thickness. *Appl. Phys. Lett.*, **78**, 1718.

41 Pan, T.M., Chen, C.L., Yeh, W.W., and Hou, S.J. (2006) Structural and electrical characteristics of thin erbium oxide gate dielectrics. *Appl. Phys. Lett.*, **89**, 222912.

42 Ono, H. and Katsumata, T. (2001) Interfacial reactions between thin rare-earth-metal oxide films and Si substrates. *Appl. Phys. Lett.*, **78**, 1832.

43 Paivasaari, J., Putkonen, M., and Niinisto, L. (2005) A comparative study on lanthanide oxide thin films grown by atomic layer deposition. *Thin Solid Films*, **472**, 275.

44 Paivasaari, J., Putkonen, M., Sajavaara, T., and Niinisto, L. (2004) Atomic layer deposition of rare earth oxides: erbium oxide thin films from β-diketonate and ozone precursors. *J. Alloy. Compd.*, **374**, 124.

45 Pan, T.M. and Huang, C.C. (2010) Effects of oxygen content and postdeposition annealing on the physical and electrical properties of thin Sm_2O_3 gate dielectrics. *Appl. Surf. Sci.*, **256**, 7186.

46 Gschneider, K.A., Jr. and Eyring, L. (1982) *Handbook on Physics and Chemistry of Rare Earths*, vol. 5.

47 Kwo, J., Hong, M., Kortan, A.R., Queeney, K.L., Chabal, Y.J., Opila, R.L., Jr., Muller, D.A., Chu, S.N.G., Sapjeta, B.J., Lay, T.S., Mannaerts, J.P., Boone, T., Krautter, H.W., Krajewski, J.J., Sergnt, A.M., and Rosamilia, J.M. (2001). Properties of high κ gate dielectrics Gd_2O_3 and Y_2O_3 for Si. *J. Appl. Phys.*, **89**, 3920.

48 Mikhelashvili, V., Eisensteini, G., and Edelmann, F. (2002) Structural properties and electrical characteristics of electron-beam gun evaporated erbium oxide films. *Appl. Phys. Lett.*, **80**, 2156.

49 Zhao, Y., Toyama, M., Kita, K., Kyuno, K., and Toriumi, A. (2006) Moisture-absorption-induced permittivity deterioration and surface roughness enhancement of lanthanum oxide films on silicon. *Appl. Phys. Lett.*, **88**, 072904.

50 van Dover, R.B. (1999) Amorphous lanthanide-doped TiO_x dielectric films. *Appl. Phys. Lett.*, **74**, 3041.

51 Zhao, Y., Kita, K., Kyuno, K., and Toriumi, A. (2006) Higher-κ $LaYO_x$ films with strong moisture resistance. *Appl. Phys. Lett.*, **89**, 252905.

52 Hoekstra, H.R. (1966) Phase relationships in the rare earth sesquioxides at high pressure. *Inorg. Chem.*, **5**, 754.

53 Hirosaki, N., Ogata, S., and Kocer, C. (2003) *Ab initio* calculation of the crystal structure of the lanthanide Ln_2O_3 sesquioxides. *J. Alloy. Compd.*, **351**, 31.

54 Mikhelashvili, V., Eisenstein, G., Edelman, F., Brener, R., Zakharov, N., and Werner, P. (2004) Structural and electrical properties of electron beam gun evaporated Er_2O_3 insulator thin films. *J. Appl. Phys.*, **95**, 613.

55 Ohmi, S., Takeda, M., Ishiwara, H., and Iwai, H. (2004) Electrical characteristics for Lu_2O_3 thin films fabricated by e-beam deposition method. *J. Electrochem. Soc.*, **151**, G279.

56 Lu, H.L., Scarel, G., Lamagna, L., Fanciulli, M., Ding, S.J., and Zhang, D.W. (2008) Effect of rapid thermal annealing on optical and interfacial properties of atomic-layer-deposited Lu_2O_3 films on Si (100). *Appl. Phys. Lett.*, **93**, 152906.

57 Ohmi, S., Kobayashi, C., Kashiwagi, I., Ohshima, C., Ishiwara, H., and Iwai, H. (2003) Characterization of La_2O_3 and Yb_2O_3 thin films for high-κ gate insulator application. *J. Electrochem. Soc.*, **150**, F134.

58 Jeon, S. and Hwang, H. (2002) Electrical and physical characteristics of $PrTi_xO_y$ for metal-oxide-semiconductor gate dielectric applications. *Appl. Phys. Lett.*, **81**, 4856.

59 Pan, T.M., Yen, L.C., and Wu, X.C. (2010) A comparative study on the structural properties and electrical characteristics of thin $HoTi_xO_y$, $TmTi_xO_y$ and $YbTi_xO_y$ dielectrics. *Semicond. Sci. Tech.*, **25**, 055015.

60 Pan, T.M., Hou, S.J., and Wang, C.H. (2008) Effects of nitrogen content on the structure and electrical properties of high-κ NdO_xN_y gate dielectrics. *J. Appl. Phys.*, **103**, 124105.

61 Green, M.L., Brasen, D., Evans-Lutterodt, K.W., Feldman, L.C., Krisch, K., Lennard, W., Tang, H.T., Manchanda, L., and Tang, M.T. (1994) Rapid thermal oxidation of silicon in N_2O between 800 and 1200 °C: incorporated nitrogen and interfacial roughness. *Appl. Phys. Lett.*, **65**, 848.

62 Botton, G.A., Gupta, J.A., Landheer, D., McCaffrey, J.P., Sproule, G.I., and Graham, M.J. (2002) Electron energy loss spectroscopy of interfacial layer formation in Gd_2O_3 films deposited directly on Si(001). *J. Appl. Phys.*, **91**, 2921.

63 Czernohorsky, M., Bugiel, E., Ostena, H.J., Fissel, A., and Kirfel, O. (2006) Impact of oxygen supply during growth on the electrical properties of crystalline Gd_2O_3 thin films on Si(001). *Appl. Phys. Lett.*, **88**, 152905.

64 Choi, C.J., Jang, M.G., Kim, Y.Y., Jun, M.S., Kim, T.Y., and Song, M.H. (2007) Electrical and structural properties of high-*k* Er-silicate gate dielectric formed by interfacial reaction between Er and SiO_2 films. *Appl. Phys. Lett.*, **91**, 012903.

65 Chen, S., Zhu, Y.Y., Xu, R., Wu, Y.Q., Yang, X.J., Fan, Y.L., Lu, F., Jiang, Z.M., and Zou, J. (2006) Superior electrical properties of crystalline Er_2O_3 films epitaxially grown on Si substrates. *Appl. Phys. Lett.*, **88**, 222902.

66 Schroeder, T., Lee, T.L., Zegenhagen, J., Wenger, C., Zaumseil, P., and Mussig, H.J. (2004) Structure and thickness-dependent lattice parameters of ultrathin epitaxial Pr_2O_3 films on Si (001). *Appl. Phys. Lett.*, **85**, 1229.

67 Dimoulas, A., Vellianities, G., Mavrou, G., Apostolopoulos, G., Travols, A., Wiemer, C., Franciulli, M., and Rittersma, Z.M. (2004) $La_2Hf_2O_7$ high-κ gate dielectric grown directly on Si (001) by molecular-beam epitaxy. *Appl. Phys. Lett.*, **85**, 3205.

68 Wei, F., Tu, H., Wang, Y., Yue, S., and Du, J. (2008) Epitaxial $La_2Hf_2O_7$ thin films on Si(001) substrates grown by pulsed laser deposition for high-κ gate dielectrics. *Appl. Phys. Lett.*, **92**, 012901.

69 Matsumoto, M., Soda, K., Ichikawa, K., Tanaka, S., Taguchi, Y., Jouda, K., Aita, O., Tezuka, Y., and Shin, S. (1994) Resonant photoemission study of CeO_2. *Phys. Rev. B*, **50**, 11340.

70 Skorodumova, N.V., Simak, S.I., Lundqvist, B.I., Abrikosov, I.A., and Johansson, B. (2002) Quantum origin of the oxygen storage capability of ceria. *Phys. Rev. Lett.*, **89**, 166601.

71 Wuilloud, E., Delley, B., Schneider, W.-D., and Baer, Y. (1984) Spectroscopic evidence for localized and extended *f*-symmetry states in CeO_2. *Phys. Rev. Lett.*, **53**, 202.

72 Khare, A., Choudhary, R.J., Bapna, K., Phase, D.M., and Sanyal, S.P. (2010) Resonance photoemission studies of (111) oriented CeO_2 thin film grown on Si (100) substrate by pulsed laser deposition. *J. Appl. Phys.*, **108**, 103712.

73 Liu, J.P., Zaumseil, P., Bugiel, E., and Osten, H. (2001) Epitaxial growth of Pr_2O_3 on Si(111) and the observation of a hexagonal to cubic phase transition during postgrowth N_2 annealing. *Appl. Phys. Lett.*, **79**, 671.

74 Schroeder, T., Zaumseil, P., Weidner, G., Wenger, C., Dabrowski, J., Mussig, H.J., and Storck, P. (2006) On the epitaxy of twin-free cubic (111) praseodymium sesquioxide films on Si(111). *J. Appl. Phys.*, **99**, 014101.

75 Weisemoeller, T., Bertram, F., Gevers, S., Greuling, A., Deiter, C., Tobergte, H., Neumann, M., Wollschlager, J., Giussani, A., and Schroeder, T. (2009) Postdeposition annealing induced transition from hexagonal Pr_2O_3 to cubic PrO_2 films on Si(111). *J. Appl. Phys.*, **105**, 124108.

76 Schroder, D.K. (2006) *Semiconductor Material and Device Characterization*, 3rd edn, John Wiley & Sons, Inc., New York.

77 Wang, J.C., Chiao, S.H., Lee, C.L., Lei, T.F., Lin, Y.M., Wang, M.F., Chen, S.C., Yu, C.H., and Liang, M.S. (2002) A physical model for the hysteresis phenomenon of the ultrathin ZrO_2 film. *J. Appl. Phys.*, **92**, 3936.

78 Zhao, C., Witters, T., Brijs, B., Bender, H., Richard, O., Caymax, M., Heeg, T., Schubert, J., Afanas'ev, V.V., Stesmans, A., and Schlom, D.G. (2005) Ternary rare-earth metal oxide high-κ layers on silicon oxide. *Appl. Phys. Lett.*, **86**, 132903.

79 Wagner, M., Heeg, T., Schubert, J., Lenk, St., Mantl, S., Zhao, C., Caymax, M., and De Gent, S. (2006) Gadolinium scandate thin films as an alternative gate dielectric prepared by electron beam evaporation. *Appl. Phys. Lett.*, **88**, 172901.

80 Lopes, J.M.J., Littmark, U., Roeckerath, M., Lenk, St., Schubert, J., Mantl, S., and Besmehn, A. (2007) Effects of annealing on the electrical and interfacial properties of amorphous lanthanum scandate high-κ films prepared by molecular beam deposition. *J. Appl. Phys.*, **101**, 104109.

81 Sun, Q.Q., Laha, A., Ding, S.J., Zhang, D.W., Osten, H.J., and Fissel, A. (2008) Effective passivation of slow interface states at the interface of single crystalline Gd_2O_3 and Si(100). *Appl. Phys. Lett.*, **92**, 152908.

82 Sun, Q.Q., Laha, A., Ding, S.J., Zhang, D.W., Osten, H.J., and Fissel, A. (2008) Observation of near interface oxide traps in single crystalline Nd_2O_3 on Si(111) by quasistatic C–V method. *Appl. Phys. Lett.*, **93**, 083509.

83 Lamb, D. (1967) *Electrical Conduction Mechanisms in Thin Insulating Films*, Methuen.

84 Lambert, M.A. and Mark, P. (1970) *Current Injection in Solids*, Academic Press.

85 Simmons, J.G. (1970) Chapter 14, in *Handbook of Thin Film Technology* (eds L. Maissel and R. Glang), McGraw-Hill.

86 O'Dwyer, J.J. (1973) *The Theory of Electrical Conduction and Breakdown in Solid Dielectrics*, Clarendon Press, Oxford.

87 Mott, N.F. and Davis, E.A. (1979) *Electronic Processes in Non-Crystalline Materials*, Oxford University Press.

88 Kao, K.C. and Hwang, W. (1981) *Electrical Transport in Solids*, Pergamon Press.

89 Hesto, P. (1986) The nature of electronic conduction in thin insulating films, in *Instabilities in Silicon Devices: Silicon*

Passivation and Related Instabilities, 1th edn (eds G. Barbottin and A. Vapaille), Elsevier Science, North Holland.

90 Hamann, C., Burghardt, H., and Frauenheim, T. (1988) *Electrical Conduction Mechanisms in Solids*, VEB Deutscher Verlag der Wissenschaften, Berlin.

91 Kao, K.C. (2004) *Dielectric Phenomena in Solids*, Academic Press.

92 Lee, J.M., Chiu, F.C., and Juan, P.C. (2009) The application of high-dielectric-constant and ferroelectric thin films in integrated circuit technology, in *Handbook of Nanoceramics and Their Based Nanodevices*, vol. 4 (eds T.Y. Tseng and H.S. Nalwa), American Scientific Publishers, California.

93 Chiu, F.C. and Lai, C.M. (2010) Optical and electrical characterizations of cerium oxide thin films. *J. Phys. D Appl. Phys.*, **43**, 075104.

94 Miranda, E., Molina, J., Kim, Y., and Iwai, H. (2006) Tunneling in sub-5nm La_2O_3 films deposited by e-beam evaporation. *J. Non-Cryst. Solids*, **352**, 92.

95 Lee, W.C. and Hu, C. (2001) Modeling CMOS tunneling currents through ultrathin gate oxide due to conduction- and valence-band electron and hole tunneling. *IEEE Trans. Electron. Dev.*, **48**, 1366.

96 Yeo, Y.C., King, T.J., and Hu, C. (2002) Direct tunneling leakage current and scalability of alternative gate dielectrics. *Appl. Phys. Lett.*, **81**, 2091.

97 Chiu, F.C., Lee, C.Y., and Pan, T.M. (2009) Current conduction mechanisms in Pr_2O_3/oxynitride laminated gate dielectrics. *J. Appl. Phys.*, **105**, 074103.

98 Chiu, F.C., Chou, H.W., and Lee, J.M. (2005) Electrical conduction mechanisms of metal/La_2O_3/Si structure. *J. Appl. Phys.*, **97**, 103503.

99 Kundu, T.K. and Lee, J.Y.M. (2000) Thickness-dependent electrical properties of Pb(Zr, Ti)O_3 thin film capacitors for memory device applications. *J. Electrochem. Soc.*, **147**, 326.

100 Schlom, D.G. and Haeni, J.H. (2002) A thermodynamic approach to selecting alternative gate dielectrics. *Mater. Res. Soc.*, **27**, 198.

101 Gutowski, M., Jaffe, J.E., Liu, C.L., Stoker, M., Hegde, R.I., Rai, R.S., and Tobin, P.J. (2002) Thermodynamic stability of high-κ dielectric metal oxides ZrO_2 and HfO_2 in contact with Si and SiO_2. *Appl. Phys. Lett.*, **80**, 1897.

102 Copel, M., Gribelyuk, M., and Gusev, E. (2000) Structure and stability of ultrathin zirconium oxide layers on Si(001). *Appl. Phys. Lett.*, **76**, 436.

103 Ma, T., Campbell, S.A., Smith, R., Hoilien, N., He, B., Gladfelter, W.L., Hobbs, C., Buchanan, D., Taylor, C., Gribelyuk, M., Tiner, M., Coppel, M., and Lee, J.J. (2001) Group IVB metal oxides high permittivity gate insulators deposited from anhydrous metal nitrates. *IEEE Trans. Electron. Dev.*, **48**, 2348.

104 Chiu, F.C., Lin, Z.H., Chang, C.W., Wang, C.C., Chuang, K.F., Huang, C.Y., Lee, J.Y., and Hwang, H.L. (2005) Electron conduction mechanism and band diagram of sputter-deposited Al/ZrO2/Si structure. *J. Appl. Phys.*, **97**, 034506.

105 Chiu, F.C., Chou, H.W., and Lee, J.Y. (2005) Electrical conduction mechanisms of metal/La_2O_3/Si structure. *J. Appl. Phys.*, **97**, 103503.

106 Chiu, F.C., Lin, S.A., and Lee, J.Y. (2005) Electrical properties of metal–HfO_2–silicon system measured from metal–insulator–semiconductor capacitors and metal–insulator–semiconductor field-effect transistors using HfO_2 gate dielectric. *Microelectron. Reliab.*, **45**, 961.

107 Kang, C.S., Cho, H.J., Onishi, K., Nieh, R., Goplan, S., Krishnan, S., and Lee, J.C. (2002) Improved thermal stability and device performance of ultra-thin (EOT < 10 Å) gate dielectric MOSFETs by using hafnium oxynitride (HfO_xN_y). Symposium on VLSI Technology, p. 146.

108 Choi, R., Kang, C.S., Lee, B.H., Onishi, K., Nieh, R., Gopalan, S., Dharmarajan, E., and Lee, J.C. (2001) High-quality ultra-thin HfO_2 gate dielectric MOSFETs with TaN electrode and nitridation surface preparation. Symposium on VLSI Technology, p. 15.

109 Lee, C.H., Luan, H.F., Bai, W.P., Lee, S.J., Jeon, T.S., Senzaki, Y., Roberts, D., and Kwong., D.L. (2000) MOS characteristics of ultra thin rapid thermal CVD ZrO_2 and Zr silicate gate dielectrics. IEEE Technical Digest of International Electron Devices Meeting, p. 27.

110 Onishi, K., Kang, L., Choi, R., Dharmarajan, E., Gopalan, S., Jeon, Y., Kang, C.S., Lee, B.H., Nieh, R., and Lee, J.C. (2001) Dopant penetration effects on polysilicon gate HfO_2 MOSFET's. Symposium on VLSI Technology, p. 131.

111 Zhu, W., Ma, T.P., Tamagawa, T., Di, Y., Kim, J., Carruthers, R., Gibson, M., and Furukawa, T. (2001) HfO_2 and HfAlO for CMOS: thermal stability and current transport. IEEE Technical Digest of International Electron Devices Meeting, p. 463.

112 Ma, Y., Ono, Y., Stecker, L., Evans, D.R., and Hsu, S.T. (1999) Zirconium oxide based gate dielectrics with equivalent oxide thickness of less than 1.0nm and performance of submicron MOSFET using a nitride gate replacement process. IEEE Technical Digest of International Electron Devices Meeting, p. 149.

113 Kang, C.S., Cho, H.J., Onishi, K., Choi, R., Kim, Y.H., Nieh, R., Han, J., Krishnan, S., Shahriar, A., and Lee, J.C. (2002) Nitrogen concentration effects and performance improvement of MOSFETs using thermally stable HfO_xN_y gate dielectrics. IEEE Technical Digest of International Electron Devices Meeting, p. 865.

114 Koyama, M., Suguro, K., Yoshiki, M., Kamimuta, Y., Koike, M., Ohse, M., Hongo, C., and Nishiyama, A. (2001) Thermally stable ultra-thin nitrogen incorporated ZrO_2 gate dielectric prepared by low temperature oxidation of ZrN. IEEE Technical Digest of International Electron Devices Meeting, p. 459.

115 Koyama, M., Kaneko, A., Ino, T., Koike, M., Kamato, Y., Iijima, R., Kamimuta, Y., Takashirma, A., Suzuki, M., Hongo, C., Inumiya, S., Takayanagi, M., and Nishiyama, A. (2002) Effects of nitrogen in HfSiON gate dielectric on the electrical and thermal characteristics. IEEE Technical Digest of International Electron Devices Meeting, p. 849.

116 Cho, H.J., Kang, C.S., Onishi, K., Gopalan, S., Nieh, R., Choi, R., Dharmarajan, E., and Lee, J.C. (2001) Novel nitrogen profile engineering for improved $TaN/HfO_2/Si$ MOSFET performance. IEEE Technical Digest of International Electron Devices Meeting, p. 655.

117 Zhu, W.J., Tamagawa, T., Gibson, M., Furukawa, T., and Ma, T.P. (2002) Effect of Al inclusion in HfO_2 on the physical and electrical properties of the dielectrics. *IEEE Electron. Device Lett.*, **23**, 649.

118 Morisaki, Y., Aoyama, T., Sugita, Y., Irino, K., Sugii, T., and Nakamura, T. (2002) Ultra-thin ($T_{eff}^{inv} = 1.7$nm) poly-Si-gated $SiN/HfO_2/SiON$ high-*k* stack dielectrics with high thermal stability (1050 °C). IEEE Technical Digest of International Electron Devices Meeting, p. 861.

119 Cheng, B., Cao, M., Rao, R., Inani, A., Voorde, P.V., Greene, W.M., Stork, J.M.C., Yu, Z., Zeitzoff, P.M., and Woo, J.C.S. (1999) The impact of high-κ gate dielectrics and metal gate electrodes on sub-100nm MOSFETs. *IEEE Trans. Electron. Dev.*, **46**, 1537.

120 Chiu, F.C., Chen, S.Y., Chen, C.H., Chen, H.W., Huang, H.S., and Hwang, H.L. (2009) Interfacial and electrical characterization in metal–oxide–semiconductor field-effect transistors with CeO_2 gate dielectric. *Jpn. J. Appl. Phys.*, **48**, 04C014.

121 Yamamoto, T., Momida, H., Uda, T., and Ohno, T. (2005) First-principles study of dielectric properties of cerium oxide. *Thin Solid Films*, **486**, 136.

122 Skorodumova, N.V., Ahuja, R., Simak, S.I., Abrikosov, I.A., Johansson, B., and Lundqvist, B.I. (2001) Electronic, bonding, and optical properties of CeO_2 and Ce_2O_3 from first principles. *Phys. Rev. B*, **64**, 115108.

123 Green, M.L., Gusev, E.P., Degraeve, R., and Garfunkel, E.L. (2001) Ultrathin

(<4nm) SiO_2 and Si−O−N gate dielectric layers for silicon microelectronics: understanding the processing, structure, and physical and electrical limits. *J. Appl. Phys.*, **90**, 2057.

124 Onishi, K., Kang, C.S., Choi, R., Cho, H.J., Gopalan, S., Nieh, R., Krishnan, S., and Lee, J.C. (2002) Effects of high-temperature forming gas anneal on HfO_2 MOSFET performance. Symposium on VLSI Technology, p. 22.

125 Choi, R., Onishi, K., Kang, C.S., Gopalan, S., Nieh, R., Kim, Y.H., Han, J.H., Krishnan, S., Cho, H.J., Shahriar, A., and Lee, J.C. (2002) Fabrication of high quality ultra-thin HfO_2 gate dielectric MOSFETs using deuterium anneal. IEEE Technical Digest of International Electron Devices Meeting, p. 613.

126 Kimizuka, N., Yamaguchi, K., Imai, K., Iizuka, T., Liu, C.T., Keller, R.C., and Horiuchi, T. (2000) NBTI enhancement by nitrogen incorporation into ultrathin gate oxide for 0.10-μm gate CMOS generation. Symposium on VLSI Technology, p. 92.

127 Hess, K., Kizilyalli, I.C., and Lyding, J.W. (1998) Giant isotope effect in hot electron degradation of metal oxide silicon devices. *IEEE Trans. Electron. Devices*, **45**, 406.

128 Robertson, J. (2002) Electronic structure and band offsets of high-dielectric-constant gate oxides. *Mater. Res. Soc.*, **27**, 217.

129 Iwai, H., Ohmi, S., Akama, S., Kikuchi, A., Kashiwagi, I., Jaguchi, J., Yamamoto, H., Tonotani, J., Kim, Y., Ueda, I., Kuriyama, A., and Yoshihara, Y. (2002). Advanced gate dielectric materials for sub-100nm CMOS. IEEE Technical Digest of International Electron Devices Meeting, p. 625.

130 Sze, S.M. (1981) *Physics of Semiconductor Devices*, John Wiley & Sons, Inc., New York.

131 Polishchuk, I., Ranade, P., King, T.J., and Hu, C. (2001) Dual work function metal gate CMOS technology using metal interdiffusion. *IEEE Electron. Device Lett.*, **22**, 444.

132 Lin, R., Lu, Q., Ranade, P., King, T.J., and Hu, C. (2002) An adjustable work function technology using Mo gate for CMOS devices. *IEEE Electron. Device Lett.*, **23**, 49.

133 Wakabayashi, H., Saito, Y., Takeuchi, K., Mogami, T., and Kunio, T. (2001) A dual-metal gate CMOS technology using nitrogen-concentration-controlled TiN_x film. *IEEE Trans. Electron. Devices*, **48**, 2363.

134 Misra, V., Lucovsky, G., and Parsons, G. (2002) Issues in high-*k* gate stack interfaces. *Mater. Res. Soc.*, **27**, 212.

135 Misra, V., Heuss, G.P., and Zhong, H. (2001) Use of metal–oxide–semiconductor capacitors to detect interactions of Hf and Zr gate electrodes with SiO_2 and ZrO_2. *Appl. Phys. Lett.*, **78**, 4166.

136 Lee, B.H., Choi, R., Kang, L., Gopalan, S., Nieh, R., Onishi, K., Jeon, Y., Qi, W.J., Kang, C., and Lee, J.C. (2000). Characteristics of TaN gate MOSFET with ultrathin hafnium oxide (8 Å–12 Å). IEEE Technical Digest of International Electron Devices Meeting, p. 39.

137 Narayanan, V., Callegari, A., McFeely, F.R., Nakamura, K., Jamison, P., Zafar, S., Cartier, E., Steegen, A., Ku, V., Nguyen, P., Milkove, K., Cabral, C., Jr., Gribelyuk, M., Wajda, C., Kawano, Y., Lacey, D., Li, Y., Sikorski, E., Duch, E., Ng, H., Wann, C., Jammy, R., Ieong, M., and Shahidi, G. (2004). Dual work function metal gate CMOS using CVD metal electrodes. Symposium on VLSI Technology, p. 192.

138 Zhong, H., Heuss, G., Misra, V., Luan, H., Lee, C.H., and Kwong, D.L. (2001) Characterization of RuO_2 electrodes on Zr silicate and ZrO_2 dielectrics. *Appl. Phys. Lett.*, **78**, 1134.

139 Zhong, H., Heuss, G., and Misra, V. (2000) Electrical properties of RuO_2 gate electrodes for dual metal gate Si-CMOS. *IEEE Electron. Device Lett.*, **21**, 593.

140 Cabral, C., Jr., Kedzierski, J., Linder, B., Zafar, S., Narayanan, V., Fang, S., Steegen, A., Kozlowski, P., Carruthers, R., and Jammy, R. (2004) Dual workfunction fully silicided metal gates. Symposium on VLSI Technology, p. 184.

141 Gusev, E.P., Cabral, C. Jr., Linder, B.P., Kim, Y.H., Maitra, K., Cartier, E., Nayfeh, H., Amos, R., Biery, G., Bojarczuk, N., Callegari, A., Carruthers, R., Cohen, S.A., Copel, M., Fang, S., Frank, M., Guha, S., Gribelyuk, M., Jamison, P., Jammy, R., Ieong, M., Kedzierski, J., Kozlowski, P., Ku, V., Lacey, D., LaTulipe, D., Narayanan, V., Ng, H., Nguyen, P., Newbury, J., Paruchuri, V., Rengarajan, R., Shahidi, G., Steegen, A., Steen, M., Zafar, S., and Zhang, Y. (2004) Advanced gate stacks with fully silicided (FUSI) gates and high-κ dielectrics: enhanced performance at reduced gate leakage. IEEE Technical Digest of International Electron Devices Meeting, p. 79.

142 Qin, M., Poon, V.M.C., and Ho, S.C.H. (2001) Investigation of polycrystalline nickel silicide films as a gate material. *J. Electrochem. Soc.*, **148**, G271.

143 Kedzierski, J., Nowak, E., Kanarsky, T., Zhang, Y., Boyd, D., Carruthers, R., Cabral, C., Amos, R., Lavoie, C., Roy, R., Newbury, J., Sullivan, E., Benedict, J., Saunders, P., Wong, K., Canaperi, D., Krishnan, M., Lee, K.L., Rainey, B.A., Fried, D., Cottrell, P., Philip Wong, H.S., Ieong, M., and Haensch, W. (2002). Metal-gate finFET and fully-depleted SOI devices using total gate silicidation. IEEE Technical Digest of International Electron Devices Meeting, p. 247.

144 Tavel, B., Skotnicki, T., Pares, G., Carriere, N., Rivoire, M., Leverd, F., Julien, C., Torres, J., and Pantel, R. (2001) Totally silicided ($CoSi_2$) polysilicon: a novel approach to very low-resistive gate ($\sim$2 Ω/ϒ without metal CMP nor etching. IEEE Technical Digest of International Electron Devices Meeting, p. 825.

145 Wolf, S. (2002) *Silicon Processing for the VLSI Era. Volume 4: Deep-Submicron Process Technology*, Lattice Press, Sunset Beach CA, p. 146, 170.

146 Takahashi, K., Manabe, K., Ikarashi, T., Ikarashi, N., Hase, T., Yoshihara, T., Watanabe, H., Tatsumi, T., and Mochizuki, Y. (2004) Dual workfunction Ni-silicide/HfSiON gate stacks by phase-controlled full-silicidation (PC-FUSI) technique for 45 nm-node LSTP and LOP devices. IEEE Technical Digest of International Electron Devices Meeting, p. 91.

147 Iijima, T., Nishiyama, A., Ushiku, Y., Ohguro, T., Kunishima, I., Suguro, K., and Iwai, H. (1992) A novel selective Ni_3Si contact plug technique for deep-submicron ULSIs. Symposium on VLSI Technology, p. 70.

148 Maszara, W.P., Krivokapic, Z., King, P., Goo, J.S., and Lin, M.R. (2002) Transistors with dual work function metal gates by single full silicidation (FUSI) of polysilicon gates. IEEE Technical Digest of International Electron Devices Meeting, p. 367.

149 Kedzierski, J., Boyd, D., Zhang, Y., Steen, M., Jamin, F.F., Ieong, M., and Haensch, W. (2003) Issues in NiSi-gated FDSOI device integration. IEEE Technical Digest of International Electron Devices Meeting, p. 441.

150 Aime, D., Froment, B., Cacho, F., Carron, V., Descombes, S., Morand, Y., Emonet, N., Wacquant, F., Farjot, T., Jullian, S., Laviron, C., Juhel, M., Pantel, R., Molins, R., Delille, D., Halimaoui, A., Bensahel, D., and Souifi, A. (2004). Work function tuning through dopant scanning and related effects in Ni fully silicided gate for sub-45nm nodes CMOS. IEEE Technical Digest of International Electron Devices Meeting, p. 87.

151 Lin, C.Y., Ma, M.W., Chin, A., Yeo, Y.C., Zhu, C., Li, M.F., and Kwong, D.L. (2003) Fully silicided NiSi gate on La_2O_3 MOSFETs. *IEEE Electron. Device Lett.*, **24**, 348.

152 Schaeffer, J.K., Capasso, C., Fonseca, L.R.C., Samavedam, S., Gilmer, D.C., Liang, Y., Kalpat, S., Adetutu, B., Tseng, H.H., Shiho, Y., Demkov, A., Hegde, R., Taylor, W.J., Gregory, R., Jiang, J., Luckowski, E., Raymond, M.V., Moore, K., Triyoso, D., Roan, D., White, Jr., B.E., and Tobin,P.J. (2004) Challenges for the integration of metal gate electrodes. IEEE

Technical Digest of International Electron Devices Meeting, p. 287.

153 Tseng, H.H., Capasso, C.C., Schaeffer, J.K., Hebert, E.A., Tobin, P.J., Gilmer, D.C., Triyoso, D., Ramon, M.E., Kalpat, S., Luckowski, E., Taylor, W.J., Yeon, Y., Adetutu, O., Hegde, R.I., Noble, R., Jahanbani, M., El Chemali, C., and While, B.E. (2004) Improved short channel device characteristics with stress relieved pre-oxide (SRPO) and a novel tantalum carbon alloy metal gate/HfO_2 stack. IEEE Technical Digest of International Electron Devices Meeting, 2004, p. 821.

154 Hou, Y.T., Yen, F.Y., Hsu, P.F., Chang, V.S., Lim, P.S., Hung, C.L., Yao, L.G., Jiang, J.C., Lin, H.J., Jin, Y., Jang, S.M., Tao, H.J., Chen, S.C., and Liang, M.S. (2005) High performance tantalum carbide metal gate stacks for nMOSFET application. IEEE Technical Digest of International Electron Devices Meeting, Session 31.

155 Michaelson, H.B. (1977) The work function of the elements and its periodicity. *J. Appl. Phys.*, **48**, 4729.

156 Yeo, Y.C., Ranade, P., King, T.J., and Hu, C. (2002) Effects of high-κ gate dielectric materials on metal and silicon gate workfunctions. *IEEE Electron. Device Lett.*, **23**, 342.

157 Yeo, Y.C., King, T.J., and Hu, C. (2002) Metal-dielectric band alignment and its implications for metal gate complementary metal-oxide-semiconductor technology. *J. Appl. Phys.*, **92**, 7266.

158 Hobbs, C., Fonseca, L., Dhandapani, V., Samavedam, S., Taylor, B., Grant, J., Dip, L., Triyoso, D., Hegde, R., Gilmer, D., Garcia, R., Roan, D., Lovejoy, L., Rai, R., Tseng, L.H., White, B., and Tobin, P. (2003). Fermi level pinning at the polySi/metal oxide interface. Symposium on VLSI Technology, p. 9.

159 Samavedam, S.B., La, L.B., Tobin, P.J., White, B., Hobbs, C., Fonseca, L.R., Demkov, A.A., Schaeffer, J., Luckowski, E., Martinez, A., Raymond, M., Triyoso, D., Roan, D., Dhandapani, V., Garcia, R., Anderson, S.G.H., Moore, K., Tseng, H.H., Capasso, C., Adettu, O., Gilmer, D.C., Taylor, W.J., Hegde, R., and Grant, J. (2003). Fermi level pinning with sub-monolayer MeO_x and metal gates. IEEE Technical Digest of International Electron Devices Meeting, p. 307.

160 Yu, H.Y., Ren, Chi., Yeo, Y.-C., Kang, J.F., Wang, X.P., Ma, H.H.H., Li, M.F., Chan, D.S.H., and Kwong, D.L. (2004) Fermi pinning-induced thermal instability of metal-gate work functions. *IEEE Electron. Device Lett.*, **25**, 337.

161 Joo, M.S., Cho, B.J., Balasubramanian, N., and Kwong, D.L. (2004) Thermal instability of effective work function in metal/high-κ stack and its material dependence. *IEEE Electron. Device Lett.*, **25**, 716.

162 Hobbs, C.C., Fonseca, L.R.C., Knizhnik, A., Dhandapani, V., Samavedam, S.B., Taylor, W.J., Grant, J.M., Dip, L.G., Triyoso, D.H., Hegde, R.I., Gilmer, D.C., Garcia, R., Roan, D., Luke Lovejoy, M., Rai, R.S., Hebert, E.A., Tseng, H.H., Anderson, S.G.H., White, B.E., and Tobin, P.J. (2004) Fermi-level pinning at the polysilicon/metal oxide interface. Part I. *IEEE Trans. Electron. Devices*, **51**, 971.

163 Schottky, W. (1940) Discrepancies in Ohm's laws in semiconductors. *Physik. Z.*, **41**, 570.

164 Bardeen, J. (1947) Surface states and rectification at a metal semi-conductor contact. *Phys. Rev.*, **71**, 717.

165 Heine, V. (1965) Theory of surface states. *Phys. Rev.*, **138**, A1689.

166 Louie, S.G. and Cohen, M.L. (1976) Electronic structure of a metal-semiconductor interface. *Phys. Rev. B*, **13**, 2461.

167 Tersoff, J. (1984) Theory of semiconductor heterojunctions: the role of quantum dipoles. *Phys. Rev. B*, **30**, 4874.

168 Monch, W. (1987) Role of virtual gap states and defects in metal-semiconductor contacts. *Phys. Rev. Lett.*, **58**, 1260.

169 Godet, C. (2002) Variable range hopping revisited: the case of an exponential distribution of localized states. *J. Non-Cryst. Solids*, **299–302**, 333.

170 Hobbs, C.C., Fonseca, L.R.C., Knizhnik, A., Dhandapani, V., Samavedam, S.B., Taylor, W.J., Grant, J.M., Dip, L.G., Triyoso, D.H., Hegde, R.I., Gilmer, D.C., Garcia, R., Roan, D., Lovejoy, M.L., Rai, R.S., Hebert, E.A., Tseng, H.-H., Anderson, S.G.H., White, B.E., and Tobin, P.J. (2004). Fermi-level pinning at the polysilicon/metal-oxide interface. Part II. *IEEE Trans. Electron. Devices,* **51**, 978.

171 Kedzierski, J., Boyd, D., Ronsheim, P., Zafar, S., Newbury, J., Ott, J., Cabral, C., Jr., Ieong, M., and Haensch, W. (2003) Threshold voltage control in NiSi-gated MOSFETs through silicidation induced impurity segregation (SIIS). IEEE Technical Digest of International Electron Devices Meeting, p. 315.

172 Nabatame, T., Kadoshima, M., Iwamoto, K., Mise, N., Migita, S., Ohno, M., Ota, H., Yasuda, N., Ogawa, A., Tominaga, K., Satake, H., and Toriumi, A. (2004). Partial silicides technology for tunable work function electrodes on high-*k* gate dielectrics – Fermi level pinning controlled $PtSi_x$ for $HfO_x(N)$ pMOSFET. IEEE Technical Digest of International Electron Devices Meeting, p. 83.

173 Anil, K.G., Veloso, A., Kubicek, S., Schram, T., Augendre, E., de Marneffe, J.F., Devriendt, K., Lauwers, A., Brus, S., Henson, K., and Biesemans, S. (2004). Demonstration of fully Ni-silicided metal gates on HfO_2 based high-*k* gate dielectrics as a candidate for low power applications. Symposium on VLSI Technology, p. 190.

174 Park, C.S., Cho, B.J., and Kwong, D.L. (2004) MOS characteristics of substituted Al gate on high-i dielectric. *IEEE Electron. Device Lett.*, **25**, 725.

175 Park, C.S., Cho, B.J., Tang, L.J., and Kwong, D.L. (2004) Substituted aluminum metal gate on high-*k* dielectric for low work-function and Fermi-level pinning free. IEEE Technical Digest of International Electron Devices Meeting, p. 299.

176 Park, C.S. and Cho, B.J. (2005) Dopant-free FUSI Pt_xSi metal gate for high work function and reduced Fermi-level pinning. *IEEE Electron. Device Lett.*, **26**, 796.

177 Datta, S., Dewey, G., Doczy, M., Doyle, B.S., Jin, B., Kavalieros, J., Kotlyar, R., Metz, M., Zelick, N., and Chau, R. (2003) High mobility Si/SiGe strained channel MOS transistors with HfO_2/TiN gate stack. IEEE Technical Digest of International Electron Devices Meeting, p. 653.

178 Ota, K., Sugihara, K., Sayama, H., Uchida, T., Oda, H., Eimori, T., Morimoto, H., and Inoue, Y. (2002) Novel locally strained channel technique for high performance 55nm CMOS. IEEE Technical Digest of International Electron Devices Meeting, p. 27.

179 Shimizu, A., Hachimine, K., Ohki, N., Ohta, H., Koguchi, M., Nonaka, Y., Sato, H., and Ootsuka, F. (2001) Local mechanical-stress control (LMC): a new technique for CMOS-performance enhancement. IEEE Technical Digest of International Electron Devices Meeting, p. 433.

180 Lee, B.H., Choi, R., Sim, J.H., Krishnan, S.A., Peterson, J.J., Brown, G.A., and Bersuker, G. (2005) Validity of constant voltage stress based reliability assessment of high-κ devices. *IEEE Trans. Device Mater. Reliab.*, **5**, 20.

181 Young, C.D., Choi, R., Sim, J.H., Lee, B.H., Zeitzoff, P., Zhao, Y., Matthews, K., Brown, G.A., and Bersuker, G. (2005) Interfacial layer dependence of $HfSi_xO_y$ gate stacks on V_t instability and charge trapping using ultra-short pulse *I–V* characterization. IEEE International Reliability Physics Symposium, p. 75.

182 Leroux, C., Mitard, J., Ghibaudo, G., Garros, X., Reimbold, G., Buillaumot, B., and Martin, F. (2004) IEEE Technical Digest of International Electron Devices Meeting, p. 737.

183 Bersuker, G., Sim, J.H., Young, C.D., Choi, R., Zeitzoff, P.M., Brown, G.A., Lee, B.H., and Murto, R.W. (2004) Effect of pre-existing defects on reliability assessment of high-*k* gate dielectrics. *Microelectron. Reliab.*, **44**, 1509.

184 Young, C.D., Bersuker, G., Brown, G.A., Lysaght, P., Zeitzoff, P., Murto1, R.W., and Huff, H.R. (2004) Charge trapping and device performance degradation in MOCVD hafnium-based gate dielectric stack structures. IEEE International Reliability Physics Symposium, p. 597.

185 Shanware, A., Visokay, M.R., Chambers, J.J., Rotondaro, A.L.P., McPherson, J., and Colombo, L. (2003) Characterization and comparison of the charge trapping in HfSiON and HfO_2 gate dielectrics. IEEE Technical Digest of International Electron Devices Meeting, p. 939.

186 Lee, B.H., Young, C.D., Choi, R., Sim, J.H., Bersuker, G., Kang, C.Y., Harris, R., Brown, G.A., Matthews, K., Song, S.C., Moumen, N., Barnett, J., Lysaght, P., Choi, K.S., Wen, H.C., Huffman, C., Alshareef, H., Majhi, P., Gopalan, S., Peterson, J., Kirsh, P., Li, H.J., Gutt, J., Gardner, M., Huff, H.R., Zeitzoff, P., Murto, R.W., Larson, L., and Ramiller, C. (2004). Intrinsic characteristics of high-k devices and implications of fast transient charging effects (FTCE). IEEE Technical Digest of International Electron Devices Meeting, p. 859.

187 Kerber, A., Cartier, E., Pantisano, L., Rosmeulen, M., Degraeve, R., Kauerauf, T., Groeseneken, G., Maes, H.E., and Schwalke, U. (2003) Characterization of the V_t-instability in SiO_2/HfO_2 gate dielectrics. IEEE International Reliability Physics Symposium, p. 41.

188 Kerber, A., Cartier, E., Pantisano, L., Degraeve, R., Kauerauf, T., Kim, Y., Hou, A., Groeseneken, G., Maes, H.E., and Schwalke, U. (2003) Origin of the threshold voltage instability in SiO_2/HfO_2 dual layer gate dielectrics. *IEEE Electron. Device Lett.*, **24**, 87.

189 Zafar, S., Callegari, A., Gusev, E., and Fischetti, M.V. (2003) Charge trapping related threshold voltage instabilities in high permittivity gate dielectric stacks. *J. Appl. Phys.*, **93**, 9298.

190 Onishi, K., Choi, R., Kang, C.S., Cho, H.J., Kim, Y.H., Nieh, R.E., Han, J., Krishnan, S.A., Akbar, M.S., and Lee, J.C. (2003) Bias-temperature instabilities of polysilicon gate HfO_2 MOSFETs. *IEEE Trans. Electron. Devices*, **50**, 1517.

191 Xu, Z., Pantisano, L., Kerber, A., Degraeve, R., Cartier, E., De Gendt, S., Heyns, M., and Groeseneken, G. (2004) A study of relaxation current in high-κ dielectric stacks. *IEEE Trans. Electron. Devices*, **51**, 402.

192 McPherson, J., Kim, J., Shanware, A., Mogul, H., and Rodriguesz, J. (2002) Proposed universal relationship between dielectric breakdown and dielectric constant. IEEE Technical Digest of International Electron Devices Meeting, p. 633.

193 McPherson, J.W. and Mogul, H.C. (1998) Underlying physics of the thermochemical E model in describing low-field time-dependent dielectric breakdown in SiO_2 thin films. *J. Appl. Phys.*, **84**, 1513.

194 McPherson, J.W., Khamankar, R.B., and Shanware, A. (2000) Complementary model for intrinsic time-dependent dielectric breakdown in SiO_2 dielectrics. *J. Appl. Phys.*, **88**, 5351.

195 Degraeve, R., Cartier, E., Kauerauf, T., Carter, R., Pantisano, L., Kerber, A., and Groeseneken, G. (2002) On the electrical characterization of high-k dielectrics. *Mater. Res. Soc.*, **27**, 222.

196 Stathis, J.H. and DiMaria, D.J. (1998) Reliability projection for ultra-thin oxides at low voltage. IEEE Technical Digest of International Electron Devices Meeting, p. 167.

197 Lee, S.J., Rhee, S.J., Clark, R., and Kwong, D.L. (2002) Reliability projection and polarity dependence of TDDB for ultra thin CVD HfO_2 gate dielectrics. Symposium on VLSI Technology, p. 78.

198 Degraeve, R., Groeseneken, G., Bellens, R., Depas, M., and Maes, H.E. (1995) A consistent model for the thickness dependence of intrinsic breakdown in ultra-thin oxides. IEEE Technical Digest of International Electron Devices Meeting, p. 863.

199 Degraeve, R., Groeseneken, G., Bellens, R., Ogier, J.L., Depas, M., Roussel, P.J., and Maes, H.E. (1998) New insights in the relation between electron trap generation and the statistical properties of oxide breakdown. *IEEE Trans. Electron. Devices*, **45**, 904.

200 Sune, J. (2001) New physics-based analytic approach to the thin-oxide breakdown statistics. *IEEE Electron. Device Lett.*, **22**, 296.

201 Chiu, F.C. (2010) Thickness and temperature dependence of dielectric reliability characteristics in cerium dioxide thin film. *IEEE Trans. Electron. Devices*, **57**, 2719.

202 Evangelou, E.K., Rahman, M.S., and Dimoulas, A. (2009) Correlation of charge buildup and stress-induced leakage current in cerium oxide films grown on Ge (100) substrates. *IEEE Trans. Electron. Devices*, **56**, 399.

203 Chatterjee, S., Kuo, Y., Lu, J., Tewg, J.Y., and Majhi, P. (2006) Electrical reliability aspects of HfO_2 high-*k* gate dielectrics with TaN metal gate electrodes under constant voltage stress. *Microelectron Reliab.*, **46**, 69.

204 Houssa, M., Stesmans, A., Naili, M., and Heyns, M.M. (2000) Charge trapping in very thin high-permittivity gate dielectric layers. *Appl. Phys. Lett.*, **77**, 1381.

205 Wu, E.Y. and Sune, J. (2005) Power-law voltage acceleration: a key element for ultra-thin gate oxide reliability. *Microelectron Reliab.*, **45**, 1809.

206 Chen, C.H., Chang, I.Y.K., Lee, J.Y.M., Chiu, F.C., Chiou, Y.K., and Wu, T.B. (2007) Reliability properties of metal-oxide-semiconductor capacitors using HfO_2 high-*k* dielectric. *Appl. Phys. Lett.*, **91**, 123507.

207 Kao, K.C. and Hwang, W. (1981) *Electrical Transport in Solids*, Pergamon, New York.

6
Hygroscopic Tolerance and Permittivity Enhancement of Lanthanum Oxide (La_2O_3) for High-*k* Gate Insulators

Yi Zhao

6.1
Introduction

Lanthanum oxide (La_2O_3) is a promising material to substitute SiO_2 as the high-*k* gate dielectric film because of its relatively high permittivity and a large bandgap of 6 eV [1–3]. However, it is well known that La_2O_3 is not very stable in air and is very hygroscopic, forming hydroxide [4, 5]. As a gate dielectric, it is inevitable to be involved with wet processes (water is used) and exposed to air in the conventional complementary metal oxide semiconductor (CMOS) process [6]. Therefore, before we consider the possibility of La_2O_3 film as a high-*k* gate dielectric, it is necessary to investigate the effects of moisture absorption on the properties of the La_2O_3 film. And if the moisture absorption can degrade the properties of La_2O_3 film as high-*k* gate dielectric, it will be very important to clarify the mechanisms of the moisture absorption to propose methods to stabilize La_2O_3 films in air by suppressing the moisture absorption. Furthermore, one very important reason for lanthanum oxide (La_2O_3) as a promising high-*k* gate dielectric to replace SiO_2 is its high permittivity. However, many low-permittivity La_2O_3 films were also reported [7–9]. The reasons for the scattering of the permittivity of La_2O_3 films are still not clear to us.

In this chapter, we first investigate the effects of the moisture absorption on the properties of La_2O_3 films in terms of XRD patterns, surface roughness, and electrical properties. The permittivity characteristics of La_2O_3 films are investigated, systematically, with and without the moisture absorption. So far, the reasons for the permittivity scattering of La_2O_3 films are not clarified yet. Therefore, it should be very important to understand the mechanisms of the moisture absorption and then find effective methods to enhance the hygroscopic tolerance of La_2O_3 films. Accordingly, the phenomenon of moisture absorption in high-permittivity (*k*) oxides was analyzed from the viewpoint of the thermodynamic process. On the base of the mechanism proposed for the moisture absorption phenomenon, methods to enhance the hygroscopic tolerance are proposed. As reported in the literatures, the permittivity of La_2O_3 film without

High-k Gate Dielectrics for CMOS Technology, First Edition. Edited by Gang He and Zhaoqi Sun.

moisture absorption still shows some low permittivity (~20). It indicates that the low permittivity should be the intrinsic property of La_2O_3 films.

6.2
Hygroscopic Phenomenon of La_2O_3 Films

La_2O_3 films were deposited on HF-last Si by sputtering the La_2O_3 target (provided by Kojundo Chemical, Japan) in argon ambient at room temperature and then were annealed at 600 °C in pure N_2 ambient for 30 s in the rapid thermal annealing (RTA) furnace. Figure 6.1 shows the sample preparation procedures. The physical thicknesses of La_2O_3 films were determined with the grazing incident X-ray reflectivity (GIXR) [10, 11] and spectroscopic ellipsometry (SE) measurements [12, 13]. The moisture absorption experiments were performed in the room air with a temperature of 24 °C and a relative humidity of 55%. The temperature and humidity were measured, synchronistically, with a sensor. As shown in Figure 6.1, two kinds of samples were prepared. One is the moisture-desorbed La_2O_3 film on silicon (type A) that was annealed at 400 °C in a high-vacuum (HV) chamber (10^{-6} Pa) to make lanthanum hydroxide decompose into La_2O_3 and H_2O [14]. The other is the moisture-absorbed sample (type B) that was exposed to air for different times (0, 6, and 12 h). Both samples were followed by 6 nm SiO_2 layer deposition before Au metal gate formation to prepare metal insulator semiconductor (MIS) capacitors to stop the moisture absorption. Type A sample was processed without any exposure to the air to prevent the moisture absorption. Figure 6.2 shows the XRD patterns of La_2O_3 films exposed to the air for different times. From the X-ray diffraction (XRD) pattern, it is found that La_2O_3 film with 0 h exposed to the air is

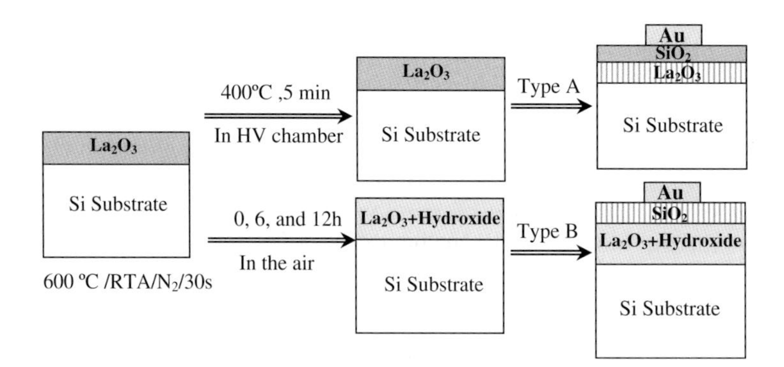

Figure 6.1 Illustration of two kinds of sample preparation procedures. One is the moisture desorbed La_2O_3 film (type A) that was annealed at 400 °C in a high-vacuum (HV) chamber (10^{-6} Pa) to make lanthanum hydroxide decompose into La_2O_3 and H_2O. The other is the moisture-absorbed sample (type B) that was exposed to the air with different period. Both samples were followed by 6 nm SiO_2 layer deposition. Type A sample was processed without any exposure to the air to prevent the moisture absorption.

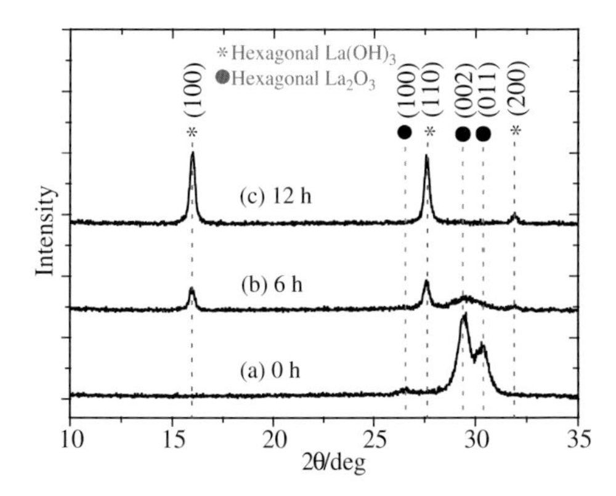

Figure 6.2 XRD patterns of La_2O_3 films on silicon after they were exposed to the air for (a) 0 h, (b) 6 h, and (c) 12 h.

polycrystallized in the hexagonal phase. After the exposure to air for 6 h, a couple of peaks attributed to $La(OH)_3$ appear, whereas the intensities of peaks attributed to the hexagonal La_2O_3 decrease. After the exposure to air for 12 h, strong $La(OH)_3$ phase peaks are found, while peaks of hexagonal La_2O_3 disappear completely. Therefore, we can conclude that the amount of hexagonal $La(OH)_3$ in the La_2O_3 film increased with the time exposed to the air.

6.2.1
Effect of Moisture Absorption on Surface Roughness of La_2O_3 Films

Figure 6.3 shows atomic force microscopy (AFM) images of La_2O_3 films after exposure to air for different times, 0, 6, and 12 h. It can be observed obviously that the surface roughness of La_2O_3 films increases with the time they are exposed to air, from 0.5 to 2.4 nm. In terms of the reason for the surface roughness enhancement after the moisture absorption, one possible reason is a nonuniform moisture absorption of the La_2O_3 film, followed by the nonuniform volume expansion of the film due to the moisture absorption. A slight film thickness increase after the moisture absorption can be observed (Figure 6.10), which indicates the volume expansion of the film. The original cause of volume expansion is the density difference between hexagonal $La(OH)_3$ and hexagonal La_2O_3. The density of hexagonal $La(OH)_3$ ($\varrho = 4.445\ g/cm^3$) [15, 16] is much smaller than that of hexagonal La_2O_3 ($\varrho = 6.565\ g/cm^3$) [17]. GIXR measurement results also indicated the film density (ϱ) decrease with the time exposed to air from the critical angle (θ_C) comparison ($\theta_{C0} > \theta_{C6} > \theta_{C12}$, θ_{C0}, θ_{C6}, and θ_{C12} are critical angles of GIXR spectra of La_2O_3 films on silicon after they were exposed to air for 0, 6, and 12 h, respectively) (Figure 6.4). The relation between the film density and the critical angle is given as follows [11]:

$$\varrho \sim \sqrt{\theta_c} \tag{6.1}$$

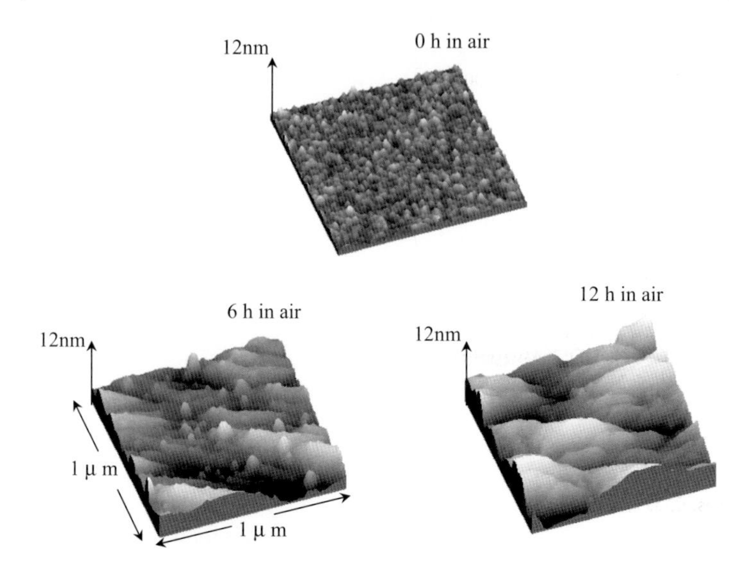

Figure 6.3 AFM images (1 µm × 1 µm) of La_2O_3 film surfaces after the exposure to the air for (a) 0 h, (b) 6 h, and (c) 12 h.

This equation indicates that a larger critical angel (θ_c) shows a larger density of the film.

6.2.2
Effect of Moisture Absorption on Electrical Properties of La_2O_3 Films

Figure 6.5 shows the *C–V* characteristics of La_2O_3 films after exposure to air for different times, 0, 3, and 7 days. In order to discuss the effect of the moisture absorption on *C–V* characteristics more conveniently, all curves were normalized to

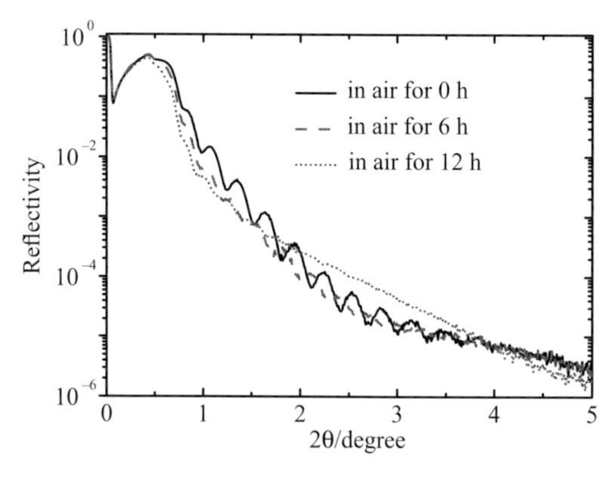

Figure 6.4 Grazing incident X-ray reflectivity spectra comparison of La_2O_3/Si sample after exposed to the air for 0, 6, and 12 h.

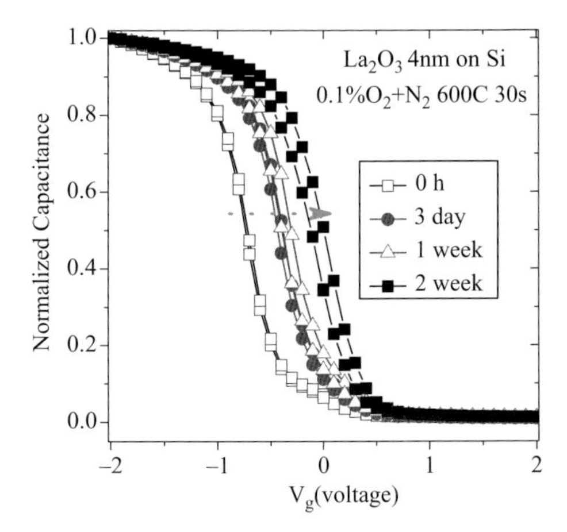

Figure 6.5 *C–V* characteristics of Au/La_2O_3/Si MIS capacitors. La_2O_3 films were exposed to the air for different times, 0, 3, 7, and 14 days.

C/C_{max}. Here, the flatband voltage (V_{FB}) and hysteresis change after the moisture absorption have been investigated.

The effect of the moisture absorption on the flatband voltage was summarized in Figure 6.6. It can be observed clearly that with the increase in the time exposure to air, the flatband voltage shifts toward the positive side. However, it was thought that moisture absorption could cause the negative flatband voltage shift [18]. The author of Ref. [8] explained it by considering the substitution of O^{-2} with OH^- after the moisture absorption. However, in our viewpoint, one O^{-2} ion could be substituted to form two OH^- ions after the moisture absorption due to the following reaction:

$$O^{2-} + H_2O \rightarrow 2OH^- \tag{6.2}$$

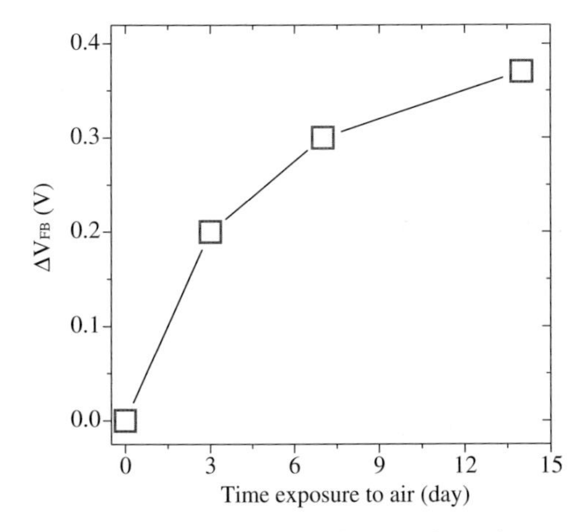

Figure 6.6 Flatband voltage shift (ΔV_{FB}) due to the moisture absorption for different times, 0, 3, 7, and 14 days.

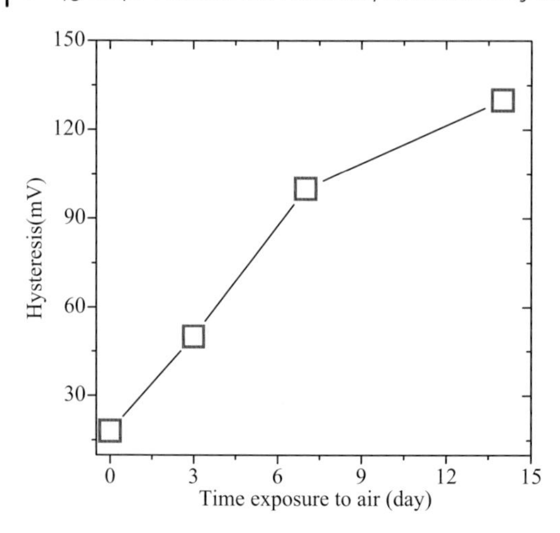

Figure 6.7 C–V curve hysteresis change due to the moisture absorption for different times, 0, 3, 7, and 14 days.

Therefore, we think that moisture absorption could not generate the positive charges and cause the negative flatband voltage shift. In our study, we observed the positive flatband voltage shift after the moisture absorption (Fig. 6.6). Although the reason is not very clear to us yet, the possible one is that the moisture absorption process just causes the formation of OH^- that contains negative charges and then induces the positive flatband voltage shift. More investigation is necessary to understand this phenomenon.

In terms of the effect of the moisture absorption on the hysteresis of *C–V* characteristic, as shown in Figure 6.7, the moisture absorption enhances the hysteresis of *C–V* curve of La_2O_3 film on silicon. Furthermore, the moisture absorption also enhances the leakage current of La_2O_3 film for several orders, as shown in Figure 6.8.

In conclusion, our experimental results show that the moisture absorption induces the formation of $La(OH)_3$ and could also enhance the surface roughness of La_2O_3 films on silicon. Thus, *in situ* gate electrode process might be needed for La_2O_3 CMOS application.

Beside of the hygroscopic phenomenon, carbon dioxide absorption phenomenon of La_2O_3 is also reported as high-*k* gate dielectric [19]. The following reaction is thought to possibly happen when La_2O_3 film is exposed to carbon dioxide or air:

$$La_2O_3 + 3CO_2 \rightarrow La_2(CO_3)_3 \tag{6.3}$$

Therefore, we also investigated carbon dioxide absorption phenomenon of La_2O_3 films. Figure 6.9 shows the XRD patterns of La_2O_3 film after exposure to CO_2 for different times. In the figure, no peaks of lanthanum carbonates appear even in CO_2 for 36 h. It indicates that compared to the moisture absorption carbon dioxide absorption is not a problem in our study.

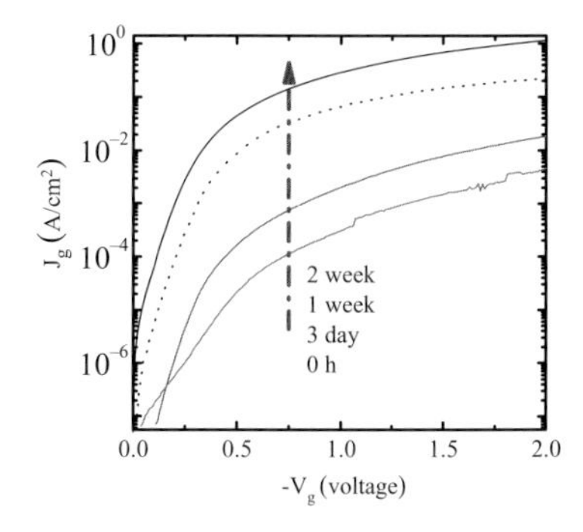

Figure 6.8 Leakage current enhancement Au/La_2O_3/Si MIS capacitors due to the moisture absorption.

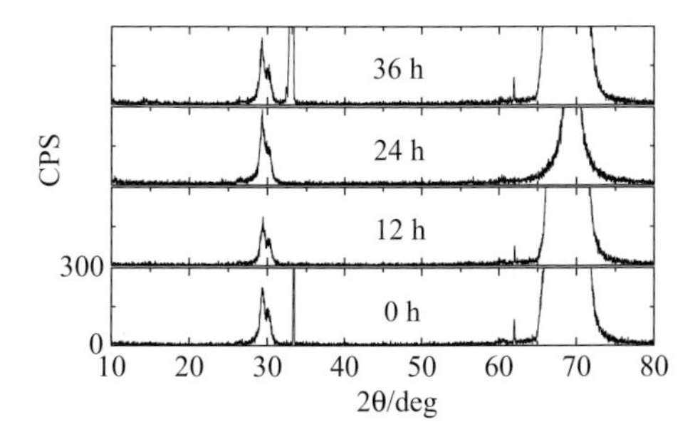

Figure 6.9 XRD patterns of La_2O_3 film on silicon with CO_2 exposure for different times, 0, 12, 24, and 36 h.

6.3
Low Permittivity Phenomenon of La_2O_3 Films

6.3.1
Moisture Absorption-Induced Permittivity Degradation of La_2O_3 Films

In terms of the reasons for the low permittivity of La_2O_3 films reported in literature, one very possible reason should be the moisture absorption due to the formation of the lanthanum hydroxide as discussed in the previous section. Figure 6.10 shows the CET (capacitance equivalent thickness) versus La_2O_3 film thickness plot for type B samples in Figure 6.1. The 0 h exposure to the air means that the sample was put in the sputtering chamber for SiO_2 layer deposition as quickly as possible after annealing in the RTA furnace. Also, the permittivity (k_{exp}, k-value obtained experimentally from the slope) of the La_2O_3 film exposed to air can be calculated from the slope of linear fitting to experimental CETs. It is observed that the permittivity of the film is degraded with the time the film is

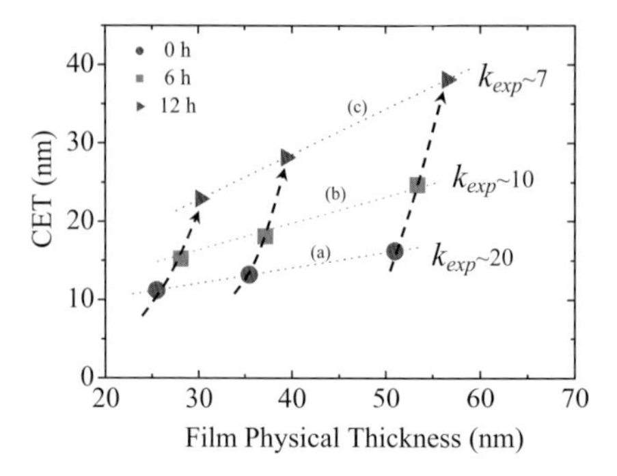

Figure 6.10 The relationship of CET to type B La_2O_3 physical thickness for $Au/SiO_2/La_2O_3/Si$ MIS capacitors. The sample was exposed to the air for (a) 0 h, (b) 6 h, and (c) 12 h before SiO_2 layer deposition.

exposed to air (Figure 6.10). The permittivity of La_2O_3 film in air for 0 h is about 20. And after the exposure to air for 12 h, the permittivity is degraded to only 7.

From the previous section, it has been concluded that the amount of hexagonal La $(OH)_3$ in the La_2O_3 film increased with the time the film is exposed to the air. Although there is no study on the permittivity of hexagonal $La(OH)_3$, we can estimate the permittivity of hexagonal $La(OH)_3$ on the basis of an additivity rule of the polarizability from Shannon's consideration [15]. From the *Clausius–Mossotti* relation, the dielectric constant is described by

$$k = (3\,V_m + 8\pi\alpha^T)/(3\,V_m - 4\pi\alpha^T), \tag{6.4}$$

where V_m and α^T denote molar volume and total polarizability, respectively. For hexagonal $La(OH)_3$, α^T is 12.81Å^3 from the Shannon's additivity rule [15] ($\alpha^T(La(OH)_3) = \alpha(La^{3+}) + 3\alpha(OH^{-1})$) and V_m is 71 Å^3 from Ref. [16]. With above values, we can estimate the permittivity of hexagonal $La(OH)_3$ that is about 10. This result indicates that hexagonal $La(OH)_3$ is with much lower permittivity compared to La_2O_3. Therefore, the effective permittivity of La_2O_3 film exposed to the air could be degraded. In fact, with the time exposed to air, Figure 6.10 shows the degradation of k_{exp} (k-value obtained experimentally from the slope), though it is necessary to take account of an inhomogeneity of the film due to the partial reaction of the La_2O_3 with the moisture as discussed below. From above discussion, the moisture absorption that causes the formation of low-permittivity lanthanum hydroxide should be a very possible reason for the scattering of the permittivity value of La_2O_3 films in reported in the literature [7–9], though details of the process are not mentioned in the literature.

From Figures 6.1 and 6.2, it could be concluded that with the time exposed to air, the amount of hexagonal $La(OH)_3$ in La_2O_3 film increases and then the density of the film is degraded. Therefore, the effect of the moisture absorption on the surface roughness and permittivity should be two big concerns in the hygroscopic

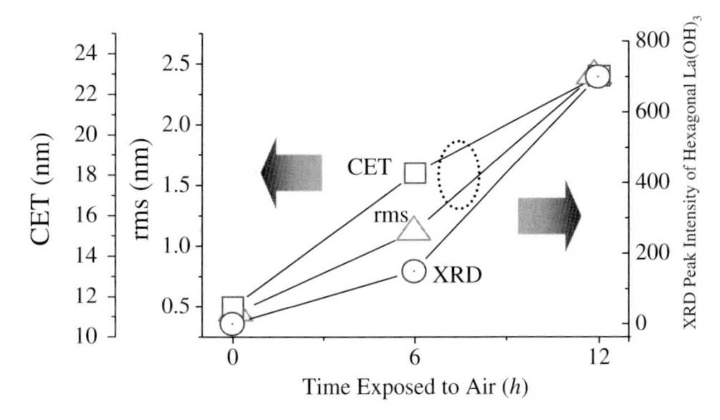

Figure 6.11 CET, rms, and hexagonal $La(OH)_3$ (100) XRD peak intensity as a function of the time exposed to the air of La_2O_3 film on silicon.

La_2O_3 film application. As summarized in Figure 6.11, all of CET, root mean square (rms) roughness, and XRD peak intensity of hexagonal $La(OH)_3$ in La_2O_3 film increase with the time the film is exposed to the air.

6.3.2 Permittivity of La_2O_3 Films without Moisture Absorption

According to the above discussion, it seems that the low permittivity of La_2O_3 reported can be attributed to the moisture absorption phenomenon. However, we also have to note that the permittivity of La_2O_3 film in air for 0 h was still a little low, about 20. This value is much lower than the reported highest one, 28, although the possibility of moisture absorption still cannot be excluded because the sample was exposed to air. To exclude the effect on the permittivity of La_2O_3 films and obtain the permittivity of La_2O_3 films without the moisture absorption, we used the *in situ* heating method in the HV chamber as shown in Figure 6.1. The La_2O_3 film (type A) was annealed at 400 °C in the HV chamber (10^{-6} Pa) to make lanthanum hydroxide decompose into La_2O_3 and H_2O and then followed by 6 nm SiO_2 layer deposition to prevent moisture absorption after it is taken out from the sputtering chamber for the electrode deposition.

Capacitance–voltage (*C*–*V*) measurements were performed for $Au/SiO_2/La_2O_3/Si/Al$ MIS capacitors with a frequency of 100 kHz. The capacitance equivalent thickness has a good linear relationship with La_2O_3 film thickness as shown in Figure 6.12, where the CET includes both La_2O_3 and SiO_2 films. Here, note that the thickness of capping SiO_2 layer was fixed (~6 nm) and the thickness of La_2O_3 film was varied. Then, the permittivity of La_2O_3 can be calculated from the slope to be about 24. This result indicates obviously that the permittivity of our La_2O_3 film is still a little low even when the film was prevented from moisture absorption. It means that the moisture absorption is not the only factor that contributes to the low permittivity of La_2O_3 films.

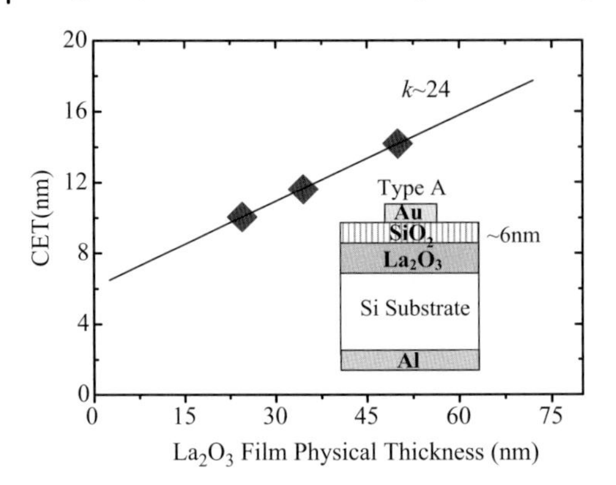

Figure 6.12 The relationship of CET to the La_2O_3 physical thickness for $Au/SiO_2/La_2O_3/Si/Al$ MIS capacitors.

6.4
Hygroscopic Tolerance Enhancement of La_2O_3 Films

6.4.1
Hygroscopic Tolerance Enhancement of La_2O_3 Films by Y_2O_3 Doping

As discussed earlier, the moisture absorption process of La_2O_3 films is related to the formation of OH ion. In the XRD pattern, peaks of hexagonal $La(OH)_3$ appeared after exposure to air for 6 h (Figure 6.2). Taking into consideration possible reactions of the moisture absorption of La_2O_3 films, one possible mechanism is the intrinsic reaction of La_2O_3 and H_2O.

Owing to the high ionicity of La_2O_3, it can react with H_2O directly as the following equations show:

$$La_2O_3 \rightarrow 2\,La^{3+} + 3\,O^{2-} \tag{6.5}$$

$$3\,H_2O + 3\,O^{2-} \rightarrow 6\,OH^- \tag{6.6}$$

This moisture absorption progress is mainly due to the small lattice energy of La_2O_3 that promotes the reaction [20]. Lattice energy (U) is the energy required to completely separate one molecule of a solid ionic compound into gaseous ions indicating the strength of the ionic bonds in an ionic lattice as shown below:

$$M_mX_n \Rightarrow mM^{n+} + nX^{m-} \tag{6.7}$$

It has been reported that the lattice energy of ionic oxides is antiproportional to the sum of the metal ion and oxygen ion radius [21]. In other words, the oxide with a larger metal ion radius shows a smaller lattice energy. In the case for rare-earth oxides, because lanthanum ion owns the largest radius, La_2O_3 shows the smallest lattice energy within rare-earth oxides [22].

Thus, to enhance the hygroscopic tolerance of La_2O_3 films, it is necessary to enhance the lattice energy of La_2O_3. It is well known that the amorphous film or poor crystallized film shows lower lattice energy than the well-crystallized film. Furthermore, the poor crystallized film is loose than the well-crystallized film. This makes water easier to diffuse into the film and react with La_2O_3. Therefore, one method to enhance the hygroscopic tolerance is to enhance the crystallinity of La_2O_3 film. As the poor crystallinity is La_2O_3's intrinsic property, to enhance the crystallinity of La_2O_3, doping of other elements or oxides is necessary. When we select oxides for doping, we have to consider the lattice energy, and larger lattice energy oxides are preferred. From the diagram of La_2O_3–Y_2O_3 system [23], a high melting point of $La_{2-x}Y_xO_3$ can be observed that indicates a low crystallization temperature of $La_{2-x}Y_xO_3$. On the other hand, Y_2O_3 shows a much lower crystallization temperature than La_2O_3, and it is very possible that $La_{2-x}Y_xO_3$ films could also have a low crystallization temperature or could be easily crystallized [24]. Furthermore, Y is in the same element group in the elements table as La and is the nearest element to La. It can be expected that $La_{2-x}Y_xO_3$ might show similar properties to La_2O_3, for example, permittivity, large bandgap, and so on.

In this study, the $La_{2-x}Y_xO_3$ films with different Y atomic concentrations (Y/La + Y = 0, 10, 40, 70, 90, and 100%) were deposited on the HF-last Si(100) substrates or thick Pt films deposited on SiO_2/Si substrates by RF cosputtering of La_2O_3 and Y_2O_3 targets (provided by Kojundo Chemical, Japan) in Ar ambient at room temperature and then were annealed at 600 °C in pure N_2 or 0.1%-O_2 + N_2 ambient for 30 s in the RTA furnace. The Y concentrations were determined by X-ray photoelectron spectroscopy (XPS) measurement. The moisture absorption experiments were performed in room air. The temperature and relative humidity of the air were 25 °C and 25%, respectively. The XRD patterns of films before and after the moisture absorption were investigated. The MIM (metal–insulator–metal) capacitors on thick Pt films deposited on SiO_2/Si substrates were prepared by depositing the Au film on the $La_{2-x}Y_xO_3$ films to evaluate the permittivities. Au was also deposited on some La_2O_3 and $La_{2-x}Y_xO_3$ films on silicon to form Au/La_2O_3 or $La_{2-x}Y_xO_3$/Si metal insulator semiconductor (MIS) capacitors. The capacitance–voltage (*C*–*V*) with the frequency of 100 kHz and the gate current density–gate voltage (*J*–*V*) measurements were performed for MIS capacitors. The physical thicknesses of films were determined with the grazing incident X-ray reflectivity and spectroscopic ellipsometry measurements as mentioned above.

Figure 6.13 shows the permittivities of all $La_{2-x}Y_xO_3$ films after their exposure to air for 0 and 24 h. No permittivity degradation of $La_{2-x}Y_xO_3$ ($x = 0.8$), $La_{2-x}Y_xO_3$ ($x = 1.4$), $La_{2-x}Y_xO_3$ ($x = 1.8$), and Y_2O_3 films is observed after films were exposed to air for 24 h. However, the permittivities of $La_{2-x}Y_xO_3$ ($x = 0.2$) film and La_2O_3 film decrease dramatically after they are exposed to air for 24 h due to the formation of low-permittivity hydroxide (Figure 6.14). The XRD patterns of all $La_{2-x}Y_xO_3$ films exposed to air for 24 h are shown in Figure 6.14. The characteristic peaks attributed to hexagonal hydroxide due to the moisture absorption appear in XRD patterns of $La_{2-x}Y_xO_3$ ($x = 0.2$) film and La_2O_3 film, while those are not found in XRD patterns of $La_{2-x}Y_xO_3$ ($x = 0.8$), $La_{2-x}Y_xO_3$($x = 1.4$), $La_{2-x}Y_xO_3$($x = 1.8$), and Y_2O_3 films. It means

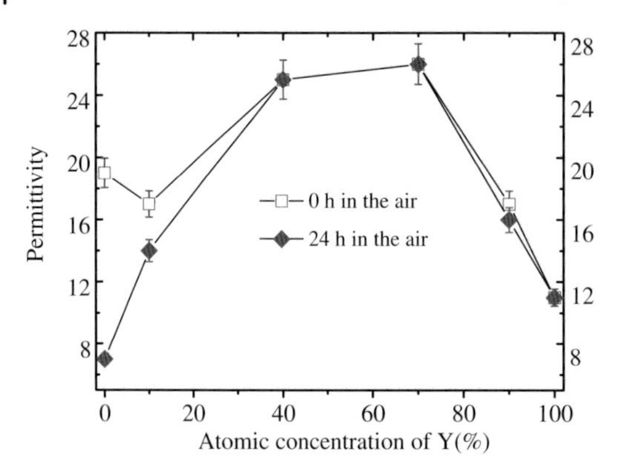

Figure 6.13 Variation in the permittivities of $La_{2-x}Y_xO_3$ films with the Y concentration. The permittivities were determined by MIM capacitors.

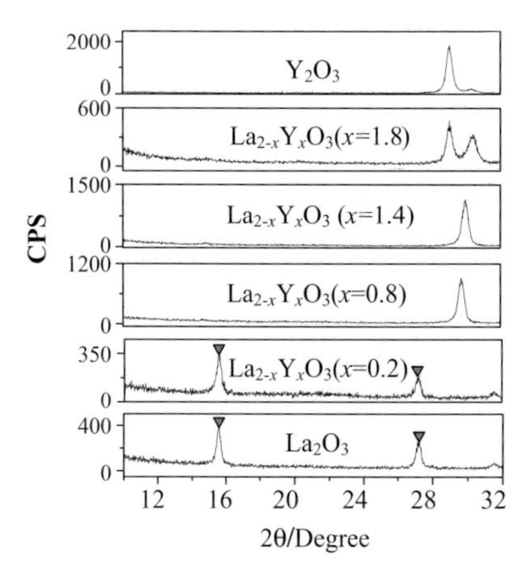

Figure 6.14 XRD patterns of $La_{2-x}Y_xO_3$ films with different Y concentrations after exposed to the air for 24 h. Temperature and relative humidity of the air are 25 °C and 50%, respectively. The films were annealed at 600 °C. (Open inverted triangle: hydroxide).

that when the Y concentration is higher than or equal to 40% ($x = 0.8$), the $La_{2-x}Y_xO_3$ film will own good moisture resistance. From the electrical properties measurement, we can also know the strong moisture resistance of $La_{2-x}Y_xO_3$ films. No degradation of *C–V* characteristics of $La_{2-x}Y_xO_3$ ($x = 1.4$) film is observed after it is exposed to air for 24 h (Figure 6.15). At the same time, the gate leakage current of Au/$La_{2-x}Y_xO_3$ ($x = 1.4$)/Si MIS capacitor shows no apparent increase after exposed to air for 24 h (Figure 6.15). On the contrary, for the La_2O_3 film after it is exposed to air for 24 h, the maximum capacitance decrease on the accumulation side of the *C–V* curve and the flatband shift are observed. The gate leakage current of Au/La_2O_3/Si MIS capacitor also increases about two orders when the La_2O_3 film is exposed to air for 24 h before Au deposition.

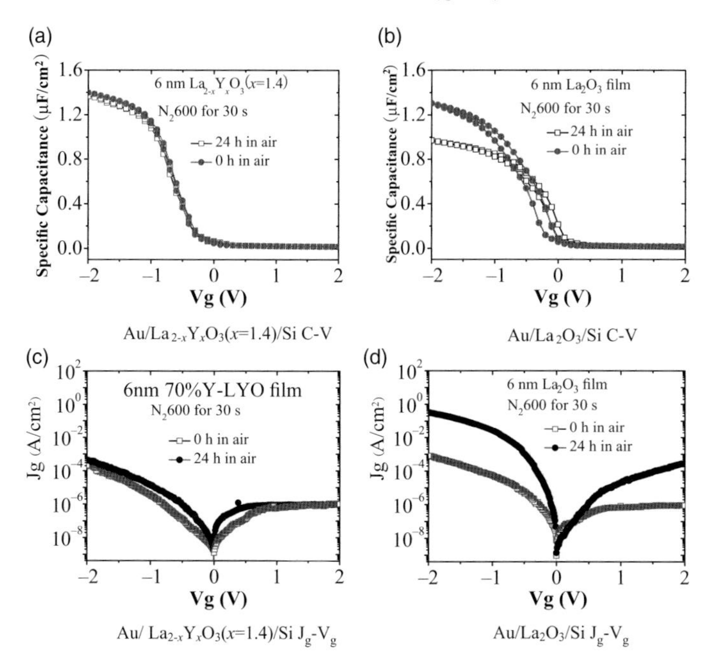

Figure 6.15 (a and b) The C–V curves at 100 kHz and (c and d) J_g–V_g curves of Au/$La_{2-x}Y_xO_3$ ($x = 1.4$)/Si and Au/La_2O_3/Si MIS capacitors after exposed to air for 0 and 24 h before Au deposition.

As already discussed, intrinsically the moisture absorption reaction is due to the small lattice energy of La_2O_3. The larger lattice energy could induce the stronger moisture resistance due to the suppression of reaction between La_2O_3 and H_2O. The well-crystallized film should own a relatively larger lattice energy than the amorphous or poorly crystallized film. In our work, La_2O_3 showed poorer crystallinity (full-width at half-maximum (FWHM) ≈1.4°) than 40%Y ($x = 0.8$) and 70%Y ($x = 1.4$) $La_{2-x}Y_xO_3$ films (FWHM ≈ 0.4°) from the XRD patterns. It indicates that the lattice energy of 40%Y ($x = 0.8$) and 70%Y ($x = 1.4$) $La_{2-x}Y_xO_3$ films should be larger than that of La_2O_3 thanks to the better crystallinity. Furthermore, Y_2O_3 owns a much larger lattice energy of 158.47 eV/mol than that of La_2O_3 (146.83 eV/mol). Therefore, Y_2O_3 doping could effectively enhance the lattice energy of La_2O_3. Furthermore, the lattice energy is related with the crystallinity of the film. One thing we should note is that 40%Y ($x = 0.8$) and 70%Y ($x = 1.4$) $La_{2-x}Y_xO_3$ films also show a much higher permittivity (~26) than La_2O_3 films in our study. The high permittivities are due to the formation of high-permittivity hexagonal phase of $La_{2-x}Y_xO_3$ films with very good crystallinity after annealing. The permittivity of lanthanum-based oxides will be discussed in more detail later in the chapter. These results indicate that $La_{2-x}Y_xO_3$ films not only show strong moisture resistance but also show a high permittivity when the Y concentration is between 40 ($x = 0.8$) and 70% ($x = 1.4$).

Therefore, due to the introduction of Y_2O_3, 40%Y ($x = 0.8$) and 70%Y ($x = 1.4$) $La_{2-x}Y_xO_3$ films after annealing at 600 °C own much larger lattice energy than La_2O_3

films, which induces the stronger hygroscopic tolerance in $La_{2-x}Y_xO_3$ films. The results also indicate that phase control is an effective method to enhance the moisture robustness of La_2O_3 films.

6.5
Hygroscopic Tolerance Enhancement of La_2O_3 Films by Ultraviolet Ozone Treatment

In our experiments, we found that the oxygen ambient-annealed La_2O_3 film shows stronger moisture resistance than nitrogen ambient-annealed La_2O_3 film, although the moisture absorption phenomenon was still observed after exposure to air for several days. So, it seems that the moisture absorption is related to oxygen vacancy in the films, partly. In other words, if the oxygen vacancies in La_2O_3 film could be eliminated, the moisture resistance could be enhanced to some degree. The most direct method is to eliminate or heal the oxygen vacancy. It has been reported that ultraviolet (UV) ozone treatment at room temperature can eliminate oxygen vacancies in oxide films [25]. Thus, the moisture absorption suppression is expected with the UV ozone posttreatment thanks to the healing of oxygen vacancies. The low temperature of UV ozone treatment is a very good merit for CMOS process that could avoid the formation of the thick interface layer. The interface layer could enhance the total EOT (equivalent oxide thickness) of gate dielectric.

La_2O_3 films were deposited on HF-last Si by sputtering the La_2O_3 target in argon ambient at room temperature and then were annealed at 600 °C in pure N_2 or 0.1%-$O_2 + N_2$ ambient for 30 s in the RTA furnace. Some samples were treated with UV ozone for 9 min at room temperature. The equipment for UV ozone treatment is illustrated in Figure 6.16. The UV lamp is a commercial Hg vapor lamp (operated at 110 W), which primarily emits light of wavelengths 185 and 254 nm. The moisture absorption experiments were performed in room air. The temperature and relative humidity of the air were 25 °C and 25%, respectively. The rms surface roughnesses and XRD patterns of films before and after the moisture absorption were investigated. Au was also deposited on some La_2O_3 films on silicon to form Au/La_2O_3/Si MIS capacitors. The capacitance–voltage with the frequency of 100 kHz and the gate current density–gate voltage (J_g–V_g) measurements were performed for MIS

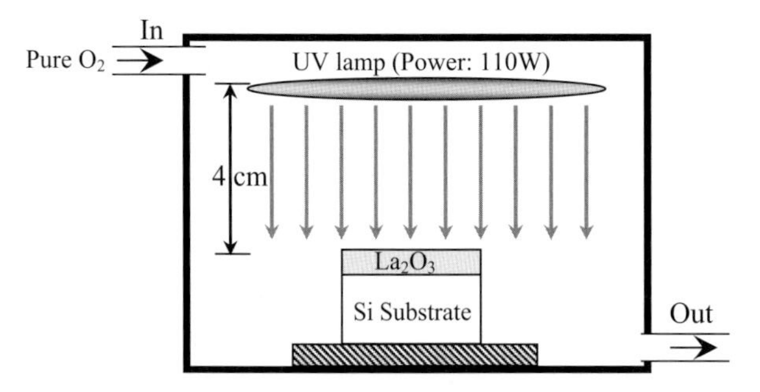

Figure 6.16 Illustration of the equipment for UV ozone posttreatment.

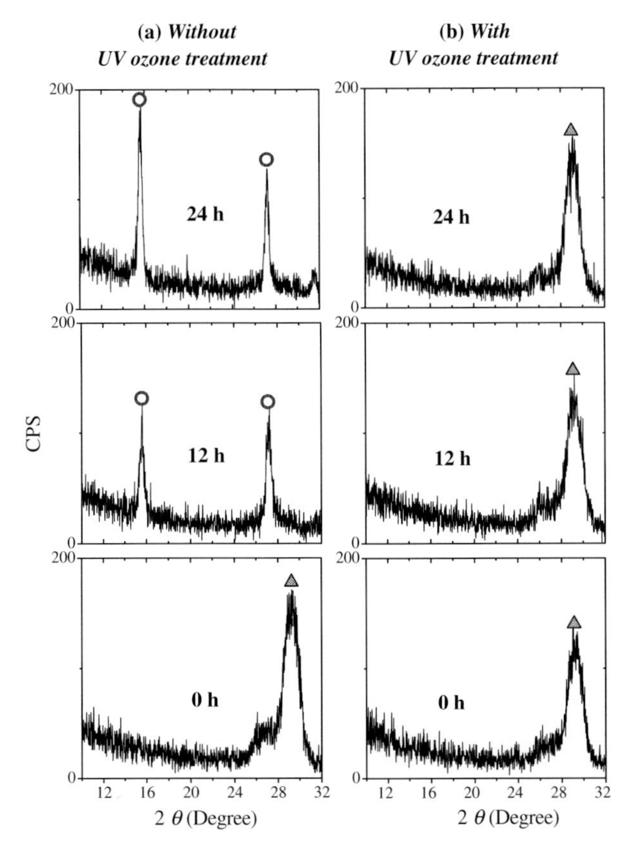

Figure 6.17 XRD patterns of La_2O_3 films (a) with and (b) without UV ozone posttreatment after N_2 annealing. Films were exposed to the air (temperature is 25 °C and relative humidity is about 25%) for different times.

capacitors. The physical thicknesses of films were determined with GIXR and SE measurements.

As it has been reported that the UV ozone treatment can eliminate the oxygen vacancies in the oxide films, the moisture absorption suppression is expected with the UV ozone posttreatment thanks to the healing of oxygen vacancies. Figure 6.17 shows the XRD patterns of La_2O_3 films with and without UV ozone posttreatment after N_2 annealing. A 0 h in air means that the sample was measured as soon as possible after annealing or UV ozone posttreatment. It is found that both are polycrystallized in the hexagonal phase when they are exposed to the air for 0 h. After exposure to air for 24 h, in the XRD pattern of the La_2O_3 film without UV ozone posttreatment after N_2 annealing, the characteristic peaks attributed to hexagonal $La(OH)_3$ due to the moisture absorption appear, while these peaks are not found in the XRD pattern of the La_2O_3 film with UV ozone posttreatment. Figure 6.18 shows AFM images of La_2O_3 films with and without UV ozone posttreatment after the films were exposed to air for different times. As shown

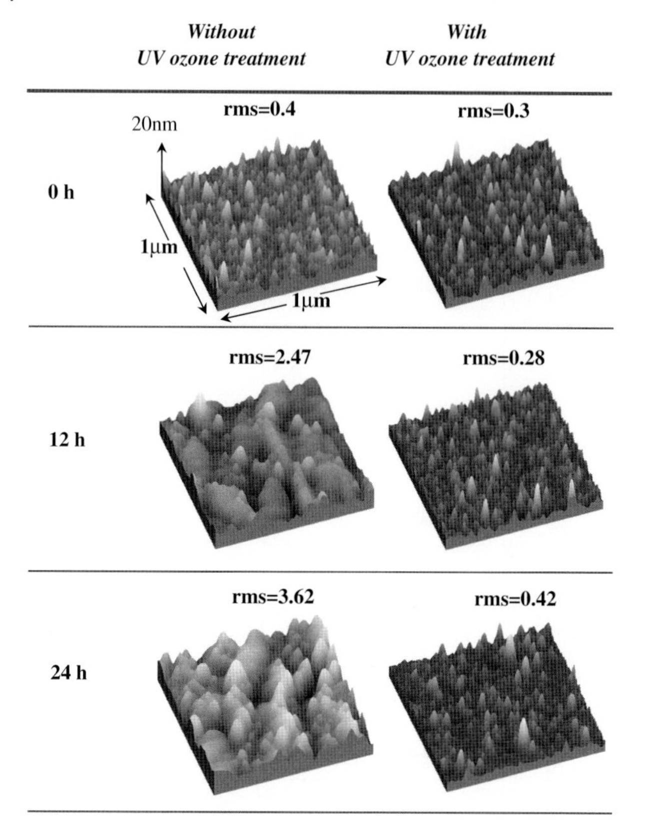

Figure 6.18 Surface AFM images (1 μm × 1 μm) of La_2O_3 films with and without UV ozone treatment after N_2 annealing at 600 °C. Films were exposed to air for different times (temperature and relative humidity of the air are 25 °C and 25%, respectively).

in Figure 6.19, the rms surface roughness of the La_2O_3 film without UV ozone posttreatment after N_2 annealing increases with the time in air due to the formation of low-density hexagonal $La(OH)_3$. Though, the surface roughness of the La_2O_3 film with UV ozone posttreatment after N_2 annealing increases very little even after the film is exposed to air for 24 h. These results suggest that UV ozone treatment can suppress the moisture absorption of La_2O_3 films.

To investigate the origin of suppression effect with UV ozone treatment, moisture resistances of La_2O_3 films with oxygen ambient (0.1%-O_2 + N_2) annealing and as-deposited La_2O_3 film (without any annealing or posttreatment) were also investigated. From Figure 6.19, it is clearly observed that the rms surface roughness of the UV ozone posttreatment film and oxygen ambient annealing film shows almost no increase with the time exposed to air. On the contrary, as-deposited and N_2 annealing films' rms surface roughnesses rapidly increase with the time exposure to air. Since UV ozone posttreatment and oxygen ambient annealing cause the same effect of healing the oxygen vacancies, it is reasonable to think that the origin of the moisture

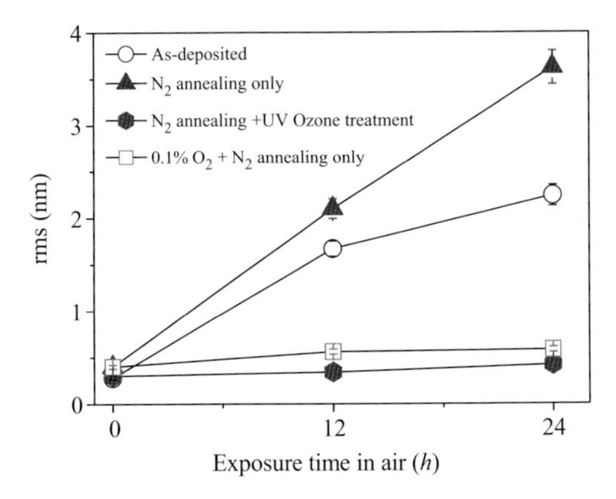

Figure 6.19 rms surface roughness as function of exposure time to air of La_2O_3 films with different treatments.

absorption suppression with the UV ozone posttreatment might be the healing of the oxygen vacancies in La_2O_3.

As discussed previously, the hygroscopic phenomenon is due to the low lattice energy of La_2O_3. Therefore, it is considered that the oxygen vacancy can decrease the lattice energy of La_2O_3. As shown in Figure 6.20, the oxygen vacancy will enlarge the charge transfer between La and O atoms and then make the La–O bond more ionic. It indicates the smaller lattice energy. Therefore, it is being considered that the lattice energy could decrease with oxygen vacancy as shown in Figure 6.21.

On the other hand, it means that if we can heal oxygen vacancies in the La_2O_3 films, moisture absorption should be suppressed to some degree. Ozone (O_3) can enhance the kinetics of oxidation (or oxygen vacancy healing) compared to the conventional thermal oxidation (oxygen ambient annealing). For the La_2O_3 films containing oxygen vacancies (La_2O_{3-x}), reaction (Equation 6.8) is possible at low temperatures, and can heal the oxygen vacancies in the La_2O_3 films during the UV ozone treatment:

$$La_2O_{3-x} + \frac{x}{2}O_3 \xrightarrow{RT} La_2O_3 + \frac{x}{2}O_2 \quad (\text{RT : room temperature}) \tag{6.8}$$

$$La_2O_{3-x} + \frac{x}{2}O_2 \xrightarrow{600^\circ} La_2O_3 \tag{6.9}$$

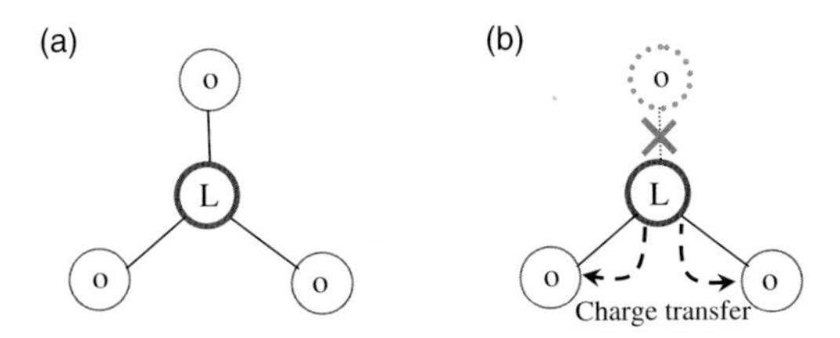

Figure 6.20 La_2O_3 (a) without and (b) with oxygen vacancy.

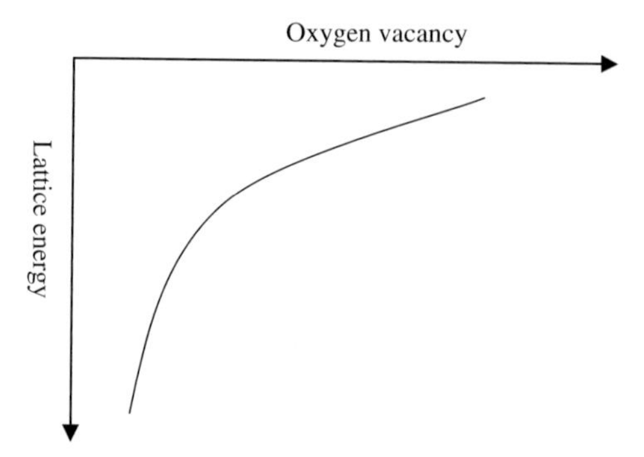

Figure 6.21 Oxygen vacancy-induced lattice energy decrease.

On the other hand, for oxygen ambient annealing to heal oxygen vacancies, a high-temperature process is generally necessary (Equation 6.9). Although the oxygen ambient annealing shows similar effect to the UV ozone treatment in terms of the moisture absorption suppression, compared to the UV ozone posttreatment, oxygen ambient annealing enhanced the CET of the film (Figure 6.22) due to the formation of thicker interface layer between silicon substrate and La_2O_3 film. Therefore, the UV ozone posttreatment is a good method to suppress the moisture absorption of La_2O_3 films with the merit of no interface layer thickness enhancement.

6.6
Thermodynamic Analysis of Moisture Absorption Phenomenon in High-*k* Gate Dielectrics

Intrinsically, the moisture absorption phenomenon in high-*k* oxides is the reaction between the solid oxide (M_mO_n) film and the gaseous state water (H_2O) in the air,

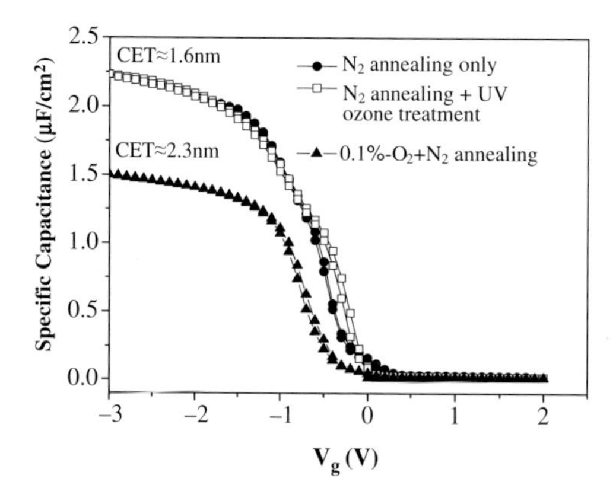

Figure 6.22 *C–V* curve (100 kHz) of $Au/La_2O_3/Si$ MIS capacitors with and without UV ozone posttreatment after N_2 annealing.

which can be expressed by Equation 6.10 as discussed above.

$$M_mO_n + H_2O(g) \leftrightarrows M(OH)_n \qquad (6.10)$$

It is well known that the speed of a chemical reaction can be indexed by the Gibbs free energy change, ΔG, of the reaction, which is given by Equation 6.11 [26].

$$\Delta G = \Delta H - T\Delta S \qquad (6.11)$$

Where, ΔH is the enthalpy change of the reaction, ΔS is the entropy change of the reaction, and T is the ambient temperature. Both ΔS and ΔH are calculated by subtracting the sum (entropy or enthalpy) on the left-hand side of the reaction equation to that of the right-hand side of the reaction equation. The entropy and enthalpy data of H_2O, $M(OH)_n$, and M_mO_n were obtained from the data base of HSC Chemistry software [27] and Ref. [28] (only for $Hf(OH)_4$). The negative ΔG, meaning the decrease in the system energy after the reaction, indicates a possibility for the occurrence of the reaction. Furthermore, when ΔG is negative, a larger absolute value of ΔG means a larger reaction speed. However, note that in a real case of high-*k* oxide films, the reaction speed is influenced by many other factors. In this chapter, we focus on the thermodynamic process of the moisture absorption reaction, which could be the main factor determining the reaction speed.

Figure 6.23 shows the calculated ΔG of the moisture absorption reactions of main high-*k* oxide candidates. For the purpose of comparison, the data of the reaction between SiO_2 and H_2O is also included in the figure. It can be obviously observed that, under standard conditions (temperature = 298.15 K; pressure = 1 atm), the moisture absorption reaction in SiO_2 could not occur since the ΔG of the reaction is positive. This fact is the chemical reason for the stable SiO_2 film in air as a gate oxide. On the other hand, a large range of ΔG values in high-*k* oxides, indicating different moisture absorption reaction speeds, could be observed. Hafnium oxide (HfO_2), the most well-studied high-*k* gate dielectric so far, shows a positive ΔG, meaning a small

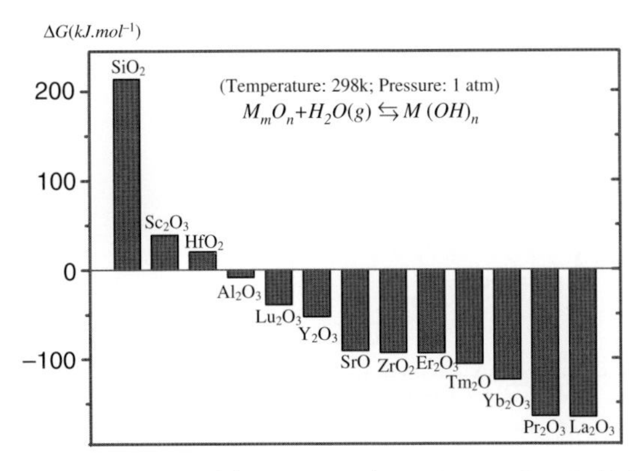

Figure 6.23 ΔG of the moisture absorption reactions in high-*k* oxides under the standard conditions. All entropy and enthalpy data of oxides, H_2O, and hydroxides are obtained from the data base of HSC chemistry software except for $Hf(OH)_4$, which is taken from Ref. [28].

moisture absorption reaction speed. This result is coincident with the experimental results since there have been few studies about the moisture absorption phenomenon in HfO_2. On the contrary, note that zirconium oxide (ZrO_2), which is also thought as a promising high-k oxide, shows a large negative ΔG. However, the moisture absorption phenomenon in ZrO_2 film as a high-k gate dielectric has not been emphasized in literatures yet. As a matter of fact, the formation of the zirconium hydroxide at the surface of ZrO_2 film has been reported [15]. Furthermore, La_2O_3 shows the most negative ΔG among all main high-k oxide candidates. This is the reason for the serious moisture absorption phenomenon we reported previously. This fact also suggests that the moisture absorption phenomenon in La_2O_3 films is the intrinsic property of La_2O_3, rather than caused by some external factors. On the other hand, it can be found from Figure 6.23 that all rare-earth oxides show a large moisture absorption reaction speed except for scandium oxide (Sc_2O_3), meaning that most of pure rare-earth oxides might not be suitable as high gate oxides although they usually show high permittivities.

Next, we discuss how to enhance the moisture resistance of rare-earth oxide, especially of La_2O_3. When considering the thermodynamic process of the moisture absorption reaction as shown in Equation 6.10, the most direct method to enhance the moisture resistance or decrease the moisture absorption reaction speed of an oxide film is doping a second oxide, which owns a stronger resistance to the moisture absorption. We have reported that the Y_2O_3-doped La_2O_3 films show much stronger moisture resistance than La_2O_3 [29, 30], which is a demonstration of this method. Figure 6.24 shows the ΔG of several La-based ternary oxides with a molecule ratio of 1: 1 between La_2O_3 and the second oxide, which are simply calculated by averaging the ΔG of the moisture absorption reaction of La_2O_3 and the second oxides. As shown in Figure 6.24, doping a second oxide is an effective method of decreasing the moisture absorption reaction speed. Furthermore, SiO_2, Sc_2O_3, HfO_2, Al_2O_3, Lu_2O_3, and Y_2O_3 are better candidates than other oxides for doping to enhance the moisture resistance of La_2O_3. On the other hand, note that the permittivity of the doped La_2O_3 has to be considered when we select the second

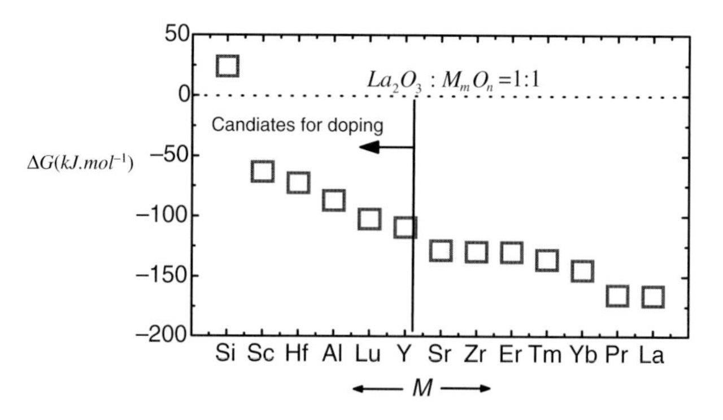

Figure 6.24 The ΔG of the moisture absorption reactions of La-based ternary oxides, which is calculated by averaging the ΔG of La_2O_3 and the second oxide. The molecule ratio between La_2O_3 and the second oxide in ternary oxides is 1 : 1.

oxide. This issue is beyond the scope of this chapter and will not be further discussed here. Furthermore, the moisture resistance of an oxide film is also affected by several external factors, like the crystallinity [29] of the film and oxygen vacancies in the film [31], as reported previously. We can understand these behaviors by a more detailed analysis of the moisture absorption reaction. The reaction in Equation 6.10 could be divided into three steps as shown in Equations (6.12), (6.13) and (6.14).

$$M_mO_n \leftrightarrows M^{+n} + O^{-2} \tag{6.12}$$

$$O^{-2} + H_2O \leftrightarrows OH^{-1} \tag{6.13}$$

$$M^{+n} + OH^{-1} \leftrightarrows M(OH)_n \tag{6.14}$$

Equations 6.12 and (6.13) are the key reactions for determining the speed of the whole moisture absorption reaction. Physically, the reaction speed of Equation 6.12 is determined by the lattice energy of the oxide, which is mainly determined by the ionicity (or electronegativity) of M ion and could also be affected by the crystallinity of the oxide in the case of a thin film. The larger electronegativity means a larger ionicity, resulting in a smaller lattice energy and a larger reaction speed of Equation 6.12. In fact, the ΔG results in Figure 6.23 are well coincident with the reported electronegativity data [32]. On the other hand, the reaction equation (6.4) is responsible for the formation of OH^{-1}, resulting in the formation of hydroxide after combined with M^{+n} (Equation 6.14). The oxygen vacancies, however, can also induce the formation of OH^{-1}, which is the reason for the more serious moisture absorption phenomenon in oxygen-deficient La_2O_3 films as reported previously. However, as shown in Figure 6.23, the thermodynamic process could be the main and intrinsic factor for determining the speed of the moisture absorption reaction.

In summary, the moisture absorption phenomenon in main high-k gate oxides has been theoretically discussed by comparing the Gibbs free energy change of the moisture absorption reactions of these oxides. The results show that the moisture absorption could occur in most high-k oxides, especially in rare-earth oxides. On the other hand, La_2O_3 shows the largest moisture absorption reaction speed among main high-k oxide candidates. To enhance the moisture resistance of La_2O_3, doping a second oxide, which has a stronger moisture resistance than La_2O_3, could be an applicable solution.

6.7
Permittivity Enhancement of La_2O_3 Films by Phase Control

One very important reason for lanthanum oxide (La_2O_3) as a promising high-k gate dielectric to replace SiO_2 is its high permittivity. However, many low-permittivity

La_2O_3 films were reported [7–9]. In terms of the reasons for the low permittivity of La_2O_3 films, two very possible ones are considered as mentioned previously. The first one is the moisture absorption that degrades the permittivity of La_2O_3 films due to the formation of the low-permittivity lanthanum hydroxide as discussed above [33]. The second one is the low density of amorphous La_2O_3 films [7]. In fact, the permittivity of La_2O_3 film without moisture absorption (0 h in air) still shows a low permittivity (~20). It indicates that the low permittivity should be the intrinsic property of La_2O_3 films. The low permittivity of La_2O_3 film can be really partly attributed to the poor cystallinity, not to the moisture absorption totally.

Therefore, in this section, we prepare well-crystallized La-based films to enhance and stabilize the permittivity of La_2O_3 films.

From the phase diagram of La_2O_3–Y_2O_3 system, a high melting point of $La_{2-x}Y_xO_3$ could be observed indicating a low crystallization temperature of $La_{2-x}Y_xO_3$. On the other hand, Y_2O_3 shows a much lower crystallization temperature than La_2O_3 (Figure 6.25), and it is very possible that $La_{2-x}Y_xO_3$ films could also have a low crystallization temperature. Furthermore, Y is in the same element group in the elements table as La and is the nearest element to La. It can be expected that $La_{2-x}Y_xO_3$ can show similar properties as La_2O_3, for example, permittivity, bandgaps, and so on, except for the moisture absorption phenomenon. On the other hand, a very common viewpoint is that amorphous film (high crystallization temperature) is better than crystallized film as high-*k* gate insulators [34]. It is believed that grain boundaries in polycrystalline films might constitute electrical leakage paths, giving rise to dramatically increased gate leakage currents. However, there are few studies on the grain boundary-induced leakage current in high-*k* gate dielectrics, and at present epitaxial (crystalline) films are technologically feasible [35, 36].

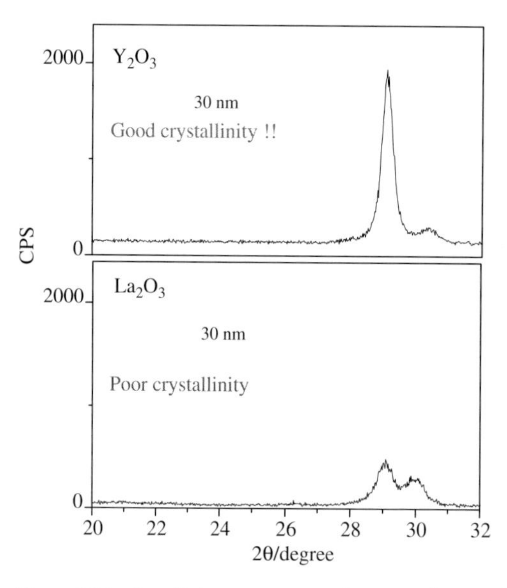

Figure 6.25 Cystallinity comparison of Y_2O_3 and La_2O_3 films.

On the other hand, among La-based high-*k* materials, $La_{1-x}Hf_xO_y$ and $LaAlO_3$ are two attractive ones because $La_{1-x}Hf_xO_y$ is a good amorphous insulator up to 900 °C [37, 38] and $LaAlO_3$ shows a high permittivity and a large bandgap [39, 40]. However, $La_{1-x}Hf_xO_y$ film crystallizes in the pyrochlore $La_2H_2fO_7$ after annealing at 1000 °C [41], while in the conventional CMOS process annealing at higher than 1000 °C is necessary to activate the source and drain dopant. In terms of $LaAlO_3$ film, as low-permittivity $LaAlO_3$ films (<20) are always reported, [7, 42], it might be very difficult to prepare high-permittivity $LaAlO_3$ films. A very possible reason for the low permittivity of $LaAlO_3$ films is their poor crystallinity that induces the low density of films. These results indicate that it is a very tough task to prepare an amorphous high-permittivity dielectric film as an alternative gate insulator. Although Ta_2O_5 film shows a high permittivity even in the amorphous state [43], owing to its very small conduction band offset with silicon, [18] it cannot be used as a high-*k* gate dielectric. It is well known that La_2O_3 has a large conduction band offset with silicon of about 2.3 eV and a high permittivity [44]. Therefore, $La_{1-x}Ta_xO_y$ film with an appropriate Ta concentration might be suitable as a gate dielectric that owns a medium conduction band offset with silicon due to introduction of La_2O_3. At the same time, a high permittivity of $La_{1-x}Ta_xO_y$ film can be expected thanks to the high permittivity of La_2O_3 and Ta_2O_5. In terms of the crystallization temperature, due to the low melting point of $La_{1-x}Ta_xO_y$ from La_2O_3–Ta_2O_5 phase diagram [45], $La_{1-x}Ta_xO_y$ films might show a high crystallization temperature. Therefore, we investigated $La_{1-x}Ta_xO_y$ films with different Ta concentrations as high-*k* gate insulators in terms of crystallization temperature, permittivity, bandgap, and electrical properties.

Therefore, theoretically and practically, both amorphous and well-crystallized high-*k* films might also be possible choices as gate insulators.

6.7.1
Experimental Procedures and Characterizations

The $La_{2-x}Y_xO_3$ and $La_{1-x}Ta_xO_y$ films with different Y or Ta atomic concentrations were deposited on the HF-last Si (100) substrates or thick Pt films deposited on SiO_2/Si substrates by RF cosputtering of La_2O_3 and Y_2O_3 or Ta_2O_5 targets (provided by Kojundo Chemical, Japan) in Ar ambient at room temperature (Figure 6.26). The Y and Ta concentrations were determined by XPS measurement. The physical thicknesses of the films were determined with SE and GIXR measurement. The crystallinity of films was investigated by X-ray diffraction measurement. The MIM capacitors on thick Pt films deposited on SiO_2/Si substrates were prepared by depositing the Au film on $La_{2-x}Y_xO_3$ or $La_{1-x}Ta_xO_y$ films to evaluate the permittivities. Au was also deposited on some $La_{2-x}Y_xO_3$ and $La_{1-x}Ta_xO_y$ films on silicon to form Au/$La_{2-x}Y_xO_3$ or $La_{1-x}Ta_xO_y$/Si MIS capacitors. The capacitance–voltage (*C*–*V*) with a frequency of 100 kHz and the gate current density–gate voltage (*J*–*V*) measurements were performed for Au/$La_{2-x}Y_xO_3$ or $La_{1-x}Ta_xO_y$/Si MIS capacitors.

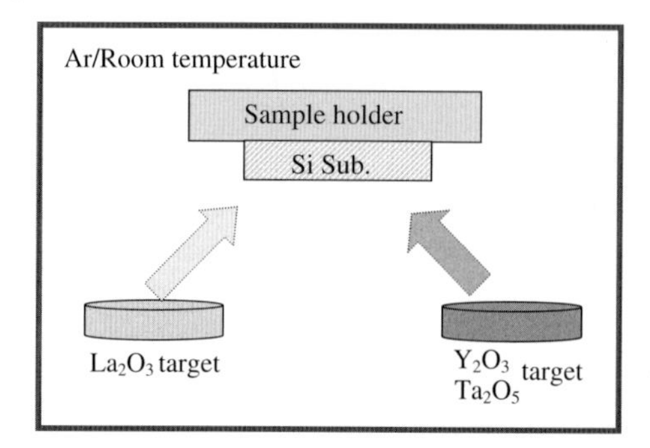

Figure 6.26 Illustration of cosputtering deposition.

6.7.2
Permittivity Enhancement by Phase Control due to Y_2O_3 Doping

Figure 6.27 shows the permittivity variation of $La_{2-x}Y_xO_3$ films annealed at 600 °C in the pure N_2 ambient with the Y concentration. It is noticed that the permittivity of La_2O_3 film is low compared to the large value of 27 reported by Chin *et al.* [1]. The low permittivity of La_2O_3 film might be attributed to the poor crystallization of the film and to the moisture absorption because we did not prevent the sample from moisture intentionally. In our study, the Y_2O_3 film is with the permittivity of 12 as reported [46]. The permittivity of $La_{2-x}Y_xO_3$ ($x = 0.2$) film is a little smaller than that of the La_2O_3 film, whereas the $La_{2-x}Y_xO_3$ ($x = 0.8$) and $La_{2-x}Y_xO_3$ ($x = 1.4$) films show much higher permittivity (~26) than La_2O_3 film in our study. This value is also very close to the high-permittivity value of La_2O_3 film as reported. When the Y concentration is as high as 90% ($x = 1.8$), the permittivity of $La_{2-x}Y_xO_3$ film decreases to 15. But this value is still higher than the permittivity of Y_2O_3.

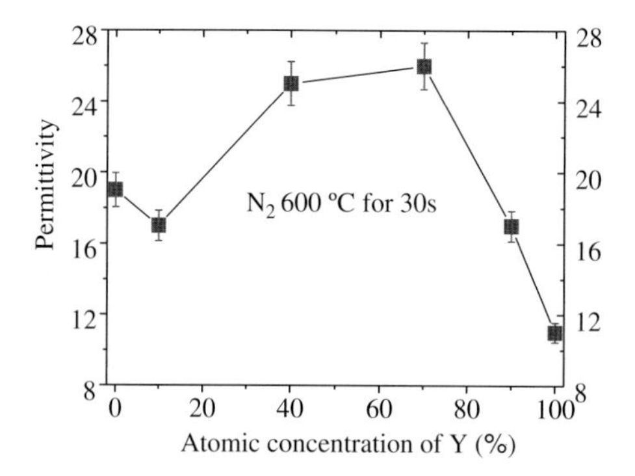

Figure 6.27 Variation in the permittivities of $La_{2-x}Y_xO_3$ films with the Y concentration. The permittivities were determined by MIM capacitors. The films were thought to be exposed to air for 0 h rather than moisture prevented because we did not prevent the films from the moisture purposely and just deposited the films with Au electrode as quickly as possible after annealing in the RTA furnace.

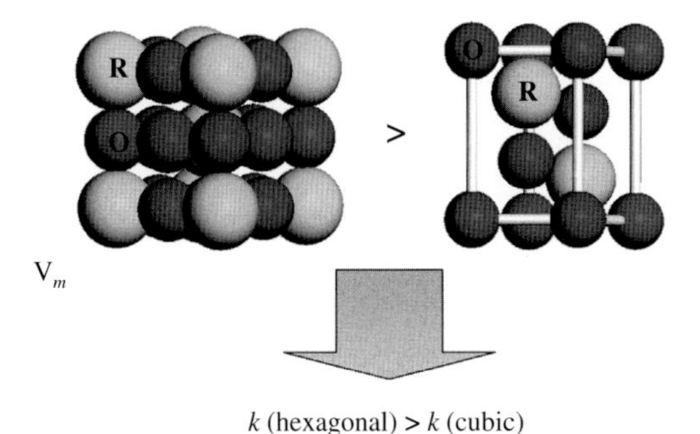

Figure 6.28 Molar volume comparison of hexagonal and cubic rare-earth oxide (R_2O_3).

To explain the reason for high permittivity of $La_{2-x}Y_xO_3$ films, it is necessary to discuss the Clausius–Mosotti equation for permittivity theory calculation [47].

Simply speaking, Clausius–Mosotti equation tells us that the permittivity of a well-crystallized film is determined by the molar volume and total polarizability given as

$$k = (3\ V_m + 8\pi\alpha^T)/(3\ V_m - 4\pi\alpha^T) \tag{6.15}$$

where V_m and α^T denote the molar volume and the total polarizability, respectively. Thus, we can understand easily that if α^T is assumed to be a constant in spite of the V_m change, the smaller V_m will induce the larger permittivity. For rare-earth oxides (R_2O_3), hexagonal phase owns much smaller molar volumes as shown in Figure 6.28. For hexagonal La_2O_3, α^T is 18.17 Å from the Shannon's additivity rule ($\alpha^T(La_2O_3) = 2\alpha(La^{3+}) + 3\alpha(O^{-2})$) and V_m is 82.7 Å^3 from Ref. [48]. With these values, we can estimate that the permittivity of hexagonal La_2O_3 is about 35, which is larger than the reported permittivities of La_2O_3 films. This difference comes from the poor crystallinity of the reported La_2O_3 films and the low-permittivity cubic phase of some La_2O_3 films. The same method is applied to hexagonal Y_2O_3 to estimate the permittivity. The V_m of hexagonal Y_2O_3 is assumed to be 90% of that of cubic Y_2O_3 [49] like the case of La_2O_3 because no XRD pattern of the hexagonal Y_2O_3 has been reported. Then, we can estimate that the permittivity of hexagonal Y_2O_3 is 22, which is much larger than the permittivity of the cubic phase Y_2O_3 in our study ($k \sim 11$). In summary, for rare-earth oxides, hexagonal phase (and well crystallized) is preferred to achieve a high permittivity. Now let me explain the reason for the high permittivity of $La_{2-x}Y_xO_3$ films. Figure 6.29 shows XRD patterns of all $La_{2-x}Y_xO_3$ films on Pt film after annealing at 600 °C. It can be observed that the La_2O_3 film is polycrystallized in the hexagonal phase. In the XRD pattern of $La_{2-x}Y_xO_3$ ($x = 0.2$) film, both peaks attributed to the cubic phase and hexagonal phase are found. Therefore, the permittivity of $La_{2-x}Y_xO_3$ ($x = 0.2$) film is smaller than that of the La_2O_3 film due to the low permittivity of cubic phase.The 40%Y ($x = 0.8$) and 70%Y ($x = 1.4$) $La_{2-x}Y_xO_3$ films are well crystallized in the hexagonal phase after annealing at 600 °C.

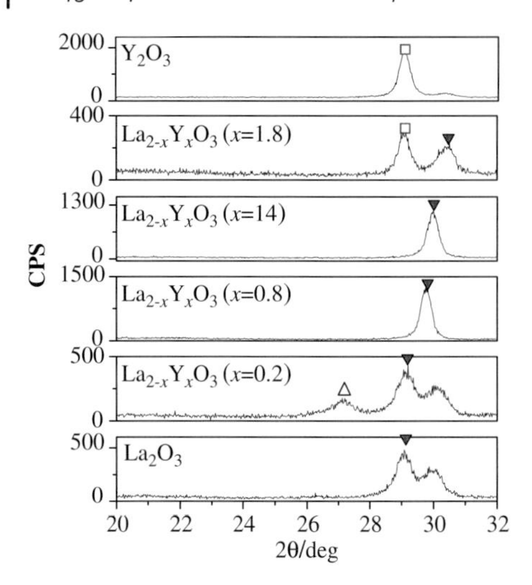

Figure 6.29 XRD patterns of $La_{2-x}Y_xO_3$ films with different Y concentrations after annealing at 600 °C. (Inverted triangle: hexagonal $La_{1-x}Y_xO_3$ (002); open square: cubic Y_2O_3 (222); open triangle: cubic La_2O_3 (222)).

The crystallinity of the film can be estimated with the FWHM of the XRD peak. The smaller FWHM indicates the better crystallinity. The FWHM of hexagonal (002) peak of 40%Y ($x=0.8$) and 70%Y ($x=1.4$) $La_{2-x}Y_xO_3$ films' XRD patterns are only 0.4°, while that of La_2O_3 film is about 1.4°. It indicates that 40%Y ($x=0.8$) and 70%Y ($x=1.4$) $La_{2-x}Y_xO_3$ films have better crystallinity than La_2O_3 film. As reported by R.A. B Devine, the permittivity of amorphous La_2O_3 is very low due to its low density. Therefore, the better crystallinity provided 40%Y ($x=0.8$) and 70%Y ($x=1.4$) $La_{2-x}Y_xO_3$ films with much higher permittivity than La_2O_3 film, even though the low-polarizability Y^{3+} ions were introduced. Another very important factor is that 40%Y ($x=0.8$) and 70%Y ($x=1.4$) $La_{2-x}Y_xO_3$ films were both crystallized in the hexagonal phase rather than in the cubic phase. As discussed above, the hexagonal rare-earth oxides show much larger permittivities than cubic rare-earth oxides as expected from the Clausius–Mossotti relation. It is reasonable that the 40%Y ($x=0.8$) and 70%Y ($x=1.4$) $La_{2-x}Y_xO_3$ films that are well crystallized in the hexagonal phase show a high permittivity of 26. In addition, the peak of hexagonal (002) $La_{2-x}Y_xO_3$ gradually shifts to a larger 2θ as Y concentration increases. This shift is attributed to the decrease in the lattice parameter due to a smaller ionic radius of Y^{3+} than that of La^{3+}. For the $La_{2-x}Y_xO_3$ ($x=1.8$) film, it is found from the XRD pattern that the film contains both the cubic and the hexagonal phases. Therefore, its permittivity is larger than that of the Y_2O_3 with cubic phase but smaller than that of 40%Y ($x=0.8$) and 70%Y ($x=1.4$) $La_{2-x}Y_xO_3$ films due to the low-polarizability Y^{3+} ion and the low-permittivity cubic phase.

As discussed earlier, a basic requirement for high-k gate dielectric is relative large bandgap and a band offset with silicon valence and conduction band larger than 1 eV, at least. Therefore, we also have to consider the bandgaps of $La_{2-x}Y_xO_3$ films.

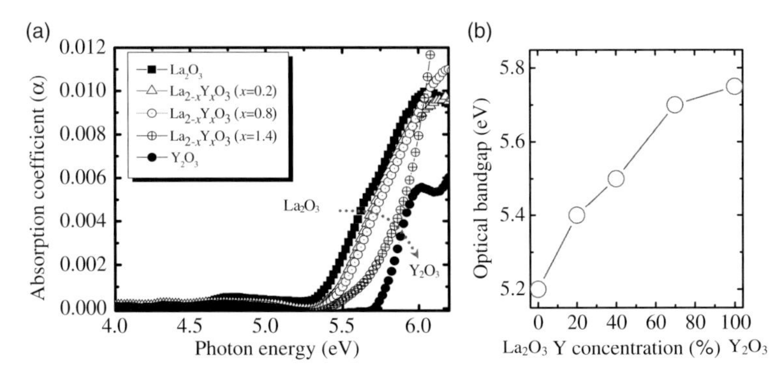

Figure 6.30 (a) Absorption coefficient versus photon energy for $La_{2\text{-}x}Y_xO_3$; (b) optical bandgaps of $La_{2\text{-}x}Y_xO_3$ films as a function of Ta atomic concentration.

Figure 6.30 shows the bandgaps of $La_{2\text{-}x}Y_xO_3$ films with different Y concentrations. From the figure, we can find that the band of $La_{2\text{-}x}Y_xO_3$ film increases with Y concentration. The bandgap of La_2O_3 is only about 5.2 eV and that of Y_2O_3 is about 5.75 eV. The 40%Y ($x = 0.8$) and 70%Y ($x = 1.4$) $La_{2\text{-}x}Y_xO_3$ films not only show a high permittivity of 26 but also show a large bandgap of 5.7 eV, which are well-crystallized in the hexagonal phase. As discussed previously, the high permittivity of 40%Y ($x = 0.8$) and 70%Y ($x = 1.4$) $La_{2\text{-}x}Y_xO_3$ films are due to the well-crystallized high-*k* hexagonal phase. Here, it is considered that the larger bandgap of 40%Y ($x = 0.8$) and 70%Y ($x = 1.4$) $La_{2\text{-}x}Y_xO_3$ films than La_2O_3 film are also attributed to the better crystallinity of these two kinds of films than La_2O_3 film. It has been reported that in the case of Hf_2O and Zr_2O, amorphous films show a much smaller bandgap.

The electrical properties of $La_{2\text{-}x}Y_xO_3$ films were also investigated. As 40%Y ($x = 0.8$) and 70%Y ($x = 1.4$) $La_{2\text{-}x}Y_xO_3$ films show much larger bandgaps, higher permittivity and stronger resistance to moisture than La_2O_3, they should also show good electrical properties. Figure 6.31 shows the *C–V* characteristics of $La_{2\text{-}x}Y_xO_3$ ($x = 1.4$) film. We can find that there is almost no frequency dependence observed. For reference, *C–V* curves with and without moisture absorption are shown in Figure 6.31. No *C–V* degradation can be observed even after exposure to air for 24 h. Figure 6.32 shows the J_g–V characteristics of $La_{2\text{-}x}Y_xO_3$ ($x = 1.4$) film with different CET. Very low leakage current density can be observed. When the CET is about 1.6 nm or EOT is about 1nm, the leakage current density is only about 10^{-5} A/cm^2. The normal criterion is smaller than1A/cm^2 when EOT is about 1 nm.

As high-*k* materials are also being considered to use as insulators for MIM capacitors due to the higher permittivity of $La_{2\text{-}x}Y_xO_3$ film and low leakage current, it might be promising as a MIM insulator. Therefore, we also investigated the *C–V* and J_g–V characteristics of Au/$La_{2\text{-}x}Y_xO_3$ ($x = 1.4$)/Pt MIM capacitors. Figure 6.33 shows the *C–V* and J_g–V characteristics. A very low leakage current when the bias is below 5 V and very high capacitance density are obtained.

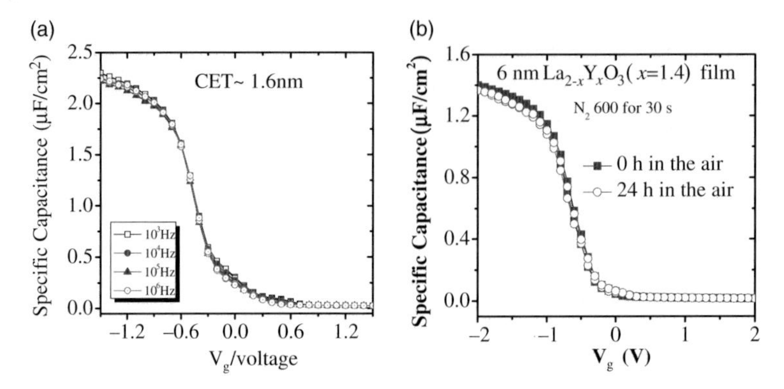

Figure 6.31 (a) Frequency and (b) moisture dependence of C–V characteristic of Au/$La_{2\text{-}x}Y_xO_3$ ($x = 1.4$)/Si MIS capacitors.

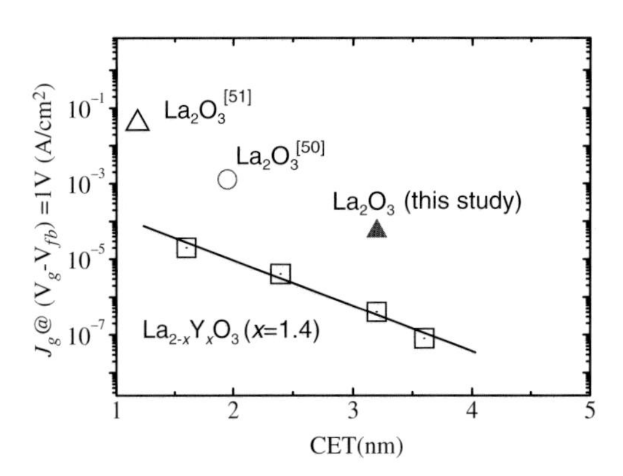

Figure 6.32 J_g–V characteristics of $La_{2\text{-}x}Y_xO_3$ ($x = 1.4$) film with different EOTs.

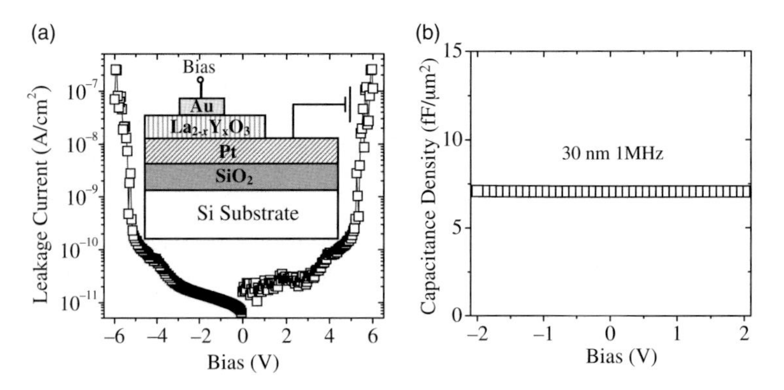

Figure 6.33 (a) J_g–V characteristic of the Au/30 nm $La_{2\text{-}x}Y_xO_3$ ($x = 1.4$)/Pt MIM capacitors; (b) C–V characteristic of Au/$La_{2\text{-}x}Y_xO_3$/Pt MIM capacitors with different film thicknesses.

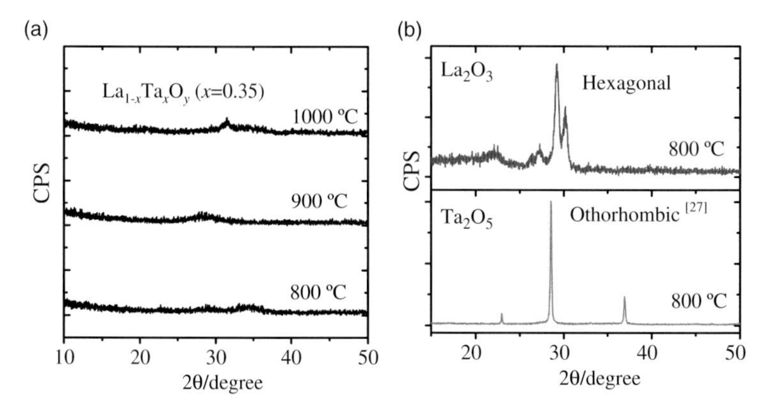

Figure 6.34 (a) XRD patterns of $La_{1-x}Ta_xO_y$ ($x = 0.35$) film annealed at 800, 900, and 1000 °C; (b) XRD patterns of La_2O_3 and Ta_2O_5 films annealed at 800 °C. The thickness of the films was about 30 nm.

6.7.3
Higher-*k* Amorphous $La_{1-x}Ta_xO_y$ Films

Figure 6.34a shows the XRD patterns of $La_{1-x}Ta_xO_y$ films with a Ta concentration of 35% ($x = 0.35$) after they were annealed at 800, 900, and 1000 °C in N_2 ambient. It can be observed that the film was still in the amorphous state even when it was annealed at 1000 °C. It indicates that the crystallization temperature of $La_{1-x}Ta_xO_y$ ($x = 0.35$) film is higher than 1000 °C. As a gate dielectric, an amorphous film is preferred rather than a polycrystallized film because grain boundaries can induce the leakage current through the dielectric [52]. As $La_{1-x}Ta_xO_y$ ($x = 0.35$) film shows a crystallization temperature higher than 1000 °C, it will be compatible with the conventional CMOS process. On the other hand, both La_2O_3 and Ta_2O_5 films crystallize after annealing at 800 °C (Figure 6.34b). Furthermore, crystallization temperatures for $La_{1-x}Ta_xO_y$ ($x = 0.35$) and $La_{1-x}Ta_xO_y$ ($x = 0.6$) films are about 800 (Figure 5.12a) and 1000 °C (Figure 6.35b), respectively. Both are higher than that of La_2O_3 [51] and Ta_2O_5 [52]. It also indicates that the crystallization temperature of $La_{1-x}Ta_xO_y$ film is sensitive to the Ta concentration, and to prepare the high crystallization temperature $La_{1-x}Ta_xO_y$ film, it is very crucial to control the Ta concentration. In the case of $La_{2-x}Y_xO_3$, it shows a very low crystallization temperature [53]. Why there is so large a difference? To answer this question, I would like to explain the crystallization mechanism of La-based ternary oxides ($La_{2-x}M_xO_3$ or $La_{1-x}M_xO_y$). We think there are two main factors that induce the high crystallization temperature. One is the size difference between La and M ions and the other is the difference of valence state of La and M ions.

It was reported that the stability of amorphous ternary metal oxide ($A_{2-x}M_xO_3$) is substantially determined by the size difference between metal A ion and metal M ion when A and M is with the same valence state [54]. The size of metal A ion close to metal M ion will induce the formation of solid solution. When there is a large size

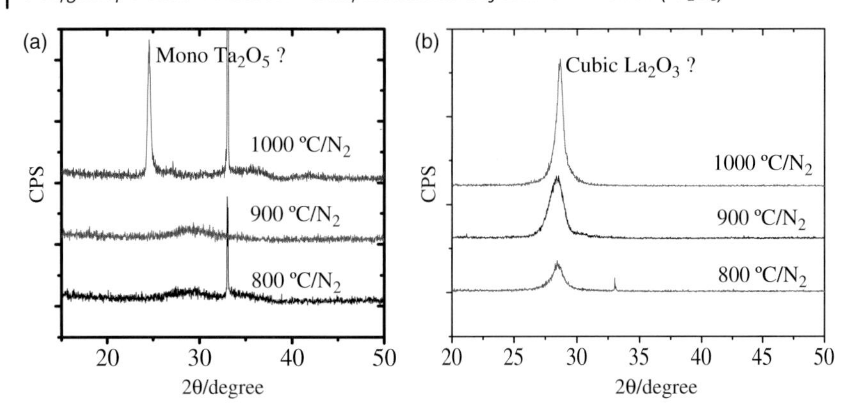

Figure 6.35 (a) XRD patterns of $La_{1-x}Ta_xO_y$ ($x = 0.6$) film annealed at 800, 900, and 1000 °C; (b) XRD patterns of $La_{1-x}Ta_xO_y$ ($x = 0.2$) film annealed at 800 °C. The thickness of the films was about 30 nm.

Table 6.1 Crystallization temperatures of $La_{1-x}M_xO_y$ ternary metal oxides in the case of M^{3+}.

Ternary oxide ($La_{2-x}M_xO_3$)	Metal ion size (La-M) (Å)	Crystallization temperature (°C)
$La_{2-x}Al_xO_3$	1.16–0.51	800
$La_{2-x}Sc_xO_3$	1.16–0.73	650
$La_{2-x}Y_xO_3$	1.16–1.02	400

difference between metal A ion and metal M ion, such solid solution is not stable. Then, the oxide can be stabilized at the amorphous state. In the case for La-based ternary oxides, the difference between La^{3+} ion and M^{3+} ion will affect the crystallization temperature of $La_{2-x}M_xO_3$. Table 6.1 shows the crystallization temperatures of $La_{2-x}Al_xO_3$ [35], $La_{2-x}Sc_xO_3$ [55], and $La_{2-x}Y_xO_3$ films [30]. Because $r(La^{3+}) > r(Y^{3+}) > r(Sc^{3+}) > r(Al^{3+})$, $La_{2-x}Al_xO_3$ is with the highest crystallization temperature and $La_{2-x}Y_xO_3$ is with the lowest one. Thus, we got the well-crystallized $La_{2-x}Y_xO_3$ film as discussed earlier.

We think that the valence state of M ion can also affect the crystallization temperature of $La_{1-x}M_xO_y$ ternary oxide because the valence state will determine the oxygen ratio in MO_y. And when we dope MO_y into La_2O_3 ($LaO_{1.5}$) and if y is larger than 1.5, there will be superfluous oxygen that will distort the oxide network and then enhance the crystallization temperature of La_2O_3. We studied several $La_{1-x}M_xO_y$ ternary metal oxides with different M (Y^{3+}, Hf^{4+}, and Ta^{5+}). Crystallization temperatures of these ternary metal oxides are shown in Table 6.2. An obvious trend can be found from the table that the ternary metal oxide that owns a larger valence state of M ion (larger y) shows a higher crystallization temperature. $La_{1-x}Ta_xO_y$ ternary oxide shows the highest crystallization temperature among these ternary metal oxides because of the difference of ion size and valence state of La and M. $La_{2-x}Y_xO_3$ films

Table 6.2 Crystallization temperatures of $La_{1-x}M_xO_y$ ternary metal oxides in the case of M^{3+}, M^{4+}, and M^{5+}.

Ternary oxide ($La_{1-x}M_xO_y$)	Metal ion size (La-M) (Å)	Crystallization temperature (°C)
$La_{1-x}Ta_xO_y$ (M^{5+})	1.16–0.74	>1000
$La_{1-x}Hf_xO_y$ (M^{4+})	1.16–0.83	1000
$La_{2-x}Y_xO_3$ (M^{3+})	1.16–1.02	400

show a low crystallization temperature due to the size of La^{3+} close to Y^{3+} and the same valence state of La^{3+} and Y^{3+}. In addition, $La_{1-x}Hf_xO_y$ film also shows a relative high crystallization temperature thanks to the large size difference between La^{3+} and Hf^{4+}. In summary, the high crystallization temperature of $La_{1-x}Ta_xO_y$ film can be attributed to the large size and valence state difference between La^{3+} and Ta^{5+}.

As a new high-*k* gate dielectric, the permittivity also should be discussed. Figure 6.36 shows the variation in permittivities of $La_{1-x}Ta_xO_y$ films with Ta concentration. The permittivities were measured with Au/$La_{1-x}Ta_xO_y$/Pt MIM capacitors. $La_{1-x}Ta_xO_y$ ($x=0.35$) film shows a high permittivity of about 30, which is comparable to the largest reported permittivity of La_2O_3 [37] and amorphous Ta_2O_5. This permittivity value is also much larger than that of amorphous $La_{1-x}Hf_xO_y$ and well-crystallized $LaAlO_3$ films. The very possible reason for the high permittivity of amorphous $La_{1-x}Ta_xO_y$ is the higher density of Ta_2O_5 [56] than La_2O_3. According to the Clausius–Mosotti relation, the permittivity is related to the density of the film [57]. The higher density will induce the higher permittivity. Therefore, the high density Ta_2O_5 doping would enhance the permittivity of La_2O_3 although the film was amorphous.

Due to the small bandgap (E_g) of Ta_2O_5, the bandgap of $La_{1-x}Ta_xO_y$ ($x=0.35$) film should be considered to confirm whether it is suitable as a gate dielectric or not.

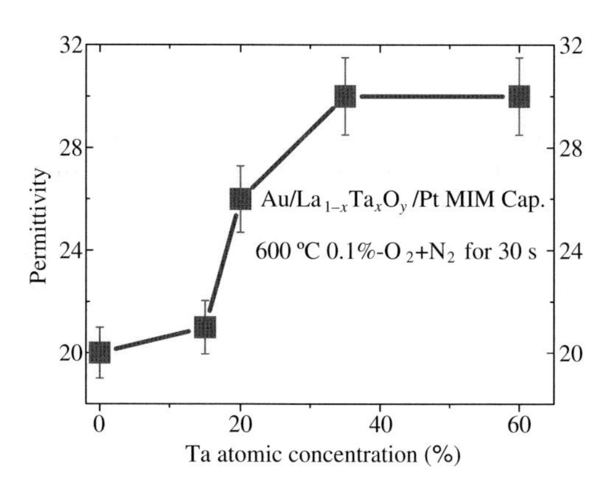

Figure 6.36 Variation in permittivities of $La_{1-x}Ta_xO_y$ films with Ta concentration.

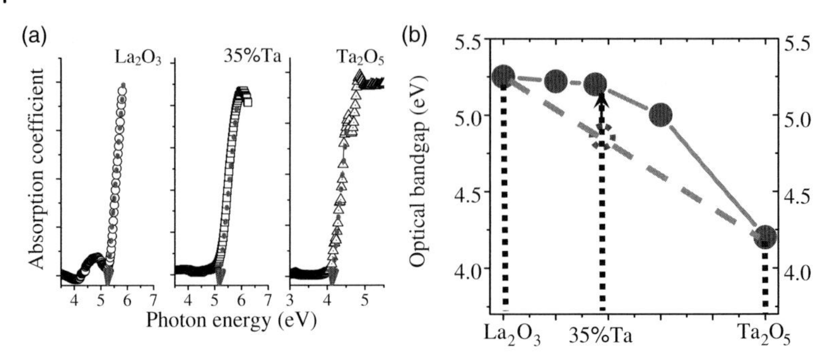

Figure 6.37 (a) absorption coefficient versus photon energy for La_2O_3, $La_{1-x}Ta_xO_y$ ($x=0.35$), and Ta_2O_5. (b) The optical bandgap of $La_{1-x}Ta_xO_y$ as function of Ta atomic concentration.

The annealed La_2O_3, $La_{1-x}Ta_xO_y$ ($x=0.35$), and Ta_2O_5 films were measured using Sorpra spectroscopic ellipsometry. Measurements were taken at three angles of incidence, 70, 72.5, and 75°, across a 3–6.24 eV spectroscopic range with a scanning step size of 0.02 eV. The point-by-point fitting algorithm was used to extract the optical constants of films [57]. The results are shown in Figure 6.37. The y-axis is absorption coefficient and x-axis is photon energy. The extracted optical bandgaps of these films are shown in Figure 6.37b. To understand the relationship between the Ta concentration and the bandgap of $La_{1-x}Ta_xO_y$ film, bandgaps of 20 and 60%Ta films are also included. Bandgaps of La_2O_3, $La_{1-x}Ta_xO_y$ ($x=0.35$), and Ta_2O_5 are about 5.25 eV, 5.2 eV, and 4.3 eV, respectively. It indicates that the bandgap of $La_{1-x}Ta_xO_y$ ($x=0.35$) is much larger than that of Ta_2O_5. We consider that the larger bandgap of $La_{1-x}Ta_xO_y$ ($x=0.35$) than that of Ta_2O_5 is attributed to the coupling effect between Ta and La 5d orbital through bonding to the same oxygen atom (Figure 6.38) [58]. As schematically shown in Figure 6.38, conduction band offsets of Ta_2O_5 and La_2O_3 with respect to silicon are determined by the energies of the unoccupied 5d-states (5d*) of Ta and La atoms, respectively. For $La_{1-x}Ta_xO_y$ film, due to the coupling effect between La 5d orbital and Ta 5d orbital through bonding to a common oxygen atom [58], the 5d* state energy of Ta atom is enhanced and then the conduction band offset of Ta with Si is increased compared to that of Ta_2O_5. Therefore, the bandgap of $La_{1-x}Ta_xO_y$ film is larger than that of Ta_2O_5 (Figure 6.38).

To understand the coupling effect more deeply, we can use Phillips's descriptions for bandgap estimation [59, 60]. The energy gap of semiconductors or insulators can be given by

$$E_{\mathrm{g}}^2 = E_{\mathrm{h}}^2 + C^2 \tag{6.16}$$

where E_{h} is equal to E_{g} in the case of a purely covalent group such as diamond or Si. The ionic or charge transfer contribution to E_{g} is represented by C.

In the case of Ta_2O_5, bandgap is determined by the difference of Ta_{5d} antibonding orbital energy and O_{2p} bonging orbital energy. According to Phillips' description, there are two parts of the bandgap, E_h and C. We can see here that C is equal to $2V_3$, which is fixed and equal to the difference of the molecular orbital energies of Ta_{5d}

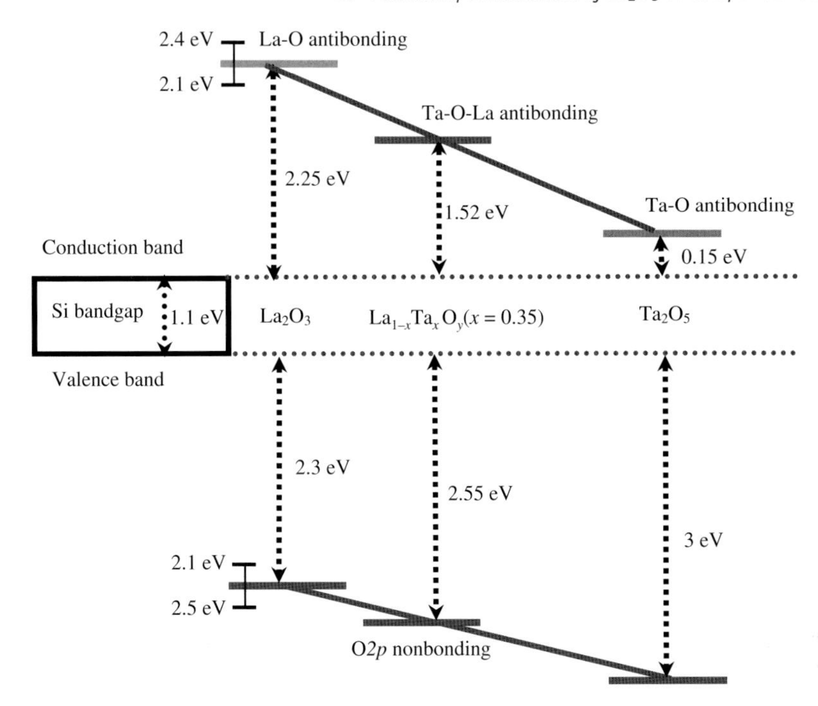

Figure 6.38 Electronic structures of La_2O_3, Ta_2O_5, and $La_{1-x}Ta_xO_y$.

and O_{2p} as shown in Figure 6.39. And E_h is equal to $2V_2$, which is due to the covalent bond contribution. In other words, V_2 is determined by the overlap of O and Ta atoms. The larger overlap will induce the larger V_2.The bandgap enlargement of $La_{1-x}Ta_xO_y$ compared to Ta_2O_5 is thought to be due to the increase in V_2 or E_h.

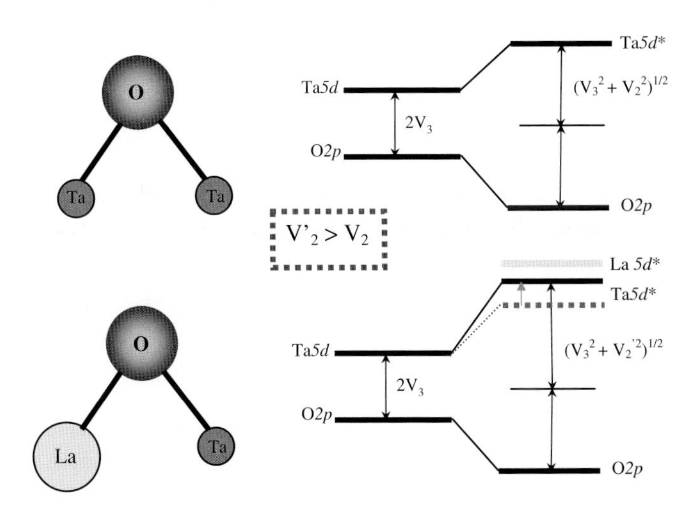

Figure 6.39 Bandgaps analysis of Ta_2O_5 and $La_{1-x}Ta_xO_y$ with Phillips theory.

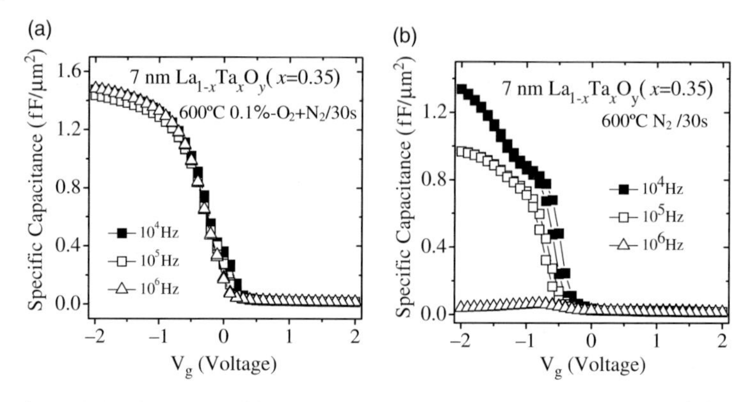

Figure 6.40 *C–V* curves of Au/$La_{1-x}Ta_xO_y$ ($x = 0.35$)/*p*-Si MIS capacitors annealed in N_2 ambient and oxygen ambient.

The ionicity of La is much larger than that of Ta. Therefore, La owns stronger ability to lose electron than Ta. And it is reasonable to think that in the La–O–Ta system, the direct charge transfer between Ta and O will be less than that in Ta–O–Ta. However, in order to form the bond, the overlap between Ta and O atoms has to increase, which will increase V_2 and then enlarge the bandgap.

Figure 6.40 shows *C–V* curves of Au/$La_{1-x}Ta_xO_y$ ($x = 0.35$)/Si MIS capacitors after annealing at 600 °C in 0.1%-O_2 + N_2 ambient and N_2 ambient. It can be observed obviously that the oxygen ambient annealing sample shows much better *C–V* characteristic than the N_2 ambient annealing sample. The *C–V* curve of oxygen ambient annealing sample is with a negligible hysteresis. This difference can attribute to the reaction of silicon substrate and $La_{1-x}Ta_xO_y$ film. As reported in the literature, at high temperature tantalum oxide can react with silicon substrate to form silicon oxide at the tantalum oxide/silicon interface [61]. This reaction can reduce the oxidation state of tantalum.

For $La_{1-x}Ta_xO_y$ film, the similar reaction will occur. In the case of N_2 ambient annealing, this reaction will consume the oxygen at the $La_{1-x}Ta_xO_y$/Si interface and drive the diffusion of oxygen in the bulk toward the $La_{1-x}Ta_xO_y$/Si interface and then induce the oxygen deficiency in the bulk $La_{1-x}Ta_xO_y$ film (Figure 6.41). The oxygen deficiency can cause defects in the film and induces the poor *C–V* characteristic. Meanwhile, in the case of oxygen ambient annealing, the oxygen in the annealing ambient can heal the oxygen deficiency in the bulk $La_{1-x}Ta_xO_y$ film during the annealing process. Therefore, oxygen ambient annealing film shows much better *C–V* characteristic than the N_2 ambient annealing film. At the same time, a much larger leakage current of the N_2 annealing film was also observed. As high-*k* materials are also being considered for use as insulators for MIM capacitors due to the higher permittivity of $La_{1-x}Ta_xO_y$, it might be promising as MIM insulator. Therefore, we also investigated the *C–V* and J_g–V_g characteristics of Au/$La_{1-x}Ta_xO_y$/Pt MIM capacitors (Figure 6.42). A very low leakage current when the bias is below 5 V and a very high capacitance density are obtained.

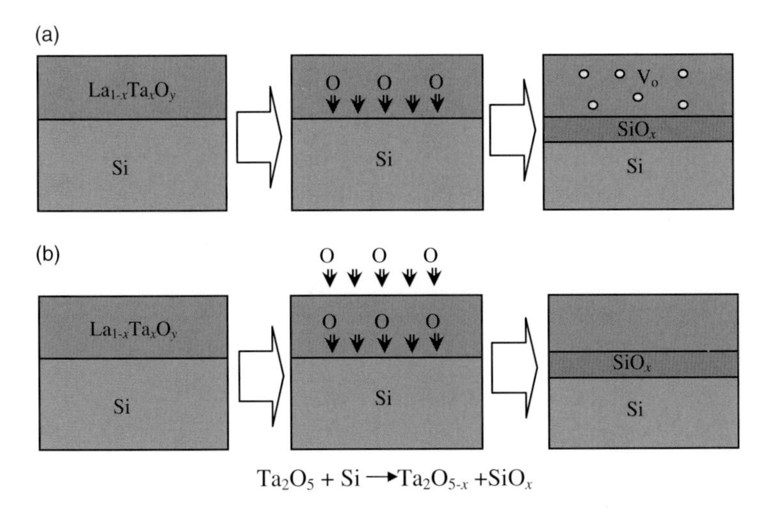

Figure 6.41 Effect of annealing ambient on the $LaTaO_x$ film quality. (a) Nitrogen ambient annealing; (b) oxygen ambient annealing.

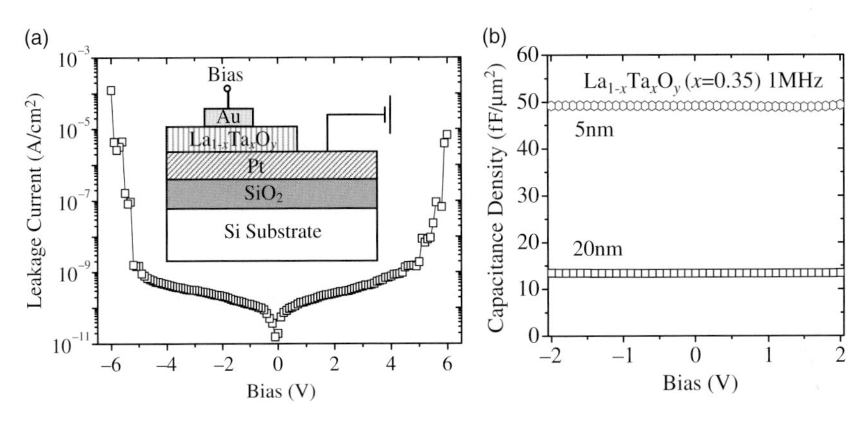

Figure 6.42 (a) J_g–V characteristic of the Au/$La_{1-x}Ta_xO_y$ (20 nm, $x = 0.35$)/Pt MIM capacitors; (b) C–V characteristic of Au/$La_{1-x}Ta_xO_y$ ($x = 0.35$)/Pt MIM capacitors with different film thicknesses.

6.8 Summary

In this chapter, two most important issues, moisture absorption phenomenon and low experimental permittivity, of La_2O_3 films as high-k gate insulators for advanced CMOS devices have been experimentally and theoretically investigated. It has been found that the moisture absorption degrades the permittivity of La_2O_3 film annealed in N_2 ambient after exposure to the air for several hours because of the formation of

$La(OH)_3$ with a lower permittivity and thus concluded that the moisture absorption should be a possible reason for the scattering *k*-values of La_2O_3 films. Furthermore, AFM results indicate that the moisture absorption also increases the surface roughness of La_2O_3 films on silicon. Thus, *in situ* gate electrode process will be needed for La_2O_3 CMOS application.

Accordingly, the moisture absorption phenomenon in main high-*k* gate oxides has been theoretically discussed by comparing the Gibbs free energy change in the moisture absorption reactions of these oxides. The results show that the moisture absorption could occur in most high-*k* oxides, especially in rare-earth oxides. On the other hand, La_2O_3 shows the largest moisture absorption reaction speed among main high-*k* oxide candidates. To enhance the moisture resistance of La_2O_3, doping a second oxide, which has a stronger moisture resistance than La_2O_3, could be an applicable solution.

The moisture absorption and associated leakage current of La_2O_3 films were suppressed by UV ozone posttreatment. The suppression effect by UV ozone treatment has been considered to come from the healing of oxygen vacancies in La_2O_3 films since the oxygen ambient annealing also shows the same suppression effect. Compared to oxygen ambient annealing, however, UV ozone posttreatment can be carried out at low temperatures, which prevents the thick interface layer formation.

With the phase control method, the permittivities and the moisture resistance of La_2O_3 films have been improved significantly. Higher-*k* well-crystallized lanthanum-based oxide films, $La_{2-x}Y_xO_3$, were prepared, which own a permittivity as high as 28 with an appropriate Y concentration due to the formation of high-permittivity hexagonal phase and also show much better resistance to moisture after annealing at 600 °C than La_2O_3 film. $La_{1-x}Ta_xO_y$ films with different Ta concentrations were investigated. The $La_{1-x}Ta_xO_y$ ($x = 0.35$) film shows not only a high crystallization temperature (>1000 °C) but also a high permittivity (~30).

Furthermore, a systematic discussion on the crystallization temperature of lanthanum-based ternary oxide has been given, which provides a possible guideline for preparing amorphous or well-crystallized lanthanum-based ternary oxides. This should also be useful for other high-*k* materials to prepare well-crystallized or amorphous films as gate insulators.

Acknowledgments

The author would like to thank Prof. Akira Toriumi, Prof. Kentaro Kyuno (now with Shibaura Institute of Technology, Japan), and Prof. Koji Kita at the University of Tokyo, Japan for their continuous supervision and support during my Ph.D study, which produced the main results reviewed in this chapter. The author also acknowledges the financial support from National Program on Key Basic Research Project (973 Program) (No. 2011CBA00607) and the Young Scientists Fund of the National Natural Science Foundation of China (No. 61106089) to continue the research topics in this chapter.

References

1 Chin, A., Yu, Y.H., Chen, S.B., Liao, C.C., and Chen, W.J. (2000) High quality La_2O_3 and Al_2O_3 gate dielectrics with equivalent oxide thickness 5–10 Å. Digest of Technical Papers Symposium on VLSI Technology, 16.

2 Wilk, G.D., Wallace, R.M., and Anthony, J.M. (2001) High-*k* gate dielectrics: current status and materials properties considerations. *J. Appl. Phys.*, **89**, 5243.

3 Robertson, J. (2004) High dielectric constant oxides. *Eur. Phys. J. Appl. Phys.*, **28**, 265.

4 Gonzales-Elipe, A.R., Espinos, J.P., Fernandez, A., and Munuera, G. (1990) XPS study of the surface carbonation/hydroxylation state of metal oxides. *Appl. Surf. Sci.*, **45**, 103.

5 Iwai, H., Ohmi, S.I., Akama, S., Ohshima, C., Kikuchi, A., Kashiwagi, I., Taguchi, J., Yamamoto, H., Tonotani, J., Kim, Y., Ueda, I., Kuriyama, A., and Yoshihara, Y. (2002) Advanced gate dielectric materials for sub-100nm CMOS. IEDM Technical Digest, 625.

6 Wolf, S. (2002) *Silicon Processing for The VLSI Era. Volume 4: Deep-Submicron Process Technology*, Lattice Press.

7 Devine, R.A.B. (2003) Infrared and electrical properties of amorphous sputtered $(La_xAl_{1-x})_2O_3$ films. *J. Appl. Phys.*, **93**, 9938.

8 Yamada, H., Shmizu, T., Kurokawa, A., Ishii, K., and Suzuki, E. (2003) MOCVD of high-dielectric-constant lanthanum oxide thin films. *J. Electrochem. Soc.*, **150** (8), G429.

9 Jin, H.J., Choi, D.J., Kim, K.H., Oh, K.Y., and Hwang, C.J. (2003) Effect of structural properties on electrical properties of lanthanum oxide thin film as a gate dielectric. *Jpn. J. Appl. Phys.*, **42**, 3519.

10 Daillant, J. and Gibaud, A. (1999) *X-Ray and Neutron Reflectivity: Principles and Applications*, Springer.

11 Tolan, M. (1999) *X-Ray Scattering from Soft-Matter Thin Films*, Springer.

12 Tompkins, H.G. and McGahan, W.A. (1999) *Spectroscopic Ellipsometry and Reflectometry*, Wiley-Interscience.

13 Shimizu, H., Kita, K., Kyuno, K., and Toriumi, A. (2005) Kinetic model of Si oxidation at HfO_2/Si interface with post deposition annealing. *Jpn. J. Appl. Phys.*, **44** (8), 6131.

14 Neumann, A. and Walter, D. (2006) The thermal transformation from lanthanum hydroxide to lanthanum hydroxide oxide. *Thermochimica Acta*, **445**, 200.

15 Shannon, R.D. (1993) Dielectric polarizabilities of ions in oxides and fluorides. *J. Appl. Phys.*, **73**, 348.

16 Koehler, W.C. and Wollan, E.O. (1953) Neutron-diffraction study of the structure of the A-form of the rare earth sesquioxides. *Acta Crystallogr.*, **6**, 741.

17 Zachariasen, W.H. (1948) Crystal chemical studies of the 5*f*-series of elements, I. New structure types. *Acta Crystallogr.*, **1**, 265.

18 Guha, S., Cartier, E., Gribelyuk, M., Bojarczuk, N., and Copel, M. (2000) Atomic beam deposition of lanthanum- and yttrium-based oxide thin films for gate dielectrics. *Appl. Phys. Lett.*, **77**, 2710.

19 Gougousi, T., Niu, D., Ashcraft, R.W., and Parsons, G.N. (2003) Carbonate formation during post-deposition ambient exposure of high-*k* dielectrics. *Appl. Phys. Lett.*, **83**, 3543.

20 Yokogawa, Y., Yoshimura, M., and Somiya, S. (1991) Lattice energy and polymorphism of rare-earth oxides. *J. Mater. Sci. Lett.*, **10**, 509.

21 Kapustinskii, A.F. (1956) Lattice energy of ionic crystals. *Q. Rev. Chem. Soc.*, **10**, 283.

22 Ohni, S., Akama, S., Kikuchi, A., Kashiwagi, I., Oshima, C., Taguchi, J., Yamamoto, H., Kobayashi, C., Sato, K., Tageda, M., Oshima, K., Ishiwara, H., and Iwai, H. (2001) Rare earth metal oxide gate thin films prepared by E-beam deposition. IWGI, 200.

23 Mizuno, M., Rouanent, A., Ymamada, T., and Noguchi, T. (1976) Phase diagram of the system La_2O_3-Y_2O_3 at high temperature. *Yogyo-Kyokai-Shi.*, **84**, 342.

24 Navrotsky, A. (2005) Thermochemical insights into refractory ceramic materials based on oxides with large tetravalent cations. *J. Mater. Chem.*, **15**, 1883.

25 Song, W.J., So, S.K., Wang, D.Y., Qiu, Y., and Cao, L.L. (2001) Angle dependent X-ray photoemission study on UV-ozone treatments of indium tin oxide. *Appl. Surf. Sci.*, **177**, 158.
26 Mortimer, R.G. (2000) *Physical Chemistry*, 2nd edn, Academic Press.
27 Vasil'ev, V.P., Lytkin, A.I., and Chernyavskaya, N.V. (1999) Thermodynamic characteristics of zirconium and hafnium hydroxides in aqueous. *J. Therm. Anal. Calorim.*, **55**, 1003.
28 Morant, C., Sanz, J.M., Galan, L., Soriano, L., and Rueda, F. (1989) The O1s X-ray absorption spectra of transition-metal oxides: the TiO_2-ZrO_2-HfO_2 and V_2O_5-Nb_2O_5-Ta_2O_5 series. *Surf. Sci.*, **218**, 331.
29 Zhao, Y., Kita, K., Kyuno, K., and Toriumi, A. (2009) Band gap enhancement and electrical properties of La_2O_3 films doped with Y_2O_3 as high-*k* gate insulators. *Appl. Phys. Lett.*, **94**, 042901.
30 Zhao, Y., Kita, K., Kyuno, K., and Toriumi, A. (2006) Higher-*k* $LaYO_x$ films with strong moisture-resistance. *Appl. Phys. Lett.*, **89**, 252905.
31 Zhao, Y., Kita, K., Kyuno, K., and Toriumi, A. (2007) Suppression of leakage current and moisture absorption of La_2O_3 films with ultraviolet ozone post treatment. *Jpn. J. Appl. Phys.*, **46**, 4189.
32 Kita, K., Kyuno, K., and Toriumi, A. (2009) Origin of electric dipoles formed at high-*k*/SiO_2 interface. *Appl. Phys. Lett.*, **94**, 132902.
33 Zhao, Y., Toyama, M., Kita, K., Kyuno, K., and Toriumi, A. (2006) Moisture-absorption-induced permittivity deterioration and surface roughness enhancement of lanthanum oxide films on silicon. *Appl. Phys. Lett.*, **88**, 072904.
34 Gusev, E.P., Narayanan, V., and Frank, M.M. (2006) Advanced high-*k* dielectric stacks with PolySi and metal gates: recent progress and current challenges. *IBM J. Res. Dev.*, **50**, 387.
35 Christen, H.M., Jellison, G.E., Jr., Okubo, I., Huang, S., Reeves, M.E., Cicerrella, E., Freeouf, J.L., Jia, Y., and Schlom, D.G. (2006) Dielectric and optical properties of epitaxial rare-earth scandate films and their crystallization behavior. *Appl. Phys. Lett.*, **88**, 262906.
36 Dimoulas, A. (2004) $La_2Hf_2O_7$ high-*k* gate dielectric grown directly on Si(001) by molecular-beam epitaxy. *Appl. Phys. Lett.*, **85**, 3205.
37 Wang, X.P., Li, M.F., Ren, C., Yu, X.F., Shen, C., Ma, H.H., Chin, A., Zhu, C.X., Ning, J., Yu, M.B., and Kwong, D.L. (2006) Tuning effective metal gate work function by a novel gate dielectric $HfLaO_x$ for nMOSFETs. *IEEE Electron. Dev. Lett.*, **27**, 31.
38 Wang, X.P., Li, M.F., Chin, A., Zhu, C.X., Shao, J., Lu, W., Shen, X.C., Yu, X.F., Chi, R., Shen, C., Huan, A.C.H., Pan, J.S., Du, A.Y., Lo, P., Chan, D.S.H., and Kwong, D.L. (2006) Physical and electrical characteristics of high-*k* gate dielectric $Hf_{(1-x)}La_xO_y$. *Solid State Electron.*, **50**, 986.
39 Edge, L.F., Schlom, D.G., Brewer, R.T., Chabal, Y.J., Williams, J.R., Chambers, S.A., Hinkle, C., Lucovsky, G., Yang, Y., Stemmer, S., Copel, M., Hollander, B., and Schubert, J. (2004) Suppression of subcutaneous oxidation during the deposition of amorphous lanthanum aluminate on silicon. *Appl. Phys. Lett.*, **84**, 4629.
40 Wilk, G.D., Wallace, R.M., and Anthony, J.M. (2001) High-*k* gate dielectrics: current status and materials properties considerations. *J. Appl. Phys.*, **89**, 5243.
41 Yamamoto, Y., Kita, K., Kyuno, K., and Toriumi, A. (2006) Structural and electrical properties of HfLaOx films for an amorphous High-*k* gate insulator. *Appl. Phys. Lett.*, **89**, 032903.
42 Vellianitis, G., Apostolopoulos, G., Mavrou, G., Argyropoulos, K., Dimoulas, A., Hooker, J.C., Conard, T., and Butcher, M. (2004) MBE lanthanum-based high-*k* gate dielectrics as candidates for SiO_2 gate oxide replacement. *Mater. Sci. Eng. B*, **109**, 85.
43 Joshi, P.C. and Cole, M.W. (1999) Influence of post-deposition annealing on the enhanced structural and electrical properties of amorphous and crystalline Ta_2O_5 thin films for dynamic random

access memory applications. *J. Appl. Phys.*, **86**, 871.
44 Robertson, J. (2000) Band offsets of wide-band-gap oxides and implications for future electronic devices. *J. Vac. Sci. Technol.*, **B18**, 1785.
45 Shishido, T., Okamura, K., and Yayima, S. (1978) Ln-M-O glasses obtained by rapid quenching using a laser beam. *J. Mater. Sci.*, **13**, 1006.
46 Kita, K., Kyuno, K., and Toriumi, A. (2005) Permittivity increase of yttrium-doped HfO_2 through structural phase transformation. *Appl. Phys. Lett.*, **86**, 102906.
47 Bottcher, C.J.F. (1973) *Theory of Electronic Polarization*, Elsevier.
48 Hirosaki, N., Ogata, S., and Kocer, C. (2003) *Ab initio* calculation of the crystal structure of the lanthanide Ln_2O_3 sesquioxides. *J. Alloy. Compd.*, **351**, 31.
49 Coutures, J., Rouanent, A., Verges, R., and Foex, E.T.M. (1976) Etude a haute temperature des systems formes par le sesquioxyde de lanthane et les sesquioxydes de lanthanides. I. Diagrammes de phases (1400 °C < T < T liquide). *J. Solid State Chem.*, **171**, 17.
50 Manchanda, L., Morris, M.D., Green, M.L., Dover, R.B., Klemens, F., Sorsch, T.W., Silverman, P.J., Wilk, G.D., Busch, B., and Aravamudhan, S. (2001) Multi-component high-*k* gate dielectrics for the silicon industry. *Microelectron. Eng.*, **59**, 351.
51 Pisecny, P., Husekova, K., Frohlich, K., Harmatha, L., Soltys, J., Machajdik, D., Espinos, J.P., Jergel, M., and Jakabovic, J. (2004) Growth of lanthanum oxide films for application as a gate dielectric in CMOS technology. *Mater. Sci. Semicond. Process.*, **7**, 231.
52 Abe, Y., Kawamura, M., and Sasaki, K. (2005) Oxidation and morphology change of Ru films caused by sputter deposition of Ta_2O_5 films. *Jpn. J. Appl. Phys.*, **44**, 1941.
53 Ushakov, S.V., Brown, C.E., and Navrotsky, A. (2004) Effect of La and Y on crystallization temperature of hafnia and zirconia. *J. Mater. Res.*, **19**, 693.
54 Yajima, S., Okayama, K., and Shishido, T. (1973) Glass formation in the Ln-Al-O system. *Chem. Lett.*, 1327.
55 Ohmi, S., Kobayashi, C., Tokumitsu, E., Ishiwara, H., and Iwai, H. (2001) Low leakage La2O3 gate insulator film with EOTs of 0.8–1.2nm. Extended Abstracts of International Conference on Solid State Device and Materials (SSDM) 496.
56 Kakio, S., Shimatai, Y., and Nakagawa, Y. (2003) Shear-horizon-type surface acoustic waves on quartz with Ta_2O_5 thin film. *Jpn. J. Appl. Phys. 1.*, **42**, 3161.
57 Li, H.J., Price, J., Gardner, M., Lu, N., and Kwong, D.L. (2006) High permittivity quaternary metal ($HfTaTiO_x$) oxide layer as an alternative high-*k* gate dielectric. *Appl. Phys. Lett.*, **89**, 103523.
58 Lucovsky, G. (2003) Electronic structure of transition metal/rare earth alternative high-*k* gate dielectrics: interfacial band alignments and intrinsic defects. *Microelectron. Reliab.*, **43**, 1417.
59 Phillips, J.C. (1970) Ionicity of the chemical bond in crystals. *Rev. Mod. Phys.*, **42**, 317.
60 Phillips, J.C. and Van Vechten, J.A. (1969) Dielectric classification of crystal structures, ionization potentials, and band structures. *Phys. Rev. Lett.*, **22**, 705.
61 Mao, A.Y., Son, K.A., Hess, D.A., Brown, L.A., White, J.M., Kwong, D.L., Roberts, D.A., and Vrtis, R.N. (1999) Annealing ultra thin Ta_2O_5 films deposited on bare and nitrogen passivated Si(100). *Thin Solid Film*, **349**, 230.

7
Characterization of High-*k* Dielectric Internal Structure by X-Ray Spectroscopy and Reflectometry: New Approaches to Interlayer Identification and Analysis

Elena O. Filatova, Andrey A. Sokolov, and Igor V. Kozhevnikov

7.1
Introduction

In order to scale down complementary metal oxide semiconductor (CMOS) devices below 0.1 μm, a thickness smaller than 2 nm of the SiO_2 gate dielectric is required. This results in leakage current arising from direct tunneling of electrons through the layer. The increase in the relative permittivity compared to that of SiO_2 ($k = 3.9$) [1, 2] permits to use physically thicker films to obtain the same effective capacitance as devices with physically thinner SiO_2 layers. The use of a physically thicker layer of an alternative "high-k" dielectric material may solve leakage and reliability problems. In many cases, properties of these layers govern the very performance of the fabricated device making mandatory strict structural and chemical control of the films and their interfaces. In particular, formation of an interlayer (IL) due to chemical reaction with a substrate may have detrimental effect on electronic properties of interfaces, for instance, precluding one from reaching the quantum confinement regime in nanoelectronic structures.

Search of suitable "high-*k*" dielectric material inevitably focused on the dielectric constants *k*, which is a function of material polarizability. As follows from Figure 7.1 [3], in the GHz frequency window, which is needed in CMOS devices, the polarizability stems from electronic and ionic dipole contribution. From the electronic contribution point of view, dielectrics with a smaller bandgap are preferable because of a greater mixture of excited electronic states compared to the ground state under an applied electric field, which results in increased polarizability and, as a result, a larger *k*-value. From the standpoint of ionic contribution, unique d orbitals with a lower symmetry of transition metal oxide initiate a greater overlap of d orbitals with p orbitals of oxygens and produce an additional electron transfer from the oxygen to the metal ion, which results in greater ion off-center displacement and, accordingly, increase in polarizability and larger *k*-value [4]. In particular, due to the open shell d-electrons with highly anisotropic electronic spatial distributions, transition metal oxide ions can provide an increased ionic and electronic component of polarizability and, as a result, a larger *k*-value. The main characteristics (bandgap E_g

High-k Gate Dielectrics for CMOS Technology, First Edition. Edited by Gang He and Zhaoqi Sun.

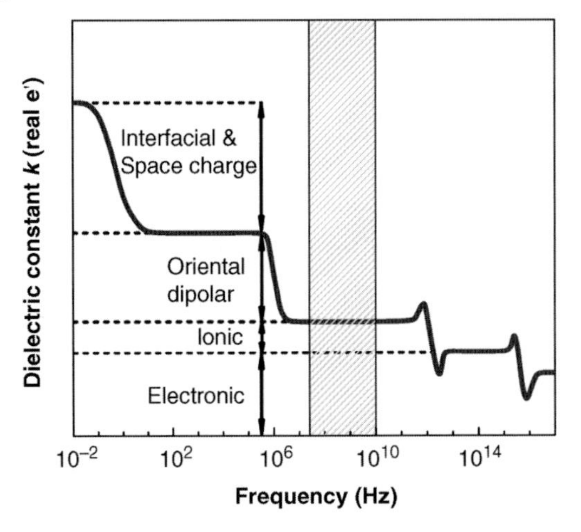

Figure 7.1 A schematic diagram illustrating the dependence of static dielectric constant on frequency (from Ref. [3] Choi, J.H., Mao, Y., and Chang, J.P. (2011) Development of hafnium based high-k materials - A review. *Materials Science and Engineering R*. 72, 97–136).

versus static dielectric constant k and electrical breakdown field E_{bd}) for representative high-k materials are shown in Figure 7.2 [5–7].

For successful implementation of high-k dielectric materials, the most important criterion is that the oxides should be stable and compatible with the crystal silicon substrate. The most important requirements include [1, 10] (Choi, J.H., Mao, Y., and Chang, J.P. (2011) Development of hafnium based high-k materials - A review. *Materials Science and Engineering R*. **72**, 97–136). (1) bandgap and band offset more than 1 eV with silicon, which guarantees the properties of an insulator; (2) thermodynamical stability with silicon substrate against crystallization and phase separation; (3) interface quality to form a good electrical interface with a silicon substrate

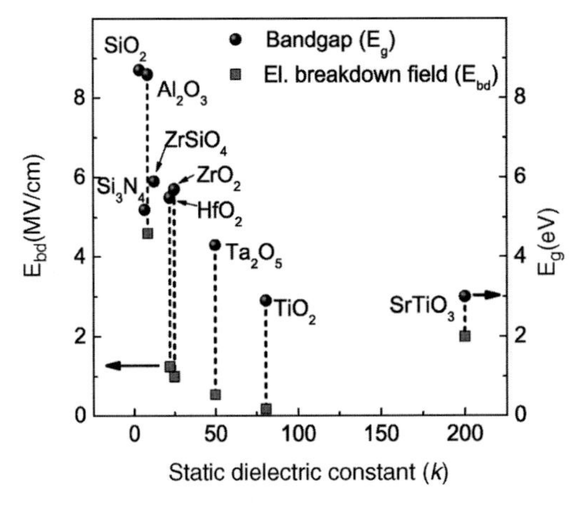

Figure 7.2 Bandgap E_g versus static dielectric constant k and electrical breakdown field E_{bd} for representative high-k materials (from Refs [5–9]).

that minimizes the fixed charge and interface-trap density; and (4) minimal bulk electrically active defects–intrinsic defects.

Our main focus is the third criterion, as the properties of the interface between Si and high-*k* dielectric material have a direct effect on the electron transport in the Si surface channel. The kinetics of the IL growth, its thickness, and its composition depend on a large number of factors including the film deposition method, substrate surface preparation, and postdeposition processing. A large number of incoming influences make it difficult to predict *a priori* IL properties. As a result, the formation of IL at the film/substrate interface needs to be closely monitored. We discuss the analytical methods suitable for both nm thin overlayers and the ILs formed at their interfaces. Potentialities of hard X-ray photoelectron spectroscopy (HAXPES) and soft X-ray (SXR) reflectometry techniques are analyzed and compared with well-known high-resolution transmission electron microscopy (HRTEM).

7.2 Chemical Bonding and Crystalline Structure of Transition Metal Dielectrics

Figure 7.3 shows the schematic molecular orbital energy-level diagram for a group IV transition metal (TM) atom bonded octahedrally to six oxygen atoms from the papers of G. Lucovsky [11, 12]. As follows from the calculations [11–13], the top of the valence band is associated with nonbonding π orbitals of oxygen atom 2p states. The first two conduction bands are assigned to transition metal atom d states with $t_{2g}(\pi^*)$ and $e_g(\sigma^*)$ symmetries, respectively, and are separated in energies. The next conduction band with $a_{1g}(\sigma^*)$ symmetry is derived from TM s states. The energy separation between the top of valence band and the $a_{1g}(\sigma^*)$ band edge defines an effective ionic bandgap

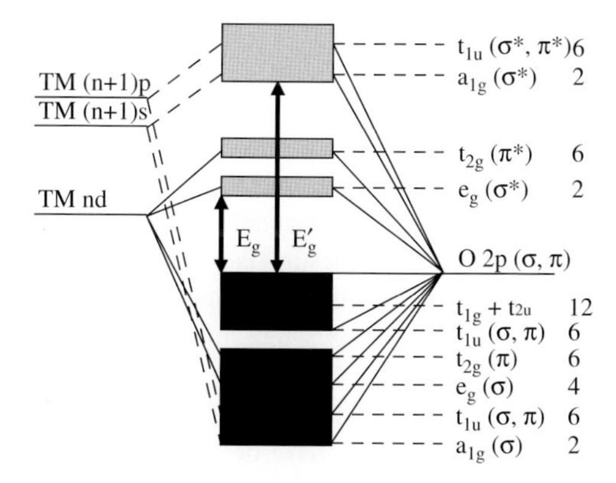

Figure 7.3 Molecular orbital energy-level diagram for a group IV transition metal in an octahedral bonding arrangement with six oxygen atom neighbors. The lowest band gap, E_g, is between the valence band oxygen 2p states and the localized $t_{2g}(\pi^*)$ states, and the ionic band gap, E'_g, is between the valence band oxigen2p states and the $a_{1g}(\sigma^*)$ band (from Refs [11, 12]).

with the value corresponding to the energy of nontransition metal insulating oxides such as MgO or Al_2O_3 (~8–9 eV) [11, 14]. Electron energy loss and X-ray spectroscopy investigations of crystalline TiO_2 confirm the ionic band energy >8 eV [15] in this compound. The analysis extended to other coordination of transition metal atoms indicates that symmetry character of the lowest conduction band is a function of the coordination and symmetry of the TM atom [12, 14, 16].

From the scaling point of view, it is the extremely important to note two aspects following from the MO calculations [11–13]: (i) the lowest antibonding d σ^* and π^* states are close in energy to the atomic nd state of TM atom; (ii) the energy separation between the nd and the n + 1s derived antibonding states is correlated with the difference between the atomic nd and the n + 1s states of the TM atoms; (iii) the larger the energy difference between the TM atomic s and d states, the closer the d state bands are to the top of valence band; (iv) bandgap E_g should scale according to the energy difference between oxygen atom p states and the TM d states.

According to the aforecited, a significant attention in the chapter has been paid to metal oxides TiO_2 and HfO_2, besides some investigations for Al_2O_3.

Depending on growth processes, TiO_2 films can be synthesized in different crystal modifications (anatase, rutile, and brookite) or in an amorphous phase. In both rutile and anatase structures, each titanium atom is coordinated to six oxygen atoms and each oxygen atom is coordinated to three titanium atoms. The essential difference between rutile and anatase lies in the secondary coordination [17, 18] and in the way the TiO_6 octahedra are joined together by sharing edges and corners. In rutile, two edges are shared and in anatase four. Following the structural information of TiO_2 polymorphs, in the rutile structure there are two types of TiO_6 octahedral with different orientations [19–21], which are joined together via two opposite edges to form a chain. The chains are linked by common corners and form an octahedral network. The $3de_g$ orbitals are directed toward the corners of the octahedral (oxygen anions), which means that they should be more sensitive to changes in the chemical state of the oxygen atoms as compared to the $3dt_{2g}$ orbitals. In the anatase structure, the octahedra are connected by edge sharing only.

TiO_2 has the very high dielectric constants (80–110) due to stronger polarizability and correspondingly the small bandgap (3.2–3.5 eV for amorphous and 3.0–3.2 eV for crystalline phase) [22]. The high dielectric constant makes it a potential gate oxide candidate, but because small bandgap and multiple oxidation states lead to various Ti–O bonds that provide leakage paths [23–28], TiO_2 application is strongly limited.

HfO_2 is one of the most attractive high-*k* dielectric materials due to its high dielectric constant ($k = 22–25$), the large bandgap (5.5–6.0), high breakdown field (3.9–6.7 MV/cm), high thermal stability, and large heat of formation (−271 kcal/mol) [29–33]. One stable monoclinic phase and four metastable phases, cubic, tetragonal, orthorhombic I, and orthorhombic II, have been identified for HfO_2 [34]. An amorphous modification can also be fabricated [35]. For a cubic HfO_2 structure, each metal ion is surrounded by eight equivalent O atoms (eight-coordinated environment of the Hf ion) and there is a four-coordinated environment of the O atom. The tetragonal HfO_2 structure has the metal ions in a distorted eightfold coordination. But the stable monoclinic phase has sevenfold coordinated metal ions (sd^3p^3 hybridization). Crystalline films often have a monoclinic structure [36], but they can also have mixed monoclinic and tetragonal [37] phases.

Al_2O_3 is considered a candidate gate oxide for the following reasons: (i) it has a rather high dielectric constant ($k=9$) [1, 10, 38, 39] and a large bandgap (8.8 eV) [1, 10, 38, 39]; (ii) it is one of the very few metal oxides that are thermodynamically stable on Si [1, 39, 40]; and (iii) the Al_2O_3/Si interface mimics, to a certain extent, the SiO_2/Si interface. Al_2O_3 is the only s–p-bonded oxide. For the MOS application, an amorphous Al_2O_3 film is preferred. The amorphous structure of Al_2O_3 consists of both (AlO_4) and (AlO_6) polyhedra dominated by slightly distorted tetrahedra, linked to each other by corners, with the Al−O−Al bond angle distribution peaked at 120°. The proportion between tetrahedra and octahedra depends strongly on preparation method [41]. This defines a network with a majority of three- and fourfold rings, where the threefold rings are planar, but where the fourfold rings present a more complex structure.

7.3
NEXAFS Investigation of Internal Structure

As was mentioned above, HfO_2 is considered one of the most promising materials to satisfy requisite properties. Most works on HfO_2 have been focused on manufacturing amorphous films to replace SiO_2, but this process is unpredictable because the microstructure of films is largely dependent on thickness, technology of synthesis, and utilizable precursors [42–44]. The technology of preparing films, that is, method and condition of synthesis (rate of deposition, temperature of substrate, etc.), as well as utilizable precursors, is an important issue for fabricating films with required properties. Another problem arises from the formation of an IL caused by oxidation of silicon during HfO_2 deposition: the formed SiO_2-like IL not only limits the gate stack downscaling but also affects electron and hole mobilities in surface channels of Si MOS devices with HfO_2 insulation [42–46]. In this chapter, we compare HfO_2/Si structures with hafnia layers grown by atomic layer deposition (ALD) and metalorganic chemical vapor deposition (MOCVD) techniques. These two growth methods allow the HfO_2 thickness control down to a monolayer but employ different thermal budgets leading to differences in properties of the IL formed between Si and HfO_2. Studied samples were prepared on the identically treated (IMEC cleaned) (100) n-type Si surfaces that result in a 0.8 nm thick initial Si oxide layer. The ALD of HfO_2 was conducted at 300 °C using $HfCl_4$ and H_2O precursors. The MOCVD growth was carried out at 485° C from tetrakisdiethylaminohafnium and O_2 precursors. To ensure that the pristine IL behavior is studied, no postdeposition processing was applied.

The O 1s reflection spectra of HfO_2 films 5, 17, and 100 nm thick synthesized by ALD and measured at a grazing incidence angle of 2° [47] are presented in Figure 7.4. In the same figure, the O 1s absorption spectra calculated from the experimental reflection spectra by means of Kramers–Kronig transform (KKT), which is described in detail in Ref. [48], are plotted. The spectral dependences of the reflectivity were

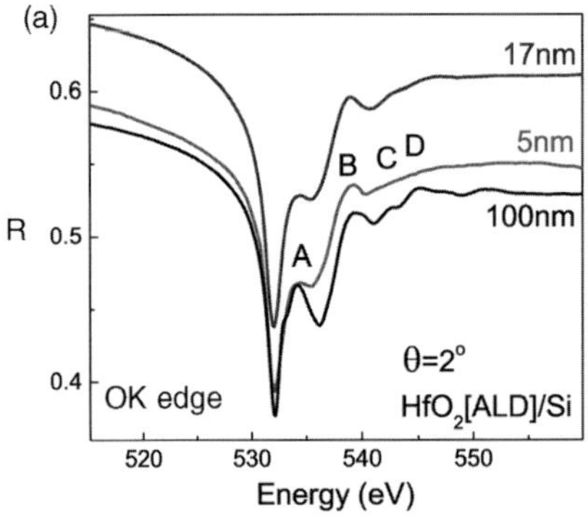

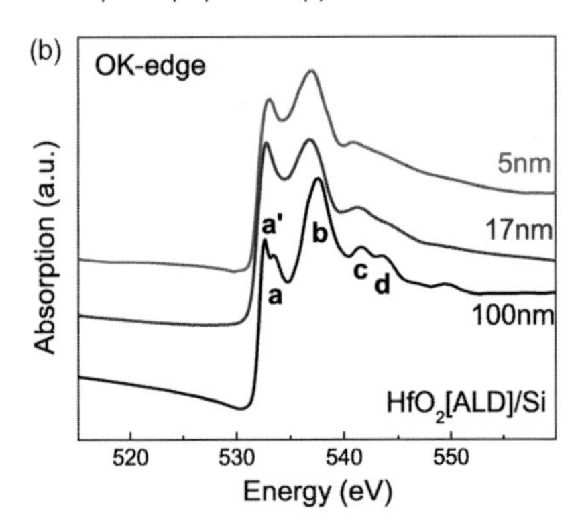

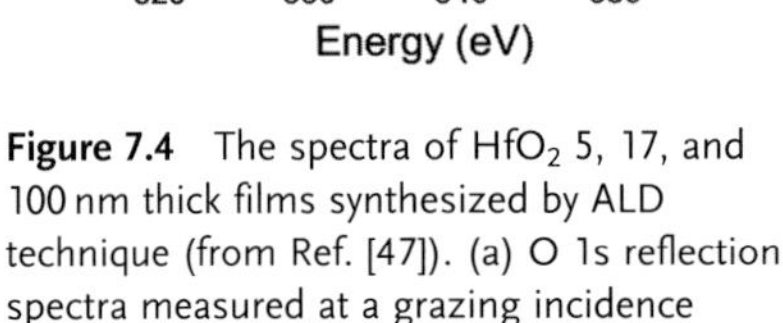

Figure 7.4 The spectra of HfO_2 5, 17, and 100 nm thick films synthesized by ALD technique (from Ref. [47]). (a) O 1s reflection spectra measured at a grazing incidence angle of 2°; (b) O 1s absorption spectra calculated from the experimental reflection spectra by means of the Kramers– Kronig transform.

measured using s-polarized synchrotron radiation in the reflectometer setup on the optics beamline (D-08-1B2) at the Berlin Synchrotron Radiation facility (BESSY II) of the Helmholtz-Zentrum Berlin [49, 50]. The accuracy of the energy scale was 10 meV. A GaAsP diode, together with a Keithley electrometer (617), was used as a detector. The accuracy of the energy scale was 10 meV. To reduce higher diffraction orders in the incident beam, a system of absorption filters was used.

As was established above, the conduction band of the HfO_2 consists of a mixture of significantly localized antibonding d σ^* and π^* states plus oxygen antibonding 2p σ and π states so that it tends to change the energy states with respect to the local bonding distortion environment. The first two conduction bands are assigned to Hf 3d states with $t_{2g}(\pi^*)$ and $e_g(\sigma^*)$ symmetries and are separated in energies. Thus, the molecular orbitals of HfO_2 derived from a linear combination of atomic orbitals (LCAOs) are characterized by four unoccupied orbitals [36]: e_g(Hf 5d + O 2pπ), t_{2g}(Hf 5d + O 2pσ), a_{1g}(Hf 6s + O 2p), and t_{1u}(Hf 6p + O 2p). In this classification, one can expect four features in the O K absorption spectrum of HfO_2 connected with one electron transitions from the O 1s orbital to the empty electronic states e_g, t_{2g}, a_{1g}, and t_{1u}. The analysis of the energy position of the peaks in the absorption spectra of the films of different thicknesses (Figure 7.4) shows that the energy separation between peaks a and b is equal to 4.0 eV (film of 5 nm thickness), 4.1 eV (film of 17 nm thickness), and 4.4 eV (film of 100 nm thickness) within the experimental accuracy, which is close to the splitting of Hf 5d states into t_{2g} and e_g components in the anisotropic field of the tetrahedral oxygen, which is equal to 4.3 eV [51, 52]. Peaks a and b, therefore, reflect core–electron transitions in the oxygen atoms to the lowest unoccupied Hf de_g and Hf dt_{2g} electronic states, which are hybridized with the 2p states of the oxygen atoms. Let us look at feature a, which is associated with doubly degenerated e_g component (Figure 7.4). Obviously, there is a broadening of this feature with the growth of the film thickness. Moreover feature a demonstrates a double structure a–a' in the absorption spectrum of the film of 100 nm thickness.

It is known that the state degeneracy can be removed due to local symmetry distortions or a decrease in coordination number. An analysis of different phases of HfO_2 points to the sevenfold coordinated metal ions in the monoclinic phase that leads to a splitting of the e_g component caused by the dynamic Jahn–Teller effect [36]. Suppression of the Jahn–Teller d state degeneracy splitting takes place, according to Ref. [36], because of grain size effects when grain size $\leq$2 nm. It might be concluded that the 100 nm thick film has the polycrystalline structure with size of the grains $\leq$2 nm. Moreover, the microstructure of the thick film involves essentially the monoclinic phase. Bands c and d in the O 1s spectrum of the 100 nm thick film have to be related to the empty electronic states with mixed (Hf 6s, 6p + O 2p) character. The probability of transitions to these states is rather low. So, the double structure *c–d* cannot be observed in the amorphous state of the film. As can be seen from Figure 7.4, there is only one wide band c in the spectra of 5 and 17 nm thick films. This result agrees well with experimental results [34, 36, 37] demonstrating the dependence of the manifestation of bands c and d, accompanied by a manifestation of Jahn–Teller effect, on the size of the grains. Even in the case of signatures of ordering on a subnanometer scale in the film, one can see a slight splitting of the features *c–d*. The best resolution of this structure is accompanied by a manifestation of the Jahn–Teller effect. One can conclude that, according to the reflection spectroscopy investigations, the 5 and 17 nm thick films are amorphous, while the film of 100 nm thickness is polycrystalline essentially in the monoclinic phase.

The films mentioned above were characterized by XRD (Figure 7.5) also. XRD data for Si(100) substrates before film synthesis have been obtained to take into account the background. Using the Gauss's approximation of the radiograph of the Si(100) substrate, the background has been constructed. As can be seen from Figure 7.5, the 100 nm thick film has a practically (100%) crystalline structure and consists of two phases (monoclinic and orthorhombic).

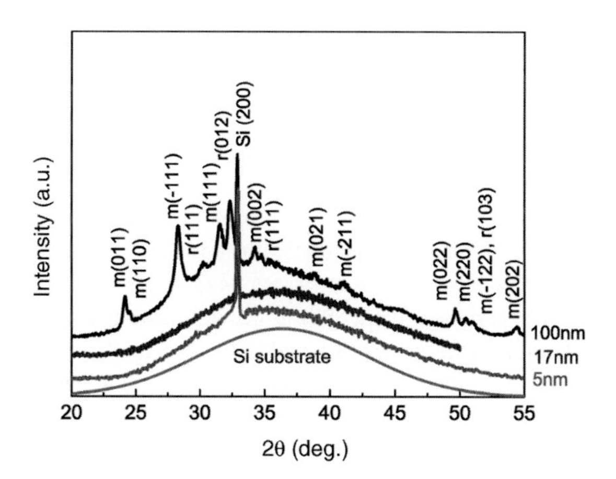

Figure 7.5 XRD data for HfO_2 5, 17, and 100 nm thick films synthesized by ALD (from Ref [47]). The letters m and r represent the monoclinic and orthorhombic structure of the HfO_2 film, respectively.

Table 7.1 XRD data for HfO_2 100 nm thick.

S. No.	2θ	Int. (meas., a.u.)	Int. (ICDD)	Phase (system)	SG	(hkl)	Peak half-width	L_{cr} (nm)
1	24.1532	515.8	15	HfO_2 (monoclinic)	$P21/c$	011	0.34	27.3
2	24.5185	160.3	12	HfO_2 (monoclinic)	$P21/c$	110	0.28	33.6
3	28.2733	1132.2	100	HfO_2 (monoclinic)	$P21/c$	−111	0.58	15.8
4	30.2837	175.8	40	HfO_2 (orthorhombic)	$Pmnb$	101	0.5	18.5
5	31.5338	719.9	74	HfO_2 (monoclinic)	$P21/c$	111	0.52	17.9
6	32.3559	1092.4	100	HfO_2 (orthorhombic)	$Pmnb$	012	0.51	18.3
7	32.8969	2443.4	—	Si (cubic)		200	0.08	—
8	34.1835	282.2	25	HfO_2 (monoclinic)	$P21/c$	002	0.32	29.8
9	49.6343	351.3	16	HfO_2 (monoclinic)	$P21/c$	022	0.40	24.8
10	50.4033	183.1	18	HfO_2 (monoclinic)	$P21/c$	220	0.39	25.6
11	54.3407	130.6	11	HfO_2 (monoclinic)	$P21/c$	202	0.38	26.7

The comparative analysis of the intensities of diffraction peaks of XRD data of the film under consideration and HfO_2 powder (data file ICDD 43–1017 and ICDD 40–1173, Table 7.1) allows us to establish that there is no preferred orientation of crystallites of both phases in the film. The crystallite sizes were calculated using the Selyakov–Scherrer formula [53]:

$$L_{cr} = \frac{\lambda \cos\theta}{\sqrt{B_{2\theta}^2 - b^2}} \tag{7.1}$$

where L_{cr} is the size of the grain, λ is the wavelength of the incident beam (0.154 178 HM), θ is the reflection angle, $B_{2\theta}$ is the half-width of the peak, and b is the broadening caused by the diffractometer (the width of the Si(200) reflection peak was used as an etalon). Table 7.1 shows the results of the calculations. As follows from the table, the sizes of the crystallites of different orientations are distinguished slightly. According to XRD data, there is no crystalline phase in HfO_2 films of thicknesses 5 and 17 nm. The results obtained from reflection spectra correlate well with XRD data.

A similar analysis of the structure of HfO_2 films of equal thickness of 5 nm (17 nm), but fabricated by two different methods (ALD and MOCVD), was carried out [47]. Figure 7.6 shows O 1s absorption spectra calculated from the experimental reflection spectra by means of the Kramers– Kronig transform [47]. Analysis of the spectra shows a good agreement between the spectra of films grown by two different methods. Obviously, bands c and d in the spectra of both films synthesized by MOCVD are clearly distinguished pointing to the presence of the signatures of ordering of the structure in the MOCVD films. Moreover, the spectrum of 17 nm thick MOCVD film correlates well with the spectrum of 100 nm thick ALD film. In both cases, the Jahn–Teller splitting of the feature *a* is traced that allows to assume that grain size in the 17 nm thick MOCVD film is >2 nm and the film is polycrystalline essentially in the monoclinic phase. Finally, it can be argued that the films

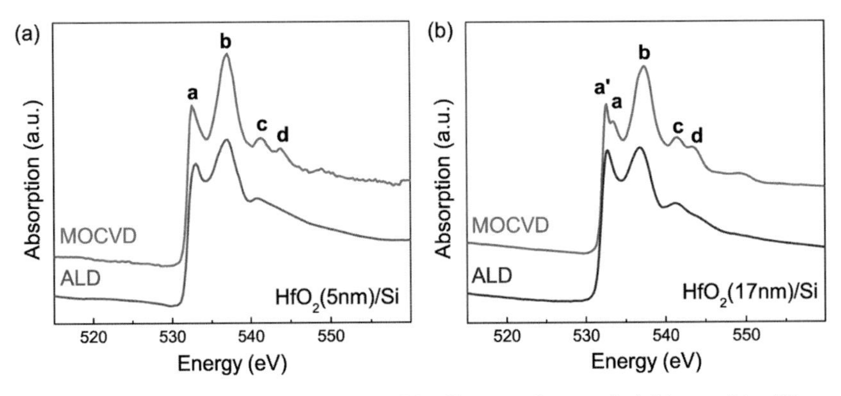

Figure 7.6 (a, b) O 1s absorption spectra of the films 5 and 17 nm thick fabricated by different methods (ALD and MOCVD) (from Ref. [47]). The spectra were calculated from experimental reflection spectra by means of the Kramers– Kronig transform.

prepared by MOCVD shows evidence for crystallization and the ALD films are amorphous.

Figure 7.7 shows Ti $L_{2,3}$ and O 1s absorption spectra of TiO_2 films, 10 and 70 nm thick, synthesized by rf magnetron sputtering onto superpolished Si(111) substrates with a 2 nm thick natural oxide on the top. Ar was used as working gas. The absorption spectra were calculated from the experimental reflection spectra [54] by means of the Kramers– Kronig transform.

According to the classical concept, the NEXAFS excitation at the Ti 2p threshold in TiO_2 occurs mainly within the octahedron of oxygen atoms and the Ti 2p absorption spectrum of TiO_2 should reflect the energies of the free Ti 3d states because it is dominated by the 2p $\rightarrow$ 3d dipole transitions in the Ti atoms [11, 12, 55]. The calculated Ti 2p absorption spectra reflect clearly the spin–orbit splitting of the Ti 2p level with energy separation close to 5.6 eV, that is, equal to the spin–orbit splitting of the Ti 2p level in TiO_2 [19, 56]. The peaks *a* and *b* in Ti $2p_{3/2}$ spectrum stem from

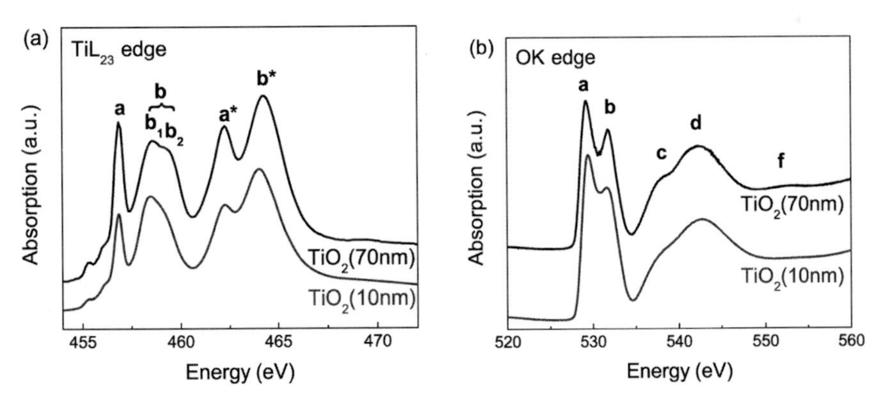

Figure 7.7 Ti $L_{2,3}$ and O 1s absorption spectra of TiO_2 films, 10 and 70 nm thick, synthesized by rf magnetron sputtering (from Ref. [54]). The spectra were calculated from experimental reflection spectra by means of the Kramers– Kronig transform. The Ti 2p1/2 structures are marked by asterisks.

dipole transitions of Ti $2p_{3/2}$ electrons to unoccupied 3d states split into $3dt_{2g}$ (peak *a*) and $3de_g$ (peak *b*) components by the octahedral crystal field created by the O ions. One can see from the figure that there is further splitting of the peak *b* into two bands b_1–b in the absorption spectrum of TiO_2 film, 70 nm thick. This doubling of the Ti 2p–$3de_g$ transitions in TiO_2 is well known [19–21], exists in absorption spectra of all crystalline modifications of TiO_2 [20, 21, 57], and depends on the crystalline structure of TiO_2. The main difference in Ti 2p spectra of different polymorphs of TiO_2 is in the relative intensities of peaks *b* and b_1. In anatase, the intensity of peak *b* is substantially stronger than peak b_1, while in rutile quite the contrary holds [19–21]. As follows from Ref. [20], the experimental XAS spectra of anatase and rutile can be solely viewed only in terms of the crystal field and longer range effects in the framework of calculations for extended clusters, consisting of five (Ti $L_{2,3}$ edge structure) or six shells (O *K* edge structure) surrounding a central (Ti or O) atom together with a damping of 0.5 eV. The intensities of components of peaks b_1 and *b* in the spectrum of the thick film (70 nm) correlates well with an anatase structure [19–21].

The molecular orbitals of TiO_2 derived from a linear combination of atomic orbitals (LCAO) are characterized by four unoccupied orbitals [36]: $2t_{2g}$(Ti 3d + O 2pπ), $3e_g$(Ti 3d + O 2pσ), $3a_{1g}$(Ti 4s + O 2p), and $4t_{1u}$(Ti 4p + O 2p). In this classification, the features *a*, *b*, *c*, and *d* in O K absorption spectra can be assigned to one-electron transitions from the O 1s orbital to the $2t_{2g}$, $3e_g$, $3a_{1g}$, and $4t_{1u}$ orbitals of TiO_2, respectively. Analysis of the energy position of the peaks shows that the energy separation between *a* and *b* peaks is equal to 2.6 eV in O K spectra within the experimental accuracy for both films that is close to the splitting of t_{2g} and e_g states in the Ti 2p absorption spectra discussed above. Peaks *a* and *b*, therefore, reflect core–electron transitions in the oxygen atoms to the lowest unoccupied Ti $3dt_{2g}$ and $3de_g$ electronic states that are hybridized with the 2p states of the ligand (oxygen) atoms. The structure in the spectrum of the TiO_2 film, 70 nm thick, is characterized by *d*, *c*, and *f* peaks and is related to transitions into the empty electronic states with mixed Ti 4s, 4p + O 2p character [20, 21]. In the spectrum of the TiO_2 film 10 nm thick, there is only one *d* peak and very weak peak *c*. Analysis of the energy positions of *c*, *d*, and *f* peaks in the spectrum of TiO_2 film 70 nm thick shows a good correlation of this structure with the structure of anatase. In that way, the analysis shows that the 70 nm thick TiO_2 film is characterized by an anatase structure. The comparison of the O K and Ti 2p spectra of both films points to the (i) absence of the splitting into b_1–b in Ti 2p spectrum of TiO_2 film 10 nm thick and (ii) absence of the peak *f* in the O K spectrum of TiO_2 film 10 nm thick. Both items indicate an amorphous state of the TiO_2 film 10 nm thick.

Finally, we would like to summarize some investigations on Al_2O_3 film 30 nm thick grown by ALD onto silicon substrate Si (100), which was preliminarily cleaned in a vapor of HF. The temperature of the substrate was 32 °C. The reactor was filled by vapors of trimethyl aluminum for a few tenths of a second, with subsequent evacuation of the reagent and admission of water vapor to the reactor, which prepared the surface for the next ALD cycle.

In the framework of quasi-atomic approach [48], the main features *a*, *b*, *c*, and *d* in Al $L_{2,3}$ absorption spectrum of Al_2O_3 from Ref. (Figure 7.8 [58]) are associated with

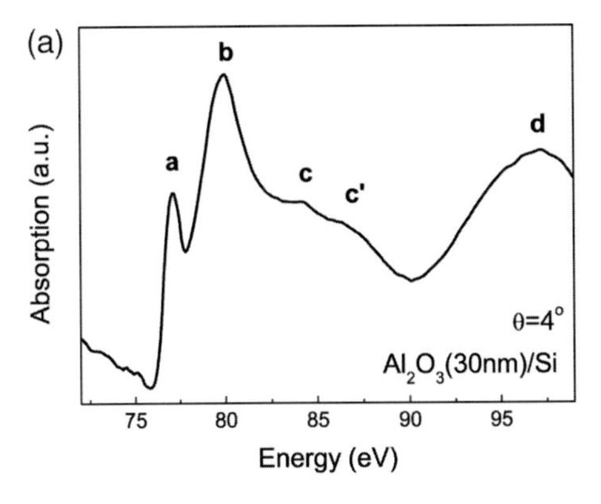

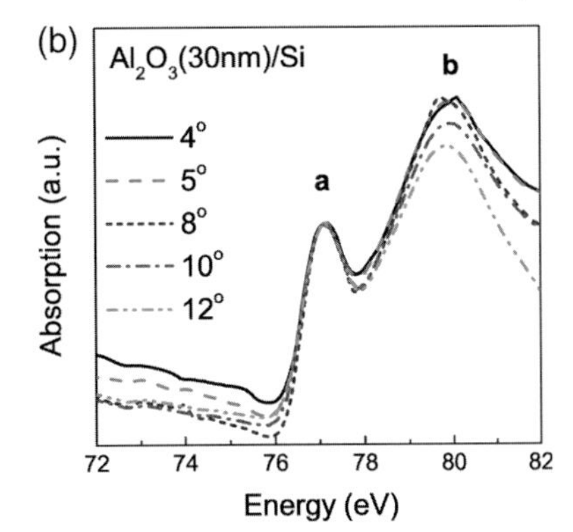

Figure 7.8 Absorption spectra of Al_2O_3 film 30 nm thick in the vicinity of the Al $L_{2,3}$ absorption edge calculated from experimental reflection spectra by means of the Kramers–Kronig transform (from Ref. [58]). (a) Reflection spectrum was measured at a glancing angle of 4°; (b) reflection spectra were measured at glancing angles between 3 and 12°.

transitions to four unoccupied orbitals: a_1(Al3s), t_2 (Al3p), e(Alεd), and e + t_2 (Alεd), correspondingly. Figure 7.8a shows the absorption spectrum calculated from the reflection spectrum measured for a glancing angle of 4°. The energy position of all peaks, the correlation of the intensities, and the value of the splitting between features *a* and *b*, that is, $\Delta E = 2.8$ eV, agree well with characteristics of amorphous Al_2O_3 [59–63]. Figure 7.8b shows the Al $L_{2,3}$ absorption spectra calculated from the reflection spectra obtained for different incident angles. The absorption spectra are normalized to the intensity of the first maximum. For clarity, we select the spectra showing the most typical changes in the fine structure as the glancing incidence angle varies, which means the depth of formation of the reflected beam varies correspondingly. The analysis of each spectrum shows that with growth of the incident angle, which was used in the calculations, (i) the shape of the first feature *a* does not change almost; (ii) the splitting between features *a* and *b* decreases from $\Delta E = 2.8$ eV (for 3°) to $\Delta E = 2.7$ eV (for 12°); (iii) the half-width and the intensity of feature *b* decreases slowly; and (iv) the value of the Al $L_{2,3}$ absorption edge changes strongly. The tendencies marked in (i–iii) are in good agreement with the basic representations of amorphous Al_2O_3 [59, 63, 64] according to which the coordination number of aluminum atom in amorphous Al_2O_3 basically is equal to 4. Taking into account that the value of the absorption edge is proportional to the concentration of the absorbed atoms in the substance and that with growth of the glancing angle the depth of the formation (probing depth) of the reflected radiation is increased, one can conclude that distribution of atoms of aluminum on depth is nonuniform, and hence, the Al_2O_3 film is inhomogeneous in depth. In view of the fact that amorphous Al_2O_3 is characterized by a mix of tetrahedrons and octahedrons, that is, each atom of Al can be surrounded 4 or 6 oxygen atoms, one can conclude that heterogeneity of an investigated film is connected with parity change between tetrahedrons and octahedrons on depth. The item (ii) points also to this conclusion.

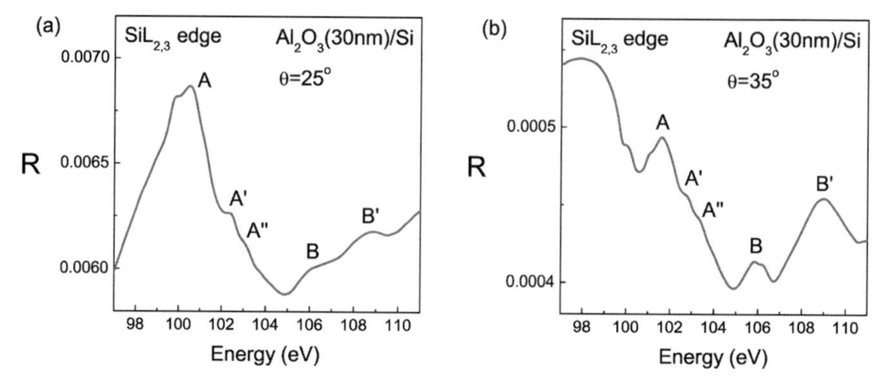

Figure 7.9 Reflection spectra in the vicinity of the Si $L_{2,3}$ absorption edge of the Al_2O_3 film, 30 nm thick, deposited on Si substrate for various glancing angles: (a) glancing angle of 25°; (b) glancing angle of 35° (from Ref. [58]).

In summary, we would like to call attention to Figure 7.9, where the reflection spectra in the vicinity of the Si $L_{2,3}$ absorption edge of Al_2O_3 for glancing angles of 25 and 35° [58] are plotted. One can see that slight structures A and B, typical for Si and SiO_2, have appeared in the energy regions 100–104 and 104–115 eV, respectively, in the reflection spectrum for 25° and are intensified with the increase in the incident angle. The favorable factor for the study of the Si–SiO_2 system is the substantial difference in the form and energy position of the fine structure components of the Si and SiO_2 reflection spectra near Si $L_{2,3}$ absorption edge [65, 66]. This fact allows to carry out the direct qualitative analysis of the reflection spectra without intermediate solution of the Kramers–Kronig transform. Distinct strengthening of features B, associated with SiO_2 structure, with growth of glancing angle specifies an increase in the contribution of a SiO_2 layer that allows to suggest the presence of an extended layer of SiO_2 on the interface Al_2O_3/Si.

7.4
Studying the Internal Structure of High-*K* Dielectric Films by Hard X-Ray Photoelectron Spectroscopy and TEM

X-ray photoelectron spectroscopy is a technique commonly used nowadays for surface analysis. The analysis of both the photoelectron energies and the intensity of each peak allows us to carry out a quantitative chemical analysis of the sample. HAXPES has an additional bonus of extended probing depth given by the large value of the photoelectron inelastic mean free path (IMPF), λ_i obtained by using hard X-rays as excitation energies. Thus, HAXPES is a nondestructive depth-sensitive method where the bulk/interface electronic properties can be accessed by systematically changing the photoelectron's emission angle or/and its kinetic energy.

Figure 7.10 shows Ti 2p photoelectron spectra from TiO_2 (sample 1) and metallic Ti (sample 2) films of about 10 nm thickness, which were deposited by rf and DC magnetron sputtering, respectively, onto superpolished Si(111) substrates with a

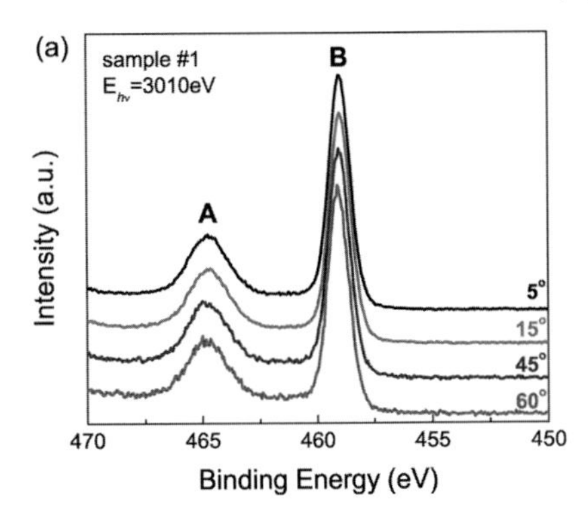

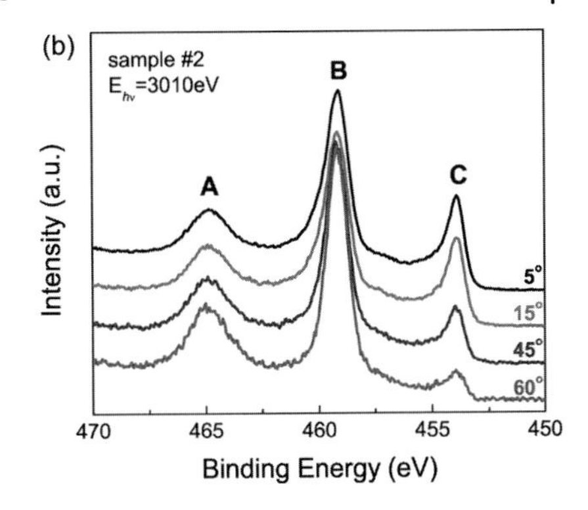

Figure 7.10 Ti 2p photoelectron spectra from TiO_2 (a) and metallic Ti (b) films of about 10 nm thickness obtained at different emission angles at an excitation energy of 3010 eV (from Filatova, E.O., Kozhevnikov, I. V., Sokolov, A.A., Ubyivovk, E.V., Yulin, S., Gorgoi, M., and Schaefers, F. (2012) Soft x-ray reflectometry, hard x-ray photoelectron spectroscopy and transmission electron microscopy investigations of the internal structure of TiO_2(Ti)/SiO_2/Si stacks. *Sci. Technol. Adv. Mater.* **13**, 015001).

2 nm thick natural oxide on the top. Ar was used as working gas (Filatova, E.O., Kozhevnikov, I.V., Sokolov, A.A., Ubyivovk, E.V., Yulin, S., Gorgoi, M., and Schaefers, F. (2012) Soft x-ray reflectometry, hard x-ray photoelectron spectroscopy and transmission electron microscopy investigations of the internal structure of TiO_2(Ti)/SiO_2/Si stacks. *Sci. Technol. Adv. Mater.* **13**, 015001). The spectra were obtained at different emission angles at excitation energy of 3010 eV. The hard X-ray photoelectron spectroscopy experiments were performed using the HIKE station at the KMC-1 beamline at the BESSY II synchrotron light source of the Helmholtz-Zentrum Berlin. All details about the experimental setup can be found in Ref [67, 68].

For clarity, all spectra were normalized to the background. The analysis of the spectra reveals the following: (i) the spectra from sample 1 measured at different emission angles are indistinguishable; (ii) the spectra from the two different samples are very similar and reveal the presence of a double peak structure A–B with a spin–orbit splitting of 5.7 eV, a 1 : 2 intensity ratio of the components and the same binding energy (BE) positions in both samples that correlate well with the characteristics of the Ti 2p peak in TiO_2 [69]; (iii) there is an additional feature C centered at 453.8 eV BE in the spectra from sample 2; the intensity of this peak decreases with increasing emission angle. Obviously, a fraction of the Ti film in sample 2 is TiO_2. Figure 7.11 shows the Ti 2p photoelectron spectra from sample 2 recorded at different excitation energies and different emission angles together with the fits. At decomposition of spectra the background of scattered electrons was subtracted with the use of function of Shirley [70]. All components were fitted by a mixed asymmetrical Gaussian/Lorentzian function in the ratio 7 : 3.

The analysis of all fitting components reveals that the peak at 453.8 eV BE (feature C of Figure 7.10b) is assigned to the Ti^0 $2p_{3/2}$ component of metallic Ti [69, 71]. Curve fitting indicates that the intensity of the Ti^0 state is dominant and exceeds other TiO_x components by a factor of 2. The intensity of the peaks

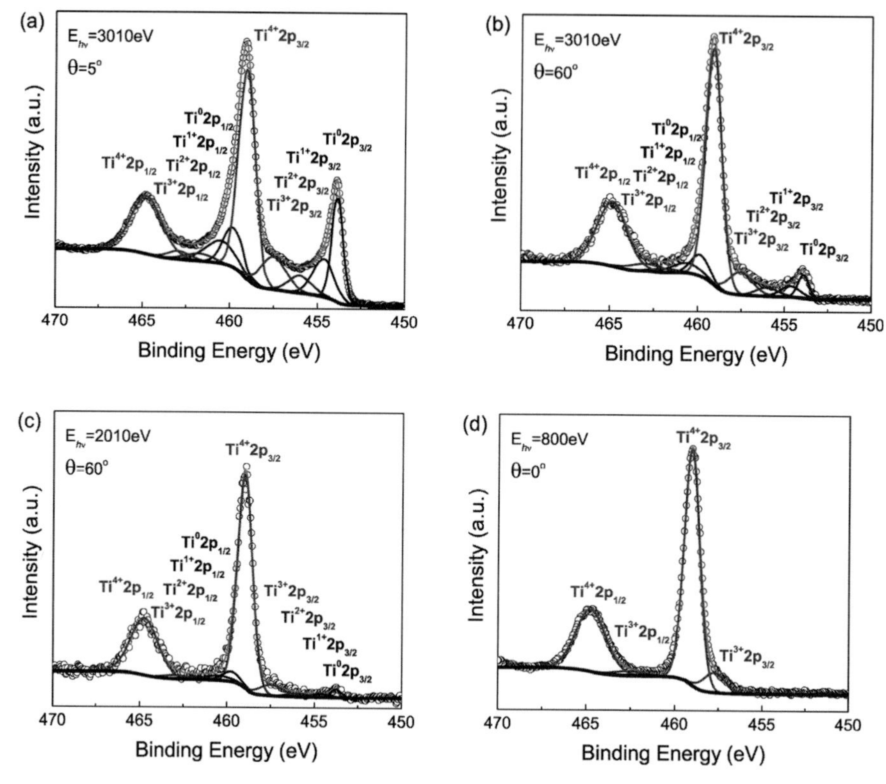

Figure 7.11 Measured and fitted Ti 2p photoelectron spectra from sample 2 recorded at different excitation energies and different emission angles (from Filatova, E.O., Kozhevnikov, I.V., Sokolov, A.A., Ubyivovk, E.V., Yulin, S., Gorgoi, M., and Schaefers, F. (2012) Soft x-ray reflectometry, hard x-ray photoelectron spectroscopy and transmission electron microscopy investigations of the internal structure of TiO_2(Ti)/SiO_2/Si stacks. *Sci. Technol. Adv. Mater.* **13**, 015001). (a) $E = 3010$ eV and 5°; (b) $E = 3010$ eV and 60°; (c) $E = 2010$ eV and 60°; and (d) $E = 800$ eV and 0°.

corresponding to metallic titanium decreases with increasing emission angle (Figure 7.11a and b) and similarly with decreasing excitation energy from 3010 to 2010 eV at 60° emission angle (Figure 7.11c). At 800 eV and normal incidence (Figure 7.11d), the peaks corresponding to metallic titanium disappear. Taking into account that larger emission angles lead to decreased probing depth, one has to conclude a metallic titanium layer between the oxide film and the substrate in sample 2. This behavior was established in all titanium core-level spectra from sample 2. One can conclude that due to the strong oxidation of metallic titanium a thick TiO_2 film is created on top of the Ti film (sample 2).

In Figure 7.12, the Si 2p spectra measured at different emission angles are presented for both samples. A prominent Si^0 peak centered at 99.3 eV BE is observed for both samples, while an additional peak at 102.8 eV BE, which according to [69, 72] reflects the Si^{4+} state, is seen for sample 1. Only a trace of SiO_2 structure can be established in sample 2 from the fitting of Si 2p spectrum at emission angle 5°

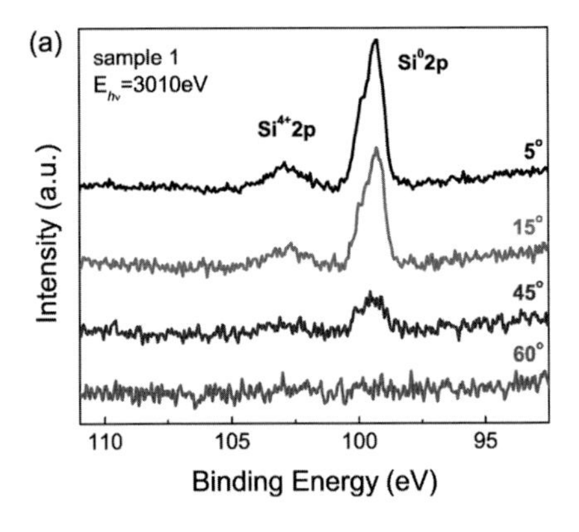

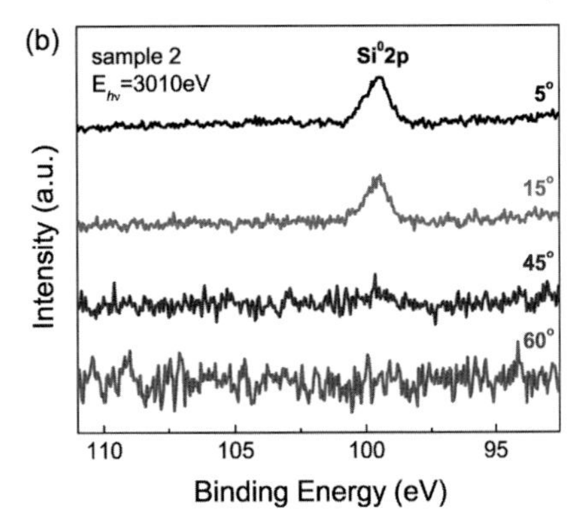

Figure 7.12 Si 2p photoelectron spectra from sample 1 (a) and sample 2 (b) obtained at different emission angles with an excitation energy of 3010 eV (from Filatova, E.O., Kozhevnikov, I.V., Sokolov, A.A., Ubyivovk, E.V., Yulin, S., Gorgoi, M., and Schaefers, F. (2012) Soft x-ray reflectometry, hard x-ray photoelectron spectroscopy and transmission electron microscopy investigations of the internal structure of TiO_2(Ti)/SiO_2/Si stacks. *Sci. Technol. Adv. Mater.* **13**, 015001).

(Figure 7.13). It is well known that metallic Ti is a very reactive material and the diffusion of oxygen from the SiO_2 into Ti is expected.

To quantitatively determine the thickness of each layer composing the investigated samples, we took into account the results [73–75] and a model was developed (Filatova, E.O., Kozhevnikov, I.V., Sokolov, A.A., Ubyivovk, E.V., Yulin, S., Gorgoi, M., and Schaefers, F. (2012) Soft x-ray reflectometry, hard x-ray photoelectron spectroscopy and transmission electron microscopy investigations of the internal structure of TiO_2(Ti)/SiO_2/Si stacks. *Sci. Technol. Adv. Mater.* **13**, 015001). Under the following assumptions: the sample is amorphous or polycrystalline, inelastic

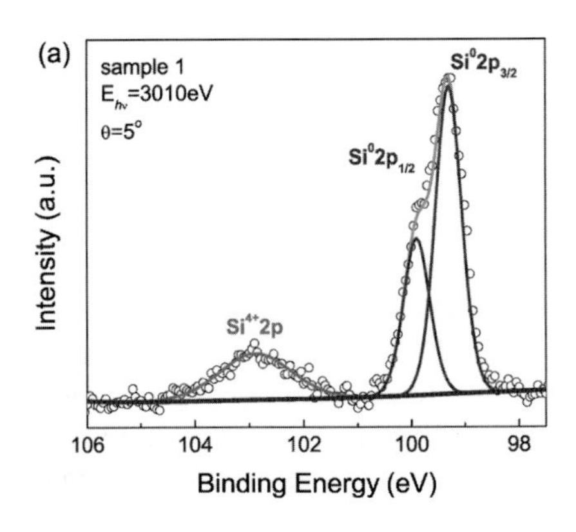

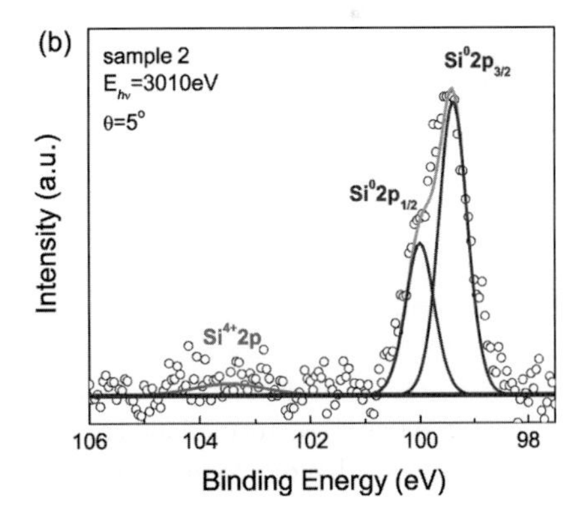

Figure 7.13 Measured and fitted Si 2p photoelectron spectra from sample 1 (a) and sample 2 (b) obtained at emission angle 5° and excitation energy of 3010 eV (from Filatova, E.O., Kozhevnikov, I.V., Sokolov, A.A., Ubyivovk, E.V., Yulin, S., Gorgoi, M., and Schaefers, F. (2012) Soft x-ray reflectometry, hard x-ray photoelectron spectroscopy and transmission electron microscopy investigations of the internal structure of TiO_2(Ti)/SiO_2/Si stacks. *Sci. Technol. Adv. Mater.* **13**, 015001).

scattering is small, refraction of escaped electrons is small, photon penetration depth does not depend on material, sample surface is smooth on the atomic scale, acceptance angle of the analyzer is small, sample is homogeneous in the XY plane, and HAXPES peak area can be estimated exactly in spite of background, asymmetry, instrumental resolution, and overlap with other peaks. In the framework of this model, the recursion formula for the intensity of a HAXPES peak from an nth layer was derived and can be written as

$$F_n(\theta) = A_{\exp}(\theta)\sigma_n c_n \lambda_n \gamma_n (1 - e^{(-d_n)/\lambda_n \cos\theta}) \prod_{i=1}^{n-1} e^{(-d_i)/\lambda_i \cos\theta} \tag{7.2}$$

where n denotes the layer number ($n = 1$ corresponds to upper layer); σ_n is the photoelectric cross section of the material of the nth layer; λ_n is inelastic mean free path in the nth layer; γ_n is an orbital angular symmetry factor of the nth layer; c_n is the atomic concentration on volume unit for the nth layer, $c(z) = \text{const.}$ for each layer; d_n is the thickness of the nth layer; $A_{\exp}(\theta) = TA\cos\theta$; T is a geometrical function defined by the parameters of the experimental equipment; and A is the analyzed part of a sample.

According to the HAXPES analysis, sample 1 can be presented as the three layers stack $TiO_2/SiO_2/Si$. Evidently in such a layer stack, the Ti^{4+}2p doublet peak corresponds only to the TiO_2 layer, Si^{4+}2p peak corresponds to the SiO_2 layer, and $Si^0$2p peak corresponds to the Si substrate. According to Equation 7.2, the intensity of these peaks can be written in the following way:

$$\begin{aligned}
F_{Ti^{4+}2p}(\theta) &= A_{\exp}(\theta)\sigma_{Ti^{4+}2p} c_{TiO_2} \lambda_{TiO_2} \gamma_{Ti^{4+}2p} (1 - e^{(-d_{TiO_2})/\lambda_{TiO_2} \cos\theta}) \\
F_{Si^{4+}2p}(\theta) &= A_{\exp}(\theta)\sigma_{Si^{4+}2p} c_{SiO_2} \lambda_{SiO_2} \gamma_{Si^{4+}2p} (1 - e^{(-d_{SiO_2})/\lambda_{SiO_2} \cos\theta}) e^{(-d_{TiO_2})/\lambda_{TiO_2} \cos\theta} \\
F_{Si^0 2p}(\theta) &= A_{\exp}(\theta)\sigma_{Si^0 2p} c_{Si} \lambda_{Si} \gamma_{Si^0 2p} e^{(-d_{TiO_2})/\lambda_{TiO_2} \cos\theta} e^{(-d_{SiO_2})/\lambda_{SiO_2} \cos\theta}
\end{aligned} \tag{7.3}$$

where σ and γ are tabulated in [76], λ can be calculated via TPP-2M formula [77], and c can be found from density of the material. To determine the λ and c, we used the value of density for the bulk materials from James and Lord database [78]. Because the density of a thin layer can differ from the density of bulk material, a fitting procedure was applied. The best agreement between calculated and measured peaks was achieved at $\varrho(TiO_2) = 4.23\ g/cm^3$, $\varrho(SiO_2) = 2.53 g/cm^3$, and $\varrho(Si) = 2.33\ g/cm^3$ that are very close to the values of bulk material densities. The value of $A_{\exp}(\theta)$ was taken as a sum of intensities of HAXPES peaks (7.3) normalized to its constants σ, c, λ and γ. The $A_{\exp}$ (θ) parameter was also fitted for each emission angle. According to the calculation carried out, sample 1 can be represented as the layer stack TiO_2 (10.1 nm)/SiO_2 (2.3 nm)/Si. The correlation between the experimental and the calculated values of the intensity of HAXPES peaks at different emission angles is shown in Figure 7.14.

According to the HAXPES analysis, sample 2 is a more complicated stack compared to sample 1. Sample 2 can be represented as a three-layer stack TiO_x/SiO_2/Si, where the first layer TiO_x is a mix of titanium oxides and metallic titanium or as a five-layer stack $TiO_2/TiO_x/Ti/SiO_2/Si$. The calculations of the thickness of layers composing the sample 2 were carried out for both layer stack models.

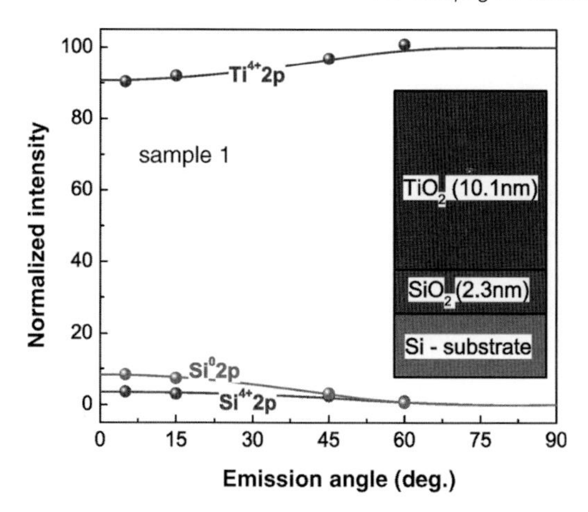

Figure 7.14 The correlation between the experimental (dots) and calculated (lines) values of HAXPES peak intensity of sample 1 obtained at different emission angles and 3010 eV excitation photon energy. For clarity, the experimental and the calculated values of the intensity of HAXPES peaks are normalized to its constants σ, c, λ, γ, and $A_{exp}(\theta)$ (from Filatova, E.O., Kozhevnikov, I. V., Sokolov, A.A., Ubyivovk, E.V., Yulin, S., Gorgoi, M., and Schaefers, F. (2012) Soft x-ray reflectometry, hard x-ray photoelectron spectroscopy and transmission electron microscopy investigations of the internal structure of TiO_2(Ti)/SiO_2/Si stacks. *Sci. Technol. Adv. Mater.* **13**, 015001).

In the framework of three-layer model, it was established that sample 2 can be denoted as the layer stack TiO_x (13.3 nm)/SiO_2 (0.5 nm)/Si. The best result in fitting was obtained for values of density $p(TiO_x) = 4.33 g/cm^3$, $p(SiO_2) = 2.53 g/cm^3$, and p (Si) = 2.33 g/cm^3. The correlation between the experimental and the calculated intensity values of HAXPES peaks at different emission angles is displayed in Figure 7.15a.

It was derived that in a five-layer model, sample 2 is the layer stack TiO_2 (3.6 nm)/TiO_x (4.4 nm)/Ti (5.8 nm)/SiO_2 (0.5 nm)/Si with the values of density $p(TiO_2) = 4.23$ g/cm^3, $p(TiO_x) = 4.33 g/cm^3$, $p(Ti) = 4.5 g/cm^3$, $p(SiO_2) = 2.53 g/cm^3$, and p(Si) = 2.33 g/cm^3. The intensity correlation of the calculated and the experimental HAXPES peaks is shown in Figure 7.15.

A comparison between results derived in three- and five-layered models shows that in five-layered models an overall thickness of the top three layers is equal to 13.8 nm value, which is larger by 0.5 nm than the total thickness of top TiO_x layer in the three-layer model. The obtained disagreement is most likely caused by the main limitation of the developed model, which assumes the sharpness of the interfaces. Moreover, the roughness of the Si substrate and the external film surface was not taken into account.

The individual thickness of layer stacks (samples 1 and 2) was also determined using the high-resolution TEM method (Filatova, E.O., Kozhevnikov, I.V., Sokolov, A.A., Ubyivovk, E.V., Yulin, S., Gorgoi, M., and Schaefers, F., (2012) "Soft x-ray reflectometry, hard x-ray photoelectron spectroscopy and transmission electron microscopy investigations of the internal structure of TiO_2(Ti)/SiO_2/Si stacks". Sci. Technol. Adv.

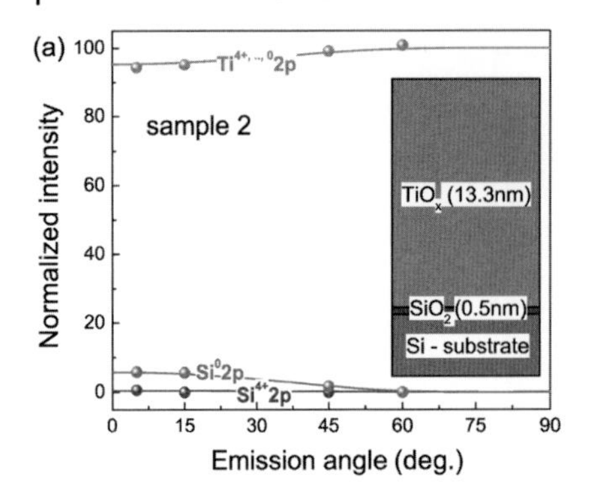

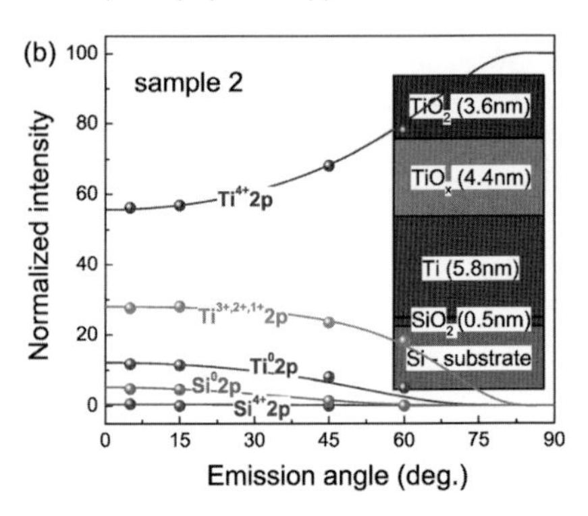

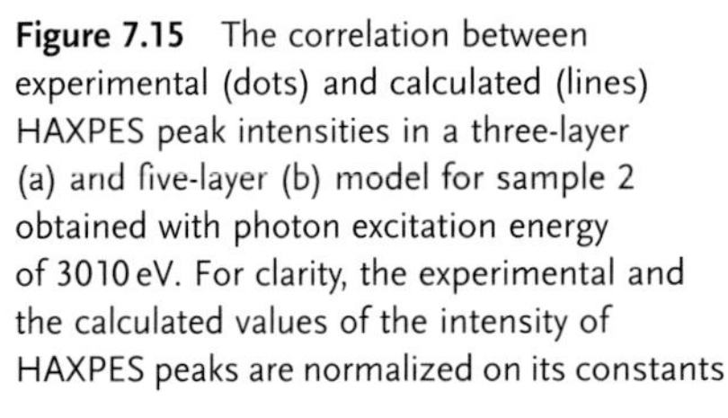
Figure 7.15 The correlation between experimental (dots) and calculated (lines) HAXPES peak intensities in a three-layer (a) and five-layer (b) model for sample 2 obtained with photon excitation energy of 3010 eV. For clarity, the experimental and the calculated values of the intensity of HAXPES peaks are normalized on its constants σ, c, λ, γ, and $A_{exp}(\theta)$ (from Filatova, E.O., Kozhevnikov, I.V., Sokolov, A.A., Ubyivovk, E.V., Yulin, S., Gorgoi, M., and Schaefers, F. (2012) Soft x-ray reflectometry, hard x-ray photoelectron spectroscopy and transmission electron microscopy investigations of the internal structure of $TiO_2(Ti)/SiO_2/Si$ stacks. *Sci. Technol. Adv. Mater.* **13**, 015001).

Mater. 13, 015001). The measurements were carried out with the Carl Zeiss Libra 200 FE 200 kEv equipment of Interdisciplinary Resource Center for Nanotechnology at St. Petersburg State University, Russia (http://nano.spbu.ru). Figure 7.16 demonstrates HRTEM results for samples 1 and 2. To reduce the possible errors associated with width gradients, the microscopic and spectral measurements were conducted at the same point on the sample. The sample was prepared in a special way for TEM measurements of the real thickness of heterostructured layers. The studies were conducted in cross-section geometry, for which the sample was cleaved into two identical parts, with the cleavage directed through the point at which the spectra were measured. Next, these sections were cemented with the front surfaces facing one

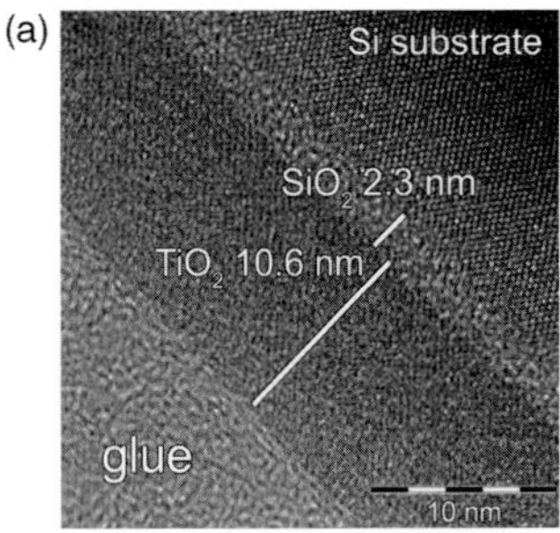

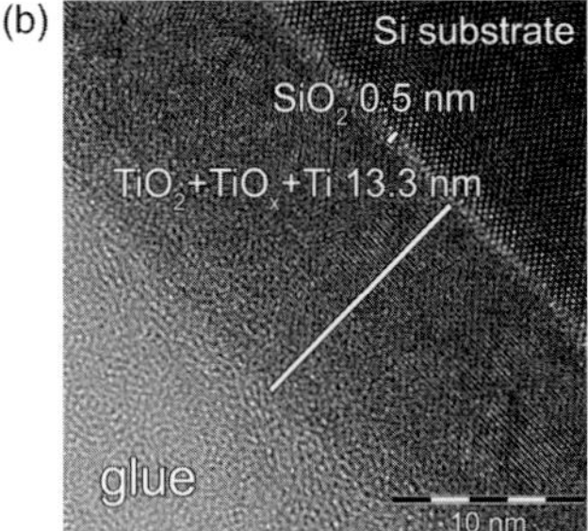

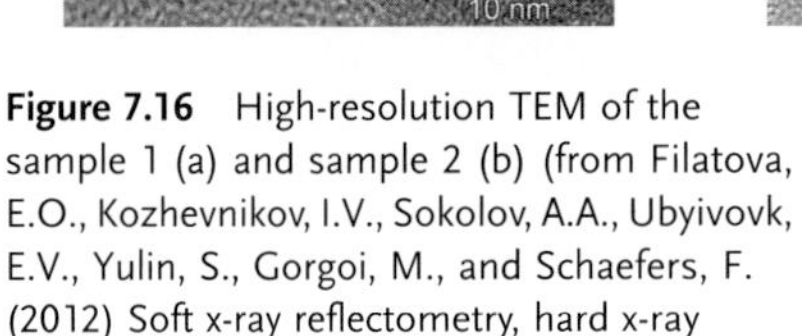
Figure 7.16 High-resolution TEM of the sample 1 (a) and sample 2 (b) (from Filatova, E.O., Kozhevnikov, I.V., Sokolov, A.A., Ubyivovk, E.V., Yulin, S., Gorgoi, M., and Schaefers, F. (2012) Soft x-ray reflectometry, hard x-ray photoelectron spectroscopy and transmission electron microscopy investigations of the internal structure of $TiO_2(Ti)/SiO_2/Si$ stacks. *Sci. Technol. Adv. Mater.* **13**, 015001).

another. After this, the central part of the sample was thinned to a thickness less than 100 Å by mechanical grinding and subsequent ion milling.

As it results from Figure 7.16, the investigated samples can be characterized as the following stacks: sample 1 is TiO_2 (10.6 nm)/SiO_2 (2.3 nm)/Si and sample 2 is (TiO_2 + TiO_x + Ti) (13.3 nm)/SiO2 (0.5 nm)/Si that correlate well with derived HAXPES data. Moreover, Figure 7.16a confirms the conclusion made above that the TiO_2 10 nm thick film is amorphous.

The developed model was applied to results obtained for HfO_2 films 5 nm thick synthesized by ALD (Figure 7.17a) and MOCVD techniques (Filatova, E.O., Kozhevnikov, I.V., Sokolov, A.A., Ubyivovk, E.V., Yulin, S., Gorgoi, M., and Schaefers, F. (2012) Soft x-ray reflectometry, hard x-ray photoelectron spectroscopy and transmission electron microscopy investigations of the internal structure of TiO_2(Ti)/SiO_2/Si stacks. *Sci. Technol. Adv. Mater.* **13**, 015001.), which were discussed above. Figure 7.17 shows the Si 2p photoelectron spectra for both samples, which were obtained by hard X-Ray photoelectron spectroscopy. At decomposition of spectra, the background of scattered electrons was subtracted with the use of function of Shirley [70]. All components were fitted by a mixed asymmetrical Gaussian/Lorentzian function in the ratio 7 : 3.

Both samples with ALD- and MOCVD-HfO_2 reveal electron photoemission from the Si substrate crystal characterized by $Si^0\ 2p_{3/2}$–$Si^0 2p_{1/2}$ spin–orbit splitting of 0.6 eV. In addition, there is second band of photoelectrons corresponding to the binding energy of 103.3 eV typical for Si dioxide, which are obviously related to the Si oxide IL formed at the interface between Si and HfO_2. Obviously, the Si^{4+} 2p line is much more intense in the MOCVD case suggesting a thicker SiO_2 IL. This result agrees well with analysis of the same samples conducted by X-ray photoelectron spectroscopy (XPS) in combination with Ar^+ ion sputtering and by a nondestructive X-ray emission spectroscopy with depth resolution (DRSXES) [79, 80]. The inferior

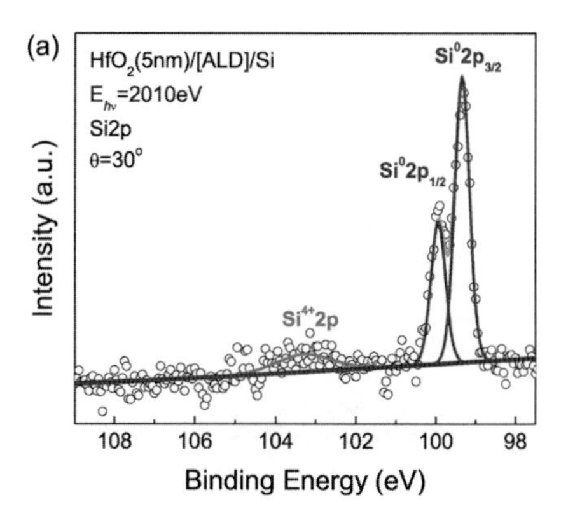

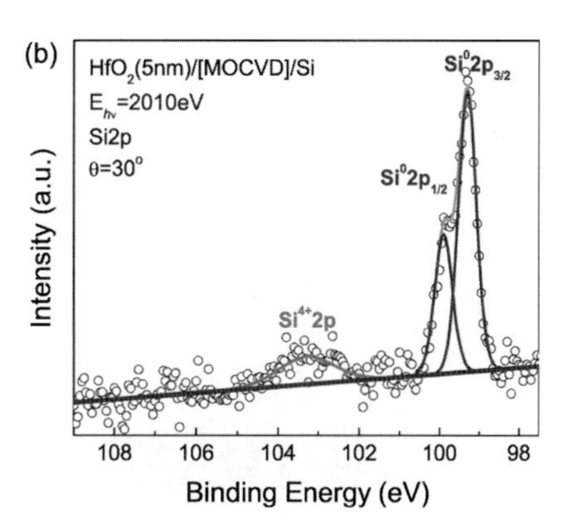

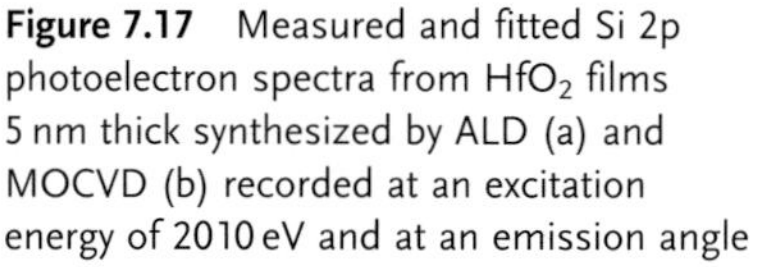

Figure 7.17 Measured and fitted Si 2p photoelectron spectra from HfO_2 films 5 nm thick synthesized by ALD (a) and MOCVD (b) recorded at an excitation energy of 2010 eV and at an emission angle of 30° (from E. O. Filatova and A. A. Sokolov (2011) X-ray photoelectron spectroscopy for nondestructive analysis of buried interfaces, *Journal of Structural Chemistry*, **52**, Supplement 1, 82–89).

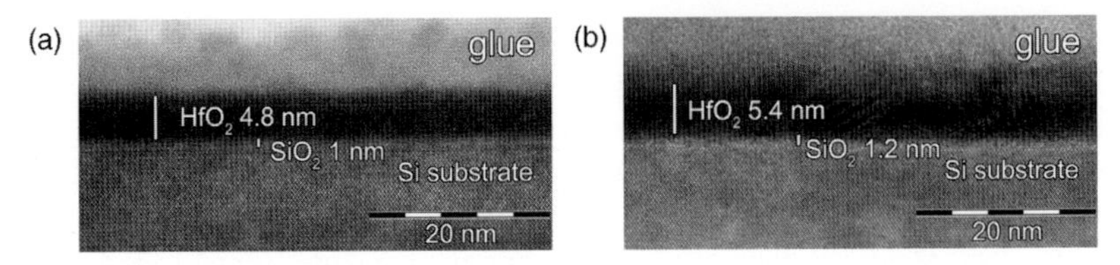

Figure 7.18 High-resolution TEM of HfO_2 films 5 nm thick synthesized by ALD (a) and MOCVD (b).

manifestation of Si^0 ($2p_{3/2}$–$Si^02p_{1/2}$) spin–orbit splitting in MOCVD-HfO_2 sample is associated with rough surface of the film in this case that will be confirmed by reflectometry studies in the next section.

To quantitatively determine the IL thickness in investigated samples, the developed model was applied. According to the calculation carried out, the samples can be characterized as a MOCVD-HfO_2 (5.1 nm)/SiO_2 (1.3 nm)/Si and ALD-HfO_2 (4.7nm)/SiO_2 (1.0nm)/Si layer stacks that correlate well with HRTEM images (Figure 7.18) for both samples. Moreover, Figure 7.18b confirms the conclusion made above that the MOCVD-HfO_2 5 nm thick film demonstrates the crystallization phase and rather rough surface.

7.5 Studying the Internal Structure of High-*K* Dielectric Films by X-ray Reflectometry

7.5.1 Reconstruction of the Dielectric Constant Profile by Hard X-Ray Reflectometry

Let us consider the wave equation for s-polarized radiation

$$\frac{d^2E}{dz^2} + \left[q^2 - k^2\chi(z)\right]E = 0; \chi(z) \equiv 1 - \varepsilon(z) \rightarrow \begin{cases} \chi_+ = \text{const.} & \text{at } z \rightarrow +\infty \\ 0 & \text{at } z \rightarrow -\infty \end{cases} \tag{7.4}$$

where $\varepsilon(z) \equiv 1 - \chi(z)$ is the dielectric permittivity varying with depth and tending to the constant values both in the vacuum and in the depth of a matter; $k \equiv \omega/c$, the wavenumber in vacuum; $q = k \sin\theta$ and θ, the grazing incidence angle of X-rays counted from the sample surface.

The inverse problem of X-ray reflectometry consists in the reconstruction of the $\chi(z)$ depth distribution (the dielectric constant profile) basing on the reflectivity curve $R(q) = |r(q)|^2$ measured in a limited range of the grazing angle $\theta \in [\theta_{min}, \theta_{max}]$, that is, at $q \in [q_{min}, q_{max}]$, where $r(q)$ is the complex amplitude reflectivity determining the asymptotic behavior of the reflected wave at $z \rightarrow -\infty$.

Notice the inverse problem as applied to the stationary Schrödinger equation is one of the classical problems in quantum mechanics, which was solved successfully in the works of Gelfand, Levitan, and Marchenko (see, for example, Ref. [81, 82]). The quantum mechanics equation is similar to (7.4), where q^2 is interpreted as the energy of a particle's striking potential $V(z) \equiv k^2\chi(z)$. The Gelfand–Levitan–Marchenko (GLM) theorem states that if the functions $\chi(z)$ and $r(q)$ obey a number of general conditions and the amplitude reflectivity $r(q)$ is known in the infinite range of the q-values $q \in (-\infty, \infty)$, then the function $\chi(z)$ can be found uniquely as a solution of the GLM equation.

At the same time, the classical proof of the GLM theorem is based essentially on the assumption that the potential tends quickly to zero in both asymptotic regions $z \to \pm\infty$ or, more correctly, the following integral exists [81, 82]:

$$\int_{-\infty}^{+\infty} \chi(z) \cdot (1 + |z|) dz < \infty \tag{7.5}$$

Evidently, this assumption is invalid in X-ray optics as a structure studied is placed typically on a substrate, so that the function $\chi(z)$ tends to the constant value in the depth of the substrate and the integral (7.5) does not exist.

In the past decades, considerable efforts were made to solve the inverse problem of potential scattering in more general definition compared to (7.5), while assuming additional information about potential to be known (see, for example, Refs [83–86] and references therein). Particularly, different proofs of the uniqueness theorem have been obtained for potentials unknown in a limited interval on the Z-axis, for example, assuming $\chi(z) \equiv 0$ at $z < z_1$ (vacuum) and $\chi(z) \equiv \text{const.}$ at $z > z_2$ (substrate). However, an analogous to the GLM equation has not been deduced for these cases and, hence, the method of practical reconstruction of the function $\chi(z)$ from real experimental data is not evident yet.

Moreover, both the classical GLM theorem and the modern approaches suggest the measurement of the complex amplitude reflectivity $r(q)$ rather than its absolute value $R(q) = |r(q)|^2$, what is an arduous task for X-ray experiments. The problem of determination of the phase of the amplitude reflectivity from its modulus is known as the phase problem of X-ray reflectometry.

There is the well-known theorem that if the absolute value of the reflectivity $R(q) = |r(q)|^2$ is known at all real $q \in (-\infty, +\infty)$ and, besides, the placement of the poles α_n and zeroes β_m of the function $r(q)$ in the upper half of the complex q-plane is known, then the amplitude reflectivity can be found using the following expression (see, for example, Ref. [82]):

$$r(q) = \sqrt{R(q)} \exp\left[\frac{1}{2\pi i} v.p. \int_{-\infty}^{+\infty} \frac{\ln R(\kappa)}{\kappa - q} d\kappa\right] \prod_n \frac{q - \alpha_n^*}{q - \alpha_n} \prod_m \frac{q - \beta_m}{q - \beta_m^*} \tag{7.6}$$

Equation 7.6 shows that, in a general case, the $\chi(z)$ distribution cannot be evaluated from the reflectivity curve $R(q)$, even though $R(q)$ is measured at all $q \in (-\infty, +\infty)$, because neither poles nor zeroes of the amplitude reflectivity are known *a priori*.

Among existing ways to solving the phase problem of X-ray reflectometry, we mention two exact approaches. The first of them has been developed by Majkrzak and Berk [87, 88] and can be explained as follows. Suppose that a sample studied can be separated into two parts: reference layer (e.g., substrate), whose amplitude reflectivity is known *a priori*, and unknown segment (e.g., layered film deposited on the substrate), whose amplitude reflectivity should be determined. Suppose that identical films of unknown internal structure were deposited on the top of three substrates of different materials. Then, measuring the absolute values of the reflectivity of all three samples, the amplitude reflectivity of the film can be found exactly. While the approach is absolutely correct from mathematical point of view, its practical application in X-ray physics is limited essentially by two facts. First, it is by no means evident that growth of a film on different substrates occurs in the same manner. Second, the proof of the statement is valid for transparent materials only (cold neutrons optics), which is not the case of X-ray radiation.

The second exact solution of the phase problem is described in Refs [89, 90] as applied to *in situ* X-ray reflectometry of a growing layered film, so that the reflectivity $R(t) = |r(t)|^2$ and its derivative dR/dt are known at time t. These two values are sufficient to find uniquely the phase of the amplitude reflectivity $r(t)$ at this point t. The approach is valid for both transparent and absorbing materials. The phase retrieval at a point t in time does not require any knowledge of the prehistory of the structure growth. Hence, there is no need to monitor the reflectivity during the whole deposition process in order to find the phase of the amplitude reflectance of a multilayer structure. It is sufficient to measure the reflectivity during a short temporal interval in the last stage of growth, the goal being only to find the derivative dR/dt at the end of deposition.

First experiments on the phase retrieval were described in Refs [89, 90], and an example is presented in Figure 7.19. *In situ* experiments on amorphous Al_2O_3 films growth were performed at the BM5 beamline of the ESRF [91]. The films were deposited by magnetron sputtering at 150 kHz pulse frequency on silicon substrates in pure (99.999%) Ar gas at the rather high working pressure of 2.66 Pa to stabilize the plasma discharge. Reflectivity versus the deposition time (Figure 7.19a, curve 1) was measured *in situ* at the fixed grazing angle of the probe beam $\theta = 0.25°$ and the X-ray energy $E = 17.5$ keV. Oscillations on the reflectivity curve are connected with the interference of waves reflected from the substrate and the external film surface. Curve 2 in Figure 7.19a shows the best fit using the simplest model of uniform Al_2O_3 placed on uniform Si substrate. Discrepancy between the simplest model and the experiment is seen clearly at initial stage of the film growth. The amplitude reflectivity phase (Figure 7.19b, curve 1) was determined from the experimental reflectivity directly without using any model of the film. For comparison, curve 2 shows the phase calculated in frames of the uniform film model. As discussed in Ref. [90], with the knowledge of the amplitude reflectivity variation in time, it is possible to determine the density profile of the film (Figure 7.19c, curve 1), even though the grazing angle of the probe beam is fixed. The density proved to be practically constant (2.29 ± 0.02 g/cm^3) regardless of the film depth and very low due to the relatively high pressure of the working gas.

The potentialities of the approach would be incomparably superior if the reflectivity curve versus the grazing angle $R(\theta, t)$ could be measured at any point in time. If this were possible, we would be able to find the phase versus the angle. Such an angular dispersive and time-resolved setup was built earlier for use with a laboratory

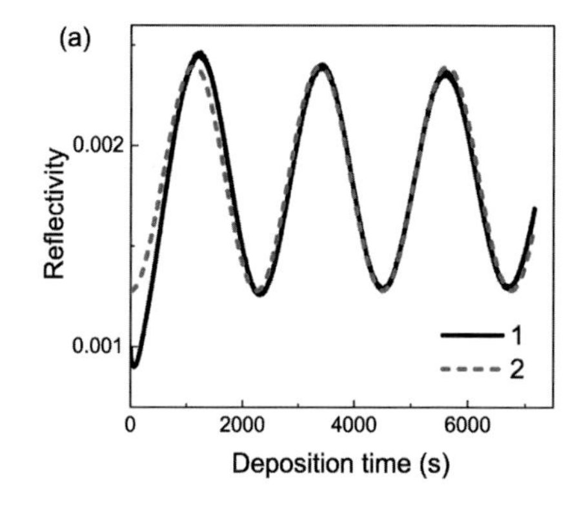

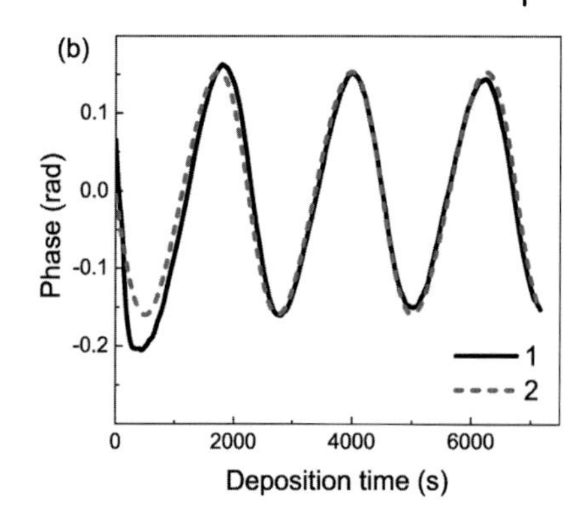

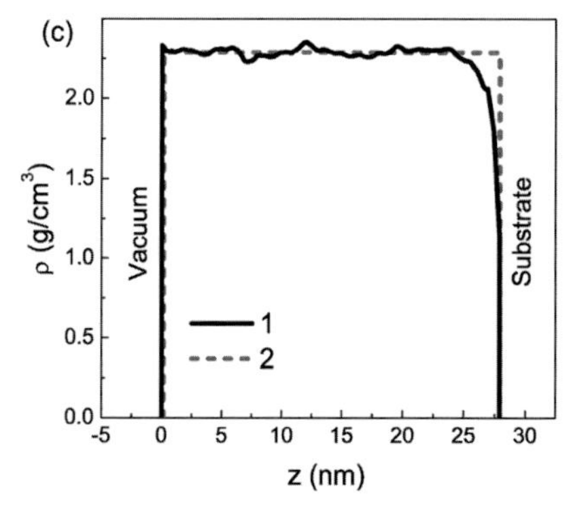

Figure 7.19 Curves 1: experimental reflectivity of growing Al_2O_3 film (a), the phase of the amplitude reflectivity (b), and the density profile (c), the phase and the density being deduced from the experimental reflectivity without using any model of the film. Curves 2: results of calculations using the simplest model of the uniform film.

X-ray source [92]. An equivalent setup combined with a powerful synchrotron source would improve the temporal sampling rate.

Therefore, as compared to the academic analysis of the inverse scattering problem in quantum mechanics, real X-ray experiments are characterized by the following points:

- While the phase retrieval methods are under development now, the absolute value of the reflectivity $R(q) = |r(q)|^2$ is measured in the vast majority of current X-ray experiments, that is, information about the amplitude reflectivity phase is unavailable.
- Contrary to quantum mechanics, where the energy of particle takes any value $q^2 \in [0, +\infty)$, X-ray measurements can be performed in a limited interval of the q-value, where $q_{\max} \leq k = 2\pi/\lambda$ in any case. Notice that in the classical definition of the inverse problem the q-value can be changed only through a variation in the incidence angle θ of X-ray beam and not through a variation in the radiation wavelength λ as in the last case the potential in (7.4) becomes a function of λ.

- In contrast to quantum mechanics or cold neutron optics, the absorption of hard X-rays can be neglected for enough thin films only. For example, absorption of X-rays at the energy $E = 8$ keV does not influence the reconstructed profile of tungsten film if the film thickness does not exceed 5 nm [93]. The film can be thicker for higher X-ray energy and much thicker for a film of light material. However, absorption can be neglected by no means in soft X-ray region. Notice no essential results were obtained in the inverse scattering problem of quantum mechanics for non-Hermitian (absorbing) potentials.
- No real surface can be considered as perfectly smooth for short-wavelength X-ray radiation. Surface and interface roughness scatters a part of the incident radiation and, hence, changes the shape of the reflectivity curve. As a result, the reconstructed dielectric constant profile can be deformed essentially and a number of artifacts can arise such as a seeming smoothening of interfaces between neighboring materials [93]. At the same time, analysis of the fine structure of interfaces is of special interest in microelectronic and optical technologies.
- Finally, the reflectivity curves are measured within some experimental error $\delta R(\theta)$.

Each of the factors listed above makes impossible an unambiguous extraction of the function $\chi(z)$ from the measured data, so that there is typically a continuum of different $\chi(z)$ distributions resulting in the same reflectivity curve $R(\theta) \pm \delta R(\theta)$ measured in the limited interval of the grazing angles $\theta \in [\theta_{\min}, \theta_{\max}]$ within the experimental error. Examples demonstrating ambiguity of the inverse problem of X-ray reflectometry are presented and discussed elsewhere [93–95]. As it was indicated in these papers, any additional knowledge about a sample is of extreme importance: even a good estimation of the overall sample thickness or understanding, where the material density is higher (on the top or on the bottom of a sample), can help in correct reconstruction of the dielectric constant profile.

Conventional approach to the inverse problem of X-ray reflectometry consists in modeling of $\chi(z)$distribution by a function with several unknown parameters, which are found by a least squares fit to experimental data. As it was mentioned in Ref. [94], the main advantage of this simplest approach is that it really works if adequate sample model is available. An example is presented in Figure 7.20 as applied to the study of the above-discussed HfO_2 films of about 5 nm and 17 nm thickness deposited on silicon substrates by the ALD technique. Reflectivities versus the grazing angle (Figure 7.20a, circles) were measured at X-ray energy $E = 8052$ eV (characteristic K_α emission line of Cu) with the use of conventional laboratory diffractometer [96]. The physically reasonable model takes into account natural oxide layer on the top of Si substrate and adhesion layer, which is always formed on any surface and consists mainly of molecules of hydrocarbons and water. For definiteness, we will suppose that the adhesion layer consists of carbon atoms. More detail discussion of an adequate choice of the sample model is given in the next section. The samples studied, thus, can be represented as three-layer films $C/HfO_2/SiO_2/Si$, where the thicknesses and the densities of all layers are unknown fitting parameters. The Si substrate density (2.42 g/cm^3) is supposed to be known.

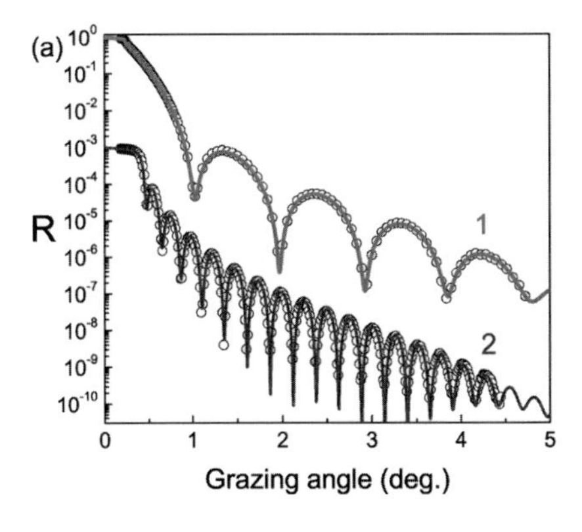

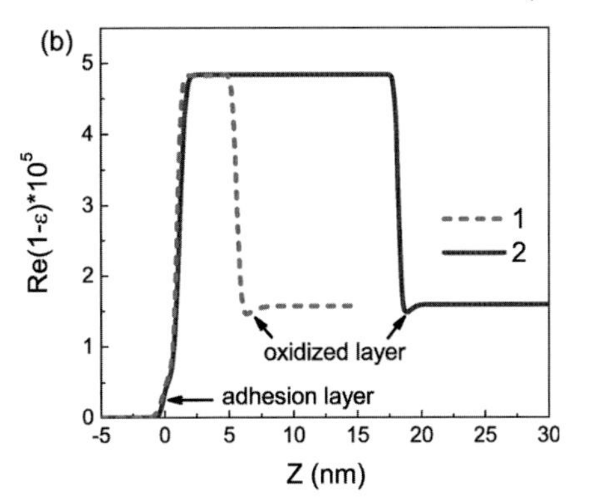

Figure 7.20 (a) The experimental reflectivity at $E = 8052$ eV (circles) versus the grazing angle of HfO_2 films of about 5 nm (1) and 17 nm (2) thickness. Solid curves are the result of fitting using the simplest three-layer model. Reflectivity of the 17 nm thick film is shifted vertically by a factor of 10^{-3} for clarity. (b) The dielectric constant profiles reconstructed with the use of the three-layer model.

In addition, it is necessary to take into account the roughness effect resulting in a quicker decrease in the reflectivity with increasing grazing angle compared to $1/\sin^4\theta$ law legitimated for a perfectly smooth sample. Notice that the roughness of a film can be typically divided into two components (see, for example, Refs [91, 97, 98]). One of them is the long-scale conformal roughness replicating the substrate one. The second component is the short-scale roughness arising due to stochastic nature of deposition process and not correlated with the substrate roughness. In the case of conformal roughness with small enough root mean-squared (rms) height σ and large enough correlation length in the interface plane, the reflectivity and the total integrated scattering (TIS) in vacuum are expressed as

$$R(\theta) = R_0(\theta)\left[1-(4\pi\sigma \sin\theta/\lambda)^2\right]; \quad \mathrm{TIS}(\theta) = R_0(\theta)(4\pi\sigma\sin\theta/\lambda)^2 \tag{7.7}$$

where R_0 is the reflectivity of the perfectly smooth sample. Notice that Equation 7.7 is valid for any layered medium (single surface, thin film, periodic, or aperiodic multilayer structure) with abrupt or smooth interfaces and for any probability density function of the roughness heights. Deducing, extended discussion, and analysis of applicability of Equation 7.7 are given in (Kozhevnikov, I.V. (2012) General Laws of X-Ray reflection from rough surfaces: II. Conformal Roughness. Crystallography Reports, 57, 417–426). An important feature of (7.7) is that the integral reflectivity $R_\Sigma(\theta) = R(\theta) + \mathrm{TIS}(\theta)$ characterizing the total intensity of radiation directed by a rough sample back into the vacuum (sum of the specularly reflected and the scattered intensities) is very close to the reflectivity of perfectly smooth sample. Therefore, the long-scale conformal roughness does not influence the detected signal if the detector entrance aperture is wide enough to collect all scattered intensity in parallel to the specularly reflected one, as it was done in our experiments.

Hence, the short-scale roughness, which is nonconformal typically, affects only the experimental reflectivity. If the roughness height is distributed in accordance with the normal law, the effect of the short-scale nonconformal roughness on the reflectivity can be described by two equivalent manners: first, introducing the Nevot–Croce factor [99] to amplitude reflectance of each interface and, second, introducing smooth variation in the dielectric constant at each interface, an "effective" transition layer formed close to a rough interface as a result of short-scale roughness averaging being written as

$$\varepsilon(z) = \frac{\varepsilon_A + \varepsilon_B}{2} - \frac{\varepsilon_A - \varepsilon_B}{2} \operatorname{erf}\left(\frac{z}{\sigma\sqrt{2}}\right) \tag{7.8}$$

where ε_A and ε_B are the dielectric constants of neighboring materials A and B, and erf(x) is the error function. If scattering of X-ray radiation by interfacial roughness is not analyzed in experiment, as in our case, it is impossible to distinguish between the effect on the reflectivity of small-scale roughness and that of real transition layer formed between neighboring materials due to interdiffusion, implantation, and chemical reactions. Thus, the σ-values found below characterize cumulative effect of both small-scale roughness and real transition layer. Notice, the interface width is approximately equal to 3σ, and the σ-values of all interfaces are considered as additional fitting parameters. Thus, the total number of fitting parameters was 10 for the HfO_2 samples studied. The merit function used under fitting had the simplest form:

$$\mathrm{MF} = \sum_j \left[\frac{R_{\mathrm{exp}}(\theta_j) - R_{\mathrm{calc}}(\theta_j)}{R_{\mathrm{exp}}(\theta_j)}\right]^2 \tag{7.9}$$

where R_{exp} and R_{calc} are the measured and the calculated reflectivity.

In spite of extreme simplicity of the merit function (7.9), there is a great variety of local minima, which is quite typical of nonlinear optimization problem. To find the solution corresponding to reality we, first, impose limitations on the possible values of the fitting parameters; for example, the layer density was supposed to not exceed the bulk material density. Second, starting from different guess values of the fitting parameters, we found solution providing the deepest minimum of the merit function.

An accuracy of fitting of experimental reflectivity curves of HfO_2 films is demonstrated in Figure 7.20a, solid curves, and the dielectric constant profiles drawn with the use of (7.8) are shown in Figure 7.20b. As seen, the simplest three-layer model resulted in a perfect agreement between the experimental and the calculated reflectivity curves as well as between the dielectric constant profiles of both films including small features in the profiles such as a fine structure of adhesion and oxidized layers. The reconstructed profiles look very realistic. In particular, the density of HfO_2 film (9.69 g/cm^3) is very close to the bulk density. The found rms roughness of the substrate (0.22–0.23 nm) and the external film surface (0.30 nm for thinner film and 0.37 nm for thicker film) agree well with the AFM data. The σ-value of SiO_2/Si interface describes a smooth variation in oxygen into the depth of the

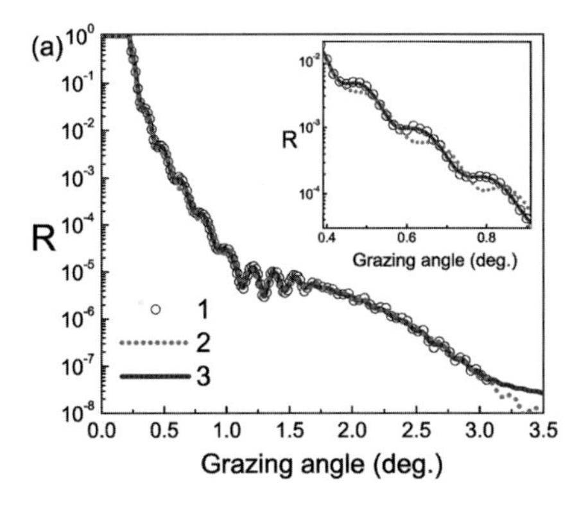

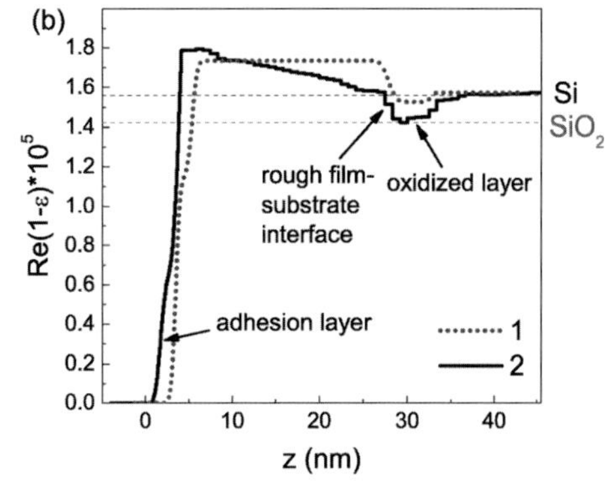

Figure 7.21 (a) The experimental reflectivity at $E = 8052$ eV (circles) versus the grazing angle of Al_2O_3 film. Curves 2 and 3 are the results of fitting using the simplest three-layer model (2) and free-form approach (3) developed by Kozhevnikov [93]. (b) The dielectric constant profiles reconstructed with the use of the three-layer model (1) and free-form approach (2). Dashed lines indicate the dielectric constant of the bulk Si (2.42 g/cm^3 density) and SiO_2 oxide (2.2 g/cm^3 density).

substrate rather than the interface roughness. The adhesion carbon layer is thin (thickness is about 1.3 nm) and loose (maximal density is on the order of 1 g/cm^3). The example considered demonstrates clearly that the simplest approach based on the physically reasonable model can be used successfully to analyze internal structure of high-quality films.

A counterexample is presented in Figure 7.21, where the results of X-ray studying Al_2O_3 film deposited on Si substrate by the ALD technique are presented. In contrast to the HfO_2 films, the reflectivity curve of Al_2O_3 film (Figure 7.21a, circles) is characterized by disappearing oscillations with increasing grazing angle. Therefore, we can conclude immediately that at least one interface of the film studied is destroyed (smoothened essentially), which results in decreasing X-ray reflectivity from this interface and, hence, decreasing amplitude of interference oscillations.

As above, the three-layer model $C/Al_2O_3/SiO_2/Si$ was used to fit the experimental reflectivity curve (Figure 7.21a, curve 2). The reconstructed dielectric constant profile is shown in Figure 7.21b, curve 1. At the first glance, an accuracy of fitting of experimental reflectivity curve is quite acceptable in logarithmic scale. However, a more careful comparison of experimental data and curve 2 in the range of small grazing angles (inset in Figure 7.21a) demonstrates clearly that oscillations in the calculated reflectivity curve are in antiphase with those in the experimental curve. Therefore, in the case considered the model-dependent fitting procedure proved to be unsuccessful and the way to improve the model is unclear.

Evidently, if we are unable to imagine an adequate model of a structure, we should use a model-independent approach to solve the inverse problem. So far, several free-form approaches to reconstruction of the polarizability profiles based on X-ray or neutron reflectivity data have been developed. Among very general approaches, we can indicate the maximum entropy method, the Bayesian spectral analysis [94], and the parameterization of $\chi(z)$ distribution using cubic B splines or sinus/cosine basis [95]. The method of maximum entropy will be discussed below in more detail as applied to analysis of soft X-ray reflectivity measurements.

One more free-form approach has been developed [93] (Kozhevnikov, I.V., Peverini, L., and Ziegler, E. (2012) Development of a self-consistent free-form approach for studying the three-dimensional morphology of a thin film. *Phys. Rev. B.*, **85**, 125439). The approach is based on the modeling of the amplitude reflectivity in the range of large q-values rather on the direct modeling of $\chi(z)$ distribution. The modeling of the amplitude reflectivity $r(q)$ at large q proves to be possible in the frame of a general model of reflecting media. A distribution $\chi(z)$ is declared as having a point of discontinuity of the nth order if $\chi(z)$ together with its n-1 derivatives $\chi'(z),\ldots,\chi^{(n-1)}(z)$ is a continuous function at point z, while its nth derivative $\varepsilon^{(n)}(z)$ suffers a step-like variation at this point. Following the approach developed in Ref. [93] (Kozhevnikov, I.V., Peverini, L., and Ziegler, E. (2012) Development of a self-consistent free-form approach for studying the three-dimensional morphology of a thin film. *Phys. Rev. B.*, **85**, 125439.), we state that a finite number of possible solutions to the inverse problem exists if the number and the order of the points of discontinuity of the dielectric function and the distances between them are known. In particular, the number of possible solutions is only four and it is independent of the number of discontinuity points if all distances between them are different. Moreover, two of these solutions result in nonphysical negative values of $\chi(z)$. A choice between the remaining two solutions can be made on the basis of available information about the sample or by performing additional experiments. In its turn, the order and the number of discontinuity points as well as the distances between them can be recognized from the Fourier analysis of the measured part of the reflectivity curve $R(q)$. The main disadvantage of the approach is neglect of X-ray absorption in the matter so that the approach can be applied only for hard X-rays and for thin enough samples. In contrary, it is necessary to determine two unknown functions $\mathrm{Re}[\chi(z)]$ and $\mathrm{Im}[\chi(z)]$ and the problem of uniqueness becomes much harder. Neglecting absorption means that, when processing, we should exclude experimental data close to the critical angle of the total external reflection, where the effect of absorption on the reflectivity is large. Evidently, then we lose some information about the dielectric constant of a substrate determining the critical angle. However, the loss of this information is not crucial because (a) the substrate influences the remaining part of the reflectivity curve, so that its dielectric permittivity can still be determined, and (b) the dielectric constant of a substrate is known *a priori*, as a rule. The roughness effect can be accurately taken into account during the dielectric constant profile reconstruction and, vice versa, the profile can be properly accounted for determination of the roughness parameters (Kozhevnikov, I.V., Peverini, L., and Ziegler, E. (2012) Development of a self-consistent free-form approach for studying the three-dimensional morphology of a thin film. *Phys. Rev. B.*, **85**, 125439). However, such a self-consistent approach demands a set of scattering distributions measured at different grazing angles of an incident X-ray beam.

First of all, we found that the reflectivity of the 5 nm thick HfO_2 film discussed above decreases in average as $1/\sin^6\theta$ with increasing grazing angle θ. It means that

$\chi(z)$ distribution can be considered as a continuous function of z, while its first derivative undergoes step-like variation in certain points z_j. After performing a Fourier analysis of the reflectivity curve of the thinner HfO_2 film, we concluded that there are three points of discontinuity, the distances between neighboring points being about 1.5 and 4.4 nm. In accordance with Kozhevnikov [93], we chose the merit function in the following form:

$$\mathrm{MF} = \sum_j \left[\frac{R_{\mathrm{exp}}(\theta_j) - R_{\mathrm{calc}}(\theta_j)}{R_{\mathrm{exp}}(\theta_j)}\right]^2 + Q \sum_{\substack{i=2,\ldots,N-1 \\ i \neq i_1, i_2, \ldots, i_m}} (\chi_{i+1} - 2\chi_i + \chi_{i-1})^2$$

$$+ Q_1 \left[\sum_{i=i_2,\ldots,i_m} (\chi_{i+1} - \chi_i)^2 + \chi_1^2\right] \tag{7.10}$$

where the dielectric constant profile was divided onto $N = 90$ subintervals, the χ-values inside them being considered as the fitting parameters.

The second sum in (7.10) is a finite difference analogue of the second derivative of function $\chi(z)$, the points of discontinuity being excluded from the sum, and the last sum is necessary to provide the continuity of function $\chi(z)$ in the points of discontinuity of the first order. Notice the second sum in the merit function (7.10) plays several roles [93]. First of all, introducing the sum into the merit function permits to find the $\chi(z)$ distribution providing the necessary asymptotic behavior of the amplitude reflectivity at large θ ($R \sim 1/\sin^6\theta$). Second, the sum results in smoothening of the merit function and in disappearance of many local minima of moderate depth. Third, the sum plays the role of a regularization function used in modern approaches to solve ill-conditioned problems. In particular, the sum results in high stability of the inverse problem solution in respect of experimental errors of the reflectivity measurements. Parameters Q and Q_1 as well as the number of subintervals should be chosen to provide the necessary asymptotic behavior of the reflectivity and the necessary accuracy of fitting of the measured part of the reflectivity curve. A comprehensive discussion of computational algorithm is given in Ref. [93].

We applied the free-form approach to the reflectivity analysis of the 5 nm thick HfO_2 film. Starting guess for the merit function minimization was uniform Si substrate. The reconstructed dielectric constant profile is presented in Figure 7.22, curve 2. For comparison, the profile found above with the use of the three-layer model is also shown in Figure 7.22, curve 1. As seen, the profiles agree well with each other including a small feature on the sample top (adhesion layer).

Similar calculations were performed for Al_2O_3 film. As it was discussed above, the three-layer model gave no way for correct fit of the measured reflectivity curve. The free-form approach resulted in excellent agreement between the experimental and the calculated reflectivity curves (Figure 7.21a, curve 3). The reconstructed dielectric constant profile shown in Figure 7.21b, curve 2, demonstrates a number of features compared to curve 1 obtained with the use of the three-layer model. First of all, the density of Al_2O_3 film was found to increase gradually with film

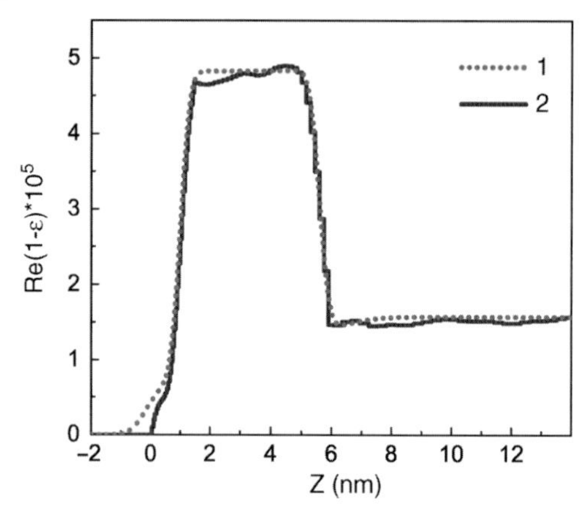

Figure 7.22 The dielectric constant profile (at $E = 8052$ eV) of HfO_2 film reconstructed with the use of the three-layer model (1) and the free-form approach (2). Curve 1 is the same as curve 1 in Figure 7.20b.

growth. In addition, a very thick (6–7 nm) oxide layer was formed on the Si substrate surface after deposition of the film, and the film–substrate interface became very rough ($\sigma \sim 0.7$ nm). All these conclusions correlate well with data obtained by NEXAFS.

7.5.2
Reconstruction of the Depth Distribution of Chemical Elements Concentration by Soft X-Ray Reflectometry

The reflectivity of a layered sample is determined entirely by the depth distribution of the matter polarizability $\chi(z, E)$ depending on the photon energy. If the sample consists of several chemical elements A, ..., B, the polarizability distribution in the soft X-ray region can be represented as [100] $1-\varepsilon(z, E) \equiv \chi(z, E) \sim E^{-2}[C_A(z) f_A(E) + \ldots + C_B(z) f_B(E)]$, where $C_j(z)$ and $f_j(E)$ are the atomic concentration profile and the complex atomic scattering factor of the jth element, the values of Ref and Imf characterizing refraction and absorption of the SXR radiation, respectively. Below, we will assume that chemical elements composing a sample are known *a priori*. The values of $f_j(E)$ were tabulated as a function of energy for all chemical elements and can be found in the Ref. [101]. The main idea of the approach consists in measurements of the reflectivity $R(\theta, E)$ versus the incident angle at different photon energies (above and below absorption edge of elements), where atomic scattering factor changes abruptly.

The variation in the atomic scattering factors Re(f) and Im(f) with the soft X-ray energy is shown in Figure 7.23 for a number of chemical elements (Ti, Si, O, and C). The figure demonstrates clearly that the contribution of chemical elements to the dielectric permittivity of matter may be quite different at different radiation wavelengths, which may result in essential variation in the reflectivity curve shape.

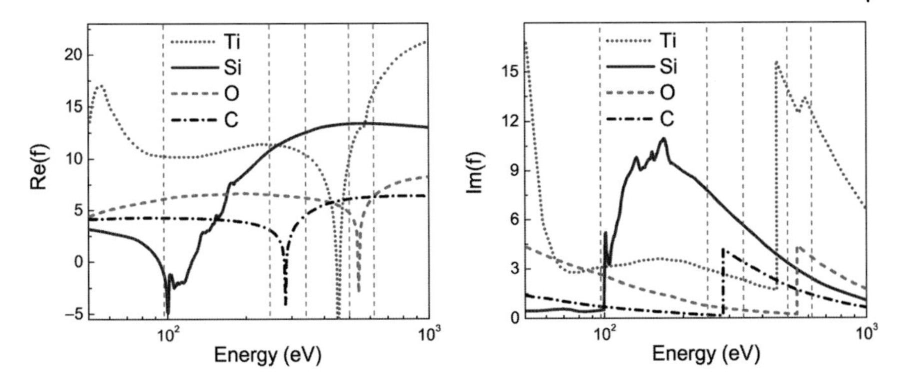

Figure 7.23 Variation in the real and the imagined parts of the atomic scattering factor Re(f) and Im (f) in the soft X-ray region for several chemical elements. Vertical dashed lines indicate SXR energies used in our experiments on the study of TiO_2 and Ti films on Si substrates.

An example is presented in Figure 7.24. Solid curves show the reflectivity of a very thin (1.5 nm thick) carbon film (2.2 g/cm^3 density) placed on top of a bulk Si substrate (2.42 g/cm^3 density). For comparison, the reflectivity of virgin Si substrate is also shown (dashed curves). As seen, the difference between the reflectivity curves of C/Si sample and Si substrate is not distinct at X-ray energy $E = 200$ eV lying below K-edge of absorption of carbon, so that a recognition of the carbon film is troublesome. However, the difference between the curves becomes essential (more than one order of magnitude) at energy $E = 300$ eV lying above the absorption edge. Moreover, the reflectivity curve shapes are quite different. Therefore, the presence of very thin carbon layer is well pronounced at this SXR energy. Evidently, quite different behavior of C/Si sample reflectivity at two radiation wavelengths is explained by a sharp variation in the phase of the amplitude reflectivity of film interfaces resulting in

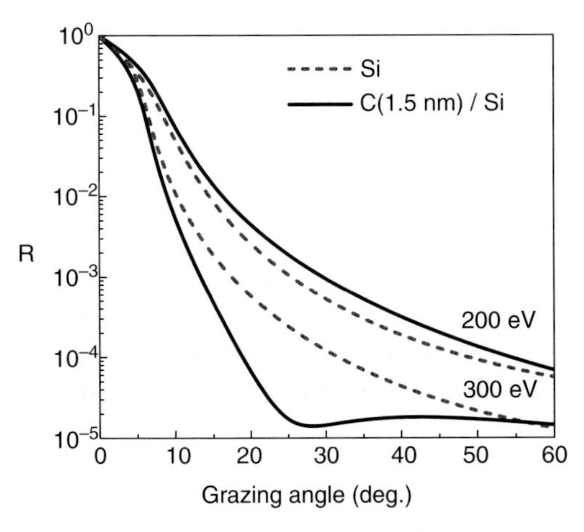

Figure 7.24 Reflectivity of Si substrate (dashed curves) and 1.5 nm thick carbon film on the Si substrate at two different SXR energies ($E = 200$ eV and 300 eV) lying below and above the K-edge of absorption of carbon.

the pronounced interference minimum at small grazing angle rather by increasing absorption in carbon, which is negligible for the very thin film considered.

Therefore, fitting simultaneously all experimental curves measured at different SXR wavelengths, we can hope to reconstruct atomic concentration profiles $C_j(z)$ of all elements composing the sample. First experiments on the reconstruction are described in Refs [47] (Filatova, E.O., Kozhevnikov, I.V., Sokolov, A.A., Ubyivovk, E.V., Yulin, S., Gorgoi, M., and Schaefers, F. (2012) Soft x-ray reflectometry, hard x-ray photoelectron spectroscopy and transmission electron microscopy investigations of the internal structure of TiO_2(Ti)/SiO_2/Si stacks. *Sci. Technol. Adv. Mater.* **13**, 015001). However, as we want to deduce several unknown functions $C_j(z)$ and absorption of radiation cannot be neglected, the problem of uniqueness becomes harder to resolve. To overcome the problem, we use an approach based on the well-known philosophical principle of Ockham's razor: when several models describe some physical phenomenon, it is preferable to use the simplest one consistent with experimental data (see also Filatova, E.O., Kozhevnikov, I.V., Sokolov, A.A., Ubyivovk, E.V., Yulin, S., Gorgoi, M., and Schaefers, F. (2012) Soft x-ray reflectometry, hard x-ray photoelectron spectroscopy and transmission electron microscopy investigations of the internal structure of TiO_2(Ti)/SiO_2/Si stacks. *Sci. Technol. Adv. Mater.* **13**, 015001 [94]). The approach is based on the use of a sequence of models, whose complexity increases progressively until the necessary accuracy of experimental data fitting is achieved.

As a practical example, let us analyze the results of SXR study of two samples: TiO_2 film deposited by rf magnetron sputtering on Si substrate with natural oxide layer on its top (sample 1) and Ti film deposited by DC magnetron sputtering on similar substrate (sample 2). Both samples were studied by HAXPES and discussed above.

The reflectivity of the studied samples was measured as a function of the incident angle at different SXR energies E. The measurements were performed using s-polarized synchrotron radiation in the reflectometer setup at the optics beamline (D-08-1B2) at the Berlin Synchrotron Radiation facility (BESSY II) of the Helmholtz-Zentrum Berlin [49, 50]. The accuracy of the energy scale was 10 meV. All curves were measured with an angular accuracy of 0.001°. A GaAsP diode, together with a Keithley electrometer (617), was used as a detector. We used a detector with a 4 mm window diameter to monitor the total intensity of the reflected beam. To reduce higher diffraction orders in the incident beam, a system of absorption filters was used.

The measured reflectivity of sample 1 versus the grazing incidence angle counted from the sample surface is shown in Figure 7.25, circles, at different SXR photon energies varying from $E = 97$ eV (the wavelength $\lambda = 12.78$ nm) to $E = 620$ eV ($\lambda = 2$ nm). All the reflectivity curves are taken into account at the same time, when processing. Notice that the energy values used in the experiment are indicated in Figure 7.23 by vertical dashed lines. As seen, they are placed between absorption edges of elements composing the films to make easier and more precisely the reconstruction of the concentration profiles.

The simplest model of sample 1 is a uniform TiO_2 film placed on a uniform Si substrate. The four fitting parameters are the thickness and the density of the film as

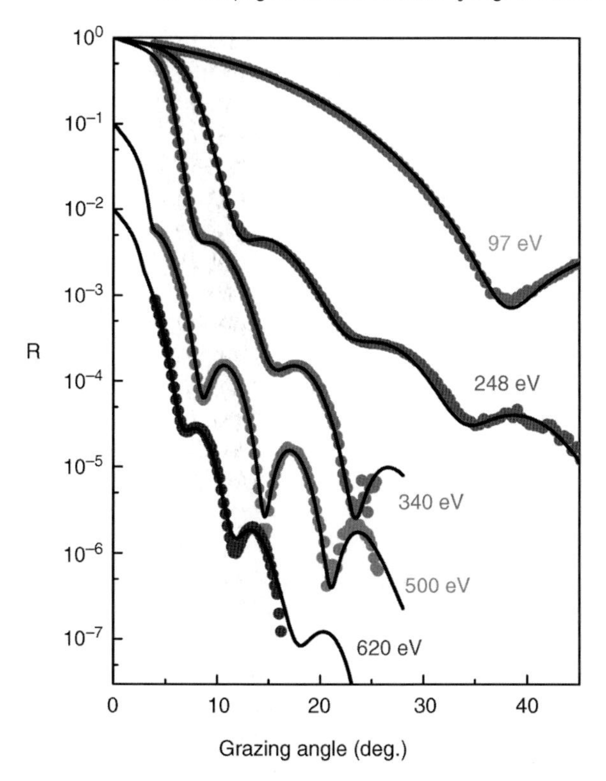

Figure 7.25 The measured reflectivity of the sample 1 (circles) versus the grazing angle at different energies of the incident beam. Solid curves show the result of fitting in the frame of the three-layer model ($C/TiO_2/SiO_2/Si$) after numerical refinement procedure with the use of the merit function (7.12). The reflectivity curves at $E = 500$ eV and 620 eV are shifted vertically for clarity (from Filatova, E.O., Kozhevnikov, I.V., Sokolov, A.A., Ubyivovk, E.V., Yulin, S., Gorgoi, M., and Schaefers, F. (2012) Soft x-ray reflectometry, hard x-ray photoelectron spectroscopy and transmission electron microscopy investigations of the internal structure of $TiO_2(Ti)/SiO_2/Si$ stacks. *Sci. Technol. Adv. Mater.* **13**, 015001).

well as the root mean-squared roughness σ of the Si substrate and the external film surface. The substrate density (2.42 g/cm^3) is supposed to be known.

Results of fitting of the reflectivity curve of sample 1 with the use of the simplest model and Equation 7.8 are shown in Figure 7.26 for two of the measured reflectivity curves. As seen, the simplest single-layer model (dashed curves) allows to describe adequately the reflectivity at $E = 97$ eV, while the calculated reflectivity differs essentially from the measured one at $E = 340$ eV.

More complex model takes into account an oxidized layer on the Si substrate surface. Therefore, at the second step we fit experimental curves using the two-layer model $TiO_2/SiO_2/Si$. As above, the fitting parameters are thickness and density of all layers as well as the rms roughness σ of all interfaces. Notice that in the model considered the σ-value of SiO_2/Si interface characterizes a gradual variation in the

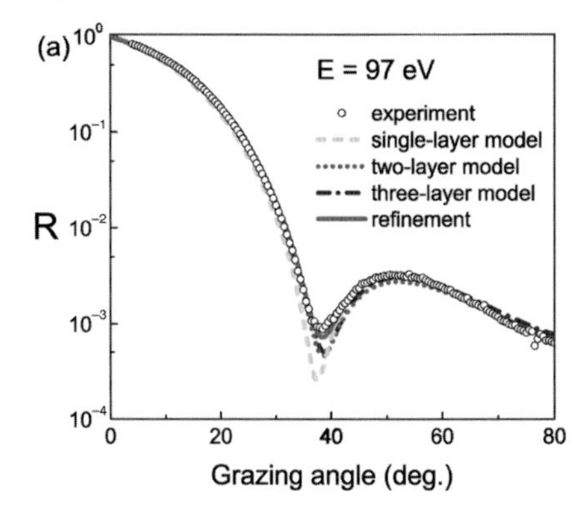

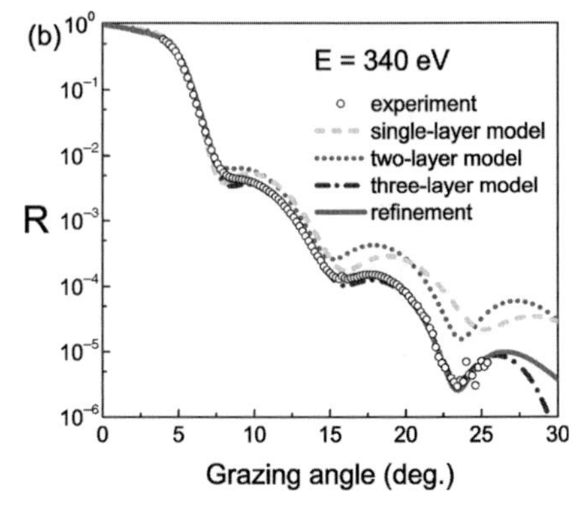

Figure 7.26 The measured reflectivities of sample 1 (circles) versus the grazing angle at two energies of the incident beam $E = 97$ eV and $E = 340$ eV. The curves demonstrate an accuracy of fitting in the frame of the single-layer (TiO_2/Si), two-layer ($TiO_2/SiO_2/Si$), and three-layer ($C/TiO_2/SiO_2/Si$) models as well as that after numerical refinement procedure with the use of the merit function (12) (from Filatova, E. O., Kozhevnikov, I.V., Sokolov, A.A., Ubyivovk, E. V., Yulin, S., Gorgoi, M., and Schaefers, F. (2012) Soft x-ray reflectometry, hard x-ray photoelectron spectroscopy and transmission electron microscopy investigations of the internal structure of $TiO_2(Ti)/SiO_2/Si$ stacks. *Sci. Technol. Adv. Mater.* **13**, 015001).

silicon and oxygen concentrations into the depth of the substrate rather than a rough boundary between the Si substrate and the oxide layer. Results presented in Figure 7.26, dotted curves, show that the accuracy of fitting is improved by a factor of 2 in the minimum of the reflectivity curve at $E = 97$ eV, while it is practically unchanged at $E = 340$ eV.

Further complication of the model implies the presence of an adhesion layer, which is always formed on any surface and consists mainly of molecules of hydrocarbons and water. For definiteness, we will suppose that the adhesion layer consists only of carbon atoms. Accuracy of fitting is illustrated in Figure 7.26, dashed–dotted curves. While introducing the adhesion layer does not influence practically the reflectivity at $E = 97$ eV, the three-layer model $C/TiO_2/SiO_2/Si$ results in a very good agreement with the measured reflectivity at the photon energy $E = 340$ eV lying just above the carbon K-edge of absorption, where its polarizability changes sharply.

The three-layer model thus allows to describe the whole set of the reflectivity curves within reasonable accuracy, while small deviations between the calculated and the experimental curves are still seen. The found depth distribution of atomic concentrations is shown in Figure 7.27a, where we used representation (7.8) to describe transition layer between neighboring materials. The thickness of TiO_2 film and SiO_2 oxide layer is found to be 9.84 and 2.25 nm, respectively. The density of TiO_2 film is 3.67 g/cm^3, which is 7–13% less than that of the bulk TiO_2 crystals, and the density of SiO_2 layer is 2.39 g/cm^3. The rms roughness of external film surface and of the substrate surface proves to be practically the same and equal to about 0.35 nm. As it was mentioned above, the σ-value of SiO_2/Si interface

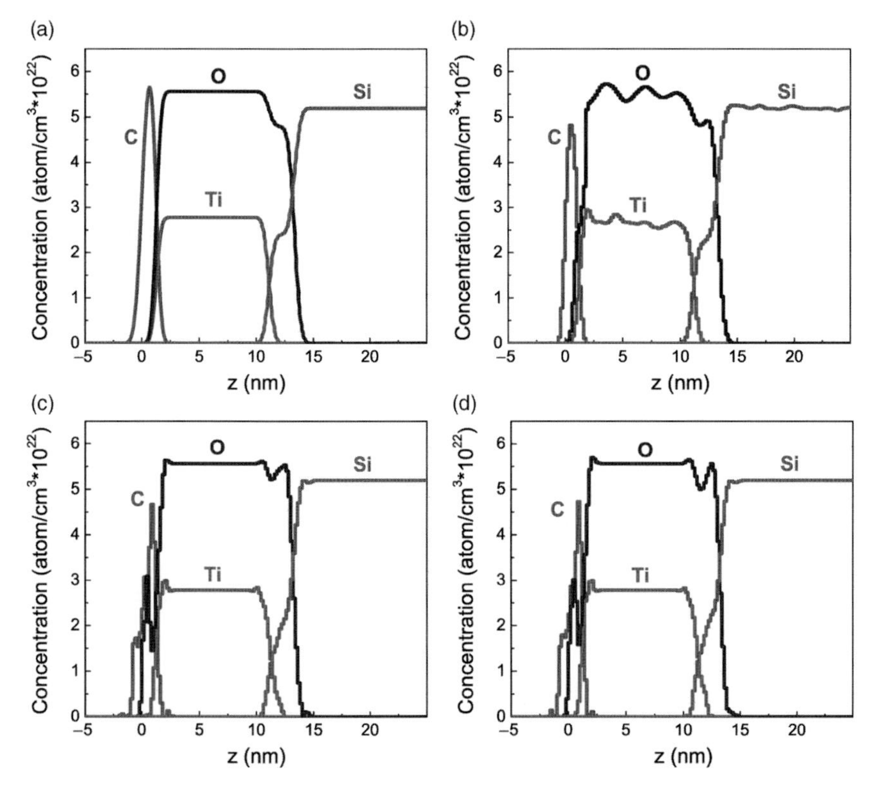

Figure 7.27 The reconstructed concentration profiles of chemical elements composed sample 1. Simulations are performed in the frame of three-layer ($C/TiO_2/SiO_2/Si$) model (a) and after numerical refinement with the use of the merit function (12) (b) or (10) at $n = 1$ (c) and $n = 2$ (d) (from Filatova, E.O., Kozhevnikov, I.V., Sokolov, A.A., Ubyivovk, E.V., Yulin, S., Gorgoi, M., and Schaefers, F. (2012) Soft x-ray reflectometry, hard x-ray photoelectron spectroscopy and transmission electron microscopy investigations of the internal structure of TiO_2(Ti)/SiO_2/Si stacks. *Sci. Technol. Adv. Mater.* **13**, 015001).

(0.41 nm) describes a smooth variation in oxygen into the depth of the substrate rather than the interface roughness. The adhesion carbon layer is thin (thickness is 1.29 nm) and loose (maximal density: 1.26 g/cm^3) according to the results obtained above by HXR reflectometry. The σ-value of the external surface of the adhesion layer (0.46 nm) describes both the roughness and the gradual decrease in the carbon density into vacuum.

The values found for the TiO_2 and the SiO_2 layer thickness as well as the SiO_2 layer density are in complete agreement with those deduced from the HAXPES measurements, while the TiO_2 layer density proved to be about 13% less. The reason of the discrepancy is not totally clear today, while it may be connected with inaccuracy in the value of inelastic mean free path of photoelectron in TiO_2 and with too simplified a model of the sample used in the HAXPES analysis (neglecting adhesion layer and interfacial roughness).

Next step of analysis consists in numerical refinement of the solution found to describe quantitatively all small features observed in the measured reflectivity curves. We used the refinement procedure based on the conception of maximum entropy, which was used in a number of studies to process hard X-ray and neutron reflectivity curves as well as HAXPES data (see, for example, Refs [94, 102] and references therein). The generalized Shannon–Jaynes entropy (negative function) is given by [103]

$$S = \sum_{i,j} \left[C_j(z_i) - C_j^{(0)}(z_i) - C_j(z_i)\log\left(C_j(z_i)/C_j^{(0)}(z_i) \right) \right] \tag{7.11}$$

where $C_j(z_i)$ is the concentration of jth element in ith pixel of the digitized reconstruction of $C_j(z)$ and $C_j^{(0)}(z)$ is a default model with respect to which the entropy is measured. In our case, $C_j^{(0)}(z)$ is the solution of the problem found in the frames of the three-layer model. If we put $C_j(z) = C_j^{(0)}(z)$, the entropy is equal to zero. The method is based on the determination of solution $C_j(z)$ providing the necessary accuracy of fitting of experimental reflectivity curves and, simultaneously, the maximum entropy. In other words, following the Ockham's razor principle, we want to find solution closest to the default one. Notice if $C_j(z)$ is close to $C_j^{(0)}(z)$, the entropy (7.11) can be written as $S \approx -\frac{1}{2}\sum_{i,j}\left[C_j(z_i) - C_j^{(0)}(z_i) \right]^2 / C_j^{(0)}(z_i)$. Therefore, we used the following merit function when fitting the measured reflectivities:

$$\mathrm{MF} = \sum_{k,m} \left[\frac{R_{\exp}(\theta_k, E_m) - R_{calc}(\theta_k, E_m)}{R_{\exp}(\theta_k, E_m)} \right]^2 + Q \sum_{i,j} \frac{\left[C_j(z_i) - C_j^{(0)}(z_i) \right]^2}{C_j^{(0)}(z_i)} \tag{7.12}$$

where Q is the parameter providing the necessary accuracy of fitting. Summations on i and j are carried out on all points z (150 in our calculations) and on all chemical elements composing the sample (four in our case). Therefore, the total number of the fitting parameters was 600. Summations on m and k are carried out on all photon energies and grazing angles of incident radiation.

Solid curves in Figures 7.25 and 7.26 show the results of fitting. As can be observed, agreement between the calculated and the measured curves is almost perfect. However, the refined concentration distributions shown in Figure 7.26b demonstrate that the merit function in the form (7.12) resulted in appearance of nonphysical oscillations. Increasing parameter Q in (7.12) smoothens these oscillations. However, an accuracy of experimental curves fitting became worse. Therefore, instead of (7.12), we tried to use the following generalization of the merit function:

$$\mathrm{MF} = \sum_{k,m} \left[\frac{R_{\exp}(\theta_k, E_m) - R_{\mathrm{calc}}(\theta_k, E_m)}{R_{\exp}(\theta_k, E_m)} \right]^2 + Q \sum_{i,j} \frac{\left[(d^n/dz^n)\left(C_i(z_j) - C_i^{(0)}(z_j) \right) \right]^2}{\alpha + \left| (d^n/dz^n) C_i^{(0)}(z_j) \right|} \tag{7.13}$$

where we added small positive number α into the denominator to overcome the problem of dividing by zero.

The measure of proximity of the solution $C_j(z)$ to the default one $C_j^{(0)}(z)$ is the difference between concentrations $C_j(z)-C_j^{(0)}(z)$ in the merit function (7.12), while it is the difference in their derivatives $d^nC_j/dz-d^nC_j^{(0)}/dz$ in the merit function (7.13).

The refined concentration distributions are shown in Figure 7.27c and d for different $n=1$ and 2, respectively. As seen, nonphysical oscillations in the profiles disappear. The distributions are very similar to each other and they are very close to the default ones shown in Figure 7.27a. Only two features were observed in the refined distributions. First, oxygen (water) appears in the adhesion layer placed on top of the sample, the carbon density being reduced. Second, concentration of oxygen increases slightly on top of Si substrate (at $z \sim 13$ nm) compared to the default distribution. These two features allow us to describe the measured reflectivity curves within the same accuracy as for the distribution shown in Figure 7.27b. Therefore, we did not draw the calculated reflectivity curves in Figure 7.25 for these cases.

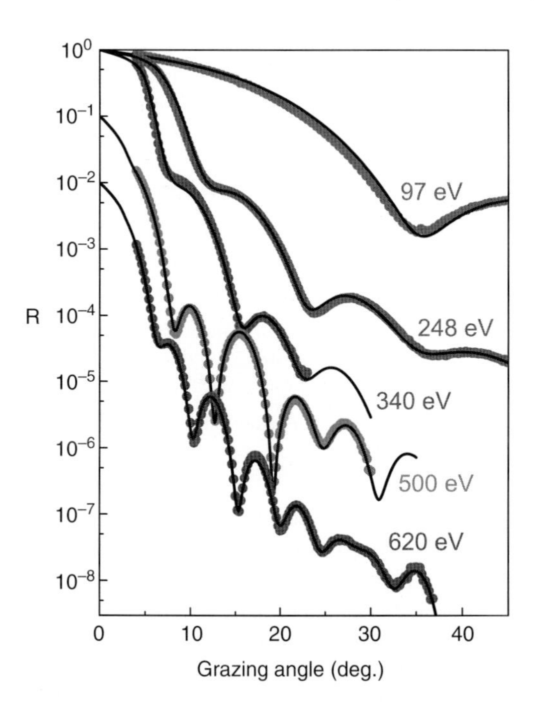

Figure 7.28 The measured reflectivity of the sample 2 (circles) versus the grazing angle at different energies of the incident beam. Solid curves show the result of fitting after numerical refinement procedure with the use of the merit function (13) at $n=1$. The reflectivity curves at $E=500$ eV and 620 eV are shifted vertically for clarity (from Filatova, E.O., Kozhevnikov, I.V., Sokolov, A.A., Ubyivovk, E.V., Yulin, S., Gorgoi, M., and Schaefers, F. (2012) Soft x-ray reflectometry, hard x-ray photoelectron spectroscopy and transmission electron microscopy investigations of the internal structure of $TiO_2(Ti)/SiO_2/Si$ stacks. *Sci. Technol. Adv. Mater.* **13**, 015001).

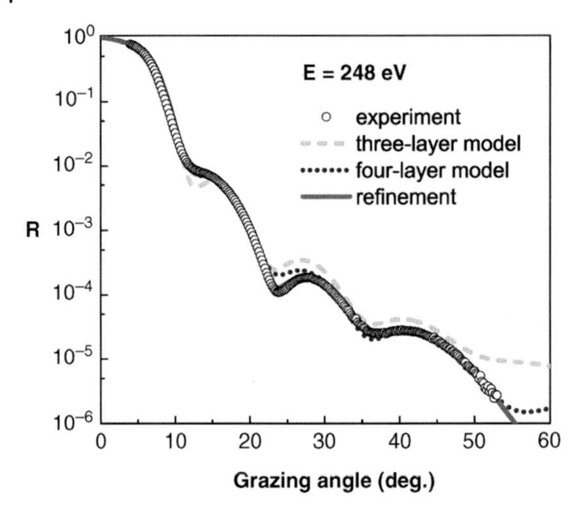

Figure 7.29 The measured reflectivity of the sample 2 (circles) versus the grazing angle at $E = 248$ eV. The curves demonstrate an accuracy of fitting in the frame of the three-layer ($C/TiO_2/SiO_2/Si$) and four-layer ($C/TiO_2/TiO_x/SiO_2/Si$) models as well as that after numerical refinement procedure with the use of the merit function (13) at $n = 1$ (from Filatova, E.O., Kozhevnikov, I.V., Sokolov, A.A., Ubyivovk, E.V., Yulin, S., Gorgoi, M., and Schaefers, F. (2012) Soft x-ray reflectometry, hard x-ray photoelectron spectroscopy and transmission electron microscopy investigations of the internal structure of $TiO_2(Ti)/SiO_2/Si$ stacks. *Sci. Technol. Adv. Mater.* **13**, 015001).

Similar analysis was performed for sample 2. The measured reflectivity curves are shown in Figure 7.28, circles. The internal structure of this sample is expected to be more complex compared to sample 1 as pure titanium can still be retained below oxidized TiO_2 layer. Actually, the dashed curve in Figure 7.29 demonstrates an accuracy of fitting at $E = 248$ eV with the use of three-layer model $C/TiO_2/SiO_2/Si$. As seen, the discrepancy between the measured and the calculated curves is well pronounced, while the same model allowed us to describe adequately all reflectivity curves of sample 1.

Let us analyze thus more complex four-layer model $C/TiO_2/TiO_x/SiO_2/Si$, where TiO_x means a mixture of titanium and its oxide and x is considered as additional fitting parameter characterizing relative concentration of oxygen inside the Ti layer. We assume the presence of TiO_x rather than pure Ti because of possible oxygen gettering from residual atmosphere of the technological chamber and/or diffusion of oxygen from the oxidized layer of the substrate due to chemical reaction with chemically active Ti. The dotted curve in Figure 7.29 shows that the accuracy of fitting improves essentially, while a small difference still persists between the measured and the calculated reflectivity curves. Therefore, as above, at the last stage of fitting we perform numerical refinement of the concentration profiles with the use of the merit function (7.13). Then, we observe almost total agreement between the calculated and the experimental reflectivities (solid curves in Figures 7.28 and 7.29).

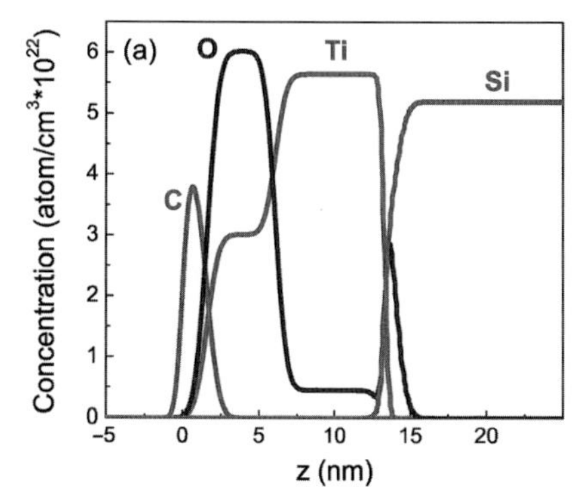

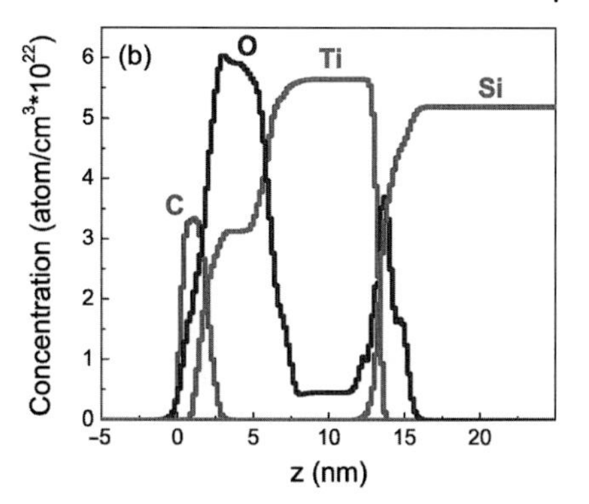

Figure 7.30 The reconstructed concentration profiles of chemical elements composed sample 2. Simulations are performed in the frame of four-layer ($C/TiO_2/TiO_x/SiO_2/Si$) model (a) and after numerical refinement with the use of the merit function (13) at $n = 1$ (b) (from Filatova, E.O., Kozhevnikov, I.V., Sokolov, A.A., Ubyivovk, E.V., Yulin, S., Gorgoi, M., and Schaefers, F. (2012) Soft x-ray reflectometry, hard x-ray photoelectron spectroscopy and transmission electron microscopy investigations of the internal structure of $TiO_2(Ti)/SiO_2/Si$ stacks. *Sci. Technol. Adv. Mater.* **13**, 015001).

The depth distributions of chemical element concentration reconstructed with the use of four-layer model are shown in Figure 7.30a. As it was expected, the upper part of the Ti layer is oxidized, TiO_2 layer being placed on the top. The oxygen concentration decreases gradually into the depth of Ti layer down to about 8% relatively (parameter $x = 0.08$). The total thickness of Ti-containing layer is 11.7 nm, while HAXPES analysis shows the 13.8 nm thickness of the layer. However, if we take into account smooth variation in material concentration near interfaces arising as a result of small-scale roughness averaging, the total thickness of Ti-containing layer proves to be about 13.5 nm (see Figure 7.30). Probably, this fact can explain the difference between results of HAXPES and SXR reflectometry. Maximal density of TiO_2 layer ($3.97\ g/cm^3$) and maximal partial density of Ti in the depth of the sample ($4.43\ g/cm^3$) are close to those of the bulk materials. The density values agree with those obtained from HAXPES measurements.

The rms roughness of the substrate surface (0.3 nm) is practically the same as for sample 1. However, the rms roughness of external sample surface (0.62 nm) is almost two times larger compared to sample 1. It is not strange because oxidation of metallic surfaces results often in essential development of small-scale roughness [104].

The most interesting feature observed in the reconstructed concentration distribution is very small thickness (0.62 nm) of the oxidized SiO_2 layer on top of the Si substrate compared to sample 1. Moreover, this layer is not seen at all in the concentration profile of Si in Figure 7.30a due to the formation of "effective" transition layer as a result of averaging of short-scale roughness in accordance with (7.8). This fact agrees well with the result of HAXPES analysis and is explained by chemical diffusion of oxygen from the top of the substrate into Ti layer.

Numerical refinement (Figure 7.30b) results in very small change in the concentration profiles. We observe appearance of oxygen in the adhesion layer, while the

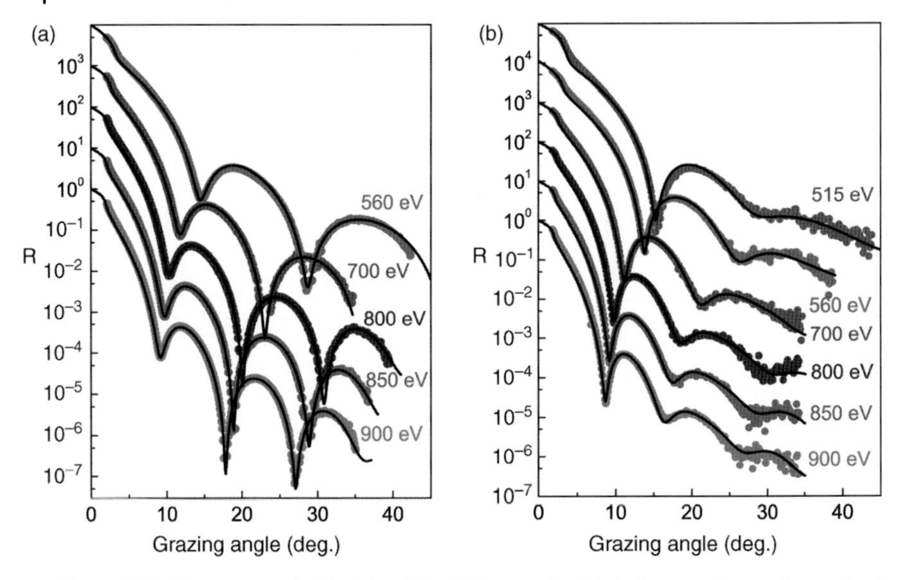

Figure 7.31 The measured reflectivity of the HfO_2 samples (circles) versus the grazing angle at different energies of the incident beam. The films were deposited by the ALD (a) and MOCVD (b) techniques. Solid curves show the result of fitting. Reflectivity curves are shifted vertically for clarity.

oxygen peak is not seen so clear as for sample 1 because of oxygen gettering by titanium film. In addition, increasing concentration of oxygen is observed in the Ti layer nearby the substrate surface ($z \sim 12.5$ nm).

Next experimental example of our approach is the study of internal structure of HfO_2 films of about 5 nm thickness deposited on Si substrates by ALD and MOCVD techniques. The reflectivity of both samples measured at different SXR energies are shown in Figure 7.31, circles. We would like to note that Hf is a very heavy chemical element, whose polarizability (both real and imaginary part) exceeds essentially the polarizability of the remaining lighter elements (O, Si, and C) composing the samples. As a result, the reflectivity curve changes only slightly when measuring below and above the edge of absorption of light elements. Therefore, we believe that the concentration profiles of light elements and, first of all, the fine structure of adhesion layer can be found within a lower accuracy compared to the Ti-containing samples discussed above.

When analyzing the reflectivity of the ALD sample, the conventional three-layer model $C/HfO_2/SiO_2/Si$ was used at the first stage of fitting. Refinement of concentration profiles was performed with the use of the merit function (7.13) at $n = 1$. The result of fitting is demonstrated in Figure 7.31a, solid curves, and the reconstructed concentration profiles are shown in Figure 7.32. As above, the refined profiles (Figure 7.32b) are very similar to initial ones (Figure 7.32a) reconstructed with the use of the three-layer model. The most pronounced difference is observed in the oxygen concentration profile and, first of all, in appearance of oxygen in the adhesion layer.

Analysis of the MOCVD sample reflectivity is not so evident. First of all, comparison of the reflectivity curves of the samples (Figure 7.31a and b) shows quick

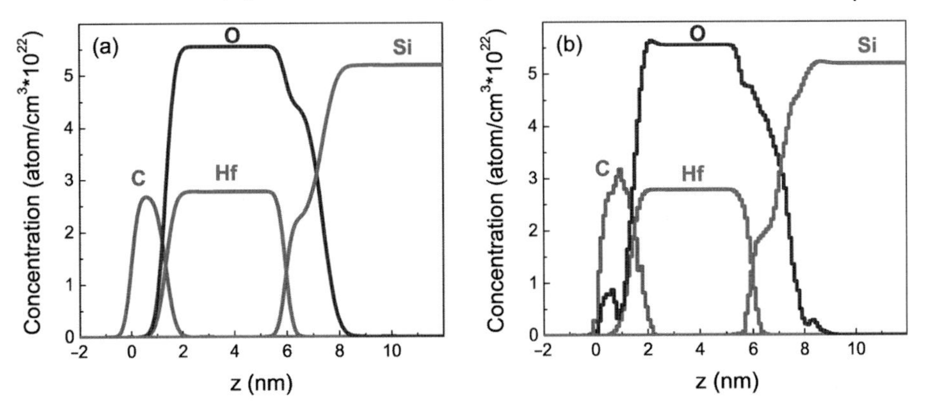

Figure 7.32 The reconstructed concentration profiles of chemical elements composed the ALD HfO_2 sample. Simulations are performed in the frame of three-layer ($C/HfO_2/SiO_2/Si$) model (a) and after numerical refinement with the use of the merit function (13) at $n = 1$ (b).

decrease in the oscillation amplitude in the reflectivity curve of the MOCVD sample with increasing grazing angle. It means that, at least, one of the interlayers of the MOCVD film is not abrupt compared to the ALD sample so that the reflectivity of this interface is considerably less at large grazing angles. The reason may be either essential development of small-scale roughness of the external film surface during deposition or deterioration of the film–substrate interface due to intermixing of HfO_2 and SiO_2 layers (see, for example, Ref. [105]).

Therefore, we tried to find out two possible solutions to the inverse problem considering two different three-layer models of the sample studied. The first model is conventional three-layer $C/HfO_2/SiO_2/Si$ film, the same as in the analysis of the ALD sample reflectivity. The second model is more complex $C/HfO_2/HfO_2^*(SiO_2)_x/Si$ film, where $HfO_2^*(SiO_2)_x$ means a mixture of HfO_2 and SiO_2 molecules, and x is one

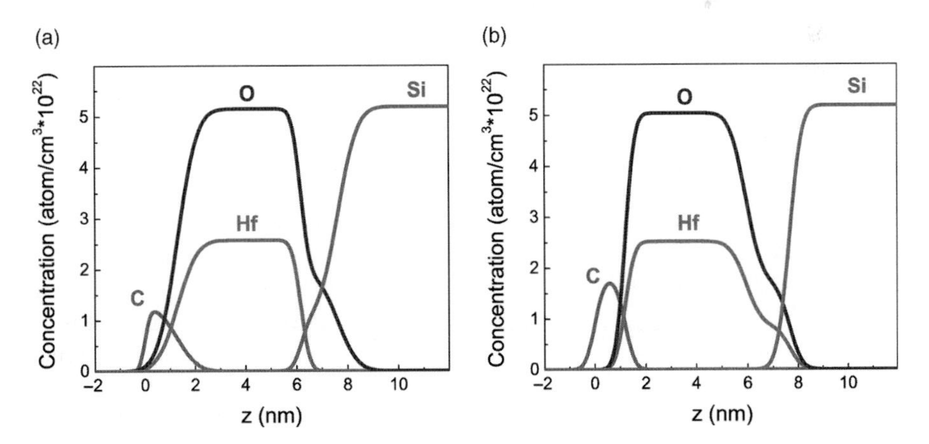

Figure 7.33 Reconstructed depth distribution of chemical elements composed the MOCVD HfO_2 sample. Simulations are performed with the use of two different three-layer models: $C/HfO_2/SiO_2/Si$ (a) and $C/HfO_2/HfO_2^*(SiO_2)_x/Si$ (b).

more fitting parameter. The depth distributions of atomic concentrations found with the use of these models are shown in Figure 7.33a and b. The main difference in the concentration profiles is observed in the depth distribution of Hf atoms. In Figure 7.33a, smooth variation in the profile is observed on the top of the sample, which can be explained by development of roughness during the film growth from $\sigma \sim 0.25$ (substrate) to $\sigma \sim 0.7$ nm. Notice the small-scale roughness increases from 0.25 to 0.3 nm only for the ALD film (Figure 7.32). In Figure 7.33b, the internal film–substrate interface smoothed essentially, which can be explained by intermixing of HfO_2 and SiO_2 layers and/or increasing small-scale roughness of the interface. In contrast to the ALD sample, in both models we observed rather thick interlayer of decreasing density between HfO_2 film and Si substrate. The interlayer thickness is about 2 nm in Figure 7.33a and about 3 nm in Figure 7.33b. Comparison of Figure 7.33 and HRTEM image (Figure 7.19b) allows us to conclude that the model shown in Figure 7.33a corresponds to reality. Refinement procedure with the use of the merit function (7.13) at $n=1$ did not change essentially the concentration profiles. The accuracy of fitting (practically the same for both models) is illustrated by Figure 7.31b, solid curves.

Acknowledgments

The authors are indebted to V.E. Asadchikov and B.S. Roshchin (Crystallography Institute, Moscow, Russia), L. Peverini and E. Ziegler (ESRF, Grenoble, France), and F. Schaefers (Helmholtz Zentrum Berlin, Germany) for help in reflectivity measurements; F. Schaefers and M.Gorgoi for help in HAXPES measurements; E.V. Ubyivovk (St. Petersburg State University, St. Petersburg, Russia) for help in TEM measurements. The authors gratefully acknowledge the assistance from the Helmholtz Zentrum Berlin (HZB). Researches were supported, in part, by the ISTC (project No. 3963). TEM experimental data presented in this chapter were obtained using the equipment of Interdisciplinary Resource Center for Nanotechnology at St. Petersburg State University, Russia (http://nano.spbu.ru).

References

1 Wilk, G.D., Wallace, R.M., and Anthony, J.M. (2001) High-κ gate dielectrics: current status and materials properties considerations. *J. Appl. Phys.*, **89**, 5243.

2 Kingon, A.I., Maria, J.-P., and Streiffer, S.K. (2000) Alternative dielectrics to silicon dioxide for memory and logic devices. *Nature*, **406**, 1032.

3 Kasap, S.O. and Kasap, S.O. (2002) *Principles of Electronic Materials and Devices*, McGraw-Hill, Boston.

4 Bersuker, G., Zeutziff, P., Brown, G., and Huff, H.R. (2004) Dielectrics for future transistors. *Mater. Today*, **7**, 26.

5 Kawamoto, A., Cho, K.J., and Dutton, R. (2001) Perspectives paper: first principles modeling of high-*k* gate dielectrics. *J. Comput. Aided Mater. Des.*, **8**, 39.

6 Gerritsen, E., Emonet, N., Caillat, C., Jourdan, N., Piazza, M., Fraboulet, D., Berthelot, A., Smith, S., and Mazoyer, P. (2005) Evolution of materials technology for stacked-capacitors in 65nm embedded-DRAM. *Solid State Electron.*, **49**, 1767.

7 Kwon, C., Jia, Q.X., Fan, Y., Hundley, M.F., and Reagor, D.W. (1998) Observation of spin-dependent transport and large magnetoresistance in $La_{0.7}Sr_{0.3}MnO_3/SrTiO_3/La_{0.7}Sr_{0.3}MnO_3$ ramp-edge junctions. *J. Appl. Phys.*, **83**, 7052.

8 Robertson, J. (2000) Band offsets of wide-band-gap oxides and implications for future electronic devices. *J. Vac. Sci. Technol. B*, **18**, 1785.

9 Robertson, J. (2002) Electronic structure and band offsets of high dielectric constant gate oxides. *Mater. Res. Soc. Bull.*, **27**, 217.

10 Robertson, J. (2006) High dielectric constant gate oxides for metal oxide Si transistors. *Rep. Prog. Phys.*, **69**, 327.

11 Lucovsky, G. and Whitten, J.L. (2004) Chapter 4 in *High-k Gate Dielectrics* (ed. Michel Houssa), IOP Publishing Ltd, Bristol.

12 Lucovsky, G. (2002) Correlations between electronic structure of transition metal atoms and performance of high-*k* gate dielectrics in advanced Si devices. *J. Non-Cryst. Solids*, **303**, 40.

13 Ballhausen, C.J. and Gray, H.B. (1964) Chapter 8, in *Molecular Orbital Theory*, Benjamin, New York.

14 Cox, P.A. (1992) Chapter 2, in *Transition Metal Oxides*, Oxford Science, Oxford.

15 Grunes, L.A., Leapman, R.D., Walker, C.D., Hoffman, R., and Kunz, A.B. (1982) Oxygen *K* near-edge fine structure: an electron-energy-loss investigation with comparisons to new theory for selected 3d transition-metal oxides. *Phys. Rev. B*, **25**, 7157.

16 Cotton, F.A. and Wilkinson, G. (1972) Chapter 4, in *Inorganic Chemistry*, 3rd edn, Interscience, New York.

17 Wyckoff, R.W.G. (1960) *Crystal Structures*, vol. 1, Wiley-Interscience, New York.

18 Ivoning, J. and Van Santen, R.A. (1983) Electrostatic potential calculations on crystalline TiO_2: the surface reducibility of rutile and anatase. *Chem. Phys. Lett.*, **101**, 541.

19 Ruus, R., Kikas, A., Saar, A., Ausmees, A., Nommiste, E., Aarik, J., Aidla, A., Uustare, T., and Martinson, I. (1997) Ti 2p and O 1s X-ray absorption of TiO_2 polymorphs. *Solid State Commun.*, **104**, 199.

20 Brydson, R., Sauer, H., Engel, W., Thomas, J.M., Zeitler, E., Kosugilland, N., and Kurodall, H. (1989) *J. Phys. Condens. Matter*, **1**, 797–812.

21 de Groot, F.M.F., Fuggle, J.C., Thole, B.T., and Sawatzky, G.A. (1990) $L_{2,3}$ X-ray-absorption edges of d^0 compounds: K^+, Ca^{2+}, Sc^{3+}, and Ti^{4+} in O_h (octahedral) symmetry. *Phys. Rev. B*, **41**, 928.

22 Campbell, S.A., Gilmer, D.C., Wang, X.C., Hsieh, M.T., Kim, H.S., Gladfelter, W.L., and Yan, J.H. (1997) MOSFET transistors fabricated with high permitivity TiO_2 dielectrics. *IEEE Trans. Electron. Dev.*, **44**, 104.

23 Smith, R.C., Ma, T.Z., Hoilien, N., Tsung, L.Y., Bevan, M.J., Colombo, L., Roberts, J., Campbell, S.A., and Gladfelter, W.L. (2000) Chemical vapour deposition of the oxides of titanium, zirconium and hafnium for use as high-*k* materials in microelectronic devices. A carbon-free precursor for the synthesis of hafnium dioxide. *Adv. Mater. Opt. Electron.*, **10**, 105.

24 Bera, M.K., Mahata, C., and Maiti, C.X. (2008) Reliability of ultra-thin titanium dioxide (TiO_2) films on strained-Si. *Thin Solid Films*, **517**, 27.

25 Goh, G.K.L., Liew, C.P.K., Kim, J., and White, T.J. (2006) Structure and optical properties of solution deposited TiO_2 films. *J. Cryst. Growth*, **291**, 94.

26 Yokota, K., Yano, Y., Nakamura, K., Ohnishi, M., and Miyashita, F. (2006) Effects of oxygen ion beam application on crystalline structures of TiO_2 films deposited on Si wafers by an ion beam assisted deposition. *Nucl. Instrum. Methods Phys. Res. B*, **242**, 393.

27 Rao, K.N. and Mohan, S. (1990) Optical properties of electron-beam evaporated TiO_2 films deposited in an ionized oxygen medium. *J. Vac. Sci. Technol. A*, **8**, 3260.

28 Ting, C.C., Chen, S.Y., and Liu, D.M. (2000) Structural evolution and optical properties of TiO_2 thin films prepared by thermal oxidation of sputtered Ti films. *J. Appl. Phys.*, **88**, 4628.

29 Nahar, R.K., Singh, V., and Sharma, A. (2007) Study of electrical and microstructure properties of high dielectric hafnium oxide thin film for MOS devices. *J. Mater. Sci. Mater. Electron.*, **18**, 615.

30 Wilk, G.D., Wallace, R.M., and Anthony, J.M. (2000) Hafnium and zirconium silicates for advanced gate dielectrics. *J. Appl. Phys.*, **87**, 484.

31 Hubbard, K.J. and Schlom, D.G. (1996) Thermodynamic stability of binary oxides in contact with silicon. *J. Mater. Res.*, **11**, 2757.

32 McPherson, J., Kim, J.Y., Shanware, A., and Mogul, H. (2003) Thermochemical description of dielectric breakdown in high dielectric constant materials. *Appl. Phys. Lett.*, **82**, 2121.

33 Takahashi, H., Toyoda, S., Okabayashi, J., Kumigashira, H., Oshima, M., Sugita, Y., Liu, G.L., Liu, Z., and Usuda, K. (2005) Chemical reaction at the interface between polycrystalline Si electrodes and HfO_2/Si gate dielectrics by annealing in ultrahigh vacuum. *Appl. Phys. Lett.*, **87**, 012903.

34 Sayan, S., Croft, M., Nguyen, N.V., Emge, T., Ehrstein, J.R., Levin, I., Suehle, J.S., Bartynski, R.A., and Garfunkel, E. (2005) The relation between crystalline phase, electronic structure and dielectric properties in high-*K* gate stacks. Proceedings of International Conference on Characterization and Metrology for ULSI Technology (Dallas) p. 92.

35 Ceresoli, D. and Vanderbilt, D. (2006) Structural and dielectric properties of amorphous ZrO_2 and HfO_2. *Phys. Rev. B*, **74**, 125108.

36 Lucovsky, G., Lee, S., Long, J.P., Seo, H., and Lüning, J. (2009) Interfacial transition regions at germanium/Hf oxide based dielectric interfaces: Qualitative differences between non-crystalline Hf Si oxynitride and nanocrystalline HfO_2 gate stacks. *Microelectronic Engineering*, **86**, 224–234.

37 Kremmer, S., Wurmbauer, H., Teichert, C., Tallarida, G., Spiga, S., Wiemer, C., and Fanciulli, M. (2005) Nanoscale morphological and electrical homogeneity of HfO_2 and ZrO_2 thin films studied by conducting atomic-force microscopy. *J. Appl. Phys.*, **97**, 074315.

38 Norton, D.P. (2004) Synthesis and properties of epitaxial electronic oxide thin-film materials. *Mater. Sci. Eng. R Rep.*, **43**, 139.

39 Manchanda, L., Morris, M.D., Green, M.L., van Dover, R.B., Klemens, F., Sorsch, T.W., Silverman, P.J., Wilk, G., Busch, B., and Aravamudhan, S. (2001) Multi-component high-*K* gate dielectrics for the silicon industry. *Microelectron. Eng.*, **59**, 351.

40 Gusev, E.P., Copel, M., Cartier, E., Baumvol, I.J.R., Krug, C., and Gribelyuk, M.A. (2000) High-resolution depth profiling in ultrathin Al_2O_3 films on Si. *Appl. Phys. Lett.*, **76**, 176.

41 Gutiérrez, G. and Johansson, B. (2002) Molecular dynamics study of structural properties of amorphous Al_2O_3. *Phys. Rev. B*, **65**, 104202.

42 Wu, W.C., Lai, C.S., Wang, Z.M., Wang, J.C., Hsu, C.W., Ma, M.W., and Chao, T.S. (2008) Current transport mechanism for HfO_2 gate dielectrics with fluorine incorporation. *Electrochem. Solid State Lett.*, **11**, H15.

43 Adelmann, C., Sriramkumar, V., Elshocht, S.V., Lehnen, P., Conard, T., and De Gendt, S. (2007) Dielectric properties of dysprosium- and scandium-doped hafnium dioxide thin films. *Appl. Phys. Lett.*, **91**, 162902.

44 Tang, C., Tuttle, B., and Ramprasad, R. (2007) Diffusion of O vacancies near Si:HfO_2 interfaces: an *ab initio* investigation. *Phys. Rev. B*, **76**, 073306.

45 Miyata, N. (2006) Two-step behavior of initial oxidation at HfO_2/Si interface. *Appl. Phys. Lett.*, **89**, 102903.

46 Jiang, R. and Li, Z. (2008) Behavior of stress induced leakage current in thin HfO_xN_y films. *Appl. Phys. Lett.*, **92**, 012919.

47 Filatova, E.O., Sokolov, A.A., Kozhevnikov, I.V., Taracheva, E.Y.,

Grunsky, O.S., Schaefers, F., and Braun, W. (2009) Investigation of the structure of thin HfO_2 films by soft X-ray reflectometry techniques. *J. Phys. Condens. Matter*, **21**, 185012.

48 Filatova, E.O. and Pavlychev, A.A. (2011) *X-Ray Optics and Inner-Shell Electronics of Hexagonal BN: Biochemistry Research Trends*, Nova Science Publishers, Inc., New York.

49 Reflectometer station, see on the BESSY websie: http://www.bessy.de/bit/upload/reflectometer.pdf.

50 Optics beamline, see on the BESSY websie: http://www.bessy.de/bit/upload/D_08_1B2.pdf.

51 Stemmer, S., Chen, Z.Q., Zhu, W.J., and Ma, T.P. (2003) Electron energy-loss spectroscopy study of thin film hafnium aluminates for novel gate dielectrics. *J. Microsc.*, **210**, 74.

52 Lucovsky, G., Hong, J.G., Fulton, C., Zou, Y., Nemanich, R.J., and Ade, H. (2004) X-ray absorption spectra for transition metal high-κ dielectrics: final state differences for intra- and inter-atomic transitions. *J. Vac. Sci. Technol. B*, **22**, 2132.

53 Rusakov, A.A. (1977) *The X-Ray Diffraction of Metals*, Atomizdat, Moscow (in Russian).

54 Filatova, E., Taracheva, E., Shevchenko, G., Sokolov, A., Kozhevnikov, I., Yulin, S., Schaefers, F., and Braun, W. (2009) Atomic ordering in TiO_2 thin films studied by X-ray reflection spectroscopy. *Phys. Status Solidi B*, **246**, 1454.

55 Fano, U. and Cooper, J.W. (1968) Spectral distribution of atomic oscillator strengths. *Rev. Mod. Phys.*, **40**, 441.

56 de Groot, F.M.F., Faber, J., Michiels, J.J.M., Czyzyk, M.T., Abbate, M., and Fuggle, J.C. (1993) Oxygen 1s X-ray absorption of tetravalent titanium oxides: a comparison with single-particle calculations. *Phys. Rev. B*, **48**, 2074.

57 Rath, S., Gracia, F., Yubero, F., Holgado, J.P., Martin, A.I., Batchelor, D., and Gonzalez-Elipe, A.R. (2003) Angle dependence of the O K edge absorption spectra of TiO_2 thin films with preferential texture. *Nucl. Instrum. Methods Phys. Res. B*, **200**, 248.

58 Filatova, E.O., Taracheva, E.Y., Sokolov, A.A., Bukin, S.V., Shulakov, A.S., Jonnard, P., André, J.-M., and Drozd, V.E. (2006) Ultrasoft X-ray reflection and emission spectroscopic analysis of Al_2O_3/Si structure synthesized by the atomic layer deposition method. *X-Ray Spectrom.*, **35**, 359.

59 Balzarotti, A., Bianconi, A., Burattini, E., Grandolfo, M., Habel, R., and Placentini, M. (1974) Core transitions from the Al 2p level in amorphous and crystalline Al_2O_3. *Phys. Status Solidi B*, **63**, 77.

60 Fomichev, V.A. (1967) Investigation of electronic structure of Al and Al_2O_3 by soft X-ray reflectometry method. *Fiz. Tverd. Tela (Leningrad)*, **8**, 2892. English transl.: *Soviet Phys. Solid State*, **8**, 2312 (1967)

61 Codling, K. and Madden, R.P. (1968) Structure in the $L_{II,III}$ absorption of aluminum and its oxides. *Phys. Rev.*, **167**, 587.

62 Swanson, N. and Powell, C.J. (1968) Excitation of L-shell electrons in Al and Al_2O_3 by 20-keV electrons. *Phys. Rev.*, **167**, 592.

63 Britov, I.A. and Romaschenko, Y.N. (1978) X-ray spectroscopic investigation of electronic structure of silicon and aluminium oxides. *Fiz. Tverd. Tela (Leningrad)*, **20**, 664; *Soviet Phys. Solid State*, **20**, 384 (1978)

64 O'Brien, W.L., Jia, J., Dong, Q.-Y., Callcott, T.A., Mueller, D.R., Ederer, D.L., and Kao, C.-C. (1993) Soft-X-ray investigation of Mg and Al oxides: evidence for atomic and bandlike features. *Phys. Rev. B*, **47**, 15482.

65 Filatova, E.O., Shulakov, A.S., and Luk'anov, V.A. (1998) Depth of formation of a reflected soft X-ray beam under conditions of specular reflection. *Phys. Solid State*, **40**, 1237.

66 Filatova, E., Stepanov, A., Blessing, C., Friedricht, J., Barchewitz, R., André, J.M., Le Guern, F., Bac, S., and Troussel, P. (1995) Total reflection and surface scattering of soft X-rays on the Si-SiO_2 system and hexagonal BN crystal. *J. Phys. Condens. Matter*, **7**, 2731.

67 Gorgoia, M., Svensson, S., Schäfers, F., Öhrwall, G., Mertin, M., Bressler, P., Karis, O., Siegbahn, H., Sandell, A., Rensmo, H., Doherty, W., Jung, C., Braun, W., and Eberhardt, W. (2009). The high kinetic energy photoelectron spectroscopy facility at BESSY progress and first results. *Nucl. Instrum. Methods Phys. Res. A*, **601**, 48.

68 Schaefers, F., Mertin, M., and Gorgoi, M. (2007) KMC-1: a high resolution and high flux soft X-ray beamline at BESSY. *Rev. Sci. Instrum.*, **78**, 123102.

69 Moulder, J.F., Stickle, W.F., Sobol, P.E., and Bomben, K.D. (1992) *Handbook of X-Ray Photoelectron Spectroscopy*, Perkin-Elmer Corp., MN, USA.

70 Shirley, D.A. (1972) High-Resolution X-Ray Photoemission Spectrum of the Valence Bands of Gold. *Phys. Rev. B*, **5**, 4709.

71 Gonbeau, D., Guimon, C., Pfister-Guillouzo, G., Levasseur, A., Meunier, G., and Dormoy, R. (1991) XPS study of thin films of titanium oxysulfides. *Surf. Sci.*, **254**, 81.

72 Himpsel, F.J., McFeely, F.R., Taleb-Ibrahimi, A., Yarmoff, J.A., and Hollinger, G. (1988) Microscopic structure of the SiO_2/Si interface. *Phys. Rev. B*, **38**, 6084.

73 Fadley, C.S. (2009) X-ray photoelectron spectroscopy: from origins to future directions. *Nucl. Instrum. Methods Phys. Res. A*, **601**, 8.

74 Fadley, C.S., Baird, R.J., Siekhaus, W., Novakov, T., and Bergström, S.Å.L. (1974) Surface analysis and angular distributions in X-ray photoelectron spectroscopy. *J. Electron. Spectrosc.*, **4**, 93.

75 Briggs, D. and Grant, J.T. (2003) *Surface Analysis*, IM Publications and Surface Spectra Ltd., Chichester.

76 Trzhaskovskaya, M.B., Nefedov, V.I., and Yarzhemsky, V.G. (2001) Photoelectron angular distribution parameters for elements Z=1 to Z=54 in the photoelectron energy range 100–5000eV. *Atom Data Nucl. Data*, **77**, 97; Photoelectron angular distribution parameters for elements Z=55 to Z=100 in the photoelectron energy range 100–5000eV, *Atom Data Nucl. Data*, **82**, 257 (2002)

77 Tanuma, S., Powell, C.J., and Penn, D.R. (1994) Calculations of electron inelastic mean free paths. V. Data for 14 organic compounds over the 50–2000eV range. *Surf. Interface Anal.*, **21**, 165.

78 James, A.M. and Lord, M.P. (1992) in *Macmillan's Chemical and Physical Data*, Macmillan, London, UK.

79 Filatova, E.O., Sokolov, A.A., Ovchinnikov, A.A., Tveryanovich, S.Y., Savinov, E.P., Marchenko, D.E., Afanas'ev, V.V., and Shulakov, A.S. (2010) XPS and depth resolved SXES study of HfO_2/Si interlayers. *J. Electron. Spectrosc.*, **181**, 206.

80 Sokolov, A.A., Filatova, E.O., Afanas'ev, V.V., Taracheva, E.Y., Brzhezinskaya, M.M., and Ovchinnikov, A.A. (2009) Interface analysis of HfO2 films on (1 0 0)Si using X-ray photoelectron spectroscopy. *J. Phys. D Appl. Phys.*, **42**, 035308.

81 Levitan, B.M. (1984) *Inverse Sturm-Liouville Problems*, Nauka, Moscow, [in Russian].

82 Ramm, A.G. (1992) *Multidimensional Inverse Scattering Problems*, Longman, New York.

83 Gesztesg, F. and Simon, B. (1999) Inverse spectral analysis with partial information on the potential. II. The case of discrete spectrum. *Trans. Am. Math. Soc.*, **352**, 2765.

84 del Rio, R., Gesztesg, F., and Simon, B. (1997) Inverse spectral analysis with partial information on the potential. III. Updating boundary conditions. *Int. Math. Res. Notices*, **15**, 751.

85 Horvath, M. (2005) Inverse spectral problems and closed exponential systems. *Ann. Math.*, **162**, 885.

86 Horvath, M. (2001) On the inverse spectral theory of Schrödinger and Dirac operators. *T. Am. Math. Soc.*, **353**, 4155.

87 Majkrzak, C.F. and Berk, N.F. (1995) Exact determination of the phase in neutron reflectometry. *Phys. Rev. B*, **52**, 10827.

88 Majkrzak, C.F. and Berk, N.F. (1998) Exact determination of the phase in neutron reflectometry by variation of the surrounding media. *Phys. Rev. B*, **58**, 15416.

89 Kozhevnikov, I., Peverini, L., and Ziegler, E. (2008) Exact determination of the phase in time resolved X-ray reflectometry. *Opt. Express*, **16**, 144.

90 Kozhevnikov, I., Peverini, L., and Ziegler, E. (2008) Exact solution of the phase problem in *in situ* X-ray reflectometry of a growing layered film. *J. Appl. Phys.*, **104**, 054914.

91 Filatova, E.O., Peverini, L., Ziegler, E., Kozhevnikov, I.V., Jonnard, P., and André, J.-M. (2010) Evolution of surface morphology at the early stage of Al_2O_3 film growth on a rough substrate. *J. Phys. Condens. Matter*, **22**, 345003.

92 Lüken, E., Ziegler, E., and Lingham, M. (1996) *In-situ* growth control of X-ray multilayers using visible light kinetic ellipsometry and grazing incidence X-ray reflectometry. Proceedings of SPIE of International Symposium on Polarization Analysis and Applications to Device Technology, *Proc.* SPIE, 2873, p. 113.

93 Kozhevnikov, I.V. (2003) Physical analysis of the inverse problem of X-ray reflectometry. *Nucl. Instrum. Methods Phys. Res. A*, **508**, 519.

94 Sivia, D.S., Hamilton, W.A., and Smith, G.S. (1991) Analysis of neutron reflectivity data: maximum entropy, Bayesian spectral analysis and speckle holography. *Phys. B*, **173**, 21.

95 Pedersen, J.S. and Hamley, I.W. (1994) Analysis of neutron and X-ray reflectivity data. II. Constrained least-squares methods. *J. Appl. Crystallogr.*, **27**, 36.

96 Asadchikov, V.E., Kozhevnikov, I.V. and Y.S. Krivonosov, Mercier, R., Metzger, T.H., Morawe, C., and Ziegler, E. (2004). Application of X-ray scattering technique to the study of supersmooth surfaces. *Nucl. Instrum. Methods Phys. Res. A*, **530** 575.

97 Peverini, L., Ziegler, E., Bigault, T., and Kozhevnikov, I.V. (2005) Roughness conformity during tungsten film growth: an *in situ* synchrotron X-ray scattering study. *Phys. Rev. B*, **72**, 045445.

98 Peverini, L., Ziegler, E., Bigault, T., and Kozhevnikov, I.V. (2007) Dynamic scaling of roughness at the early stage of tungsten film growth. *Phys. Rev. B*, **76**, 045411.

99 Névot, L. and Croce, P. (1980) Caractérisation des surfaces par réflexion rasante de rayons X. Application à l'étude du polissage de quelques verres silicates. *Rev. Phys. Appl.*, **15**, 761.

100 Attwood, D. (1999) *Soft X-Rays and Extreme Ultraviolet Radiation. Principles and Applications*, Cambridge University Press, Cambridge.

101 Henke, B.L., Gullikson, E.M., and Davis, J.C. (1993) X-ray interactions: photoabsorption, scattering, transmission, and reflection at $E = 50$–30000 eV, $Z = 1$–92, *Atomic Data and Nuclear Data Tables*, **54**, 181–342.

102 Livesey, A.K. and Smith, G.S. (1994) The determination of depth profiles from angle-dependent XPS using maximum entropy data analysis. *J. Electron. Spectrosc.*, **67**, 439.

103 Skilling, J. (1989) *Maximum Entropy and Bayesian Methods: Cambridge 1988* (ed. J. Skilling), Kluwer, Amsterdam.

104 Underwood, J.H., Gullikson, E.M., Ng, W., Ray-Chaudhuri, A., and Cerrino, F. (1995) *OSA Proc. on Extreme Ultraviolet Lithography, 1994*, vol. 23 (eds F. Zernike and D.T. Attwood), Optical Society of America, pp. 61–66.

105 Smirnova, T.P., Kaichev, V.V., Yakovkina, L.V., Kosyakov, V.I., Beloshapkin, S.A., Kuznetsov, F.A., Lebedev, M.S., and Gritsenko, V.A. (2008) Composition and structure of hafnia films on silicon. *Inorg. Mater.*, **44**, 965.

8
High-*k* Insulating Films on Semiconductors and Metals: General Trends in Electron Band Alignment

Valeri V. Afanas'ev, Michel Houssa, and Andre Stesmans

8.1
Introduction

Wide-bandgap oxide layers on semiconductors and metals are used as insulators in a broad variety of integrated electron devices including the important areas of logic, memory, and high-frequency applications. To meet the scaling requirements of the future generations of electron devices in terms of specific capacitance and gate leakage current, the search for suitable dielectrics capable of replacing the traditional SiO_2 insulator touches more and more exotic material systems [1–3]. The two most important parameters commonly used to evaluate the potential of an insulating material for applications are the dielectric permittivity (κ) and the bandgap width (E_g) [4–7]. However, the electron and hole injection rates at the interfaces are determined by the conduction and valence band offsets, respectively, rather than by the oxide bandgap width *per se* [8], thus stimulating the analysis of the electrode effects on the interface band alignments. Up to now, most of the barrier heights and band offsets have been calculated theoretically [4, 6] because reproducible fabrication of metal/oxide and oxide/semiconductor samples of high quality is a formidable challenge given the complexity of the processing technology. On the other hand, reliable quantification of the barrier height at the semiconductor/insulator or metal/insulator interfaces poses a significant experimental challenge by itself as the frequently used photoelectron spectroscopy methods are known to suffer from the attendant insulator charging effects [9–13]. The latter, unless adequately corrected for [14–16], will lead to a systematic error in the measured band offset value.

In this chapter, we overview the internal photoemission (IPE) and photoconductivity (PC) experiments [17, 18] used to characterize barriers for electrons and holes at interfaces of silicon and high-mobility semiconductors (GaAs, GaSb, $In_xGa_{1-x}As$, Ge, $Si_{1-x}Ge_x$, and InP) with a broad variety of metal oxide insulators. Three general issues are addressed: (1) the impact of the oxide crystallinity on the interface band diagram compared to an amorphous insulator, (2) the evolution of band offsets with oxide bandgap width as a function of ionic radius (r_i) of the oxide cation, and (3) electrostatic dipoles at the metal/oxide interfaces. It is found that in the case of

High-k Gate Dielectrics for CMOS Technology, First Edition. Edited by Gang He and Zhaoqi Sun.

compact cations ($r_i < 0.07$ nm) such as Al, Mg, and Sc, the oxide crystallization results in widening of the gap, predominantly associated with the downshift of the O2p-derived electron states at the top of the oxide valence band (VB). In contrast, in the case of larger cations ($r_i > 0.07$ nm), no difference between the gaps of amorphous or (poly)crystalline oxides is found. The top of the VB preserves its energy in these oxides, the differences in bandgap being mostly reflected in the energy of the cation-derived electron states at the bottom of the conduction band (CB). This observation suggests that O2p states at the VB top can be used as reliable reference level to evaluate the intrinsic band offsets both between semiconductors and oxides, and between two oxides if r_i exceeds 0.07 nm.

The comparison of energy barriers at interfaces between different semiconductors and oxide insulators indicates that the bulk density of electron states (DOS) represents the major factor determining the band alignment, with little influence of the structure- or processing-sensitive dipoles. However, in the case of metal/insulator contacts, interfacial dipoles are found to cause significant barrier height variation as influenced by sample processing. We suggest that the different behavior of semiconductor/oxide and metal/oxide barriers is related to the fact that interface charges induce an *extended* space charge layer in semiconductors resulting in an insignificant variation of the electrostatic potential across the interfacial region. In the case of metal electrodes, the small thickness of the polarization layer at the metal surface leads to a much higher strength of the electric field at the interface, giving rise to a significant dipole.

8.2
Band Offsets and IPE Spectroscopy

In most of the experiments discussed here, metal/insulator and semiconductor/insulator structures were prepared by depositing an insulating oxide on a metal or semiconductor surface. In addition, in some cases metal electrodes were deposited on top of the insulator as it is done when fabricating the control gate.

While amorphous oxide layers can easily be deposited by a variety of techniques [2, 3], fabrication of crystalline oxide films is more challenging. To obtain crystalline oxide phases, we employed two approaches. First, thanks to the thermal stability of silicon, the amorphous oxide layers can be crystallized into polycrystalline films by using a postdeposition anneal. In some cases, temperatures as high as 1000 °C are required, for example, for γ-Al_2O_3/Si [19]. Second, if the lattice mismatch between the oxide and the substrate is not prohibitively large, the oxide layer can be grown epitaxially on (100) or (111) silicon surfaces because of the possibility to conduct this growth at a sufficiently high temperature (>650 °C). The crystallinity of the oxide layer was routinely assessed by X-ray diffractometry (XRD) or reflection high-energy electron diffraction (RHEED). The annealing approach enables fabrication of relatively thick (>20 nm) crystallized oxide layers that usually are difficult to grow epitaxially because of the limited pseudomorphic growth thickness achievable. The Si-related component of the interface band diagram remains unchanged, thus providing a reliable energy reference that allows reliable monitoring of the oxide bandgap edge energies as a function of oxide composition and structure.

Similar structures were also fabricated on a number of high-mobility semiconductor materials by deposition of an insulating metal oxide to analyze the behavior of the interface barriers as affected by the bulk DOS of the semiconductor and the crystallographic orientation of its surface plane. As these semiconductors lack the thermal stability of Si crystals, most of the oxide layers are amorphous, but some of them, for instance, HfO_2 and ZrO_2, show polycrystalline features if deposited at temperatures exceeding 350 °C. Nevertheless, as will be shown later on, one can obtain a meaningful comparison of the band offsets to those observed in the case of the Si substrate because, for cations of large ionic radius r_i, the oxide crystallinity has no measurable effect on the interface band alignment.

Band alignment at semiconductor interfaces can be determined in a variety of ways by observing electron transport across the interface [18]. However, in the case of interfaces with insulators, application of the classical methods becomes problematic because of two major reasons. First, the equilibrium injection rate at room temperature becomes negligible if the interface barrier height or the band offset exceeds 2 eV. Under these circumstances, only charge carriers excited to a sufficient energy can be transported across the interface. Second, wide-gap insulators usually contain a considerable density of defects with energy levels inside the gap that would thus act as charge traps. If the insulating layer is subjected to charge injection or exposed to excitation by UV radiation, X-rays, electron, or particle beams, these traps will accumulate charge, leading to a global shift of the energy levels in the insulator compared to the uncharged initial state [9, 10], sometimes referred to as "the differential charging effect" [11]. This effect is illustrated in Figure 8.1 showing a semiconductor/insulator interface band diagram prior to (a) and after trapping of a positive charge in the bulk of the insulator (b). In theory, in a neutral sample (Figure 8.1a), the VB offset can be measured by observing simultaneously the energy distribution of electrons emitted from the VBs of the semiconductor and insulator, or by comparing two core-level energies (the Kraut's method [20, 21]). However, in the presence of the oxide charge (Figure 8.1b), all electron states in the outer layer of the insulator, which provide a dominant contribution to electron photoemission, will be shifted in energy by ΔE_{ch} because of the electric field induced by the trapped charge. This global electrostatic energy shift will result in a systematic error in the measured band offset value. To account for this radiation-induced measurement artifact, a correction based on the surface potential monitoring, for example, the shift of the C1s core level of adsorbed carbon atoms, can be used [14]. Nevertheless, the accuracy of this approximation may be questioned as it is based on the assumption that all electron states across the insulator are shifted by the same energy as those at the surface.

In order to avoid the above-discussed ambiguities, we measured the interface band offsets using a combination of electron/hole IPE and PC spectroscopies on metal oxide semiconductor (MOS) or metal–insulator–metal (MIM) capacitors obtained by deposition of a semitransparent metal electrode on top of the oxide [17, 18, 22]. The physical idea behind the technique is illustrated by the interface band diagram

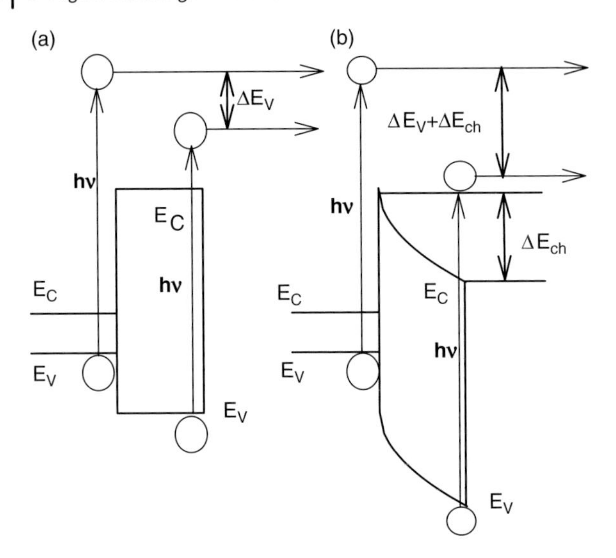

Figure 8.1 Schematic of the VB offset (ΔE_V) determination from an external photoemission experiment: while in the ideal case of the uncharged insulator (a) the energy difference between electrons emitted from the VB tops of the semiconductor and insulator corresponds to the VB offset, the X-ray-induced charging of the insulator introduces (b) an energy shift ΔE_{ch} due to variation in the electrostatic potential across the insulating layer leading to a systematic error in the measured band offset value.

shown in Figure 8.2: when the sample is illuminated with photons of energy ($h\nu$) exceeding the energy barrier for electrons (Φ_e) under positive gate bias (Figure 8.2a) or the barrier for holes (Φ_h) under negative gate bias (Figure 8.2b), charge carriers may be injected into the oxide producing a photocurrent as these drift toward the opposite metal electrode. Thanks to the electric field in the oxide induced by biasing the top electrode, the measured barrier height is determined by the relative energy position of the band edges in the semiconductor and in the insulator taken at a distance of the mean photoelectron escape depth and at the image barrier top position, respectively. With both spacial values being of the order of 1 nm, the potential impact of oxide charges becomes negligible. Moreover, the IPE spectra are usually measured for different metal bias values and extrapolated to zero field yielding the barrier height unaffected by any built-in charge [18].

It is also worth mentioning here that IPE experiments usually require photons of much lower energy (typically below 5 eV) than applied in photoelectron spectroscopy measurements. This prevents generation of electron–hole pairs in the insulator, thus minimizing the potential charging-related phenomena. In addition, by simply extending the spectral range to higher photon energies, one can also determine the oxide bandgap E_g(ins) from the spectral threshold of the intrinsic PC. This oxide gap value can be used either to calculate both the CB and the VB offset at the interface if only one IPE barrier is measured experimentally, or to provide the internal proof of consistency by comparing the experimental E_g(ins) value with that calculated from the electron and hole IPE barriers $E_g(\text{ins}) = \Phi_e + \Phi_h - E_g(\text{sc})$ (cf. Figure 8.2).

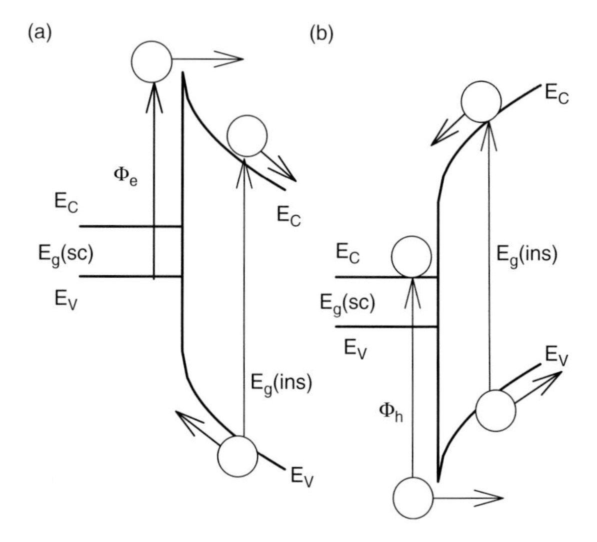

Figure 8.2 Schematic semiconductor/insulator energy band diagram for positive (a) and negative (b) bias applied to the top metal electrode (not shown), indicating the electron transitions in the case of electron IPE (threshold Φ_e) and hole IPE (threshold Φ_h) from the semiconductor into the insulator, as well as the insulator PC (threshold E_g(ins)). Thanks to the externally applied electric field, the thresholds Φ_e and Φ_h correspond to the energy position of the insulator band edges at the interface, not influenced by the charges distributed over the remaining thickness of the insulating layer. E_g(sc) represents the semiconductor bandgap.

More details on IPE measurements and examples of the IPE/PC spectra can be found elsewhere [17, 18].

8.3 Silicon/Insulator Band Offsets

To illustrate the application of the IPE spectroscopy for interface barrier height inference, we consider here the impact of crystallization on the band alignment between Si and Al_2O_3. While amorphous alumina (a-Al_2O_3) films exhibit only a 6.1–6.2 eV wide gap [23], its annealing-induced crystallization into cubic γ-Al_2O_3 eliminates the low-energy PC [19], pointing to an increase in the gap to 8.7 eV value as reported before for this Al_2O_3 phase [24]. The most interesting aspect here is the evolution of the VB and CB edges caused by the transition from amorphous to cubic Al_2O_3. Electron IPE spectra, such as shown in Figure 8.3a for the as-deposited a-Al_2O_3 and polycrystalline γ-Al_2O_3 on (100)Si obtained after 1000 °C annealing, indicate a ~0.5 eV upshift of the oxide CB relative to the reference level of the Si VB top. This shift is consistent with measurements performed on single-crystal γ-Al_2O_3 epitaxially grown on (111)Si at 775 °C as illustrated in Figure 8.3b, showing the field dependence (the Schottky plot) of the electron IPE thresholds in all samples, that is, a-Al_2O_3/(100) Si (open circle), polycrystalline γ-Al_2O_3/(100)Si (open square), and single-crystal

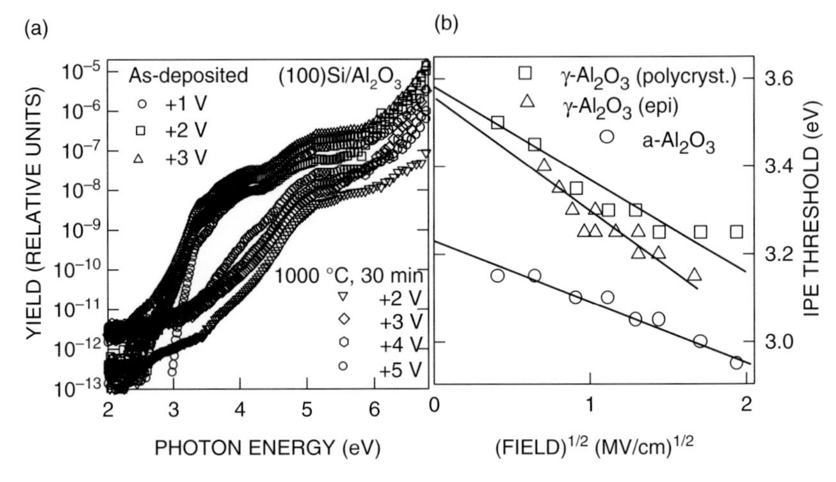

Figure 8.3 (a) Logarithmic plot of the IPE/PC yield spectra of (100)Si/Al_2O_3(12 nm)/Au capacitors measured under different bias values applied to the Au contact for the as-deposited (a-Al_2O_3) layer and after annealing-induced crystallization to cubic γ-Al_2O_3; (b) Schottky plot of the spectral thresholds of electron IPE from the silicon VB into the CB of a-Al_2O_3 on (100)Si (open circle), polycrystalline γ-Al_2O_3 on (100)Si (open square) obtained by crystallization of the amorphous alumina film by annealing at 1000 °C, and single-crystal γ-Al_2O_3 (open triangle) epitaxially grown on the (111)Si substrate. Lines illustrate linear fits used to find the zero-field barrier heights.

γ-Al_2O_3/(111)Si (open triangle). This result would mean that the gap widening from 6.2 to 8.7 eV occurs for about 80% through a downshift of the Al_2O_3 VB top. In other oxides of light metals such as MgO [25] or Sc_2O_3 [26], the transition from the amorphous to the crystalline cubic phase is found to occur without any measurable change in the electron IPE threshold, leaving the oxide VB shift entirely responsible for the crystallization-induced bandgap widening. For the oxides of heavier metals, including the rare-earth ones and complex oxides (aluminates, scandates, and hafnates), no difference in the bandgap width or in the band offsets with Si were found between amorphous and crystalline oxide layers [17].

Although it will not be discussed in depth here, oxide crystallization has another effect on the electron structure of interfaces with silicon [17, 26]. Both the epitaxial and the crystallized oxides exhibit much less “band tailing” in the PC spectra than the amorphous layers of the same composition. Comparison of electron and hole IPE spectra of amorphous and crystalline oxide phases reveals that emergence of the “tails” is associated mainly with the smearing out of the oxide CB edge while VB tailing is barely detectable. Keeping in mind that in the insulating oxides the top of the VB is derived from lone-pair states of O2p anions and the bottom of the CB is predominantly constituted of the unoccupied states of metal cations [27], this result suggests that the disorder-induced band “tailing” is mostly caused by the electron states derived from cation atomic shells. As a possible physical mechanism for the CB “tailing,” one may consider fluctuations in the local electrostatic potential at the site of a metal ion, caused, for instance, by a change in the number of the nearest-neighbor oxygen ions.

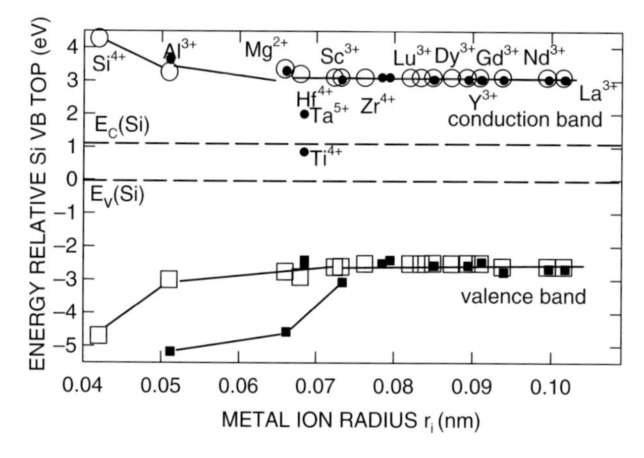

Figure 8.4 Compilation of the experimental IPE results on the energies of the VB top and CB bottom in different oxides as a function of metal cation radius referenced to the band edges of the Si crystal. Open and filled symbols correspond to amorphous and crystalline oxide layers, respectively.

In addition to electron IPE and PC, at many Si/oxide interfaces it also appears possible to detect hole IPE [17, 18], which delivers the VB offset value in the most straightforward way. The corresponding results on the CB and VB band offsets at interfaces of silicon with various oxides are summarized in Figure 8.4. The energy positions of the oxide VB and CB edges are shown as a function of the metal cation radius r_i, with the silicon VB top chosen as the origin of the energy scale. Some important trends revealed include the following:

- In the case of oxides of light metals (Mg, Al, and Sc), layers crystallized in a cubic structure exhibit a significantly wider gap than the corresponding amorphous films. As already mentioned above, this bandgap widening is predominantly associated with a shift of the O2p-derived top of the oxide valence band and may be correlated with decreasing the metal cation radius as evident from Figure 8.4. Upon crystallization, a CB shift is seen only in Al_2O_3, a trend that might further be extrapolated to the largest CB offset (3.15 eV) encountered at the a-SiO_2/Si interface for the smallest Si^{4+} cation, although no results for crystalline SiO_2 are available yet.
- For larger cations, including not only the cases of 2+ and 3+ metal ions but also 4+ ones (Zr and Hf) and Ti [28] as well as Ta^{5+}, the top of the oxide VB is universally found at around 2.5 eV below the Si VB top. The same appears to be true for more complex oxide systems for which r_i is estimated as the weighted average of the constituent cation radii.
- Neither variations in the oxide layer growth methods nor the change in the Si surface plane orientation from (100) to (111) is found to affect the interface barriers [29] mapped in Figure 8.4. This suggests a negligible (<0.1 eV) influence of structure-sensitive interface dipoles on the interface band diagram. Therefore, the interface band alignment appears to be predominantly determined by the bulk DOS of silicon and the bulk DOS of the oxide.

8.4 Band Alignment at Interfaces of High-Mobility Semiconductors

Also, in the case of high-mobility semiconductor materials such as GaAs, GaSb, $In_xGa_{1-x}As$, InP, Ge, and $Si_{1-x}Ge_x$, no dipoles sensitive to the crystallographic surface orientation or its predeposition treatment are encountered at the interfaces with insulating oxides [17, 30]. This observation supports the previous suggestion that the band diagram of semiconductor/oxide insulator interfaces is primarily determined by the *bulk* DOS of the solids in contact. The most striking example of a high-mobility semiconductor–insulator barrier, in terms of insensitivity to the particular structural aspects of the interface, is presented in Figure 8.5 that shows the spectra of electron IPE from the VB of a (100)Ge crystal into an 8 nm thick HfO_2 insulator with several monolayers (MLs) of silicon inserted between Ge and the oxide. Even with 6 MLs of silicon on top of the Ge, no measurable shift of the spectral threshold Φ_e(Ge-HfO_2) is seen, indicating that the bulk DOS of Ge and HfO_2 are the crucial factors determining the band alignment.

Pertinently, however, at the interfaces of germanium with insulating oxides, formation of a narrow-gap interlayer (IL) caused by Ge oxidation is revealed as an influential factor affecting the interface barriers [31]. By adding its own DOS to this band diagram, the IL-related component of the electron spectrum effectively reduces the interface barrier (see Φ_e(Ge-GeO_x) in Figure 8.5), thus facilitating low-field injection of electrons and leading to charge instability of the insulating stack. A similar effect is also found at the interfaces of $A_{III}B_V$ semiconductors (GaAs, InGaAs, and InP) with Al_2O_3 and HfO_2 [32–34] suggesting IL formation as the major obstacle in creating charge-free insulating stacks on these semiconductors.

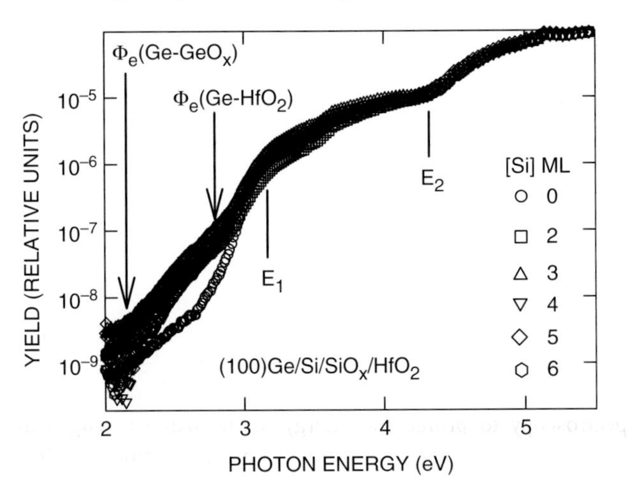

Figure 8.5 Logarithmic plot of the IPE yield spectra of (100)Ge Si/SiOx/HfO_2 (8 nm)/Au capacitors with different thickness of the passivating silicon layer (in MLs) grown on Ge prior to HfO_2 deposition. Vertical arrows indicate the threshold energies of electron IPE from the VB of germanium both into the CB of HfO_2 and into the CB of the GeO_x IL formed during oxide deposition. The vertical lines E_1 and E_2 mark the energies of direct optical transitions in the Ge crystal.

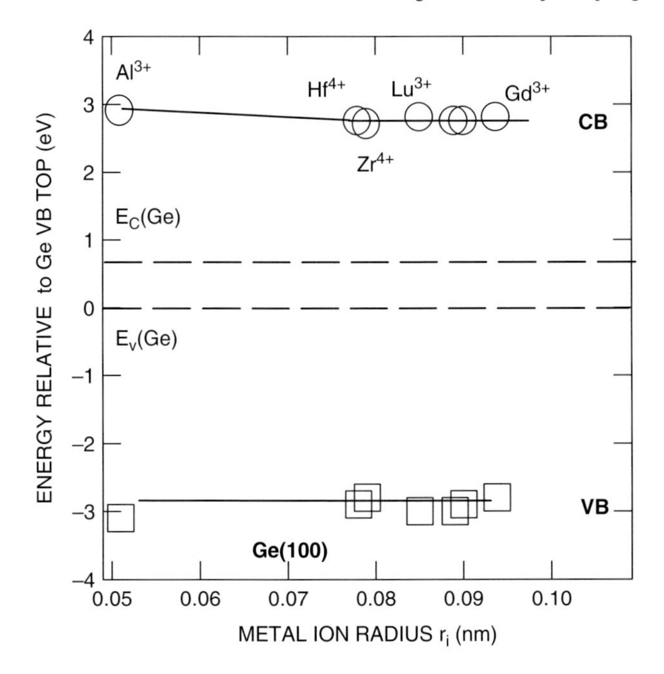

Figure 8.6 Experimental results on the energies of the VB top and CB bottom in different amorphous oxides as a function of cation ionic radius referenced to the band edges for the Ge crystal. The top of the Ge VB is at $\delta E_V = 0.4$ eV above the silicon VB as measured relative to the common Al_2O_3 CB bottom reference level.

Being inferior to that of Si, the thermal stability of Ge and $A_{III}B_V$ compounds precluded a reliable analysis of the annealing-induced crystallization effects. Yet for these semiconductors we still can check for the above-inferred universality of the band offset evolution with the insulator cation radius at the interfaces of silicon with various oxides on top. The results of the CB and VB band offset measurements are summarized in Figures 8.6 and 8.7 for Ge and GaAs, respectively. Remarkably, taking into account that the top of Ge VB is energetically located at 0.4 eV above the VB top in Si [35–37] and that in GaAs the VB edge is at 0.2 eV below its counterpart in Si [32], the O2p-derived VB top of the various oxides addressed remains at the same energy value within an accuracy of ± 0.2 eV, notwithstanding the largely different chemical properties and electronic structure of the substrate crystals.

The absence of significant structure- or composition-sensitive dipoles also allowed us to use the IPE spectroscopy to profile the energy of the bandgap edges in semiconductor alloys as a function of composition. Comparative study of $Si_{1-x}Ge_x/SiO_2$ [38] and $In_xGa_{1-x}As/Al_2O_3$ [33] interfaces reveals opposite trends in the band diagram evolution: While in the former case the shift of the semiconductor VB appears to be responsible for the gap narrowing with increasing Ge concentration, as illustrated in Figure 8.8, the semiconductor VB top in the $In_xGa_{1-x}As/Al_2O_3$ structures remains unaffected by increasing In content (Figure 8.9), suggesting a predominant variation in the CB edge energy. Interestingly, when changing the anion

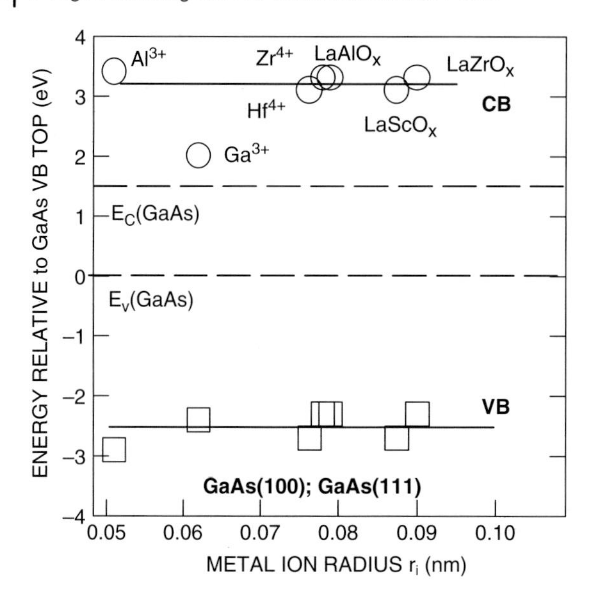

Figure 8.7 Experimental results on the energies of the VB top and CB bottom in different amorphous oxides as a function of cation ionic radius referenced to the band edges for GaAs crystals. The top of the GaAs VB is at $\delta E_V = 0.2$ eV below the silicon VB as measured relative to the common Al_2O_3 CB bottom reference level.

in the $A_{III}B_V$ compound from As in $In_xGa_{1-x}As$ to P in InP, a considerable VB downshift (~1 eV) is found, while in GaSb the VB top is shifted up by 0.4 eV as compared to GaAs, as illustrated in Figure 8.9. These results suggest a wealth of possibilities for the fabrication of functional quantum well channel structures through engineering the composition-dependent semiconductor band edge profiles.

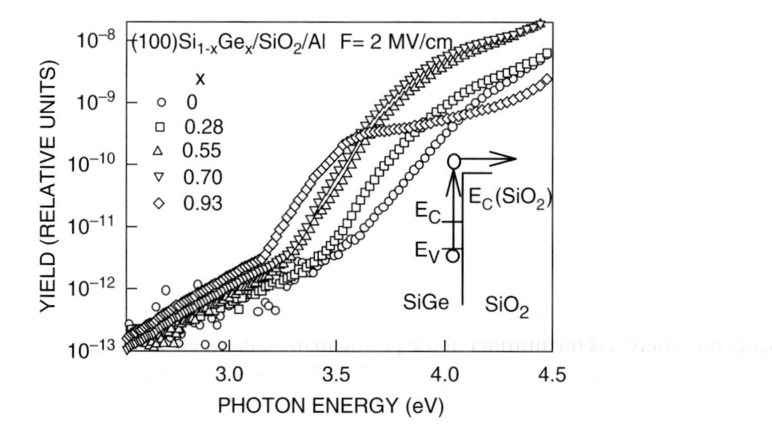

Figure 8.8 Logarithmic plots of the IPE yield spectra of (100)$Si_{1-x}Ge_x/SiO_2$/Al capacitors with different concentration of germanium. The inset illustrates the scheme of the observed electron IPE from the semiconductor VB into the CB of oxide insulator. Measurements are carried out under an applied electric field strength of $F = 2$ MV/cm over the SiO_2 insulating layer.

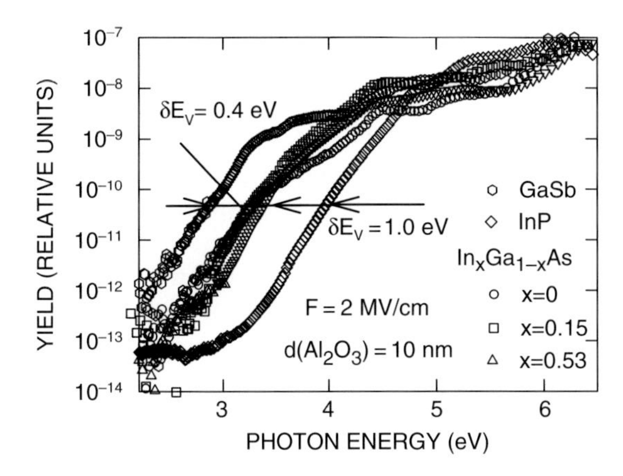

Figure 8.9 Comparison of the yield spectra for (100) faces of GaSb, GaAs, $In_xGa_{1-x}As$ ($x = 0.15, 0.53$) and InP in contact with a 10 nm thick a-Al_2O_3 insulating film measured under equal strength of electric field in the oxide (positive bias applied to the opposite metal electrode). The inferred variations in the semiconductor VB top energies (δE_V) relative to the common reference level of Al_2O_3 CB edge are indicated by arrows.

At least two aspects of the revealed trends in the semiconductor/oxide interface band offset evolution merit further discussion. First, there is the empirical observation that there is no significant difference (within ±0.2 eV accuracy) in the energy of the O2p-derived electron states near the VB top between oxides of different composition prepared on various substrates by a broad variety of deposition methods. This result suggests that this electron state represents a convenient and reliable reference level that would allow one to infer the oxide CB edge energy simply from its known bandgap width. This simple picture is valid for both amorphous and crystallized oxides with cation radii exceeding approximately 0.07 nm and, for amorphous oxides, even over the range $r_i > 0.05$ nm (cf. Figures 8.4 ,8.6 and 8.7). This range of ionic radii would imply that the suggested approach is applicable to the majority of the high-permittivity and ferroelectric oxides considered nowadays for device implementation. In particular, the difference in the bandgap width between two insulating oxides, for instance, $SrTiO_3$ and $LaAlO_3$, will be reflected mainly in the CB offset, with the O2p states at the VB top being almost aligned. In this way, the intrinsic – that is, unaffected by charged defects – band alignment between oxide insulators can be estimated simply from the known difference in bandgap width. An obvious condition for the applicability of this approach consists in the absence of additional states associated with partially occupied metal d- or f-shells inside the O2p metal (*ns*, *np*, *nd*, where *n* is the number of the period in the table of elements) gap as encountered in semiconducting and magnetic oxides. These partially occupied metal shells give rise to an additional DOS, narrowing down the gap and changing the character of the VB top states, as is observed in, e.g., Nd_2O_3 [39] and NiO [40].

Second, the band alignment between an insulating oxide and a semiconductor can be estimated with sufficient accuracy (0.2 eV or even better) if the energy position of the semiconductor bandgap edge relative to the reference O2p VB top is known (cf. Figures 8.4, 8.6 and 8.7). In this way, oxides with a (nearly) equal bandgap will deliver close band offsets at interfaces with the same semiconductor, as was already noticed earlier [17, 18]. This, again, allows one to evaluate the interface band alignment on the basis of the oxide gap width. To be noted here is that this result explains the success of the E_g(ins) versus dielectric permittivity diagram [4–7] in predicting the insulating behavior of an oxide, as the change in gap in this plot directly reflects the variation in the oxide CB bottom energy measured relative to the *common* reference level of the oxide VB top. For instance, in the case of silicon (Figure 8.4), oxides with a bandgap smaller than 3.6 eV can provide no barrier for electrons in the silicon CB. Another example is provided by semiconducting silicon carbide SiC: In this group of materials, the VB top preserves its energy at 1.75 eV below the VB of (100)Si irrespective of the SiC surface plane orientation and polytype, as referenced to the CB edge of the common native insulating oxide SiO_2 [41], while the SiC polytype-related bandgap changes are reflected in variations in the CB bottom energy. By using the results shown in Figure 8.4, one can estimate that, despite sufficiently high CB offsets, the VB offset between SiC and high-permittivity metal oxides (except for crystalline MgO and Al_2O_3) will be less than 1 eV [42, 43] leading to a high hole injection rate. To avoid this undesirable phenomenon, one may insert between SiC and the high-permittivity metal oxide a wide-gap barrier layer of SiO_2 that, thanks to its high VB offset with SiC [44], effectively blocks the hole injection [45].

With the above trends successfully inferred, a question may arise regarding the physical factors that affect the O2p VB top position in the case of crystallized compact oxides, for example, MgO and Al_2O_3. As one may notice from Figure 8.4, the energy of the VB top for the light metal oxides is significantly reduced by crystallization, pointing to some attractive interaction between the lone-pair 2p electron states of neighboring oxygen anions in the densified insulator lattice. In the less dense amorphous network or in the case of larger cations ($r_i > 0.07$ nm), this interaction seems to vanish, resulting in the "universal" O2p VB top energy position. As the Coulomb interaction between two electron clouds is repulsive, it obviously cannot account for the observed downshift of the VB states in energy upon crystallization. A possible candidate for the observed effect is the attractive exchange interaction. It is likely that not only the oxygen-to-oxygen distance but also the number of the neighboring oxygen ions available for electron exchange are of importance. Another possible factor may entail the mutual orientation of the lone-pair 2p states of the neighboring oxygens. Yet, either confirmation or disproof of this hypothesis must await in-depth theoretical verification on the basis of *ab initio* quantum mechanical calculations.

8.5
Metal/Insulator Barriers

While the above-overviewed results regarding interface band alignment between semiconductors and insulating oxides reveal remarkable universality of the barrier height, the electron barrier between the metal Fermi level and the insulator CB edge at metal/oxide interfaces exhibits a more complicated behavior. Early compilations of the measured electron barrier heights indicated [8, 17, 46], besides a large spread in the data, a considerable sensitivity to the sample processing conditions, for instance, to the oxide growth temperature and subsequent thermal treatments [47]. Furthermore, unlike the semiconductor/oxide barriers, inserting an interlayer of another oxide between the metal and the high-permittivity oxide may significantly alter the barrier for electrons as well as the work function difference between the metal and the semiconductor electrodes in a MOS capacitor [48]. These observations suggest that, in addition to the bulk DOS spectra of a metal and an insulating oxide, there must be an additional factor influencing the metal/oxide barrier heights. In this section, we will try to reveal the origin of the metal/oxide barrier variation based on the IPE spectroscopy method.

An example of the annealing-induced metal/oxide barrier variation is illustrated in Figure 8.10 that shows the Fowler plot of electron IPE spectra from TaN_x (open symbols) and TiN_x (filled symbols) into the HfO_2 insulator as affected by the temperature of the anneal performed after depositing the metal layer on the oxide [49]. While TiN_x exhibits a stable barrier, an annealing-induced shift of about 1 eV in the spectral curves is seen in the case of TaN_x. Pertinently, this kind of barrier instability is also observed for Ta carbides of different composition (TaC and Ta_2C) deposited on HfO_2 [50]. In the case of TaN_x/HfO_2, analysis of the field-dependent IPE indicates that the barrier height increase observed after annealing is caused by the creation of a layer of negatively charged defects in the oxide near the interface with the metal [50] rather than by annealing-induced changes in the DOS of the metal.

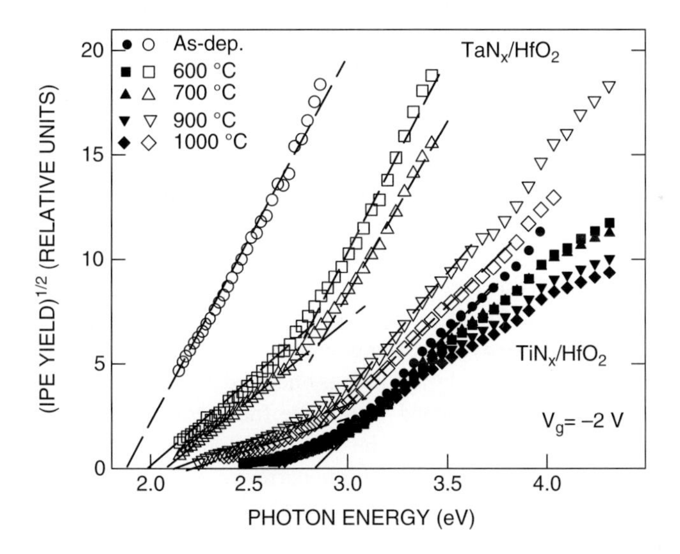

Figure 8.10 Fowler plots of electron IPE yield spectra from TaN_x (open symbols) and TiN_x (filled symbols) into 10 nm thick HfO_2 measured under a negative bias of $V_g = -2$ V on the as-deposited (100)Si/HfO_2/TaN_x (open circle) and (100)Si/HfO_2/TiN_x (filled circle) samples and on the samples after 10 s annealing in N_2 at the indicated temperature. Lines illustrate the spectral threshold determination procedure by means of the $Y^{1/2}$–$h\nu$ plot (the Fowler plot).

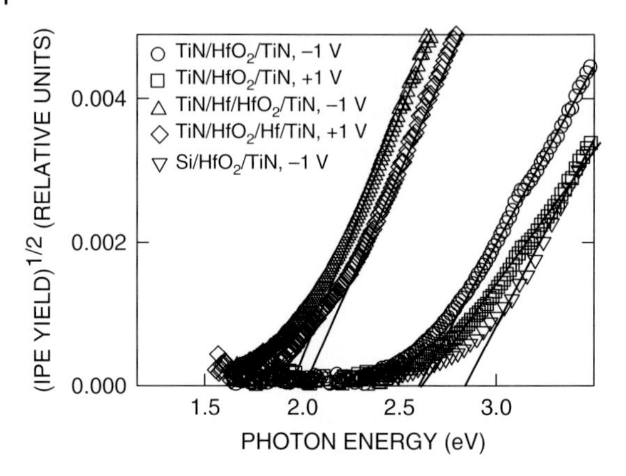

Figure 8.11 Fowler plots illustrating the determination of the electron IPE spectral thresholds at the top (open circle) and bottom (open square) TiN/HfO_2 interfaces in a $Si/TiN/HfO_2/TiN$ structure, the top TiN/HfO_2 interface (open triangle) in a $Si/TiN/Hf/HfO_2/TiN$ sample, the top TiN/HfO_2 interface (inverted open triangle) in $Si/HfO_2/TiN$, and the bottom TiN/HfO_2 interface (open diamond) in $Si/TiN/HfO_2/Hf/TiN$. The applied bias to the top TiN electrode is either +1 V or −1 V, which enables observation of electron IPE either from the bottom or from the top electrode. Lines illustrate extrapolations to zero yield used to infer the spectral thresholds of the IPE.

An even more striking example of defect-induced metal/oxide barrier variations is demonstrated in Figure 8.11 that compares the spectra of electron IPE from TiN into 10 nm thick HfO_2 layers as influenced by the kind of material of the opposite electrode [51]: A high barrier is observed at the chemically inert Si (passivated with a thin SiO_2 layer) substrate (inverted open triangle), while, relative to this value, the barrier decreases slightly (by ~0.2 eV) in the $TiN/HfO_2/TiN$ stack for both top (open circle) and bottom (open square) interface injection. A lower IPE quantum yield from the bottom TiN/HfO_2 interface suggests an attenuation of the electron flux by some IL that appears to be oxidized Ti. However, by inserting a 5 nm thick metallic Hf layer between the bottom TiN (open triangle) and HfO_2 or under the top TiN electrode (open diamond), the electron IPE threshold at the opposite TiN/HfO_2 interface can be shifted by nearly 1 eV below that found in the $Si/HfO_2/TiN$ stack (inverted open triangle). These results suggest that oxygen scavenging from HfO_2 by metallic Hf can lower the barrier between TiN and HfO_2 by as much as one-third of its original value. Obviously, this effect cannot be understood within the framework of metal- or interface-induced gap states derived from the bulk DOS of the contacting solids [52–54]. As PC measurements indicate that the narrowest bandgap of HfO_2, associated with the monoclinic phase, remains uninfluenced by the electrode material (not shown), it is likely that oxygen scavenging introduces near-interface positively charged defects that, together with electrons from the metal electrode, form an interface dipole responsible for the barrier lowering demonstrated in Figure 8.11.

One may wonder why the effect of near-interface insulator charges appears to be so different at the interfaces of oxides with metals and semiconductors. To shed some light on this query, one may compare the interface properties for the transition from semiconductor to metallic behavior by increasing the doping concentration in the

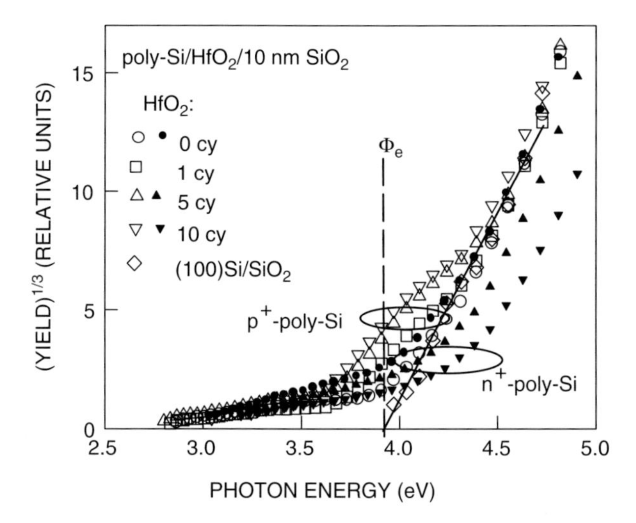

Figure 8.12 Cube root of the electron IPE yield as a function of photon energy at the interfaces of the n^+- (filled symbols) and p^+-doped (open symbols) poly-Si with 10 nm SiO_2 for different amounts (cycles) of HfO_2 (1 cy $\approx$ 0.1 monolayer) IL inserted. The curve for the low-doped p-type Si(100)/SiO_2 interface is added for comparison (open diamond). The average strength of the applied electric field in the oxide is 1 MV/cm. The solid line illustrates the determination of the electron IPE spectral threshold Φ_e (marked by dotted line) from the (yield)$^{1/3}$–$h\nu$ plot shown.

semiconductor [55]. In Figure 8.12 are shown spectra of electron IPE from the VB of heavily p^+-(open symbols) and n^+- (filled symbols) doped (n_a,$n_d > 10^{19}$ cm^{-3}) polycrystalline silicon (poly-Si) into the CB of SiO_2 as affected by the number of ($HfCl_4$ + H_2O) cycles (cy's) used to introduce a HfO_2 IL prior to poly-Si deposition. Compared to the low-doped (100)Si/SiO_2 interface results, shown on the same plot (open diamond), without HfO_2 deposition no difference is encountered either for p^+-(open circle) or for the n^+-(solid circle) poly-Si electrode. However, introduction of the HfO_2 IL between poly-Si and SiO_2 causes a shift of the IPE threshold in a direction dependent on the poly-Si doping type: in the p^+ -poly-Si samples the threshold becomes lower (open square, open triangle, and open inverted triangle), while in the n^+ ones the threshold increases (solid triangle and solid inverted triangle).

By stepping to the Schottky plot of the IPE spectral thresholds as inferred from the results shown in Figure 8.12, it is possible to reveal the physical picture of the observed interface barrier variations [18]. As can be seen from Figure 8.13, the barrier reduction found in the p^+-doped poly-Si samples is caused by a strong field-induced lowering that increases when more HfO_2 is added to the poly-Si/SiO_2 interface. The slope of the Schottky plot, usually determined by the image-force dielectric constant (ε_i), nearly doubles in the low field range $F < 1$ MV/cm compared to the case of the low-doped Si case (grey circles) pointing to barrier perturbation by a Coulombic attractive potential of positively charged defects located close to the Si/oxide interface plane [18]. This picture is consistent with the fact that the screening depth in Si is still larger than the photoelectron escape depth ($\lambda_e \approx 1.3$ nm [56]). However, in the case of a metallic emitter, the near-interface charges will create a dipole because the metal polarization layer depth rarely exceeds 0.1 nm, thus leading to the global barrier lowering observed at the TiN/HfO_2 interfaces (Figure 8.11).

In the case of n^+- poly-Si, the observed increase in barrier height would be consistent with the introduction of negative charged defects into the Si oxide when adding the HfO_2 IL (solid squares in Figure 8.13). The repulsive potential of these centers will scatter photoelectrons preventing them from entering the SiO_2 layer. Then, the photocurrent detected in the IPE experiment will predominantly originate from the interface areas *between* the negative charge positions, resulting in a barrier field dependence resembling that of the uncharged case (solid circle) but now with the barrier height increased by the repulsive electric fields of the charged defects. Again, in the case of a metal photoemitter the small thickness of the surface polarization layer would lead to a dipole-like contribution with "tailing" of the IPE spectra toward the initial (without charges) barrier value because of the strong electrostatic screening expected in the case of metal electrodes. This is exactly the kind of behavior observed in the case of TaN_x and TaC_x electrodes on HfO_2 [49, 50]. Thus, the observed additional dipole component of the metal/oxide interface barriers can be associated with dipole layers induced by near-interface charged centers (structural defects or impurities) in the insulating layer.

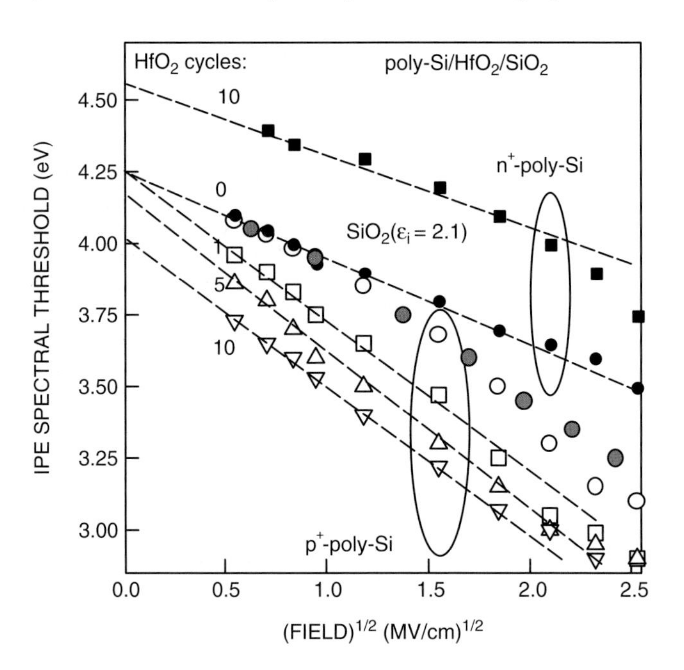

Figure 8.13 Schottky plot of the spectral threshold of electron IPE from the VB of nominally undoped, and p^+- and n^+-doped poly-Si electrodes (gray, open, and filled symbols, respectively) into the CB of 10 nm thick thermal SiO_2 grown on (100)Si. The number of atomic layer deposition ($HfCl_4 + H_2O$) cycles used to grow HfO_2 on SiO_2 prior to the poly-Si growth is indicated for each sample. The abscissa (field) denotes the square root of the applied electric field in the Si oxide. Lines represent linear fits to the data used to infer the zero-field barrier height.

8.6
Conclusions

Using the IPE/PC spectroscopy, general trends have been revealed regarding the evolution of electron band offsets at the interfaces of insulating oxides with semiconductors and metals. With the bulk electron DOS determining the semiconductor/oxide interface band alignment most of the variations in the interface barriers are expected to originate from the variations in the bandgap width and electron affinity of the materials in contact.

On the oxide side, the energy of the O2p-derived VB top can serve as the convenient reference as its value is found to be almost invariant (± 0.2 eV) both in the oxides with large sized ($r_i > 0.07$ nm) cations and in the amorphous oxide layers down to even smaller cation size ($r_i > 0.05$ nm). In this way, the band offsets can be evaluated on the basis of the known oxide bandgap values and energies of semiconductor bandgap edges.

The gap widening in the oxides of compact ($r_i < 0.07$ nm) cations upon observed crystallization appears to be predominantly caused by the downshift of the O2p-derived VB states, suggesting some attractive interaction between them.

The considerable barrier height variability generally encountered at the metal/oxide interfaces is considered to be related to the formation of interface dipole layers between charged defects in the insulating film and a polarization layer at the metal surface. The corresponding contribution to the barrier height amounts to one-third of the original barrier height. This effect may be invoked to tune the metal/oxide barrier height as well as the built-in electric field in the insulator by generating or passivating interface oxide defects.

References

1 Wilk, G.D., Wallace, R.M., and Anthony, J.M. (2001) High-κ gate dielectrics: current status and materials properties considerations. *J. Appl. Phys.*, **89**, 5243.

2 Houssa, M. (ed.) (2004) *High-κ Gate Dielectrics*, IoP Publishing, Bristol.

3 Huff, H.R. and Gilmer, D.C. (eds) (2005) *High Dielectric Constant Materials*, Springer, Berlin.

4 Robertson, J. (2000) Band offsets of wide-band-gap oxides and implications for future electronic devices. *J. Vac. Sci. Technol. B*, **18**, 1785.

5 Schlom, D.J. and Haeni, J.H. (2002) A thermodynamic approach to selecting alternative gate dielectrics. *MRS Bull.*, **27**, 198.

6 Robertson, J. (2006) High dielectric constant gate oxides for metal oxide Si transistors. *Rep. Prog. Phys.*, **69**, 327.

7 Schlom, D.G., Datta, S., and Guha, S. (2008) Gate oxides beyond SiO_2. *MRS Bull.*, **33**, 1017.

8 Afanas'ev, V.V., Houssa, M., Stesmans, A., Adriaenssens, G.J., and Heyns, M.M. (2002) Band alignment at the interfaces of Al_2O_3 and ZrO_2-based insulators with metals and Si. *J. Non-Cryst. Solids*, **303**, 69.

9 Lau, W.M. (1989) Use of surface charging in X-ray photoelectron spectroscopic studies of ultrathin dielectric films on semiconductors. *Appl. Phys. Lett.*, **54**, 338.

10 Lau, W.M. (1989) Effects of a depth-dependent specimen potential on X-ray photoelectron spectroscopic data. *J. Appl. Phys.*, **65**, 2047.

11 Alay, J.L. and Hirose, M. (1997) The valence band alignment at ultrathin SiO_2/Si interfaces. *Appl. Phys. Lett.*, **81**, 1606.
12 Toyoda, S., Okabayashi, J., Kumigashira, H., Oshima, M., Liu, G.L., Liu, Z., Ikeda, K., and Usuda, K. (2005) Precise determination of band offsets and chemical states in SiN/Si studied by photoemission spectroscopy and X-ray absorption spectroscopy. *Appl. Phys. Lett.*, **87**, 102901.
13 Tanimura, T., Toyoda, S., Kamada, H., Kumigashira, H., Oshima, M., Sukegawa, T., Liu, G.L., and Liu, Z. (2010) Photoinduced charge-trapping phenomena in metal/high-*k* gate stack structures studied by synchrotron radiation photoemission spectroscopy. *Appl. Phys. Lett.*, **96**, 162902.
14 Nohira, H., Tsai, W., Besling, W., Young, E., Petry, J., Conard, T., Vandervorst, W., DeGendt, S., Heyns, M., Maes, J., and Tuominen, M. (2002). Characterization of ALCVD-Al_2O_3 and ZrO_2 layer using X-ray photoelectron spectroscopy. *J. Non-Cryst. Solids*, **303**, 83.
15 Perego, M., Seguini, G., and Fanciulli, M. (2008) XPS and IPE analysis of HfO_2 band alignment with high-mobility semiconductors. *Mater. Sci. Semicond. Process.*, **11**, 221.
16 Bersch, E., Di, M., Consiglio, S., Clark, R.D., Leusnik, G.J., and Diebold, A.C. (2010) Complete band offset characterization of the HfO_2/SiO_2/Si stack using charge corrected X-ray photoelectron spectroscopy. *J. Appl. Phys.*, **107**, 043702.
17 Afanas'ev, V.V. and Stesmans, A. (2007) Internal photoemission at interfaces of high-κ insulators with semiconductors and metals. *J. Appl. Phys.*, **102**, 081301.
18 Afanas'ev, V.V. (2008) *Internal Photoemission Spectroscopy: Principles and Applications*, Elsevier, Oxford.
19 Afanas'ev, V.V., Stesmans, A., Mrstik, B.J., and Zhao, C. (2002) Impact of annealing-induced densification on electronic properties of atomic-layer-deposited Al_2O_3. *Appl. Phys. Lett.*, **81**, 1678.
20 Kraut, E.A. (1984) Heterojunction band off-sets – variation with ionization potential compared to experiment. *J. Vac. Sci. Technol. B*, **2**, 486.
21 Waldrop, G.R., Grant, G.W., Kowalszyk, S.P., and Kraut, E.A. (1985) Measurement of semiconductor heterojunction band discontinuities by X-ray photoemission spectroscopy. *J. Vac. Sci. Technol. B*, **3**, 835.
22 Adamchuk, V.K. and Afanas'ev, V.V. (1992) Internal photoemission spectroscopy of semiconductor-insulator interfaces. *Prog. Surf. Sci.*, **41**, 111.
23 Afanas'ev, V.V., Houssa, M., Stesmans, A., and Heyns, M.M. (2002) Band alignments in metal-oxide-silicon structures with atomic-layer deposited Al_2O_3 and ZrO_2. *J. Appl. Phys.*, **91**, 3079.
24 Ealet, B., Elyakhloufi, M.H., Gillet, E., and Ricci, M. (1994) Electronic and crystallographic structure of gamma-alumina thin-films. *Thin Solid Films*, **250**, 92.
25 Afanas'ev, V.V., Stesmans, A., Cherkaoui, K., and Hurley, P.K. (2010) Electron energy band alignment at the (100)Si/MgO interface. *Appl. Phys. Lett.*, **96**, 052103.
26 Afanas'ev, V.V., Shamuilia, S., Badylevich, M., Stesmans, A., Edge, L.F., Tian, W., Schlom, D.G., Lopes, J.M.J., Roeckrath, M., and Schubert, J. (2007) Electronic structure of silicon interfaces with amorphous and epitaxial insulating oxides: Sc_2O_3, Lu_2O_3, $LaLuO_3$. *Microelectron. Eng.*, **84**, 2278.
27 Lucovsky, G. (2002) Correlations between electronic structure of transition metal atoms and performance of high-*k* gate dielectrics in advanced Si devices. *J. Non-Cryst. Solids*, **303**, 40.
28 This follows not only from IPE work but also from photoelectron spectroscopy data, see, for example, Amy, F., Wan, A.S., Kahn, A., Walker, F.J., and McKee, R.A. (2004) Band offsets at heterojunctions between $SrTiO_3$ and $BaTiO_3$ and Si(100). *J. Appl. Phys.*, **96**, 1635. The oxide charging effect in the latter Ti case is unlikely to be of significance because of a negative CB offset with Si.
29 Badylevich, M., Shamuilia, S., Afanas'ev, V.V., Stesmans, A., Laha, A.,

Osten, H.J., and Fissel, A. (2007) Investigation of the electronic structure at interfaces of crystalline and amorphous Gd_2O_3 thin layers with silicon substrates of different orientation. *Appl. Phys. Lett.*, **90**, 252101.

30 Afanas'ev, V.V. and Stesmans, A. (2009) Barrier characterization at interfaces of high-mobility semiconductors with oxide insulators. *ECS Trans.*, **25** (6), 95.

31 Afanas'ev, V.V., Stesmans, A., Delabie, A., Bellenger, F., Houssa, M., and Meuris, M. (2008) Electronic structure of GeO_2-passivated interfaces of (100)Ge with Al_2O_3 and HfO_2. *Appl. Phys. Lett.*, **92**, 022109.

32 Afanas'ev, V.V., Badylevich, M., Stesmans, A., Brammertz, G., Delabie, A., Sionke, S., O'Mahony, A., Povey, I.M., Pemble, M.E., O'Connor, E., Hurley, P.K., and Newcomb, S.B. (2008). Energy barriers at interfaces of (100)GaAs with atomic layer deposited Al_2O_3 and HfO_2. *Appl. Phys. Lett.*, **93**, 212104.

33 Afanas'ev, V.V., Stesmans, A., Brammertz, G., Delabie, A., Sionke, S., O'Mahony, A., Povey, I.M., Pemble, M.E., O'Connor, E., Hurley, P.K., and Newcomb, S.B. (2009) Energy barriers at interfaces between (100)$In_xGa_{1-x}As$ ($0 \leq x \leq 0.53$) and atomic-layer deposited Al_2O_3 and HfO_2. *Appl. Phys. Lett.*, **94**, 202110.

34 Chou, H.-Y., Afanas'ev, V.V., Stesmans, A., Lin, H.C., Hurley, P.K., and Newcomb, S.B. (2010) Electron band alignment between (100)InP and atomic-layer deposited Al_2O_3. *Appl. Phys. Lett.*, **97**, 132112.

35 Afanas'ev, V.V. and Stesmans, A. (2006) Spectroscopy of electron states at interfaces of (100)Ge with high-κ insulators. *Mater. Sci. Semicond. Process.*, **9**, 764.

36 Afanas'ev, V.V. and Stesmans, A. (2004) Energy band alignment at the (100)Ge/HfO_2 interface. *Appl. Phys. Lett.*, **84**, 2319.

37 Afanas'ev, V.V., Shamuilia, S., Stesmans, A., Dimoulas, A., Panayiotatos, Y., Sotiropoulos, A., Houssa, M., and Brunco, D.P. (2006) Electron energy band alignment at interfaces of (100)Ge with rare-earth oxide insulators. *Appl. Phys. Lett.*, **88**, 132111.

38 Afanas'ev, V.V., Stesmans, A., Souriau, L., Loo, R., and Meuris, M. (2009) Valence band energy in confined $Si_{1-x}Ge_x$ ($0.28 < x < 0.93$) layers. *Appl. Phys. Lett.*, **94**, 172106.

39 Afanas'ev, V.V., Badylevich, M., Stesmans, A., Laha, A., Fissel, A., Osten, H.J., Tian, W., Edge, L.F., and Schlom, D.G. (2008) Band offsets between Si and epitaxial rare earth sesquioxides (RE_2O_3, RE = La, Nd, Gd, Lu): effect of 4*f*-shell occupancy. *Appl. Phys. Lett.*, **93**, 192105.

40 Afanas'ev, V.V., Badylevich, M., Houssa, M., Stesmans, A., Aggrawal, G., and Campbell, S.A. (2010) Electron energy band alignment at the NiO/SiO_2 interface. *Appl. Phys. Lett.*, **96**, 172105.

41 Afanas'ev, V.V., Bassler, M., Pensl, G., Schulz, M., and Stein von Kamienski, E. (1996) Band offsets and electronic structure of SiC/SiO_2 interfaces. *J. Appl. Phys.*, **79**, 3108.

42 Afanas'ev, V.V., Ciobanu, F., Dimitrijev, S., Pensl, G., and Stesmans, A. (2004) Band alignment and defect states at SiC/oxide interfaces. *J. Phys. Condens. Matter*, **16**, S1839.

43 Afanas'ev, V.V., Ciobanu, F., Pensl, G., and Stesmans, A. (2004) Contributions to the density of interface states in SiC MOS structures, in *Silicon Carbide: Recent Major Advances* (eds W.J. Choyke, H. Matsunami, and G. Pensl), Springer, Berlin, pp. 344–371.

44 Afanas'ev, V.V. and Stesmans, A. (2000) Valence band offset and hole injection at the 4H-, 6H-SiC/SiO_2 interfaces. *Appl. Phys. Lett.*, **77**, 2024.

45 Afanas'ev, V.V., Stesmans, A., Chen, F., Campbell, S.A., and Smith, R. (2003) HfO_2-based insulating stacks on 4H-SiC (0001). *Appl. Phys. Lett.*, **82**, 922.

46 Afanas'ev, V.V. and Stesmans, A. (2003) Band alignment at the interfaces of Si and metals with high-permittivity insulating oxides, in *High-κ Gate Dielectrics* (ed. M. Houssa), IoP Publishing, Bristol, pp. 217–250.

47 Pantisano, L., Schram, T., Li, Z., Lisoni, J.G., Pourtois, G., De Gendt, S., Brunco, D.P., Akheyar, A., Afanas'ev, V.V., Shamuilia, S., and Stesmans, A. (2006)

Ruthenium gate electrodes on SiO_2 and HfO_2: sensitivity to hydrogen and oxygen anneals. *Appl. Phys. Lett.*, **88**, 243514.

48 Pantisano, L., Schram, T., O'Sullivan, B., Conard, T., De Gendt, S., Groeseneken, G., Zimmerman, P., Akheyar, A., Heyns, M.M., Shamuilia, S., Afanas'ev, V.V., and Stesmans, A. (2006). Effective work function modulation by controlled dielectric monolayer deposition. *Appl. Phys. Lett.*, **89**, 113505.

49 Afanas'ev, V.V., Stesmans, A., Pantisano, L., and Schram, T. (2005) Electron photoemission from conducting nitrides (TiN_x, TaN_x) into SiO_2 and HfO_2. *Appl. Phys. Lett.*, **86**, 232902.

50 Shamuilia, S., Afanas'ev, V.V., Stesmans, A., Schram, T., and Pantisano, L. (2008) Internal photoemission of electrons from Ta-based conductors into SiO_2 and HfO_2 insulators. *J. Appl. Phys.*, **104**, 073722.

51 Afanas'ev, V.V., Stesmans, A., Pantisano, L., Cimino, S., Adelmann, C., Goux, L., Chen, Y.Y., Kittl, J.A., Wouters, D., and Jurczak, M. (2011) TiN_x/HfO_2 interface dipoles induced by oxygen scavenging. *Appl. Phys. Lett.*, **98**, 132901.

52 Tersoff, J. (1984) Schottky-barrier heights and the continuum of gap states. *Phys. Rev. Lett.*, **52**, 465.

53 Tersoff, J. (1984) Theory of semiconductor heterojunctions – the role of quantum dipoles. *Phys. Rev. B*, **30**, 4874.

54 Monch, W. (1996) Empirical tight-binding calculation of the branch-point energy of the continuum of the interface-induced gap states. *J. Appl. Phys.*, **80**, 5076.

55 Afanas'ev, V.V., Stesmans, A., Pantisano, L., and Chen, P.J. (2005) Electrostatic potential perturbation at the polycrystalline Si/HfO_2 interface. *Appl. Phys. Lett.*, **86**, 072107.

56 Sebenne, C., Bolmont, D., Guichar, G., and Balkanski, M. (1975) Surface states from photoemission threshold measurements on a clean, cleaved, Si(111) surface. *Phys. Rev. B*, **12**, 3280.

Part Three
Challenge in Interface Engineering and Electrode

High-k Gate Dielectrics for CMOS Technology, First Edition. Edited by Gang He and Zhaoqi Sun.

9
Interface Engineering in the High-*k* Dielectric Gate Stacks

Shijie Wang, Yuanping Feng, and Alfred C.H. Huan

9.1
Introduction

The continual downscaling of semiconductor devices into the "nano" era requires not only the replacement of silicon dioxide (SiO_2) gate dielectric by high dielectric constant (high-*k*) materials but also the replacement of polycrystalline Si (poly-Si) gate electrodes by metal gates. In order to integrate the metal gate material and the high-*k* dielectric with current semiconductor technology, it is essential to understand and control two interfaces at the atomic scale, the interface between metal gate and high-*k* dielectric and the interface between high-*k* dielectric and Si. Band discontinuity is one of the most important parameters for semiconductor interfaces [1, 2]. The transport properties in heterojunction devices are controlled by the electronic band profiles at the interfaces, namely, the valence and conduction band offsets (VBO and CBO) in the case of semiconductor heterojunctions and the n-type (for electrons) and p-type (for holes) Schottky barrier heights (SBHs) in the case of metal/semiconductor interfaces (see Figure 9.1). Concerning the high-*k* dielectric applications, the conduction and valence band offset between high-*k* oxide and Si channel should be larger than 1.0 eV, while the n-type SBH (for electrons) or effective work function of metal gate on high-*k* dielectrics between high-*k* oxide and metal gate electrodes (or effective work function of metal gate on high-*k* dielectrics) should be tunable to a range of $\sim$1.1 eV (the magnitude of Si bandgap) to meet the requirements for both of p- and n-type metal oxide semiconductors (MOS) devices.

9.2
High-*k* Oxide/Si Interfaces

The conduction (valence) band offset between the dielectric and the Si channel acts as an electrical barrier for the electrons (holes) in Si channel and it should be large enough to minimize carrier injection into the conduction band states of the dielectric. Normally, a minimum requirement of 1.0 eV for CBO and VBO is needed and the high-*k* gate dielectric candidates with CBO or VBO smaller than

High-k Gate Dielectrics for CMOS Technology, First Edition. Edited by Gang He and Zhaoqi Sun.

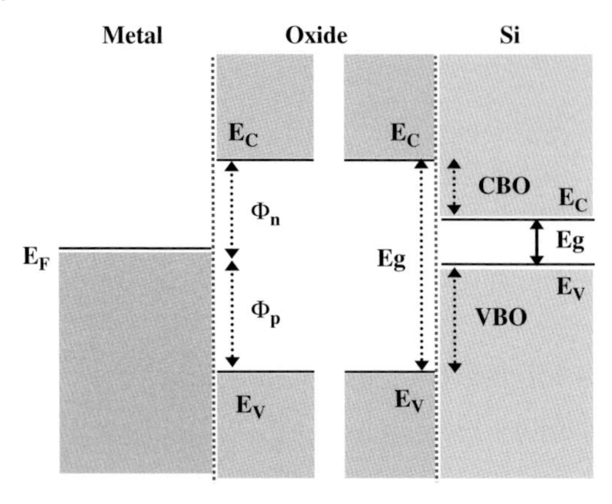

Figure 9.1 Schematic band diagrams for metal/oxide/Si stacks. Definitions of band offsets (VBO and CBO) and of Schottky barrier heights (Φ_n and Φ_p) are shown.

1.0 eV are not considered for further applications. However, many oxides with large dielectric constant have bandgaps much smaller than 8.8 eV of silicon dioxide. How the oxide and silicon bands line up critically affects the capacitance–voltage characteristics and transport properties of the MOS devices. So, it is very important to accurately determine and control the band offsets for high-*k* dielectric materials on Si substrate.

X-ray photoemission spectroscopy (XPS) is a powerful probe technique that has been used to determine reliable band offsets for numerous heterojunction systems in the past decades. Especially, an XPS core level-based method proposed by Kraut *et al.* [3] exploits the longer sampling depth of XPS (up to ~10 nm) while overcoming its limited experimental energy resolution compared to synchrotron radiation photoemission techniques. Kraut's method has been widely used to accurately determine the valence band offset between ultrathin high-*k* dielectric film and Si substrate in the past few years. Chambers *et al.* [4] used the XPS core level-based method to directly measure the valence band offset at epitaxial $SrTiO_3$/Si interfaces. They found that the entire band discontinuity lies at the valence band edge and the conduction band offset is very small, far lower than the minimum requirement. Their band offset measurement immediately precludes the possibility of applying $SrTiO_3$ as a high-*k* dielectric material if there were no further atomic-level bandgap engineering methods at the interface to equalize the valence and conduction band offsets.

The asymmetric distribution of band offsets (with the majority of discontinuity occurring at the valence band edges) is the main characteristic of most high-*k* dielectric materials. For ultrathin ZrO_2 film on Si interfaces, Wang *et al.* [5, 6] used X-ray photoemission method to measure the VBO to be ~3.0 eV, while the CBO is as small as ~1.7 eV. This result is close to that obtained by Miyazaki *et al.* [7]. Puthenkovilakam and Chang [8] reported a more asymmetric distribution of band offsets for ultrathin ZrO_2 films grown by atomic layer deposition (ALD) method, with the magnitude of ~3.7 eV for VBO and only ~0.8 eV for CBO. Such low CBO ($<$ 1.0 eV) indicates the limitation of ZrO_2 grown by the ALD method as an alternative high-*k* gate dielectrics. For ultrathin HfO_2 film on Si interfaces, the situation is similar because of the similar character of Zr and Hf. Using high-resolution angle-resolved photoelectron spectroscopy, Oshima *et al.* [9] determined the valence band offset for HfO_2/Si interfaces to be ~3.0 eV, versus ~1.7 eV for the conduction band offset. Alloying ultrathin Al and Hf oxide films (Hf–Al–O) on Si substrate strengthens the asymmetric distribution. For Hf-Al-O/Si interfaces, Jin *et al.* [10] obtained a valence band offset of 3.6 eV, versus ~1.3 eV for the CBO.

While the conduction band offsets for ZrO_2 and HfO_2 on Si seem large enough for high-*k* gate dielectric applications, the strong asymmetric distribution of band offsets for $SrTiO_3$/Si interfaces brings forward an important task, that is, how to fulfill the bandgap engineering at the interfaces to obtain a positive and large enough conduction band offset for $SrTiO_3$/Si interfaces. Theoretically, Först *et al.* [11] demonstrated that atomic control of interfacial structures can improve the electronic properties of the interface to meet technological requirements. They increased the CBO of the $SrTiO_3$/Si interface from near zero to ~1.0 eV by controlling the chemical environment. This sizable change in CBO results from a local interface dipole formed by the interfacial polarized bonds. However, we noticed there may be a possible pitfall in Först's first-principles calculations, that is, the calculated VBOs did not include any many-body correction, which can be as high as 1.0 eV for metal oxide/Si heterostructures. If such many-body corrections were counted, the achieved increase in CBO would be nearly canceled out, and the resulting CBO (near-zero eV) is in good agreement with the XPS results from Chambers *et al.* [4]. Nevertheless, the demonstrated ability of tuning the band alignments between high-*k* oxide and Si by Först *et al.* is still very encouraging and has wide applications in engineering high-*k* dielectric/Si interfaces and other semiconductor heterostructures.

9.2.1 Growth of Crystalline High-*k* Oxide on Semiconductors

Epitaxial yttrium-stabilized ZrO_2 (YSZ) has been prepared on Si by using pulsed laser deposition and its band alignment has been measured by using high-resolution XPS. Cross-sectional TEM has been performed to characterize the YSZ-Si heterostructures. Figure 9.2a is a typical cross-sectional HRTEM image of epitaxial YSZ thin film on Si(100) substrate taken with the incident electron beam parallel to Si[110]. It is very clear that the epitaxial relationship between YSZ and Si is a high-symmetry cube-on-cube structure. The electron diffraction pattern of YSZ/Si interface (Figure 9.2d) shows that the orientation relationship between YSZ and Si is YSZ(001)//Si(001) and YSZ(011)//Si(011). The most obvious point in Figure 9.2a is that the YSZ thin film grows epitaxially on Si(100) substrate, without the formation of interfacial amorphous oxide layer at the interface in this case.

We proposed a possible interface structure formed by *c*-ZrO2 and Si with a –Si–Si–O–Zr–O– interfacial atomic structure, as shown in Figure 9.2c. Based on the interface structure model shown in Figure 9.2c, the simulated HRTEM image

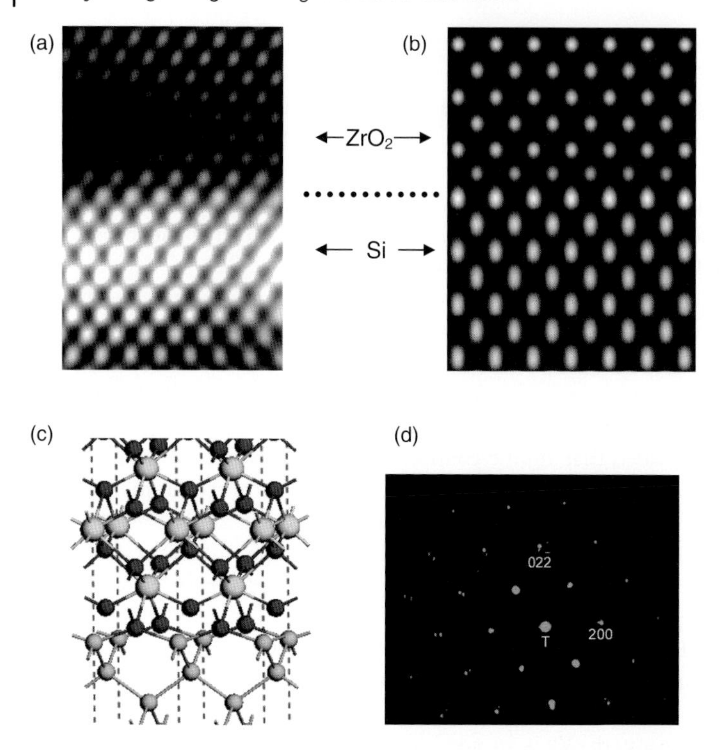

Figure 9.2 (a) HRTEM image of the interface between YSZ and Si(001) substrate, showing the high-symmetry cube-on-cube epitaxial relationship with YSZ(001)//Si(001) and YSZ [011]//Si[011]; (b) simulated HRTEM image by JEMS program; (c) proposed ZrO_2/Si interface structure; and (d) the electron diffraction pattern for YSZ on Si(001).

of crystalline YSZ on silicon (in Figure 9.2b) is in good agreement with experimental HRTEM image (Figure 9.2a), indicating that the film on silicon is single-crystal cubic YSZ film.

9.2.2
Measurement of Band Alignment at High-*k* Oxide/Si Interfaces

In order to measure the band alignment at the epitaxial ZrO_2/Si interface, a YSZ film with a thickness of 3.0 nm was grown on silicon(001). As the sampling depth of X-ray photoelectron with energy of 1486.6 eV can be up to ~10 nm, the information from both the film and the substrate can be obtained from photoemission spectra simultaneously.

We applied the photoemission-based method of Kraut *et al.* [12] to determine the valence and conduction band discontinuities. The procedure to measure the band offsets is shown in Figure 9.3. First, the energy differences between the valence band maximum (VBM) (E_v) and the core-level centroids were measured in respective bulk materials (Δ_1 for bulk-Si and Δ_2 for bulk-ZrO_2). Then, the core levels (Si $2p_{3/2}$ and

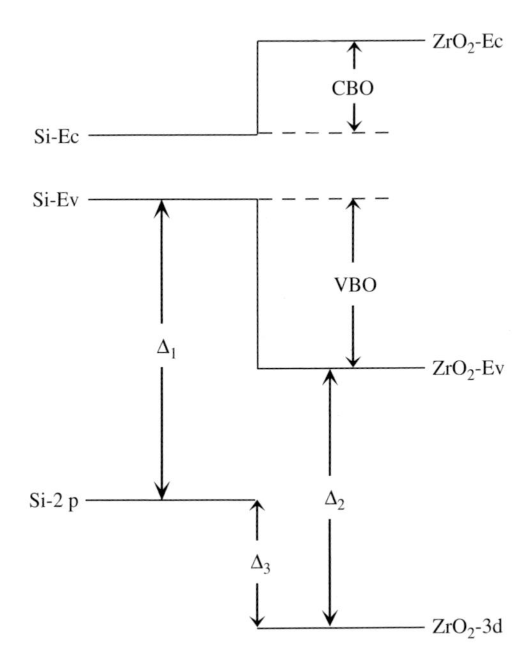

Figure 9.3 Schematic representation of a band alignment problem at heterojunctions.

Zr $3d_{5/2}$ for Si substrate and YSZ thin film, respectively) for the thin film and substrate were simultaneously measured from the ZrO_2/Si contact. These core levels from the interface were used to align the energy bands on the two sides. In this procedure, there is an important assumption, that is, the energy differences (Δ_1 for bulk-Si and Δ_2 for bulk-ZrO_2) are kept constant for a given material. This assumption is essentially right, which has been checked carefully for various materials and serves as the basis for this band alignment method for heterostructures.

In using the above procedure to measure valence band offset, it is essential to accurately determine the valence band maximum for semiconductor or oxide from XPS valence band spectra. One simple and accurate method is the linear method by finding the intersection of two straight line segments, of which one fits to the linear portion of the valence band leading edge and one fits to the background channels between the VBM and the Fermi level. In Figure 9.4, we show how the VBMs for Si (001) and YSZ(001) surfaces can be determined accurately by this linear method from XPS valence band spectra.

In Figure 9.5, we show the core-level and valence band spectra for bulk Si(001), bulk YSZ(001), and YSZ/Si(001) heterojunctions used to determine the band offsets. The energy bands in Figure 9.5 have been aligned using Zr $3d_{5/2}$ and Si $2p_{3/2}$ core levels as the reference levels. In addition, an important feature of Si 2p spectra is the absence of the bonding peak corresponding to Si in SiO_2 or silicate, which is normally in a higher energy position (100–104 eV) than that for element Si (~99.5 eV). There is only one Si 2p spin–orbital doublets, which originates from pure Si substrate. This finding confirms the result obtained from the HRTEM images shown in

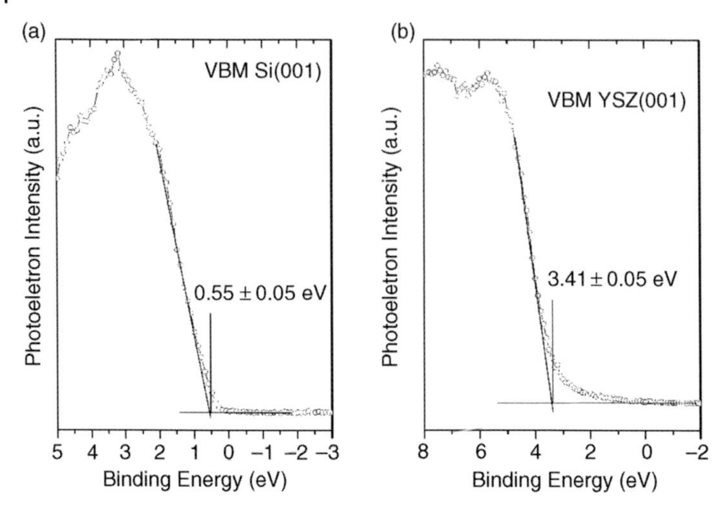

Figure 9.4 Valence band spectra for (a) Si(001) and (b) YSZ(001).

Figure 9.2a, that is, there is no interfacial SiO_x or silicate layer between the 3.0 nm thick YSZ ultrathin film and Si substrate.

In Figure 9.5, the energy difference between the valence band edges and the energy position of Si $2p_{3/2}$ peak was measured to be 98.70 ± 0.05 eV for Si(001) substrate, which is in good agreement with the value we obtained previously [13]. It is worth noticing that this value is slightly lower than that measured by Yu *et al.* [14] (98.95 eV for n-Si) and that obtained by Chambers *et al.* [15] (98.90 eV, for n-Si and 98.98 eV, for p-Si). They measured the energy difference between the valence band edges and the

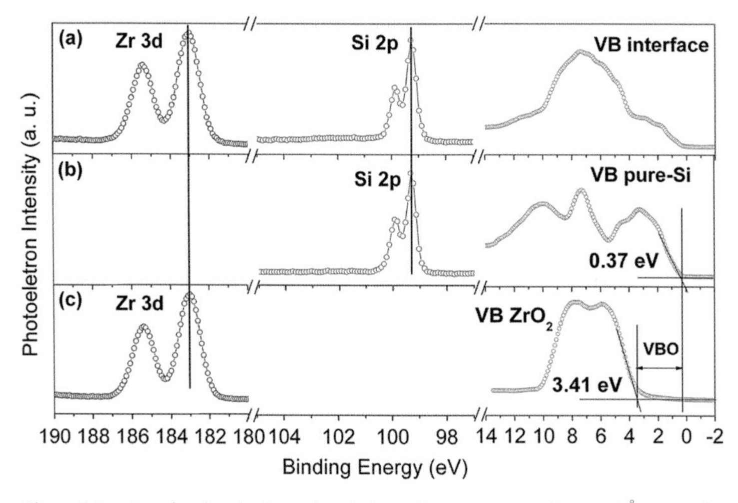

Figure 9.5 Core-level and valence band photoelectron spectra for (a) 30 Å epitaxial YSZ on Si(001), (b) H-terminated Si (001) surface, and (c) YSZ(001) surface. The energy bands have been aligned using Zr 3d and Si 2p core levels as the reference levels.

centroids of Si 2p, instead of Si $2p_{3/2}$. The resolution of XPS in our experiment is high (up to 0.45 eV) enough to distinguish the spin–orbital doublets of Si 2p even for the buried Si substrate (the energy separation between Si $2p_{1/2}$ and Si $2p_{3/2}$ is only 0.6 eV). Thus, we used the energy position of Si $2p_{3/2}$ peak instead of the centroids of the whole Si 2p as the reference level. The energy difference between Zr $3d_{5/2}$ and the VBM of ZrO_2 was measured to be 179.53 ± 0.05 eV for YSZ(001) surface and the energy difference between Zr $3d_{5/2}$ and Si $2p_{3/2}$ was determined to be 83.67 ± 0.05 eV for 30 Å epitaxial YSZ on Si(001). Then, the valence band offset was determined to be 3.04 ± 0.05 eV by using the equation

$$\mathrm{VBO} = (E_{\mathrm{Zr}\ 3d_{5/2}} - E_{\mathrm{Si}\ 2p_{3/2}})_{\mathrm{ZrO_2/Si(001)}} - (E_{\mathrm{Zr}\ 3d_{5/2}} - E_{\mathrm{v}})_{\mathrm{ZrO_2(001)}} + (E_{\mathrm{Si}\ 2p_{3/2}} - E_{\mathrm{v}})_{\mathrm{Si(001)}} \tag{9.1}$$

where $E_{\mathrm{Zr}\ 3d_{5/2}}$ denotes the energy position of Zr $3d_{5/2}$, $E_{\mathrm{Si}\ 2p_{3/2}}$ is the energy position of Si $2p_{3/2}$, and E_V is the valence band edge.

To check the above core level-based method, we simulated the valence band spectra of the 3.0 nm YSZ/Si interface by summing the appropriately weighted spectra for clean Si(001) and YSZ(001) surface in their respective energy positions shown in Figure 9.5. The result of this check is shown in Figure 9.6. The simulated valence spectra (YSZ + Si) are in good agreement with that for the actual 3.0 nm YSZ/Si interface, which demonstrates that the above core level-based method can accurately determine the valence band offset for heterostructures.

With the measured valence band offset, the conduction band offset (CBO) can be obtained by the simple equation

$$\mathrm{CBO} = (E_{\mathrm{g-ZrO_2}} - E_{\mathrm{g-Si}}) - \mathrm{VBO} \tag{9.2}$$

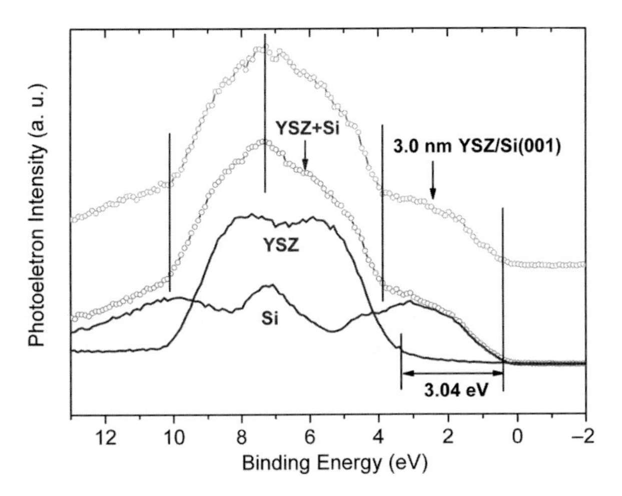

Figure 9.6 Comparison between the valence band for the actual 3.0 nm YSZ/Si interface and the simulated valence spectra (YSZ + Si). The latter was obtained by summing the valence band for Si(001) and YSZ(001) surface in their respective energy positions aligned by the core level-based method.

where E_{g-ZrO_2} and E_{g-Si} are bandgaps for ZrO_2 and Si, respectively. The bandgap of ZrO_2 was determined to be ~5.82 eV using the ultraviolet adsorption, while that for Si is 1.12 eV. Therefore, the CBO for ZrO_2/Si interface is 1.66 eV.

The valence band offset (3.04 ± 0.05 eV) we obtained here for epitaxial ZrO_2/Si interface is in good agreement with the value (3.15 eV) measured by Miyazaki *et al.* [7] for thermal evaporation-deposited ultrathin amorphous ZrO_2 on Si(001) interface. For ZrO_2/SiO_x (~0.5 nm)/Si stacks with an ultrathin SiO_2 buffer layer between ZrO_2 and Si, Puthenkovilakam *et al.* [16] obtained a relatively larger value of 3.35 eV. Fulton *et al.* [17] argued that the charge layer in the SiO_x buffer layer would give rise to large changes in the band alignment. They found an astonishing valence band maximum shift for ZrO_2 on Si with SiO_2 buffer layer between them, as large as 2.0 eV. The VBO was shifted from 0.95 eV for the as-grown ZrO_2 on Si to 3.0 eV for the postannealing sample. The band alignment is so dependent on the interface structures for the amorphous ZrO_2 on Si stacks that it inspires us to consider what are the situations for epitaxial ZrO_2/Si interfaces where there is no silicon oxide or silicate buffer layer between them.

Theoretical approach based on the first-principles density functional theory (DFT) has been used to study interfacial properties of metal oxide dielectric/Si interface on atomic level [18–21]. The general bonding rules proposed by Robertson and Peacock [22] are very instructive in understanding the atomic structure of oxide/Si interfaces, and such rules have been tested for several model interfaces of ZrO_2/Si in free-standing mode, in which both oxide and Si were relaxed simultaneously in DFT calculations. Puthenkovilakam *et al.* [16] studied the detailed atomic and electronic structures of ZrO_2/Si interfaces using first-principles calculations. The valence band offset evaluated from their interface model agrees with their XPS experimental results (3.65 eV), which indicated that the conduction band offset is smaller than the critical value for high-*k* dielectric applications. However, there have been other reported experimental values for the valence band offsets, for example, 3.0 eV and 3.15 eV. Such variations indicate that band offsets depend on deposition process. Thus, further study is required in order to clarify the dependence of band offset on interface structure.

In our group, atomic structure and electronic properties of various ZrO_2/Si interface models have been systematically studied [23]. The stabilities of these interfaces have been compared to explore the possibility of atomic control of interface structure by altering the chemical environment (oxygen chemical potential). The valence band offsets for the various interface structures are calculated to determine the dependence of band offset on interface structures, thus to provide information for band offset engineering with these materials. It was found that the valence band offset strongly depends on the interface structure or the interface chemical composition strongly modifies the lineup term of VBO, as is the case in CaF_2/Si heterostructures [24]. The variation in the band offset between different structures is supposed to occur due to the change in interface net dipole. The change in the interface dipole forces a rigid shift of the energy position of the materials on both sides of the interface. For the structures with different interface atomic configurations, the difference is as large as 2.5 eV. Such large variations with interface structures bring some technological difficulties, that is, the chemical environment should be well controlled to achieve reproducible band offset. Such a conclusion drawn from the model interfaces relaxed in the Si substrate mode is different from that in the free-standing mode, where a relatively constant band offset has been found for O-terminated models. Such differences can be due to two effects: the structural (uniaxial deformation) and chemical (interface chemical composition) effects. The former affects both the bulk and the lineup terms of band offsets, while the latter affects only the lineup term. Our results can offer some explanation for the different experimental results that vary from 3.0 to 3.6 eV. This difference may be due to different interface structures. However, it is difficult to make quantitative comparison directly because the samples used in the experiments are inhomogeneous and far from perfect epitaxial crystal ZrO_2 on Si structure.

9.3 Metal Gate/High-*k* Dielectric Interfaces

For metal gates replacing conventional poly-Si as the gate electrodes, a major stumbling block is the difficulty in identifying gate materials that can locate the dielectric/Si interface Fermi energy near the Si conduction (or valence) band edges for NMOS (or PMOS) devices. Here, an effective work function of metal gates ($\Phi_{m,eff}$) is defined as the location of the energy position of metal gate Fermi level in the bandgap of Si channel, as shown in Figure 9.7. Notice the vacuum discontinuity

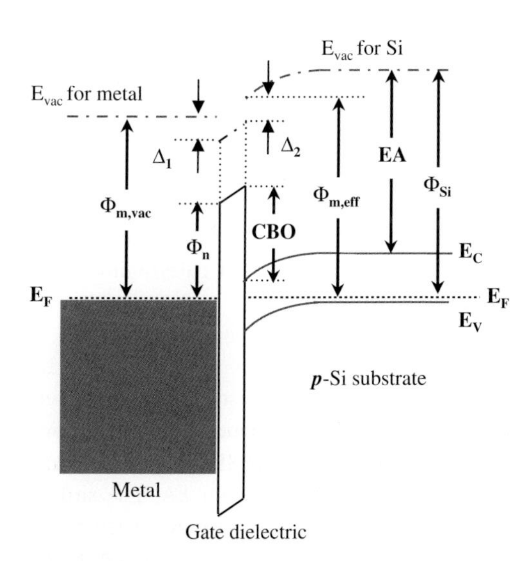

Figure 9.7 Energy band diagram of a MOS diode in thermal equilibrium. E_{vac} is the vacuum level, $\Phi_{m,vac}$ is the vacuum work function of metal gate, $\Phi_{m,eff}$ is the effective work function of metal gate on gate dielectric, Φ_{Si} is the work function of p-Si substrate, Φ_n is the barrier height between metal and gate dielectric, CBO is the conduction band offset between gate dielectric and Si, EA is the electron affinity of Si (~4.05 eV), and Δ_1 and Δ_2 denote the vacuum-level discontinuity at the two interfaces.

(Δ_1 and Δ_2) at the two interfaces. In typical semiconductor text books, the band diagrams of a MOS diode in thermal equilibrium are often simplified and the energy bands are aligned only using the electron affinity (EA) of the materials on the two sides [25]. Thus, the vacuum levels there are continuous. As we will discuss below, this corresponds to the Schottky limits, not the general situation.

For real MOS capacitors, it is not easy to evaluate the effective work function accurately because it is determined by the band alignments at two interfaces that are very complex. The most common way to extract the effective work function of metal gates is to vary the thickness of the dielectric layer and plot the capacitor flatband voltage (V_{FB}) versus the effective oxide thickness (EOT) of the dielectric layer. The flatband condition means no charge induced in the semiconductor and V_{FB} is expressed as $V_{FB} = (\Phi_{eff.m} - \Phi_{Si}) + V_{oxide}$, where V_{oxide} is the potential across the oxide dielectric layer. By assuming that V_{oxide} originates only from the fixed charge at the dielectric/Si interface (Q_f, can be positive or negative, depending on the specific interface.), V_{FB} can be reexpressed as $V_{FB} = (\Phi_{eff.m} - \Phi_{Si}) - Q_f \times \text{EOT}/\varepsilon_{SiO_2}$. The thickness series (EOT) of dielectric layer allows one to extract the effective work function and fixed charge at the dielectric/Si interface. This capacitance-based method needs no detailed information concerning the band alignments at the two interfaces, for example, barrier height for metal/dielectric interface and dielectric/silicon interface. However, appropriate interfacial structures at the dielectrics/Si interfaces and the corresponding charge distributions should be carefully considered.

The effective work function can also be obtained by measuring the barrier height for metal/dielectric interface and dielectric/silicon interface separately, expressed by $\Phi_{m,eff} = \Phi_n - \text{CBO} + \text{EA}$, where Φ_n is the n-type Schottky barrier height between metal gate electrode and high-k gate dielectric, CBO is the conduction band offset between high-k gate dielectric and Si, and EA is the electron affinity of Si (4.05 eV), which is the energy difference between the vacuum level and the Si conduction band maximum (CBM). We note that the above definition is for a flatband condition, where the surface charge-induced band bending is negligible within the interface-specific region (several nanometers). For a given high-k oxide/Si interface, the effective work function is determined only by the band alignment at metal gate/high-k oxide interface, or Φ_n.

The question what determines the magnitude of Schottky barrier height Φ_n at metal/semiconductor interfaces has troubled scientists for decades [26]. The firsta ttempts to understand the rectifying properties of metal/semiconductor interfaces are the views attributed to Schottky himself and to Mott [27]. The famous Schottky–Mott relationship, $\Phi_n = \Phi_{m,vac} - \text{EA}$, states that the SBH depends on the vacuum work function of metal ($\Phi_{m,vac}$) and the semiconductor EA. The simple Schottky–Mott picture did not take into account the local electronic structures at the interface but considered only the "pseudo-bulk" properties, $\Phi_{m,vac}$ and EA. Here, "pseudo-bulk" means that although based on the bulk electronic structures of the two materials, they are measured by performing experiments on their surfaces and thus are surface sensitive.

In real metal gate/high-k dielectric interfaces, the simple Schottky–Mott relationship is rarely realized. Instead, the dependence of SBH on metal vacuum work function was found not as sensitive as predicted by Schottky–Mott relationship. A term, "Fermi-level pinning," has often been used to describe the insensitivity of the experimental SBH to the metal vacuum work function. One popular model proposes that the SBH is modulated by the metal-induced gap states (MIGS) [28]. MIGS are the evanescent states of metal wavefunctions tunneling into the forbidden bandgap of semiconductor. Tersoff [29] claimed that if the MIGS decay length is large, the variation in metal Fermi level would be screened efficiently within several atomic layers on the semiconductor side at the interface. This MIGS screening-induced Fermi-level pinning effect is described by [30]

$$\Phi_n = S(\Phi_{m,vac} - \Phi_{CNL}) + (\Phi_{CNL} - EA) \quad (9.3)$$

where S is the so-called Schottky pinning parameter that is a characteristic of semiconductor, with $S = 1$ describing the Schottky–Mott picture and $S = 0$ describing the Barding limit of strong pinning, and Φ_{CNL} is the energy position of charge neutrality level (CNL) of the semiconductor with respect to the vacuum level. CNL is defined as the energy above which the surface states are empty. One may wonder how MIGS are related to CNL. Notice that MIGS are actually Bloch states of the bulk semiconductor with complex wave vector. The most important assumption in the MIGS or CNL model is that the MIGS (the exponentially decaying bulk states) of the semiconductor are the main component of the surface (or interface) states, irrespective of the type of metal. In MIGS theory, both pinning parameter S and the energy position of CNL can be estimated from the bulk properties of the semiconductor only. Mönch [31] found that pinning parameter S can be estimated from the empirical formula

$$S = [1 + 0.1(\varepsilon_\infty - 1)^2]^{-1} \quad (9.4)$$

where ε_∞ is the electronic component of the dielectric constant of semiconductor. A wide-gap material would tend to have a smaller ε_∞ and thus pin less. To estimate the energy position of charge neutrality level, Robertson used Tersoff's idea [32] and associated it with the branch points of the complex band structure of the dielectric. The branch point is located by calculating the zero of Green functions of the band structure, taken over the Brillion zone. However, this method to find the CNL using direct integration of density of states is in general not in good agreement with the result from the calculation of actual complex band structures [33]. Thus, how to evaluate the energy position of CNL in the fundamental bandgap of high-k material is still an open question.

The above MIGS model is appealing because it offers a simple scheme to predict SBH from the bulk properties of the two components (metal and semiconductor) only, with no need for the detailed interface information, for example, interface structures or chemical compositions. However, the MIGS model reflects only one side of the coin [34]. In addition to the contribution of MIGS for the SBH, there is another contribution, a short-range part, the polarized interface bonding [35]. The polarized interface bonds form interface dipole, which will shift the energy bands on both sides of the interface and change the SBH. Then, there is a competition between

MIGS and interface bonding. If MIGS is large enough, the electrostatic potential from interface bonds or the metal–semiconductor electronegativity difference would be screened efficiently and strong Fermi-level pinning is formed. This is the situation that has been demonstrated for metal on covalent or small-gap semiconductors (Si, Ge, and GaAs) [36]. However, high-*k* dielectric materials are mostly large-bandgap ionic semiconductors or insulators, where the decay length of MIGS is rather short. It is expected that interface bonding instead of MIGS would play a dominant role in deciding the SBH of metal gate/high-*k* oxide interfaces.

To understand the formation mechanism of SBH at metal gate/high-*k* dielectrics, one must measure the effective work function of metal gates ($\Phi_{m,eff}$) or Schottky barrier height accurately. As mentioned above, a widely used method is to extract the values of effective work function from the relations of flatband voltage versus equivalent oxide thickness (EOT) for metal gate/high-*k* dielectrics/Si capacitors [37]. In this capacitance-based method, appropriate interfacial structures at the high-*k* dielectrics/Si interfaces and the corresponding charge distributions should be carefully considered [38]. Alternatively, p-type SBHs for metal gate/dielectric interfaces can be directly measured by X-ray photoemission method, which can accurately determine the barrier height between metal/high-*k* oxide interfaces.

We found that the measured p-type SBHs for Ni/ZrO_2 interfaces either depend on the initial structure and chemical composition of the ZrO_2 surface or depend on the interface structure. A possible explanation for this interface structure-dependent SBH at Ni/ZrO_2 interfaces is that the local interface dipole is created mainly by charge transfer between the metal and the oxide. This will be discussed first in the framework of MIGS model the next. Then, a detailed consideration based on first-principles calculations will be given.

In the conventional MIGS model for the formation of Schottky barrier height at metal/semiconductor interfaces, charge neutrality level serves as the “Fermi level” on the semiconductor side to align the energy bands at the interface. If the energy position of CNL in the semiconductor is higher than that of Fermi level in the metal, electrons will flow from the semiconductor into the metal and vice versa. To offer a simple scheme to predict SBH from the bulk properties of the two components (metal and semiconductor) only, with no need for the detailed interface information, one often assumes CNL as a bulk property with no relation to the surface states of the semiconductor. However, if one wants to characterize the Schottky barrier height in the framework of MIGS model only, one would better admit that CNL includes both intrinsic and extrinsic parts [39]. For Schottky barriers, the semiconductor CNL may depend on the geometry of the interface, which has been demonstrated widely in various metal/semiconductor interfaces. It was proposed to distinguish the intrinsic and extrinsic charge neutrality levels: the intrinsic one is associated with an ideal, broad, and structureless metal-induced density of states, while the extrinsic one depends on the specific characteristics of each interface. Obviously, intrinsic CNL depends only on the bulk properties of the semiconductor. It is associated with the branch points of the complex band structure of the semiconductor, and can be evaluated from the calculation of actual complex band structures. The extrinsic CNL depends on the specific characteristics of each interface, and thus different surface

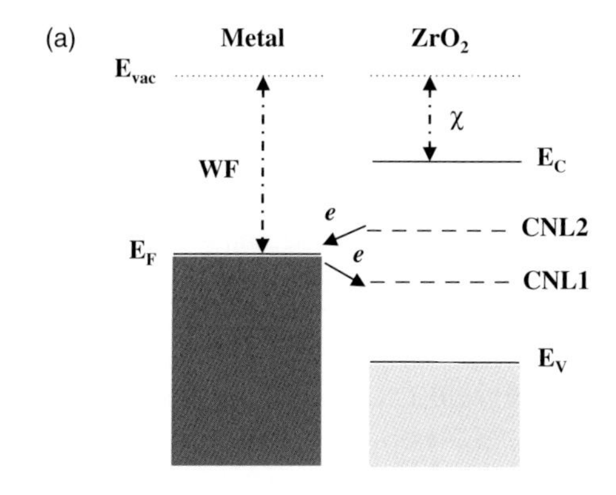

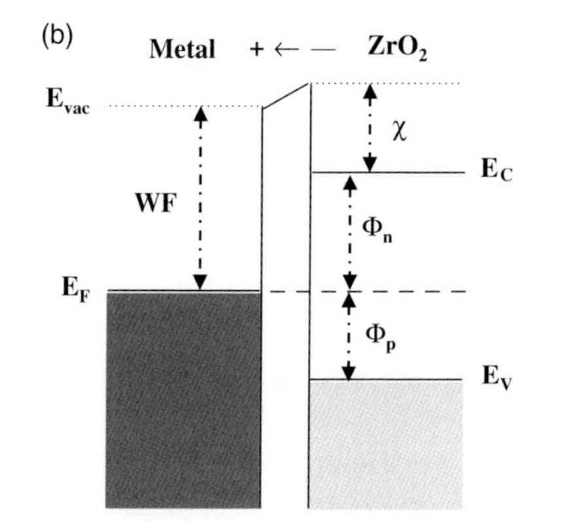

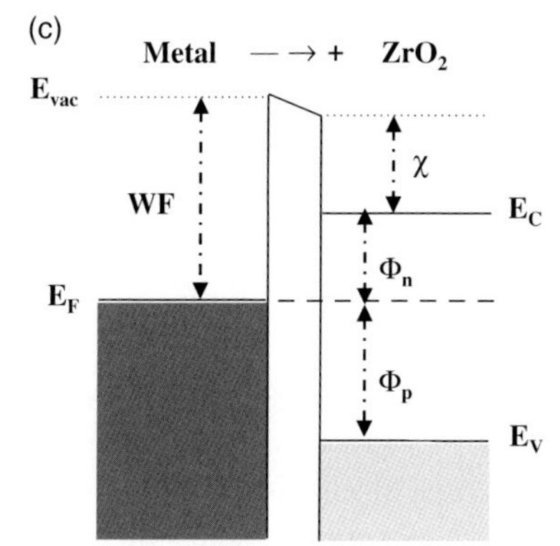

Figure 9.8 Energy band diagram for (a) separated Ni(001) surface and ZrO_2(001) surfaces with different surface status; (b) Ni–YSZ interface (oxygen rich); and (c) Ni–S–YSZ interface (oxygen deficient). WF is the vacuum work function of Ni (001) with a value of 5.22 eV, χ denotes the electronegativity of ZrO_2 with an estimated value of 2.5 eV, and Φ_p and Φ_n denote the p- and n-type SBH.

status will strongly modify the energy position of the charge neutrality level. For narrow bandgap semiconductors, such as Si and GaAs, the allowed range of modification for CNL is within several tenths eV. Superficially, the variation range of CNL for Si or GaAs is restricted by its bandgap. The physical mechanism behind it is the strong MIGS screening in such narrow bandgap semiconductors and intrinsic CNL dominates the formation of Schottky barrier for these semiconductors. However, for wide-bandgap semiconductors or insulators, the variation range of CNL modification can be as large as several eV. Without the consideration of the influence of specific interface characteristics, appropriate conclusions cannot be drawn. Metal gate/high-*k* dielectric interfaces belong to the latter situation.

Here, we use Ni/ZrO_2 interface as an example to explain this theory. Figure 9.8a shows the isolated energy diagrams of metal and oxide immediately before they combine to form interfaces. Figure 9.8b and c illustrate the different charge transfer at the two kinds of Ni/ZrO_2 interfaces, oxygen-rich (Ni-YSZ) and oxygen-deficient

(Ni-S-YSZ) interfaces. Charge neutrality level, defined as the energy above which the surface states are empty for a neutral surface, serves as "Fermi level" for oxide surfaces. Generally, the energy positions of the metal Fermi level and the oxide CNL are not at the same height, which will incur charge transfer after the metal and oxide contact to form interface.

According to the above discussion, the charge neutrality level not only depends on the bulk properties of the material (intrinsic CNL) but also is strongly related to the surface status of the material (extrinsic CNL). Therefore, it is expected that the ideal YSZ(001) surface and the defect-rich YSZ(001) surface have totally different energy positions of "Fermi levels" or CNL, with CNL1 for the former and CNL2 for the latter, as shown in Figure 9.8a. For Ni–YSZ interface, the energy position of CNL1 is lower than that of the metal Fermi level, which leads electrons to transfer from metal side to oxide side when the two sides combine to form interface. When equilibrium is reached, the energy band diagram is shown in Figure 9.8b, where an interface dipole forms and points toward the metal side. For Ni–S–YSZ interface, the charge transfer is in an inverse direction, from oxide to metal. Thus, the interface dipole points toward the oxide side, as shown in Figure 9.8c. It is the different interface dipole that shifts the energy bands on the two sides and induces the variation in band alignment.

9.4 Chemical Tuning of Band Alignments for Metal Gate/High-*k* Oxide Interfaces

In an actual metal oxide semiconductor field-effect transistor (MOSFET) device, it is the effective work function of metal gate ($\Phi_{m,eff}$), the energy position of metal Fermi level in the Si channel, that determines the effective confinement of carriers and the gate threshold voltage. The integration of metal gate with high-*k* gate dielectric requires the effective metal work functions ($\Phi_{m,eff}$) to be within ± 0.1 eV of the Si valence and conduction band edges for p- and n-channel MOSFETs, respectively [40]. However, to find two metals with suitable work functions and to integrate them with current semiconductor technology remain a challenge. In addition, the selection of metal gate materials is not straightforward because the effective work function of a metal depends on the underlying gate dielectrics and could differ appreciably from the metal work function in vacuum ($\Phi_{m,vac}$) [41]. In order to identify the right metal gate material, understanding and controlling the interfaces between metal gate electrode and high-κ dielectric at the atomic scale are essential.

In our research, we propose a method for atomic-level chemical tuning of band alignment for metal gate on high-*k* dielectric, which aims to provide not only a practical way of modifying the band alignments for metal gate/high-*k* oxide interfaces to satisfy the engineering requirement for the metal gate technology but also a fundamental understanding of the metal–dielectric interfaces at atomic scale [42].

Instead of growing the metal directly on oxide, we proposed to include a layer of heterovalent metal m (m = Au, Pt, Ru, Mo, Al, V, Zr, Ti, and W) between them. We consider two models, one with a full monolayer of heterovalent atoms (Figure 9.9a) and the other with half monolayer (Figure 9.9b) between an oxygen-terminated

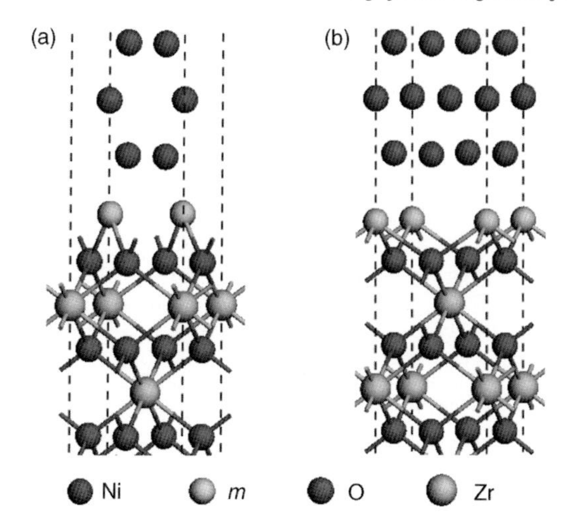

Figure 9.9 Supercells for the Ni-m-ZrO_2 interfaces, (a) with one monolayer metal m (m = Ni, V, and Al) and (b) with half monolayer metal m (m = Au, Pt, Ni, Ru, Mo, Al, V, Zr, and W). The interface is formed using *c*-ZrO_2(001) and fcc Ni(001) surfaces, with either half or one monolayer of heterovalent metal (m) between them.

ZrO_2(001) surface and Ni(001). Due to different electronegativity of the heterovalent metal atoms, the interlayer is expected to change the interface electric dipole moment and result in changes in the band alignment at the metal–oxide interfaces. Chemical tuning of Schottky barrier height and effective metal work function can be achieved by using different metals or different coverage for the interlayer.

The p-type Schottky barrier height (p-SBH) Φ_p was determined using the standard bulk-plus-lineup approach [43], with the average electrostatic potential at the ion core (V_{core}) in the "bulk" region as reference energy,

$$\Phi_p = \Delta E_b + \Delta V \tag{9.5}$$

where ΔE_b is the difference between the Fermi energy of Ni and the energy of the VBM of oxide, each measured relative to V_{core} of the corresponding "bulk" ions, and ΔV is the lineup of V_{core} through the interface.

The average electrostatic potential V_{core} of Zr ions in the oxide and that of Ni atoms in the Ni-m-ZrO_2 interfaces show that V_{core} of the Ni ions near the interface was perturbed by the interface dipole, but quickly recovered its value in bulk region. On the oxide side, V_{core} for Zr ions has converged to its bulk value at the second Zr layer, which is consistent with the short decay length of MIGS in ZrO_2. ΔV was evaluated from the difference in V_{core} of the Zr and Ni ions in their respective bulk regions.

The p-SBHs (Φ_p) were calculated for various Ni-m-ZrO_2 interfaces using the calculated $\Delta V$and the quasi-particle [44], and spin–orbital-corrected ΔE_b. The n-SBH (Φ_n) was derived from $\Phi_n = E_g - \Phi_p$, where E_g is the energy gap of the dielectric. The experimental bandgap of 5.82 eV was used here instead of the DFT-GGA result because of the well-known underestimation of the latter. The effective metal work function ($\Phi_{m,eff}$) on oxide was evaluated using the formula $\Phi_{m,eff} = \Phi_n - \text{CBO} + \text{EA}$,

Table 9.1 Interface metal coverage (θ), electronegativity of metal atom m (χ, Mulliken scale, in eV), work function of metal m in vacuum ($\Phi_{m,vac}$, in eV), Mulliken charge (Q_m, in e), n-SBHs (Φ_n, in eV) for Ni-m-ZrO_2 interfaces, and effective metal work functions ($\Phi_{m,eff}$, in eV) for Ni-m-ZrO_2-Si capacitors.

m	θ	χ	$\Phi_{m,vac}$	Q_m	Φ_n	$\Phi_{m,eff}$
Au	0.50	5.77	5.10	0.16	4.62	6.92
Pt	0.50	5.60	5.65	0.16	3.84	6.14
Ni	0.50	4.40	5.15	0.37	2.76	5.06
Ru	0.50	4.50	4.71	0.27	2.76	5.06
Mo	0.50	3.90	4.60	0.51	2.38	4.68
Al	0.50	3.23	4.28	1.06	2.18	4.48
V	0.50	3.60	4.30	0.69	2.09	4.39
Zr	0.50	3.64	4.05	1.01	1.96	4.26
Ti	0.50	3.45	4.33	0.80	1.95	4.25
W	0.50	4.40	4.55	0.15	1.80	4.10
Ni	1	4.40	5.15	0.24	3.63	5.93
V	1	3.60	4.30	0.44	2.65	4.95
Al	1	3.23	4.28	0.63	1.82	4.12

where CBO is the conduction band offset between the oxide and the Si substrate (1.75 eV), and EA is the electron affinity of Si (4.05 eV).

The calculated n-SBHs and effective metal work functions are listed in Table 9.1 for the Ni-m-ZrO_2 interfaces with various interlayer metals. It can be seen that a tunability as wide as 2.8 eV for Φ_n ($\Phi_{m,eff}$) can be achieved for Ni-m-ZrO_2 interfaces by simply introducing half or one monolayer of heterovalent metal m between Ni and ZrO_2. Furthermore, we found that n-SBHs and effective metal work functions follow a crude chemical tuning trend: for a given metal m coverage (half or one monolayer), the n-SBH (effective work function) increases linearly with the electronegativity (χ) of the interlayer metal atom m. The simple linear relationship between $\Phi_{m,eff}$ and χ is obeyed by most transition metals except Al and W. The exceptional case of Al is probably due to its simple metal character (s and p valence electrons), in contrast to the transition metals (including d valence electrons). But the reason is not clear for the exceptional behavior of tungsten.

The above linear chemical tuning trend is certainly due to the localized interfacial dipole formed by different interfacial chemical bonds. Because of the short decay length (0.9 Å) of MIGS in ZrO_2, the localized interface dipole formed by interfacial polarized bonds plays a dominating role compared to MIGS in the determination of Schottky barrier heights for the metal–dielectric oxide interfaces. We may regard the interface region as a large "molecule" that connects the Ni bulk reservoir on the one side and ZrO_2 bulk reservoir on the other side, so that the Ni-m-ZrO_2 interface takes the form of (Ni-Bulk)−Ni−m−O−(ZrO_2-bulk), with −Ni−m−O− being the interface specific region, as shown in Figure 9.10. The interface dipole comprises two parts, one from the ionic m−O bonds and the other from the positive charged metal layer and its image charge on the metal (Ni) side. The former raises all energy levels

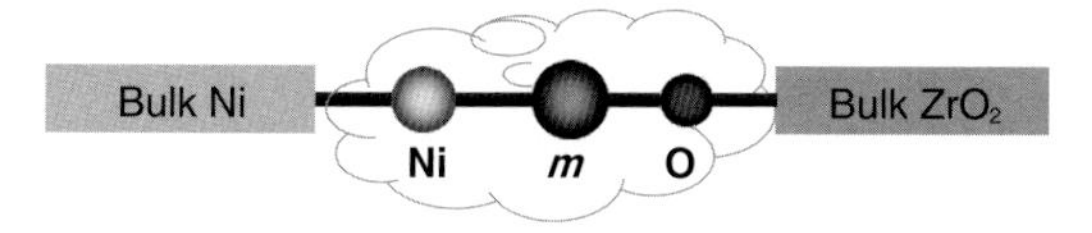

Figure 9.10 Carton of Ni-m-ZrO_2 interfaces illustrating the interface bonding.

on the oxide side with respect to the values in the metal, thus increasing n-SBH, while the latter decreases n-SBH. With decreasing electronegativity of m, both types of dipole increase. But n-SBH decreases indicating that the net interface dipole is pointing from the metal to the oxide, and the dipole formed by the m cation layer and its image charge plays the dominant role in the determination of SBH. This may be the basis for observed chemical tuning trend. The above argument is based only on a phenomenal electrostatic model. We note that without a detailed knowledge of the charge redistribution profile, and the microscopic dielectric constant at the interface, the shift of SBH cannot be derived exactly from electrostatic models, although such shift has been obtained self-consistently in our first-principles calculations.

Experimental studies have confirmed our predictions above. The vapor deposition of metal film on well-defined oxide surfaces under the clean conditions of ultrahigh vacuum (UHV) provides a controlled method for studying fundamental details concerning metal/oxide interfaces [45]. To experimentally demonstrate the chemical tuning effects of SBH between Ni and ZrO_2 interfaces, one stringent requirement is to keep the submonolayer of heterovalent metal at the Ni–ZrO_2 interface in a two-dimensional (2D) geometry instead of aggregating to form 3D clusters. Whether the submonolayer of metal can "wet" the oxide surface or not depends on the competition of adatom–oxide interaction and intraadatom interaction. If the former is strong enough, all the metal adatoms can have bonds with the oxide surface and a 2D geometry is formed; otherwise, 3D cluster geometry is preferred for metal adatoms.

Aluminum has high affinities for oxygen (i.e., large negative heat of metal oxide formation, ~550 kJ per mol O). Thus, it is expected that the adsorption energy is high for the adsorption of aluminum on ZrO_2(001) surface, which is terminated by a layer of oxygen ions. Actually, for submonolayer of adatoms, the adsorption energies are determined by the strength of local chemical bonds formed at the interface, Al–O bonds for the case of submonolayer Al on ZrO_2 (001) surface. First-principles calculations were carried out to calculate the adsorption of 0.5 ML Al on *c*-ZrO_2(001) surfaces. The results show that Al–O bonding is rather strong, with the adsorption energy of 8.89 eV/Al-adatom or 10.71 J/m^2. Notice that the reported work of separation for the Nb–Al_2O_3 interface [46], which is known to be a system with an extremely good adherence, is 9.80 J/m^2. It is thus safe to expect that 0.5 ML of Al can "wet" the ZrO_2(001) surface and be appropriately included between Ni thin film and ZrO_2 substrate.

Figures 9.11 and 9.12 show the survey spectra for Ni-Al(0.5 ML)–YSZ(001) and Ni–YSZ(001) stacks, respectively. The small peak around 118.0 eV corresponds to the Al 2s, as indicated by the arrow in Figure 9.11. The peak adjacent to Al 2s in the lower energy of 111.0 eV is for Ni 3s, as shown in the same energy position in Figure 9.12. The intensity of Al 2s peak is so weak that it almost pushes to the resolution limit of

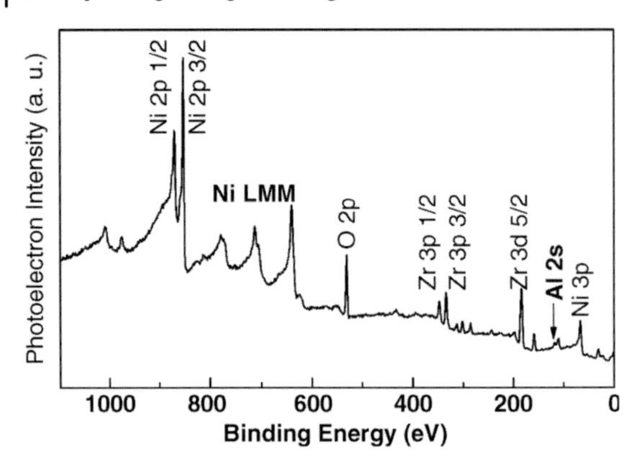

Figure 9.11 Survey spectra for 4 nm thick Ni-Al (0.5 ML)-YSZ(001) interfaces.

XPS instrument. Therefore, the chemical state (metallic or oxidized) of the interfacial Al atoms cannot be identified from the XPS spectra. The weak intensity of Al 2s spectra is due to the small amount of Al and the attenuation of the signal through the upper Ni layer. For a comparison, the survey spectra for Ni–YSZ(001) interface without interfacial layer is shown in Figure 9.12. The two spectra are identical except the additional Al 2s spectra for Ni/Al(0.5 ML)/ZrO_2 stacks in Figure 9.11. The existence of Al at the interface is obvious.

The p-type Schottky barrier height was evaluated by the core level-based method. One recalls that the energy difference between the energy position of Zr $3d_{5/2}$ peak and the leading edge of the valence bands ($\Delta E = E_{\mathrm{Zr\ 3d5/2}} - E_{\mathrm{VBM}}$, 179.62 ± 0.05 eV in this case of Ni on ZrO_2 substrate) can be viewed as a bulk property with no relation to the surface status or the overlayer on the ZrO_2 surface. Then, the p(n)-type Schottky barrier height $\Phi_p(\Phi_n)$ can be obtained using the simple equations

$$\Phi_{\mathrm{p}} = (E_{\mathrm{Zr\ 3d5/2}})_{\mathrm{Ni/ZrO_2}} - \Delta E \tag{9.6}$$

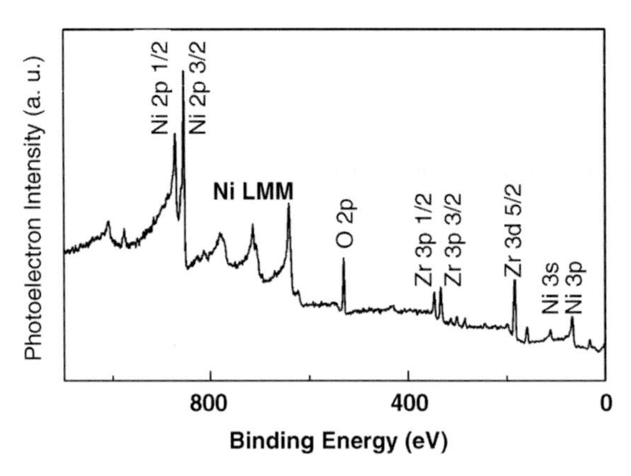

Figure 9.12 Survey spectra for 4 nm thick Ni on YSZ(001) interfaces.

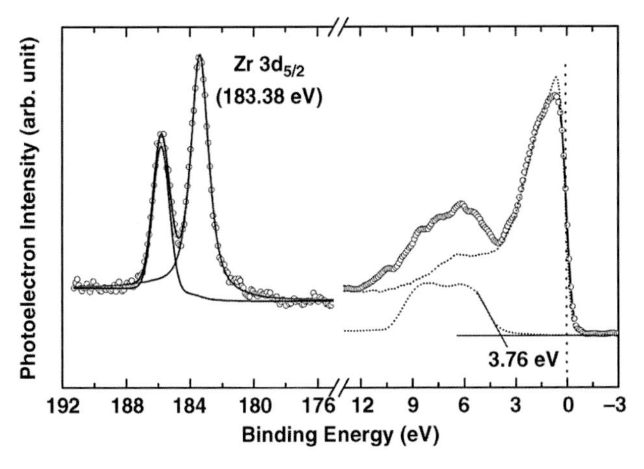

Figure 9.13 XPS spectra for Zr 3d and valence band for 4 nm thick Ni-Al (0.5 ML)-YSZ(001) interfaces. The dotted lines show the valence bands for Ni bulk and YSZ(001) substrate, respectively. The Fermi level is at energy zero.

and

$$\Phi_n = E_g - \Phi_p \tag{9.7}$$

where $(E_{Zr\ 3d5/2})_{Ni/ZrO_2}$ is the peak position of Zr 3d core-level spectra in Ni/m/ZrO_2 stack and E_g is the bandgap of ZrO_2 (5.80 eV for the ZrO_2 sample in our experiment).

Figure 9.13 shows the Zr 3d and valence band spectra for the 4 nm Ni-Al(0.5 ML)–YSZ(001) interfaces. The dotted lines show the valence bands for Ni bulk and YSZ (001) substrate, respectively. The Fermi level is at energy zero. One can see the valence band spectra for Ni bulk is slightly sharper than that for Ni ultrathin film on YSZ substrate. This is due to the small contribution from the interfacial Al layer. The p-type SBH was evaluated to be 3.76 ± 0.05 eV and was checked by decomposing the valence band of the stack to that of Ni and ZrO_2 components, as shown in Figure 9.13. The inclusion of half monolayer of Al between Ni and ZrO_2 dramatically changes the band alignment between metal and oxide, from 2.60 eV for Ni/ZrO_2 interface to 3.76 eV for Ni/Al(0.5 ML)/ZrO_2 interface. Furthermore, the measured SBH (3.76 eV) for Ni/Al (0.5 ML)/ZrO_2 interface is in good agreement with the calculated value (3.62 eV). Considering the accuracy of the experimental and calculation methods (both around 0.1 eV), this consistency is rather satisfying and serves as a strong evidence that the band alignment between high-*k* dielectric and metal gate can be tuned chemically by including only half monolayer of heterovalent metal at the interface.

Table 9.2 summarizes the experimental and calculated values of SBHs for Ni–ZrO_2(001) interfaces with different interfacial chemical compositions and interface coverage. For Ni–ZrO_2(001) interface with no heterovalent interfacial metal, the experimental SBH takes a value between that of one monolayer coverage of Ni and that of half monolayer. One possible reason is that Ni–Ni(0.5 ML)–ZrO_2 and Ni-Ni (1 ML)–ZrO_2 have similar interface formation energies so that the actual interface structure takes a coverage between 0.5 and 1. So the resultant SBH takes an average.

Table 9.2 Comparison of calculated (DFT-GGA) and XPS experimental values of Schottky barrier heights for Ni-*m*-ZrO_2 (m = Ni or Al) interfaces.

Structure	Coverage	Method	Φ_p (eV)	Φ_n (eV)
	0.50	DFT-GGA	3.02	2.76
Ni-ZrO_2	1	DFT-GGA	2.17	3.63
	0.50–1	XPS	2.60	3.20
	0.50	DFT-GGA	3.62	2.18
Ni-Al-ZrO_2	1	DFT-GGA	4.98	1.82
	~0.50	XPS	3.76	2.04

For the chemical tuning effects, the most important feature here is that the inclusion of a layer of low work function metal (Al) between high work function metal (Ni) and gate dielectric could tune the effective work function of metal gate or SBH up to ~1.2 eV. This experimentally and theoretically proven tuning ability is of great importance. It provides a practical way of modifying the band alignments for metal gate/high-κ oxide interfaces to satisfy the engineering requirement for the metal gate technology but also a fundamental understanding of the metal gate/high-*k* dielectric interfaces on the atomic scale.

9.5 Summary and Discussion

Although intensive studies have been carried out on the alternative high-*k* gate dielectric and metal gate materials, there are still many challenges in their application for the future generations of Si-based MOSFET technologies, for example, growth of high-quality interfaces between Si and high-*k* oxides and tuning the magnitude of the effective work functions for metal gate on high-*k* dielectrics up to a range of 1.1 eV. In order to identify the right alternative high-*k* gate dielectric and metal gate materials, it is essential to understand and control the two interfaces, the front interface between metal gate electrode and high-*k* dielectric and the back one between high-*k* dielectric and Si substrate, on the atomic scale.

The main objective of this chapter is to review how the atomic-level interface structures affect the properties of metal gate/high-*k* dielectric oxide/Si stacks or how the macroscopic properties (e.g., band alignments for metal gate/high-*k* dielectric oxide and high-*k* dielectric/Si interfaces) are related to the electronic structures on small length scale (e.g., interface bonding). A combined approach of (i) characterization techniques, for example, transmission electron microscopy (TEM) and XPS, and (ii) first-principles calculations based on DFT is applied. Characterization studies (TEM and XPS) provide a complete picture for the metal/oxide and oxide/Si interfaces on the atomic scale. The first-principles calculations would offer explanations for the related experimental results and provide insight into the physical mechanism behind the formation of metal/high-*k* oxide and high-*k* oxide/Si interfaces.

It is recognized that high-k oxide in amorphous or in single crystalline can both be used as the alternative dielectric material to replace SiO_2. In this chapter, we focus the study on crystalline materials and choose the single-crystalline ZrO_2, one of the most promising candidates, as the prototype of high-k dielectric material because it has a well-defined surface and simple cubic structure, which leads us to focus on interface effects without perturbation from the bulk contribution. In addition, crystalline structure is very appropriate for our first-principles calculations where supercell structures are used to mimic the infinite crystalline material.

We readily acknowledge that our study is based only on prototypes of metal gate and high-k oxide materials, for example, crystalline interfaces at metal gate/high-k oxide/silicon stacks, which represent ideal situations compared to the much more complicated situations in real metal gate and high-k material applications. Future work would be needed to identify to what extent the results (e.g., the dependence of band alignments on interface structures) obtained here can be applied to amorphous metal gate and high-k oxide materials. For example, as the first approximation, the nonpolar interfaces between the most abundant high-k oxide and metal surfaces, such as Ni(111) and $ZrO_2(111)$ surface, could be studied. Our preliminary studies [47] have shown that nonpolar Ni(111)/$HfO_2(111)$ interface has much weaker interface tension (at least five times smaller) than that for polar Ni(001)/$HfO_2(001)$ interface, which already indicates different types of interface interaction for polar and nonpolar metal gate/high-k interfaces. This would have significant effects on the interface electronic structures and properties. Further studies to explore the chemical tuning effects on nonpolar metal gate/high-k interfaces are highly recommended.

Further studies may also extend the present work for defect-free interfaces to study the interfaces incorporating defects, for example, oxygen vacancies [48, 49], hydrogen-related defects [50], and their complex. It is well known that high-k dielectric material have much more defects than SiO_2 and the defects at high-k oxide/silicon interfaces heavily degrade the channel mobility [51]. But the physical mechanism behind this mobility degrading is not clear yet. In addition, to couple the oxygen vacancies at high-k oxide and preclude the substantial formation of interfacial silicon oxide layer, nitridation of high-k oxide is being intensively studied [52, 53]. Therefore, further studies to explore the role of incorporation of nitrogen at high-k dielectric/silicon interfaces would also be an interesting subject.

References

1 Franciosi, A. and van de Walle, C. (1996) Heterojunction band offset engineering. *Surf. Sci. Rep.*, **25**, 1.

2 Peressi, M., Binggeli, N., and Baldereschi, A. (1998) Band engineering at interfaces: theory and numerical experiments. *J. Phys. D Appl. Phys.*, **31**, 1273.

3 Kraut, E.A., Grant, R.W., Waldrop, J.R., and Kowalczyk, S.P. (1980) Precise determination of the valence-band edge in X-ray photoemission spectra: application to measurement of semiconductor interface potentials. *Phys. Rev. Lett.*, **44**, 1620.

4 Chambers, S.A., Liang, Y., Yu, Z., Droopad, R., and Ramdani, J. (2001)

Band offset and structure of $SrTiO_3$/Si (001) heterojunctions. *J. Vac. Sci. Technol. A*, **19**, 934.

5 Wang, S.J., Huan, A.C.H., Foo, Y.L., Chai, J.W., Pan, J.S., Li, Q., Dong, Y.F., Feng, Y.P., and Ong, C.K. (2004) Energy-band alignments at ZrO_2/Si, SiGe, and Ge interfaces. *Appl. Phys. Lett.*, **85**, 4418.

6 Wang, S.J., Dong, Y.F., Huan, A.C.H., Feng, Y.P., and Ong, C.K. (2005) The epitaxial ZrO_2 on silicon as alternative gate dielectric: film growth, characterization and electronic structure calculations. *Mater. Sci. Eng. B*, **118**, 122.

7 Miyazaki, S. (2001) Photoemission study of energy-band alignments and gap-state density distributions for high-*k* gate dielectrics. *J. Vac. Sci. Technol. B*, **19**, 2212.

8 Puthenkovilakam, R. and Chang, J.P. (2004) Valence band structure and band alignment at the ZrO_2/Si interface. *Appl. Phys. Lett.*, **84**, 1353.

9 Oshima, M., Toyoda, S., Okumura, T., Okabayashi, J., Kumigashira, H., Ono, K., Niwa, M., Usuda, K., and Hirashita, N. (2003) Chemistry and band offsets of HfO_2 thin films for gate insulators. *Appl. Phys. Lett.*, **83**, 2172.

10 Jin, H., Kang, H.J., Lee, S.W., Lee, Y.S., and Cho, M.-H. (2005) Band alignment in ultrathin Hf-Al-O/Si interfaces. *Appl. Phys. Lett.*, **87**, 212902.

11 Först, C.J., Ashman, C.R., Schwarz, K., and Blochl, P.E. (2004) The interface between silicon and a high-*k* oxide. *Nature (London)*, **427**, 53.

12 Kraut, E.A., Grant, R.W., Waldrop, J.R., and Kowalczyk, S.P. (1983) Semiconductor core-level to valence-band maximum binding-energy differences: precise determination by X-ray photoelectron spectroscopy. *Phys. Rev. B*, **28**, 1965.

13 Li, Q., Wang, S.J., Li, K.B., Huan, A.C.H., Chai, J.W., Pan, J.S., and Ong, C.K. (2004) Photoemission study of energy-band alignment for RuO_x/HfO_2/Si system. *Appl. Phys. Lett.*, **85**, 6155.

14 Yu, E.T., Croke, E.T., McGill, T.C., and Miles, R.H. (1990) Measurement of the valence-band offset in strained Si/Ge (100) heterojunctions by X-ray photoelectron spectroscopy. *Appl. Phys. Lett.*, **56**, 569.

15 Chambers, S.A., Liang, Y., Yu, Z., Droopad, R., Ramdani, J., and Eisenbeiser, K. (2000) Band discontinuities at epitaxial $SrTiO_3$/Si(001) heterojunctions. *Appl. Phys. Lett.*, **77**, 1662.

16 Puthenkovilakam, R., Carter, E.A., and Chang, J.P. (2004) Electronic structure of ZrO_2/Si and $ZrSiO_4$/Si interfaces. *Phys. Rev. B*, **69**, 155329.

17 Fulton, C.C., Lucovsky, G., and Nemanich, R.J. (2004) Process-dependent band structure changes of transition-metal (Ti,Zr,Hf) oxides on Si (100). *Appl. Phys. Lett.*, **84**, 580.

18 Fiorentini, V. and Gulleri, G. (2002) Theoretical evaluation of zirconia and hafnia as gate oxides for Si microelectronics. *Phys. Rev. Lett.*, **89**, 266101.

19 Zhang, X., Demkov, A.A., Li, H., Hu, X., and Wei, Y. (2003) Atomic and electronic structure of the SiÕSrTiO3 interface. *Phys. Rev. B*, **68** 125323.

20 Peacock, P.W. and Robertson, J. (2004) Bonding, energies, and band offsets of Si-ZrO_2 and HfO_2 gate oxide interfaces. *Phys. Rev. Lett.*, **92**, 057601.

21 Fissel, A., Dabrowski, J., and Osten, H.J. (2002) Photoemission and *ab initio* theoretical study of interface and film formation during epitaxial growth and annealing of praseodymium oxide on Si (001). *J. Appl. Phys.*, **91**, 8986.

22 Peacock, P.W. and Robertson, J. (2003) Structure, bonding, and band offsets of (100)$SrTiO_3$–silicon interfaces. *Appl. Phys. Lett.*, **83**, 5497.

23 Dong, Y.F., Feng, Y.P., Wang, S.J., and Huan, A.C.H. (2005) First-principles study of ZrO_2/Si interfaces: energetics and band offsets. *Phys. Rev. B*, **72**, 045327.

24 Satpathy, S. and Martin, R.M. (1989) Energetics and valence-band offset of the CaF2/Si insulator-on semiconductor interface. *Phys. Rev. B*, **39**, 8494.

25 Sze, S.M. (2001) *Semiconductor Devices: Physics and Technology*, John Wiley & Sons, Inc.

26 Brillson, L.J. (1979) Chemical mechanisms of Schottky barrier formation. *J. Vac. Sci. Technol.*, **16**, 1137.

27 Schottky, W., Stormer, R., and Waibel, F. (1931) Rectifying Action at the Boundary

between CuProus Oxide and Applied Metal Electrodes. *Z. Hoch Frequentztechnik*, **37**, 162.

28 Louie, S.G., Chelikowsky, J.R., and Cohen, M.L. (1977) Ionicity and the theory of Schottky barriers. *Phys. Rev. B*, **15**, 2154.

29 Tersoff, J. (1984) Schottky barrier heights and the continuum of gap states. *Phys. Rev. Lett.*, **52**, 465.

30 Robertson, J. and Chen, C.W. (1999) Schottky barrier heights of tantalum oxide, barium strontium titanate, lead titanate, and strontium bismuth tantalite. *Appl. Phys. Lett.*, **74**, 1168.

31 Mönch, W. (1987) Role of virtual gap states and defects in metal–semiconductor contacts. *Phys. Rev. Lett.*, **58**, 1260.

32 Tersoff, J. (1984) Theory of semiconductor heterojunctions: the role of quantum dipoles. *Phys. Rev. B*, **30**, 4874.

33 Demkov, A.A., Fonseca, L.R.C., Verret, E., Tomfohr, J., and Sankey, O.F. (2005) Complex band structure and the band alignment problem at the Si–high-*k* dielectric interface. *Phys. Rev. B*, **71**, 195306.

34 Tung, R.T. (2000) Chemical bonding and Fermi level pinning at metal–semiconductor interfaces. *Phys. Rev. Lett.*, **84**, 6078.

35 Chang, S., Brillson, L.J., Kime, Y.J., Rioux, D.S., Kirchner, P.D., Pettit, G.D., and Woodal, J.M. (1990) Orientation-dependent chemistry and Schottky-barrier formation at metal–GaAs interfaces. *Phys. Rev. Lett.*, **64**, 2551.

36 Mönch, W. (1999) Barrier heights of real Schottky contacts explained by metal-induced gap states and lateral inhomogeneities. *J. Vac. Sci. Technol. B*, **17**, 1867.

37 Jha, R., Gurganos, J., Kim, Y.H., Choi, R., Jack, L., and Veena, M. (2004) A capacitance-based methodology for work function extraction of metals on high-κ. *IEEE Electron. Device Lett.*, **25**, 420.

38 Yang, H., Son, Y., Baek, S., Hwanga, H., Lim, H., and Jung, H.S. (2005) Ti gate compatible with atomic-layer-deposited HfO_2 for n-type metal-oxide-semiconductor devices. *Appl. Phys. Lett.*, **86**, 092107.

39 Flores, F., Muñoz, A., and Durán, J.C. (1989) Semiconductor interface formation: the role of the induced density of interface state. *Appl. Surf. Sci.*, **41/42**, 144.

40 De, I., Johri, D., Srivastava, A., and Osburn, C.M. (2000) Impact of gate workfunction on device performance at the 50nm technology node. *Solid State Electron.*, **44**, 1077.

41 Knizhnik, A.A., Iskandarova, I.M., Bagatur'yants, A.A., Potapkin, B.V., and Fonseca, L.R.C. (2005) Impact of oxygen on the work functions of Mo in vacuum and on ZrO_2. *J. Appl. Phys.*, **97**, 064911.

42 Dong, Y.F., Wang, S.J., Feng, Y.P., and Huan, A.C.H. (2006) Chemical tuning of band alignments for metal gate/high-κ oxide interfaces. *Phys. Rev. B*, **73**, 045302.

43 Peressi, M., Binggeli, N., and Baldereschi, A. (1998) Band engineering at interfaces: theory and numerical experiments. *J. Phys. D*, **31**, 1273.

44 Králik, B., Chang, E.K., and Louie, S.G. (1998) Structural properties and quasiparticle band structure of zirconia. *Phys. Rev. B*, **57**, 7027.

45 Henrich, V.E. and Cox, P.A. (1994) *The Surface Science of Metal Oxides*, Cambridge University Press, Cambridge.

46 Zhukovskii, Y.F., Kotomin, E.A., Jacobs, P.W.M., and Stoneham, A.M. (2000) *Ab initio* modeling of metal adhesion on oxide surfaces with defects. *Phys. Rev. Lett.*, **84**, 1256.

47 Li, Q., Dong, Y.F., Wang, S.J., Huan, A.C.H., Feng, Y.P., and Ong, C.K. (2006) Evolution of Schottky barrier heights at Ni/HfO_2 interfaces. *Appl. Phys. Lett.*, **88**, 222102.

48 Foster, A.S., Lopez Gejo, F., Shluger, A.L., and Nieminen, R.M. (2002) Vacancy and interstitial defects in hafnia. *Phys. Rev. B*, **65**, 174117.

49 Feng, Y.P., Lim, A.T.L., and Li, M.F. (2005) Negative-U property of oxygen vacancy in cubic HfO_2. *Appl. Phys. Lett.*, **87**, 062105.

50 Kang, J.G., Lee, E.-C., Chang, K.J., and Jin, Y.-G. (2004) H-related defect complexes in HfO_2: a model for positive fixed charge defects. *Appl. Phys. Lett.*, **84**, 3894.

51 Robertson, J. (2004) High dielectric constant oxides. *Eur. Phys. J. Appl. Phys.*, **28**, 265.

52 Sayan, S., Nguyen, N.V., and Ehrstein, J. (2005) Effect of nitrogen on band alignment in HfSiON gate dielectrics. *Appl. Phys. Lett.*, **87**, 212905.

53 Toyoda, S., Okabayashi, J., Takahashi, H., Oshima, M., Lee, D.I., Sun, S., Sun, S., Pianetta, P.A., Ando, T., and Fukuda, S. (2005) Nitrogen doping and thermal stability in $HfSiO_xN_y$ studied by photoemission and X-ray absorption spectroscopy. *Appl. Phys. Lett.*, **87**, 182908.

10
Interfacial Dipole Effects on High-*k* Gate Stacks

Li Qiang Zhu

10.1
Introduction

The thickness of SiO_2-based gate dielectrics needs to be reduced down to below ~1 nm as the feature size of MOS transistor is scaled down to smaller dimensions. However, as the gate dielectric thicknesses decrease, a number of key dielectric parameters vital for high-performance device operation, which can be ignored in bigger devices, must be taken into account, such as gate leakage current, oxide breakdown, channel mobility, and so on [1]. Moreover, the continual device shrinking also makes SiO_2 run out of atoms for further scaling. At the very beginning, the consideration of replacing SiO_2 by high-*k* gate dielectrics was thought to be a direct solution to overcome the intrinsic problems in polysilicon/SiO_2/Si substrate gate stacks. Since high-*k* insulators can be grown physically thicker at the same (or thinner) equivalent electrical oxide thickness (EOT), the usage of high-*k* dielectrics would offer a significant advantages in gate leakage reduction while maintaining other dielectric characteristics similar to SiO_2, thus offering many potential advantages. However, it is becoming evident that only replacing the gate insulator, without changing electrode materials at the same time, may not be sufficient for the scaled devices with the improved device performance. When introducing the polysilicon/high-*k* dielectric stacks as a candidate for polysilicon/SiO_2 stacks, several problems might be encountered for high-performance CMOS applications.

First, polysilicon gate electrodes tend to get degraded. Negative effects of polysilicon gate depletion on the device performance will get more and more significant with decreasing dielectric thickness, resulting in the reduced inversion layer charge density and the degradation of transconductance [2]. In fact, polysilicon depletion existing in polysilicon gate electrode, which is typically ~0.2–0.5 nm region, is becoming a larger and larger fraction of the equivalent oxide thickness (EOT) and will limit the gate capacitance the stacks can obtain [3]. Such effects cannot be ignored for sub-2 nm gate dielectric thickness. With the presence of the depletion region, the voltage drop across the gate dielectrics will be reduced because part of the gate voltage

High-k Gate Dielectrics for CMOS Technology, First Edition. Edited by Gang He and Zhaoqi Sun.

will be dropped across the depletion region within the polysilicon gate, which means that the effective gate voltage will be reduced.

It is reported that there is decreased ratio between the inversion gate capacitance (C_{inv}) and the accumulation gate capacitance (C_{acc}), that is, C_{inv}/C_{acc}, as the gate length is scaled down from 0.4 to 0.12 μm and the potential drop (ΔV_p) in polysilicon gate increases as the gate length is scaled down [4]. Furthermore, with decrease in the gate length, the C_{inv}/C_{acc} value is rapidly decreased and the potential drop (ΔV_p) is rapidly increased as the MOS device size scales down due to the rapidly increased depletion effects, implying that the device performance deterioration by the polysilicon depletion effects will get more significant as the devices continue to scale down. The schematic illustrations of the depletion region for MOS devices as a function of their size are shown in Figure 10.1. Due to the polysilicon gate depletion effects, an additional potential drop will be experienced at the depletion region.

Second, as a method to decrease the negative effects on the device performance, the polysilicon gate should be doped heavily to reduce the polysilicon depletion effects and the relative high resistance in the short gate electrodes. The doping level in the polysilicon electrodes will arrive at around 10^{20} dopant atoms per cm^3. Heavily doped polysilicon is also the most widely used gate material in Si MOSFETs. However, even at the high doping concentrations, the gate depletion is not entirely avoidable and more parasitic charge centers will be introduced at the polysilicon/gate

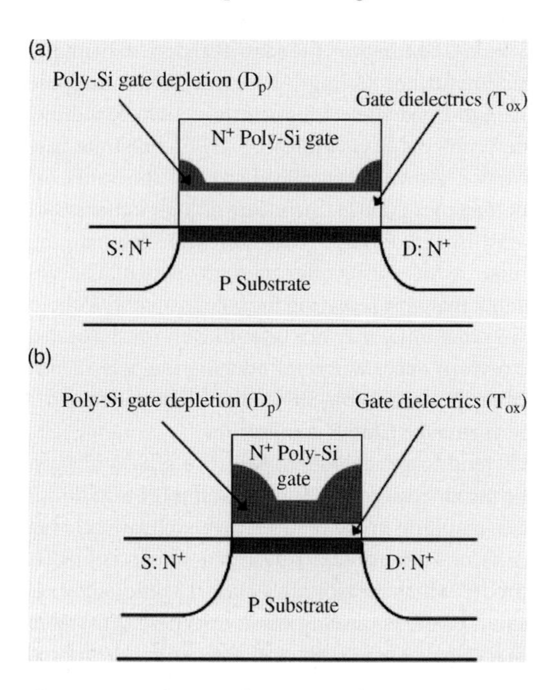

Figure 10.1 Schematic illustrations of the depletion region for polysilicon-gated MOS devices as a function of MOS device size. (a) MOS with bigger size and (b) MOS with scaled size. Effective depletion width is getting wider for the scaled MOS size, leading to an additional potential drop, ΔV_p.

dielectric interface. Due to such parasitic charges in the gate stacks, the charge carriers in the channel will be affected by the remote Coulomb scattering, degrading the device performance in terms of drain current because of the mobility degradation. These effects are getting more and more significant with the thinner dielectrics. Moreover, there is relatively increased resistance of the gate electrode due to the reduction in geometry for scaling requirements. Furthermore, polysilicon gate electrode is usually annealed at high temperature ($> 1000\,^{\circ}$C) in order to activate dopants in gate electrode for the heavily doped n^+ (or p^+) gate CMOS process. The annealing process is also limited by the diffusion of dopants, especially for p-MOS where boron is very likely to penetrate from the gate, through the ultra thin gate oxide barrier, and into the silicon channel, which will degrade the performance of the transistor. Moreover, the step from conventional SiO_2 to high-k materials will also induce the critical requirements of thermal stabilities for high-k materials in contact with the polysilicon gate electrodes, making it a less favorable material for continual use as a gate electrode. The concerns will also arise when the polysilicon gate is not doped heavily enough. On the other hand, to prevent the dopants in polysilicon gate from diffusing into the thin dielectrics, implant dose with a relatively low energy is preferred, resulting in the nonuniform, graded doping profiles in the polysilicon gate due to the implant and constrained annealing conditions, while maintaining the required source/drain junction depths [5]. This nonetheless gives rise to an insufficient doping in the polysilicon gate near the oxide; therefore, the contribution of the depletion capacitance will get significant.

Third, high-k dielectrics and polysilicon are incompatible due to the Fermi-level pinning occurring at the polysilicon/high-k interface because of which the effective work function of a polysilicon gate electrode could not be modulated easily for polysilicon/high-k gate stacks by doping level in the polysilicon gates [6]. Therefore, the threshold voltages in polysilicon/high-k CMOS devices, especially for p-MOS, are always unacceptably high, resulting in degraded transistor performance.

Researchers have tried to solve the problem of polysilicon depletion effects and reduce their negative effects on the device performance. Improved polysilicon processes have been developed. For example, polysilicon gate depletion effects have been reduced by using pulsed excimer laser annealing (ELA), without degrading the gate oxides, which is promising for applications in advanced Si CMOS technologies [7]. These new processing techniques present new opportunities for polysilicon gate electrodes. However, from the point of view of the materials' intrinsic limitations for the polysilicon gate electrodes, we must find new alternative conductive materials to act as gate electrodes.

10.2 Metal Gate Consideration

To overcome the fundamental limitations in polysilicon gate with shrinking MOS device feature size, such as the unavoidable polysilicon gate depletion, degradation of the channel performance by penetrating dopants from the polysilicon gate, the relatively high gate resistance when the polysilicon gate is not doped heavily enough, and the Fermi-level pinning effects experienced at polysilicon/high-*k* interfaces, metallic gate electrode is deemed as the most effective solution. Metallic gate electrode shows no depletion effects, shows a reduced sheet resistance, and is free of additional doping process because of the existence of the abundant electrons in the metal. Therefore, the applications of the metallic gate electrode avoid the dopants penetrating the gate dielectrics and the channels, which is expected in the polysilicon gate process. Moreover, metal gate electrodes will have potential advantages in improving the channel mobility than the polysilicon gate electrode because the metal gate electrode illustrates more effective screening effects for the high-*k* surface optical phonons from coupling to the channel under inversion conditions [8]. In terms of the processing temperature, there is an obvious advantage that the application of the metal gate may decrease the required thermal budget, while the high-temperature thermal step is unavoidable in the case of the polysilicon gate stacks because of the requirements of the dopant activation annealing. Therefore, the application of the metal gate electrode opens a new door for the advanced CMOS technologies.

Though metallic gates possess several advantages, the technical innovation from the polysilicon gate process to metallic gate process also faces several fundamental challenges. The polysilicon gate possesses one advantage in high-performance MOS transistor that the desired threshold voltage (V_{th}) for both n-MOS and p-MOS applications can be easily realized by the way of heavily n^+-type and p^+-type doped polysilicon gate, obtaining the work function of ~4.1 eV and ~5.0 eV for n-MOS and p-MOS application, respectively. However, in metal gate technology, obtaining desirable threshold voltage is not so direct due to the lack of certain metal possessing both the high work function and the low work function. Therefore, effectively controlling suitable gate work function is quite important to maintain the low threshold voltage. Moreover, the thermal/chemical interface stability and processing compatibility with the underlying dielectric are also the main issues in the metal gate applications. Up to now, several attempts have been made for the application of metal electrodes in CMOS flow.

The first direct approach was thought to be the introduction of a single metal with mid-gap work function into CMOS flow for both n-MOS and p-MOS applications, as shown in Figure 10.2a, where the gate Fermi level is located at the mid-gap position of

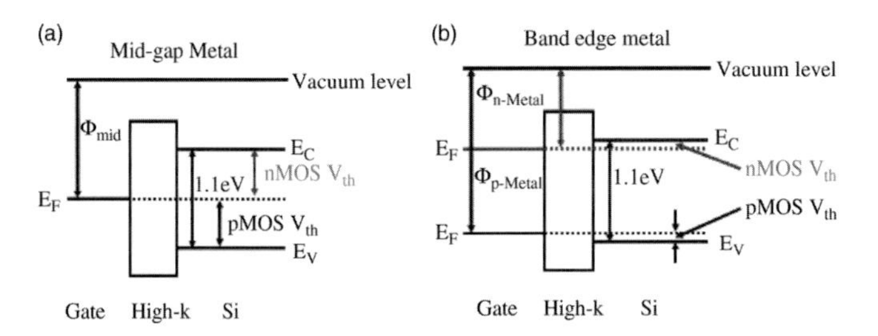

Figure 10.2 Simplified band alignments for n-MOS and p-MOS devices using (a) mid-gap metal gate and (b) band edge metal gates.

Si substrate [9]. Under this consideration, the employed mid-gap metal will result in the symmetrical V_{th} values for n-MOS and p-MOS. The simple CMOS processing scheme is expected by using only one mask and one gate electrode. However, there are limitations. Since the bandgap of Si substrate is $\sim$1.1 eV, the high threshold voltage of $\sim$0.5 V is expected for any mid-gap metal on Si for both n-MOS and p-MOS. Such big threshold voltage imposes big concerns on high-performance devices, making it difficult to turn on the device, unsuitable for low-voltage operations.

The second scheme would be the introduction of the band edge metal gate, in other words, dual metal dual work function electrodes, one for n-MOS and one for p-MOS devices, respectively, as shown in Figure 10.2b. In order to obtain high-performance CMOS logic applications on bulk Si, the candidate metal gate stacks require an n^+-type metal and a p^+-type metal with the right work functions on high-*k* dielectrics of $\sim$4.1 and $\sim$5.0 eV, respectively, which means that each metal should line up their Fermi level favorably with the conduction band (CB) and valence band (VB) of Si substrate, respectively, resulting in the threshold voltage of below $\sim$0.2 V for both n-MOS and p-MOS applications.

One of the main challenges in metal/high-*k* CMOS technologies may be the correct selection of gate work functions for both n-MOS and p-MOS applications. Their compatibilities with CMOS process should also be considered. The first idea comes from the relationship of the metal work function as a function of atomic number. On the one hand, the low work function metals are generally the reactive elements and tend to be unstable on high-*k* dielectrics due to their high free energy of formation [10], while the high work function metals will suffer from process difficulties and adhesion issues on dielectrics because they are not reactive with the gate dielectrics and might result in the discontinuous films. On the other hand, most of the metal gate electrodes suffer from high-temperature instabilities on high-*k* dielectrics resulting in a degraded interface. Therefore, the changes in work function and equivalent oxide thickness may be unavoidable, which poses significant concerns for the high-performance CMOS applications. Another difficulty in the realization of the metal vacuum work function is the modifications of the work function by the Fermi-level pinning effects at metal/high-*k* interface, especially for p-MOS devices. Moreover, it is also reported that there are dipole effects on high-*k* stacks at both metal/high-*k* interface and high-*k*/SiO_2 (or Si) interface, adding another uncertainty to the effective work function controlling in MOS applications. Furthermore, there are many more metal candidates for n-MOS devices; however, the metal candidates for p-MOS devices are relatively few. Only limited elements (Ir, Pt, etc.) in the periodic table have the required work function above 5.2 eV, if allowing enough space for the additional work function lowering because of the Fermi-level pinning effects.

To date, different schemes have been considered to obtain different work functions. The first consideration is to modulate the gate vacuum work function directly. Since single metal electrode in the periodic table could not give us processing-friendly properties, the disadvantages in the single metal might be released and further avoided by alloying other metals or nonmetal elements. At the same time, the advantages of each element in the alloyed metal might be magnified. Thereafter, the compound metal materials have been widely studied and integrated with CMOS flow. Such compound metal materials include metal fully silicide materials (NiSi, TiSi, and so on, in other words, FUSI), conducting metal nitrides (WN_x, TiN_x, MoN_x, HfN_x, etc.), alloyed metals (Hf_xRu_{1-x}, Ta_xMo_{1-x}, etc.), nitride alloyed metals, alloyed FUSI, and so on. Though the work function modulation by these processes is very important in CMOS industry, a detailed discussion of the metal gate electrode is out of the scope of this chapter.

10.3 Interfacial Dipole Effects in High-*k* Gate Stacks

The previous section simply overviews the applications of metal gate electrode and the ways to modulate the gate work function through the vacuum work function tuning process. However, the metal gate effective work function will be different from their vacuum work function when they are integrated with the CMOS devices, resulting in deviation of the threshold voltage from the desired value. It has been found that metal gates with additional thermal annealing at high temperature will result in mid-gap work functions for almost all the metal candidates because of the Fermi-level pining effects occurring at gate/high-*k* interface. Therefore, special attention should be paid to the Fermi-level pinning effects in order to select suitable metal gate and the high-*k* materials for advanced CMOS technologies. Moreover, when the high-*k* gate stacks are proceeded under the high-temperature thermal budgets, there might be the separation of the positive and negative charges across the gate/high-*k* interface or across the high-*k*/substrate interfaces, that is, the interfacial dipole effects, which results in the modification of the high-*k* band alignments and therefore the deviation of the gate effective work function from the desired value. Though the existence of the interfacial dipole effects affords us a new possibility to modify the threshold voltage, the control of the interfacial dipole is not easy. First, the physical microscopic origins for such dipole effects are not clear yet, and there are several models to explain these effects. At the same time, each model is not universal enough to explain all the examples. Therefore, the dipole model is still needed to be upgraded. Second, the observation of the dipole effects is not so direct, which might affect the establishment of the interfacial dipole model in the high-*k* gate stacks. In this section, an overview of the typical interfacial dipole model developed in the literature will be made. Then, the developed methods to observe the interfacial dipole effects will be discussed.

10.3.1 Modification of the Gate Work Function by the Interfacial Dipole

When a metal (or a semiconductor) and a dielectric form an interface, the barrier height between the metal Fermi level (or semiconductor conduction band) and the dielectric conduction band depends on the metal work function (or semiconductor electron affinity) and dielectric electron affinity in the Schottky limit model with no charge transfer (Figure 10.3) [11, 12]. However, in general, this rule is not obeyed, and

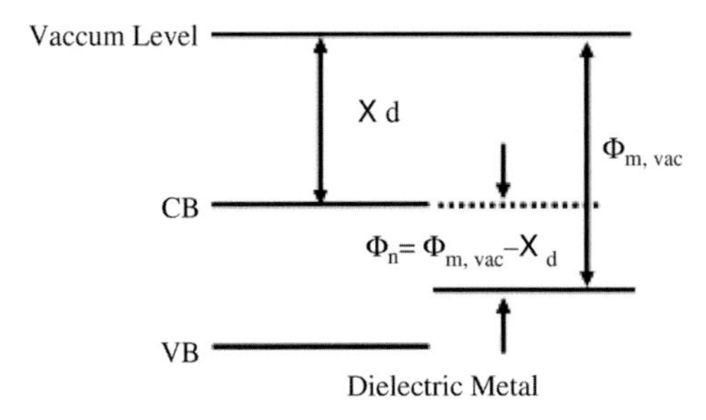

Figure 10.3 Schottky limit model: the band barrier between metal and dielectric depends on the metal work function and dielectric electron affinity.

the charge transfer is always expected at such interfaces. During the research conductivities performed on high-*k* materials for CMOS applications, charge transfer at either gate/high-*k* interface or high-*k*/substrate interface has been observed. Such charge transfer is well known and termed as the interfacial dipole effect, which not only depends on the dielectrics and the gate materials but also depends on the process the gate stacks have been proceeded with. The presence of the interfacial dipole in high-*k* gate stacks, located at either gate/high-*k* interface or high-*k*/substrate interface, will modulate the band alignments in high-*k* gate stacks. Therefore, the effective work functions of metal gates on high-*k* gate dielectrics will deviate from their vacuum ones, which poses significant concerns in terms of the controlling of threshold voltage in MOSFETs.

On the one hand, the presence of the interfacial dipole in high-*k* stacks will be detrimental to threshold voltage stabilities. However, on the other hand, we should take the advantage of such dipole effects by getting a good understanding of them. The presence of the interface dipole at either gate/high-*k* interface or high-*k*/substrate interface should positively offer us new possibilities for the modification of gate effective work functions in CMOS applications. Figure 10.4 illustrates the simplified band diagrams modulated by the interfacial dipole located at either gate/high-*k* (Figure 10.4a) or high-*k*/substrate interface (Figure 10.4b). The dipole

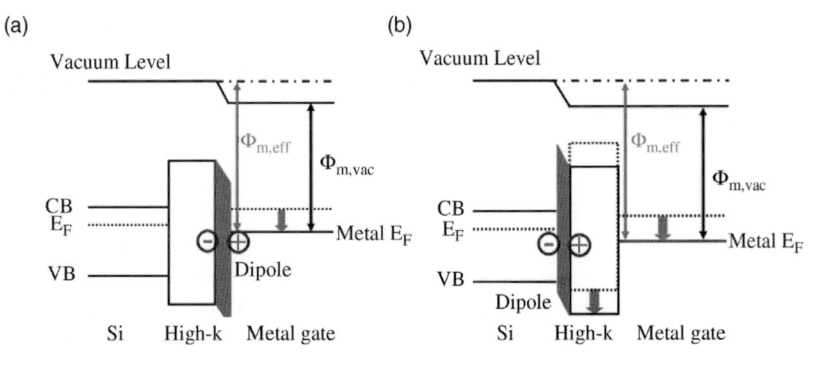

Figure 10.4 Simplified band diagrams modulated by the interfacial dipole located at (a) gate/high-*k* interface and (b) high-*k*/substrate interface.

orientations depend on the neighboring materials and the process the gate stacks have been proceeded with. The dipole located at metal/dielectric interface is normally due to the presence of the intrinsic interface states that normally will drive the gate effective work function to the mid-gap position, that is, the Fermi-level pinning. Such pinning effects normally will get significant for high work function metals [13, 14]. As an effective method to overcome such effects, researchers have put a capping layer on high-*k* to release the Fermi-level pinning effects. In some cases, the atoms in the capping layer will also diffuse to the high-*k*/substrate interfaces, forming a dipole layer at high-*k*/substrate interfaces, which also helps to pull the gate effective work function to either n-edge or p-edge work function. More directly, researchers also integrate the dipole atoms with high-*k* layer or just insert the dipole atoms at high-*k*/substrate interface. Doing so, the desired gate effective work function both at n-edge and at p-edge could be realized.

10.3.2
Fermi-Level Pinning Effects at Gate/High-*k* Interfaces

Fermi-level pinning effects are experienced by the way of separation of the positive charge and negative charge across gate/high-*k* interface by different mechanisms that have been observed at both polysilicon/high-*k* interface and metal gate/high-*k* interface, and both will result in the modification of the high-*k* band alignments on substrate and therefore the gate effective work functions.

For polysilicon/SiO_2 gate stacks, the threshold voltage could be changed easily by the polysilicon doping. However, in polysilicon/high-*k* gate stacks, the effective work function of a polysilicon gate could not be altered easily. Hobbs *et al.* studied the high-frequency *C–V* curves for HfO_2 and SiO_2 gated with polysilicon electrode for both p-MOS and n-MOS [15]. Through the polysilicon doping, the polysilicon Fermi-level shifts from the conduction band edge to the valence band edge, separated by ~1.1 eV, the flatband voltage (V_{fb}) difference between p^+-polysilicon and n^+-polysilicon gated SiO_2 gate stacks is found to be ~1 V. However, for n^+- or p^+-gated HfO_2 MOS structure, the *C–V* curves shift only by ~0.13 V and 0.24 V for n-MOS and p-MOS, respectively. Such difference is mainly related to the Fermi-level pinning effects experienced at polysilicon/Hf-based high-*k* interface. In general, V_{th} shifts reported for polysilicon-gated CMOS devices by using Hf-based high-*k* materials follow the same trend. For p-MOS applications, V_{th} is shifted in the negative direction by as much as ~0.6 V compared to that of SiO_2, while for n-MOS applications, V_{th} is closer to that of SiO_2.

Shiraishi *et al.* developed a model that very well explained the Fermi-level pinning effects occurring at polysilicon/HfO_2 interface [13]. For polysilicon/HfO_2 interface, there is partial oxidation by formation of Si–O–Si bonds by O atoms when pulled out from HfO_2. Accordingly, two additional electrons are generated after one O atom is pulled out. If these electrons remain inside HfO_2, they will occupy the V_o level in HfO_2. However, since HfO_2 is in contact with the polysilicon gate, electrons have to be transferred into the gate electrode depending on where the polysilicon gate Fermi level is located.

The total energy gain (ΔG_{total}) in this process can be given by the following reaction equation [14]:

$$(HfO_2) + \frac{1}{2}Si \rightarrow Vo^{2+} + 2e^- + (HfO_2) + \frac{1}{2}SiO_2 + \Delta G_{total} \tag{10.1}$$

Therefore, Fermi level is elevated to the polysilicon pinning position (E_p) where $\Delta G_{total} = 0$ is satisfied. Oxygen vacancies related to Fermi-level pinning effects can also be found in other p-metal gate electrode on HfO_2. In the case of p-metal/HfO_2/Si stacks, high-temperature annealing would block O transfer from high-*k* to electrode because of the inert properties for almost all the p-metals. Therefore, oxygen has to be transferred from high-*k* to the substrate and subsequently electron transfer to the electrode has to take place. Since Equation 10.1 will be approved without including the thermodynamic parameters of p-metal, the p-metal pinning effect is inevitable in p-metal gated Hf-based high-*k* gate stacks after high-temperature treatment.

Except for the oxygen vacancies related to Fermi-level pinning effects for Hf-based dielectrics, the presence of the Si–Hf bonds at polysilicon/HfO_2 interface is also thought to be the possible origin for the Fermi-level pinning effects [6, 15, 16]. On the basis of the element electronegativity considerations, the interfacial dipole created by the charge transfer between Si and Hf decreases the effective work function of p^+-polysilicon gate and pins the Fermi level at the position below the polysilicon conduction band. Similar effects can also be found on polysilicon/Al_2O_3 system [6]. Since Al is a p-type dopant in Si, it is likely that the interfacial Al atoms behave like dopants and that there is no pinning due to Si–Al bonds, while Fermi-level pinning effects occur only due to charge transfer across Si–O–Al bonds.

From the above discussions, two methods to decrease the Fermi-level pinning effects have been considered. Since the Fermi-level pinning in gate/HfO_2 stacks is related to the oxygen vacancy production in HfO_2, the first consideration to decrease the Fermi-level pinning effects is to modify the Hf-based dielectric composite intentionally and prevent the formation of oxygen vacancies in HfO_2. The introduction of Si into Hf-based dielectric layer is considered [13]. In $HfSiO_x$ films, two types of oxygen coexist. One is an O atom surrounded by Si atoms, while the other is surrounded by Hf atoms. It is easier for the Hf-surrounded O to form the oxygen vacancies than that for the Si-surrounded one. Therefore, the decreased oxygen vacancy density is expected for $HfSiO_x$ films with increased Si percentages, which would help decrease the Fermi-level pinning effects. Second, the Fermi-level pinning effect is also related to the oxygen and charge diffusion across the interface. Therefore, the second scheme to decrease the Fermi-level pinning effects is the introduction of a barrier layer at gate/high-*k* interface, such as SiO_2 and SiN_x. However, in these stacks, the oxygen transfer to substrate remains and the electron transfer to gate still exists. Moreover, the introduction of low-*k* layer in high-*k* stacks also imposes significant concerns for the EOT scalability. The introduction of scaled Al_2O_3 or AlN cap layer sandwiched between polysilicon gate and high-*k* stack would get more effective in improving the V_{th} value. On the one hand, the introduction of

Al will introduce an interfacial dipole at high-*k*/bottom interface compensating the Fermi-level pinning [17]. On the other hand, the introduction of Si−O−Al bond configurations at top gate/high-*k* interface would also compensate Si−Hf bond-related charge transfer, releasing the Fermi-level pinning effects.

Although these efforts continue, the resulting V_{th} are always not competitive with or better than SiO_2 or SiON/polysilicon stacks for high-performance applications. The difference in V_{th} improvements between different capping layers might originate from the differences in the ability of the cap layer to block O atom diffusion from the Hf-related high-*k* dielectrics to the polysilicon gate. The Fermi-level pinning effects cannot be suppressed without blocking the oxygen transfer both to the electrode and to the substrate. Another method would be the introduction of another interfacial dipole at gate/high-*k* interface or high-*k*/substrate interface, which will compensate the Fermi-level shifting upward for p-metal; therefore, the desired work function would be obtained.

10.3.3
Micromodels for the Interfacial Dipole in High-*k* Stacks

It has been shown that putting an additional capping layer at metal gate/high-*k* interface or inserting an interlayer at bottom high-*k*/Si interface by introducing an interfacial dipole can modify the high-*k* band alignments, therefore obtaining the band edge work function suitable for both p-MOS and n-MOS applications [18–20]. In some cases, the introduction of the interfacial dipole is realized by incorporating the dipole atoms directly into high-*k* layers, which results in the dipole formation at high-*k*/substrate interface. Such methods would be desirable for the concerns of the EOT scalability and processing complications for the additional layer incorporation in high-*k* stacks. However, though there have been successful applications of the dipole effects in the gate effective work function modifications, the microphysical origin behind them have not been well established yet. Several possible mechanisms have been proposed both for metal gate/high-*k* interface and for high-*k*/SiO_2 interface, including Fermi-level pinning, oxygen vacancy model, electronegativity and group electronegativity model, area oxygen density model, and so on.

10.3.3.1 Fermi-Level Pinning

Origins of the Fermi-level pinning effects could be classified into at least two groups, the extrinsic and the intrinsic interface gate states.

Extrinsically, the interface between high-*k* and metal is often experienced by a chemical reaction that produces a lot of interface defects or oxygen vacancies inside high *k*. Then, the gate Fermi level will be pinned by charge transfer because of the chemical reactions. The oxygen vacancy formation discussed in the previous section would be extrinsic in origin for Fermi-level pinning effects.

Intrinsically, when high *k* is in contact with a metal, the intrinsic states are presented in the energy gap of the dielectrics. These states are sometimes called "metal-induced gap states" (MIGS) at the metal/semiconductor interface and can be thought either as the dangling bond states of the broken surface bonds dispersed

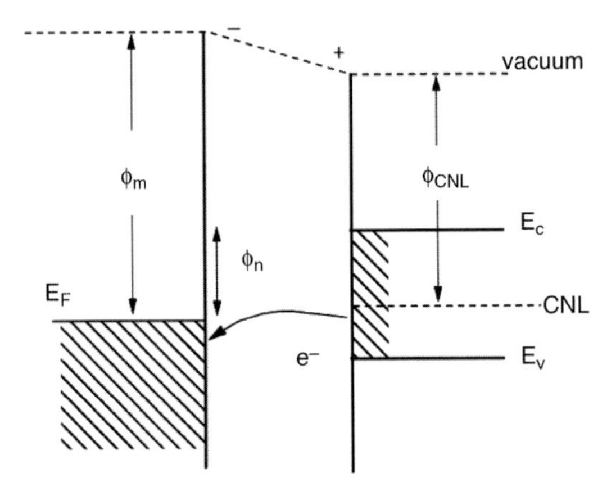

Figure 10.5 Schematic of charge transfer at a Schottky barrier. Reproduced from Ref. [22]. Copyright 2006, Elsevier.

across the bandgap of the semiconductor or as the evanescent states of the metal wave functions continuing into the forbidden energy gap [12]. Moreover, the intrinsic interface states are predominantly donor-like close to the valence band and acceptor-like close to the conduction band [21]. Changes in these interface states will create a dipole, driving the band alignments, and the metal Fermi level $E_{F,m}$ will move toward $E_{CNL,d}$ (charge neutrality level). Therefore, the barrier will deviate from that of Schottky model. As shown in Figure 10.5, the barrier height between metal Fermi level and the dielectric conduction band depends on the pinning parameter and is given by $\Phi_n = S(\Phi_{m,vac} - \Phi_P) + (\Phi_P - \chi_d)$, where S is the pinning parameter and Φ_P is the pinning energy of the semiconductor surface measured from the vacuum level. Therefore, the effective work function $\Phi_{m,eff}$ will differ from the vacuum metal work function $\Phi_{m,vac}$, which is given by $\Phi_{m,eff} = \Phi_P + S(\Phi_{m,vac} - \Phi_P)$. Similarly, there is also charge transfer between the dielectrics and the semiconductors when they form an interface; therefore, the barrier height can be defined as $\Phi_n = (\chi_s - \Phi_{P,s}) - (\chi_d - \Phi_{P,d}) + S(\Phi_{P,s} - \Phi_{P,d})$.

10.3.3.2 **Oxygen Vacancy Model**

Oxygen vacancies are very useful concepts in explaining the interfacial dipole effects in high-k stacks, especially in explaining the high work function metal gate (p-gate) Fermi-level pinning effects. As discussed above, the formation of oxygen vacancies in Hf-based high-k stacks plays an important role in the Fermi-level pinning effects. For the p^+-polysilicon-gated or p-metal-gated high k, the oxidation of Si substrate or the oxidation of polysilicon would result in the reduced high k due to the oxygen out-diffusion, which introduces a neutral oxygen vacancy with two electrons left at the sites where the oxygen is diffused out. Because of the lower Fermi level for the p-gate electrode, the electrons will move out of the sites leaving the positive charged oxygen vacancies (Vo^{2+} or Vo^+) and the Fermi level for the p-gate electrode will shift upward due to the electrons transferring in, which results in decreased gate effective work function.

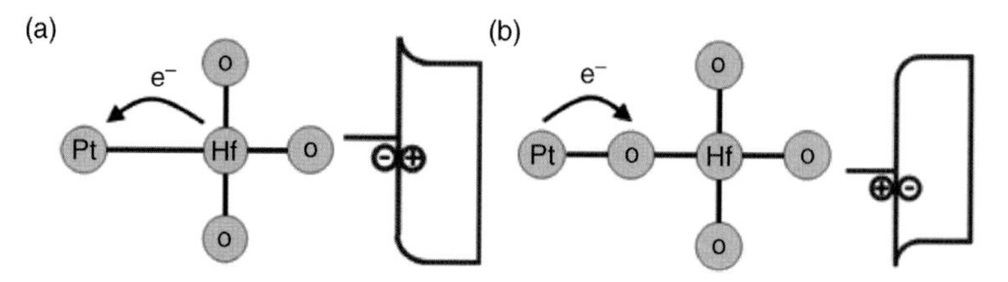

Figure 10.6 Schematic of the interface dipole formation for (a) O-deficient interface: Pt–[V_o]–Hf and (b) O-rich interface: Pt–O–Hf.

The interface oxygen behavior is also important for the interfacial dipole modulation considering the interface bond configurations. The oxygen-rich and oxygen-deficient interface would modulate the work function, obviously. As an effective method, the modulation of the oxygen densities at the interface has been realized through changing the annealing ambience [23]. For the forming gas annealed Pt/HfO_2 stacks, an O-deficient Pt/HfO_2 interface is produced. Due to the fact that Pt is more electronegative than hafnium (~2.28 versus ~1.3), an interface bonding configuration of Pt–Hf or Pt–[V_o]–Hf could result in electron transfer from HfO_2 side to Pt, which drives the Pt Fermi-level shifting upward, therefore decreasing the gate effective work function, as shown in Figure 10.6. For the O_2 ambient annealed Pt/HfO_2 stacks, an O-rich Pt–HfO_2 interface is produced. Because oxygen is more electronegative than Pt (~3.44 versus ~2.28), interface bonding configuration of Pt–O–Hf could result in the electron transfer out from Pt to HfO_2 side, which will increase the gate effective work function.

Except for the oxygen vacancy formation within high *k*, the oxygen-deficient interface at metal/high-*k* interface will also result in the change in the gate effective work function. In order to obtain a high effective work function, putting a capping layer on high-*k* dielectrics to modulate the oxygen density at the interface is an effective solution. Taking the advantages that W can catch oxygen atoms, a W monolayer (ML) introduced at Ru/HfO_2 interface will create a thin WO_x layer, which promotes the formation of Ru–O bonding resulting in a high effective work function [24]. Another reliable solution would be the incorporation of La into HfO_2. The lower V_o densities in HfLaO lead to the decreased amounts of electron transfer in the gate stacks and the partially released Fermi-level pinning effects at metal/high-*k* interfaces [25].

10.3.3.3 Pauling Electronegativity Model

Oxygen vacancy model fails to explain the dipole effects by the incorporation of rare-earth (RE) elements into Hf-based stacks. For example, La substitution of Hf results in negative charged defects La_{Hf}' and the formation of positive charged oxygen vacancy defects: $La_2O_3 = 2La_{Hf}' + 3O_o^x + V_o$ [26]. Assuming that oxygen vacancies will be mobile, there will be charge separation across the dielectrics, which will result in the threshold voltage shifts. Therefore, the same amount of threshold voltage would be expected for different RE elements. However, RE element-dependent threshold voltage shifts are observed in high-*k* stacks.

Since there are bond configuration differences at the interface in high-*k* stacks, the electronegativity (EN) differences among the elements forming the interface bonds are also thought to be the possible origin of the interface dipole effects in high-*k* stacks. This model can well explain the flatband voltage shifts modulated by RE elements in n-MOS applications. When there are rare-earth elements at the HfSiON/SiO_2 interface, they tend to form silicates. Hf–O–Si–N matrix will separate from RE–O–Si or Si–O matrix. Therefore, the interfacial Hf–O–RE configuration is formed across the interface, which results in a charge transfer determining the magnitude of the dipole [27]. The dipole magnitude (μ) is proportional to the charge on the pole (Q) and the separate distance (d) between two poles (+ versus −): $\mu = Qd$. Q is inversely proportional to electronegativity. The larger Qd values will possess a larger dipole strength.

Considering the bond configurations at the interface, a more reliable model to explain the dipole effects is to carefully number the valence differences. For example, since the valences of La and Hf are 3 and 4, respectively, overall dipole moments at the La_2O_3/Si interface and the HfO_2/Si interface will be $2/3\ \mu_{\text{La–O}} - 1/2\ \mu_{\text{Si–O}}$ and $1/2\ \mu_{\text{Hf–O}} - 1/2\ \mu_{\text{Si–O}}$, respectively [28], while for La_2O_3/SiO_2 or HfO_2/SiO_2 interface, an additional dipole moment of Δμ should be considered because of the effects of second nearest-neighbors that are estimated to be the same between the La_2O_3/SiO_2 and the HfO_2/SiO_2 interface because EN value of both La (EN = 1.1) and Hf (EN = 1.3) is much lower than that of O (EN = 3.44).

The application of the group electronegativity concepts improves their effectiveness in explaining the interfacial dipole effects in high-*k* stacks. For the dielectric M_xN_y, the geometric mean electronegativity is defined according to Sanderson criterion [29]:

$$\text{EN}_{\text{mean}} = (\text{EN(M)}^x \text{EN(N)}^y)^{1/(x+y)} \tag{10.2}$$

Therefore, the group electronegativity of Al_2O_3, HfO_2, and MgO is estimated to be ~2.54, ~2.49, and ~2.11, respectively. Since the group electronegativity of HfO_2 is between that of Al_2O_3 and MgO, both p-MOS and n-MOS are realized by capping HfO_2 with Al_2O_3 or MgO, respectively, resulting in the formation of a relative dipole that helps the realization of the positive or negatively shift of the gate effective work function with respect to that of HfO_2/SiO_2 stacks [29]. It is interesting to note that Mg and Hf have the identical electronegativities; however, the Mg-incorporated stacks have resulted in a very different flatband voltage compared to that of Hf-only stacks.

Notwithstanding these achievements, difficulties still remain. Simply speaking, almost all high-*k* materials exhibit the smaller group electronegativities compared to that of SiO_2; therefore, electronegativity model will fail to explain the opposite directions of interfacial dipole at high-*k*/SiO_2 interface between different high-*k* materials.

10.3.3.4 **Area Oxygen Density Model [30]**

As discussed above, oxygen behavior is very important for the gate effective work function in Pt/HfO_2 interface, which modulates the interfacial dipole magnitude significantly. It is also appropriate for high-*k* dielectrics on SiO_2. Oxygen diffusion

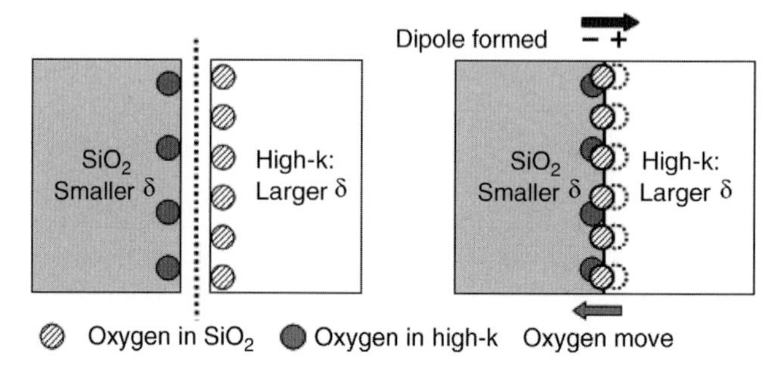

Figure 10.7 Formation of the dipole at the high-k/SiO_2 interface driven by the different oxygen area densities between high-k and SiO_2. The larger δ value of high-k oxide than that of SiO_2 results in the dipole with negative charge on the SiO_2 side and positive charge on the high-k side.

across the interface would well modulate the interfacial dipole at high-k/SiO_2 interface as a result of the charge transfer as long as the oxygen diffusion. Generally speaking, the interface formed by two oxides coming into contact with each other will be in a substable state. When the stacks are activated under annealing process, there will be redistribution of the atoms at the interface region, especially for oxygen atoms. Driven by the interface free energy reduction requirements, there will be the oxygen diffusion across the interface between high-k and SiO_2 due to the different oxygen area density (δ) between high-k dielectrics and SiO_2. Moreover, the movement of oxygen across the interface will result in the charge imbalance and therefore the dipole is formed across the interface, as is illustrated schematically in Figure 10.7.

One of the big achievements for this model is that it successfully explains both the dipole strength and the opposite dipole directions at the high-k/SiO_2 interface. Since the larger the $|\delta_{\text{high } k} - \delta_{SiO_2}|$ value is, the stronger the interfacial dipole is expected, and the sign of $\delta_{\text{high } k} - \delta_{SiO_2}$ will determine the interfacial dipole directions; it is quite important for this model in terms of both engineering and understanding of effective work function tuning in high-k gate stacks.

To date, though several achievements have been made in improving the interfacial dipole models in high-k stacks, the developed models can be explained only in the relatively limited examples. In fact, the dipole formation is a complicated process, not limited to only one mechanism in some cases. So far, though the dipole formation has been successfully applied in the gate effective work function modifications, there still is the debate on the dipole formation mechanisms, and the microorigin of the interfacial dipole formation in high-k stacks is still unclear. More work is obviously needed to establish a universal model to explain the dipole effects in high-k gate stacks.

10.4 Observation of the Interfacial Dipole in High-*k* Stacks

The dipole characterization via capacitance–voltage (C–V) measurements has some limitations, including the difficulties in estimation of V_{FB} shifts due to the traps and interfacial states (D_{it}) at the high-k/SiO_2 interface, the difficulty to differentiate the interfacial dipole effects (a kind of intrinsic property of high-k stacks) from the process-induced extrinsic charges. On the other hand, the reasons of the flatband voltage (V_{FB}) roll-off effects observed in the low EOT region are still not clear yet [31–33]. The possible origins are the charge effects, the dipole strength changes, and so on, which means that the announced dipole models are not well established and need further upgrades. Application of new techniques to observe and study the dipole effects would help us to understand the dipole behavior. By applying X-ray photoelectron spectroscopy (XPS), the direct evidence of the dipole effects in high-k stacks is expected to be obtained [34–38]. Application of the internal photoemission and Kelvin probe based on the analysis of the band offset and Kelvin potential, respectively, would also result in the observation of the interfacial dipole effects. Successful application of such new techniques would help understand and establish the mechanisms and the microscopic origin of the interfacial dipole more clearly.

10.4.1
Flatband Voltage Shifts in Capacitance–Voltage Measurements

For metal/high-k/SiO_2/Si stacks, a schematic model of the possible charge locations and interfacial dipoles is illustrated in Figure 10.8. In electrical characterization, the flatband voltage (V_{FB}) can be expressed as a function of the fixed charges and dipoles presented in MOS stacks as follows:

$$\begin{aligned} V_{FB} &= \frac{\Phi_{ms}}{q} - \frac{\varrho_1 T_{\text{high }k}^2}{2\varepsilon_{\text{high }k}} - \frac{Q_1 T_{\text{high }k}}{\varepsilon_{\text{high }k}} - \frac{\varrho_0 T_{SiO_2}^2}{2\varepsilon_{SiO_2}} \\ &\quad - \frac{Q_0 \text{EOT}}{\varepsilon_{SiO_2}} + \Delta_2 + \Delta_1 + \Delta_0 \end{aligned} \tag{10.3}$$

where Δ_2, Δ_1, and Δ_0 are the interfacial dipole at metal/high-k, high-k/SiO_2, and SiO_2/Si interface, respectively. Generally speaking, the bulk charge density inside

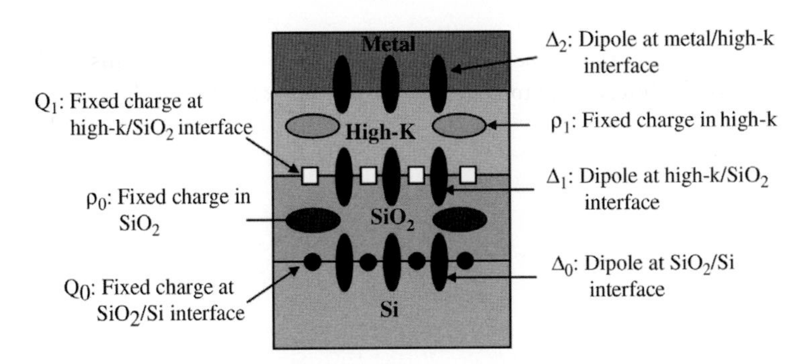

Figure 10.8 Schematic model of possible fixed charges and dipole locations in high-k gate stacks.

high-*k* and SiO_2 is relatively low compared to the interface fixed charges and it is believed that there is no dipole at SiO_2/Si interface; therefore, V_{FB} is simplified to be

$$V_{FB} = \frac{\Phi_{ms}}{q} - \frac{Q_1 T_{high\ k}}{\varepsilon_{high\ k}} - \frac{Q_0 \mathrm{EOT}}{\varepsilon_{SiO_2}} + \Delta_2 + \Delta_1 \tag{10.4}$$

From this expression, a linear relation between V_{FB} and high-*k* thickness has been found. In fact, the interfacial dipole will get saturated for high-*k* thickness above ~1 nm. Therefore, the interfacial dipole-induced V_{FB} shift is independent of the high-*k* thickness, while the charge-induced shift is proportional to the thickness. These two effects could be differentiated from the V_{FB} behavior. After the *C–V* measurements, the flatband voltage is extracted from the *C–V* curves. The V_{FB} versus EOT (or high-*k* thickness) curves would be drawn. The linear relations could be found in almost all of the high-*k* stacks. The slope is related to the interface fixed charges. Through extrapolating the curves to the point with EOT (or $T_{high\ k}$) equal to zero, the gate effective work function could be extracted. The deviation of the metal gate effective work function would be related to the interfacial dipole at the interfaces. Figure 10.9 illustrates V_{FB} versus EOT plot for the fabricated W/HfO_2/SiO_2/Si and W/La_2O_3/SiO_2/Si stacks with W/SiO_2/Si stack as a reference [28]. Through linear extrapolating V_{FB} versus EOT curves to the point with EOT equal to zero, the intercept tells the different ($\Delta_2 + \Delta_1$) values between these two stacks of ~0.36 eV, indicating the relative dipole magnitude of ~0.36 eV between these two stacks.

As shown in Figure 10.8, there are two possible interface dipole locations, that is, metal/high-*k* interface and high-*k*/SiO_2 interface. In order to clarify the possible dipole locations, a specific structure is needed. Here is an example for Au/$HfLaO_x$/SiO_2/Si stacks as illustrated in Figure 10.10 [39]. After Au deposition, no additional annealing is proceeded. From these stacks, the dipole locations could

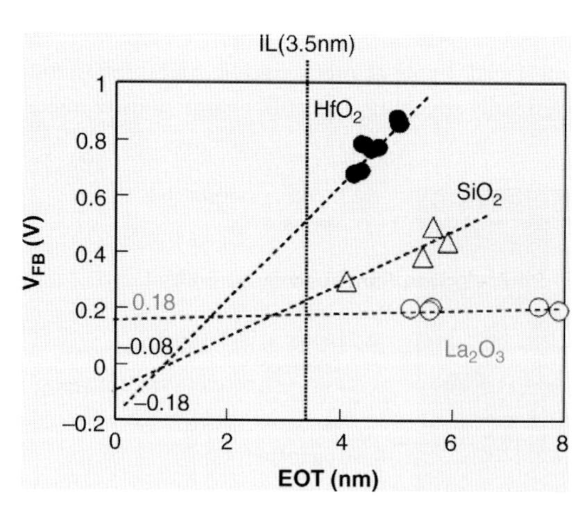

Figure 10.9 V_{FB} versus EOT plot for the fabricated W/HfO_2/SiO_2/Si and W/La_2O_3/SiO_2/Si stacks with W/SiO_2/Si stack as a reference. Reproduced from Ref. [28]. Copyright 2008, Elsevier.

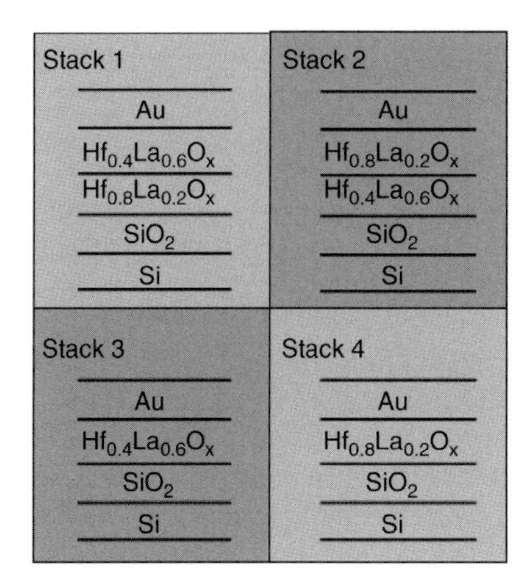

Figure 10.10 Schematic descriptions of MOS capacitors with four kinds of La concentration profiles in $HfLaO_x$ films.

be extracted. If the dipole is located at the $Au/HfLaO_x$ interface, stack 1 and stack 3 should exhibit the same flatband voltage value since there is the same $Au/HfLaO_x$ interface with the same La percentage of 60%. Similarly, stack 2 and stack 4 should also exhibit the same flatband voltage value because of the same $Au/HfLaO_x$ interface with the same La percentage of 20%. However, there is the same flatband voltage value between stacks 1 and 4 and similarly the same flatband voltage value between stacks 2 and 3. Considering the fact that there is the same $HfLaO_x/SiO_2$ interface with the same La percentage of 20% between stack 1 and stack 4 and the same La percentage of 60% between stack 2 and stack 3, it could be concluded that the interfacial dipole is located at the $HfLaO_x/SiO_2$ interface rather than at the $Au/HfLaO_x$ interface.

Though the dipole direction and strength have been estimated typically through *C–V* measurements, other techniques, especially for the physical insights, are still needed to study the interfacial dipoles in detail in terms of the limitations of each technique. The techniques to detect the interfacial dipole effects in high-*k* gate stacks include the photoelectron spectroscopy (PES, XPS, and UPS), internal photoemission (IPE), and Kelvin probe.

10.4.2 Core-Level Binding Energy Shift in Photoelectron Spectroscopy

Photoemission spectroscopy (PES), also known as photoelectron spectroscopy, is a kind of technique to measure the energy of electrons emitted from materials through the photoelectric effects, which can determine the binding energies of electrons in a substance. Depending on the light source, the photoelectron spectroscopy can be divided into XPS and ultraviolet photoelectron spectroscopy (UPS). Since the interfacial dipole at the interface across the heterostructures would result in the relative shift in the core-level binding energies between two materials, the photoelectron spectroscopy should play an important role in studying such effects.

10.4.2.1 **Band Discontinuities and Schottky Barrier Analysis in Heterostructures**

Application of the photoemission spectroscopy has made a substantial progress in understanding the nature of parameters for heterostructures. There are a large number of studies on the band discontinuities in semiconductor junctions or at insulator/semiconductor interfaces.

In the case of the semiconductor band discontinuity modifications, photoelectron spectroscopy was used to study their dependence on interlayer thickness. A homojunction with Al–P double interlayer was stacked between Si substrate and a Si overlayer, as shown in Figure 10.11 [40]. The creation and the magnitude of the band discontinuities were deduced from core-level photoemission spectra of the two sides of the homojunction, that is, Si $2p^{sub}$ versus Si $2p^{over}$. For the two-stacked double layers of 2(Al–P), the shift in Si 2p is ~0.5 eV, while for the two-stacked double layers of 2(P–Al), the shift in Si 2p is ~ -0.5 eV. When the three-stacked double layers are introduced, the shift in Si 2p is saturated, which means that the band discontinuities, that is, the interfacial dipole at Si/Si, have already saturated for the two-stacked double layers.

In 1986, Perfetti *et al.* studied the dipole-induced changes in the band discontinuities at SiO_2/Si interface [41]. They introduced an intralayer of two different materials with very different electronegativity, cesium (Cs) and hydrogen (H), at SiO_2/Si interface. A giant change in the valence band discontinuities (ΔE_v) was observed to shift in the opposite direction when taking the SiO_2/Si interface as a reference. ΔE_v changes from 4.9 eV for the intralayer-free SiO_2/Si interface to 5.15 eV for the Cs intralayer, while to 4.4 eV for the hydrogen intralayer. Such giant intralayer-induced changes in ΔE_v are attributed to the modifications of the interface dipoles, which play a major role in the band lineup modifications.

Photoemission spectroscopy has also been widely applied to determine the Schottky barrier profile in metal/semiconductor interface. Based on photoemission spectroscopy, systematic studies on the initial stage band bending of metal/GaAs (100) stacks as a function of metal coverage have been conducted [42]. When Al is deposited on p-GaAs(110) surface, the Ga 3d or As 3d core-level binding energy will

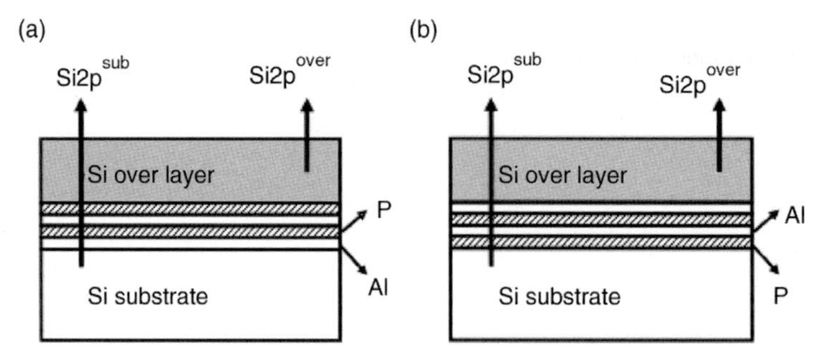

Figure 10.11 Schematic structures for (a) Si 2(Al–P)–Si and (b) Si 2(P–Al)–Si.

shift as a function of the Al coverage at room temperature. When there is 0.001 monolayer of Al deposited on GaAs, both Ga 3d and As 3d peaks start to shift toward the high binding energy. Moreover, both Ga 3d and As 3d peaks (as well as the valence band maximum, VBM) shift in the same direction by the same amount of magnitude at a given Al coverage, indicating that the shifts are related to the band bending occurring at the GaAs surface.

10.4.2.2 **Interfacial Charge Investigation**

More recently, photoelectron spectroscopy has also been applied in investigating the charges inside high-*k* stacks. The Si 2p spectra for thin Zr-silicates on Si before and after annealing in forming gas (90%N_2:10%H_2) at 400 °C for 30 min was studied [43]. The binding energies have been referenced to the C 1s (binding energy) peak for adventitious carbon, taken to be at 285.0 eV. Though there is no shift for the Si 2p core-level binding energy originating from silicate (Si $2p^{silicate}$), there is a big shift in the Si 2p core-level binding energy originating from Si substrate (Si $2p^{Si\_sub}$). The annealing process in forming gas ambient increases the Si $2p^{Si\_sub}$ binding energy for Si substrate component, which indicates the changes in the fixed charges inside Zr-silicate and therefore the changes in the Si band bending strength. For the as-deposited sample, the position of Si 2p component from the substrate (Si $2p^{Si\_sub}$) deviates substantially from the typically observed energy at 99.6 eV for elemental Si, and the Si $2p^{Si\_sub}$ core level is located at 98.5 eV, which indicates that there are a large number of positive fixed charges at Zr-silicate side and that the band bending occurs on the Si substrate side because of the negative charges at Si side. The positive charge density is estimated to be $\sim 1 \times 10^{13}/\text{cm}^2$. After annealing, Si $2p^{Si\_sub}$ core-level binding energy shifts approximately 0.5 eV to higher binding energy. During additional annealing in the forming gas, the positive charge on the Zr-silicate side decreases; hence the decreased band bending on the Si side because of the decreased negative charge.

Similar investigations have also been done for $HfO_2/SiO_2/Si$ stacks [44]. The band alignments of $HfO_2/SiO_2/Si$ stacks and the corresponding core-level shifts have been studied. The shifts in both Si $2p^{Si\_sub}$ core level and Hf 4f core level are attributed to the band bending occurring in Si substrate and the positive charges at HfO_2/SiO_2 interface.

10.4.2.3 **Band Alignment Determination**

Heterojunctions have been widely employed to fabricate the electrical and optical devices, taking advantage of the difference in bandgaps between the two materials of the junction. In such a structure, valence band offset (VBO) and the conduction band offset (CBO) are the important parameters to determine their performance. It is also right for the gate dielectrics. Except for the bandgap (E_g), both the VBO and the CBO to Si substrate would be the most important factors in terms of controlling the gate leakage current and assessing the high-*k* gate dielectrics in the future CMOS device technologies. Considering the importance of the band alignments in the device performance, the photoelectron spectroscopy has been widely used to

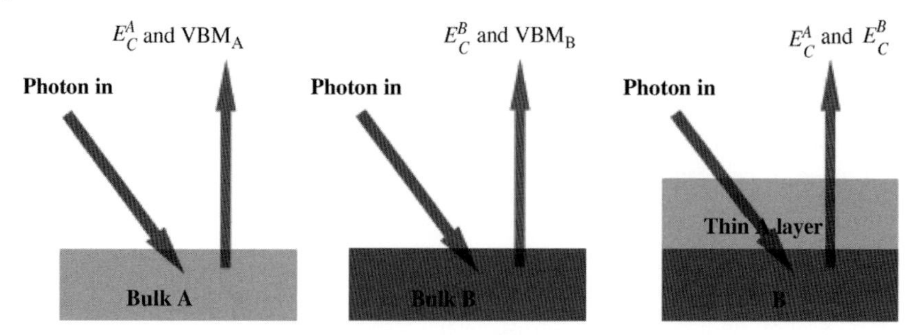

Figure 10.12 Samples for determining the band offset at A/B interface.

determine the VBO and the CBO in both semiconductor/semiconductor interface and insulator/semiconductor interface.

In a conventional method to determine the band offsets at A/B interface, three samples are required, that is, bulk A sample, bulk B sample, and sample with thin A layer on bulk B, as illustrated in Figure 10.12. Then, the valence band offset $\Delta E_V(A/B)$ at A/B structure could be obtained through the following relation:

$$\Delta E_V(A/B) = \Delta E_{V,C}^{B} - \Delta E_{V,C}^{A} + \Delta E_C^{A/B} \tag{10.5}$$

where

$$\Delta E_{V,C}^{A} = VBM_A - E_C^A \tag{10.6}$$

$$\Delta E_{V,C}^{B} = VBM_B - E_C^B \tag{10.7}$$

denote the difference between core level (E_C) and valence band maximum energy for pure A and pure B, respectively, while

$$\Delta E_C^{A/B} = E_C^B - E_C^A \tag{10.8}$$

is the core-level binding energy difference between A and B at A/B interface.

In a more direct method, the valence band offset can be obtained by measuring only the valence band maximum difference between A in A/B structure and bare B substrate [45]. Figure 10.13a illustrates the XPS valence band spectra for as-deposited ZrO_xN_y on (100)Si substrate with different N/Zr atomic ratio and clean Si(100) substrate sample [46]. The valence band maximum value of each sample is determined by extrapolating the leading edge of valence band spectra to the base line. The separation of valence band maximum of ZrO_xN_y from that of clean Si(100) substrate tells the valence band offsets for ZrO_xN_y/Si structures. When combining the bandgap (E_g) of both Si substrate and the thin ZrO_xN_y layer, both the valence band offset and the conduction band offset could be extracted. Both the valence band offset and the conduction band offset are observed to decrease with increased N/Zr atomic ratio, as shown in Figure 10.13b. Since O 2p and N 2p form the valence band, the incorporation of nitrogen inside ZrO_2 films would result in the upshifting of the valence band maximum value due to the higher N 2p states than O 2p states.

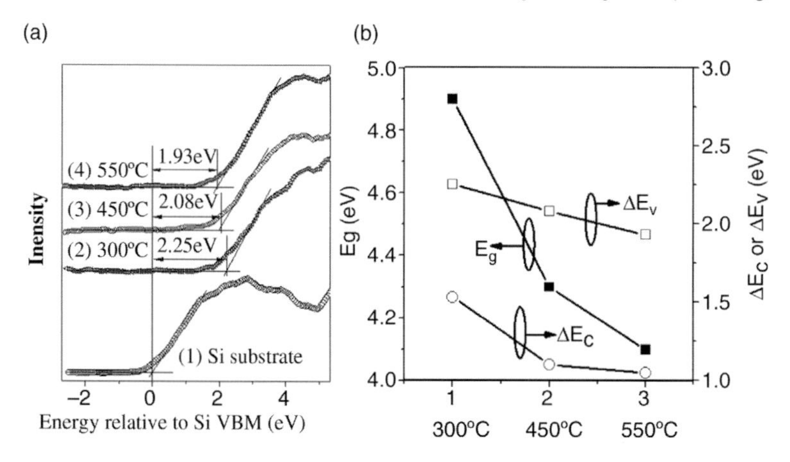

Figure 10.13 (a) XPS valence band spectra taken from (1) clean Si(100) substrates and as-deposited ZrO_xN_y films on Si with different N/Zr atomic ratio deposited at different temperatures, (2) 0.07 at 300 °C, (3) 0.10 at 450 °C, and (4) 0.12 at 550 °C. The cross point from each spectrum denotes the VBM for that specific sample. (b) Bandgap (E_g) values, zero-field valence band offsets (ΔE_C), and conduction band offsets (ΔE_C) of as-deposited ZrO_xN_y films on Si at different temperatures.

The antibonding d-states of zirconium form the lowest conduction band states in ZrO_2; therefore, the decreased ΔE_C values are expected due to the formation of Zr−N bonds and Zr−O−N bonds in ZrO_xN_y. The band alignment results indicate the decreased valence band maximum and the lowered conduction band minimum for ZrO_xN_y on Si substrate with the increased N incorporation.

Since the interfacial dipole existing in high-*k* stacks would result in not only the vacuum level shifts but also the shifts of the conduction band and valence band, the measurements of the valence band offsets (ΔE_C) and conduction band offsets (ΔE_C) would result in the estimation of the interfacial dipole magnitude in high-*k* stacks.

10.4.2.4 **Interfacial Dipole Measurement by Photoelectron Spectroscopy**

As already discussed, the measurements of the band discontinuities at the semiconductor/semiconductor interface and the Schottky barrier modifications have been performed by using photoelectron spectroscopy. The interfacial fixed charges have also been investigated through the relative core-level shifts in the high-*k* stacks. Moreover, the band alignments in high-*k* stacks have also been studied by using photoelectron spectroscopy. When combining the bandgaps for the stacks, both the valence band offsets and the conduction band offsets could be extracted. There are some clues for the interfacial dipole effects on the heterostructures. However, only limited studies on the interfacial dipole measurements in high-*k* gate stacks have been performed. In fact, the interfacial dipole is a kind of positive charge and negative charge pair across the interface, which introduces the band bending on both high-*k* side and SiO_2 side. As illustrated in Figure 10.14, both conduction band and valence band as well as the high-*k* core-level ($CL^{high\ k}$) binding

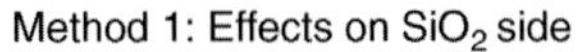

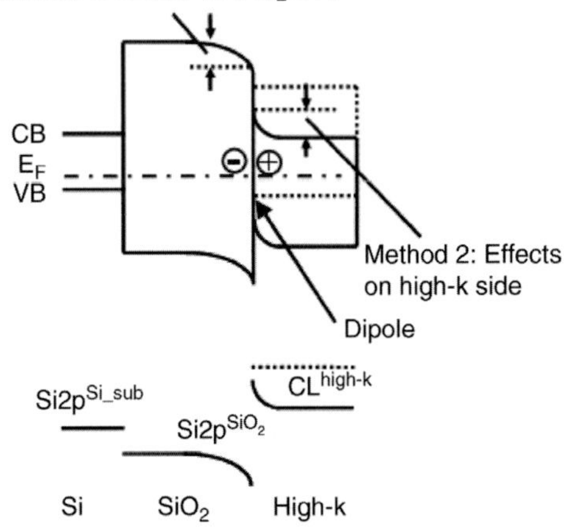

Figure 10.14 Simplified band diagrams for the high-k/SiO_2/Si stack when there is a dipole at the high-k/SiO_2 interface. The diagram with dashed lines is for the case assuming no dipole at the high-k/SiO_2 interface. Because of the positive charge and negative charge pairs, there will be band bending at both SiO_2 side and high-k side. In fact, it is also possible that the opposite dipole could exist at the high-k/SiO_2 interface with negative charge at high-k side and positive charge at SiO_2 side, depending on the high-k materials.

energy will also shift the same way as the vacuum level as a result of the interfacial dipole at the high-k/SiO_2 interface. Therefore, we can also observe the dipole effects in high-k stacks by analyzing the core-level binding energy shifts and the valence band maximum.

In an attempt, Zhu *et al.* applied both XPS and UPS to observe the interfacial dipole effects in high-k stacks [36]. Atomic layer deposition (ALD) HfO_2/Al_2O_3 stacks have been prepared on a p-type Si substrate with ~0.8 nm SiO_2. The Al_2O_3 thickness was controlled to be 0.4, 0.8, and 1.2 nm. HfO_2 thickness was controlled to be ~1 nm. A 1 nm HfO_2 layer was also deposited under the same conditions on SiO_2/Si as a control sample. The surface carbon contamination was carefully sputtered off using Ar^+ ions. Postdeposition annealing was conducted at 500 °C for 15 min in ultrahigh vacuum.

Figure 10.15 illustrates the definitions of the UPS measured work function. The decrease in the UPS measured work function means a downward shift in high-k band alignments above the dipole layer, increasing the effective gate work function. The work function Φ_W can be obtained by using the following relation:

$$\Phi_W = h\nu + E_{cutoff} - E_{Fermi} \tag{10.9}$$

where E_{cutoff} and E_{Fermi} are the kinetic energies of the secondary electron cutoff and the Fermi level in the UPS spectra, respectively. The secondary electron cutoff at low kinetic energies is given by inelastically scattered electrons that have just enough energy to reach the vacuum level, while the Fermi level at high kinetic energies represents the most energetic electrons excited from the Au foil in contact with the

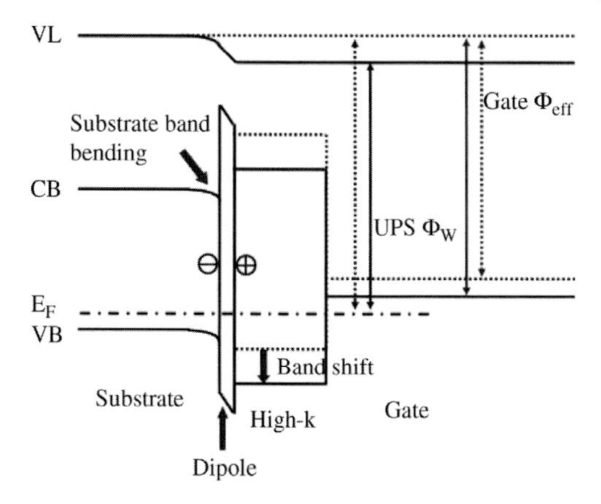

Figure 10.15 The band diagrams for high-*k* gate stacks when there is a dipole at the high-*k*/substrate interface. The diagram with dotted lines is for the case assuming no dipole at the high-*k*/substrate interface and no band bending at the substrate.

substrate. The obtained work function and valence band maximum as a function of Al_2O_3 thickness are illustrated in Table 10.1. A linear dependence of both the work function and the VBM on Al_2O_3 thickness is observed. Following the band diagram shown in Figure 10.15, the decrease in the work function (or increase in the VBM) with increase in the Al_2O_3 thickness is related to the introduction of an additional interfacial dipole in high-*k* stacks.

The application of XPS spectra would help to shed light on the origin of the work function shifts as a function of Al_2O_3 thickness and yield both the magnitude of substrate band bending and the high-*k* core-level shifts. As shown in Table 10.1, with increase in Al_2O_3 thickness, Si 2p substrate emission changes from 99.2 eV for Al_2O_3 thickness of 0 nm, through 99.27 and 99.32 eV for Al_2O_3 thickness of 0.4 and 0.8 nm, to 99.4 eV for Al_2O_3 thickness of 1.2 nm. These observations mean that there is band bending occurring in Si substrate, which will also shift the vacuum level and the band alignments for high-*k* stacks on substrate. Therefore, such band bending should be considered in the vacuum level shifts to tell the dipole magnitude.

The evidences of the dipole effects could be obtained from Hf 4f spectra and Al 2p spectra. In Hf 4f spectra, HfO_2 component can be extracted. For $HfO_2/SiO_2/Si$ stacks, the emission from HfO_2 can be extracted at binding energies of 18.02 eV. With the increase in Al_2O_3 layer from 0.4 to 1.2 nm, a gradual increase in Hf 4f core-level binding energy for HfO_2 component from 18.16 to 18.36 eV is observed, following closely the work function shifts. Such shifts are related to the presence of an interfacial dipole in the gate stack. By compensating the band bending, the relative dipole strength is estimated to be 0.14 eV compared to the $HfO_2/SiO_2/Si$ stack. Similar conclusion can also be drawn from Al 2p spectra. Al 2p component originated from Al_2O_3 can be extracted, which is shifting with increase in Al_2O_3 thickness, also following closely the work function shifts. After compensating the substrate band bending, the interfacial dipole strength increases by 0.07 eV, going from sample with 0.4 nm Al_2O_3 to sample with 1.2 nm Al_2O_3. Such shifts are the same as that deduced

Table 10.1 The UPS measured work function and VBM, and XPS core-level binding energy for Si 2p, Hf 4f, and Al 2p from Si substrate component, HfO_2 component, and Al_2O_3 component as a function of Al_2O_3 thickness.

Al_2O_3 thickness (nm)	Φ (eV)	VBM (eV)	Si 2p (eV)	Hf 4f (eV)	Al 2p (eV)
0	4.43	4.24	99.20	18.02	—
0.40	4.34	4.35	99.27	18.16	75.36
0.80	4.26	4.55	99.32	18.24	75.45
1.20	4.14	4.60	99.4	18.36	75.56

from Hf 4f core level, indicating that the shifts are correlated with the presence of a dipole situated below Al_2O_3 layer, in other words, a dipole related to the Al–silicate interface layer.

In this study, the interfacial dipole effects have been observed in high-*k* stacks with low SiO_2 thickness. However, there may be the possible interactions through the charge transfer between high-*k* layer and Si substrate for high-*k* layer on ultrathin SiO_2. In order to see closely the dipole effects just across the high-*k*/SiO_2 interface, thick SiO_2 is needed in high-*k* stacks to avoid the possible charge transfer between high-*k* and Si substrate.

It is needed to note that the dipole formed at high-*k*/SiO_2 interface has two effects on the band diagram as illustrated in Figure 10.14. On the one hand, the dipole at high-*k*/SiO_2 interface should induce a field at SiO_2 side at the high-*k*/SiO_2 interface, resulting in SiO_2 band bending. Therefore, the first method is presented [48]. The dipole effects at SiO_2 side can be extracted by investigating $\Delta(\text{Si } 2p^{SiO2}\text{-Si } 2p^{Si\_sub})$ behaviors related to SiO_2 band bending for stacks with and without high-*k* on SiO_2/Si as a function of SiO_2 thickness. On the other hand, the dipole will also induce a field in high-*k* side at the high-*k*/SiO_2 interface, resulting in high-*k* band bending. Therefore, for the second method, the dipole effects at high-*k* side can be extracted by investigating $\Delta(\text{Si } 2p^{SiO2}\text{-CL}^{high\ k})$ behavior related to high-*k* dipole conversion layer band bending in high-*k*/high-*k* dipole conversion layer/SiO_2 stacks [49].

Figure 10.16 illustrates $\Delta(\text{Si2p}^{SiO2}\text{-Si2p}^{Si\_sub})$ values as a function of SiO_2 thickness for bare SiO_2/Si and high-*k*/SiO_2/Si stacks (high $k = HfO_2$ or Y_2O_3) [48]. In thicker SiO_2 region, $\Delta(\text{Si } 2p^{SiO2}\text{-Si } 2p^{Si\_sub})$ values tend to saturate. Moreover, $\Delta(\text{Si } 2p^{SiO2}\text{-Si } 2p^{Si\_sub})$ value for the stacks with a high-*k* layer on SiO_2/Si is different from that for the bare SiO_2/Si when SiO_2 is thicker than ~5 nm. $\Delta(\text{Si } 2p^{SiO2}\text{-Si } 2p^{Si\_sub})$ at the saturated region for the HfO_2/SiO_2/Si stacks is larger than that for the bare SiO_2/Si stacks, while on the contrary, $\Delta(\text{Si } 2p^{SiO2}\text{-Si } 2p^{Si\_sub})$ at the saturated region for the Y_2O_3/SiO_2/Si stacks is smaller than that for the bare SiO_2/Si stacks.

In HfO_2/SiO_2/Si stacks, a dipole layer formed at the HfO_2/SiO_2 interface introduces an inner electric field in SiO_2 and causes SiO_2 band bending down at HfO_2/SiO_2 interface because of the negative dipole charges at SiO_2 side (Figure 10.17a). Therefore, $\Delta(\text{Si } 2p^{SiO2}\text{-Si } 2p^{Si\_sub})$ value is increased compared to that for bare SiO_2/Si stack (Figure 10.17b). In Y_2O_3/SiO_2/Si stacks, a dipole layer formed at the

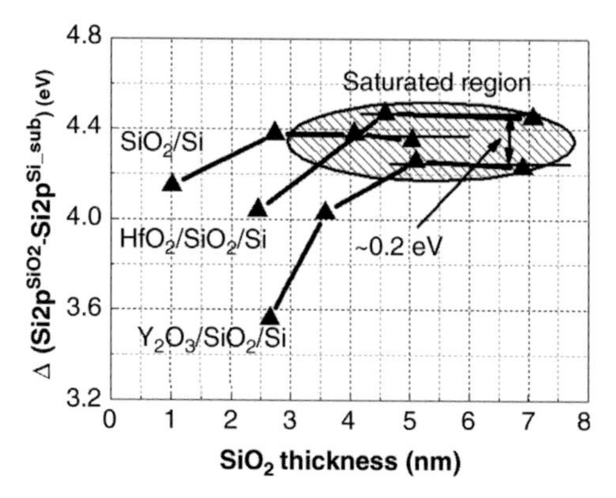

Figure 10.16 Δ(Si $2p^{SiO2}$-Si $2p^{Si\_sub}$) values as a function of SiO_2 thickness for SiO_2/Si, HfO_2/SiO_2/Si, and Y_2O_3/SiO_2/Si stacks. For SiO_2 thicker than ~5 nm, Δ values for SiO_2/Si are between Δ for HfO_2/SiO_2/Si and Δ for Y_2O_3/SiO_2/Si.

Y_2O_3/SiO_2 interface introduces an inner electric field in SiO_2 and causes SiO_2 band bending up at the Y_2O_3/SiO_2 interface because of the positive dipole charges at SiO_2 side (Figure 10.17c). Therefore, Δ(Si $2p^{SiO2}$-Si $2p^{Si\_sub}$) value is decreased compared to that for bare SiO_2/Si stack (Figure 10.17b). The dipole direction considered above

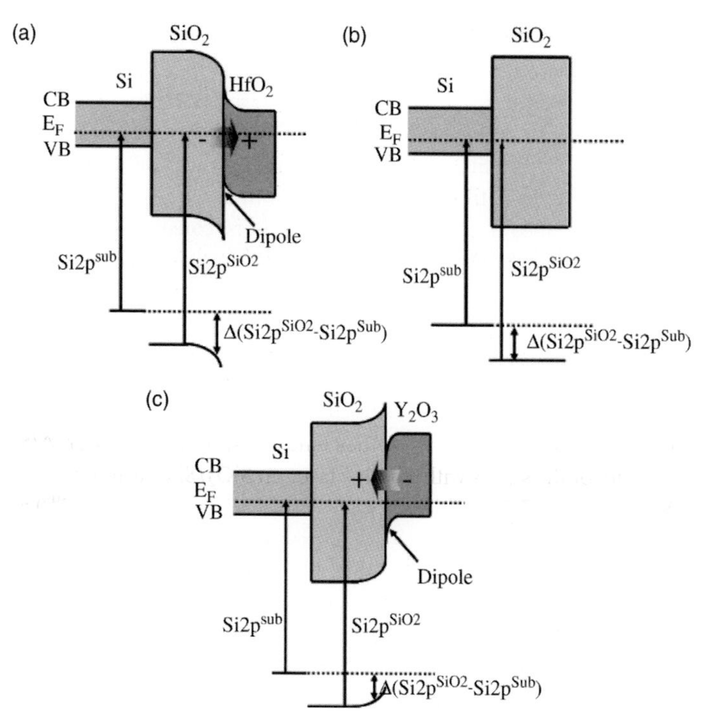

Figure 10.17 Schematic pictures of the band diagrams for (a) HfO_2/SiO_2/Si, (b) SiO_2/Si, and (c) Y_2O_3/SiO_2/Si stacks, where dipole layer with opposite directions are considered for (a) and (c).

for both cases is consistent with *C–V* results and strongly suggests dipole layer formation at the high-k/SiO_2 interface [20]. The difference in $\Delta(\text{Si }2p^{SiO2}\text{-Si }2p^{Si\_sub})$ value between $HfO_2/SiO_2/Si$ stacks and $Y_2O_3/SiO_2/Si$ stacks at the saturated region is ~0.2 eV, indicating that the observed relative SiO_2 band bending difference is ~0.2 eV between these two stacks.

The previous method only tells the dipole directions at the high-k/SiO_2 interface. In order to tell the relative dipole magnitude, studies on the band bending are needed both on SiO_2 side and on high-*k* side. However, in the previous method, we studied only the band bending occurring on the SiO_2 side. In the next method, the band bending on the high-*k* side will be studied. Through tracking the shifts in chemical difference between Si 2p in SiO_2 and Hf 4f in HfO_2 as a function of dipole conversion layer of $(HfO_2)_x(Y_2O_3)_{1-x}$ between HfO_2 and SiO_2, the relative dipole magnitude at HfO_2/SiO_2 interface and at Y_2O_3/SiO_2 interface can also be addressed. Figure 10.18 illustrates $\Delta(\text{Si }2p^{SiO2}\text{-Hf }4f^{HfO2})$ values as a function of *x*-values in $HfO_2/(HfO_2)_x(Y_2O_3)_{1-x}/SiO_2/Si$ stacks. $\Delta(\text{Si }2p^{SiO2}\text{-Hf }4f^{HfO2})$ is observed to increase as Y/Hf ratio increases. The overall increase in $\Delta(\text{Si }2p^{SiO2}\text{-Hf }4f^{HfO2})$ is ~0.4 eV from $HfO_2/SiO_2/Si$ stack to $HfO_2/Y_2O_3/SiO_2/Si$ stack. Such results mean that the relative band bending shift at high-*k* side is ~0.4 eV from HfO_2/SiO_2 interface to Y_2O_3/SiO_2 interface.

For the above two methods, it is easy to conclude the relative dipole magnitude at HfO_2/SiO_2 interface and Y_2O_3/SiO_2 interface. Though the band bending in SiO_2 is underestimated, while that of high-*k* is overestimated, such effects on the estimation of relative dipole magnitude are automatically canceled. The relative difference in the dipole magnitude at HfO_2/SiO_2 interface and at Y_2O_3/SiO_2 interface is estimated to be ~0.6 eV. The corresponding *C–V* curves for MOS capacitors of $Au/HfO_2/SiO_2/Si/Al$, $Au/HfO_2/(HfO_2)_{0.63}(Y_2O_3)_{0.37}/SiO_2/Si/Al$, and $Au/HfO_2/Y_2O_3/SiO_2/Si/Al$ are illustrated in Figure 10.19. The relative flatband voltage shift is ~0.5 V between $Au/HfO_2/SiO_2/Si/Al$ and $Au/HfO_2/Y_2O_3/SiO_2/Si/Al$, which is consistent with the relative difference in dipole magnitude obtained from XPS measurements.

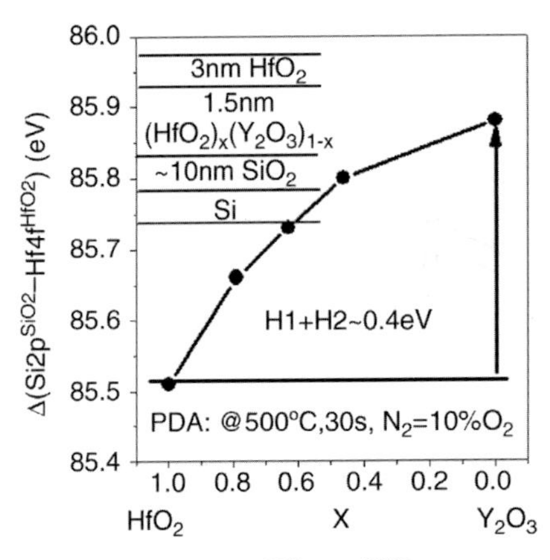

Figure 10.18 $\Delta(\text{Si }2p^{SiO2}\text{-Hf }4f^{HfO2})$ values as a function of inserted dipole conversion layer ($(HfO_2)_x(Y_2O_3)_{1-x}$) in $HfO_2/(HfO_2)_x(Y_2O_3)_{1-x}/SiO_2/Si$ stacks.

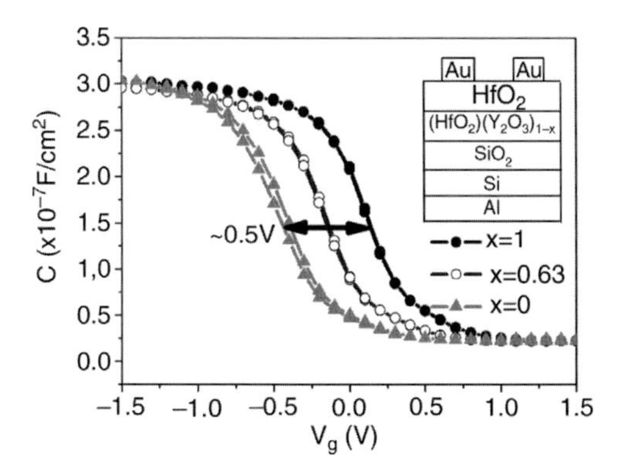

Figure 10.19 *C–V* curves obtained for MOS capacitors of $Au/HfO_2/SiO_2/Si/Al$, $Au/HfO_2/(HfO_2)0.63(Y_2O_3)_{0.37}/SiO_2/Si/Al$, and $Au/HfO_2/Y_2O_3/SiO_2/Si/Al$.

Successful applications of the photoelectron spectroscopy in interfacial dipole characterization suggest that it would be a useful technique in providing the physical insights into the dipole formations at the high-k/SiO_2 interfaces.

10.4.3 Band Alignments Measured by Using Internal Photoemission

Internal photoemission is also a well-known technique to determine the band alignments in semiconductor heterojunctions. It has been widely used in high-*k* gate stacks to tell both the valence band offset and the conduction band offset, as well as the barrier at the metal/high-*k* interfaces. In the internal photoemission measurements, the IPE yield (Y) is usually defined as the photocurrent (I) normalized to the incident photon flux (I_0), that is, $Y = I/I_0$. When the metal gate is positively biased (Figure 10.20a), the current across the high-*k* layer is related to electrons emitted from the Si substrate. In this case, there is a relationship shown as follows:

$$Y = A\,(h\nu - \phi_{\mathrm{IPE}})^3 \tag{10.10}$$

where A is a constant and ϕ_{IPE} is the spectral threshold of electron emission from the valence band of silicon substrate to the high-*k* conduction band determined

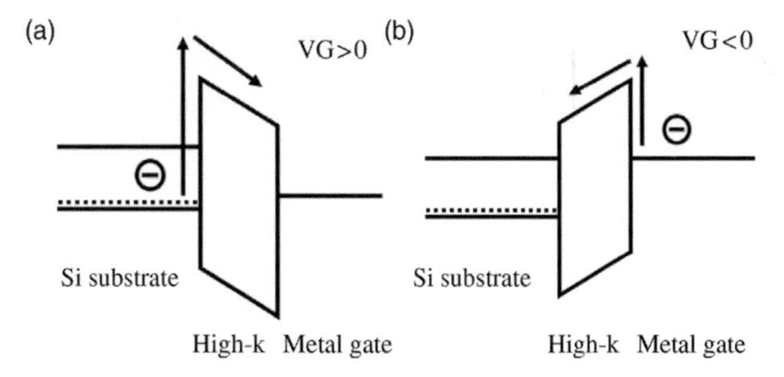

Figure 10.20 Electron emitted from Si substrate or metal gate when metal gate is (a) positive or (b) negative biased.

by linear extrapolation in the $Y^{1/3}$ versus $h\nu$ plot. However, when the metal gate is negative biased (Figure 10.20b), the current across the high-k layer is related to electrons emitted from the metal gate electrode. In this case, there is a relationship shown as follows:

$$Y = A\,(h\nu - \phi_{\mathrm{IPE}})^2 \tag{10.11}$$

where ϕ_{IPE} is the spectral threshold of electron emission from the metal gate electrode to the high-k CB determined by linear extrapolation in the $Y^{1/2}$ versus $h\nu$ plot. However, in some cases, there will be the image force barrier reduction at metal/dielectric interface or semiconductor/dielectric interface, the obtained IPE spectral threshold is not the true barrier. So, in order to obtain a more accurate barrier value, the electric field-dependent spectral threshold ($\phi_{\mathrm{IPE}}(E)$) could be obtained. By drawing Schottky plot (i.e., $\phi_{\mathrm{IPE}}(E)$ versus $E^{1/2}$), the zero-field barrier height could be obtained by extrapolating $\phi_{\mathrm{IPE}}(E)$ versus $E^{1/2}$ curves to zero electric field.

IPE techniques can also work under the photoconductivity mode in MOS structures. In this case, the current is related to electrons emitted from the valence band to conduction band. Then, the insulator bandgap could be extracted. Therefore, the band alignments of high-k gate stacks could be obtained by using the IPE techniques. Since the band alignments will be modulated by the interfacial dipole formation in the high-k gate stacks, such as the VB barriers or CB barriers, both the interfacial dipole directions and the magnitude could be analyzed by using IPE techniques [50–52].

The study on IPE spectral thresholds for $Al/HfO_2/SiO_2/Si$ and $Al/SiO_2/Si$ stacks tells the dipole effects at HfO_2/SiO_2 interface [52]. Zero-field barrier energies could be obtained by extrapolating IPE spectra threshold $\phi_{\mathrm{IPE}}(E)$ to zero field in the corresponding Schottky plot with Al gate positive or negative biased. The resulting barrier between Si valence band and HfO_2 conduction band is determined to be ~3 eV, that is, the conduction band offset between Si and HfO_2 is ~1.9 eV. The resulting barrier between Al Fermi level and HfO_2 conduction band is determined to be ~2.3 eV, while the resulting barrier between Al Fermi level and SiO_2 conduction band is determined to be ~3.25 eV. Since there is no Fermi-level pinning effect for low-temperature thermal budgets, considering the band alignment parameters for SiO_2/Si stacks, the Al effective work function for $Al/SiO_2/Si$ stacks is determined to be ~4.15 eV. For the $Al/HfO_2/SiO_2/Si$ stacks, the Al effective work function is determined to be ~4.45 eV. Figure 10.21 illustrates the band alignments for $Al/HfO_2/SiO_2/Si$ and $Al/SiO_2/Si$ stacks. Therefore, there is a dipole formation in $Al/HfO_2/SiO_2/Si$ stacks. Considering the low-temperature thermal budget, which is not enough to result in the Fermi-level pinning effects at Al/HfO_2 interface, the dipole is determined to be ~0.3 eV located at HfO_2/SiO_2 interface.

10.4.4
Potential Shifts in Kelvin Probe Measurements

As a noncontact, nondestructive vibrating capacitor device, Kelvin probe force microscopy (KPFM) is a powerful tool to measure the work function of conducting

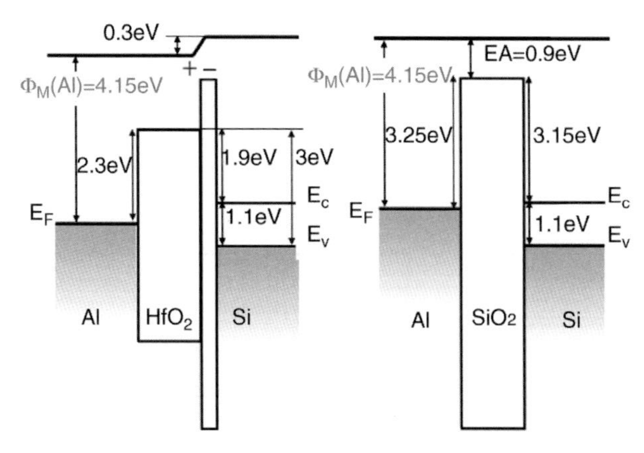

Figure 10.21 Electron energy band diagram for $Al/HfO_2/SiO_2/Si$ and $Al/SiO_2/Si$ MOS structures. Reproduced from Ref. [52]. Copyright 2008, The Japan Society of Applied Physics.

materials and the surface potential of semiconductor or insulator down to a nanometer-scale resolution, without need to touch the sample surface. An electrical contact is made with another part of the sample or sample holder. The measurements are based on detection of electrostatic interactions, arising from the difference in the work function between a conducting specimen and a vibrating tip, that is, the so-called contact potential difference (CPD), as shown in Figure 10.22 [53]. The vibrating tip and the sample form a capacitor, having an ideal or parallel plate geometry. During the measurements, DC offset bias V_B is applied between the tip and the sample. As the tip vibrates periodically, electric charge is pushed around the external detection circuit, resulting in a current $i(t)$. If V_B is equal to the work function difference between the tip and the sample, the vacuum levels of the tip and the sample become aligned, and the specific properties could be found in $i(t)$. Therefore, we can obtain the intended potential by adjusting DC offset bias V_B to obtain CPD. Since the interfacial dipole in high-k stacks would result in the vacuum-level shifts, Kelvin probe can also be utilized to estimate the interfacial dipole effects inside high-k stacks [54, 55].

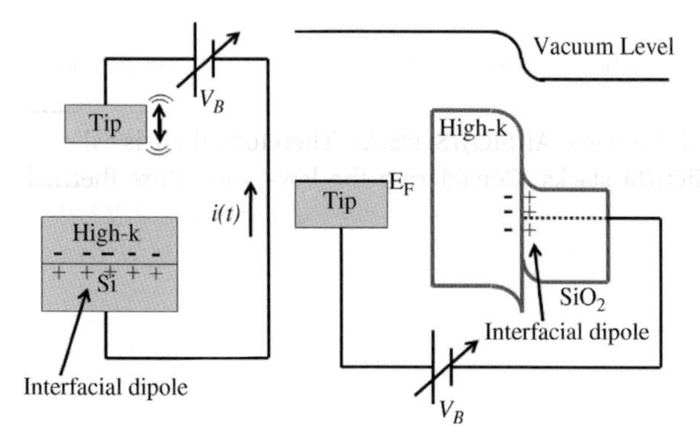

Figure 10.22 Schematic diagrams of Kelvin probe. DC offset bias V_B is applied between the tip and the sample during the measurements.

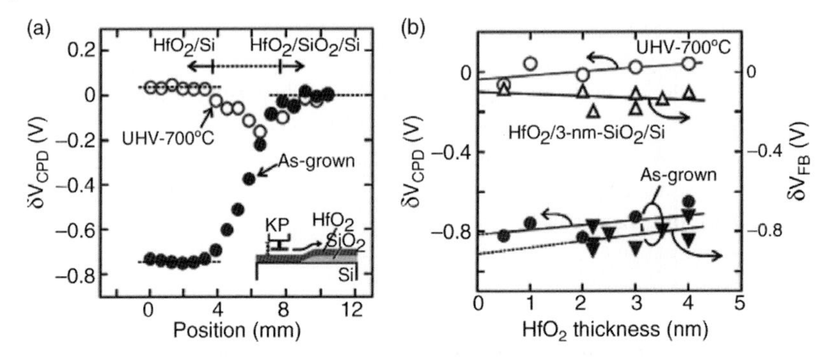

Figure 10.23 An example of the interfacial dipole measurements by using Kelvin probe technique. (a) Contact potential difference voltage (V_{CPD}) measured on the $HfO_2/SiO_2/Si$ stack structures with graded SiO_2 thickness. (b) HfO_2 thickness dependence of δV_{CPD} and δV_{FB} for as-grown and UHV-annealed samples. Reproduced from Ref. [54]. Copyright 2010, The Japan Society of Applied Physics.

Figure 10.23a illustrates an example of the interfacial dipole measured by using Kelvin probe technique [54]. The graded SiO_2/Si with SiO_2 thickness from 0 to 3 nm worked as the substrate. HfO_2 was deposited by using an ultrahigh vacuum electron beam deposition system, followed by postdeposition annealing at 400 °C in 2×10^{-6} Torr O_2 for 1 min (as-grown) as illustrated in the inset in Figure 10.23. The sample was also annealed in the same UHV chamber at 700 °C for 3 min. By using the following relation, the effects of the surface contamination could be avoided:

$$\delta V_{CPD} = V_{CPD} - V'_{CPD} \tag{10.12}$$

where V'_{CPD} is the average V_{CPD} value for $HfO_2/SiO_2/Si$ stacks with SiO_2 thickness of 3 nm. For the as-grown sample, V_{CPD} value for abrupt structure HfO_2/Si is much lower than that for $HfO_2/SiO_2/Si$ stacks with SiO_2 thickness of 3 nm, which means that the effective work function of the abrupt structure HfO_2/Si is about 0.8 eV larger than that of $HfO_2/SiO_2/Si$ stacks with SiO_2 thickness of 3 nm. V_{FB} was also obtained on the Ir gated stacks. Figure 10.23b illustrates the HfO_2 thickness dependence of δV_{CPD} and δV_{FB} ($= V_{FB} - V'_{FB}$, where V'_{FB} is the ideal V_{FB} estimated by assuming the Ir work function of 5.3 eV). From flatband voltage shifts, it could be concluded that there is a dipole at HfO_2/Si interface with the magnitude of ~0.9 eV, which is consistent with the Kelvin probe results of ~0.8 eV shifts in the work function. After annealing in UHV chamber at 700 °C, δV_{CPD} value tends to change not so much, which means the relative interfacial dipole decreases dramatically.

10.5 Summary

To solve the fundamental limitations in polysilicon gate with the shrinking of the MOS device feature size, such as the unavoidable polysilicon gate depletion, penetration of dopants from the polysilicon gate degrading the channel performance, the relatively high gate resistance when the polysilicon gate is not doped heavily enough, and the Fermi-level pinning effects experienced between the polysilicon and the high-*k* dielectrics, metallic gate electrode is deemed as the most effective solution. One of the main challenges in the metal/high-*k* gate stacks applications is to achieve the gate effective work functions of ~4.1 and ~5.0 eV on high-*k* for both n-MOS and p-MOS applications, respectively.

The conventional method to modulate the work function is to directly modulate the gate vacuum work function. However, the gate work function is also affected by both the metal/high-*k* interface and the high-*k*/SiO_2 interface because of the existence of interfacial dipoles at such interfaces. Though there have been successful applications of the interface dipoles in the gate effective work function modifications, the microphysical origin behind them has not been well established yet. An overview on the possible mechanisms proposed both for the metal/high-*k* interface and for the high-*k*/SiO_2 interface has been provided in this chapter, including Fermi-level pinning, oxygen vacancy model, electronegativity and group electronegativity model, and area oxygen density model. However, these reported dipole models are not well established to explain all the examples. More work is obviously needed to establish a universal model to explain the dipole effects in high-*k* gate stacks.

The observation of the dipole effects is not so direct, which might affect the establishment of the interfacial dipole model for the high-*k* gate stacks. Though the interfacial dipole is typically observed by *C–V* measurements based on the study of the flatband voltage shifts, the application of new techniques to observe and study the dipole effects would help us to understand the dipole behavior. By using XPS, the direct evidence of the dipole effects in high-*k* stacks is expected to be obtained. The applications of internal photoemission and Kelvin probe based on the analysis of the band offset and Kelvin potential, respectively, would also result in the observation of the interfacial dipole effects. The successful application of such new techniques would help us to understand the microscopic origin of the dipole effects more clearly.

References

1 Green, M.L., Gusev, E.P., Degraeve, R., and Garfunkel, E.L. (2001) Ultrathin (<4nm) SiO_2 and Si–O–N gate dielectric layers for silicon microelectronics: understanding the processing, structure, and physical and electrical limits. *J. Appl. Phys.*, **90**, 2057.

2 Rios, R., Arora, N.D., and Huang, C.L. (1994) An analytic polysilicon depletion effect model for MOSFET's. *IEEE Electron. Device Lett.*, **15**, 129.

3 Josse, E. and Skotnicki, T. (1999) Polysilicon gate with depletion – or – metallic gate with buried channel: what evil worse? IEEE International Electron Devices Meeting Technical Digest, p. 661.

4 Choi, C.H., Chidambaram, P.R., Khamankar, R., Machala, C.F., Yu, Z., and Dutton, R.W. (2002) Gate length dependent polysilicon depletion effects. *IEEE Electron. Device Lett.*, **23**, 224.

5 Arora, N.D., Rios, R., and Huang, C.L. (1995) Modeling the polysilicon depletion effect and its impact on submicrometer

CMOS circuit performance. *IEEE Trans. Electron. Dev.*, **42**, 935.

6 Hobbs, C., Fonseca, L., Dhandapani, V., Samavedam, S., Taylor, B., Grant, J., Dip, L., Triyoso, D., Hegde, R., Gilmer, D., Garcia, R., Roan, D., Lovejoy, L., Rai, R., Hebert, L., Tseng, H., White, B., and Tobin, P. (2003) Fermi level pinning at the polySi/metal oxide interface. IEEE Symposium on VLSI Technology, p. 9.

7 Wong, H.Y., Takeuchi, H., King, T.J., Ameen, M., and Agarwal, A. (2005) Elimination of poly-Si gate depletion for sub-65-nm CMOS technologies by excimer laser annealing. *IEEE Electron. Device Lett.*, **26**, 234.

8 Shah, R. and Souza, M.M.D. (2007) Impact of a nonideal metal gate on surface optical phonon-limited mobility in high-*k* gated MOSFETs. *IEEE Trans. Electron. Dev.*, **54**, 2991.

9 Wilk, G.D., Wallace, R.M., and Anthony, J.M. (2001) High-*k* gate dielectrics: current status and materials properties considerations. *J. Appl. Phys.*, **89**, 5243.

10 Misra, V., Heuss, G.P., and Zhong, H. (2001) Use of metal-oxide-semiconductor capacitors to detect interactions of Hf and Zr gate electrodes with SiO_2 and ZrO_2. *Appl. Phys. Lett.*, **78**, 4116.

11 Schottky, W. (1940) Abweichungen von ohm'schen gesetz in halbleitern. *Physikalische Z.*, **41**, 570.

12 Robertson, J. (2000) Band offsets of wide-band-gap oxides and implications for future electronic devices band offsets of wide-band-gap oxides and implications for future electronic devices. *J. Vac. Sci. Technol. B*, **18**, 1785.

13 Shiraishi, K., Yamada, K., Torii, K., Akasaka, Y., Nakajima, K., Konno, M., Chikyow, T., Kitajama, H., Arikado, T., and Nara, Y. (2006) Oxygen-vacancy-induced threshold voltage shifts in Hf-related high-*k* gate stacks. *Thin Solid Films*, **508**, 305.

14 Akasaka, Y., Nakamura, G., Shiraishi, K., Umezawa, N., Yamabe, K., Ogawa, O., Lee, M., Amiaka, T., Kasuya, T., Watanabe, H., Chikyow, T., Oktsuka, F., Nara, Y., and Nakamura, K. (2006) Modified oxygen vacancy induced Fermi level pinning model extendable to p-metal pinning. *Jpn. J. Appl. Phys.*, **45**, L1289.

15 Hobbs, C.C., Fonseca, L.R.C., Knizhnik, A., Dhandapani, V., Samavedam, S.B., Taylor, W.J., Grant, J.M., Dip, L.G., Triyoso, D.H., Hegde, R.I., Gilmer, D.C., Garcia, R., Roan, D., Lovejoy, M.L., Rai, R.S., Hebert, E.A., Tseng, H.H., Aderson, S.G.H., White, B.E., and Tobin, P.J. (2004) Fermi level pinning at the polysilicon/metal oxide interface-part I. *IEEE Trans. Electron. Dev.*, **51**, 971.

16 Xiong, K., Peacock, P.W., and Robertson, J. (2005) Fermi level pinning and Hf–Si bonds at HfO_2: polycrystalline silicon gate electrode interfaces. *Appl. Phys. Lett.*, **86**, 012904.

17 Wen, H.C., Song, S.C., Park, C.S., Burham, C., Bersuker, G., Choi, K., Quevedo-Lopez, M.A., Ju, B.S., Alshareef, H.N., Niimi, H., Park, H.B., Lysaght, P.S., Majhi, P., Lee, B.H., and Jammy, R. (2007) Gate first metal-aluminum-nitride PMOS electrodes for 32nm low standby power applications. IEEE Symposium on VLSI Technology, 160.

18 Alshareef, H.N., Quevedo-Lopez, M., Wen, H.C., Harris, R., Kirsch, P., Majhi, P., Lee, B.H., Jammy, R., Lichtenwalner, D.J., Jur, J.S., and Kingon, A.I. (2006) Work function engineering using lanthanum oxide interfacial layers. *Appl. Phys. Lett.*, **89**, 232103.

19 Abe, Y., Miyata, N., Shiraki, Y., and Yasuda, T. (2007) Dipole formation at direct-contact HfO_2/Si interface. *Appl. Phys. Lett.*, **90**, 172906.

20 Kamimuta, Y., Iwamoto, K., Nunoshige, Y., Hirano, A., Mizubayashi, W., Watanabe, Y., Migita, S., Ogawa, A., Ota, H., Nabatame, T., and Toriumi, A. (2007) Comprehensive study of V_{FB} shift in high-*k* CMOS dipole formation, Fermi-level pinning and oxygen vacancy effect. IEEE International Electron Devices Meeting Technical Digest, p. 341.

21 Yeo, Y.C., Ranade, P., King, T.J., and Hu, C. (2002) Effects of high-*k* gate dielectric

materials on metal and silicon gate workfunctions. *IEEE Electron. Device Lett.*, **23**, 342.

22 Robertson, J. and Falabretti, B. (2006) Band offsets of high *k* gate oxides on high mobility semiconductors. *Mater. Sci. Eng. B*, **135**, 267.

23 Schaeffer, J.K., Fonseca, L.R.C., Samavedam, S.B., Liang, Y., Tobin, P.J., and White, B.E. (2004) Contributions to the effective work function of platinum on hafnium oxide. *Appl. Phys. Lett.*, **85**, 1826.

24 Jha, R., Lee, B., Chen, B., Novak, S., Majhi, P., and Misra, V. (2005) Dependence of PMOS metal work functions on surface conditions of high-*k* gate dielectrics. IEEE International Electron Devices Meeting Technical Digest, p. 43.

25 Wang, X.P., Yu, H.Y., Li, M.F., Zhu, C.X., Biesemans, S., Chin, A., Sun, Y.Y., Feng, Y.P., Lim, A., Yeo, Y.C., Loh, W.Y., Lo, G.Q., and Kwong, D.L. (2007) Wide V_{fb} and V_{th} tunability for metal-gated MOS devices with HfLaO gate dielectrics. *IEEE Electron. Device Lett.*, **28**, 258.

26 Guha, S., Paruchuri, V.K., Copel, M., Narayanan, V., Wang, Y.Y., Batson, P.E., Bojarczuk, N.A., Linder, B., and Doris, B. (2007) Examination of flatband and threshold voltage tuning of HfO_2/TiN field effect transistors by dielectric cap layers. *Appl. Phys. Lett.*, **90**, 092902.

27 Kirsch, P.D., Sivasubramani, P., Huang, J., Young, C.D., Quevedo-Lopez, M.A., Wen, H.C., Alshareef, H., Choi, K., Park, C.S., Freeman, K., Hussain, M.M., Bersuker, G., Harris, H.R., Majhi, P., Choi, R., Lysaght, P., Lee, B.H., Tseng, H.H., Jammy, R., Boscke, T.S., Lichtenwalner, D.J., Jur, J.S., and Kingon, A.I. (2008) Dipole model explaining high-*k*/metal gate field effect transistor threshold voltage tuning. *Appl. Phys. Lett.*, **92**, 092901.

28 Kakushima, K., Okamoto, K., Adachi, M., Tachi, K., Ahmet, P., Tsutsui, K., Sugii, N., Hattori, T., and Iwai, H. (2008) Origin of flat band voltage shift in HfO_2 gate dielectric with La_2O_3 insertion. *Solid State Electron.*, **52**, 1280.

29 Schaeffer, J.K., Gilmer, D.C., Capasso, C., Kalpat, S., Taylor, B., Raymond, M.V., Triyoso, D., Hegde, R., Samavedam, S.B., and White, Jr., B.E. (2007) Application of group electronegativity concepts to the effective work functions of metal gate electrodes on high-*k* gate oxides. *Microelectron. Eng.*, **84**, 2196.

30 Kita, K. and Toriumi, A. (2009) Origin of electric dipoles formed at high-*k*/SiO_2 interface. *Appl. Phys. Lett.*, **94**, 132902.

31 Song, S.C., Park, C.S., Price, J., Burham, C., Choi, R., Wen, H.C., Choi, K., Tseng, H.H., Lee, B.H., and Jammy, R. (2007) Mechanism of V_{FB} roll-off with high work function metal gate and low temperature oxygen incorporation to achieve PMOS band edge work function. IEEE International Electron Devices Meeting Technical Digest, p. 337.

32 Choi, K., Wen, H.C., Bersuker, G., Harris, R., and Lee, B.H. (2008) Mechanism of flatband voltage roll-off studied with Al_2O_3 film deposited on terraced oxide. *Appl. Phys. Lett.*, **93**, 133506.

33 Akiyama, K., Wang, W., Mizubayashi, W., Ikeda, M., Ota, H., Nabatame, T., and Toriumi, A. (2008) Roles of oxygen vacancy in HfO_2/ultra-thin SiO_2 gate stacks comprehensive understanding of V_{FB} roll-off. IEEE Symposium on VLSI Technology, p. 80.

34 Barrett, N.T., Renault, O., Besson, P., Tiec, Y.L., and Martin, F. (2006) Band offsets of nitrided ultrathin hafnium silicate films. *Appl. Phys. Lett.*, **88**, 162906.

35 Kakushima, K., Okamoto, K., Tachi, K., Song, J., Sato, S., Kawanago, T., Tsutsui, K., Sugii, N., Ahmet, P., Hattori, T., and Iwai, H. (2008) Observation of band bending of metal/high-*k* Si capacitor with high energy X-ray photoemission spectroscopy and its application to interface dipole measurement. *J. Appl. Phys.*, **104**, 104908.

36 Zhu, L.Q., Barrett, N., Jégou, P., Martin, F., Leroux, C., Grampeix, H., Renault, O., and Chabli, A. (2009) X-ray photoelectron spectroscopy and ultraviolet photoelectron spectroscopy investigation of Al-related dipole at the HfO_2/Si interface. *J. Appl. Phys.*, **105**, 024102.

37 Zhu, L.Q., Kita, K., Nishimura, T., Nagashio, K., Wang, S.K., and Toriumi, A. (2009) X-ray photoelectron spectroscopy study of dipole effects at HfO_2/SiO_2/Si stacks. Extended Abstracts of the International Conference on Solid State Devices and Materials, p. 40.

38 Mori, T., Ohta, A., Murakami, H., Higashi, S., and Miyazaki, S. (2009) Evaluation of effective work function of Pt on Bi-layer high-k/SiO_2 stack structure using by backside X-ray photoelectron spectroscopy. Extended Abstracts of the International Conference on Solid State Devices and Materials, p. 44.

39 Yamamoto, Y., Kita, K., Kyuno, K., and Toriumi, A. (2007) Study of La-induced flat band voltage shift in metal/$HfLaO_x/SiO_2$/Si capacitors. *Jpn. J. Appl. Phys.*, **46**, 7251.

40 Marsi, M., Stasio, G.D., and Margaritondo, G. (1992) Local nature of artificial homojunction band discontinuities. *J. Appl. Phys.*, **72**, 1443.

41 Perfetti, P., Quaresima, C., Coluzza, C., Fortunato, C., and Margaritondo, G. (1986) Dipole-induced changes of the band discontinuities at the SiO_2-Si interface. *Phys. Rev. Lett.*, **57**, 2065.

42 Cao, R., Miyano, K., Kendelewicz, T., Chin, K.K., Lindau, I., and Spicer, W.E. (1987) Kinetics study of initial stage band bending at metal GaAs(110) interfaces. *J. Vac. Sci. Technol. B*, **5**, 998.

43 Opila, R.L., Wilk, G.D., Alam, M.A., Dover, R.B.V., and Busch, B.W. (2002) Photoemission study of Zr- and Hf-silicates for use as high-*k* oxides: role of second nearest neighbors and interface charge. *Appl. Phys. Lett.*, **81**, 1788.

44 Sayan, S., Emge, T., Garfunkel, E., Zhao, X., Wielunski, L., Bartynski, R.A., Vanderbilt, D., Suehle, J.S., Suzer, S., and Banaszak-Holl, M. (2004) Band alignment issues related to HfO_2/SiO_2/p-Si gate stacks. *J. Appl. Phys.*, **96**, 7485.

45 Yu, H.Y., Li, M.F., Cho, B.J., Yeo, C.C., Joo, M.S., Kwong, D.L., Pan, J.S., Ang, C.H., Zheng, J.Z., and Ramanathan, S. (2002) Energy gap and band alignment for $(HfO_2)_x(Al_2O_3)_{1-x}$ on (100) Si. *Appl. Phys. Lett.*, **81**, 376.

46 Zhu, L.Q., Fang, Q., Wang, X.J., Zhang, J.P., Liu, M., He, G., and Zhang, L.D. (2008) Structural, optical properties and band gap alignments of ZrO_xN_y thin films on Si (100) by radio frequency sputtering at different deposition temperatures. *Appl. Surf. Sci.*, **254**, 5439.

47 Sharia, O., Demkov, A.A., Bersuker, G., and Lee, B.H. (2008) Effects of aluminium incorporation on band alignment at the SiO_2/HfO_2 interface. *Phys. Rev. B*, **77**, 085326.

48 Zhu, L.Q., Kita, K., Nishimura, T., Nagashio, K., Wang, S.K., and Toriumi, A. (2010) Observation of dipole layer formed at high-*k* dielectrics/SiO_2 interface with X-ray photoelectron spectroscopy. *Appl. Phys. Express*, **3**, 061501.

49 Zhu, L.Q., Kita, K., Nishimura, T., Nagashio, K., Wang, S.K., and Toriumi, A. (2011) Interfacial dipole at high-*k* dielectric/SiO_2 interface: X-ray photoelectron spectroscopy characteristics. *Jpn J. Appl. Phys.*, **50**, 031502.

50 Dell'Orto, T., Almeida, J., Coluzza, C., Baldereschi, A., Margaritondo, G., Canlile, M., Yildirim, S., Sorba, L., and Franciosi, A. (1994) Internal photoemission studies of artificial band discontinuities at buried GaAs(100)/GaAs(100) homojunctions. *Appl. Phys. Lett.*, **64**, 2111.

51 Widiez, J., Kita, K., Nishimura, T., and Toriumi, A. (2007) Advanced characterization of high-*k* gate stack by internal photo emission (IPE): interfacial dipole and band diagram in Al/Hf(Si)O_2/Si MOS structure. Extended Abstracts of the International Conference on Solid State Devices and Materials, p. 1028.

52 Widiez, J., Kita, K., Tomida, K., Nishimura, T., and Toriumi, A. (2008) Internal photoemission over HfO_2 and $Hf_{(1-x)}Si_xO_2$ high-*k* insulating barriers: band offset and interfacial dipole characterization. *Jpn. J. Appl. Phys.*, **47**, 2410.

53 Nonnenmacher, M., O'Boyle, M.P., and Wickramasinghe, H.K. (1991) Kelvin probe force microscopy. *Appl. Phys. Lett.*, **58**, 2921.

54 Miyata, N., Yasuda, T., and Abe, Y. (2010) Kelvin probe study of dipole formation and annihilation at the HfO_2/Si interface. *Appl. Phys. Express*, **3**, 054101.

55 Miyata, N., Abe, Y., and Yasuda, T. (2008) Chemical bonding-induced dipole at the HfO_2/Si interface. Extended Abstracts of the International Conference on Solid State Devices and Materials, p. 682.

11
Metal Gate Electrode for Advanced CMOS Application

Wenwu Wang, Xiaolei Wang, and Kai Han

11.1
The Scaling and Improved Performance of MOSFET Devices

In 1930, Lilienfeld patented the basic concept of the field-effect transistor (FET) [1]. Thirty years later, in 1960, it was finally reduced to practice in Si–SiO_2 by Kahng and Attala [2]. Since then, it has been incorporated into integrated circuits and has grown to be the most important device in the electronics industry. Especially, the rapid progress of integrated circuit industry since the late 1980s has enabled the Si-based microelectronics industry to simultaneously meet several technological requirements to an expanding commercial market [3]. These requirements include performance (speed), low static (off-state) power, and a wide range of power supply and output voltages [4]. This has been accomplished by developing the ability to perform a calculated reduction of the dimensions of the fundamental active device in the circuit: the FET – a practice termed "scaling" [5–7]. The scaling of minimum feature sizes in the MOSFET has been the major driving force for improving circuit speed, reducing power dissipation, and increasing packing density [8].

To meet the requirements of MOSFET shrinking, various scaling rules have been proposed including constant field scaling, constant voltage scaling, quasi-constant voltage scaling, and generalized scaling [5, 6, 9]. In commonly used constant field scaling, the maximum magnitude and shape of the internal electric field remain constant, and thus the shorter device is just a scaled version of the larger device. The scaling rules for the constant field scaling are listed in Table 11.1 [9]. A schematic illustration of the scaling of MOSFET device is provided in Figure 11.1: a large FET is scaled down by a factor K to produce a smaller FET with similar behavior [10]. When all of the voltages and dimensions are reduced by the scaling factor K and the doping and charge densities are increased by the same factor, the electric field configuration inside the FET remains the same as it was in the original device. This results in circuit speed increasing in proportion to the factor K and circuit density increasing as K^2. The limitation with this type of scaling is that the fraction of voltage required to turn the device off is very high since weak inversion region width does not scale. Also, the external chip interface may require the voltage level to be maintained

High-k Gate Dielectrics for CMOS Technology, First Edition. Edited by Gang He and Zhaoqi Sun.

Table 11.1 Scaling rules for constant field scaling and constant voltage scaling.

Scaling factor	Electric field constant	Gate voltage constant
Gate length (L)	$1/K$	$1/K$
Gate width (W)	$1/K$	$1/K$
Junction depth(X_j)	$1/K$	$1/K$
Gate area (S)	$1/K^2$	$1/K^2$
Gate oxide thickness (T_{ox})	$1/K$	$1/K$
Doping density (N_A)	K	K^2
Voltage (V)	$1/K$	1
Electric field (E)	1	K
Current ($I = (W/L)(1/T_{ox})V^2$)	$1/K$	K
Gate capacitance ($C = eS/T_{ox}$)	$1/K$	$1/K$
Gate delay (VC/I)	$1/K$	$1/K^2$
Power dissipation (P)	$1/K^2$	K
Gate delay × power dissipation	$1/K^3$	$1/K$
Current density (J)	K	K^3
Power density (VI/S)	1	K^3

K is the scaling factor.

constant, so that sometimes voltages are not scaled and are maintained constant. This type of scaling is called constant voltage scaling, as shown in Table 11.1. However, in constant voltage scaling, the reduced device dimensions can increase the electric fields. To minimize the field, quasi-constant voltage scaling is often used wherein

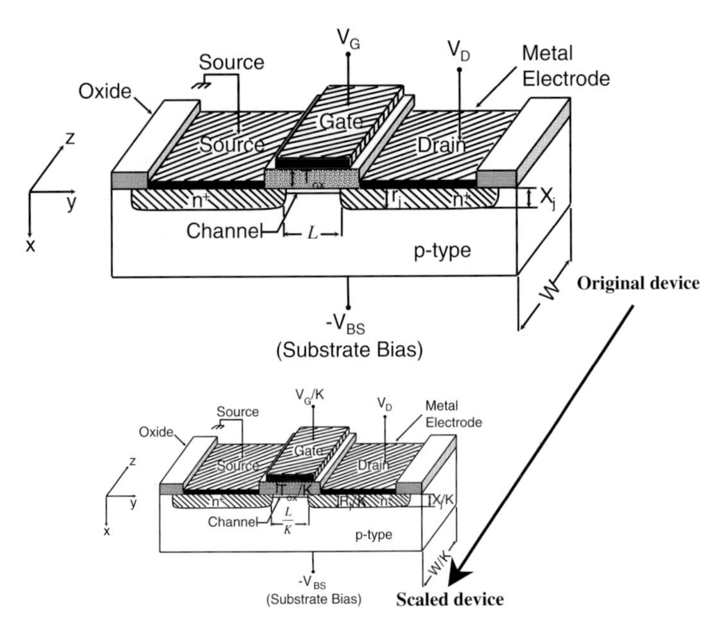

Figure 11.1 Schematic illustration of the scaling of MOSFET device by factor K [10].

Table 11.2 Scaling rules for different types of scaling ($K > U > 1$).

Scaling factor	Electric field constant	Gate voltage constant	Quasi-constant voltage scaling	Generalized scaling
W, L	$1/K$	$1/K$	$1/K$	$1/K$
T_{ox}	$1/K$	$1/K$	$1/U$	$1/K$
N_A	K	K^2	K	K^2/U
V, V_T	$1/K$	1	$1/U$	$1/U$

K and U are scaling factors.

device dimensions are scaled according to constant field scaling rules but voltages are scaled less dramatically. However, in that case, the depletion region widths do not scale in the same ratio as the device dimensions. This problem is avoided by using a different scaling factor for the substrate doping. This is called generalized scaling. The different scaling rules are summarized in Table 11.2 [10].

In a groundbreaking paper written in 1965, Intel cofounder Gordon E. Moore did forecast an exponential growth in the number of transistors per integrated circuit and predicted this trend would continue [11]. His prediction, popularly known as "Moore's law," states that the number of transistors on integrated circuits doubles approximately every 24 months, resulting in higher performance at lower cost. This simple but profound statement is the foundation of semiconductor and computing industries. It is the basis for the exponential growth of computing power, component integration that has stimulated the emergence of generation after generation of PCs and intelligent devices. This observation about silicon integration, made a reality by Intel, has fueled the worldwide technology revolution. Schematic depiction of the transistor count growth for Intel processors (dots) and Moore's law (line) is shown in Figure 11.2. To realize increased integration density, device

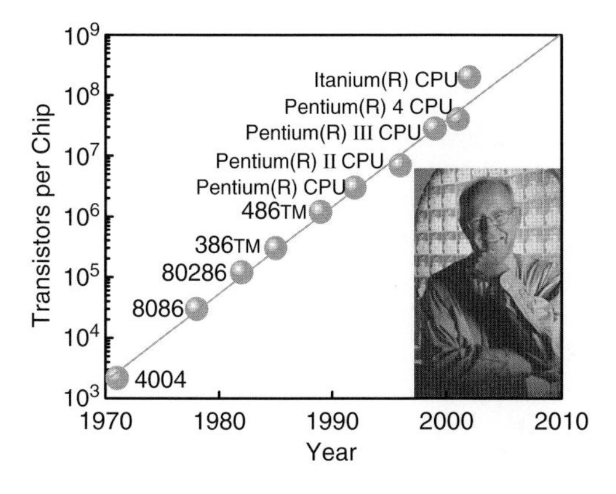

Figure 11.2 Growth of transistor counts for Intel processors (dots) and Moore's law (line). Moore's law means more performance and decreasing cost. (For a color version of this figure, please see the color plate at the beginning of this book.)

dimensions have shrunk drastically over the past several decades and will continue if technology predictions of the International Technology Roadmap for Semiconductors (ITRS) can be realized [12].

Regarding the improved performance associated with the scaling of logic device dimensions, it can be seen by considering a simple model for the drive current associated with a FET [2]. The drive current can be written (using the gradual channel approximation) as

$$I_D = \frac{W}{L}\mu C_{inv}\left(V_G - V_T - \frac{V_D}{2}\right)V_D \tag{11.1}$$

where W is the width of the transistor channel, L is the channel length, μ is the channel carrier mobility (assumed constant here), and C_{inv} is the capacitance density associated with the gate dielectric when the underlying channel is in the inverted state. V_G and V_D are the voltages applied to the transistor gate and drain, respectively. The threshold voltage is denoted by V_T. It can be seen that in this approximation, the drain current is proportional to the average charge across the channel (with a potential $V_D/2$) and the average electric field (V_D/L) along the channel direction. Initially, I_D increases linearly with V_D and then eventually saturates to a maximum when $V_{D,sat} = V_G - V_T$ to yield

$$I_{D,sat} = \frac{W}{L}\mu C_{inv}\frac{(V_G - V_T)^2}{2} \tag{11.2}$$

The term ($V_G - V_T$) is limited in range due to reliability and room-temperature operation constraints since too large a V_G would create an undesirable, high electric field across the oxide. Furthermore, V_T cannot be easily reduced below about 200 mV because it should be larger than kT value, which is about 25 mV at room temperature. Typical specification temperatures ($\leq$100 °C) could therefore cause statistical fluctuations in thermal energy, which would adversely affect the desired V_T value. Thus, even in this simplified approximation, a reduction in the channel length or an increase in the gate dielectric capacitance will result in an increased $I_{D,sat}$.

From a CMOS circuit performance point of view, a performance metric considers the dynamic response (i.e., charging and discharging) of transistors, which should be associated with a specific circuit element and the supply voltage provided to the element at a representative (clock) frequency. A common element employed to examine such switching time effects is a CMOS inverter [2]. This circuit element is shown in Figure 11.3, where the input signal is attached to the gates and the output signal is connected to both the n-type MOS (n-MOS) and the p-type MOS (p-MOS) transistors associated with the CMOS stack. The switching time is limited by both the fall time required to discharge the load capacitance by the n-FET drive current and the rise time required to charge the load capacitance by the p-FET drive current. That is, the switching response time is given by [2]

$$\tau = \frac{C_{Load}V_{DD}}{I_D} \tag{11.3}$$

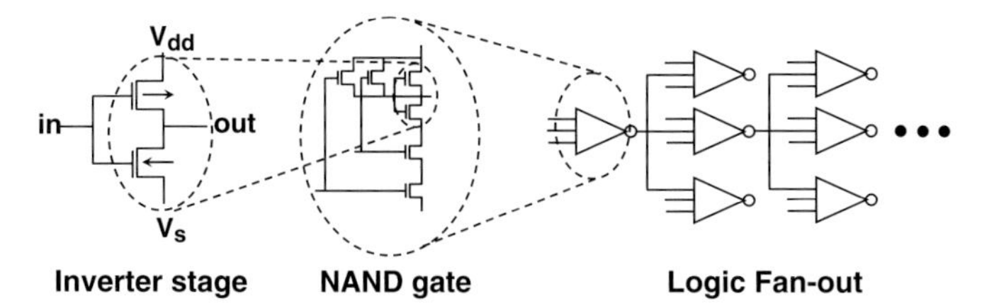

Figure 11.3 Components used to test a CMOS FET technology. V_{DD} and V_S serve as the source and drain voltages, respectively, and are common to the NAND gates shown. Each NAND gate is connected to three resulting in a fan-out of 3.

where $C_{Load} = FC_{Gate} + C_j + C_i$, and C_j and C_i are parasitic junction and local interconnection capacitances, respectively. The "fan-out" for interconnected devices is given by factor "F". Ignoring delay in gate electrode response, as $\tau_{Gate} \ll \tau_{n,p}$, the average switching time is therefore

$$\bar{\tau} = \frac{\tau_p + \tau_n}{2} = C_{Load} V_{DD} \left\{ \frac{1}{I_D^n + I_D^p} \right\} \tag{11.4}$$

The load capacitance in the case of a single CMOS inverter is simply the gate capacitance if one ignores parasitic contributions such as junction and interconnection capacitance. Hence, an increase in I_D is desirable to reduce switching speeds. Considering Equation 11.1 described above, Equation 11.3 could also be simplified to

$$\tau = \frac{C_{Load} V_{DD}}{I_D} = \frac{C_{ox} WLV_{GS}}{\frac{W}{L}\mu C_{ox}(V_{GS} - V_T)} \approx \frac{L^2}{\mu} \tag{11.5}$$

It is obvious that decreasing the channel length is the most effective way to improve the device speed. With the aggressive downscaling of CMOS devices, short-channel effects, such as (1) threshold voltage lowering, (2) channel length modulation, (3) carrier velocity saturation, (4) drain-induced barrier lowering (DIBL), (5) punch-through, (6) breakdown, and (7) hot carrier generation, have to be controlled. First, threshold voltage lowering is of serious concern for switching properties and off-state power consumption. With decreasing channel length, the source and drain depletion regions become a more significant part of the channel. Thus, due to this shared charge from the source and drain depletion regions, lower charges induced by the gate are needed to invert the transistor. This results in lowering of the transistor threshold voltage. Second, channel length modulation is the effect of reduction in the length of the channel, as the pinch-off region becomes a significant part of the channel. Thus, the channel length varies from the "physical" channel length with the applied voltages. This results in dependence of I_{DS} on V_{DS} even in saturation region. Carrier velocity saturation refers to a situation where, beyond a critical electric field, carrier velocity becomes constant and no longer depends on the applied electric field. This essentially leads to lower mobility, and hence, reduced current. DIBL refers to drain-induced barrier lowering, where the barrier for the carriers to come to the channel from the source is reduced due to field lines penetrating from the drain to the source. This effect is enhanced at higher drain bias. As the channel length reduces, the depletion regions of source and drain approach each other. If the substrate doping is not high enough eventually these two depletion regions could touch each other, a situation called punch-through. Due to the high field near drain region, the carriers gain large velocities and may cause impact ionization. These newly generated carriers may be accelerated in the high fields and generate more carriers. This can result in an avalanche. This effect is called hot carrier generation and can lead to a rapid increase in current and, eventually, to a transistor breakdown.

11.2 Urgent Issues about MOS Gate Materials for Sub-0.1 μm Device Gate Stack

The industry's demand for greater integrated circuit functionality and performance at lower cost requires an increased circuit density, which has translated into a higher density of transistors on a wafer. This rapid shrinking of the transistor feature size has forced the channel length and gate dielectric thickness to also decrease rapidly, as listed in Table 11.3. As a result, a series of problems associated with gate materials occurred, and the current CMOS gate dielectric silicon dioxide (SiO_2) and polysilicon gate electrode will have to be phased out in the near future according to the prediction of the ITRS [12]. In other words, "how much longer can Moore's law continue?" is an emerging question for the current integrated circuit industry.

11.2.1 SiO_2 Gate Dielectric

The traditional MOS gate stack mainly consists of heavily doped polysilicon electrode, SiO_2 dielectric and silicon substrate, as schematically denoted in Figure 11.4. SiO_2 dielectric has served for more than four decades as an excellent gate insulator responsible for blocking current in insulated gate field-effect transistor (IGFET) channels from the gate electrode in CMOS devices. To improve the device performance, SiO_2 has been scaled aggressively to invert the surface to a sufficient sheet charge density to obtain the desired current for the given supply voltage and to avoid short-channel behavior. However, with the rapid scaling down of devices, SiO_2 is scaled from a thickness of 100 nm 30 years ago to a mere 1.2 at 90 nm process node. This represents an oxide layer composed of only four atom thick. This tendency will cause a problem as the SiO_2 layer gets thinner. The rate of gate leakage tunneling exponentially goes up. Current leakage contributes to power dissipation and heat.

Table 11.3 EOT versus technology node (2009 ITRS).

Year	2009	2010	2011	2012	2013	2014	2016
Generation	52	45	40	36	32	28	22
T_{ox} (nm)	1–1.3	0.95–1.3	0.88–1.2	0.75–1	0.65–1	0.55–0.95	0.5–0.9

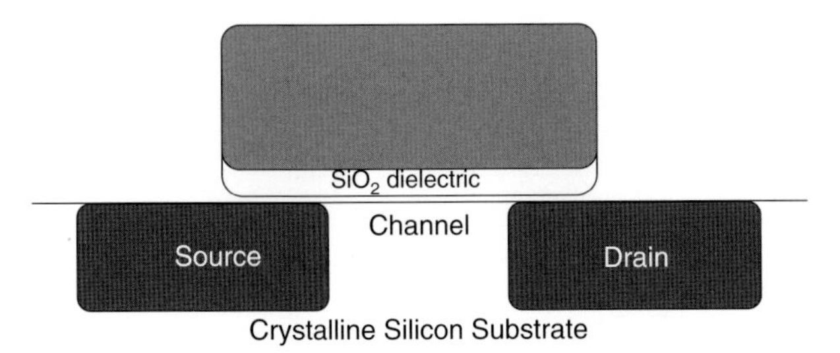

Figure 11.4 Basic CMOS transistor. Below 90 nm process node, the thickness of the SiO_2 gate dielectric will be lower than 1.2 nm (only about four atomic layers). Current leakage due to tunneling imposes significant power dissipation. (For a color version of this figure, please see the color plate at the beginning of this book.)

The relation of tunneling leakage current with SiO_2 thickness can be obtained by the following equation [3, 13]:

$$J_g = \frac{A}{T_{ox}^2} e^{-2T_{ox}\sqrt{(2m^* q/\hbar^2)\{\Phi_B - (V_{ox}/2)\}}} \tag{11.6}$$

where A is an experimentally constant, T_{ox} is the physical thickness of SiO_2 dielectric, Φ_B is the potential barrier height between the metal and the SiO_2, V_{ox} is the voltage drop across the dielectric, and m^* is the electron effective mass in the dielectric. For highly defective films that have electron trap energy levels in the SiO_2 bandgap, electron transport will instead be governed by a trap-assisted mechanism such as Frenkel–Poole emission or hopping conduction [3]. The dependence of leakage current on SiO_2 physical thickness is shown in Figure 11.5 [14]. Consequently, this

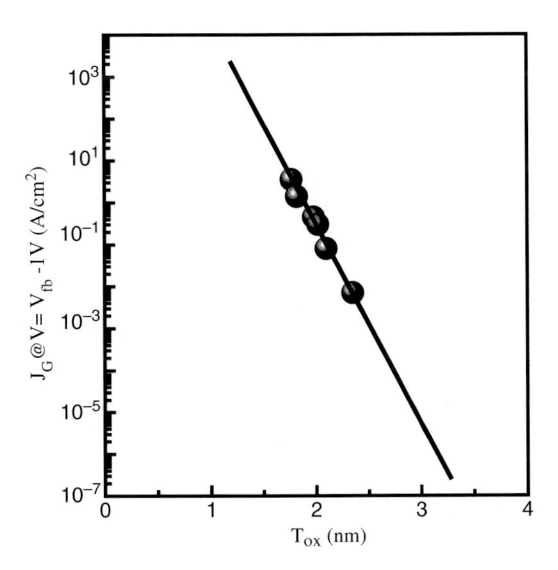

Figure 11.5 Gate leakage current density plot as a function of SiO_2 physical thickness [14].

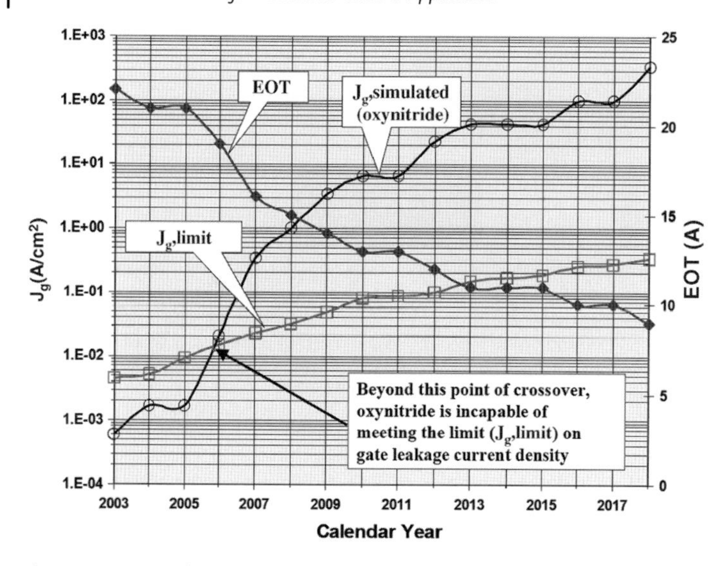

Figure 11.6 LSTP logic scaling up of gate leakage current density limit and of simulated gate leakage due to direct tunneling [12]. (For a color version of this figure, please see the color plate at the beginning of this book.)

conduction problem causes the transistor to stray from its purely "on" and "off" state and into an "on" and "leaky off" behavior. In addition, the ITRS also gave the gate leakage current density limit and simulated gate leakage due to direct tunneling for various logic technology requirements, for example, low standby power logic, low operating power logic, and high performance logic, as shown in Figures 11.6–11.8 [12]. The curves for maximum allowed gate leakage current density

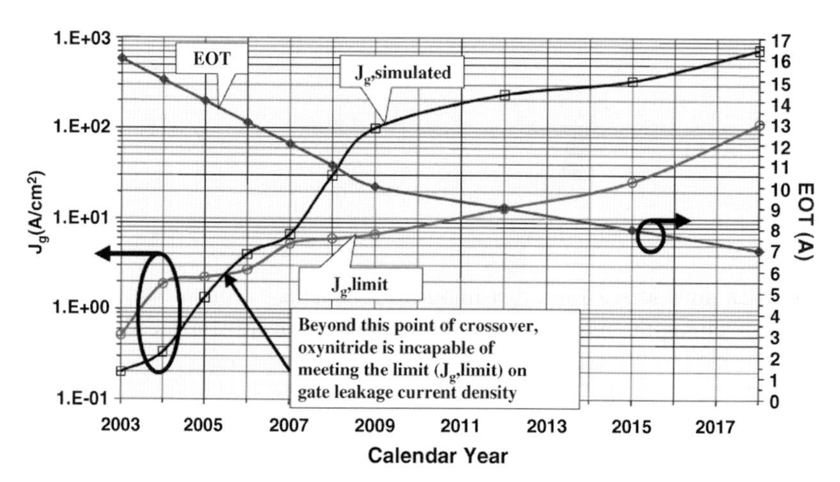

Figure 11.7 LOP logic scaling up of gate leakage current density limit and of simulated gate leakage due to direct tunneling [12]. (For a color version of this figure, please see the color plate at the beginning of this book.)

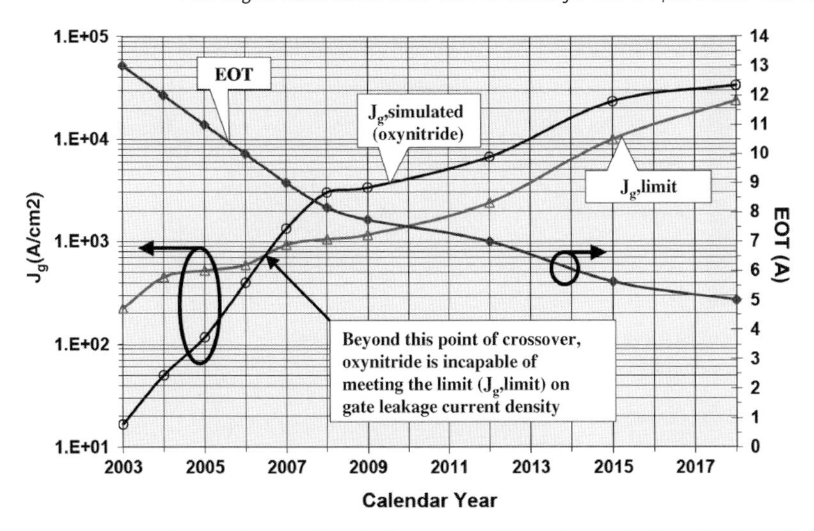

Figure 11.8 HP logic scaling up of gate leakage current density limit and of simulated gate leakage due to direct tunneling [12]. (For a color version of this figure, please see the color plate at the beginning of this book.)

are labeled with "J_g, limit." Obviously, silicon oxide or oxynitride gate dielectrics are projected to be unable to meet the gate leakage current limit by 2006. Without a new dielectric material with increased physical thickness and a higher-κ value, Moore's law would inevitably hit a wall. The mission of next-generation technology development is to break down the barriers and keep Moore's law rolling forward. Solving the gate dielectric problem is a critical issue for the industry.

11.2.2
Polysilicon Electrode

Besides the tunneling leakage problem associated with the scaled SiO_2 dielectric, the issues related to heavily doped polysilicon gate electrode are also becoming severe, such as polysilicon depletion effect and boron (B) penetration [15–19]. The process requirements of traditional n^+/p^+ dual-gate CMOS technologies result in a compromise in the achievable electronically active impurity concentration in the polysilicon gate [20]. The implant and annealing conditions for the polysilicon doping must be carefully selected to avoid impurity penetration through the gate oxide, while maintaining the required source/drain junction depth and lateral diffusion length dictated by scaling rules. The reduced active dopant levels in the polysilicon gate give rise to the formation of a depletion layer near the polysilicon/oxide interface when the device is biased in strong inversion, which in turn results in degraded device characteristics (*the so-called polysilicon depletion or poly-depletion effect*), as schematically shown in Figure 11.9. This effect adds ~3–4 Å to the effective dielectric thickness, which accentuates a significant effect of polysilicon depletion on the sub-0.1 μm technology nodes [21]. It seems that the electronically active

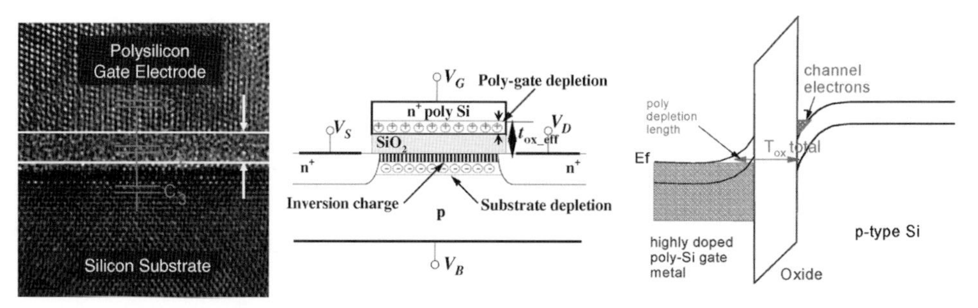

Figure 11.9 Schematic diagram of polysilicon depletion effect.(For a color version of this figure, please see the color plate at the beginning of this book.)

doping density is difficult to be above $10^{20}/\text{cm}^3$ for n^+ polysilicon and above the mid-$10^{19}/\text{cm}^3$ for p^+ polysilicon electrode. This implies an inherently limited capability of improving device performance. In addition, regarding the boron penetration problem, the drive to add more B to the heavily doped p^+ polysilicon gate to minimize depletion, together with the thinning of the gate dielectric, results in increased B diffusion through the gate dielectric into the channel of the p-MOS device than does the n^+ dopant used for n-MOS. Thus, the out-diffused B dopant accumulates in the n-Si substrate where it can change the threshold voltage and reduce dielectric reliability, thereby again degrading the intended device characteristics in an uncontrollable and unacceptable way [22]. Finally, it is anticipated that polysilicon will not be stable on most high-κ dielectric materials since it can react with high-κ dielectric to form silicides, as schematically depicted in Figure 11.10. For CMOS scaling in the longer term, current roadmap predicts that polysilicon gate technology will likely be phased out beyond the 65 nm node, after which the use of stable metal gate electrodes will be required.

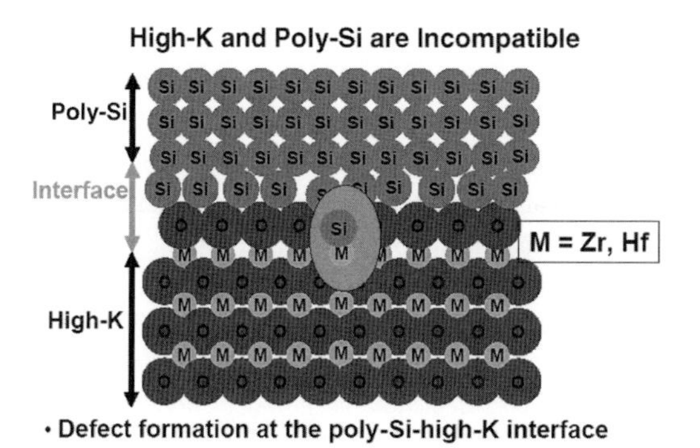

Figure 11.10 Instability of polysilicon electrode on high-κ dielectrics. (For a color version of this figure, please see the color plate at the beginning of this book.)

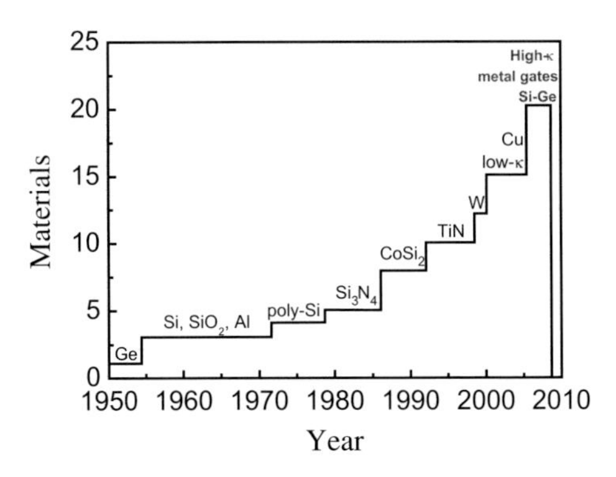

Figure 11.11 Schematic depiction for the requirements of material development in IC technology [23, 24].

11.3 New Requirements of MOS Gate Materials for Sub-0.1 μm Device Gate Stack

To meet the aggressive progress of MOS devices, large efforts are devoted to use new gate materials to solve the unendurable integrating problems involved in producing chips at the 45 nm, 32 nm, and even smaller levels. This technology will postpone the industry meeting the limits of Moore's law. Consequently, high-κ gate dielectric and metal gate electrode materials are proposed. Figure 11.11 schematically summarizes the requirements for material development [23, 24].

11.3.1 High-κ Gate Dielectric

In order to realize the continued scaling of MOS devices, a suitable replacement for SiO_2 should soon be developed. After more than 10 years of intensive effort, a new material known as "high-κ" dielectric has been identified. "High κ" stands for high dielectric constant, which is a measure of how much charge a material can hold. Different materials similarly have different abilities to hold charge. As an alternative to SiO_2, high-κ metal oxides can provide a substantially thicker (physical thickness) dielectric for reduced leakage and improved gate capacitance.

As for the gate capacitance issue, capacitance of a parallel plate capacitor can be expressed as (ignoring quantum mechanical and depletion effects from a Si substrate and gate) [25]

$$C = \frac{\kappa \varepsilon_0 A}{t} \tag{11.7}$$

where κ is the dielectric constant (also referred to as the relative permittivity) of the material, ε_0 is the permittivity of free space ($= 8.85 \times 10^{-3}$ fF/μm), A is the area of the capacitor, and t is the thickness of the dielectric. This expression for C can be rewritten in terms of t_{eq} (i.e., equivalent oxide thickness, EOT) and κ_{ox}

(= 3.9, dielectric constant of SiO_2) of the capacitor. The term t_{eq} represents the theoretical thickness of SiO_2 that would be required to achieve the same capacitance density as the dielectric (ignoring issues such as leakage current and reliability). For example, if the capacitor dielectric is SiO_2, $t_{eq} = 3.9\varepsilon_0(A/C)$, and a capacitance density of $C/A = 34.5\ \mathrm{fF}/\mu\mathrm{m}^2$ corresponds to $t_{eq} = 1$ nm. Thus, the physical thickness of an alternative dielectric employed to achieve the equivalent capacitance density of $t_{eq} = 1$ nm can be obtained from the following expression:

$$\frac{t_{eq}}{\kappa_{ox}} = \frac{t_{\mathrm{high}\ \kappa}}{\kappa_{\mathrm{high}\ \kappa}} \tag{11.8}$$

For example, a dielectric with a relative permittivity of 16 therefore affords a physical thickness of ~4 nm to obtain $t_{eq} = 1$ nm (as noted above, the actual performance of a CMOS gate stack does not scale directly with the dielectric due to possible quantum mechanical and depletion effects) [25]. Reminding the relation of leakage current with the physical thickness of oxides described in Equation 11.6, the increased physical thickness in high-κ films would effectively decrease the tunneling leakage, while keeping the equivalent capacitance density constant, that is, drive current constant. This means that the higher "κ" increases the transistor capacitance so that the transistor can switch properly between "on" and "off" states, and has very low drain current when off, yet very high drain current when on. However, despite the encouraging results from the potential application of high-κ dielectrics in reducing current leakage, there also are several issues associated with it, such as decreased bandgap, degraded thermal stability, compatibility with gate electrode, unexpected band misalignment, and low channel mobility [26–29]. The bandgap of oxide dielectric, or more importantly the barrier height, tends to decrease with increasing dielectric constant [26]. The decrease in bandgap would cause an increase in tunneling leakage at a particular bias, and this could then offset the reduction in leakage current caused by the increase in physical thickness of the high-κ dielectrics; in addition, some interfacial oxide between the high-κ dielectrics and the Si substrates could be formed during gate stack preparation. Consequently, compared to the case without interlayer, the tunneling current through the stacks increased, even with the same t_{eq}, because of significantly reduced barrier height related to the interface region [28]; the interface quality of high-κ dielectric and Si underneath is also a critical issue for achieving high channel mobility. It is still ambiguous that low interface state densities, low fixed charge, and smooth surface could be achieved by using any high-κ dielectric material other than SiO2. Furthermore, even in the case of high-κ dielectric, there still exist some inherent limitations that could seriously threaten the continued scaling of all gate dielectrics, regardless of the material [3, 30]. First, the electrical thickness of any dielectric is given by the distance between the centroids of charge in the gate and the substrate. This thickness, as denoted by t_{eq}, therefore includes the effective thickness of the charge sheet in the gate and the inversion layer in the substrate (channel). These effects can add significantly to the expected t_{eq} derived from the physical thickness of the dielectric alone [25]. Depletion in the polysilicon gate electrode arises from the depletion of mobile charge carriers in the polysilicon near the gate dielectric interface, particularly in the gate bias polarity required to invert the channel, as described above. Consequently, about 3–4 Å in the polysilicon electrode nearest to the gate dielectric interface essentially behaves like intrinsic Si, which adds about 3–4 Å to the effective dielectric thickness (rather than acting as a metal with a Fermi sea of electrons right up to the dielectric interface); the nature of the inversion charge layer in the Si substrate (or channel, for transistors) also contributes about 3–6 Å to the effective t_{eq} value, thus even for ideal, degenerately doped poly-Si gates, it is difficult to realize an overall $t_{eq} < 10$ Å in MOSFETs using current process technology. Second, for a constant electric field scaling, the ideal case is that the operating voltage and transistor dimensions are reduced by the same factor, while in practice, the feature dimensions have been reduced more rapidly than the operating voltage, therefore causing a rapidly increasing electric field across the gate dielectric. The continually decreasing t_{eq} value for scaling CMOS also increases the effective electric field in the channel region. As a result, this increased electric field pulls the carriers in the channel closer against the dielectric interface, which causes increased phonon scattering of more confined carriers and thereby decreases the channel carrier mobility. At very high electric fields in the channel, such as established by $t_{eq} < 10$ Å, interface roughness scattering further reduces carrier mobility.

As discussed above, replacing the SiO_2 with a material having a higher dielectric constant is not as simple as it may seem. There is a wide variety of films with higher κ values than SiO_2, ranging from Si_3N_4 with a κ value of about 7 to Pb-La-Ti (PLT) oxide with a κ value of 1400 [31, 32]. Unfortunately, many of these films are not thermodynamically stable with respect to Si, which makes it easy to react with Si, or are lacking in other properties. The alternative high-κ materials should have high breakdown voltage, low diffusion coefficients of the elements withstanding process temperatures to avoid mixed oxide formation, high band offset (at least 1 eV) restricting leakage current due to thermal emission or tunneling, good adhesion, low deposition temperature, low defect density and charge states on Si, and ability to be patterned easily. Table 11.4 and Figures 11.12 and 11.13 summarize the properties of some of the studied high-κ dielectrics [26, 33]. Considering these objectives, recent research on high-κ dielectrics has primarily focused on metal oxides and their silicates. Among these, the group IVB transition metals, Hf and Zr, and rare-earth metals have generated a substantial amount of investigations [34–38]. Transistors based on these films have shown excellent overall performance presenting possible solutions to the need for thinner t_{eq} with low leakage current.

11.3.2 Metal Gate Electrode

Besides the exploration for alternative high-κ dielectrics, to resolve boron penetration, polysilicon depletion, and compatibility issues with high-κ dielectrics, new metal electrodes are required to replace the traditional polysilicon gate electrode used in n-MOS and p-MOS transistors.

Metal gate electrodes could offer a possible solution to the gate depletion problem, but the addition of 3–6 Å to the t_{eq} value from the Si channel will remain. In addition,

Table 11.4 Experimental bandgaps, relative dielectric constants, and conduction band offset on Si for high-κ candidates.

Dielectrics	Dielectric constant	Bandgap E_g (eV)	Conduction band offset (eV)
Si	11.9	1.1	
SiO_2	3.9	8.9	3.5
Si_3N_4	7.9	5.3	2.4
Al_2O_3	9.5–12	8.8	2.8
ZrO_2	12–16	5.7–5.8	1.4–15
Al_2O_3	15	5.6	2.3
$ZrSiO_4$	10–12	~6	1.5
HfO_2	16–30	4.5–6	1.5
$HfSiO_4$	~10	~6	1.5
La_2O_3	20.8	~6	2.3
Ta_2O_5	25	4.4	0.36
TiO_2	80–170	3.05	~0

it has been reported that Fermi-level pinning at the high-κ/polysilicon interface causes high threshold voltages in MOSFET transistors [39]. High-κ/polysilicon transistor exhibits severely degraded channel mobility due to the coupling of low-energy surface optical (SO) phonon modes arising from the polarization of the high-κ to the inversion channel charge carriers. Metal gate may be more effective in screening the high-κ SO phonons from coupling to the channel under inversion conditions [40]. In addition, the use of metal gates in a replacement gate process can lower the required thermal budget by eliminating the need for dopant activation anneal in the polysilicon electrode.

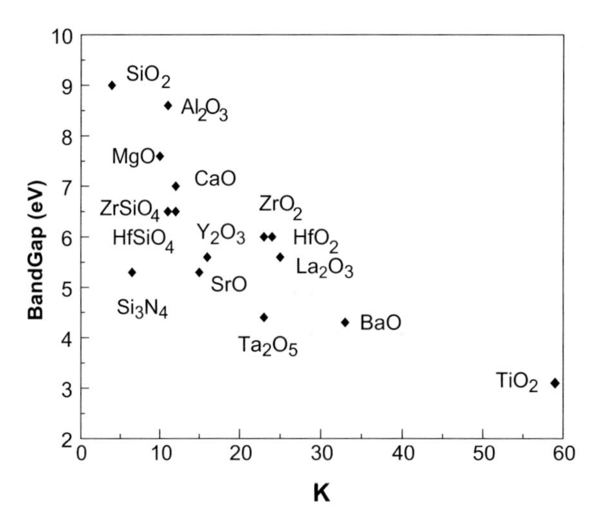

Figure 11.12 Dependence of bandgap on dielectric constant for various high-κ candidates.

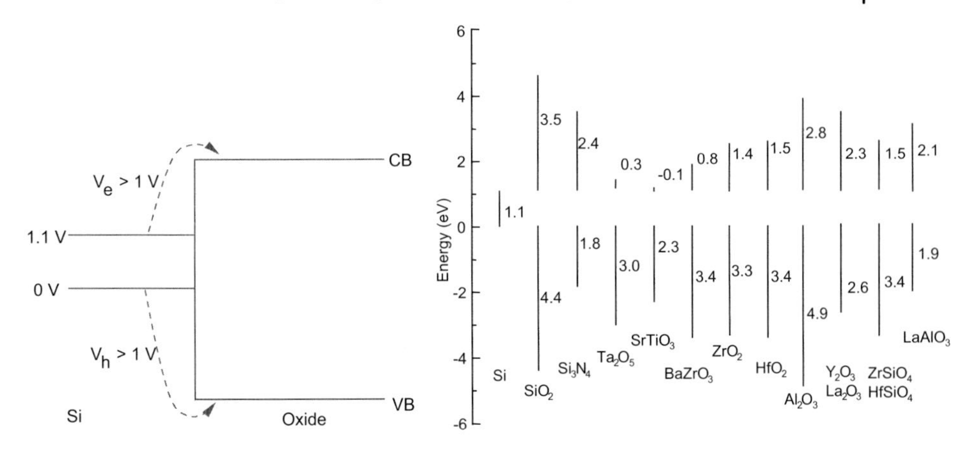

Figure 11.13 Schematic diagram of band offset, and the calculated conduction and valence band offsets for a number of potential high-κ candidates on Si.

Over the years, the use of polysilicon gate electrodes has become nearly as deeply rooted in Si device technology as has the SiO_2–Si interface. The necessity of finding new gate electrode materials and processes at the time for implementing new high-κ gate dielectrics has caused great concern in the industry, which has tried to maintain a baseline of familiar, controllable processes in the face of continued scaling challenges. Therefore, the first impulse in searching for a metal gate replacement for polysilicon was to try to find a single metal that would work for both n-MOS and p-MOS devices in a circuit by selecting one with a Fermi level aligned near the mid-gap position between the valence and the conduction bands in the Si substrate, as depicted in Figure 11.14a [3]. This would maintain the nearly equal but opposite polarity n-MOS and p-MOS threshold voltage (V_t) values required for circuit operation, although these would be higher than those for dual-doped polysilicon gates because of the need to overcome the approximate 0.5 V of band bending (half the bandgap; the energy difference between the mid-gap and the band edges) required to charge the minority carrier bands. It is thought that this might be

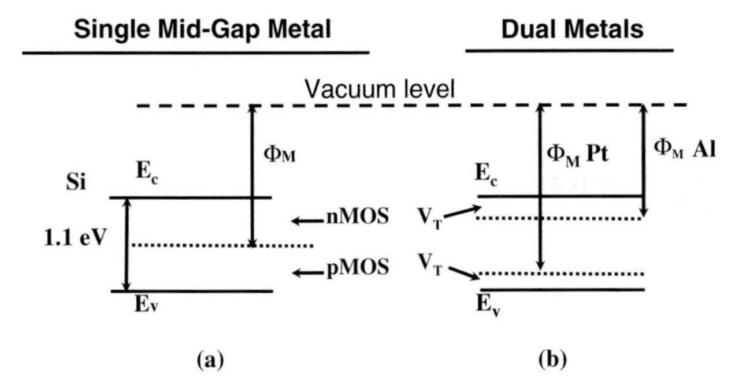

Figure 11.14 Energy diagrams of threshold voltages for n-MOS and p-MOS devices using (a) midgap metal gates and (b) dual metal gates.

ameliorated to some extent by channel doping of the devices. The main advantage of employing a mid-gap metal arises from a symmetrical V_T value for both n-MOS and p-MOS. This affords a simpler CMOS processing scheme since only one mask and one metal would be required for the gate electrode (no ion implantation step would be required). However, in view of device performance and scaling requirements, it seems that this approach might not be a feasible alternative for current planar bulk CMOS devices due to the large threshold voltage ($\sim$0.5 V), while it may be a promising candidate for nonclassical CMOS structures such as ultrathin body (UTB) fully depleted silicon-on-insulator (FDSOI) initially and later some type of multiple gate UTB MOSFETs, which are expected to be utilized to deal with the difficulties encountered in current bulk CMOS technology [41]. For high-frequency circuits (about 5 GHz and above), capacitive coupling to the silicon substrate in bulk CMOS devices limits the switching frequency. Also, leakage into the substrate from the small devices can cause extra power dissipation. These problems would not be sufficiently avoided just by simply scaling the existing device structures to these extremely short-channel lengths. Thereby, several new types of device structures, for example, SOI and multiple gate UTB MOSFETs, are being investigated by making circuits on insulating substrates (either sapphire or silicon dioxide) that have a thin, approximately 100 nm, layer of crystalline silicon, in which the MOSFETs are fabricated. For this case, because "intrinsic" channels (or assumed undoped channels) could be used for both n-MOS and p-MOS devices, a gate material with a mid-gap work function is desirable [42], at which the V_T can also be set to the desired value of several tenths of a volt for n-MOSFETs and similar negative values for p-MOSFETs, the gate electrode work function must be near mid-gap (quasi-mid-gap), for example, several tenths of a volt above the Si mid-gap position for n-MOSFETs and several tenths of a volt below it for p-MOSFETS. In a word, different gate electrodes can be utilized for different devices.

Naturally, for the current bulk CMOS devices, a dual-metal gate strategy would be needed [3, 43]. As shown in Figure 11.14b, two metals could be chosen by their work functions, ϕ_M, such that their Fermi levels line up favorably with the conduction and the valence bands of Si, respectively. Device simulation taking into account quantum effect on device drive current suggests that the ϕ_M for n-MOS and p-MOS gate electrodes must be around 4 and 5 eV, respectively [43]. In the ideal case depicted in Figure 11.14b, the ϕ_M value of Al could achieve V_T $\sim$0.2 V for n-MOS, while the higher ϕ_M value of Pt could achieve V_T $\sim$0.2 V for p-MOS. In practice, Al is not a feasible electrode metal because it will reduce nearly any oxide gate dielectric to form an Al_2O_3-containing interface layer. Similarly for p-MOS, Pt is not a practical choice for the gate metal since it is not easily processed, does not adhere well to most dielectrics, and is expensive.

To search for metallic materials with suitable work functions that might be compatible with Si process technology, some relatively unfamiliar candidates such as In, Sn, Os, Ru, Ir, Zn, Mo, Re, and their oxides have been uncovered [4, 44–47]. More commonly used materials such as Ta, V, Zr, Hf, and Ti were also investigated as the metal electrode with low work functions for n-MOS devices, and Mo, W, Co, and Au were examined for p-MOS devices with high work functions. In addition, some

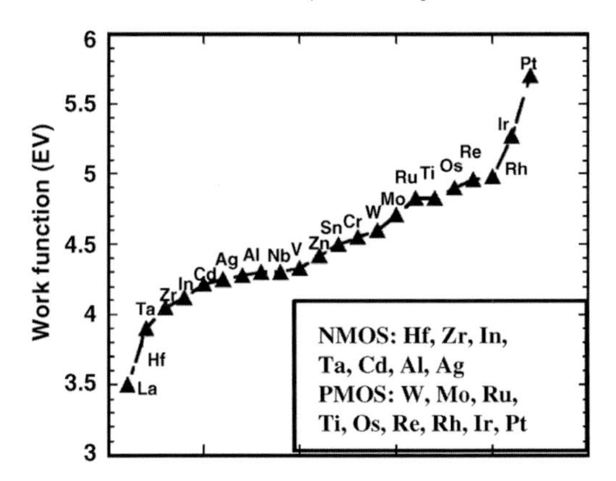

Figure 11.15 Work functions of some potential metals for CMOS devices [48].

metal nitrides and metal alloys are also being considered such as WN_x, TiN_x MoN_x, TaN_x, $TaSi_xN_y$, Ru-Ta, Ru-Zr, Pt-Hf, Pt-Ti, Co-Ni, and Ti-Ta. The work functions of some potential metals for CMOS devices are summarized in Figure 11.15 [48]. The real challenge is that a suitable work function is the first requirement for a dual-metal technology, but it is only one of the many equally important requirements such as processing ambient and temperature used in Si technology, thermal stability, and compatibility with many other materials and deposition techniques.

First, the straightforward issue is the sequential deposition and selective etching of both metals for a dual-metal gate technology. In a sequential dual-metal process, the etching of the first metal from top of the gate oxide will cause potential damage, particularly during the overetching required to assure complete removal of the first metal layer. To avoid exposure of the gate dielectric to etching environment, several approaches have been proposed to modify or "tune" the work function of metal gate systems. In general, these methods depend upon the specific properties of the materials chosen. For example, a metal layer with a work function appropriate for one of the devices, the n-MOS, for instance, could be first deposited. After that, the second metal chosen to form an alloy with a work function, appropriate for the p-MOS device, was deposited and subjected to thermal treatment [49]. Prior to the thermal treatment, the second metal film is etched off from the n-MOS devices, leaving the first metal film in place. The specific materials proposed in this approach are Ti for the first metal layer and Ni for the second film. Another technique depends upon the fact that compounds of some metals, for instance, their nitrides, can have metallic properties with work functions different from that of the parent material. As for this process, after deposition of the parent metal layer, appropriate devices are masked photolithographically so that those remaining exposed can be implanted with N_2. The nitride is then formed postannealing [50]. In addition, in the study of high-κ dielectric, by employing silicates of Hf and Zr, which may be considered as metal-doped SiO_2, the beneficial properties of the SiO_2-Si system could be possibly maintained. A similar philosophy may be applied to the case of dual-work function gated electrodes by using fully silicided metal gates (FUSI) [51]. For more advanced devices, such as dual-gate or 'fin' field-effect transistors (FinFETs) [52], it is found that only limited tuning of the work function near the mid-gap value is required. Both pure metal systems [53], such as W, and FUSI structures have been proposed for this application [54].

Thermal and chemical stability under high-temperature process is also a critical issue for those potential metal electrode candidates. The highest temperatures used in device fabrication with gate-first process are for the activation of dopant atoms in the source, drain, and gate regions of the transistor. Typical activation thermal budgets include rapid thermal annealing between 900 and 1000 °C (a few seconds to a minute). Metal gates are expected to be stable with respect to the underlying gate dielectric during high-temperature process. As described above, alternative metal gates include elemental metals, metal oxides, metal nitrides, metal silicides, or other metallic alloys; many metal gate electrodes, however, are unstable both on SiO_2 and on high-κ dielectrics, such as Ti, Zr, and Hf [55–57]. The performance of CMOS devices could be degraded by the reaction or intermixing at the interface of metal electrode and dielectric due to any possible change in work function, EOT, or other parameters. Typically, low work function metals are not thermodynamically stable on SiO_2 or high κ, while high work function metals tend to be inert and naturally immune to oxidation. Figure 11.16 shows the thermal stability of various gate metal candidates in contact with SiO_2. The thermodynamic heats of formation for some elements on SiO_2 are also summarized in Table 11.5.

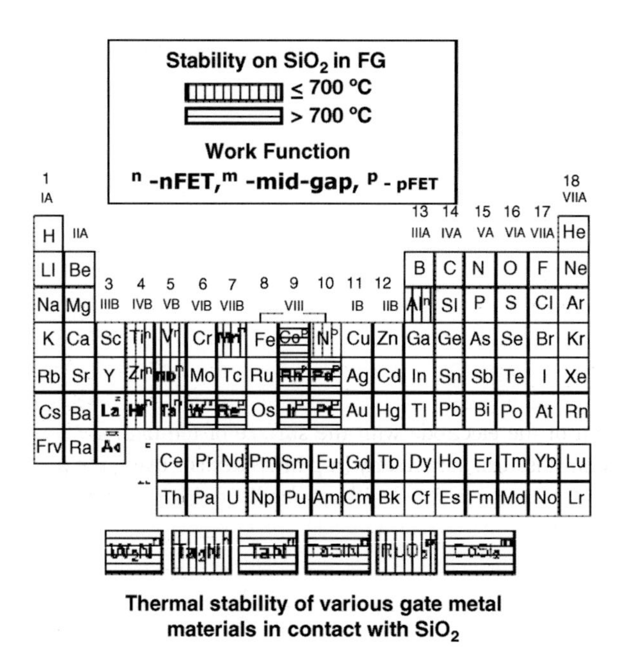

Figure 11.16 Thermal stability of various gate metal candidates in contact with SiO_2 [3].

Table 11.5 Heat of formation of some elements toward oxygen.

Heat of formation of oxide ($-\Delta H_f$ kJ/mole)	Elements
0–50	Au, Au, Pt
50–100	Pd, Rh
150–200	Ru, Cu
200–250	Re, Co, Ni
250–300	Na, Fe, Mo, Sn, Ge, W
300–350	Rb, Cs, Zn
350–400	K, Cr, Nb, Mn
400–450	V
450–500	Si
500–550	Ti, U, Ba, Zr, Hf
550–600	Al, Sr, La, Y, Ce

The properties of metal gate electrodes are also affected by the deposition technique used in the material fabrication such as the morphology of the gate electrode, and the interface quality of the gate electrode and the dielectric underneath. In general, a deposition technique that creates sharp interfaces is expected. Various methods such as physical vapor deposition (PVD) and chemical vapor deposition (CVD) techniques have been used for the deposition of thin films. Most metal thin films are deposited using PVD method (sputtering or evaporation). Unfortunately, most of these techniques involve the use of energetic particles (ions, electrons) and are likely to introduce physical damage to the gate dielectric (rough interfaces, metal ions penetrating into the dielectric) and result in degraded gate dielectric reliability [58–62]. Compared to PVD methods, CVD techniques have advantages for film fabrication of IC process owing to its good step coverage and compatibility with large-scale processing. Especially, it would lead to minimal damage to the underlying dielectric and provide a number of variables such as temperature, pressure, or gas flow to control the film microstructure. However, apart from a few exceptions (W and Ta), investigations for refractory metal CVD have not been very well identified or characterized. This is a challenging area that can be expected to become even more important as CMOS technology continues to be scaled down below sub-0.1 μm technology.

Another issue related to the development of metal gate systems for high-κ dielectrics is the interaction of the electrode with the surface of the high-κ films [63–68]. Unexpected V_T control issues, particularly in p-MOS devices using the p^+ polysilicon–HfO_2 system, have been associated with a polysilicon–HfO_2 interaction resulting in the pinning of the Fermi level at the interface at a value different from that predicted by the doping of the polysilicon [39]. Such Fermi-level pinning effects have also been predicted and described for metal gates in conjunction with high-κ dielectrics [63]. This rather general effect may result in the need to identify metal gate candidates with work functions even smaller for n-MOS devices and larger for p-MOS devices than those required for a SiO_2 gate dielectric. An important method to overcome the instability of the metal gate with high-κ dielectric during the high-temperature fabrication process is the gate-last process, which is demonstrated by the Intel Corporation [69–71]. After the deposition of the high-κ dielectric, poly-Si is deposited as the replacement gate, which goes through the following anneal activation of the source and drain regions. Then, the poly-Si gate is removed and the appropriate metal gates with typical work functions for n-MOS and p-MOS performance are deposited. Thus, the reaction of the metal gate and high-k dielectric during the high-temperature process can be suppressed. The reliability such as the V_T instability can be improved. In addition, using a replacement metal gate flow enables stress enhancement techniques to be in place before removing the poly-Si gate from the transistor. It has been shown that this can further enhance strain and is a key benefit of this flow [72, 73].

More work should be done to better understand alternative metal electrodes, both for dual-metal and for mid-gap metal approaches, as a means of alleviating potentially limiting properties of highly doped polysilicon.

11.4 Summary

The introduction of high-κ dielectric and metal gate electrode is demonstrated based on the downscaling requirements of MOS devices. Enough high-κ value, thermal stability of high-κ dielectric with Si substrate, compatibility of high-κ dielectric with gate electrode, the interfacial characteristics between high-κ dielectric and Si substrate, band alignment of band offset, and high channel mobility have to be considered carefully for the introduction of high-κ dielectric. The combination of high-κ dielectric and metal gate electrode must be simultaneously investigated as integration. The tuning of flatband voltage for MOS devices with high-κ/metal gate stack is the main challenge especially for gate-first process. Thermal and chemical stability, film deposition technique, and interaction of metal gate electrode with high-κ film also affect the electrical performance of MOS devices.

References

1 Lilienfeld, J.E. (1930) Method and apparatus for controlling electric currents. U.S. Patent 1 745 175.

2 Kahng, D. and Atalla, M.M. (1960) Silicon–silicon dioxide field induced surface devices. IRE Solid-State Device Research Conference, Pittsburgh, PA, June.

3 Wilk, G.D., Wallace, R.M., and Anthony, J.M. (2001) High-κ gate dielectrics: current status and materials properties considerations. *J. Appl. Phys.*, **89**, 5243.

4 Hori, T. (1997) *Gate Dielectrics and MOS ULSIs*, Springer, New York.

5 Dennard, R.H., Gaensslen, F.H., Yu, H.-N., Ideout, V.L., Bassous, E., and LeBlanc, A.R. (1974) Design of ion-implanted MOSFETs with very small physical dimensions. *IEEE J. Solid State Circ.*, **9**, 256.

6 Baccarani, G., Wordeman, M.R., and Denard, R.H. (1984) Generalized scaling theory and its application to a $^1/_4$ micrometer MOSFET design. *IEEE Trans. Electron. Dev.*, **31**, 452.

7 Packan, P.A. (1999) Pushing the limits. *Science*, **285**, 2079.
8 Critchlow, D.L. (1999) MOSFET scaling: the driver of VLSI technology. *Proc. IEEE*, **87**, 659.
9 Tsividis, Y. (1999) *Operation and Modeling of The MOS Transistor*, 2nd edn.
10 Sze, S.M. (1981) *Physics of Semiconductor Devices*, Murray Hill, New Jersey.
11 Moore, G.E. (1965) Cramming more components onto integrated circuits. *Electronics*, **38**, 8.
12 Semiconductor Industry Association (2004) The international Technology Roadmap for Semiconductor, Semiconductor Industry Association.
13 Kim, N., Austin, T., Blaauw, D., Mudge, T., Flautner, K., Hu, J., Irwin, M., Kandemir, M., and Narayanan, V. (2003) Designing computer architecture research workloads. *IEEE Comput.*, **36**, 65.
14 Hokazono, A., Ohuchi, k., Takayanagi, M., Watanabe, Y., Magoshi, S., Kato, Y., Shimizu, T., Mori, S., Oguma, H., Sasaki, T., Yoshimura, H., and Miyano, K. (2002). 14nm gate length CMOSFETs utilizing low thermal budget process with poly-SiGe and Ni salicide. IEEE International Electron Devices Meeting Technical Digest, p. 639.
15 Wakabyashi, H., Saito, Y., Takeuchi, K., Mogami, T., and Kunio, T. (1999) A novel W/TiN_x metal gate CMOS technology using nitrogen-concentration-controlled TiN_x film. IEEE International Electron Devices Meeting Technical Digest, p. 253.
16 Stewart, E.J., Carroll, M.S., and Sturm, J.C. (2001) Suppression of boron penetration in p-channel MOSFETs using polycrystalline $Si_{1-x-y}Ge_xC_y$ gate layers. *IEEE Electron. Dev. Lett.*, **22**, 574.
17 Choi, C.H., Chidambaram, P.R., Khamankar, R., Machala, C.F., Yu, Z.P., and Dutton, R.W. (2002) Gate length dependent polysilicon depletion effects. *IEEE Electron. Dev. Lett.*, **23**, 224.
18 Vasileska, D. (1999) The influence of space-quantization effects and poly-gate depletion on the threshold voltage, inversion layer and total gate capacitances in scaled Si-MOSFETs. *J. Model. Simul. Microsyst.*, **1**, 49.
19 Kim, S.D., Park, C.M., and Woo, J.C.S. (2000) Advanced model and analysis for series resistance in sub-100nm CMOS including poly depletion and overlap doping gradient effect. IEEE International Electron Devices Meeting Technical Digest, p. 723.
20 Pfiester, J.R., Baker, F., Mele, T., Tseng, H., Tohin, P., Hayden, J., Miller, J., Gunderson, C., and Panillo, L. (1990) The effects of boron penetration on p^+ polysilicon gated PMOS devices. *IEEE Trans. Electron. Dev.*, **37**, 1842.
21 Huang, C.L., Arora, N.D., Nasr, A., and Bell, D. (1993) Effect of polysilicon depletion on MOSFET I–V characteristics. *Electron. Lett.*, **29**, 1208.
22 Wu, E., Nowack, E., Han, L., Dufresne, D., and Abadeer, W. (1999) Nonlinear characteristics of Weibull breakdown distributions and its impact on reliability projection for ultra-thin oxides. IEEE International Electron Devices Meeting Technical Digest, p. 441.
23 Robertson, J.,New Materials for Electronics, Session 1, 2nd series of 2004 Horizon Seminars.
24 Braun, A.E. (2001) Photostrip faces 300mm, copper and low-κ convergence. *Semiconduct. Int.*, **23** (10), 78–90.
25 Rios, R. and Arora, N.D. (1994) Determination of ultra-thin gate oxide thicknesses for CMOS structures using quantum effects. IEEE International Electron Devices Meeting Technical Digest, p. 613.
26 Robertson, J. (2000) Band offsets of wide-band-gap oxides and implications for future electronic devices. *J. Vac. Sci. Technol. B*, **18**, 1785.
27 Brar, B., Wilk, G.D., and Seabaugh, A.C. (1996) Direct extraction of the electron tunneling effective mass in ultrathin SiO_2. *Appl. Phys. Lett.*, **69**, 2728.
28 Vogel, E.M., Ahmed, K.Z., Hornung, B., Henson, W.K., Mclarty, P.K., Lucovsky, G., Hauser, J.R., and Wortman, J.J. (1998) Modeled tunnel currents for high dielectric constant dielectrics. *IEEE Trans. Electron. Dev.*, **45**, 1350.
29 Hubbard, H.J. and Schlom, D.G. (1996) Thermodynamic stability of binary oxides in contact with silicon. *J. Mater. Res.*, **11**, 2757.
30 Iwai, H., Momose, H.S., and Ohmi, S. (2000) The physics and chemistry of SiO_2

and the Si-SiO_2 interface. Proceedings of the Electrochemical Society, 3.

31 Ma, T.P. (1998) Making silicon nitride film a viable gate dielectric. *IEEE Trans. Electron. Dev.*, **45**, 680.

32 Dey, S.K. and Lee, J.J. (1992) Cubic paraelectric (nonferroelectric) perovskite PLT thin films with high permittivity for ULSI DRAMs and decoupling capacitors. *IEEE Trans. Electron. Dev.*, **39**, 1607.

33 Robertson, J. and Chen, C.W. (1999) Schottky barrier heights of tantalum oxide, barium strontium titanate, lead titanate, and strontium bismuth tantalite. *Appl. Phys. Lett.*, **74**, 1168.

34 Takeuchi, H. and King, T.J. (2003) Scaling limits of hafnium-silicate films for gate-dielectric applications. *Appl. Phys. Lett.*, **83**, 788.

35 Ding, S.J., Zhu, C.X., Li, M.F., and Zhang, D.W. (2005) Atomic-layer-deposited Al_2O_3–HfO_2–Al_2O_3 dielectrics for metal-insulator-metal capacitor applications. *Appl. Phys. Lett.*, **87**, 053501.

36 Seong, N.J., Yoon, S.G., Yeom, S.J., Woo, H.Y., Kil, D.S., Roh, J.S., and Sohn, H.C. (2005) Effect of nitrogen incorporation on improvement of leakage properties in high-k HfO_2 capacitors treated by N_2-plasma. *Appl. Phys. Lett.*, **87**, 132903.

37 Zhao, C., Witters, T., Brijs, B., Bender, H., Richard, O., and Caymax, M. (2005) Ternary rare-earth metal oxide high-k layers on silicon oxide. *Appl. Phys. Lett.*, **86**, 132903.

38 Barlage, D., Arghavani, R., Dewey, G., Doczy, M., Doyle, B., Kavalieros, J., Murthy, A., Roberds, B., Stokley, P., and Chau, R. (2001) High-frequency response of 100nm integrated CMOS transistors with high-κ gate dielectrics. IEEE International Electron Devices Meeting Technical Digest, p. 231.

39 Hobbs, C., Fonseca, L., Dhandapani, V., Samavedam, S., Taylor, B., Grant, J., Dip, L., Triyoso, D., Hegde, R., Gilmer, D., Garcia, R., Roan, D., Lovejoy, L., Rai, R., Hebert, L., Tseng, H., White, B., and Tobin, T. (2003). Fermi level pinning at the polySi/metal oxide interface. IEEE Symposium on VLSI Technology, p. 9.

40 Fischetti, M.V., neumayer, D.A., and Cartier, E.A. (2008) Effective electron mobility in Si inversion layers in metal–oxide–semiconductor systems with a high-κ insulator: the role of remote phonon scattering. *J. Appl. Phys.*, **90**, 4587.

41 Brown, G.A., Zeitzoff, P.M., Bersuker, G., and Huff, H.R. (2004) Scaling CMOS: materials and devices. *Mater. Today*, **7**, 20.

42 Buchanan, D.A., McFeely, F.R., and Yurkas, J.J. (2005) Fabrication of midgap metal gates compatible with ultrathin dielectrics. *Appl. Phys. Lett.*, **73**, 1676.

43 De, I., Johri, D., Srivastava, A., and Osburn, C.M. (2000) Impact of gate workfunction on device performance at the 50nm technology node. *Solid State Electron.*, **44**, 1077.

44 Thomas, M., Kroemer, H., Blank, H.R., and Wong, K.C. (1998) Induced superconductivity and residual resistance in InAs quantum wells contacted with superconducting Nb electrodes. *Physica E*, **2**, 894.

45 Chen, T., Li, X.M., and Zhang, X. (2004) Epitaxial growth of atomic-scale smooth Ir electrode films on MgO buffered Si(100) substrates by PLD. *J. Cryst. Growth*, **267**, 80.

46 Zhong, H., Heuss, G., and Misra, V. (2000) Electrical properties of RuO_2 gate electrodes for dual metal gate Si-CMOS. *IEEE Electron. Dev. Lett.*, **21**, 593.

47 Zhong, H., Heuss, G., Misra, V., Lee, C.H., and Kwong, D.L. (2005) Characterization of RuO_2 electrodes on Zr silicate and ZrO_2 dielectrics. *Appl. Phys. Lett.*, **78**, 1134.

48 Misra, V. (2003) Dual metal gate selection issues. 6th Annual Topical Research Conference on Reliability, Session 4, p. 15.

49 Polishchun, I., Ranade, P., King, T.J., and Hu, C.M. (2002) Dual work function metal gate CMOS transistors by Ni-Ti interdiffusion. *IEEE Electron. Dev. Lett.*, **23**, 200.

50 Lu, Q., Yeo, Y.C., Yang, K.J., Polishchun, I., King, T.J., and Hu, C. (2001) Metal gate work function adjustment for future CMOS technology. IEEE Symposium on VLSI Technology, p. 45.

51 Maszara, W.P., Krivokapic, Z., King, P., Goo, J.S., and Lin, M.R. (2002) Transistors

with dual work function metal gates by single full silicidation (FUSI) of polysilicon gates. IEEE International Electron Devices Meeting Technical Digest, p. 367.

52 Hisamoto, D., Lee, W.C., Kedzierski, J., Takeuchi, H., Kuo, C., King, T.J., Bokor, J., and Hu, C. (1998) A folded-channel MOSFET for deep-sub-tenth micron era. IEEE International Electron Devices Meeting Technical Digest, p. 1032.

53 Buchanan, D.A., McFeely, F.R., and Yurkas, J.J. (2005) Fabrication of midgap metal gates compatible with ultrathin dielectrics. *Appl. Phys. Lett.*, **73**, 1676.

54 Kedzierski, J., Nowak, E., Kanarsky, T., Zhang, Y., Boyd, D., Carrruthers, R., Cabral, C., Amos, R., Lavoie, C., Roy, R., Newbury, J., Sullivan, E., Benedict, J., Saunders, P., Wong, K., Canaperi, D., Krishnan, M., Lee, K.L., Rainey, B.A., Fried, D., Cottrell, P., Wong, H.-S.P., Leong, M.K., and Haensch, W. (2002). Metal-gate FinFET and fully-depleted SOI devices using total gate silicidation. IEEE International Electron Devices Meeting Technical Digest, p. 247.

55 Wang, S.Q. and Mayer, J.W. (1988) Reactions of Zr thin films with SiO_2 substrates. *J. Appl. Phys.*, **64**, 4711.

56 Misra, V., Heuss, G.P., and Zhong, H.C. (2001) Use of metal-oxide-semiconductor capacitors to detect interactions of Hf and Zr gate electrodes with SiO_2 and ZrO_2. *Appl. Phys. Lett.*, **78**, 4166.

57 Luan, H.F., Lee, S.J., Lee, C.H., Song, S.C., Mao, Y.L., Senzaki, Y., Roberts, D., and Kwong, D.L. (1999) High quality Ta_2O_5 gate dielectrics with $T_{ox\text{-}eq} < 10$ Å. IEEE International Electron Devices Meeting Technical Digest, p. 141.

58 Yang, H., Brown, G.A., Hu, J.C., Lu, J.P., Kraft, R., Rotondaro, A.L.P., Hattangady, S.V., Chen, I.C., Luttmer, J.D., Chapman, R.A., Chen, P.J., Tsai, H.L., Amirhekmat, B., and Magel, L.K. (1997). A comparison of TiN processes for CVD W/TiN gate electrode on 3nm gate oxide. IEEE International Electron Devices Meeting Technical Digest, p. 459.

59 Amazawa, T. and Oikawa, H. (1998) Surface state generation of Mo gate metal oxide semiconductor devices caused by mo penetration into gate oxide. *J. Electrochem. Soc.*, **145**, 1297.

60 Lundgren, P. (1999) Impact of the gate material on the interface state density of metal-oxide-silicon devices with an ultrathin oxide layer. *J. Appl. Phys.*, **85**, 2229.

61 Ushiki, T., Yu, M.-C., Kawai, K., Shinohara, T., Ino, K., Morita, M., and Ohmi, T. (1999) Gate oxide reliability concerns in gate-metal sputtering deposition process: an effect of low-energy large-mass ion bombardment. *Microelectron. Reliab.*, **39**, 327.

62 Park, D.G., Cbo, H.J., Lim, K.Y., Cba, T.H., Yeo, I.S., and Park, J.W. (2001) Effects of TiN deposition on the characteristics of W/TiN/SiO_2/Si metal oxide semiconductor capacitors. *J. Electrochem. Soc.*, **148**, F189.

63 Yeo, Y.C., Ranade, P., King, T.J., and Hu, C. (2002) Effects of high-*k* gate dielectric materials on metal and silicon gate workfunctions. *IEEE Electron. Dev. Lett.*, **23**, 342.

64 Shinkai, S. and Sasaki, K. (1999) Influence of sputtering parameters on the formation process of high-quality and low-resistivity HfN thin film. *Jpn. J. Appl. Phys.*, **38**, 2097.

65 Gotoh, Y., Tsuji, H., and Ishikawa, J. (2003) Measurement of work function of transition metal nitride and carbide thin films. *J. Vac. Sci. Technol. B*, **21**, 1607.

66 Yu, H.Y., Lim, H.F., Chen, J.H., Chen, J.H., Li, M.F., Zhu, C.X., Tung, C.H., Du, A.Y., Wang, W.D., Chi, D.Z., and Kwong, D.L. (2002). Physical and electrical characteristics of HfN gate electrode for advanced MOS devices. *IEEE Electron. Dev. Lett.*, **24**, 230.

67 Yu, H.Y., Li, M.F., and Kwong, D.L. (2004) Thermally robust HfN metal as a promising gate electrode for advanced MOS device applications. *IEEE Trans. Electron. Dev.*, **51**, 609.

68 Osburn, C.M., Kim, I., Han, S.K., De, I., Yee, K.F., Gannavaram, S., Lee, S.J., Luo, Z.J., Zhu, W., Hauser, J.R., Kwong, D.L., Lucovsky, G., Ma, T.P., and Öztürk, M.C. (2002). Vertically scaled MOSFET gate stacks and junctions: how far are we likely to go? *IBM J. Res. Dev.*, **46**, 299.

69 Natarajan, S., Armstrong, M., Bost, M., Brain, R., Brazier, M., Chang, C.-H., Chikarmane, V., Childs, M., Deshpande, H., Dev, K., Ding, G., Ghani, T., Golonzka, O., Han, W., He, J., Heussner, R., James, R., Jin, I., Kenyon, C., Klopcic, S., Lee, S.H., Liu, M., Lodha, S., McFadden, B., Murthy, A., Neiberg, L., Neirynck, J., Packan, P., Pae, S., Parker, C., Pelto, C., Pipes, L., Sebastian, J., Seiple, J., Sell, B., Sivakumar, S., Song, B., Tone, K., Troeger, T., Weber, C., Yang, M., Yeoh, A., and Zhang, K. (2008). A 32nm logic technology featuring 2nd-generation high-κ + metal-gate transistors enhanced channel strain and 0.171 μm^2 SRAM cell size in a 291 Mb array. IEEE International Electron Devices Meeting Technical Digest, p. 941.

70 Mistry, K., Allen, C., Auth, C., Beattie, B., Bergstrom, D., Bost, M., Brazier, M., Buehler, M., Cappellani, A., Chau, R., Choi, C.H., Ding, G., Fischer, K., Ghani, T., Grover, R., Han, W., Hanken, D., Hattendorf, M., He, J., Hicks, J., Huessner, R., Ingerly, D., Jain, P., James, R., Jong, L., Joshi, S., Kenyon, C., Kuhn, K., Lee, K., Liu, H., Maiz, J., McIntyre, B., Moon, P., Neirynck, J., Pae, S., Parker, C., Parsons, D., Prasad, C., Pipes, L., Prince, M., Ranade, P., Reynolds, T., Sandford, J., Shifren, L., Sebastian, J., Seiple, J., Simon, D., Sivakumar, S., Smith, P., Thomas, C., Troeger, T., Vandervoorn, P., Williams, S., and Zawadzki, K. (2007). A 45nm logic technology with high-κ + metal gate transistors, strained silicon, 9 Cu interconnect layers, 193nm dry patterning, and 100% Pb-free packaging. IEEE International Electron Devices Meeting Technical Digest, p. 247.

71 Packan, P., Akbar, S., Armstrong, M., Bergstrom, D., Brazier, M., Deshpande, H., Dev, K., Ding, G., Ghani, T., Golonzka, O., Han, W., He, J., Heussner, R., James, R., Jopling, J., Kenyon, C., Lee, S.-H., Liu, M., Lodha, S., Mattis, B., Murthy, A., Neiberg, L., Neirynck, J., Pae, S., Parker, C., Pipes, L., Sebastian, J., Seiple, J., Sell, B., Sharma, A., Sivakumar, S., Song, B., Amour, A.St., Tone, K., Troeger, T., Weber, C., Zhang, K., and Luo, Y., and Natarajan, S. (2009). High performance 32nm logic technology featuring 2nd generation high-κ + metal gate transistors. IEEE International Electron Devices Meeting Technical Digest, p. 659.

72 Auth, C., Cappellani, A., Chun, J.-S., Dalis, A., Davis, A., Ghani, T., Glass, G., Glassman, T., Harper, M., Hattendorf, M., Hentges, P., Jaloviar, S., Joshi, S., Klaus, J., Kuhn, K., Lavric, D., Lu, M., Mariappan, H., Mistry, K., Norris, B., Rahhal-orabi, N., Ranade, P., Sandford, J., Shifren, L., Souw, V., Tone, K., Tambwe, F., Thompson, A., Towner, D., Troeger, T., Vandervoorn, P., Wallace, C., Wiedemer, J., and Wiegand, C. (2008). 45nm high-κ + metal gate strain-enhanced transistors. IEEE Symposium on VLSI Technology, p. 128.

73 Wang, J., Tateshita, Y., Yamakawa, S., Nagano, K., Hirano, T., Kikuchi, Y., Miyanami, Y., Yamaguchi, S., Tai, K., Yamamoto, R., Kanda, S., Kimura, T., Kugimiya, K., Tsukamoto, M., Wakabayashi, H., Tagawa, Y., Iwamoto, H., Ohno, T., Saito, M., Kadomura, S., and Nagashima, N. (2007). Novel channel-stress enhancement technology with eSiGe S/D and recessed channel on damascene gate process. IEEE Symposium on VLSI Technology, p. 46.

Part Four
Development in non-Si-based CMOS technology

High-k Gate Dielectrics for CMOS Technology, First Edition. Edited by Gang He and Zhaoqi Sun.

12
Metal Gate/High-κ CMOS Evolution from Si to Ge Platform

Albert Achin

12.1
Introduction

The technology trend for integrated circuit (IC) is to reach lower cost, higher speed, higher density, and lower power consumption. The important circuit operation frequency (f) is related to the drive current (I_d) of metal oxide semiconductor field-effect transistor (MOSFET) by

$$f = I_d/2\pi C_{load} V_s \tag{12.1}$$

Here, the C_{load} and V_s are the load capacitance and maximum voltage swing, respectively. The present advanced IC technology of 2011 uses the 28 nm node complementary MOS (CMOS), with a transistor gate length as small as 28 nm. I_d in such highly scaled sub-100 nm MOSFET can be further expressed as

$$I_d = W\, v_{eff}\, C_{inv}(V_g - V_t) \tag{12.2}$$

The W, v_{eff}, and C_{inv} are gate width, effective channel velocity, and inversion capacitance, respectively. Figure 12.1 shows the schematic diagram of an n-type MOSFET (n-MOSFET), where the device was fabricated in counterdoped p-Si well to decrease the source–drain leakage current. The transistor was isolated by shallow trench isolation with filled SiO_2 and chemical–mechanical polishing (CMP) for planarization. Ion implantation of arsenic (As) atoms was used to form the highly doped n^+ regions at source–drain. Such n^+ dopants are needed to lower the series and contact resistances of MOSFET.

The transistor dimension scales 0.7X for every technology generation to reach an area reduction of half or double the transistor numbers in the wafer. The typical time span for each technology generation is 2–3 years. It is expected that the transistor size may scale to a 10 nm regime around 2020. Besides the horizontal area scaling, the vertical scaling of gate dielectric and source–drain n^+ regions are also required to suppress the short channel effect. In the past, ultrathin SiO_2 gate dielectric with 1.2 nm thickness has been used for 90 nm node CMOS since 2003, with a gate length of 50 nm. However, at such thin SiO_2 thickness, the gate leakage current is

High-k Gate Dielectrics for CMOS Technology, First Edition. Edited by Gang He and Zhaoqi Sun.

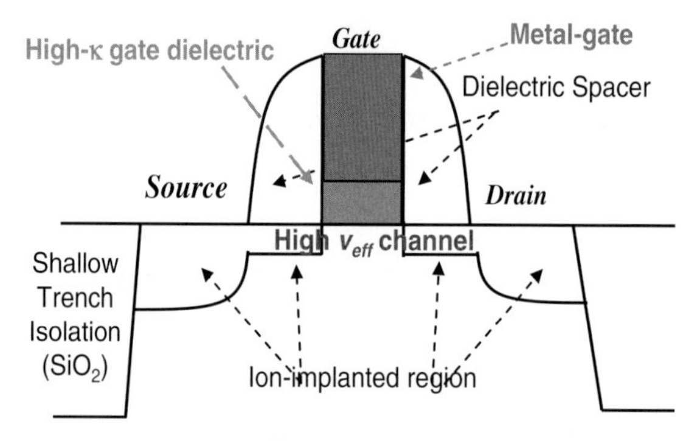

Figure 12.1 Schematic device structure of an n-MOSFET.

10^2–10^3 A/cm^2 from direct quantum mechanical tunneling mechanism. Further downscaling the gate oxide thickness (t_{ox}) will increase the gate current exponentially. For a typical 1 G transistor high-performance IC with 50 nm gate length and 200 nm gate width, the leakage current of IC is as high as 10–100 A! Such large DC power consumption is the fundamental limitation of IC. To continue the scaling trend, high dielectric constant (κ) gate dielectric [1–21] must be used to provide higher capacitance density (C_{ox}):

$$C_{ox} = \varepsilon_0 \kappa_{ox}/t_{ox} = \varepsilon_0 \kappa_{SiO_2}/\mathrm{EOT} \tag{12.3}$$

The κ_{ox} and κ_{SiO2} are dielectric constant of high-κ dielectric and SiO_2, respectively. The EOT is the equivalent oxide thickness that equals $t_{ox} \times \kappa_{SiO_2}/\kappa_{ox}$.

Figure 12.2 shows the schematic energy band diagram of an n^+-poly-Si/SiO_2/Si n-MOS structure. The C_{inv} is smaller than C_{ox} due to extra gate poly-Si depletion (t_{poly}) and distance from gate dielectric/semiconductor interface to peak quantum mechanical carrier inversion (t_{QM}). To increase C_{inv} with zero t_{poly}, metal gate must

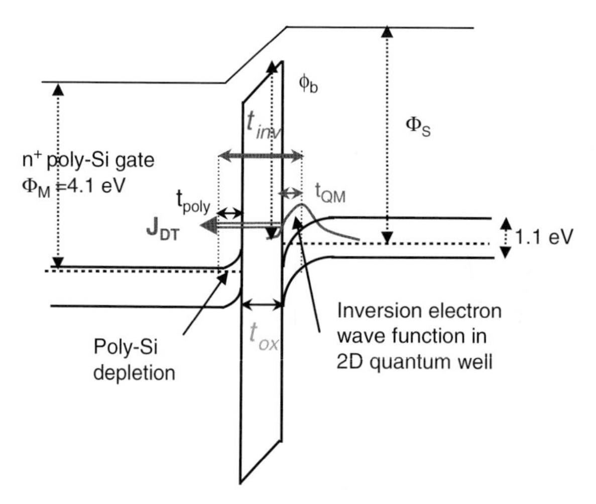

Figure 12.2 Schematic energy band diagram of an n^+-poly-Si/SiO_2/Si n-MOSFET.

be used to replace poly-Si gate that has been used nearly for half century. It is important to notice that t_{QM} is significantly larger for III–V InGaAs with small electron effective mass (m_e^*) than that of Si, which in turn lowers the needed high C_{inv}. According to Intel Cofounder Gordon Moore and Moore's law inventor: "The implementation of high-κ and metal materials marks the biggest change in transistor technology since the introduction of polysilicon gate MOS transistors in the late 1960s." Therefore, the high κ and metal gate are the enabling technologies for continuous downscaling the CMOS-based ICs with low DC power consumption.

To replace the poly-Si gate electrode, the metal gate needs to have close metal work function (Φ_M) with poly-Si for low threshold voltage (V_t) MOSFET. This will require metal gate with Φ_M of 4.1 eV for n-MOSFET as shown in Figure 12.2. For low V_t p-MOSFET, an additional 1.1 eV of Si bandgap energy is necessary for metal gate leading to a high Φ_M of 5.2 eV. Besides, the high-κ/Si interface reaction increases V_t in both n- and p-MOSFETs, where metal gates with $\Phi_M < 4.1$ and > 5.2 eV are required to compensate the interface reaction effect. The iridium (Ir) and platinum (Pt) in the periodic table have the desired higher $\Phi_M > 5.2$ eV, but cause the p-MOSFET failure due to metal diffusion into gate dielectric during 1000 °C rapid thermal annealing (RTA). Such high-temperature RTA is needed for source–drain dopant activation after ion implantation. Thus, there is no metal or metal compound available in periodic table for low V_t p-MOSFET.

One solution to reach low V_t MOSFET, pioneered by us as early as 2000, is to tune the flatband voltage (V_{fb}) by using oxide charge (Q_{ox}) instead of Φ_M:

$$V_t = V_{fb} + 2\phi_F + Q_{dpl}/C_{ox}; \quad V_{fb} = (\Phi_M - \Phi_S) - Q_f/C_{ox} - Q_{ox}/C_{ox} \qquad (12.4)$$

Here, the $2\phi_F$, Q_{dpl}, Q_f, and Φ_S are the surface inversion potential, depletion charge, fixed oxide charge, and Si work function, respectively. Figure 12.3 shows the unique negative V_{fb} in high-κ La_2O_3 for low V_t n-MOSFET [4, 5]. Alternatively,

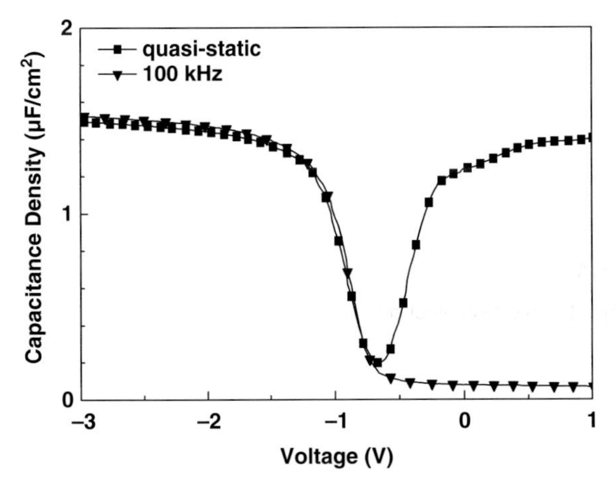

Figure 12.3 *C–V* characteristics of first La_2O_3 capping on high-κ gate dielectric to form La_2O_3/high-κ/Si n-MOSFET, with unique $V_{fb} < 0$ for low V_t n-MOSFET.

Table 12.1 Major process flow chart for gate-first and gate-last metal gate/high-κ MOSFETs.

Gate first	Gate last
1. Shallow trench isolation	1. Shallow trench isolation
2. High-κ gate dielectric deposition and postdeposition annealing (PDA)	2. High-κ gate dielectric deposition and PDA
3. Metal gate deposition and patterning	3. Poly-Si deposition and gate patterning
4. Ion implantation into source–drain extension	4. Ion implantation into source–drain extension
5. Spacer formation	5. Spacer formation
6. Ion implantation into source–drain contacts	6. Ion implantation into source–drain contacts
7. Source–drain activation by RTA at 1000°C	7. Source–drain activation by RTA at 1000 °C
	8. Isolation oxide deposition
	9. CMP planarization
	10. Mask alignment and opening contact window to poly-gate
	11. Poly-gate removal
	12. Metal gate deposition and CMP planarization

the high-κ Al_2O_3 has positive V_{fb} and can be used for low V_t p-MOSFET [1, 2, 20]. Our pioneered La_2O_3 and Al_2O_3 gate dielectrics on HfSiO have been successfully implemented for 32 nm node low V_t gate-first n- and p-MOSFETs, respectively.

Alternatively, low V_t MOSFET can also be obtained by changing the device fabrication process using the *replacement-gate and gate-last process* [17]. Table 12.1 summarizes the comparison of conventional *gate-first* with *gate-last process* for metal-gate/high-κ CMOS. This gate-last MOSFET fabricates the metal gate last, after source–drain ion implantation and 1000 °C RTA. Since the metal gate will not face the high temperature, Φ_M tuning for low V_t CMOS is possible and free from metal diffusion through high-κ gate dielectric. This method requires additional mask and process steps to fabricate the dummy poly-Si gate and high-κ MOSFET first, isolation oxide deposition, CMP planarization, mask alignment, opening contact window to poly-gate, poly-gate removal, and metal gate replacement. The replacement gate process is similar to the VLSI back-end metal damascene process via contact. One version of this gate-last process is to use the dummy poly-Si/SiO_2 and replace it by the metal gate/high-κ gate stack. Since both high-κ gate dielectric and metal gate are formed after source–drain RTA, no high-temperature process will be applied to gate stack. This can lower the high-κ/Si interface reaction [16] and is important to reach low V_t CMOS. The gate-last processed metal gate/high-κ CMOS has been implemented in Intel's 45 nm node CMOS [17] with a scaled EOT of 1.0 nm and 25X gate leakage reduction.

To increase the important I_d, the increasing v_{eff} is the other method in addition to increasing C_{inv} using high-κ gate dielectric. Here, the v_{eff} is related to the mobility at high effective field since the MOSFET is biased at saturation for higher I_d. In the past, strained Si has been successfully implemented into IC manufacture [22]. However, further v_{eff} improvement and lower V_d are needed for high-performance and low-power operation. The Ge has several times higher electron and hole mobilities than Si that is an excellent candidate for CMOS beyond strain-Si [23–36]. Although III–V $In_{0.7}Ga_{0.3}As$ has higher v_{eff}, the Ge p-MOS and III–V n-MOS architecture is very complicated for manufacture. Besides, there is no known method to lower the dislocation density in both III–V n-MOS and Ge p-MOS platform on Si, where low density < 1 dislocation/cm^2 is required for 12 in. Si wafer used for IC manufacture [24]. The large source to drain off-state leakage current (I_{OFF}) is the other challenge for small bandgap $In_{0.7}Ga_{0.3}As$ and Ge. This high I_{OFF} is also the basic limitation to continue scaling MOSFET into sub-10 nm.

To address theses issues, we invented the first dislocation-free low leakage Ge-on-insulator (GeOI) [24]. The GeOI can be formed by wafer bonding of Ge/oxide and oxide/Si and thinning down or smart cut [24–26]. The low-temperature Ge/oxide–oxide/Si wafer bonding is a challenge since the Ge has much lower melting temperature than Si. To enhance the bonding strength, the oxygen plasma bombardment into oxide surface is applied. The smart cut process has been used for Si-on-insulator (SOI) wafer fabrication that includes the proton ion implantation, thermal cut at elevated temperature, and surface smoothness steps. The smooth Ge/oxide bottom interface allows the formation of ultrathin body (UTB) GeOI.

The modern IC requires logic, memory, and communication functions in the same chip for system-on-chip (SoC). The memory function becomes more important since the logic IC operates at a much higher speed than memory. Thus, the memory IC block is placed near the logic core to lower the communication delays. The logic and memory blocks should be processed using as many shared masks as possible, in order to decrease the exorbitant high mask cost. This is shown by the very high price of extreme ultraviolet (EUV) lithography system and relatively slow throughput, although the small 13.5 nm EUV wavelength is capable for 10 nm node IC manufacture. In comparison with the embedded DRAM in back end, the fabrication of nonvolatile memory (NVM) on Ge logic platform is more difficult.

The present flash memory cell stores charges in poly-Si floating gate (FG), but the conductive poly-Si can cause all the charges to leak out via a single oxide defect. The conductive poly-Si also causes the FG-FG capacitive coupling in nearby flash cells and requires a nonplanar structure for coupling shielding. By replacing the poly-Si with insulating Si_3N_4, this charge trapping (CT) flash [37–45] has better scalability and simpler planar structure due to the discrete charge storage in zero-dimensional (0D) quantum traps within Si_3N_4. However, the smaller conduction band offset (ΔE_C) to barrier oxide is the drawback that causes stored charge leakage. We pioneered the use of deep E_C high-κ AlGaN CT flash [38, 39] shown schematically in Figure 12.4, where a large extrapolated 10 year retention memory window of 3.3 V is obtained at 85°C at low 11 V 100 μs program/-11 V 1 ms erase [39]. Such low-voltage operation is important for low-power Green memory. Our high-κ AlGaN CT flash is also affirmed by another group at a major flash company [41], with >5X better retention than conventional Si_3N_4 CT flash. The high-κ trapping layer is the enabling technology to continue downscaling the CT flash, as listed in the *International Technology Roadmap for Semiconductors* (*ITRS*) [45]. At present, scaling trapping layer thickness to 3.6 nm

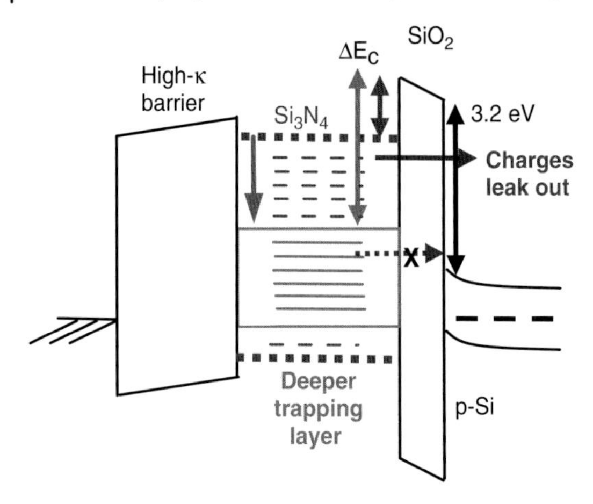

Figure 12.4 Schematic energy band diagram of high-κ trapping CT Flash.

is reached, with still large 3.1 V 10 year extrapolated retention window at 125 °C and excellent 10^6 endurance at a fast 100 μs and ±16 V program/erase (P/E) [44].

However, the fabrication of CT flash on Ge CMOS platform is very challenging, owing to a maximum allowed 600 °C process temperature to prevent high-κ/Ge interface degradation. Unfortunately, the gate oxide quality of flash memory cell degrades largely at lower temperature [43]. Therefore, new flash device [42, 43] or novel NVM memory such as resistive RAM (RRAM) [46–48] needs to be developed. The cross-point RRAM shows high potential to downscaling beyond CT flash; however, the high set and reset currents are the technology bottleneck for low-power and high-density operation. For array and 3D integration, the unipolar mode with one-resistor-1-diode (1R1D) structure is preferred, but the poor endurance, small high- to low- resistance state (HRS/LRS) window, high current requiring large driver transistor and high forming power are the challenges [46]. The better CT flash and ultralow energy (ULE) RRAM [47] will be discussed for embedded NVM for Ge logic platform in the following sections.

12.2
High-κ/Si CMOSFETs

As mentioned previously, the lack of high Φ_M metal gate with good thermal stability is the basic challenge for low V_t gate-first metal gate/high-κ p-MOSFET. To improve the thermal stability, an ultrathin amorphous Si is added between Ir and high-κ gate dielectric, which forms the Ir-rich silicide (Ir_3Si) metal gate [15] after 1000 °C RTA. Surface nitridation was also applied to high-κ gate dielectric to decrease the metal diffusion. Figure 12.5 shows the *C–V* characteristics of Ir_3Si/HfLaON/Si p-MOSFETs. Using high Φ_M Ir_3Si metal gate, the needed positive V_{fb} and low V_t is obtained at 1.6 nm EOT. However, V_{fb} is reduced for the same Ir_3Si/HfLaON p-MOSFET, as EOT scales to 1.2 nm [16]. Here, the EOT values were determined by quantum

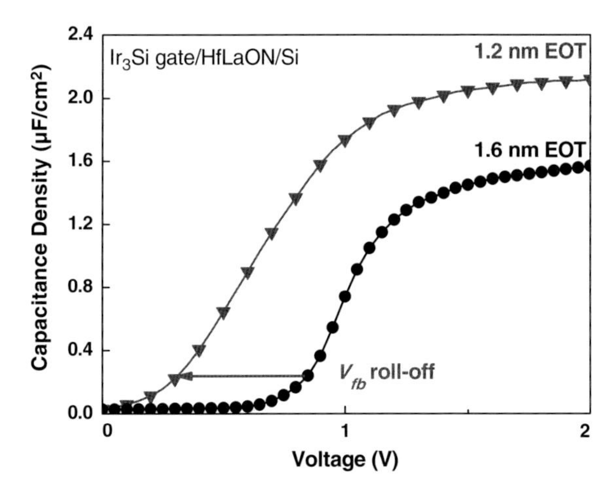

Figure 12.5 *C–V* characteristics of Ir_3Si/HfLaON p-MOSFETs at 1.6 and 1.2 nm EOT.

mechanical *C–V* simulation. Such V_{fb} roll-off at smaller EOT is a difficult challenge for low V_t MOSFET.

12.2.1 Potential Interface Reaction Mechanism

To study the V_{fb} roll-off effect, we have examined the V_{fb} dependence. From Equation 12.4, the V_{fb} can be affected only by the change in charges (Q_{ox} and Q_f). This is because the same Ir_3Si on HfLaON was used, the V_{fb} roll-off at thinner EOT is not due to the Φ_M–Φ_S. Here, the Φ_M is defined as the minimum energy to remove an electron from a metal outside the surface that is obviously independent of the source–drain RTA. Similarly, the Φ_S is unrelated to RTA thermal cycle but depends only on the dopant concentration in the channel. From the cross-sectional transmission electron microscopy (TEM) and secondary ion mass spectrometry (SIMS) analysis, the high-κ/Si interface layer was formed during RTA. The interface layer thickness also increases monotonically with increasing RTA temperature.

To further investigate the relation of interface reaction with Q_{ox}, we have plotted bond enthalpy of various high-κ dielectrics shown in Figure 12.6.

Here, the bond enthalpy is the energy required to break 1 mol of molecules into their individual atoms. The close bond enthalpy of high-κ metal oxide and Si-O allows the interface reaction at a high temperature:

$$Si + HfO_2 \xrightarrow[\Delta]{} SiO_x + HfO_{2-x} \tag{12.5}$$

Such interface reaction can generate charged oxygen vacancy and dangling bonds in the form of nonstoichiometric SiO_x and HfO_{2-x} ($x < 2$) [16]. The interface reaction can be formed by diffusion through an ultrathin SiO_2 layer during 1000 °C RTA. This interface reaction and V_{fb} roll-off effect are the bottleneck for EOT scaling.

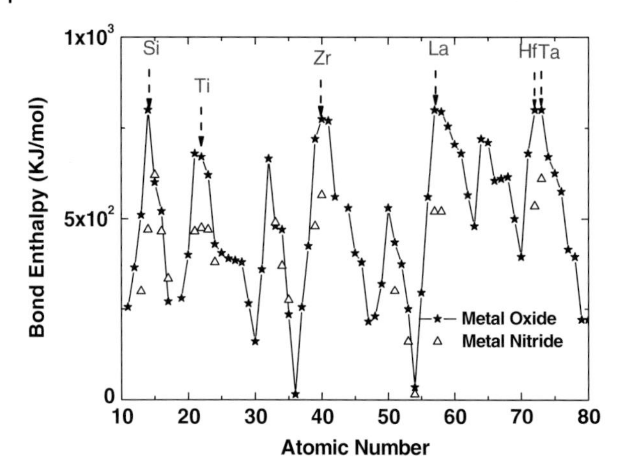

Figure 12.6 Bond enthalpy of binary oxides versus atomic number in the periodic table.

12.2.2
Inserting an Ultrathin SiON

One method of lowering interface reaction is to insert an interfacial oxide [20]. Figure 12.7 shows the *C–V* characteristics of MoN/HfAlO/SiON p-MOSFET, where a 1.5 nm thick SiON was inserted between HfAlO and Si channel. The SiON was formed by standard thermal oxide growth and surface nitridation, which is relatively thick to lower the high-κ and Si interdiffusion and interface reaction. Adding Al_2O_3 to HfO_2 improves the high-temperature thermal stability. In addition, Al_2O_3 provides the unique positive V_{fb} for low V_t p-MOSFET. The high Φ_M MoN gate further increases the needed positive V_{fb}. These combined effects lead to a low V_t of

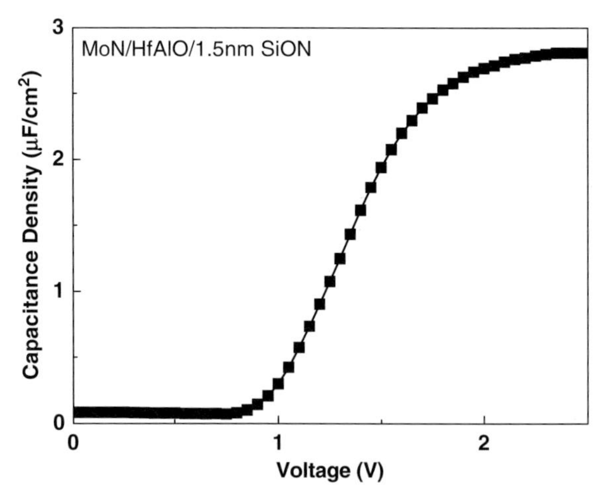

Figure 12.7 *C–V* characteristics of MoN/HfAlO/SiON/Si p-MOSFET after 1000 °C RTA.

only −0.1 V and near the ideal 4 kT/q value for the ultimate V_t scaling [18]. After 1000 °C RTA, a small EOT of 0.85 nm was obtained that is smaller than the physical thickness of interfacial SiON. This is due to the high-κ diffusion into the SiON from SIMS analysis.

To further analyze the fabricated MoN/HfAlO/SiON p-MOSFET with a small 0.85 nm EOT, we have plotted the mobility versus vertical effective field. As shown in Figure 12.8, the hole mobility is lower than the universal mobility. This is not due to the SiON since the control MoN/SiON p-MOSFET at 1.65 nm EOT shows close values with universal mobility. Thus, the limited EOT scaling by interfacial oxide and degraded mobility are the major issues of metal gate/high-κ/SiON/Si CMOS. These are part of the reasons for slower EOT scaling from 45 to 32 nm node CMOS [17, 21].

12.2.3
Low-Temperature Process

The other solution for interface reaction is to lower the process temperature since the interface reaction in formula (12.5) follows the Arrhenius temperature dependence. However, the traditional CMOS process requires a 1000 °C RTA to activate the ion-implanted dopants at source–drain, where an interfacial SiON or SiO_2 is needed to lower the interface reaction and V_{fb} roll-off. One method is to use gate-last process developed by Intel [17], which requires extra mask and process steps.

Alternatively, we have used a new source–drain doping technique to replace the ion implantation and high-temperature RTA. The shallow junction was formed by using low-temperature silicide-induced doping [16], where the dopants were driven into source–drain by silicidation process at 600–700 °C RTA. Figure 12.9 shows the I_d–V_g characteristics of high-κ LaTiO n- and p-MOSFETs using dual TaN and Ir metal gates, respectively. Low V_t of 0.13 and −0.16 V were measured at very small EOT of 0.66

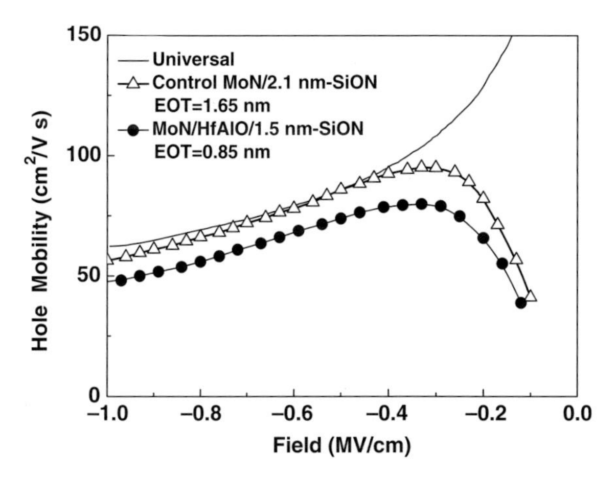

Figure 12.8 Hole mobility of MoN/HfAlO/SiON/Si p-MOSFET at 0.85 nm EOT and control MoN/SiON/Si p-MOSFET at 1.65 nm EOT.

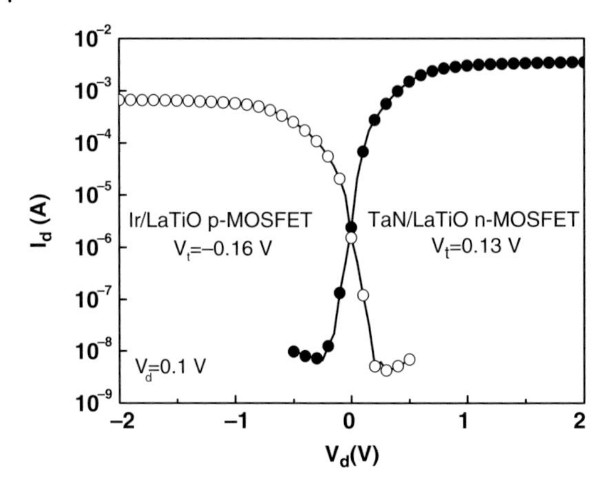

Figure 12.9 I_d–V_g characteristics of [TaN-Ir]/LaTiO p- and n-MOSFETs at small EOT of 0.66 and 0.59 nm EOT, respectively.

and 0.59 nm, respectively, from measured *C–V* characteristics and quantum mechanical *C–V* simulations shown in Figure 12.10.

However, it was reported that the mobility degrades monotonically with decreasing EOT, which is due to the closer carrier wave function to high-κ/Si interface by remote phonon scattering from high-κ gate dielectric. The degraded mobility is the major limitation at small EOT.

Another problem to limit EOT scaling is the smaller C_{inv} improvement at smaller EOT. As shown in Figure 12.11, the C_{inv} in a TaN/LaTiO n-MOSFET is noticeably lower than the accumulation capacitance (C_{acc}) at an EOT of 0.63 nm. Such smaller C_{inv} than C_{acc} is well predicted by quantum mechanical *C–V* simulation, also plotted in Figure 12.11, in good agreement with the measured *C–V* data.

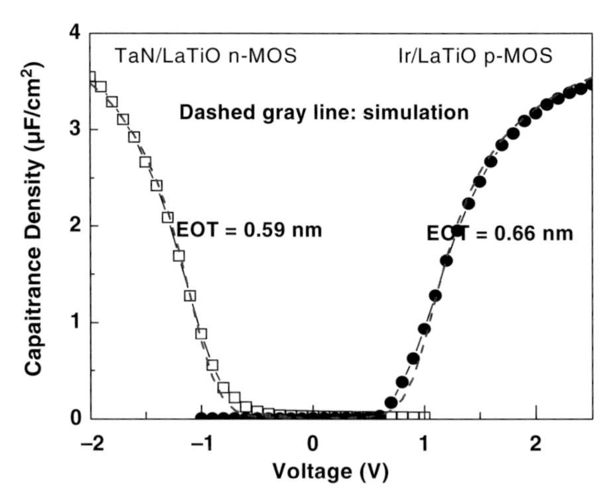

Figure 12.10 *C–V* characteristics of [TaN-Ir]/LaTiO p- and n-MOSFETs.

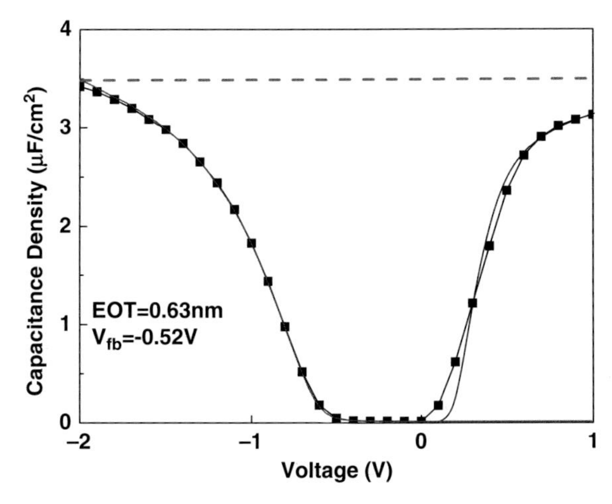

Figure 12.11 *C–V* characteristics of TaN/LaTiO n-MOSFET measured from accumulation to inversion at a small 0.63 nm EOT. The line is the simulation results.

We further plot the accumulation and inversion energy band diagrams of metal gate/high-κ n-MOSFET in Figure 12.12, to explain such significant lower C_{inv} than C_{acc}. At a small EOT < 1 nm, the inversion electrons have wider wave function spreading compared to that of accumulation holes. This is from the much lighter effective mass of m_e^* than m_h^*. This fundamental physics limits further C_{inv} improvement by EOT scaling or using higher ν_{eff} III–V material, which has an even smaller m_e^* and a larger t_{QM}. The trade-off between higher ν_{eff} III–V material and smaller C_{inv} should be considered, in addition to the higher process cost.

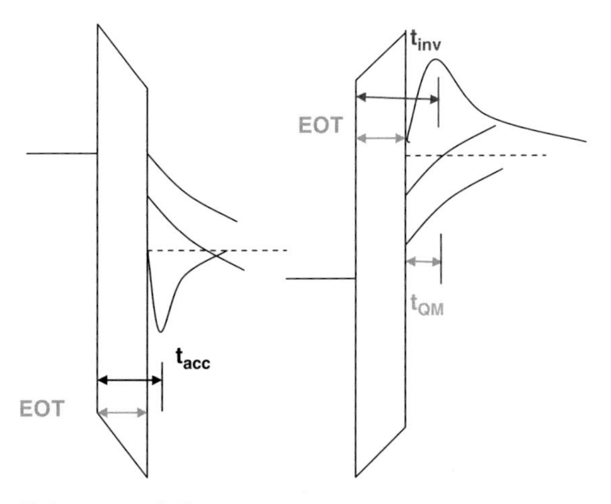

Figure 12.12 Electron and hole wave function distributions of metal gate/high-κ n-MOSFET at accumulation and inversion with a small EOT <1 nm.

12.3 High-κ/Ge CMOSFETs

As mentioned in the previous section, I_d improvement by C_{inv} is limited by the wide wave function spreading at small EOT < 1 nm. Since the V_g decreases with device scaling and V_t is limited to 4 kT/q thermal energy, the only method to improve I_d shown in Equation 12.2 is to use higher v_{eff} channel. However, this kind of new material, typical III–V InGaAs, also has a smaller m_e^* and a larger t_{QM} that leads to a smaller C_{inv}. Such higher v_{eff} and smaller C_{inv} must be traded off in III–V materials, in addition to the much higher dislocation density and process cost to grow InGaAs on Si. Alternatively, the Ge material has both higher electron and hole mobility than those of Si that is the ideal channel material for CMOS [23–36]. However, the challenges are integration of defect-free Ge on Si, poor high-κ/Ge interface, low n-type dopant activation and Fermi-level pinning close to valance band. The defect-free Ge on Si is essential to prevent large leakage current via dislocations and improve the yield. The typical dislocation density of epitaxial Ge on Si is as high as $>10^6$ dislocation/cm^2 – several orders of magnitude larger than the required < 1 dislocation/cm^2 specification for 12 in. Si wafer. The above issues make the metal gate/high-κ/Ge n-MOSFET more difficult than p-MOSFET. This is one of the reasons why III–V InGaAs was proposed to replace the Ge n-MOS, but the III–V n-MOS and Ge p-MOS architecture is very complicated for manufacture, in addition to the much higher process cost and lower yield. To improve the Ge n-MOSFET, a novel technology will be introduced in the following section that has led to the record high-performance Ge transistor to date. The other merit of metal gate/high-κ/Ge CMOS is the small energy bandgap of 0.67 eV, even smaller than the 0.75 eV of $In_{0.53}Ga_{0.47}As$ lattice-matched to InP substrate. The small energy bandgap and high mobility are essential to realize the low-voltage operation Green transistor.

12.3.1 Defect-Free Ge-on-Insulator

One difficult challenge for Ge CMOS is integration with Si substrate, where such integration is necessary to use existing Si CMOS VLSI platform. However, the direct growth of Ge on Si will cause high defect density of $>10^6$ dislocation/cm^2 that in turn will increase the leakage current and lower the device yield. Besides, the very high source–drain leakage current is the other physical limitation of a small energy bandgap material like Ge.

To address these two issues at the same time, we invented the first defect-free GeOI technology [24]. Figure 12.13 shows the cross-sectional TEM of GeOI on Si substrate, which was formed by low-temperature chemical vapor deposition (CVD) of SiO_2 on both Si and Ge substrates, O_2-plasma treatment to SiO_2 surface, Ge/SiO_2-SiO_2/Si wafer bonding, and finally thinning down the top Ge by polishing. Such O_2-plasma treatment is an important step to enhance wafer bonding at a low temperature < 600 °C that is necessary for low melting temperature Ge. The thin-body GeOI can be formed by H^+ implantation into SiO_2/Ge, wafer bonding, and smart cut

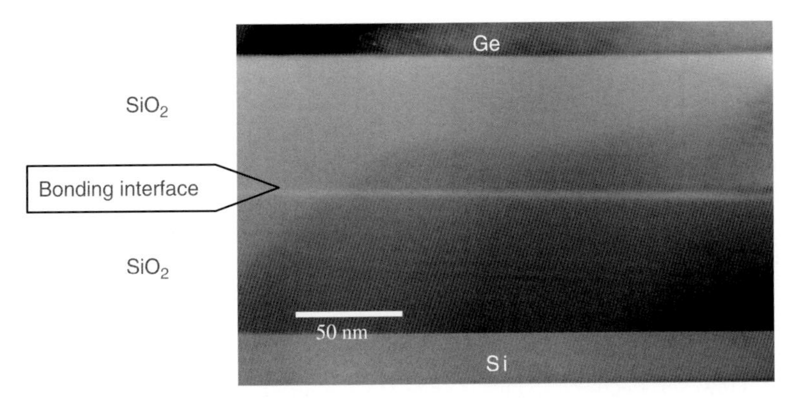

Figure 12.13 Cross-sectional TEM image of GeOI, with dislocation-free top Ge and smooth Ge/SiO_2 interface.

separation by forming H_2 gas and expansion cleavage at an elevated temperature [26]. As shown in Figure 12.13, the formed GeOI has no dislocation and the smooth bottom Ge/SiO_2 interface allows further thinning down the Ge body. Besides, the low transistor I_{OFF} was directly measured and verified by device simulation [28], where the leakage current decreases monotonically with decreasing the Ge body thickness. Such UTB structure [27] is needed to suppress the source–drain leakage current for small energy bandgap Ge and highly scaled sub-15 nm CMOS.

Based on the good GeOI quality, we further fabricated the high-κ $LaAlO_3$ p-MOSFETs on GeOI with IrO_2 metal gate. Figure 12.14 shows the hole mobilities as a function of vertical effective field, at a small 1.4 nm EOT. The control IrO_2/$LaAlO_3$ p-MOSFET on Si substrate shows a lower mobility than universal mobility due to the high-κ/Si interface scattering.

In sharp contrast, the same metal gate/high-κ p-MOS on GeOI shows significantly higher hole mobility than that of control Si and 2.2X higher mobility than SiO_2/Si universal mobility at 1 MV/cm. Such better high-field mobility is important since the

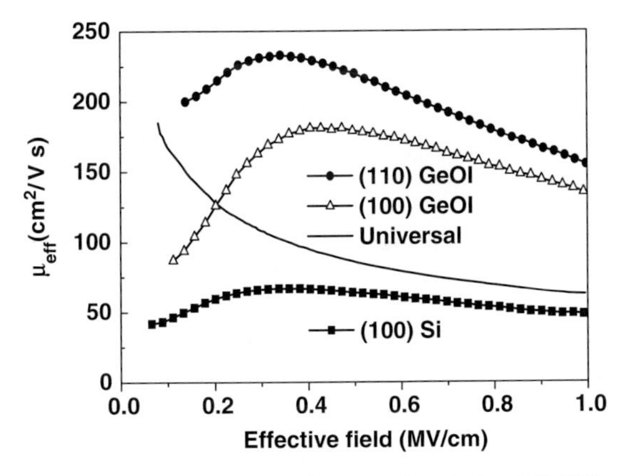

Figure 12.14 Hole mobility of IrO_2/$LaAlO_3$/GeOI p-MOSFETs at 1.4 nm EOT.

MOSFET is operated at saturation bias rather than at peak mobility with a small V_g. The hole mobility can further be improved using (110)-oriented GeOI with 2.5X higher hole mobility than SiO_2/Si universal mobility, at a high field of 1 MV/cm and a small EOT of 1.4 nm. Such higher hole mobility of Ge p-MOS is due to the smaller m_h^* than that of Si.

12.3.2
The Challenge for Ge n-MOS

Although high-performance Ge p-MOSFET with high hole mobility and small EOT has been reported as early as 2004, there is little progress on high-performance Ge n-MOSFET. However, electron mobility is higher than hole mobility due to the smaller m_e^* at conduction band. The Ge n-MOSFET suffers from the low n-type source–drain dopant activation, high surface defect-related Fermi-level pinning close to valance band, and poor high-κ/Ge interface. Figure 12.15 shows the *C–V* characteristics of TaN metal gate and high-κ TiLaO gate dielectric on 5 nm thick epitaxial Ge grown on 6 in. Si substrate. To lower the interface reaction, an extra ~0.8 nm GeO_2 was added between high-κ and Ge.

For control Ge n-MOS without the ultrathin GeO_2, strong interface reaction is evident from the large V_{fb} degradation and the decrease in capacitance density at 450–550 °C RTA. Such interface reaction originates from the high-κ and Ge interdiffusion as observed by SIMS [30], which is due to the weaker cohesive energy of Ge (372 kJ/mol) than Si (446 kJ/mol) to separate single atom from crystal lattice. In sharp contrast, the device with ultrathin GeO_2 shows much improved V_{fb} degradation from 450 to 550 °C RTA, with the needed negative V_{fb} value for low V_t Ge n-MOSFET. However, the lower gate capacitance after increasing RTA temperature to 550 °C indicates the interface layer growth in the presence of ultrathin GeO_2.

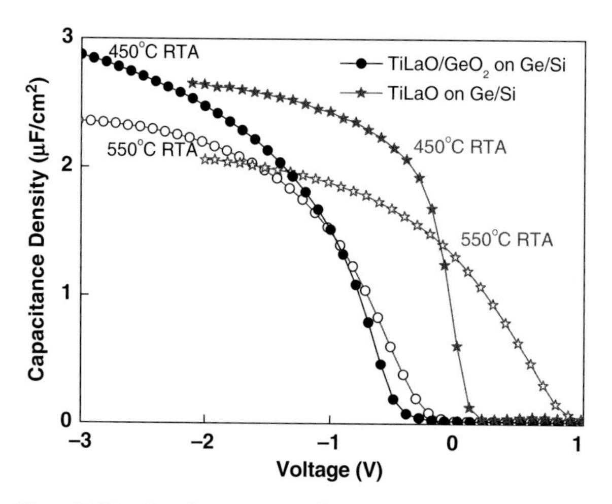

Figure 12.15 *C–V* characteristics of TaN/TiLaO/Ge/Si n-MOSFETs with and without the ultrathin GeO_2 interfacial layer.

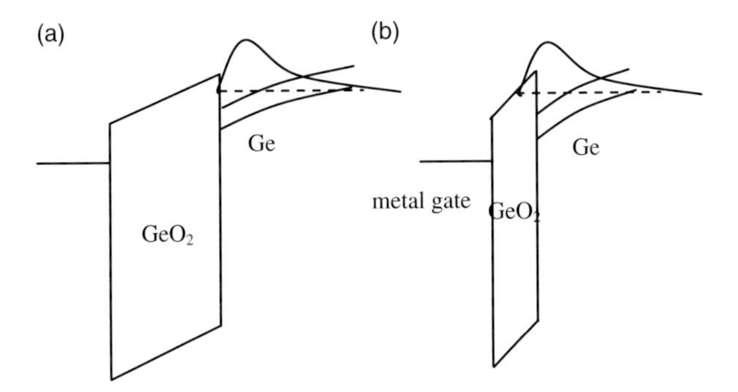

Figure 12.16 Schematic energy band diagram of metal gate/GeO_2/Ge n-MOSFETs.

The thermally grown GeO_2 on Ge has been used to improve the interface and mobility of Ge n-MOSFET, similar to SiO_2 on Si. High peak mobility at low effective field was reported for GeO_2/Ge n-MOSFET at large EOT [31–34]. Unfortunately, the downscaling EOT of GeO_2/Ge n-MOS is unsuccessful due to the theoretical limitation shown in the energy band diagram in Figure 12.16. Here, the GeO_2/Ge has a very small ΔE_C of ~0.8 eV [35]. At large EOT or low effective field, the electron wave function can be confined to GeO_2/Ge interface that gives the high peak electron mobility (Figure 12.16a). However, the electron wave function can easily spread out into GeO_2 at small EOT and high effective field shown in Figure 12.16b.

Since the m_e^* is much higher in GeO_2 than in Ge, the overall electron mobility degrades rapidly at the small EOT and high effective field. Besides, the small ΔE_C at GeO_2/Ge prohibits scaling EOT to 1 nm due to the large leakage current of inversion electrons. These difficult challenges are part of the reason why III–V InGaAs was proposed to replace Ge for n-MOSFET; however, the III–V InGaAs n-MOS and Ge p-MOS architecture is very complicated to manufacture, in addition to the high process cost and low yield by high-density dislocations. Therefore, better high-κ, interfacial layer, and novel process must be developed to obtain high-performance Ge n-MOSFET.

12.3.3 High-Mobility Ge n-MOS Using Novel Technology

To overcome the interface reaction, we have used SiO_2 instead of GeO_2. SiO_2 has larger ΔE_C than GeO_2 to prevent electron wave function from spreading into gate dielectric. Besides, SiO_2 is much more thermally robust than GeO_2 that can dissociate and form volatile GeO even at a low temperature of 589–758 K [30]:

$$GeO_{2(s)} + Ge_{(s)} \rightarrow 2\,GeO_{(g)} \tag{12.6}$$

The above reaction also forms the vacancies in GeO_2 layer to cause high-κ and Ge interdiffusion and interface reaction.

To further lower the interface reaction and improve the high-κ/SiO_2/Ge interface, we have applied the ultrafast laser annealing on gate dielectric [36]. Figure 12.17

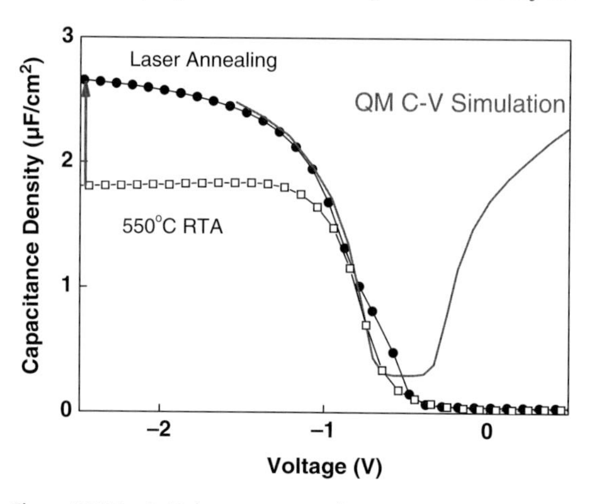

Figure 12.17 *C–V* characteristics of $TaN/ZrO_2/La_2O_3/SiO_2/Ge$ n-MOS devices with laser annealing or conventional RTA.

shows the *C–V* characteristics of $TaN/ZrO_2/La_2O_3/SiO_2/Ge$ n-MOS devices with or without laser annealing. Large improvement of capacitance density was reached after laser annealing, which is due to the ZrO_2 crystallization to higher κ (111) orientation. An EOT of 0.95 nm is obtained after laser annealing, from quantum mechanical *C–V* simulation using Ge parameters. This EOT value is significantly smaller than the reported data of metal gate/GeO_2/Ge n-MOS [31–34], where the later cases are limited by the small ΔE_C in GeO_2/Ge. Small interface reaction in high-κ/Ge can also be reached due to the 30 ns pulsed KrF excimer laser. At the same time, the laser energy of 5.0 eV (248 nm wavelength) is absorbed by surface Ge and in turn heats up the high-κ/SiO_2 to improve the quality. This is evident from the small *C–V* hysteresis of only 21 mV and much better than the 73 mV in control 550 °C RTA device [36]. The higher κ value after LA may also solve the EOT scaling issue of metal gate/high-κ/Si CMOS without changing to a new high-κ material.

Another challenge of Ge n-MOSFET is the poor ion-implanted dopant activation using conventional RTA. We have further applied laser annealing for source–drain dopant activation. Using the laser annealing on P^+-implanted Ge, a low sheet resistance of 73 Ω/sq and small 1.10 ideality factor of n^+/p Ge junction were obtained, while still maintaining the low reverse leakage current. Such excellent n^+/p junction characteristics, after ion implantation and laser annealing, are important to reach high-performance Ge n-MOS. Figure 12.18 shows the electron mobility versus gate oxide effective field of $TaN/ZrO_2/La_2O_3/SiO_2/Ge$ n-MOSFET. Although the mobility at low effective field is lower than reported data [31–34], the high-field mobility at 1 MV/cm is the highest value among all publications, with the additional merit of the lowest EOT <1 nm. It is important to notice that the transistor is biased at high field for high I_d rather than at low field with a small V_g. Such excellent results suggest the good interface property using both ultrathin SiO_2 interfacial layer and pulsed laser annealing. The high-field mobility at 1 MV/cm is 1.6X higher than universal SiO_2/Si data, indicating the importance of Ge n-MOS beyond strained Si.

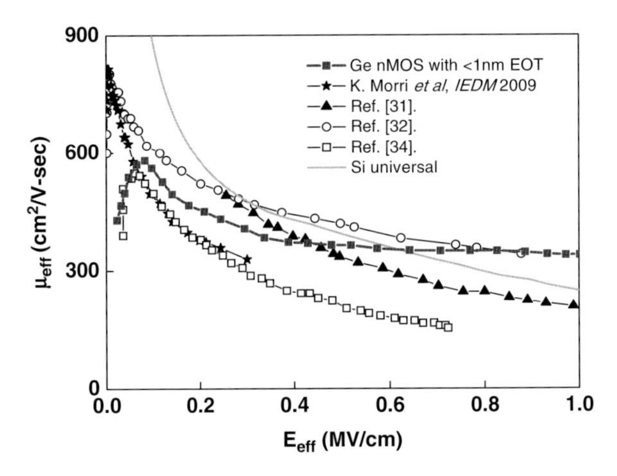

Figure 12.18 Comparison of mobility of laser-annealed $TaN/ZrO_2/La_2O_3/SiO_2/Ge$ n-MOSFET with various published data [31–34].

12.4 Ge Platform

Based on the excellent performance of both Ge p- and n-MOSFETs discussed above, it is highly possible to form the Ge logic ICs. Although no 12-in. Ge substrate is available now, the integration of defect-free Ge with Si can be realized by selective area wafer bonding and separation to form GeOI. Such GeOI structure with thin Ge is similar to UTB SOI, which is essential to lower the source–drain leakage current of small energy bandgap Ge at sub-15 nm node. The low-temperature processed GeOI can be fabricated on existing Si IC and forms the 3D integration [26]. Such device-level 3D IC is quite different from the wafer-level 3D IC with an inherent merit of a much higher interconnect density, a unique advantage of low-temperature processed Ge CMOS.

However, the integration of memory function into Ge logic platform is required for SoC to use as many shared masks as possible. This is especially important to lower the prohibitive mask cost of 13.5 nm EUV lithography system with relatively slow throughput. Such logic and memory IC integration is also necessary to improve the performance of SoC since the logic IC operates at a much higher speed than memory. However, the challenge is to fabricate high-performance flash memory on Ge, where the gate dielectric quality of flash memory degrades largely at a lower temperature of maximum 600 °C. The important logic–memory integration scheme and novel 3D application based on Ge platform will be discussed in then following sections.

12.4.1 Logic and Memory Integration

To improve the memory device integrity, a CTE flash [42] is used to compensate the degrade oxide quality at $<600\,°C$ process for Ge logic. Figure 12.19 shows the

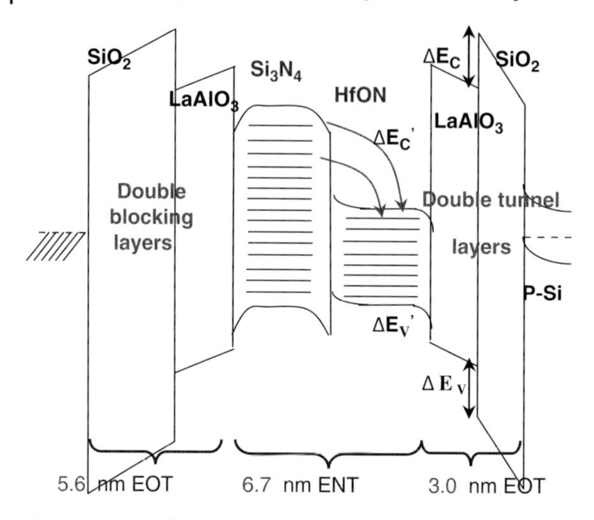

Figure 12.19 Schematic energy band diagram of CTE Flash.

schematic energy band diagram of CTE flash. The double tunnel and double blocking layers have a high-energy bandgap SiO_2 and a high-κ layer. The larger physical thickness improves the retention, even though the high-κ/SiO_2 gate dielectrics were formed at a low temperature for Ge logic. The existing ΔE_C and valance band offset (ΔE_V) in the high-κ/SiO_2 tunnel layers allow the electron and hole tunneling at lower P/E voltage and improves the endurance. The small bandgap high-κ trapping layer provides extra $\Delta E_C{}'$ and $\Delta E_V{}'$ for better carrier confinement. The relatively low trapping capability of high-κmetal–oxide–nitride (MON) is improved by adding Si_3N_4 to form the double trapping layers.

We have implemented the CTE flash on Si first at conventional 1000 °C. The TaN-[SiO_2/$LaAlO_3$]-[Si_3N_4/HfON]-[$LaAlO_3$/SiO_2]-Si MONOS CTE flash has a large 3.3 V 10 year extrapolated memory window at 150 °C to allow 4-bits per cell multilevel cell (MLC) operation [42]. The fast 100 μs and low ± 16 V P/E and good 10^5 or even 10^6 cycling endurance [44] are due to the extra ΔE_C and ΔE_V in double-tunnel oxide layers.

We further fabricated the CT flash of TaN-[20 nm-HfLaO]-[20 nm-HfON]-[6 nm-HfLaO]-InGaZnO on display glass substrate with a low temperature of only 400 °C – compared to the 600 °C for Ge CMOS logic. Such low process temperature is necessary for high-mobility InGaZnO thin-film transistors (TFT). As shown in Figure 12.20, retention can last only 10 s, degrading fast with time, which is due to the much degraded oxide quality at the low process temperature. Besides, endurance can last for only 100 P/E cycles. To improve the memory device integrity, we also fabricated the CTE flash of TaN-[2.5 nm-SiO_2/12 nm-HfLaO]-[5nm-Si_3N_4/5nm-HfON]-[4 nm-HfLaO/2.5 nm-SiO_2]-InGaZnO on glass again at 400 °C [43]. As shown in Figure 12.20, largely improved retention to 10^5 s is obtained, with very small decay rate and large memory window. Besides, improved endurance to 5×10^3 P/E cycles is also obtained. Such large improvement is due to the unique merit of bandgap

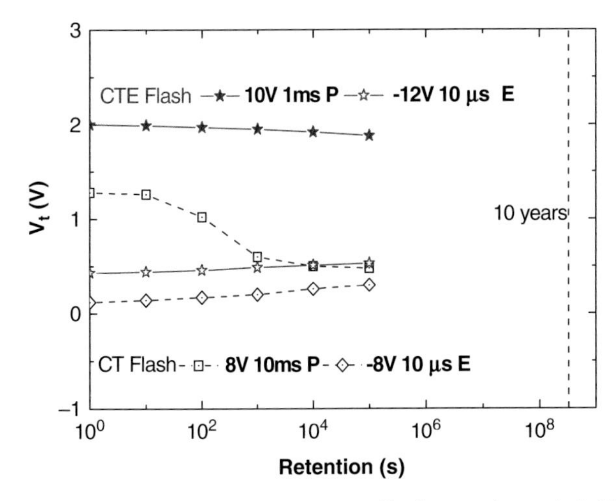

Figure 12.20 Retention characteristics of both CT and CTE Flash fabricated on glass at 400 °C.

engineering explained above. The good memory device performance processed at 400 °C on glass ensures high-performance CTE flash fabricated on Ge platform at a higher 600 °C temperature. Such low-temperature processed CT or CTE flash is especially important for vertical 3D NVM, where the complicated metal shielding for FG-FG coupling is not possible due to the vertical 3D structure.

Alternatively, the RRAM shows high potential to downscale beyond CT flash with even lower cost and simpler process for 3D NVM. However, high switching power is the major drawback for high-density memory IC. To address this issue, we invented the ULE RRAM [46–48] using the unique hopping conduction mechanism [49]. Figure 12.21 shows the switching *I–V* and endurance characteristics of Ni/GeO_x/$SrTiO_3$/TaN RRAM. Very low SET current of −3.5 μA at −1.1 V (4 μW), RESET current of 0.12 nA at 0.13 V (16 pW), very large 5×10^5 HRS/LRS memory window, and good SET/RESET endurance $>10^5$ cycles are obtained at the

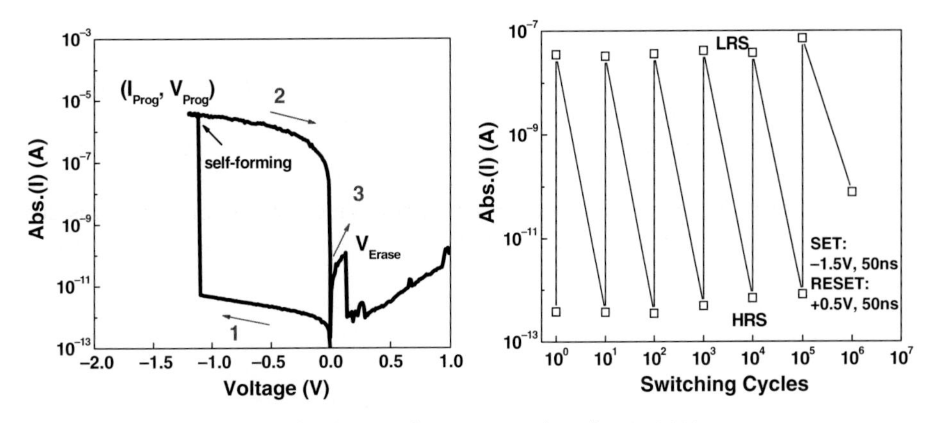

Figure 12.21 Switching *I–V* and endurance characteristics of novel ULE RRAM.

same time. Switching energy as small as 6 fJ in 20 ns (10^6X faster than flash), large 125 °C retention window, and 10^6 endurance were reached [47]. Although these are early demonstration of ULE RRAM using novel hopping conduction mechanism, this device shows a high potential to downscaling memory cell beyond CT flash. Besides, such < 400 °C process temperature is ideal for embedded NVM and Ge logic–memory platform integration, although further improvement of endurance, variation, and array performance will be required to understand the full potential of ULE RRAM. At present, the hopping conduction is the only known mechanism to obtain ULE RRAM, which is also demonstrated by a major flash supplier [50].

12.4.2
3D GeOI/Si IC

The metal gate/high-κ/Si CMOS provides the ~25X lower DC power consumption due to the lower gate leakage current using high-κ gate dielectric. The metal gate/high-κ/Ge CMOS can further lower V_d due to the small energy bandgap, higher v_{eff}, and better high-field mobility that is ideal for low DC power Green transistor. The CTE flash also provides the lower P/E voltage for low DC power Green memory.

In addition to the DC power consumption, the AC switching power becomes more important due to the increasing transistor density and parasitic capacitance in highly scaled sub-45 nm ICs. The dynamic AC power (P_{ac}) can be expressed as

$$P_{ac} = C_{load}\, V_s^2 f \tag{12.7}$$

Here, the C_{load}, V_s, and f are defined in Equation 12.1. The higher f is the technology trend for IC and unable to lower the P_{ac}. The V_s scaling is limited to the minimum V_d of ~0.5 V in the IC. Thus, the effective method to lower P_{ac} is to decrease the C_{load} that can be obtained only by using 3D integration first proposed by us [26]. To simulate the complicated back-end interconnect lines, two parallel lines were used on VLSI standard Si substrate. Figure 12.22 shows the P_{ac} of 1 mm long parallel lines with wide 10 μm spacing calculated by electromagnetic simulation. The 3 dB P_{ac} loss limits the maximum f to 17 GHz that can be largely increased to 38 GHz by using one-layer 3D integration or much higher by using two-layer 3D integration.

The maximum 17 GHz limitation is close to the existing high-density CPU chip with frequency less than 10 GHz, which is the reason to use a multicore architecture. We further demonstrated the first device-level 3D integration. Figure 12.23a shows the 3D GeOI on 1-poly-6-metal (1P6M) 0.18 μm Si CMOS. The dark area is the selectively bonded Ge using smart cut separation; the schematic diagram of 3D GeOI/0.18 μm Si is shown in Figure 12.23b. The fabricated metal gate/high-κ GeOI CMOS on selective Ge area shows good device performance with higher mobility than that of Si. The device-level 3D integration is quite different from the wafer-level 3D using bonding and thinning technique, where the former provides the high-density interconnect similar to neurons in biosystem (Figure 12.23c). Such 3D high-density interconnect is the only method to lower the AC switching power based on Equation 12.7.

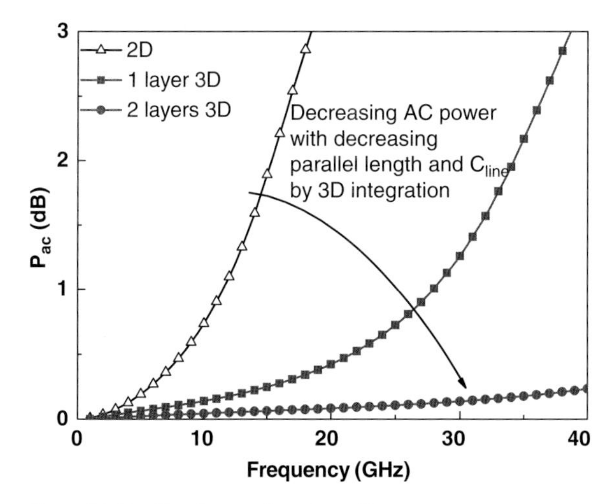

Figure 12.22 The P_{ac} as a function of frequency.

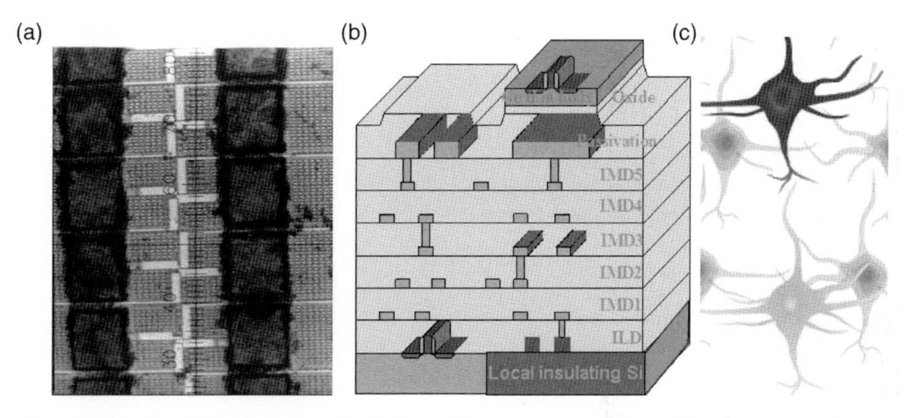

Figure 12.23 The 3D IC is the only method to lower AC power consumption and first demonstrated by us in 3D GeOI on 1P6M 0.18 μm Si CMOS.

12.5 Conclusions

The high-κ and metal gate are the enabling technologies to obtain low DC power CMOS and ICs. However, further C_{inv} improvement with EOT much less than 1 nm is limited by the electron wave function spreading and mobility degradation. Better transistor performance can be obtained by using high v_{eff} (or high-field mobility) Ge CMOS, where the 1 MV/cm high-field mobility shows 2.5X and 1.6 X higher values than SiO_2/Si universal mobility for p- and n-MOSFETs, respectively. The small energy bandgap and high v_{eff} Ge allow lower V_d operation for low DC power Green transistor. The AC switching power becomes more important with increasing operation frequency and integration density. The AC power can be lowered using only device-level 3D integration and demonstrated in the 3D GeOI on sub-μm Si

CMOS. The integration of memory function into Ge logic platform can be realized by using advanced CTE flash device structure and ULE RRAM, where both can reach low switching energy on the order of 10 fJ per cell. Such CTE flash and ULE RRAM are also ideal for 3D NVM application. Finally, the device-level 3D integration is the only method to decrease the performance gap of semiconductor logic–memory chip with organic neurons that may be realized by engineers in the near future.

Acknowledgments

The support from the National Nano Project of National Science Council, Taiwan ROC, is highly appreciated.

References

1 Liao, C.C., Chin, A., and Tsai, C. (1998) Electrical characterization of Al_2O_3 on Si from MBE-grown AlAs and Al. 10th Intl. Molecular Beam Epitaxy (MBE) Conference Technical Digest, p. 652; *J. Cryst. Growth*, **201/202**, 652 (1999)

2 Chin, A., Liao, C.C., Lu, C.H., Chen, W.J., and Tsai, C. (1999) Device and reliability of high-*k* Al_2O_3 gate dielectric with good mobility and low D_{it}. Symposium on VLSI Technical Digest, p. 135.

3 Lee, B.H., Kang, L., Qi, W.-J., Nieh, R., Jeon, Y.J., Onishi, K., and Lee, J.C. (1999) Ultrathin hafnium oxide with low leakage and excellent reliability for alternative gate dielectric application. IEEE International Electron Devices Meeting Technical Digest, p. 133.

4 Chin, A., Wu, Y.H., Chen, S.B., Liao, C.C., and Chen, W.J. (2000) High quality La_2O_3 and Al_2O_3 gate dielectrics with equivalent oxide thickness 5–10 Å. Symposium on VLSI Technical Digest, p. 16.

5 Wu, Y.H., Yang, M.Y., Chin, A., and Chen, W.J. (2000) Electrical characteristics of high quality La_2O_3 dielectric with equivalent oxide thickness of 5 Å. *IEEE Electron. Dev. Lett.*, **21**, 341.

6 Wilk, G., Wallace, R., and Anthony, J. (2001) High-*k* gate dielectrics: current status and materials properties considerations. *J. Appl. Phys.*, **89**, 5243.

7 Iwai, H., Ohmi, S., Akama, S., Ohshima, C., Kikuchi, A., Kashiwagi, I., Taguchi, J., Yamamoto, H., Tonotani, J., Kim, Y., Ueda, I., Kuriyama, A., and Yoshihara, Y. (2002). Advanced gate dielectric materials for sub-100nm CMOS. IEEE International Electron Devices Meeting Technical Digest, p. 625.

8 Datta, S., Dewey, G., Doczy, M., Doyle, B.S., Jin, B., Kavalieros, J., Kotlyar, R., Metz, M., Zelick, N., and Chau, R. (2003) High mobility Si/SiGe strained channel MOS transistors with HfO_2/TiN gate stack. IEEE International Electron Devices Meeting Technical Digest, p. 653.

9 Nabatame, T., Kadoshima, M., Iwamoto, K., Mise, N., Migita, S., Ohno, M., Ota, H., Yasuda, N., Ogawa, A., Tominaga, K., Satake, H., and Toriumi, A. (2004). Partial silicides technology for tunable work function electrodes on high-*k* gate dielectrics: Fermi level pinning controlled $PtSi_x$ for HfO_x(N) p-MOSFET. IEEE International Electron Devices Meeting Technical Digest, p. 83.

10 Tseng, H.-H., Capasso, C.C., Schaeffer, J.K., Hebert, E.A., Tobin, P.J., Gilmer, D.C., Triyoso, D., Ramón, M.E., Kalpat, S., Luckowski, E., Taylor, W.J., Jeon, Y., Adetutu, O., Hegde, R.I., Noble, R., Jahanbani, M., El Chemali, C., and White, B.E. (2004). Improved short channel device characteristics with stress relieved pre-oxide (SRPO) and a novel tantalum carbon alloy metal gate/HfO_2

stack. IEEE International Electron Devices Meeting Technical Digest, p. 821.

11 Yamamoto, Y., Kita, K., Kyuno, K., and Toriumi, A. (2005) A new Hf-based dielectric member, $HfLaO_x$, for amorphous high-κ gate insulators in advanced CMOS. Solid State Devices and Materials Technical Digest, p. 252.

12 Xiong, K., Robertson, J., Gibson, M.C., and Clark, S.J. (2005) Defect energy levels in HfO_2 high-dielectric-constant gate oxide. *Appl. Phys. Lett.*, **87**, 183505.

13 Yu, D.S., Chin, A., Wu, C.H., Li, M.-F., Zhu, C., Wang, S.J., Yoo, W.J., Hung, B.F., and McAlister, S.P. (2005) Lanthanide and Ir-based dual metal-gate/HfAlON CMOS with large work-function difference. IEEE International Electron Devices Meeting Technical Digest, p. 649.

14 Wang, X.P., Shen, C., Li, M.-F., Yu, H.Y., Sun, Y., Feng, Y.P., Lim, A., Sik, H.W., Chin, A., Yeo, Y.C., Lo, P., and Kwong, D.L. (2006). Dual metal gates with band-edge work functions on novel HfLaO high-κ gate dielectric. Symposium on VLSI Technical Digest, p. 12.

15 Wu, C.H., Hung, B.F., Chin, A., Wang, S.J., Chen, W.J., Wang, X.P., Li, M.-F., Zhu, C., Jin, Y., Tao, H.J., Chen, S.C., and Liang, M.S. (2006). High temperature stable [Ir_3Si-TaN]/HfLaON CMOS with large work-function difference. IEEE International Electron Devices Meeting Technical Digest, p. 617.

16 Cheng, C.F., Wu, C.H., Su, N.C., Wang, S.J., McAlister, S.P., and Chin, A. (2007) Very low V_t [Ir-Hf]/HfLaO CMOS using novel self-aligned low temperature shallow junctions. IEEE International Electron Devices Meeting Technical Digest, p. 333.

17 Mistry, K., Allen, C., Auth, C., Beattie, B., Bergstrom, D., Bost, M., Brazier, M., Buehler, M., Cappellani, A., Chau, R., Choi, C.-H., Ding, G., Fischer, K., Ghani, T., Grover, R., Han, W., Hanken, D., Hattendorf, M., He, J., Hicks, J., Huessner, R., Ingerly, D., Jain, P., James, R., Jong, L., Joshi, S., Kenyon, C., Kuhn, K., Lee, K., Liu, H., Maiz, J., McIntyre, B., Moon, P., Neirynck, J., Pae, S., Parker, C., Parsons, D., Prasad, C., Pipes, L., Prince, M., Ranade, P., Reynolds, T., Sandford, J., Shifren, L., Sebastian, J., Seiple, J., Simon, D., Sivakumar, S., Smith, P., Thomas, C., Troeger, T., Vandervoorn, P., Williams, S., and Zawadzki, K. (2007). A 45 nm logic technology with high-k + metal gate transistors, strained silicon, 9 Cu interconnect layers, 193 nm dry patterning, and 100% Pb-free packaging. IEEE International Electron Devices Meeting Technical Digest, p. 247.

18 Liao, C.C., Chin, A., Su, N.C., Li, M.-F., and Wang, S.J. (2008) Low V_t gate-first Al/TaN/[Ir_3Si-$HfSi_{2-x}$]/HfLaON CMOS using simple laser annealing/reflection. Symposium on VLSI Technical Digest, pp. 190–191.

19 Fonseca, L.R.C., Liu, D., and Robertson, J. (2008) P-type Fermi level pinning at a Si:Al_2O_3 model interface. *Appl. Phys. Lett.*, **93**, 122905.

20 Chang, M.F., Lee, P.T., and Chin, A. (2009) Low threshold voltage MoN/HfAlO/SiON p-MOSFETs with 0.85-nm EOT. *IEEE Electron. Device Lett.*, **30**, 861.

21 Jan, C.-H., Agostinelli, M., Buehler, M., Chen, Z.-P., Choi, S.-J., Curello, G., Deshpande, H., Gannavaram, S., Hafez, W., Jalan, U., Kang, M., Kolar, P., Komeyli, K., Landau, B., Lake, A., Lazo, N., Lee, S.-H., Leo, T., Lin, J., Lindert, N., Ma, S., McGill, L., Meining, C., Paliwal, A., Park, J., Phoa, K., Post, I., Pradhan, N., Prince, M., Rahman, A., Rizk, J., Rockford, L., Sacks, G., Schmitz, A., Tashiro, H., Tsai, C., Vandervoorn, P., Xu, J., Yang, L., Yeh, J.-Y., Yip, J., Zhang, K., Zhang, Y., and Bai, P. (2009). A 32 nm SoC platform technology with 2nd generation high-k/metal gate transistors optimized for ultra low power, high performance, and high density product applications. IEEE International Electron Devices Meeting Technical Digest, p. 647.

22 Ghani, T., Armstrong, M., Auth, C., Bost, M., Charvat, P., Glass, G., Hoffmann, T., Johnson, K., Kenyon, C., Klaus, J., McIntyre, B., Mistry, K., Murthy, A., Sandford, J., Silberstein, M., Sivakumar, S., Smith, P., Zawadzki, K., Thompson, S., and Bohr, M. (2003).

A 90 nm high volume manufacturing logic technology featuring novel 45 nm gate length strained silicon CMOS transistors. IEEE International Electron Devices Meeting Technical Digest, p. 978.

23 Chui, C., Kim, H., Chi, D., Triplett, B., Mcintyre, P., and Saraswat, K. (2002) A sub-400 °C germanium MOSFET technology with high-*k* dielectric and metal gate. IEEE International Electron Devices Meeting Technical Digest, p. 437.

24 Huang, C.H., Yang, M.Y., Chin, A., Chen, W.J., Zhu, C.X., Cho, B.J., Li, M.-F., and Kwong, D.L. (2003) Very low defects and high performance Ge-On-Insulator p-MOSFETs with Al_2O_3 gate dielectrics. Symposium on VLSI Technical Digest, p. 119.

25 Huang, C.H., Yu, D.S., Chin, A., Chen, W.J., Zhu, C.X., Li, M.-F., Cho, B.J., and Kwong, D.L. (2003) Fully silicided NiSi and germanided NiGe dual gates on SiO_2/Si and Al_2O_3/Ge-on-insulator MOSFETs. IEEE International Electron Devices Meeting Technical Digest, p. 319.

26 Yu, D.S., Chin, A., Liao, C.C., Lee, C.F., Cheng, C.F., Chen, W.J., Zhu, C., Li, M., Yoo, W.J., McAlister, S.P., and Kwong, D.L. (2004) 3D GOI CMOSFETs with novel IrO_2(Hf) dual gates and high-κ dielectric on 1P6M-0.18 μm-CMOS. IEEE International Electron Devices Meeting Technical Digest, p. 181.

27 Low, T., Li, M.F., Fan, W.J., Tyam, N.S., Yeo, Y.-C., Zhu, C., Chin, A., Chan, L., and Kwong, D.L. (2004). Impact of surface roughness on silicon and germanium ultra-thin-body MOSFETs. IEEE International Electron Devices Meeting Technical Digest, p. 151.

28 Chin, A., Kao, H.L., Tseng, Y.Y., Yu, D.S., Chen, C.C., McAlister, S.P., and Chi, C.C. (2005) Physics and modeling of Ge-on-insulator MOSFETs. European Solid State Device Research Conference Technical Digest, p. 285.

29 Zhang, Q., Huang, J., Wu, N., Chen, G., Hong, M., Bera, L.K., and Zhu, C. (2006) Drive-current enhancement in Ge n-channel MOSFET using laser annealing for source/drain activation. *IEEE Electron. Device Lett.*, **27**, 728.

30 Chen, W.B. and Chin, A. (2009) Interfacial layer dependence on device property of high-κ TiLaO Ge/Si n-type metal-oxide-semiconductor capacitors at small equivalent-oxide thickness. *Appl. Phys. Lett.*, **95**, 212105.

31 Kuzum, D., Krishnamohan, T., Nainani, A., Sun, Y., Pianetta, P.A., Wong, H.S.-P., and Saraswat, K.C. (2009) Experimental demonstration of high mobility Ge N-MOS. IEEE International Electron Devices Meeting Technical Digest, p. 453.

32 Lee, C.H., Nishimura, T., Saido, N., Nagashio, K., Kita, K., and Toriumi, A. (2009) Record-high electron mobility in Ge n-MOSFETs exceeding Si universality. IEEE International Electron Devices Meeting Technical Digest, p. 457.

33 Morii, K., Iwasaki, T., Nakane, R., Takenaka, M., and Takagi, S. (2009) High performance GeO_2/Ge n-MOSFETs with source/drain junctions formed by gas phase doping. IEEE International Electron Devices Meeting Technical Digest, p. 681.

34 Yu, H.Y., Kobayashi, M., Jung, W.S., Okyay, A.K., Nishi, Y., and Saraswat, K.C. (2009) High performance n-MOSFETs with novel source/drain on selectively grown Ge on Si for monolithic integration. IEEE International Electron Devices Meeting Technical Digest, p. 685.

35 Lin, L., Xiong, K., and Robertson, J. (2010) Atomic structure, electronic structure, and band offsets at Ge:GeO:GeO_2 interfaces. *Appl. Phys. Lett.*, **97**, 242902.

36 Chen, W.B., Shie, B.S., Cheng, C.H., Hsu, K.C., Chi, C.C., and Chin, A. (2010) Higher κ metal-gate/high-κ/Ge n-MOSFETs with $<$1nm EOT using laser annealing. IEEE International Electron Devices Meeting Technical Digest, p. 420.

37 Lee, C.H., Choi, K.I., Cho, M.K., Song, Y.H., Park, K.C., and Kim, K. (2003) A novel SONOS structure of SiO_2/SiN/Al_2O_3 with TaN metal gate for multi-giga bit flash memories. IEEE International Electron Devices Meeting Technical Digest, p. 613.

38 Lai, C.H., Chin, A., Chiang, K.C., Yoo, W.J., Cheng, C.F., McAlister, S.P., Chi, C.C., and Wu, P. (2005) Novel SiO_2/AlN/HfAlO/IrO_2 memory with fast

erase, large ΔV_{th} and good retention. Symposium on VLSI Technical Digest, p. 210.

39 Chin, A., Laio, C.C., Chiang, K.C., Yu, D.S., Yoo, W.J., Samudra, G.S., McAlister, S.P., and Chi, C.C. (2005) Low voltage high speed SiO_2/AlGaN/$AlLaO_3$/TaN memory with good retention. IEEE International Electron Devices Meeting Technical Digest, p. 165.

40 Lai, C.H., Chin, A., Kao, H.L., Chen, K.M., Hong, M., Kwo, J., and Chi, C.C. (2006) Very Low voltage SiO_2/HfON/HfAlO/TaN memory with fast speed and good retention. Symposium on VLSI Technical Digest, p. 54.

41 Joo, K.H., Moon, C.R., Lee, S.N., Wang, X., Yang, J.K., Yeo, I.S., Lee, D., Nam, O., Chung, U.I., Moon, J.T., and Ryu, B.I. (2006) Novel charge trap devices with NCBO trap layers for NVM or image sensor. IEEE International Electron Devices Meeting Technical Digest, p. 979.

42 Lin, S.H., Chin, A., Yeh, F.S., and McAlister, S.P. (2008). Good 150 °C retention and fast erase charge-trapping-engineered memory with scaled Si_3N_4. IEEE International Electron Devices Meeting Technical Digest, 843.

43 Su, N.C., Wang, S.J., and Chin, A. (2010) A nonvolatile InGaZnO charge trapping engineered flash memory with good retention characteristics. *IEEE Electron. Device Lett.*, **31**, 201.

44 Tsai, C.Y., Lee, T.H., Wang, H., and Chin, A. (2010) Highly-scaled 3.6-nm ENT trapping layer MONOS device with good retention and endurance. IEEE International Electron Devices Meeting Technical Digest, p. 110.

45 ITRS (2009) International Technology Roadmap for Semiconductors (ITRS) 2009 Edition ((http://www.itrs.net/).

46 Cheng, C.H., Chin, A., and Yeh, F.S. (2010) Novel ultra-low power RRAM with good endurance and retention. Symposium on VLSI Technical Digest, p. 85.

47 Cheng, C.H., Chin, A., and Yeh, F.S. (2010) High performance ultra-low energy RRAM with good retention and endurance. IEEE International Electron Devices Meeting Technical Digest, p. 448.

48 Cheng, C.H., Chin, A., and Yeh, F.S. (2010) Very high performance non-volatile memory on flexible plastic substrate. IEEE International Electron Devices Meeting Technical Digest, p. 512.

49 Chin, A., Lee, K., Lin, B.C., and Horng, S. (1996) Picosecond photoresponse of carriers in Si ion-implanted Si. *Appl. Phys. Lett.*, **69**, 653.

50 Kim, M.J., Baek, I.G., Ha, Y.H., Baik, S.J., Kim, J.H., Seong, D.J., Kim, S.J., Kwon, Y.H., Lim, C.R., Park, H.K., Gilmer, D., Kirsch, P., Jammy, R., Shin, Y.G., Choi, S., and Chung, C. (2010). Low power operating bipolar TMO ReRAM for sub 10 nm era. IEEE International Electron Devices Meeting Technical Digest, p. 444.

13
Theoretical Progress on GaAs (001) Surface and GaAs/high-κ Interface

Weichao Wang, Ka Xiong, Robert M. Wallace, and Kyeongjae Cho

13.1
Introduction

The recent resurgence of interest in III–V transistor channel materials has come about due to the scaling requirement of Si-based metal oxide semiconductor (MOS) devices [1]. III–V channel materials have much higher bulk electron mobilities than Si (see Table 13.1) and thus expected to provide high channel injection velocities. Among the III–V compounds, gallium arsenide (GaAs) is a prototype material because of its many advantages, such as, 5× higher electron mobility than Si, the availability of semiinsulating GaAs substrates, and a higher breakdown field over Si to achieve GaAs-based high-speed devices.

Nevertheless, III–V materials suffer from a lack of thermodynamically stable gate dielectrics [2], with an acceptable low density of III–V/dielectric interface states, which makes it very difficult to be implemented into future CMOS-type devices. The interfacial density of states (D_{it}) leads to many problems such as Fermi-level pinning, reduced electron mobility, and instability of device operations. The Fermi-level pinning results in the loss of the modulation of the carrier concentration at the oxide/III–V interface by the gate bias. For oxide/III–V interfaces, the interfacial group III- or V-dangling bonds are partially saturated [3] and they can hardly be fully saturated to obtain a superior high-*k*/III–V interface with low D_{it}. Furthermore, the partially saturated bonds can induce mid-gap states that lead to Fermi-level pinning. This pinning is most likely due to the presence of III(V)–O bonds and the resultant structural disorders. Therefore, understanding the oxidation and passivation mechanisms of the III–V surface are very useful to gain an important insight to control the quality of high-*k*/III–V interfaces [4–8].

When high-*k* oxides are deposited upon III–V materials, the III–V/high-*k* oxide interfaces are much more complicated than the ideal III–V surface alone. The low electronic quality of an III–V/dielectric interface can be attributed to several reasons.

High-k Gate Dielectrics for CMOS Technology, First Edition. Edited by Gang He and Zhaoqi Sun.

Table 13.1 Basic parameters of semiconductors.

	Bandgap (eV)	Electron mobility ($cm^2 V^{-1} s^{-1}$)	Hole mobility ($cm^2 V^{-1} s^{-1}$)	Lattice constant (Å)
Si	1.12	1400	450	5.43
GaAs	1.42	8500	400	5.65
InAs	0.37	4×10^4	500	6.03
$Ga_{0.53}In_{0.47}As$	0.85	1.2×10^4	850	5.87
InSb	0.17	7.7×10^4	300–400	6.48
InP	0.35	5400	200	5.87

First, it is the intrinsic complexity of the III–V/high-k interface compared to the Si/SiO interface. The Si/SiO_2 interface has low gap states due to the nature of tetrahedral covalent bonding of Si and SiO_2 in which dangling Si bonds are easily passivated by H [9]. But high-k oxides such as HfO_2 and Al_2O_3 have ionic bonding without a fixed coordination number, resulting in poor interface bonding quality because of the intrinsic complexity of high k/III–V. Unlike the Si/SiO_2 interface in which only Si—O bonds dominate, the high-k/III–V interface, for example, HfO_2/GaAs, could have Ga—O, As—O, In—O, In—As, Hf—Ga, and Hf—As interfacial bonds. Each type of bonding introduces various compounds at the interface, and the presence of multiple compounds makes the interface not be well defined. In addition, the realistic interface is inevitably inducing interfacial defects, such as dangling bonds, Ga(As) dimers, and so on. These imperfections at interface structures essentially introduce gap states. Second, each interfacial Si dangling bond at the Si/SiO_2 interface contains one electron so that it can be saturated by Si—O or Si—H bonds resulting in a high-quality interface. However, for the III–V/high-k interface, interfacial III (V), dangling bonds contain 0.75 (1.25) electrons. Theoretically, to form an insulating interface, the electron counting rule must be satisfied. The compensation of these dangling bonds with partial charges becomes more difficult than that of Si. Third, it is known that the oxygen content at the interface strongly depends on the ambient oxygen pressure and the stoichiometry of the deposited high-k materials. Without properly controlling the ambient oxygen pressure, mutual interdiffusion (i.e., metallic ions from high-k oxides diffused into GaAs and Ga (As) ions into high-k oxides) at the interface of high k/GaAs may occur during the thermal annealing process. Diffused Hf atoms may go into O vacancy sites at the interface, forming Hf—Ga and Hf—As bonds that generate metallic interface states. Consequently, varying oxygen content at the interface will result in different interfacial defects that act as charge trapping centers. Such traps may result in a high Coulomb scattering rate and hence degrade the channel mobility. Therefore, in principle, the interface states could be reduced by varying the interfacial oxygen content through the ambient oxygen pressure control. Finally, there are other scattering mechanisms, such as phonon scattering [10], surface roughness scattering [11], and so on, to influence the carrier transport.

At present, there are limited theoretical studies on the high-k/III–V interface [12, 13]. In this chapter, we primarily review the GaAs surface and interface with HfO_2 from a theoretical point of view.

13.2 Computational Method

The state-of-art supercomputer provides a great opportunity to explore the atomic and electronic structures of molecules, bulk materials, surface, and interfaces, without actually doing experiment. Today, computational science has become the third main branch in the research field besides experiment and analytical theory. Especially, in the field of computational material science, density functional theory (DFT) produces a single-particle method to solve the complicated many-body Schrödinger equation without inputs of any arbitrary parameters. In this chapter, the theoretical work uses DFT-based methods to calculate the ground state energies as implemented in the plane-wave basis code VASP [14]. The trajectories of the atoms are computed using the ab initio molecular dynamics (MD) scheme [15]. The pseudo-potential is described by projector-augmented wave (PAW) method [16].

13.3 GaAs Surface Oxidation and Passivation [17]

13.3.1 Clean GaAs Surface

The GaAs(001) surface is the most intensively studied system among all III–V materials [18–20]. It consists of alternating planes of Ga and As that are separated by 1.41 Å [19]. Both Ga-terminated and As-terminated GaAs (001) surfaces are observed to reconstruct forming As–As dimers or Ga–Ga dimers on the surface [21]. Several GaAs(001) reconstructed surfaces, namely, (2×4), (4×2), $\alpha(2 \times 4)$, $\alpha(4 \times 2)$, $\beta(2 \times 4)$, $\beta(4 \times 2)$, $\beta2(2 \times 4)$, $\beta2(4 \times 2)$, $\gamma(2 \times 4)$, and $\gamma(4 \times 2)$ have been studied previously [22]. The (2×4) reconstruction ((4×2) reconstruction) corresponds to four As (Ga) dimers on top of the GaAs(001) surface in an As (Ga) rich condition. The $\alpha(2 \times 4)$ reconstruction ($\alpha(4 \times 2)$) represents two As (Ga) dimers missing compared to the (2×4) ((4×2) reconstruction) in the As (Ga) rich condition. The $\beta(2 \times 4)$ ($\beta(4 \times 2)$) has one As (Ga) dimer missing at the top surface compared to (2×4) ((4×2)). The $\beta2$ (2×4) ($\beta2(4 \times 2)$) has two less As (Ga) dimers at the top surface and one Ga(As) dimer in the second top layer compared to the (2×4) ((4×2)). To form $\gamma(2 \times 4)$ ($\gamma(4 \times 2)$), one of the (2×4) ((4×2)) As dimers leaves its original position and relocates at the center of the top two As (Ga) dimers. The new As (Ga) dimer is perpendicular to the rest of the three As (Ga) dimers, however, parallel to the GaAs(001) surface.

To find the relative stabilities among these different reconstructed surface structures, formation energies per unit area are calculated by using $E_F = \frac{1}{A}(U^{tot}_{slab} - N_{Ga}\mu_{Ga} - N_{As}\mu_{As})$, where U^{tot}_{slab} is the total energy of the slab, N_i is the number of atoms of type i, A is the surface area, and μ_i is the corresponding chemical potential for species i (Ga or As) in the slab. The Ga and As chemical potentials are constrained by $\mu_{Ga} + \mu_{As} = \mu_{GaAs}$, where μ_{GaAs} is the chemical potential of bulk GaAs. The As chemical potential is bounded by $\mu_{GaAs} < \mu_{As} < 0$. In the Ga-rich condition, the Ga chemical potential is the same as in bulk Ga that is zero, and the As chemical potential simply equals to GaAs chemical potential. As a result, E_F depends only on one variable, that is, μ_{As}, which varies from -0.78 eV to 0. In As-rich limit, E_F only depends on μ_{Ga}. Figure 13.1 shows the formation energies of different GaAs(001) surface reconstructions as a function of μ_{As}. It is clear that GaAs(001)-β2(2 × 4) (see Figure 13.2) is the most stable surface over a wide range of μ_{As}, consistent with other reports [23].

The calculated band structure and density of states (DOS) of GaAs(001)-β2(2 × 4) are shown in Figure 13.3. The projected band structure of bulk GaAs is presented with dots, together with the bound surface states for GaAs(001)-β2(2 × 4) in the energy region of the bandgap. In this work, the dotted bulk region in Figure 13.3 is produced by two Ga and As layers at the bottom of the supercell that show GaAs bulk behavior. For this surface, the gap is essentially free of surface states. Five valence bands (VB) (labeled as V1–V5 in Figure 13.3) appear above the bulk band edge at the K point in this work rather than four valence bands in Schmidt's work [24]. Slightly above the bulk valence band edge at the K point, we find the two highest occupied states V1 and V2 lie at 0.41 and 0.32 eV, respectively. They correspond to the combinations of antibonding π^* and p_z orbitals of the As dimers located at the third layer (V1) and top layer (V2). The π bonding of As dimers at the third layer gives rise to V3 and V4. V5 is composed of the combinations of bonding σ and p orbitals for Ga and top As atoms located at the center of the second top layer. Compared to other GaAs(001) surface reconstructions, it is found that the states localized at the top-layer

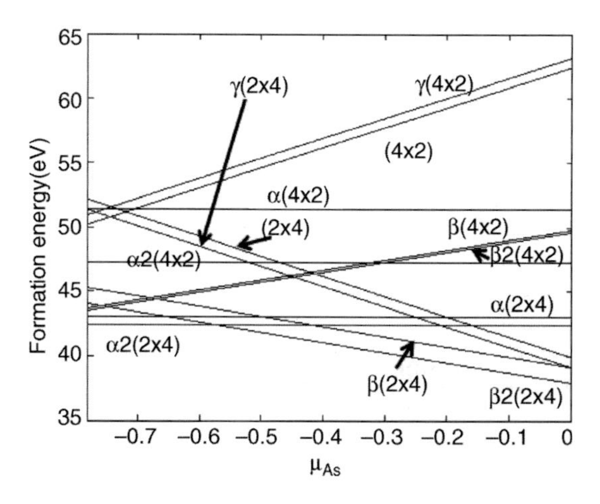

Figure 13.1 Formation energies of different GaAs(001) surface reconstructions versus As chemical potentials. Maximum As chemical potential corresponds to As-rich conditions, on the contrary, minimum As chemical potential indicates Ga-rich conditions. (Reproduced with permission from Ref. [17].)

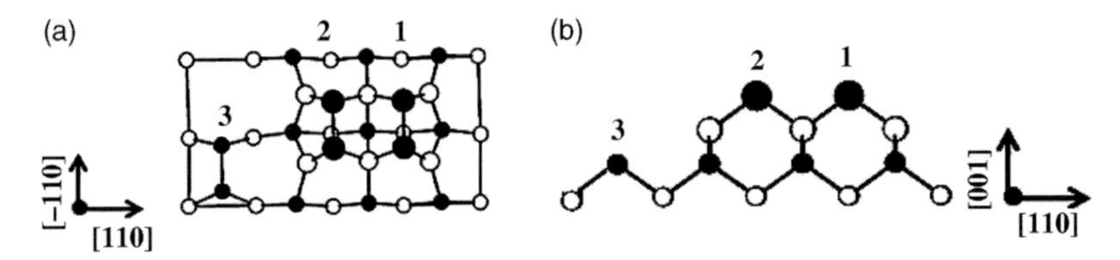

Figure 13.2 Top (a) and side (b) view of GaAs(001)-β2(2 × 4) surface. Large (small) filled circles indicate top- (third-) layer As atoms, whereas large (small) empty circles represent second- (fourth-) layer Ga atoms. (Reproduced with permission from Ref. [17].)

dimers show nearly identical charge distribution due to the symmetry of surface geometry. The lowest unoccupied state C1 is a combination of antibonding σ^* and in-plane p orbitals of the top-layer As dimers. C2 is related to the threefold-coordinated Ga atoms located at the second layer. This state is almost entirely localized at the Ga atoms on one side of the dimer block (close to the third-layer As dimer). Based on the analysis, all the As dangling bonds are fully saturated and attributed to valence band edge states. Four completely empty Ga dangling bonds (threefold coordinated second-layer Ga atoms) contribute conduction band (CB) edge states.

From the charge transfer point of view, the threefold coordinated Ga atoms transfer three electrons to six As dimer atoms so that all the As dangling bonds are fully saturated by forming lone pair states, thus opening a surface gap.

13.3.2
GaAs Surface Oxidation

The oxidation of a GaAs (001) surface is an important issue for GaAs-based device fabrication. Previous work [25–27] studied diverse surface oxidation models, and their results showed that a possible mechanism of Fermi-level pinning is not due to the intrinsic properties of GaAs (001) surface, but due to the specific bonding geometries resulting from the oxidation. Recently, based on scanning tunneling

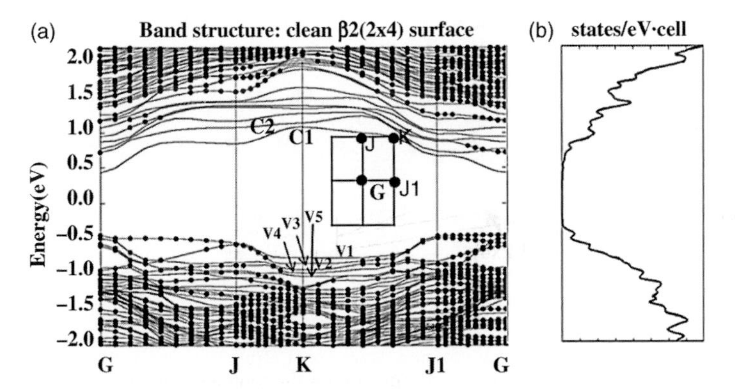

Figure 13.3 (a) Band structure (bound states) for GaAs(001)-β2(2 × 4) surface plotted over the projected bulk band structure (dotted regions). (b) The corresponding total DOS of the clean surface. (Reproduced with permission from Ref. [17].)

spectroscopy (STS) and DFT studies for different adsorbates on the GaAs (001) surface, Winn et al. [27] have proposed that the Fermi-level pinning mechanisms could be either direct or indirect. The direct Fermi-level pinning is due to the gap states induced by the adsorbate, while the indirect Fermi-level pinning is due to the gap states induced by the secondary effects, such as the generation of undimerized As atoms. The mechanisms on how excess As and As oxide induce surface gap states are of interest with regard to Fermi-level pinning.

In this section, we review the interaction between oxygen atoms and GaAs (001)-β2 (2 × 4) clean surface. For the interaction between a single oxygen atom and the surface, we consider two different oxide structures, namely, adsorption and replacement. The single oxygen adsorption cases are shown in Figure 13.4, where an oxygen atom adsorbed on the bridge site of two Ga atoms (Figure 13.4, 1a) shows its instability compared to the other three structures (1(b), 1(c), and 1(d)) from the formation energy shown in Figure 13.5 (the previous section describes how to calculate the formation energy). Oxygen adsorption on the back-bond site shown in Figure 13.4, 1(b) exhibits a high stability based on its low formation energy. The trough site As dimer (1(d) site) is slightly more stable than the 1(c) structure. Four possible configurations for the replacement of one surface atom by oxygen atom are also shown in Figure 13.4. In Figure 13.4, 2(a), one of the second (top) layer Ga is substituted by one oxygen atom forming two As—O bonds. From the high formation energy shown in Figure 13.5, we find that this is not an energetically favorable structure. In Figure 13.4, 2(b), oxygen replaces one of the row site As dimer atoms and forms three bonds, that is, two Ga—O bonds and one O–As (dimer) bond. In Figure 13.4, 2(c), a row site As atom is replaced by one oxygen atom, but the O

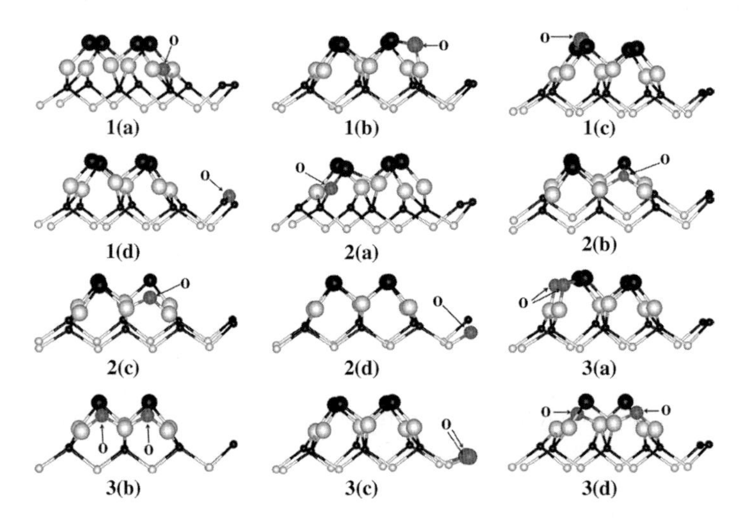

Figure 13.4 Side view of one oxygen atom adsorption (1(a)–1(d) in the first row) and replacement (2(a)–2(d) in the second row) on GaAs(001))-β2(2 × 4) surface. 3(a)–3(d) indicate the two oxygen atom adsorption 3(a) and replacement (3(b), 3(c), and 3(d)) on GaAs(001)-β2(2 × 4) surface. Black ball indicates As atom, white ball Ga atoms, and gray ball O atom. (Reproduced with permission from Ref. [17].)

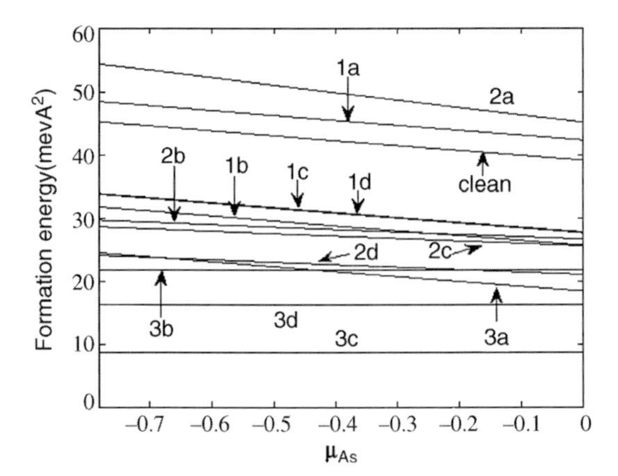

Figure 13.5 Different oxidized surface formation energies versus As chemical potentials. Maximum As chemical potential corresponds to As-rich conditions, on the contrary, minimum As chemical indicates Ga-rich conditions. The notation of each line corresponds to that in Figure 13.4. (Reproduced with permission from Ref. [17].)

atom forms only two Ga−O bonds compared to three bonds in 2(b). The total energies of structure 2(b), 2(c), and 2(d) can be directly compared since they have the same number of Ga, As, and O atoms. Our calculation exhibits the energy difference of 0.27 eV between 2(b) and 2(c), and the structure 2(d) is 1.43 eV and 1.16 eV more stable than the structure 2(b) and 2(c), respectively. Among the four possible replacement sites, the structure 2(d) has the lowest formation energy. In this case, one of the trench site As dimer atoms was replaced by one oxygen atom. Comparison of these scenarios finds that one oxygen tends to replace surface As atoms rather than Ga atoms.

For two oxygen atoms interacting with the surface shown in Figure 13.4, 3(a–d), a similar study was done to determine the possible stable structures. Two oxygen atoms prefer to stay in the two back-bond sites (see Figure 13.4, 3(a)). We find that for two oxygen replacement cases, the formation energy is lower than that of Figure 13.4, 3 (a), and this lower energy indicates that two oxygen atoms prefer to replace surface As atoms rather than adsorb on the surface. This finding is similar to the case of single oxygen atom adsorption discussed above. In the case of replacement, two oxygen atoms prefer to replace the trench sites As dimer atoms and form two Ga−O−Ga bonds (Figure 13.4, 3(c)). Nevertheless, the O row structure shown in Figure 13.4, 3 (b) is consistent with the adsorption site according to the STM study by Hale et al. [25], which was a motivation to study the Fermi energy pinning mechanism. As shown in Figure 13.4, 3(b), two As dimer atoms are replaced by two oxygen atoms and two dangling bonds associated with the top As atoms are present. Other possible structures for two O atoms replacing surface As atoms, such as the structure shown in Figure 13.4, 3(d), were also studied. The energies of such structures are only 5 meV lower than that of the structure 3(b).

From these results, we can make a general statement that the clean surface is easily oxidized since most of the oxidized surface formation energies are lower than that of the clean surface. In addition, the replacement cases always show higher stabilities than the adsorption cases.

To further analyze the surface properties and possible Fermi-level pinning mechanisms of GaAs(001) surfaces, the electronic structures of oxidized GaAs (001) surfaces are investigated. For the electronic structure of an oxygen atom adsorption on the back-bond site on GaAs-β2(2 × 4) surface shown in Figure 13.4, 1(b), the corresponding band structure (Figure 13.6a) indicates no defect states in the bandgap region. Considering the charge distribution, the specific back bonds Ga–O–As are saturated: 0.75 electrons and 1.25 electrons transfer from adjacent Ga atom and As atoms to this oxygen, respectively. At the K point of this specific oxidized surface Brillouin zone, surface bands are formed by the empty bonds of four three-coordinated Ga atoms at the second layer. The oxygen replacement of a As dimer atom in the trough site shown in Figure 13.4, 2(d) behaves differently from the oxygen adsorption on the back-bond site of Figure 13.4, 1(b). There is one band crossing the Fermi level that leads to Fermi-level pinning according to Tersoff's pinning model [28]. This specific band contributes one peak in the total density of state gap region shown in Figure 13.6b, right panel. Further electronic analysis shows that this specific band corresponds to the As half-saturated dangling bond in the remaining undimerized As atom, and this specific As atom p orbital mainly contributes to the gap states.

For the two oxygen atoms interacting with the GaAs(001)-β2(2 × 4) surface, the corresponding band structure shown in Figure 13.6c also indicates no gap states.

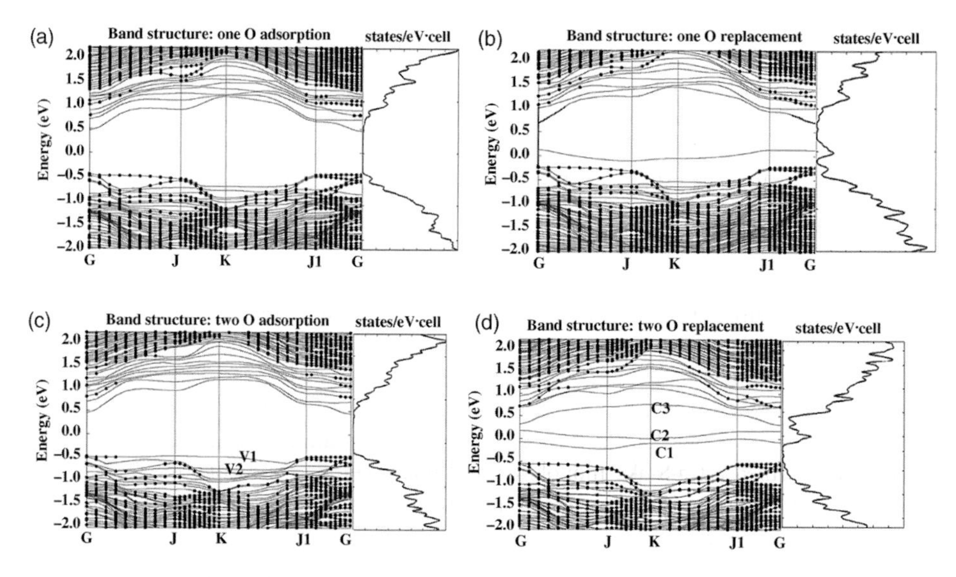

Figure 13.6 (a–d) Left panels represent band structures for one oxygen adsorption, one oxygen replacement, two oxygen adsorption, and two oxygen replacement shown as Figure 13.4, 1(b), 2(c), 3(a), and 3(c), respectively. The right panel indicates the corresponding oxidized total DOS. The dotted region indicates the projected GaAs bulk bands. (Reproduced with permission from Ref. [17].).

At 0.30 and 0.07 eV above the top of valance band of the clean surface at the K point, there are two valence bands V1 and V2 shown in Figure 13.6c. These two bands primarily originate from two As–O bonds similar to single oxygen atom adsorption at the back-bond sites. V1 indicates the combinations of antibonding π^* and p orbitals of top As dimer atoms, and V2 is related to the oxygen contribution. Two configurations for the replacement type are studied as shown in Figure 13.4, 3(b) and 3(c). The more stable surface (Figure 13.4 3(c)) exhibits no defect level in the gap region since two As atoms of the As dimer are replaced by two oxygen atoms and formed two Ga–O bonds without any partially occupied As dangling bonds. The structure 3(b) in Figure 13.4 shows three bands in the gap region, which is 0.75 eV, 0.82 eV, and 1.57 eV above the top of valance band clean surface band edge at the K point. These three bands correspond to one main peak and several shoulder peaks in the total density of state gap region (Figure 13.6d, right panel). In fact, C1 and C2 are combinations of antibonding σ^* and p orbitals of O–As bonds. In case of C3, combination of p orbitals of Ga and O gives rise to C3. Recently, Winn et al. [27] revealed that the Fermi energy pinning is due to the undimerized As atom, which agrees well with the results here.

The atomic and electronic structures for multiple oxygen atoms corresponding to extensive oxidation, that is, three, four, five, and six oxygen atoms, interactions with GaAs(001)-$\beta2(2 \times 4)$ surface have also investigated. Surfaces oxidized with multiple O atoms are found to be the combination of one and two oxidized surfaces, and reveal that surface states are from unsaturated surface As atoms rather than other mechanisms.

13.3.3 Passivation of the Oxidized GaAs Surface

To passivate the oxidized GaAs surface and remove the surface gap states, several species such as sulfur and gallium oxide are applied to oxidized GaAs (001) surfaces. Traub et al. [29] used chlorine to passivate the oxidized surface, and the associated X-ray photoelectron spectroscopy (XPS) revealed that the GaAs(001) surface with Cl could be effectively passivated with wet chemical methods. Winn et al. [27] proposed that H could be used to passivate the undimerized As atoms induced by the adsorbates. With H passivation, the As-induced gap states are suppressed into the valence band edge region, yielding a clean unpinned surface. The interaction of S with the GaAs(001) surface has been studied by Szucs et al. [30] with several model geometries of the GaAs(001) surface by DFT. For the Ga-rich GaAs surface, the top layer formed a Ga-S-like monolayer, and under certain conditions the S-passivated surface can also reconstruct forming a S–S dimer pair. Jiang et al. [31] also studied S passivation of GaAs(100) surfaces through ab initio molecular orbital calculations. Their results showed that S pasivation results in the opening of surface Ga dimers, which in turn lowers (raises) the highest occupied (lowest unoccupied) surface states. In addition to the use of extrinsic elements to passivate GaAs(001) surface, native oxides are also studied. The origin of Ga_2O_3 passivation mechanisms for reconstructed GaAs(001) surface has been investigated by using $Ga_7As_7O_2H_{20}$

cluster model [32]. The simulation showed that the reduction in the density of surface states located within the bulk energy gap derives from initial near-bridge-bonded O atoms. However, this cluster model could overestimate energy gap due to the strong quantum size effect of the GaAs cluster. Although several diverse experimental and theoretical works are done to investigate the passivation mechanisms for GaAs (001) surface, the systematic theoretical atomic-level interpretation of favored passivation structures and properties may lead to a more complete understanding of the mechanisms that would be very useful for optimization of GaAs-based MOSFET devices.

In this section, the passivation of oxidized GaAs(001)-β2(2 × 4) surface is examined by applying several candidates including H, Cl, F, S, and GaO to the oxidized surfaces shown in Figure 13.4, 2(c) and Figure 13.4, 3(b). We first consider the atomic structures of H, Cl, F, S, and GaO-passivated oxidized GaAs surface. Figure 13.7a presents the energetically favorable H position on Figure 13.4, 2(c) structure obtained after examining possible adsorption sites. H prefers to stay 0.28 Å above the top As atomic layer to form a H–As bond of 1.55 Å. For the F case shown in Figure 13.7b, the favorable site is 0.90 Å above the surface top layer with a F–As bond length of 1.82 Å. A similar result is obtained for a Cl interaction with the specific oxidized surface (Figure 13.7c) and the As–Cl bond length is 2.25 Å. To compare the stabilities of these three bonds, a binding energy analysis was conducted.

The binding energy of passivation is defined as $E_b = E_{oxd} + E_{pasv} - E_{tot}$, where E_b is the binding energy, E_{tot} is the total energy of H, F, Cl, and S interacting with the oxidized surface, E_{oxd} is the energy of oxidized surface. In the case of H, Cl, F, and S passivation, E_{pasv} may be defined as $E_{pasv} = (E_{mol} - n \times E_{atm})/n$, where E_{mol} are total energies for H_2, Cl_2, F_2, and S_8 [33], E_{atm} are the energies of the isolated H, F, Cl, and S atom. N represents the number of atoms in each molecule. In the case of GaO, E_{pasv} was obtained by calculating the total energy of an isolated GaO molecule. This calculation results in a H_2 bond energy of 4.50 eV compared to the experimental value of 4.53 eV [34]. Based on the parameters, we found $E_b = 2.15$ eV for H passivation. F, Cl, S, 2S, and GaO binding energies are 3.88 eV and 2.22 eV, 0.03 eV, 2.52 eV, and 3.84 eV, respectively.

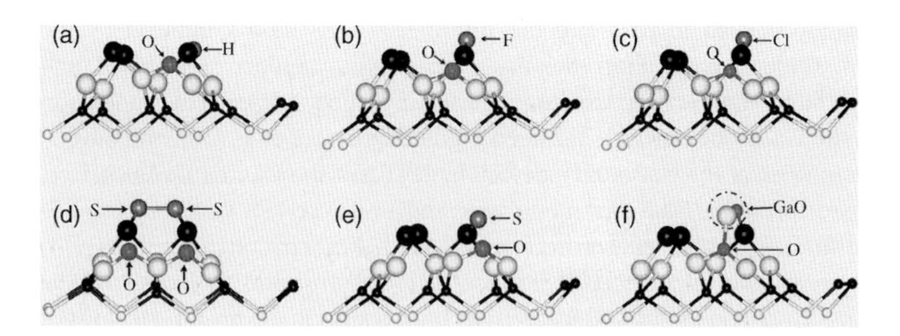

Figure 13.7 Side view of H, Cl, F, S and GaO adsorption on oxidized GaAs(001)-β2(2 × 4) surface(see Figure 13.4, 2(c). For the 2S case, the corresponding structure is for Figure 13.4, 3 (b)). Black ball indicates As atom, white ball Ga atoms, gray ball O atoms; F, Cl, H, and S are indicated in the figure. (Reproduced with permission from Ref. [17].)

For S passivation, two different oxidized surface configurations are considered. Figure 13.7d shows that two S atoms form a S–S dimer pair with a dimer bond length of 2.11 Å and that the dimer forms two bonds with two dangling As atoms. With the presence of S–S dimer pair and S–As bonds, this surface model is consistent with experimental observations [9]. Figure 13.7e indicates one S atom interacting with a specific As atom with a dangling bond. It sits 0.73 Å above the top layer of the oxidized surface. Figure 13.7f displays the adsorption of molecular species cluster GaO on the oxidized GaAs(001)-β2(2 × 4) surface. The GaO (see Figure 13.7f) acts as a bridge to connect the dangling As atom and oxygen atom.

Now, we consider the electronic structures of these passivated GaAs surfaces. Hydrogen is known to be very effective at passivating silicon [35, 36]. In the case of GaAs, the H atom transfers 0.16 electrons to the specific adjacent As atom based on a Bader charge analysis [37] so that the dangling As bond is partially saturated. The gap states induced by the dangling bond are suppressed into the valence band region, thus opening a surface gap like the clean surface. This finding agrees with that of Winn's work [27]. For the F and Cl cases, F and Cl obtain 0.61 and 0.46 electrons from the adjacent undimerized As atom, respectively. Therefore, F and Cl help to compress the gap bands induced by specific As dangling bonds into the conduction band. This F and Cl could unpin the Fermi level. In these two cases, p orbital surface states move to valence bands region and form bulk-like new p orbital states since the As half-saturated dangling bond becomes saturated. Comparing F passivating Si-based gate stacks [38], F is also effective in passivating the GaAs-based device due to the strong F–As bond and the F-induced unpinned oxidized GaAs surface.

Sulfur plays dual roles in its interaction with the oxidized GaAs(001)-β2(2 × 4) surface. One S atom forms a S–As bond (see Figure 13.7e) with the single oxygen atom oxidized surface. The band structure (Figure 13.8a) shows one band crossing the Fermi level. This specific S band contributes one peak to the total density of states gap region, as shown in Figure 13.8a. Moreover, 0.57 eV above the bulk band at the K point, there is a second band. From a charge transfer perspective, there is only one extra electron in the oxidized surface compared to the clean GaAs(001)-β2(2 × 4) surface. Nevertheless, the S atom needs two more electrons to form a closed electron shell when S interacts with oxidized surface. Therefore, the S–As bond is not fully

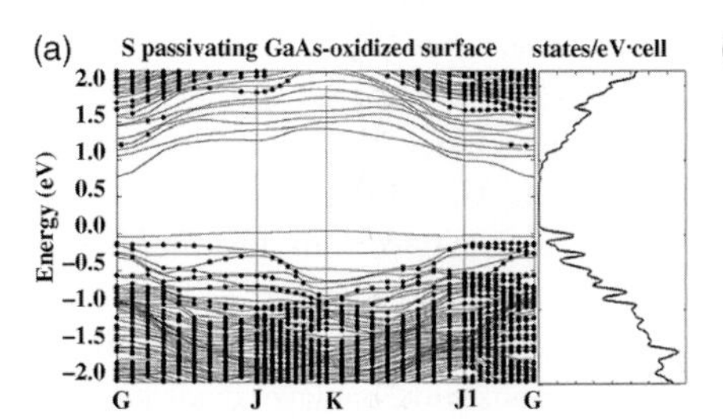

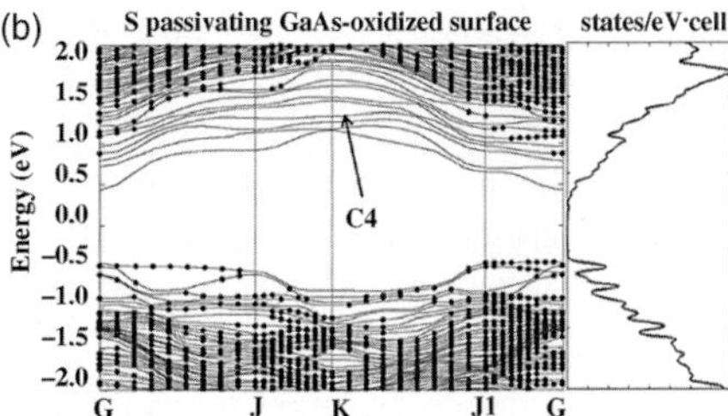

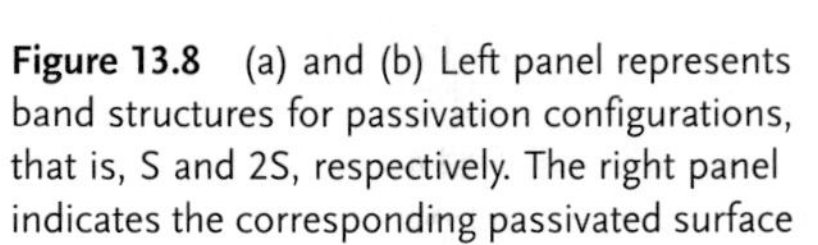
Figure 13.8 (a) and (b) Left panel represents band structures for passivation configurations, that is, S and 2S, respectively. The right panel indicates the corresponding passivated surface total density of states. The dotted region indicates the projected GaAs bulk bands. (Reproduced with permission from Ref. [17].)

saturated and leads to high density of surface states. When two S atoms interact with two undimerized As surface, a charge transfer becomes more complex than the single S atom case. Among the three oxidized gap bands shown in Figure 13.6d (i.e., C1, C2, and C3), C1 and C2 are related to undimerized As bonds, and C3 is induced by O–Ga–O unsaturated bonds. Each of the two S atoms obtains 0.17 electrons from its own adjacent As atom and forms a S–S dimer pair. Meanwhile, 0.2 more electrons from Ga connected two oxygen atoms transfer to As–S bonds. Finally, C1, C2, and C3 are pushed into conduction band region and open a clean gap shown in Figure 13.8b.

In the case of the adsorption of oxide molecule GaO on the oxidized GaAs(001)-β2 (2 × 4) surface, a clean gap is obtained, indicating As dangling bonds are fully saturated by obtaining one electron from the GaO species.

According to the present analysis, H, Cl, F, and GaO could be used to passivate any system that has undimerized As atoms. However, once these candidates have been used to passivate the undimerized As atoms, any states left in the bandgap region should be from the extrinsic adsorbate bonding with the GaAs(001)-β2(2 × 4) surface. For the S case, if S bonds are saturated, it would also help to eliminate the surface state density. On the other hand, if there remain S dangling bonds, the generation of new states within the gap region occurs, leading to Fermi-level pinning.

13.3.4
Initial oxidation of the GaAs(001)-β2(2 × 4) Surface

As the above study is based on static calculations, first-principles molecular dynamics was performed recently by Scarrozza et al. [39] to investigate the initial oxidation of the GaAs(001) surface and an atomistic model of the oxidation at its early stages was provided [39].

The reconstructed GaAs(001)-β2(2 × 4) surface (see Figure 13.2) is gradually loaded with O, placing four atoms per step until the total O number increasing up to 12 (∼1.5 ML of oxide). The simulations at this stage were done at 680 K for a simulation time ∼4.0 ps. Subsequently, the temperature was increased to a target 1400 K, for a total time of 3.0 ps. This O-loading annealing cycle is repeated and for a longer simulation time in order to increase the oxide thickness: six O atoms are added at 1400 K for ∼9.5 ps. The GaAs/oxide structure is finally quenched at room temperature with 1.0 ps followed by a structural relaxation.

On the basis of the observation of the atomic-scale events occurring during the MD simulations, Scarrozza et al [39] derive a qualitative model of the surface oxidation and the growth of a few monolayer thick amorphous $GaAsO_x$, which is outlined in the following four-step scenario (see Figure 13.9): (a) the O atoms are adsorbed on the surface by breaking the Ga–As bonds and forming bridging Ga–O–As units. The adsorption proceeds with an increasing saturation of the bonding sites available at the surface. During this initial phase, the amorphous submonolayer oxide tends to expand vertically in order to minimize the surface stress induced by the adsorption. This seems to be the most favorable way of releasing the stress at this stage of the oxidation because the surface atoms can move upward freely. (b) The reaction of oxidation proceeds into the layer underneath through a mechanism of successive

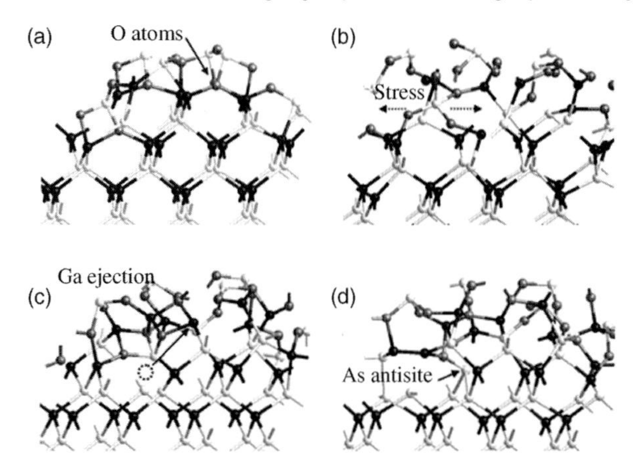

Figure 13.9 Mechanism of oxidation of the GaAs(001) surface: (a) initial oxidation, vertical growth of the oxide layer; (b) oxidation of the inner layer: stress accumulation at the interface; (c) stress release by a Ga atom ejection into the oxide layer with a vacancy generated at the interface; (d) filling of the vacancy by a neighboring atom and formation of an As antisite defect (As, Ga, and O atoms are depicted by white, black, and gray balls, respectively). (Reproduced with permission from Ref. [39].)

Ga–As bond breaking, deformation and reformation of the Ga–O–As bonding units. The incorporation of the O atoms into the substrate leads inevitably to the formation of a compressed amorphous $GaAsO_x$ layer relative to the crystalline substrate, with a resulting accumulation of stress at the formed interface. (c) As the stress at the interface reaches a critical threshold, an atomistic mechanism of stress release characterized by the ejection of Ga atoms into the amorphous $GaAsO_x$ layer above occurs, which generates naturally a vacancy at the interface. (d) In a subsequent stage, as the oxidation proceeds, a neighboring As atom fills the vacancy, leading to the formation of an As antisite defect near the surface. Such antisite defects are thought to result in mid-gap states.

13.4
Origin of Gap States at the High-*k*/GaAs Interface and Interface Passivation

As the heart of a MOSFET, the high-*k*/GaAs interface has drawn great attention experimentally during the past decade [40]. Only recently, there have been conducted several important theoretical studies on this interface [12]. In this section, we summarize the latest achievements in the theoretical understanding of the high-*k*/GaAs interface.

13.4.1
Strained HfO_2/GaAs Interface [42]

Because of the computational cost of DFT calculations, GaAs/HfO_2 interfaces with large strain have been investigated [41, 42]. A GaAs(001)/HfO_2(001) interface, with a resultant lattice mismatch of 10%, is considered as a model system suitable for DFT computations. The application of a 10% planar strain on the HfO_2 layer reduces its energy gap by 1.3 eV but does *not* induce any gap states. Moreover, the static dielectric constant becomes larger with increasing lateral lattice constant [43]. Since it is interesting to investigate what factors cause gap states and how to remove such states, the planar strain associated with this model does not qualitatively affect the electronic structure analysis, and similar methods have been applied with success for interfaces such as HfO_2/Si [44, 45]. Moreover, this strain changes only the band offset [46]. Four possible ideal HfO_2/GaAs interfaces could form: the interface terminated by (1) O–Ga bonds (denoted as "O4"), (2) O–As bonds ("O/As"), (3) Hf–Ga bonds ("Hf/Ga"), and (4) Hf–As bonds ("Hf/As"). The calculation shows the formation energy of the O4 interface is 0.23 eV/Å^2, 0.17 eV/Å^2, and 0.09 eV/Å^2 lower than those of the O/As, Hf/Ga, and Hf/As interfaces, respectively. The higher formation energy of the Hf/Ga interface than that of Hf/As is related to metal–metal bonds energetically unfavorable due to strong metallic repulsive interactions. Therefore, it is important to focus on investigating the O4 interface.

Figure 13.10 shows the electronic structure and partial charge distribution of O4 interface. The interface states arise (see Figure 13.10a and b) from the interfacial charge mismatch, mainly due to charge loss of the interfacial As (see Figure 13.10c) compared to interfacial Ga and Hf. An interface (O2) with neutralized interfacial charge, satisfying electron counting rule [47], tends to have much less interface states because interfacial As lose fewer electrons compared to that of the polarized interface (O4). Compared to the stable interface O4, it is found that F and H may be used to

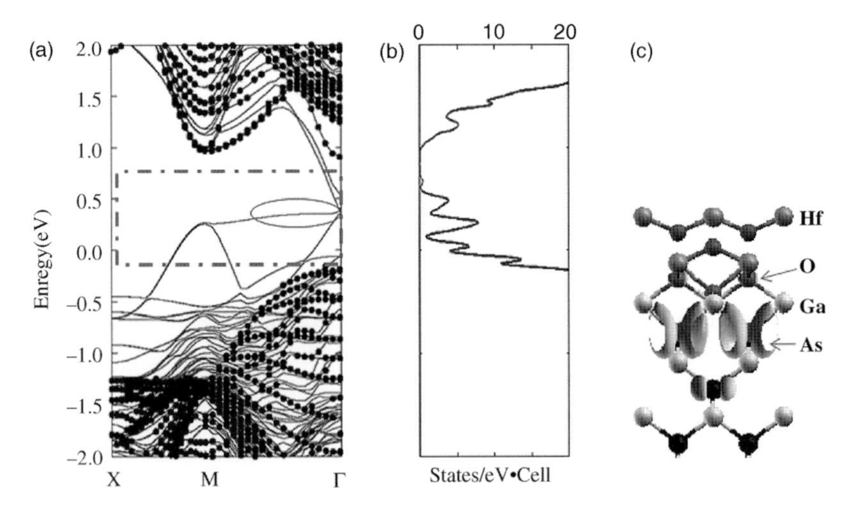

Figure 13.10 (a) The band structure of the O4 interface. The dotted region indicates the projected GaAs bulk bands. D_{it} is 1.35 × 10^{13}/ev cm^2. The dashed square frame covers the interface states and one flatband is highlighted within the ellipse. (b) Total density of states of the O4 interface. (c) The charge density plot of the interfacial bands. Charge is in solid grey color. The electron density is 2 × 10^{-2} eÅ^{-3}. (Reproduced with permission from [42].)

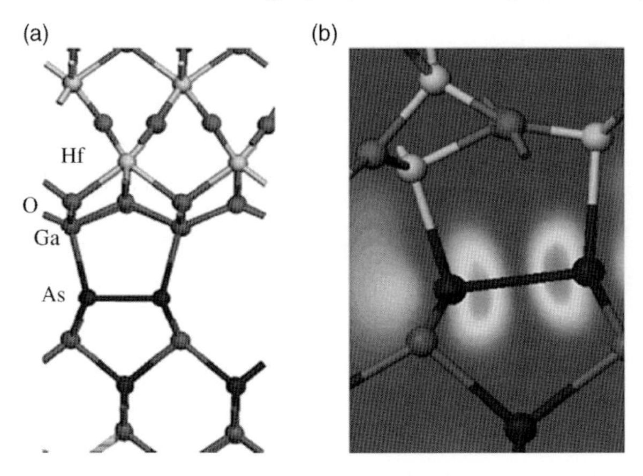

Figure 13.11 (a) An As–As dimer bond at the GaAs:HfO_2 interface. (b) Charge density of the σ^* state. (Reproduced with permission from Ref. [13].)

effectively passivate the unsaturated interfacial bonds and hence reduce the interface states.

Lin and Robertson [13] proposes a different optimized interface with the existence of the As–As dimer at the interface, as shown in Figure 13.11a. The HfO_2 lattice is shifted to 1/4a(100) to accommodate the As–As bond. The interfacial O sites become threefold coordinated. The structure is relaxed, and the local density of states is calculated. The As–As bonding (σ) state lies well into the valence band (Figure 13.11b). The As–As antibonding (σ^*) state lies in the bandgap, just below the conduction band edge, E_c. Its wave function is strongly localized on the As–As bond.

In Lin's work [13], the As–As bond appears to be a prime candidate for the gap states below E_c observed in the experimental interface state density D_{it} [48]. Calculations were carried out on interfaces with InP, InAs, or InGaAs. Nevertheless, the chemical trends are clearer from such defects in amorphous III–V semiconductors [49]. Whereas the As–As σ^* state lies just below E_c in GaAs, it lies above E_c in InAs, and the analogous P–P σ^* lies above E_c in InP. This is because the bandgap narrows from GaAs to InAs as a downshift of E_c, while the As–As σ^* state lies at the same energy in both arsenides, above E_v in InGaAs. For InP, the P–P bond is stronger, so the P–P σ^* state lies higher above E_v in InP than in GaAs, well above E_c.

13.4.2 Strain-Free GaAs/HfO_2 Interfaces [12]

In the previous section, the study of the strained interface shows that the gap states arises from the As atom (or As–As bond) at the interfacial sublayer. However, the interface contains ~10% interfacial planar strain that significantly changes the band edges of HfO_2 and the distribution of the gap states. Thus, the distribution of the density of states within the GaAs bandgap does not match well the experimental observations [50]. Moreover, the calculated valence band offset (VBO) is much lower

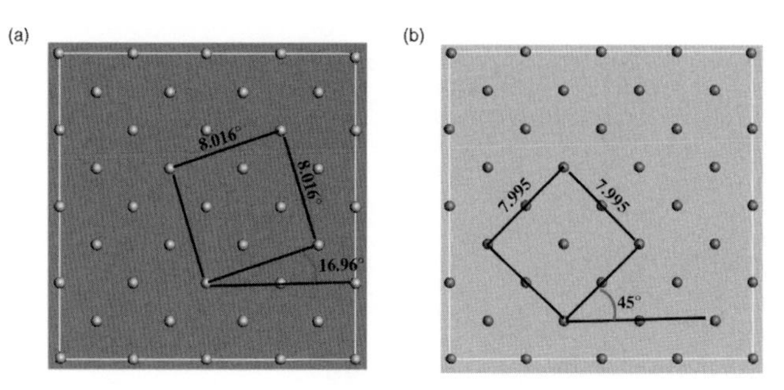

Figure 13.12 (a) Hf monolayer for HfO_2 bulk normal to (001) direction and (b) Ga monolayer for GaAs(001) bulk normal to (001) direction. (Reproduced with permission from Ref. [12].)

than the experimental data, which indicates the limitation of the strained interface. To overcome the limit, an alternative HfO_2/GaAs interface was examined with minimal strain (within ~0.3%) to perform a more accurate investigation.

To build a model interface HfO_2/GaAs with ~0.3% planar strain, the (001)-oriented HfO_2 surface is compressed by ~0.3% and rotated counterclockwise by 28.04° to match the GaAs(001) surface (see Figure 13.12). Each atomic layer has ten O, five Hf, four Ga, and four As atoms, respectively. They use a periodic slab model with Ga–O bonds at the interface (formed by Ga-terminated GaAs and O-terminated HfO_2 surfaces), which is supported by experimental results [51, 52]. A 10 Å vacuum region is used to avoid the interactions between top and bottom atoms in the periodic slab images. The bottom Ga atoms are passivated by pseudo-hydrogen (with 1.25 valence electron) to mimic As bulk bonds. Meanwhile, the top layer of HfO_2 is initially terminated by 10 oxygen atoms in the unit cell, and half of them are removed to generate an insulating HfO_2 surface without surface states.

At the interface, the average Ga–O bond length is 1.97 Å, which is close to that in bulk Ga_2O_3 (2.01 Å) [53]. To determine the oxidation state of interfacial Ga, the Bader charge decomposition method [37] (i.e., zero flux surfaces to divide atoms) is utilized to compute the atomic charge distribution. We applied the method to calculate charge states of interfacial Ga atoms. The average charge state of the interfacial Ga is + 1.45, comparable to + 1.70 of bulk Ga_2O_3 rather than + 0.52 of bulk Ga_2O. Thus, the interfacial Ga–O bonds have similar character to those in Ga_2O_3 that is consistent with what is found experimentally. In these experimental works, only Ga suboxides (e.g., Ga_2O_3) exist and As related oxides are consumed by the atomic layer deposition (ALD) via a "self-cleaning" reduction reaction effect. The relaxed structure (Figure 13.12a) shows the reconstructed interface in which one Ga–As bond between the second (As) and the third top layers (Ga) of GaAs is broken and leaves a Ga dangling bond. Meanwhile, two As–As dimer pairs are formed resulting from the charge loss of the interfacial As atoms.

The left and central panels in Figure 13.13b are the band structure and the total DOS of O9 interface, respectively. The Fermi level is shifted to 0 eV. At the Γ point in the band structure, there are six bands within the bulk bandgap of GaAs. In the lower half of the GaAs bandgap, there is a flatband that produces a high DOS (refer to the

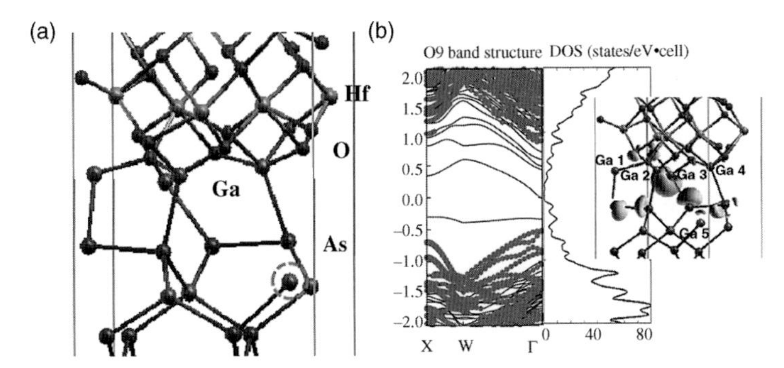

Figure 13.13 (a) Side view of GaAs:HfO_2 interface model (O9). (b) Left and central panels represent the band structures and total DOS of interface O9, respectively. The dotted region indicates the projected GaAs bulk bands. The right inset in (b) shows the partial charge distribution within the bandgap of bulk GaAs. Charge is in solid grey color. The electron density is 4.0×10^{-2} e Å^{-3}. Fermi level is set at 0 eV. (Reproduced with permission from Ref. [12].).

central panel of Figure 13.13b) which can lead to Fermi-level pinning. This flatband is induced by the interfacial Ga atom bonded to As and oxygen (e.g., Ga3). The gap states located in the upper half of GaAs bandgap show lower DOS than that in the lower half region and are induced by As–As dimers and Ga dangling bond (Ga5 in Figure 13.13b inset). The right inset in Figure 13.13b describes partial charge distribution within the gap region of bulk GaAs, which clearly indicates that gap states arise from partial Ga (Ga3) oxidation, two As–As dimers, and a Ga dangling bond. Specifically, Ga1, Ga2, Ga3, and Ga4 all contribute gap states due to their parital oxidation. However, Ga1, Ga2, and Ga4's contribution to the gap states is much smaller than that of Ga3 so that only the Ga3 contribution is clearly visible in Figure 13.13b inset. In general, one should note that to obtain a high-quality interface, interfacial As must be restored to its near-bulk charge state in GaAs. In addition, the Ga dangling bonds and the Ga partial oxidation (e.g., Ga^{3+}) should also be avoided. These considerations provide a clue on how to remove the gap states by different interface passivation strategies.

13.4.3
Si Passivation of HfO_2/GaAs Interface [54]

Based on the experimental and theoretical understanding on the origin of gap states, some interfacial control layers such as Si, Ge, and Si/Ge are used in experimental studies to passivate GaAs related interfaces [55–57]. Using a Si passivation layer, Hinkle *et al.* [52] found that the high oxidation state of interfacial Ga species (Ga^{3+}) was reduced through the formation of SiO_x. The reduction in the interfacial Ga^{3+} species reflects the effect of Si passivation, as the frequency dispersion of capacitance reduced substantially when Ga_2O_3 (Ga^{3+}) is removed from the interface. However, the mechanism of this unpinning remains unexplained, and the impact of other structural disorders is not clearly understood as well. Consequently, an understanding on the interfacial atomic bonding and the mechanism of Si passivation at the GaAs/HfO_2 interface is needed to elucidate the interface atomic and electronic properties. For this purpose, a first-principles DFT study of GaAs–HfO_2 interface was performed with a controlled silicon passivation at the interface.

In this study [54], two kinds of local bonding are considered, Si interstitial (labeled as Si_i) and replacement (Si_{Hf}) of interfacial Hf. For each case, only the most stable interface is reviewed here. Figure 13.14a shows the band structure (left panel) and the corresponding total DOS (central panel) of Si_{Hf} interface, which is the most stable interface in our model study. The flatband (originating from Ga3 interface atom) in the lower half region of GaAs gap in O9 (Figure 13.13b) is now suppressed into the valence band of bulk GaAs. Meanwhile, the gap states (originating from As–As dimers and Ga dangling bond) in the upper half of the bandgap are partially pushed into the GaAs conduction band. Tail states right below the CB edge still remain, and this partial passivation effect could be explained by partial charge distribution at the interface (see Figure 13.14a right inset). Specifically, Si atoms have a charge of ~0.89e compared to ~1.67e of Hf, and this different charge can stabilize the charge state of the interfacial Ga ("Ga3" in Figure 13.13b). Consequently, the corresponding interface gap states (in the lower half region of bulk GaAs gap) are suppressed into valence band region. In addition, according to the Bader charge analysis, As–As dimers gain limited charge compensation from both Si and "Ga3" so that As–As related gap states are less stable and correspondingly partially pushed upward to the CB region.

The electronic structure of metastable Si_i interface is shown in Figure 13.14b. The flatband located in the lower half region of GaAs gap in O9 was removed by silicon interstitial atoms in the $4Si_i$ interface. In the upper half of the GaAs gap, there are three tail bands (Figure 13.14b), and the charge density plot of the gap states

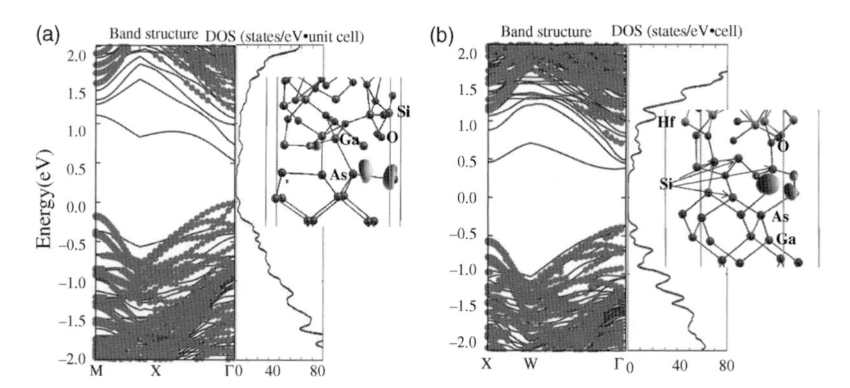

Figure 13.14 (a) and (b) Left and right panels represent the band structures and DOS of the Si_{Hf} and Si_i interface, respectively. The dotted region indicates the projected GaAs bulk band. The inset figures of (a) and (b) depict the partial charge distribution plot of interfacial bands for the Si_{Hf} and Si_i interface, respectively. The electron density is 0.5×10^{-2} e Å^{-3} and 3.0×10^{-2} e Å^{-3}, respectively. Charge is in solid grey color. Fermi level is set at 0 eV. (Reproduced with permission from Ref. [54].)

reveals that these three tail bands originating from the interstitial Si and Ga atoms. We find that the Ga atom has an excess charge of 0.22*e* compared to that in bulk GaAs. This excess charge may adjust the local bonding and produce interfacial gap states. The Si_i atoms behave like a bulk Si since interfacial Si–Si bonds remain 2.35 Å (comparable to that of bulk Si, 2.37 Å). Considering that Si has a smaller bandgap (1.12 eV) than GaAs (1.42 eV), the interfacial Si can contribute to the CB edge tail states, as shown in Figure 13.14b.

For all the Si-passivated interfaces, the gap states close to the E_v maximum are removed, and the gap states within the upper half region is slightly shifted upward to E_c. This DOS analysis clearly shows a partial passivation effect of silicon, but the gap states are not completely removed by increasing the Si-passivation density. This indicates that increasing Si amount at the interface is not helpful in removing the gap states.

Experimentally, the frequency dispersion of capacitance versus gate voltage curves was observed with the detection of Ga oxidation states corresponding to Ga_2O_3-like species at the interface. This frequency dispersion could be eliminated by removing Ga atoms in the Ga_2O_3 state using the Si passivation interlayer so that only Ga atoms corresponding to Ga_2O state would be present. This experimental observation can be explained by these Si-passivated Si_i and Si_{Hf} interface models. With the presence of Si interstitials or Si_{Hf} defects at the interface, the charge state of Ga decreases from + 1.24 to + 0.59 or + 0.73–0.80, and this charge state change corresponds to the trend of experimentally detected reaction from Ga_2O_3 to Ga_2O at the interface. Therefore, this model (Ref. [12]) is consistent with the experimental findings [20, 21] and provides an important insight into the nature of Si-passivation at high-*k*/III–V interfaces.

13.4.4
Sulfur Passivation Effect on HfO_2/GaAs Interface

Recently, Aguirre-Tostado *et al.* [58] observed that the GaAs surface treated with $(NH4)_2S$ showed a decreased band bending due to the reduction of interfacial space charge with respect to NH_4OH. This bending was attributed to a binding energy shift to higher energies for S-passivated surfaces. However, fundamental systematic knowledge of how S atomically changes the interface bonding and thus passivates the HfO_2/GaAs interface states remains unknown. In previous studies [59], the gap states have been assigned to As–As dimers, Ga partial oxidation and Ga-dangling bonds and Ga^{3+}-like oxidation states. However, due to the well-known limitations of GGA/DFT in predicting bandgap, the quantitative location of the gap states has not been fully identified. In this section, the Heyd–Scuseria–Ernerhof (HSE) [60] hybrid functional was applied to improve the accuracy of the bandgap and location of the gap states. Furthermore, the S passivation effect on the interface is investigated and provides an atomic level insight into gap state removal mechanism [59].

In order to study how S can passivate the interfacial gap states, various S substitutions and interstitials (up to four S atoms at the interface with different configurations) are examined. Experimentally, XPS data of Aguirre-Tostado *et al.* [58]

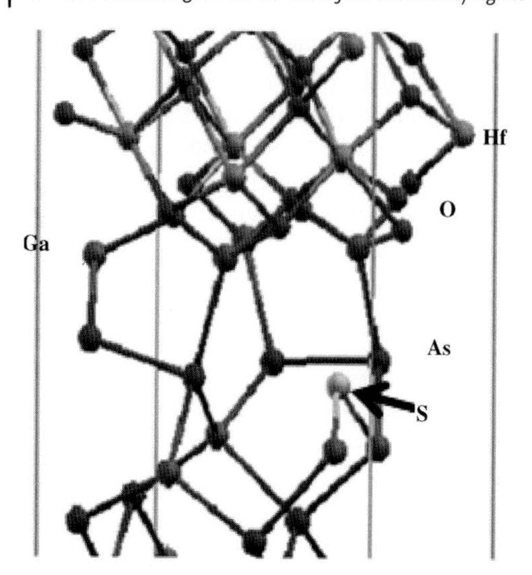

Figure 13.15 Side view of GaAs:HfO_2 interfaces (O9/S/Ga4).

showed that S bonding peak was under the XPS detection limit, indicating a low amount of S at the interface. Moreover, Chen *et al.* [38] observed Ga−S bonding at HfO_2/GaAs interface. Compared to the experimental information, the highest stability of the interface with one S interstitial is consistent, and we chose a model with one S interstitial located between Ga and As, forming a Ga–S and As−S bond. Figure 13.15 shows the final relaxed configuration with only Ga−S bond formed without breaking As–As dimers, which agrees well with XPS observations [58].

Figure 13.16 illustrates the D_{it} of the interface models O9/Ga4 and O9/S/Ga4. Valence band edge of bulk GaAs is used as a reference to line up these two D_{it} distributions. By performing HSE correction, a bulk GaAs bandgap of 1.40 eV was obtained, which agrees well with experimental value of 1.42 eV [61]. In the case

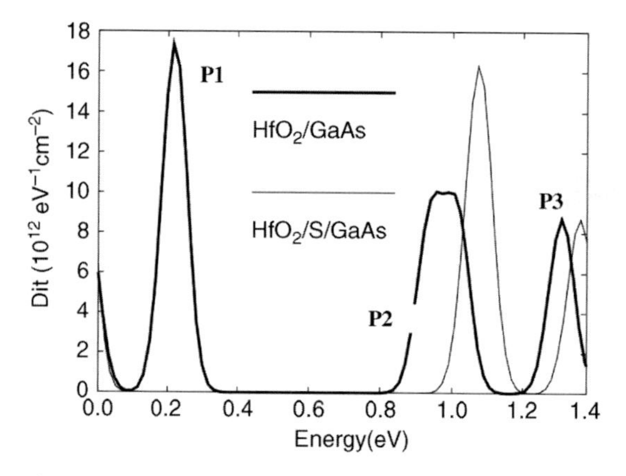

Figure 13.16 D_{it} of interface HfO_2/GaAs and HfO_2/S/GaAs. VBM (CBM) is valence (conducting) band edge.

of O9/Ga4, within the GaAs gap region three peaks labeled as P1, P2, and P3 are found. The charge density distribution of the bandgap states is plotted to explore the origin of these gap states. Figure 13.13b clearly shows the gap states arising from As–As dimers, Ga (Ga5) dangling bonds, and Ga (Ga3) partial oxidation. Further partial charge density analysis indicates P1, P2, and P3 correspond to Ga partial and Ga^{3+} like oxidation (Ga3), Ga (Ga5) dangling bonds and As–As dimers, respectively. According to the Bader charge [37] calculation, Ga1, Ga2, Ga3, and Ga4 contain 1.64*e*, 1.65*e*, 2.19*e*, and 1.55*e*, respectively. Ga1, Ga2, and Ga4 oxidation states show Ga^{3+} state because their charge state is close to Ga charge of 1.30*e* in bulk Ga_2O_3. However, Ga3 has 2.19*e* that is neither close to that of Ga_2O_3 nor close to that of GaAs, and this state corresponds to a partial oxidation state leading to gap states. The contribution of Ga1, Ga2, and Ga4 to the gap states is less than that of Ga(3), which is not shown in Figure 13.17 because of the high electron density value used to emphasize Ga(3)'s contribution. As a result, we conclude that Ga^{3+} and Ga partial charge states are responsible of the gap state P1.

In the presence of S at the interface, Ga3 (bonded to three O and two As atoms) charge state is reduced from 2.19*e* to 1.81*e* while Ga1 (bonded to three O and one As atoms) charge state is increased from 1.64*e* to 1.69*e* without changing Ga—O bonding environment. Ga dangling bond is removed by forming Ga—S bond, but the Ga atom (with 2.23*e*) is not fully saturated by S. Moreover, the As–As dimers survive after S being introduced into the interface. Consequently, gap states located in the lower half of the gap region are removed (P1 in Figure 13.16), and two upper half trap peaks (P2 and P3) slightly move upward the conducting band. This partial passivation effect is also observed in experimental studies [50]. Therefore, the main effect of S passivation is to remove the P1 gap state near the valance band edge. Under the S passivation, newly formed Ga (Ga1) partial oxidation states and Ga—S bonding contribute to the

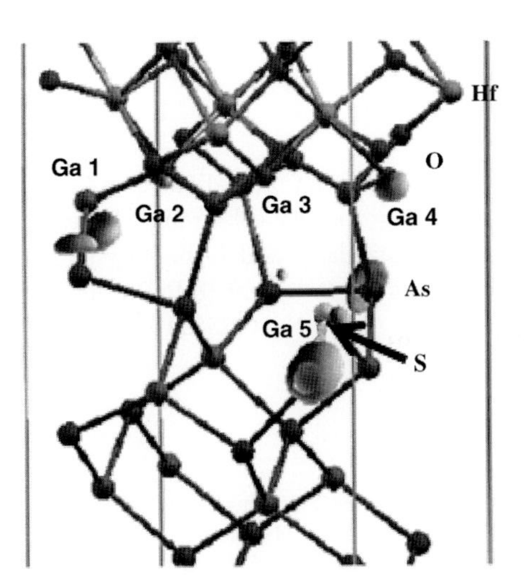

Figure 13.17 Side view of partial charge density of interface O9/S/Ga4. The Ga, As, Hf, and O atoms are depicted by gray, purple, light blue, and red balls, respectively. Yellow color indicates partial charge distribution. The electron density is $6.5 \times 10^{-2}\ e\,Å^{-3}$.

gap states in the upper half of the gap region. As S cannot fully eliminate As–As dimers and recover Ga partial oxidized states, S has a limited passivation capability on HfO_2/GaAs interface.

VBO is accurately predicted by reference potential method [62, 63]. In the interface O9/Ga4, VBO is 1.81 eV compared to experimental data of 2.00 [64], 2.10 [65], and 2.85 eV [66]. The variation in VBOs indicates experimental samples' diversity, and the VBO difference between theoretical prediction and experimental measurement results from the variety of the interfacial oxygen concentration. At the O9/S/Ga4 interface, they obtain a VBO of 1.98 eV. This finding confirms large enough tunneling barriers between HfO_2 and GaAs.

13.5
Conclusions

We have reviewed first-principles studies of the GaAs surface oxidation and passivation, and the high-k/GaAs interface. For the GaAs surface, GaAs(001)-β2(2 × 4) surface shows the most stability and this surface tends to be oxidized. The surface states originate from the dangling As bonds with half charge filled. H, Cl, F, S, and GaO stand out as effective candidates to remove the surface states. In the case of the high-k/GaAs interface, a strained interface predicts that the interfacial gap states arise from the interfacial As–As bonds but fails to provide right band offsets due to the large planar strain at the interface. The strain-free interface indicates that the gap states are distributed as three parts and are attributed to the Ga^{3+} like and partial oxidation, As–As dimer, and Ga-dangling bond. Moreover, these calculations show that the interfacial oxygen essentially controls the band offsets of the interface. In order to passivate the interface, Si and S appear to be promising in partially removing the gap states and conduction band tails survive.

Acknowledgments

This work was supported by the Semiconductor Research Corporation Focus Center on Materials Structures and Devices (MSD), by the Nanoelectronics Research Initiative (NRI) with the National Institute for Standards and Technology (NIST) through the Midwest Institute for Nanoelectronics Discovery (MIND), and the National Science Foundation (NSF)under ECCS Award No. 0925844

References

1 Oktyabrsky, S. and Ye, P.D. (2010) Chapter 1, in *Non-Silicon MOSFET Technology: A Long Time Coming, A Volume of Fundamentals of III–V Semiconductor MOSFETs*, Springer, New York, p. 1.

2 Schlom, D.G., Chen, L., Eom, C., Rabe, Karin M., Streiffer, S.K., and Triscone, J. (2007) Strain tuning of ferroelectric thin films. *Annu. Rev. Mater. Sci.*, **37**, 589.

3 Baraff, G.A., Appelbaum, J.A., and Hamann, D.R. (1977) Electronic structure at an abrupt GaAs–Ge interface. *J. Vac. Sci. Technol.*, **14**, 999.

4 Shen, J., Chagarov, E.A., Feldwinn, D.L., Melitz, W., Santagata, N.M., Kummel, A.C., Droopad, R., and Passlack, M. (2010) Scanning tunneling microscopy/spectroscopy study of atomic and electronic structures of In_2O on InAs and $In_{0.53}Ga_{0.47}As(001)$-(4×2) surfaces. *J. Chem. Phys.*, **133**, 164704.

5 Dorn, R., Luth, H., and Russel, G.J. (1974) Adsorption of oxygen on clean cleaved (110) gallium-arsenide surfaces. *Phys. Rev. B*, **10**, 5049.

6 Modine, N.A. and Kaxiras, E. (1999) Theory of the (3×2) reconstruction of the GaAs(001) surface. *Mater. Sci. Eng. B*, **67**, 1; J.E. Northrup and S. Froyen (1993) Energetics of GaAs(100)-(2 × 4) and –(4×2) reconstructions. *Phys. Rev. Lett.*, **71**, 2276.

7 Bone, P.A., Ripalda, J.M., Bell, G.R., and Jones, T.S. (2006) Surface reconstructions of InGaAs alloys. *Surf. Sci.*, **600**, 973.

8 Bracker, A.S., Yang, M.J., Bennett, B.R., Culbertson, J.C., and Moore, W.J. (2000) Surface reconstruction phase diagrams for InAs, AlSb, and GaSb. *J. Cryst. Growth*, **220**, 384.

9 Paget, D., Bonnet, J.E., Berkovits, V.L., Chiaradia, P., and Avila, J. (1996) Sulfide-passivated GaAs(001). I. Chemistry analysis by photoemission and reflectance anisotropy spectroscopies. *Phys. Rev. B*, **53**, 4604.

10 Fischetti, M., Neumayer, D.A., and Cartier, E.A. (1996) Effective electron mobility in Si inversion layers in metal–oxide–semiconductor systems with a high-κ insulator: the role of remote phonon scattering. *J. Appl. Phys.*, **90**, 4587.

11 Leitz, C.W., Gurrie, M.T., Lee, M.L., Cheng, Z.Y., Antoniadis, D.A., and Fitzgerald, E.A. (2002) Hole mobility enhancements and alloy scattering-limited mobility in tensile strained Si/SiGe surface channel metal–oxide–semiconductor field-effect transistors. *J. Appl. Phys.*, **92**, 3745.

12 Wang, W., Xiong, K., Wallace, R.M., and Cho, K. (2010) Impact of interfacial oxygen content on bonding, stability, band offsets, and interface states of GaAs:HfO_2 interfaces. *J. Phys. Chem. C*, **114**, 22610.

13 Lin, L. and Robertson, J. (2011) Defect states at III–V semiconductor oxide interfaces. *Appl. Phys. Lett.*, **98**, 082903.

14 Kresse, G. and Furthmüller, J. (1996) Efficiency of *ab-initio* total energy calculations for metals and semiconductors using a plane-wave basis set. *Comp. Mater. Sci.*, **6**, 15.

15 Car, R. and Parrinello, M. (1985) Unified approach for molecular dynamics and density-functional theory. *Phys. Rev. Lett.*, **55**, 2471.

16 Blochl, P.E. (1994) Projector augmented-wave method. *Phys. Rev. B.*, **50**, 17953.

17 Wang, W., Lee, G., Huang, M., Wallace, R.M., and Cho, K. (2010) First-principles study of GaAs(001)-β2(2×4) surface oxidation and passivation with H, Cl, S, F, and GaO. *J. Appl. Phys.*, **107**, 103720.

18 Goringe, C.M., Clark, L.J., Lee, M.H., Payne, M.C., Stich, I., White, J.A., Gillan, M.J., and Sutton, A.P. (1997) The GaAs(001)-(2 × 4) surface: structure, chemistry, and adsorbates. *J. Phys. Chem. B.*, **101**, 1498.

19 Schmidt, W.G., Bechstedt, F., Fleischer, K., Cobet, C., Esser, N., Richter, W., Bernholc, J., and Onida, G. (2001) GaAs (001): surface structure and optical properties. *Phys. Status Solidi A*, **188**, 1401.

20 LaBella, V.P., Krause, M.R., Ding, Z., and Thibado, P.M. (2005) Arsenic-rich GaAs(001) surface structure. *Surf. Sci. Rep.*, **60**, 1.

21 Cho, A.Y. (1971) GaAs epitaxy by a molecular beam method: observations of surface structure on the (001) face. *J. Appl. Phys.*, **42**, 2074.

22 Qian, G., Martin, X.R., and Chadi, D. (1988) Stoichiometry and surface reconstruction: an *ab initio* study of GaAs(100) surfaces. *Phys. Rev. Lett.*, **60**, 1962.

23 Northrup, J.E. and Froyen, S. (1993) Energetics of GaAs(100)-(2×4) and -(4×2) reconstructions. *Phys. Rev. Lett.*, **71**, 2276.

24 Schmidt, W.G. and Bechstedt, F. (1996) Geometry and electronic structure of GaAs(001)(2×4) reconstructions. *Phys. Rev. B*, **54**, 16742.

25 Hale, M., Yi, S., Sexton, J., Kummel, A., and Passlack, M. (2003) Scanning tunneling miscroscopy and spectroscopy of gallium oxide deposition and oxidation on GaAs(001)-c(2×8)/(2 × 4). *J. Chem. Phys.*, **119**, 6719.

26 Kruse, P., McLean, J.G., and Kummel, A.C. (2000) Relative reactivity of arsenic and gallium dimers and backbonds during the adsorption of molecular oxygen on GaAs(100)(6×6). *J. Chem. Phys.*, **113**, 9217.

27 Winn, D.L., Hale, M.J., Grassman, T.J., Kummel, A.C., Droopad, R., and Passlack, M. (2007) Direct and indirect causes of Fermi level pinning at the SiO/GaAs interface. *J. Chem. Phys.*, **126**, 084703.

28 Tersoff, J. (1984) Schottky barrier heights and the continuum of gap states. *Phys. Rev. Lett.*, **52**, 465.

29 Traub, M.C., Biteen, J.S., Brunschwig, B.S., and Lewis, N.S. (2008) Passivation of GaAs nanocrystals by chemical functionalization. *J. Am. Chem. Soc.*, **130** (3), 955.

30 Szucs, B., Hajnal, Z., Frauenheim, Th., González, C., Ortega, J., Pérez, R., and Flores, F. (2003). Chalcogen passivation of GaAs(1 0 0) surfaces: a theoretical study. *Appl. Surf. Sci*, **212** 861.

31 Jiang, G. and Harry, R. (1996) The origin of Ga_2O_3 passivation for reconstructed GaAs (001) surfaces. *J. Appl. Phys.*, **83**, 5880.

32 Jiang, G. and Harry, R. (1996) *Ab initio* studies of S chemisorption on GaAs(100). *J. Appl. Phys.*, **79**, 3758.

33 Meyer, B. (1964) Solid allotropes of sulfur. *Chem. Rev.*, **64** (4), 429–451.

34 Lide, D.R. (2001) *Handbook of Chemistry & Physics*, 82nd edn, CRC, Boca Raton, pp. 9–15.

35 Lüdemann, R. (1999) Hydrogen passivation of multicrystalline silicon solar cells. *Mater. Sci. Eng. B*, **58**, 86.

36 Capasso, F. and Williams, G.F. (1982) A proposed hydrogenation/nitridization passivation mechanism for GaAs and other III–V semiconductor devices, including InGaAs long wavelength photodetectors. *J. Electrochem. Soc.*, **129**, 821.

37 Henkelman, G., Arnaldsson, A., and Jónsson, H. (2006) A fast and robust algorithm for Bader decomposition of charge density. *Comp. Mater. Sci.*, **36**, 354.

38 Chen, Y.-T., Zhao, H., Yum, J.H., Wang, Y., Xue, F., Zhou, F., and Lee, J.C. (2009) Improved electrical characteristics of TaN/Al_2O_3/$In_{0.53}Ga_{0.47}As$ metal–oxide–semiconductor field-effect transistors by fluorine incorporation. *Appl. Phys. Lett.*, **95**, 013501.

39 Scarrozza, M., Pourtois, G., Houssa, M., Caymax, M., Stesmans, A., Meuris, M., and Heyns, M.M. (2009) A theoretical study of the initial oxidation of the GaAs (001)-*(*2(2×4) surface. *Appl. Phys. Lett.*, **95**, 253504.

40 Wallace, R.M., McIntyre, P.C., Kim, J., and Nishi, Y. (2009) Atomic layer deposition of dielectrics on Ge and III–V materials for ultrahigh performance transistors. *MRS Bull.*, **34**, 493.

41 Ok, I., Kim, H., Zhang, M., Zhu, F., Park, S., Yum, J., Zhao, H., and Lee, J.C. (2007) Temperature effects of Si interface passivation layer deposition on high-*k* III–V metal–oxide–semiconductor characteristics. *Appl. Phys. Lett.*, **91**, 132104.

42 Wang, W., Xiong, K., Wallace, R.M., and Cho, K. (2010) Origin of HfO_2/GaAs interface states and interface passivation: a first principles study. *Appl. Surf. Sci.*, **256**, 6569.

43 Sadakazu, W., Jun, N., and Akiko, N. (2009) In-plane strain effects on dielectric properties of the HfO_2 thin film. *J. Vac. Sci. Technol. B*, **27**, 2020.

44 Dong, Y.F., Feng, Y.P., Wang, S.J., and Huan, A.C.H. (2005) First-principles study of ZrO_2/Si interfaces: energetics and band offsets. *Phys. Rev. B*, **72**, 045327.

45 Peacock, P.W., Xiong, K., Tse, K., and Robertson, J. (2006) Bonding and interface states of Si:HfO_2 and Si:ZrO_2 interfaces. *Phys. Rev. B*, **73**, 075328.

46 Franceschetti, A., Wei, S.-H., and Zunger, A. (1994) Absolute deformation potentials of Al, Si, and NaCl. *Phys. Rev. B*, **50**, 17797.

47 Pashley, M.D. (1989) Electron counting model and its application to island structures on molecular-beam epitaxy grown GaAs(001) and ZnSe(001). *Phys. Rev. B*, **40**, 10481.

48 Passlack, M., Droopad, R., Fejes, P., and Wang, L. (2009) Electrical properties of Ga_2O_3/GaAs interfaces and GdGaO dielectrics in GaAs-based MOSFETs. *IEEE Electron. Device Lett.*, **30**, 2.
49 O'Reilli, E.P. and Robertson, J. (1986) Electronic structure of amorphous III–V and II–VI compound semiconductors and their defects. *Phys. Rev. B*, **34**, 8684.
50 Caymax, M., Brammertz, G., Delabie, A., Sioncke, S., Lin, D., Scarrozza, M., Pourtois, G., Wang, W., Meuris, M., and Heyns, M. (2009). Interfaces of high-*k* dielectrics on GaAs: their common features and the relationship with Fermi level pinning. *Microelectron. Eng.*, **86**, 1529.
51 Hinkle, C.L., Sonnet, A.M., Vogel, E.M., McDonnell, S., Hughes, G.J., Milojevic, M., Lee, B., Aguirre-Tostado, F.S., Choi, K., Kim, H.C., Kim, J., and Wallace, R.M. (2008). GaAs interfacial self-cleaning by atomic layer deposition. *Appl. Phys. Lett.*, **92**, 071901.
52 Hinkle, C.L., Milojevic, M., Brennan, B., Sonnet, A.M., Aguirre-Tostado, F.S., Hughes, G.J., Vogel, E.M., and Wallace, R.M. (2009) Detection of Ga suboxides and their impact on III–V passivation and Fermi-level pinning. *Appl. Phys. Lett.*, **94**, 162101.
53 Yoshioka, S., Hayashi, H., Kuwabara, A., Oba, F., Matsunaga, K., and Tanaka, I. (2007) Structures and energetics of Ga_2O_3 polymorphs. *J. Phys. Condens. Matter*, **19**, 346211.
54 Wang, W., Xiong, K., Gong, C., Wallace, R., and Cho, K. (2011) Si passivation effects on atomic bonding and electronic properties at HfO_2/GaAs interface: a first-principles study. *J. Appl. Phys.*, **109**, 063704.
55 Kim, H., Ok, I., Zhang, M., Lee, T., Zhu, F., Yu, L., and Lee, J.C. (2006) Metal gate-HfO_2 metal–oxide–semiconductor capacitors on n-GaAs substrate with silicon/germanium interfacial passivation layers. *Appl. Phys. Lett.*, **89**, 222903.
56 Tiwari, S., Wright, S.L., and Batey, J. (1988) Unpinned GaAs MOS capacitors and transistors. *IEEE Electron. Device Lett.*, **9**, 488.
57 Kim, H., Ok, I., Zhang, M., Zhu, F., Park, S., Yum, J., Zhao, H., Lee, J.C., Oh, J., and Majhi, P. (2008) Flatband voltage instability characteristics of HfO_2-based GaAs metal–oxide–semiconductor capacitors with a thin Ge layer. *Appl. Phys. Lett.*, **92**, 102904.
58 Aguirre-Tostado, F.S., Milojevic, M., Choi, K.J., Kim, H.C., Hinkle, C.L., Vogel, E.M., Kim, J., Yang, T., Xuan, Y., Ye, P.D., and Wallace, R.M. (2008) S passivation of GaAs and band bending reduction upon atomic layer deposition of HfO_2/Al_2O_3 nanolaminates. *Appl. Phys. Lett.*, **93**, 061907.
59 Wang, W., Hinkle, C.L., Vogel, E.M., Cho, K., and Wallace, R.M. (2011) Is interfacial chemistry correlated to gap states for high-k/III–V interfaces? *Microelect. Eng.*, **88**, 1061.
60 Perdew, J.P., Ernzerhof, M., and Burke, K. (1996) Rationale for mixing exact exchange with density functional approximations. *J. Chem. Phys.*, **105**, 9982.
61 GaAs Band struture and carrier concentration, http://www.ioffe.rssi.ru/SVA/NSM/Semicond/GaAs/bandstr.html. Accessed 02-April-2012.
62 Van de Walle, C.G. and Martin, R.M. (1987) Theoretical study of band offsets at semiconductor interfaces. *Phys. Rev. B*, **35**, 8154.
63 Al-Allak, H.M. and Clark, S.J. (2001) Valence-band offset of the lattice-matched β-$FeSi_2$(100)/Si(001) heterostructure. *Phys. Rev. B*, **63**, 033311.
64 Afanas'ev, V.V., Badylevich, M., Stesmans, A., Brammertz, G., Delabie, A., Sionke, S., O'Mahony, A., Povey, I.M., Pemble, M.E., O' Connor, E., Hurley, P.K., and Newcomb, S.B. (2008). Energy barriers at interfaces of (100)GaAs with atomic layer deposited Al_2O_3 and HfO_2. *Appl. Phys. Lett.*, **93**, 212104.
65 Seguini, G., Perego, M., Spiga, S., Fanciulli, M., and Dimoulas, A. (2007) Conduction band offset of HfO_2 on GaAs. *Appl. Phys. Lett.*, **91**, 192902.
66 Dalapati, G.K., Oh, H., Lee, S.J., Sridhara, A., See, A., Wong, W., and Chi, D. (2008) Energy-band alignments of HfO_2 on p-GaAs substrates. *Appl. Phys. Lett.*, **92**, 042120.

14
III–V MOSFETs with ALD High-κ Gate Dielectrics

Jack C. Lee and Han Zhao

14.1
Introduction

To achieve higher density and performance and lower power consumption, silicon complementary metal–oxide–semiconductor (CMOS) devices have been scaled for more than 30 years. Transistor delay times decrease by more than 30% per technology generation, resulting in doubling of microprocessor performance every 2 years [1]. The key enabler for the exponential growth of the transistor density on a chip is the scaling of the metal–oxide–semiconductor field effect transistor (MOSFET) gate length by a factor of 0.7 per technology node. In addition, other transistor dimensions have been scaled according to Dennard's scaling theory [2].

The half-pitch of the first metal layer and the physical gate length of the transistors for high-performance logic circuits are going to be 29 and 16 nm, respectively, in 2014 according to 2008 International Technology Roadmap for Semiconductors (IRTS) [3]. This is approaching the optical limit of the photolithography. Due to the physical limit (~14 Å thickness [4, 5]) and high gate leakage of SiO_2 gate oxide, Intel Inc. has used high-κ oxide in their 45 nm node products [1]. All these constraints and limits are calling for a new material and/or a new device structure to continue improving semiconductor products. The 2007 edition of ITRS [6] states the structures and technology innovation from 65 nm node onward: employing a group of technology boosters such as stained Si [7, 8], high-κ/metal gate [4, 9–11], and high-mobility channels [12–14] can provide performance benefit. The following part of this section will focus on one of the potential solutions: high-mobility channels with high-κ gate oxide.

From device characteristics point of view, a high on current (I_{on}) for MOSFETs can provide smaller transistor switching time and thus improved transistor performance. Power consumption for a CMOS circuit consists of active power and standby power. To reduce the power consumption, both operating voltage (V_{dd}) and off current (I_{off}) have to be reduced. I_{off} consists of subthreshold leakage current, gate-induced drain leakage (GIDL) current, and gate leakage current [15, 16]. In summary, to obtain a

1) http://www.intel.com/technology/45nm/index.htm.

High-k Gate Dielectrics for CMOS Technology, First Edition. Edited by Gang He and Zhaoqi Sun.

semiconductor device with low power consumption and high performance, a high I_{on}, a low I_{off}, and a small V_{dd} are required.

Table 14.1 shows the material properties of various semiconductors. III–V materials have much higher electron mobility compared to Si; thus, they can potentially provide higher on current at lower operating voltages, thereby providing both power and performance benefit.

Applying high-κ oxides instead of SiO_2 as the gate oxides on III–V materials can reduce the gate leakage current at the same equivalent oxide thickness (EOT), thus reducing the power consumption. The requirements of the high-κ dielectrics [4, 17, 18] include (1) a large bandgap for high barrier heights to both electrons and holes, (2) thermodynamic stability and low interface trap density with substrate, (3) compatibility with gate electrode, and (4) compatibility with conventional planar CMOS processing. The material properties of various gate oxides are summarized in Table 14.2.

The key challenge for III–V MOSFETs with high-κ oxides is the lack of high-quality, thermodynamically stable insulators that passivate the gate oxide/III–V interface. Recently, surface channel inversion-mode III–V MOSFETs with atomic layer deposited (ALD) Al_2O_3, HfO_2, ZrO_2, or $LaAlO_3$ dielectrics [19–22], molecular beam epitaxy (MBE) $Ga_2O_3(Gd_2O_3)$ dielectrics [23–25], and Si, Ge, Si_xN_y, Ge_xN_y, or Al_xN_y interfacial passivation layer (IPL) and high-κ gate stacks [26–30] show promising results. Various interface treatment techniques such as sulfur compounds [31], N_2 plasma [29], HBr solution [32], PH_3 passivation [33], and fluorine treatment [34, 35] have also been demonstrated with improved device characteristics. On the other hand, buried channel III–V MOSFETs with InAlAs barrier layer and Si interfacial passivation layer [36, 37], or with single InP barrier layer or InP/InAlAs double-barrier layer using *ex situ* ALD oxide [38, 39], flatband InGaAs MOSFETs with GaAs/AlGaAs barrier layer and Si δ-doping using *in situ* MBE $GaGdO_x$ gate oxide [40, 41], or MOS high electron mobility transistors (MOS-HEMTs) [42] demonstrate much higher channel mobility (e.g., $\mu_{eff} > 3800\,cm^2/(V\,s)$ [36–42]) compared to surface channel MOSFETs (e.g., $\mu_{eff} < 2000\,cm^2/(V\,s)$ [19–22, 26–35]). Moreover, the gate leakage current density of buried channel InGaAs MOSFETs [36–38] can be several orders of magnitude lower than that of HEMTs [43, 44]. There are also other challenges for III–V MOSFETs [45] such as small electron mass-induced quantum capacitance, low density of states, low Γ–L valley separation, low hole mobility, and their process integration issues with Si industry. Moreover, the reported transistor characteristics, including the drive capability (saturation current and transconductance), electrostatic integrity (subthreshold swing (SS) and drain-induced barrier lowering (DIBL)), and the channel mobility, are still far from the satisfactory level. There is still a long way to go on successfully implementing III–V MOSFETs.

The motivation of this work is to explore the possibility of combining the high-κ gate dielectrics with the III–V substrates to implement high-performance MOSFETs for digital applications in the post-silicon era. On one hand, the device performance of surface channel $In_{0.53}Ga_{0.47}$ As MOSFETs is improved by proper process, and substrate and interface engineering techniques. Effects of gate-first and gate-last processes on interface quality are compared. It has been found that applying gate-last

Table 14.1 Material properties of various semiconductors.

	Si	Ge	GaAs	InP	$In_{0.53}Ga_{0.47}As$	$In_{0.7}Ga_{0.3}As$	InAs
Lattice constant (Å)	5.431	5.658	5.653	5.869	5.869	5.937	6.058
Electron effective mass (m^*/m_0)	0.19	0.082	0.067	0.077	0.041	0.034	0.023
Electron affinity (eV)	4.05	4	4.07	4.38	4.5	4.65	4.9
Bandgap (eV)	1.12	0.66	1.42	1.35	0.74	0.58	0.35
n_i at RT (cm^{-3})	1×10^{10}	2×10^{13}	2.1×10^{6}	1.3×10^{7}	6.3×10^{11}	1.1×10^{13}	1.0×10^{15}
N_c (cm^{-3})	3.2×10^{19}	1×10^{19}	4.7×10^{17}	5.7×10^{17}	2.1×10^{17}	1.5×10^{17}	8.7×10^{16}
N_v (cm^{-3})	1.8×10^{19}	5×10^{18}	9.0×10^{18}	1.1×10^{19}	7.7×10^{18}	7.5×10^{18}	6.6×10^{18}
Electron mobility ($cm^2/(V\ s)$)	1500	3900	8500	4600	12 000	20 000	33 000
Hole mobility ($cm^2/(V\ s)$)	450	1900	400	150	300	400	460
Saturation velocity (cm/s)	1×10^{7}	6×10^{6}	2.1×10^{7}	2.5×10^{7}	3.1×10^{7}	6.1×10^{7}	7.7×10^{7}

Table 14.2 Material properties of various gate oxides.

Properties	SiO_2	Al_2O_3	HfO_2	ZrO_2	La_2O_3
Dielectric constant	3.9	8–9	18–25	18–30	20–36
Bandgap (eV)	9	8.8	6.0	5.8	4.3
Band offset for electrons (eV)	3.5	2.8	1.5	1.4	2.3
Band offset for holes (eV)	4.4	4.9	3.4	3.3	0.9

process provides significant capacitance–voltage (C–V) frequency dispersion reduction and interface trap density (D_{it}) reduction for InGaAs devices compared to gate-first process. By investigating the dependence of channel doping concentration and channel thickness on device performance for $In_{0.53}Ga_{0.47}As$ MOSFETs with ALD Al_2O_3 dielectrics, it has been found that undoped channel provides the highest drive current compared to p-doped channel. With proper substrate doping concentration, reasonable subthreshold swing can be achieved. Thinner InGaAs channel provides lower I_{off} but also lower I_{on}. Among ALD Al_2O_3, HfO_2, and $LaAlO_3$ gate oxides, Al_2O_3 exhibits the best interface quality with InGaAs, HfO_2 exhibits the thinnest EOT, while $LaAlO_3$ gives better thickness scalability than Al_2O_3 and better interface quality than HfO_2. On the other hand, buried channel InGaAs MOSFETs with single InP barrier layer with different thicknesses and $InP/In_{0.52}Al_{0.48}As$ double-barrier layer were investigated. InP barrier layer was found to provide MOSFETs with higher transconductance for both $In_{0.7}Ga_{0.3}As$ and $In_{0.53}Ga_{0.47}As$ MOSFETs, especially for $In_{0.7}Ga_{0.3}As$. $In_{0.7}Ga_{0.3}As$ MOSFETs with InP barrier layer show much higher transconductance and peak mobility, and lower subthreshold swing than those without barrier. Devices using $InP/In_{0.52}Al_{0.48}As$ double barrier achieve mobility enhancement at both low and high fields compared to those without barrier.

14.2 Surface Channel InGaAs MOSFETs with ALD Gate Oxides

14.2.1 Effects of Gate-First and Gate-Last Processes on Interface Quality of $In_{0.53}Ga_{0.47}As$ MOSCAPs Using ALD Al_2O_3 and HfO_2

Source and drain (S/D) usually need to be formed by ion implantation and thermal activation for surface channel inversion-mode MOSFETs. However, dielectric/semiconductor interface quality may degrade during this high-temperature S/D activation annealing process for III–V MOS structure. In this section, we applied S/D activation process on metal–oxide–semiconductor capacitors (MOSCAPs) to investigate the effect of this process on oxide/III–V interface quality. We compared the interface quality of $In_{0.53}Ga_{0.47}As$ MOSCAPs with ALD Al_2O_3 and HfO_2 under three process conditions: (a) only postdeposition annealing (PDA-only), no S/D activation, (b) gate-first process (S/D activated after gate stack deposition) [46], and (c) gate-last process

(S/D activated before gate stack deposition) [47]. It has been found that MOSCAPs with gate-last process can maintain similar interface trap density (D_{it}) to PDA-only samples, while those with gate-first process have much larger D_{it}. This suggests that gate-last process is more promising for surface channel inversion-type III–V MOSFETs. Moreover, D_{it} is higher for MOSCAPs with HfO_2 than those with Al_2O_3. X-ray photoelectron spectroscopy (XPS) indicates that gate-first process results in a larger amount of In–O, Ga–O, and As–As bonds on InGaAs surface, while gate-last process maintains similar surface chemical bonding condition to PDA-only process. MOSCAPs with HfO_2 exhibit more Ga–O bonds than those with Al_2O_3 and similar In or As bonding condition to those with Al_2O_3. Transmission electron microscopy (TEM) and electron energy loss spectroscopy (EELS) analyses show that MOSCAPs with HfO_2 using gate-first process exhibit thicker interfacial layer and more intermixing between oxide and substrate than those using gate-last process.

$In_{0.53}Ga_{0.47}As$ MOSCAPs were fabricated on 400 nm n-type $In_{0.53}Ga_{0.47}As$ (Si-doped, 5×10^{16} cm^{-3}) epitaxially grown on n-InP substrate. Three different process conditions were applied: (a) PDA-only, (b) gate-first process (G-first), and (c) gate-last process (G-last). Table 14.3 shows the process flow chart for these three processes. 6 nm Al_2O_3 (capacitance equivalent thickness (CET) = 4.2 nm) or 7 nm HfO_2 (CET = 2.1 nm) was deposited on different samples for gate dielectrics.

Figure 14.1 illustrates typical C–V characteristics of TaN/Al_2O_3/InGaAs and TaN/HfO_2/InGaAs MOSCAPs using PDA-only process and gate-first process. The C–V curves of MOSCAPs using gate-last process are the same as those for PDA-only process (data not shown), indicating similar interface quality to PDA-only process. Gate-first process exposes gate stacks on InGaAs at high temperature and degrades interface quality, thus causing higher frequency dispersion on C–V curves. Larger frequency dispersion of MOSCAPs with HfO_2 compared to those with Al_2O_3 indicates higher D_{it} for HfO_2 samples. D_{it} at upper half of the

Table 14.3 Process flow chart.

PDA-only	G-first	G-last
1. Wafer cleaning and S passivation	1. Wafer cleaning and S passivation	1. Wafer cleaning and S passivation
2. ALD gate dielectrics	2. ALD gate dielectrics	2. 10 nm ALD Al_2O_3 capping layer
3. PDA at 500 °C, 90 s in N_2	3. PDA at 500 °C, 90 s in N_2	3. S/D activation at 700 °C, 10 s in N_2
4. TaN gate metal and back-side metal deposition	4. TaN gate metal and back-side metal deposition	4. Remove capping layer
	5. S/D activation at 700 °C, 10 s in N_2	5. Wafer cleaning and S passivation
		6. ALD gate dielectrics
		7. PDA at 500 °C, 90 s
		8. TaN gate metal and back-side metal

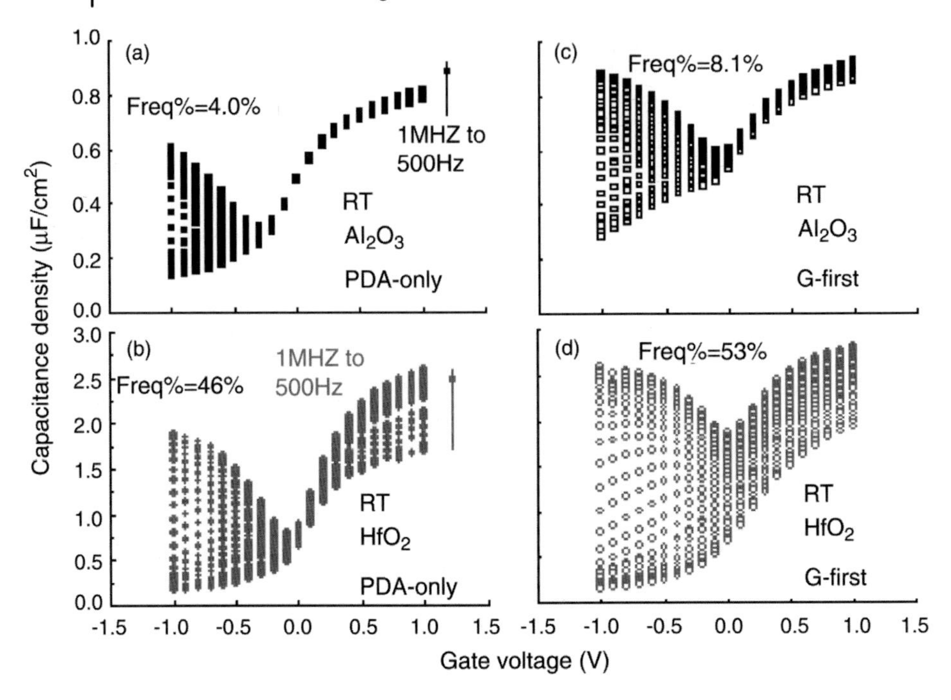

Figure 14.1 *C–V* characteristics of TaN/Al_2O_3/InGaAs and TaN/HfO_2/InGaAs MOSCAPs as a function of frequencies from 1 MHz to 500 Hz at room temperature using PDA-only process and gate-first process (freq% = $(C_{500\ Hz} - C_{1\ MHz})/C_{1\ MHz}$, at $V_g = 1$ V).

bandgap of MOSCAPs with different processes was measured using conductance method (Figure 14.2). Low-temperature (150 K) measurement allows detecting D_{it} close to band edge [48, 49]. It is clearly seen that MOSCAPs using gate-last process have similar D_{it} to PDA-only samples, while gate-first process results in

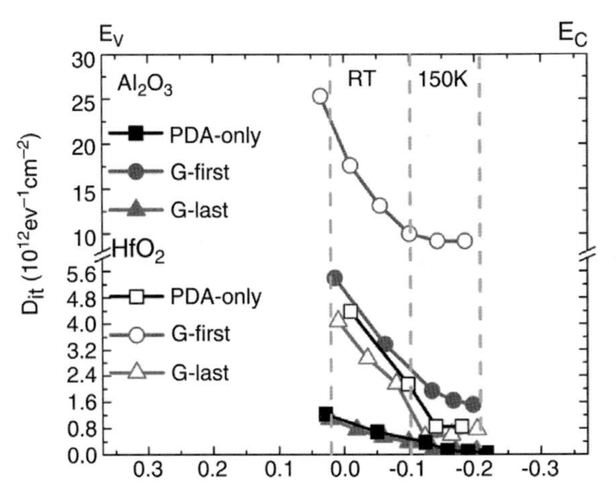

Figure 14.2 D_{it} versus energy position at bandgap for InGaAs MOSCAPs with Al_2O_3 and HfO_2 using PDA-only, gate-first, and gate-last processes.

much larger value of D_{it}. MOSCAPs with Al_2O_3 have smaller D_{it} value than those with HfO_2.

XPS was measured on different samples using processes including PDA-only, G-first, and G-last (Figures 14.3 and 14.4). Ga^{3+}–O bond is fitted at about 1.1 eV above Ga–As bond, and Ga^{+}–O bond and Ga–S bond [50, 51] are under the detection limit; this might be because we have comparably thick Al_2O_3 or HfO_2 (25–30 Å) for our *ex situ* XPS measurement. There is no As–O bond detected for both Al_2O_3 and HfO_2 using gate-last process and the substrate oxide self-cleaning effect during ALD oxide deposition is considered to be the reason [52–54]. Some As^{3+}–O bonds were detected for HfO_2 with gate-first process. The line shape for the In spectra, even for a surface that is oxygen free, is asymmetric [55]. This makes the deconvolution of In spectra exceedingly difficult; thus, we only pointed out the difference in In spectra between gate-first and gate-last processes. For gate-first samples, it is obvious that the excessive In oxide growth has changed the shape of In spectra dramatically. For InGaAs MOSCAPs with Al_2O_3 gate dielectrics (Figure 14.3), samples with gate-last process exhibit similar amount of surface oxide components to PDA-only samples, indicating no interface reaction at Al_2O_3/InGaAs interface. On the other hand, samples with gate-first process show excessive interfacial oxidation and increased In–O, Ga–O, and As–As bonds on the substrate surface. This explains why the samples using gate-last process can maintain similar D_{it} to those using PDA-only process while samples using gate-first process exhibit increased D_{it}. InGaAs MOSCAPs with HfO_2 gate dielectrics (Figure 14.4) show similar trend of surface oxide condition for different processes to those with Al_2O_3, except that

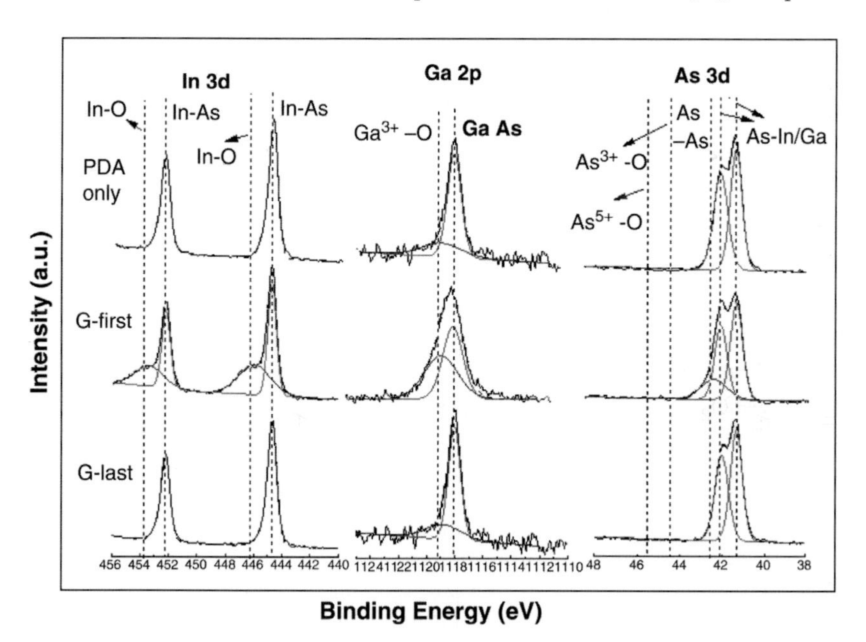

Figure 14.3 XPS spectra of In 3d, Ga 2p, and As 3d after applying PDA-only, gate-first, and gate-last processes for Al_2O_3/InGaAs structure.

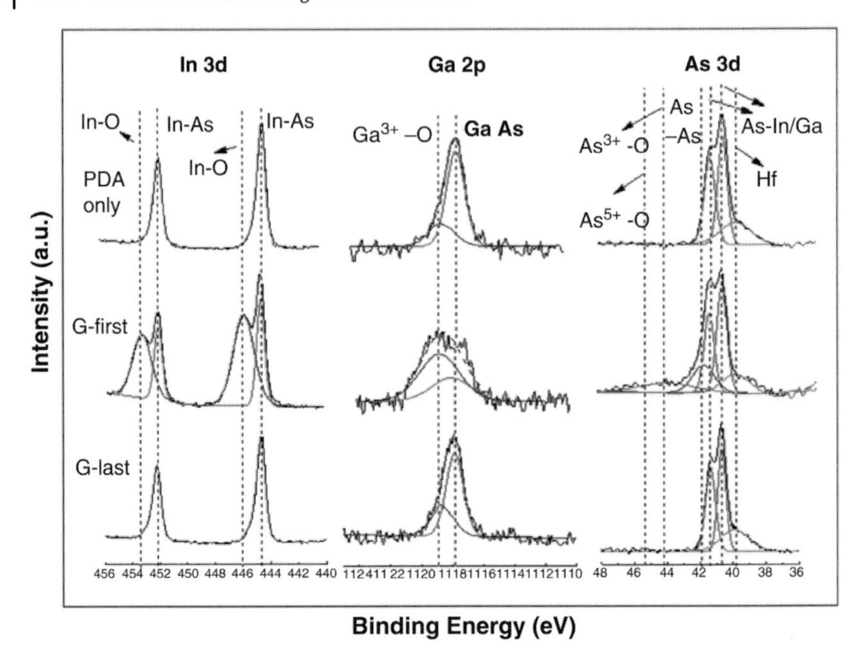

Figure 14.4 XPS spectra of In 3d, Ga 2p, and As 3d after applying PDA-only, gate-first, and gate-last processes for HfO_2/InGaAs structure.

MOSCAPs with HfO_2 using gate-first process show even much larger amount of surface oxides than those with Al_2O_3 and those with HfO_2 using gate-last process show larger amount of Ga–O component than those with Al_2O_3, which might be responsible for the higher D_{it} for MOSCAPs with HfO_2.

Figure 14.5 illustrates the high-resolution bright-field and dark-field scanning TEM and EELS analyses of MOSCAPs with HfO_2 using gate-first and gate-last processes. There is evidence of an interfacial layer approximately 0.5 nm thick right at the interface between InGaAs and HfO_2 using gate-first process in dark-field TEM. The interfacial oxide is still undetectable using gate-last process with dark-field TEM. The EELS analysis shows clearly that the samples with gate-first process have larger oxide/substrate interfacial mixing than those with gate-last process.

In conclusion, we have fabricated $In_{0.53}Ga_{0.47}As$ MOSCAPs with ALD Al_2O_3 and HfO_2 by applying three different processes including PDA-only process, gate-first process, and gate-last process. Samples using HfO_2 show larger D_{it} than those with Al_2O_3, which results from more Ga–O bonds on HfO_2/InGaAs samples than Al_2O_3/InGaAs samples indicated by XPS spectra. MOSCAPs with gate-first process have much larger D_{it} than those with PDA-only process while those with gate-last process have similar D_{it} value to PDA-only samples. TEM and EELS results indicate that MOSCAPs with HfO_2 using gate-first process exhibit thicker interfacial layer and more intermixing between oxide and substrate than those using gate-last process. These results suggest that gate-last process is preferable over gate-first process for surface channel inversion-type III–V MOSFETs.

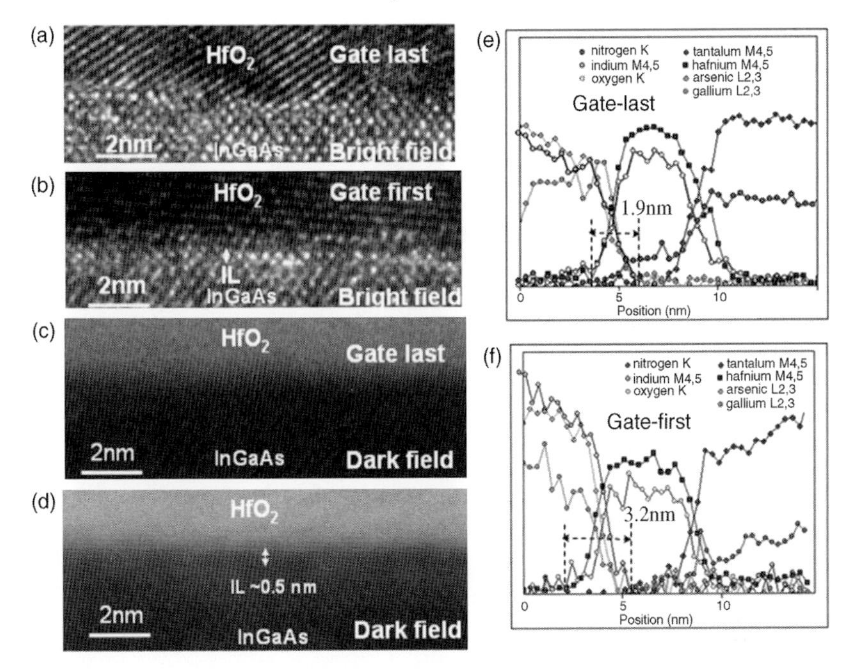

Figure 14.5 High-resolution bright-field TEM (a, b), dark-field TEM (c, d), and EELS (e, f) of MOSCAPs with HfO_2 using gate-first and gate-last processes.

14.2.2
Effect of Channel Doping Concentration and Thickness on Device Performance for $In_{0.53}Ga_{0.47}$ As MOSFETs with ALD Al_2O_3 Dielectrics

High drive current density of 400 mA/mm for 0.5 μm $In_{0.53}Ga_{0.47}As$ channel [56] and 1 A/mm for 0.4 μm $In_{0.65}G_{0.35}As$ channel [19] using ALD Al_2O_3 gate dielectric and 0.9 A/mm for 1 μm $In_{0.53}Ga_{0.47}As$ channel using MBE $Ga_2O_3(Gd_2O_3)$ gate dielectric [57] have been reported recently. The drive current is comparable to 65 nm strained Si channel technology ($I_{on} = 1.6$ mA/μm, subthreshold swing = 105 mV/dec) [58] even with much longer gate length and thicker EOT. However, InGaAs surface channel n-MOSFETs usually exhibit fairly high off-current density (e.g., 5×10^{-4} mA/mm [19]) and large subthreshold swing (e.g., 179 mV/dec [59], 240 mV/dec [56], and 330 mV/dec [19]). In our work, we investigate the effects of $In_{0.53}Ga_{0.47}As$ doping concentration and thickness on the MOSFET device performance. By carefully engineering the channel doping concentration and thickness with ALD Al_2O_3 gate dielectrics, reasonable subthreshold swing of 104 mV/dec and low off-current density of 4.0×10^{-6} mA/mm have been obtained.

$In_{0.53}Ga_{0.47}As$ MOSFETs were fabricated by gate-last process on 200 nm undoped $In_{0.53}Ga_{0.47}As$ (sample (a)), 300 nm p-type $In_{0.53}Ga_{0.47}As$ (Be-doped, $2 \times 10^{16}\,cm^{-3}$, sample (b)), 30 nm p-type $In_{0.53}Ga_{0.47}As$ ($2 \times 10^{16}\,cm^{-3}$, sample (c)), or 300 nm p-type $In_{0.53}Ga_{0.47}As$ ($5 \times 10^{16}\,cm^{-3}$, sample (d)). Note that sample (a) was grown on SI-InP substrate and an undoped InAlAs buffer layer was grown before InGaAs layer.

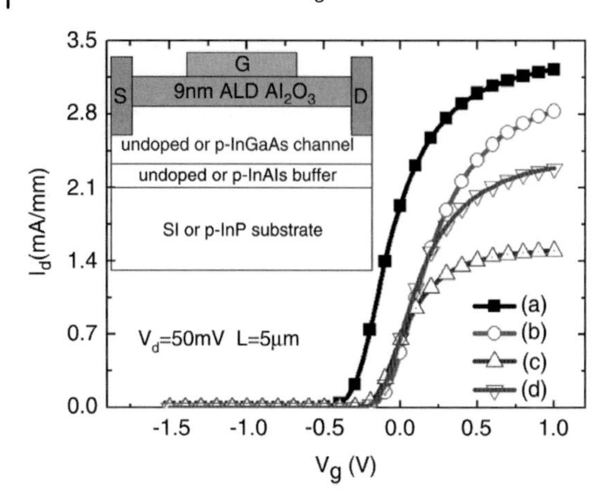

Figure 14.6 I_d–V_g characteristics at $V_d = 50$ mV for samples (a)–(d) with gate width (W) of 600 μm and gate length (L) of 5 μm: (a) 200 nm undoped $In_{0.53}Ga_{0.47}As$ channel; (b) 300 nm p-type $In_{0.53}Ga_{0.47}As$ channel with 2×10^{16} cm^{-3} doping concentration; (c) 30 nm p-type $In_{0.53}Ga_{0.47}As$ with 2×10^{16} cm^{-3} doping concentration; and (d) 300 nm p-type $In_{0.53}Ga_{0.47}As$ with 5×10^{16} cm^{-3} doping concentration. Inset shows cross-sectional structures of samples (a)–(d).

Samples (b), (c), and (d) were all grown on p-InP substrates and a p-InAlAs buffer layer was grown before p-InGaAs. The surface oxides of InGaAs were removed with diluted hydrofluoric acid (HF) cleaning, and then 100 Å Al_2O_3 capping layer was deposited by ALD. After 35 keV, 2×10^{14} cm^{-2} Si ion implantation at the S/D region, S/D activation annealing was performed at 700 °C for 10 s. The Al_2O_3 capping layer was removed using buffered oxide etch (BOE). Gate oxide (90 Å Al_2O_3) was then deposited by ALD (EOT = 4.7 nm) after HF cleaning and sulfur passivation of the surface. After 500 °C postdeposition annealing, TaN gate electrode was deposited by PVD and AuGe/Ni/Au S/D ohmic contact was deposited by e-beam evaporation. For p-type substrates, Cr/Au was used for back contact. The inset in Figure 14.6 shows the cross-sectional structures of samples (a)–(d).

Table 14.4 summarizes the device performance for samples (a)–(d). Figures 14.6 and 14.7 compare the I_d–V_g curves at $V_d = 50$ mV and I_d–V_d curves at different V_g values from V_{th} to $V_{th} + 2$ V for samples (a)–(d). The gate length is 5 μm and gate width is 600 μm. Undoped InGaAs channel (sample (a)) shows the lowest threshold voltage (−0.31 V) and the highest drive current density (125 mA/mm at $V_g - V_{th} = 2$ V). Sample (d) (300 nm p-InGaAs, 5×10^{16} cm^{-3}) exhibits slightly lower drive current (80 mA/mm versus 83 mA/mm) than sample (b) (300 nm p-InGaAs, 2×10^{16} cm^{-3}), and sample (c) (30 nm p-InGaAs) shows the lowest drive current density (55 mA/mm). The extrinsic transconductance at $V_d = 50$ mV (Figure 14.8) shows a similar trend as the drive current density for samples (a)–(d). The maximum extrinsic transconductance is 6.5 mS/mm at $V_d = 0.05$ V and 70.5 mS/mm at $V_d = 2$ V for sample (a). In addition to reduced ionized impurity scattering from undoped InGaAs, the different carrier distribution is another important reason for the better current

Table 14.4 Device performances for samples (a)–(d).

Samples ($L = 5$ μm, $W = 600$ μm)	V_{th} (V) (non uniformity ±0.05 V)	I_d @ $V_g - V_{th} = 2$ V (mA/mm)	g_{mmax} @ $V_d = 50$ mV (mS/mm)	SS (mV/dec)	I_{off} @ $V_g = -1$ V (mA/mm)	Maximum μ_{eff} (cm²/(V s))
(a) Undoped InGaAs (200 nm)	−0.31	125	6.5	147	1.7×10^{-4}	1964
(b) p-InGaAs (2×10^{16} cm^{-3}, 300 nm)	−0.10	83	5.3	121	4.8×10^{-4}	1120
(c) p-InGaAs (2×10^{16} cm^{-3}, 30 nm)	−0.17	55	3.7	116	4.0×10^{-6}	847
(d) p-InGaAs (5×10^{16} cm^{-3}, 300 nm)	−0.14	80	5.0	104	6.1×10^{-5}	1066

driving capability on undoped InGaAs channel. Since the "undoped" InGaAs is very lightly n-type doped (10^{14} cm^{-3}) in reality, MOSFETs on undoped InGaAs are actually buried channel transistors. The channel of MOSFETs on p-type InGaAs (surface channel MOSFETs) is closer to the InGaAs/Al_2O_3 interface, which will be degraded more by the interface roughness scattering and interface states.

Figure 14.8 also shows the log(I_d)–V_g at $V_d = 50$ mV for samples (a)–(d). For undoped InGaAs (sample (a)), the off-current density is 1.7×10^{-4} mA/mm at $V_g = -1$ V and the subthreshold swing is 147 mV/dec. Sample (d) (300 nm p-InGaAs, 5×10^{16} cm^{-3}) shows the minimum subthreshold swing of 104 mV/dec, while sample (b) (300 nm p-InGaAs, 2×10^{16} cm^{-3}) exhibits 121 mV/dec. The off-current densities are 4.8×10^{-4} and 6.1×10^{-5} mA/mm, respectively, at $V_g = -1$ V for

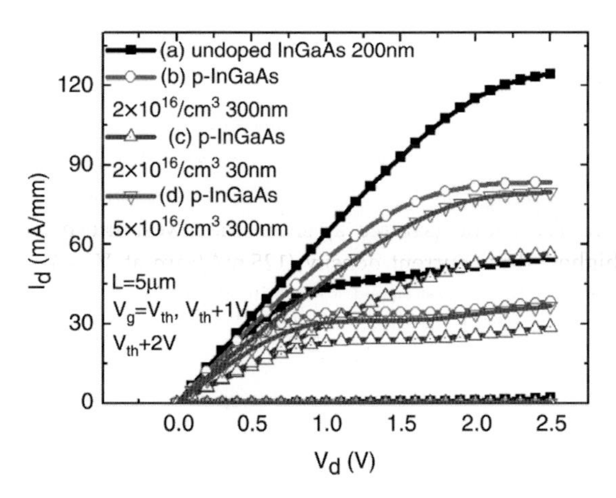

Figure 14.7 I_d–V_d characteristics at $V_g = V_{th}$, $V_g = V_{th} + 1$ V, and $V_g = V_{th} + 2$ V for samples (a)–(d) ($W = 600$ μm, $L = 5$ μm).

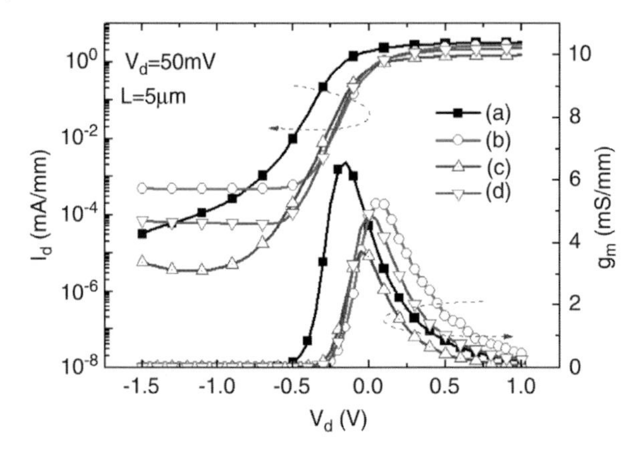

Figure 14.8 Log-scale I_d–V_g and extrinsic transconductance g_m versus V_g at $V_d = 50$ mV for samples (a)–(d) ($W = 600\,\mu$m, $L = 5\,\mu$m).

samples (b) and (d). Sample (c) (30 nm p-InGaAs, $2 \times 10^{16}\,\mathrm{cm}^{-3}$) exhibits the lowest off current of 4.0×10^{-6} mA/mm. The off current is believed to be due to S/D junction leakage, and for thinner InGaAs channel sample, S/D ($x_j = 500$ Å) diffuses into the larger bandgap InAlAs buffer region, thus resulting in lower I_{off}. Sample (d) has smaller depletion width than sample (b) due to its higher doping concentration, which leads to smaller off-current density. The high I_{off} of sample (b) degrades its subthreshold swing.

From split-CV measurement, the frequency dispersion is less than 3% at $V_g = 2$ V from 1 MHz to 10 kHz for all four samples. The maximum mobility is 1964, 1120, 847, and 1066 $\mathrm{cm}^2/(\mathrm{V\,s})$ for samples (a)–(d), respectively, calculated from 1 MHz split-CV (see Figure 14.9; inset shows the 1 MHz split-CV for samples (a)–(d)). Lower

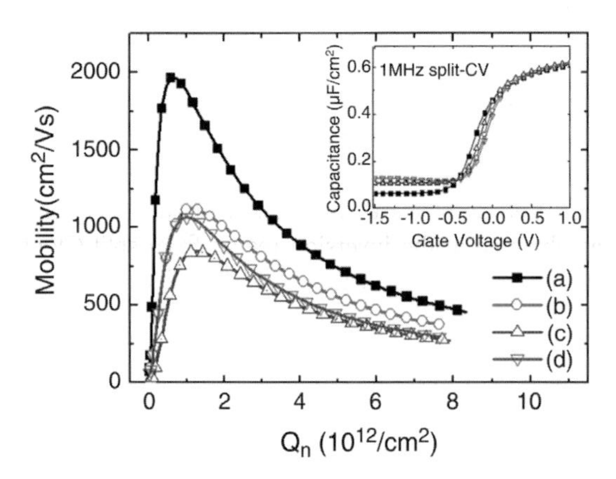

Figure 14.9 Effective channel mobility versus inversion charge density for samples (a)–(d) ($W = 600\,\mu$m, $L = 20\,\mu$m). Inset shows 1 MHz split-CV of samples (a)–(d).

ionized impurity scattering and reduced interface scattering are believed to be responsible for the higher mobility of the undoped InGaAs samples.

In summary, the impact of $In_{0.53}Ga_{0.47}As$ channel doping concentration and thickness on device performance has been studied. The undoped channel provides the highest drive current but relatively poor subthreshold swing. With proper substrate doping concentration ($5 \times 10^{16}\,cm^{-3}$), small subthreshold swing can be achieved. Thinner InGaAs channel exhibits lower off-current density but also relatively low drive current.

14.2.3
$In_{0.53}Ga_{0.47}$ As n-MOSFETs with ALD Al_2O_3, HfO_2, and $LaAlO_3$ Gate Dielectrics

Various high-κ gate dielectrics have been demonstrated on III–V MOSFETs with high drive current density. However, MOS device performance, EOT scalability, and high-κ dielectric/III–V interface quality using different gate dielectrics on III–V substrate have not been fairly compared. This section systematically compares the characteristics of $In_{0.53}Ga_{0.47}As$ n-MOSFETs with different gate dielectrics (Al_2O_3, HfO_2, and $LaAlO_3$) deposited by ALD. HfO_2 is demonstrated to have the best EOT scalability, while Al_2O_3 exhibits the best interface quality with InGaAs substrates. By using $LaAlO_3$, transistors can achieve smaller EOT than those using Al_2O_3 and accordingly smaller subthreshold swing. Al_2O_3 on $In_{0.53}Ga_{0.47}As$ shows minimum interface trap density D_{it} of $1.17 \times 10^{12}\,cm^{-2}$, MOSFETs with HfO_2 dielectric demonstrate the minimum EOT of 1 nm with drive current of 133.3 mA/mm for 5 μm gate length, and MOSFETs with $LaAlO_3$ gate dielectric have obtained subthreshold swing of 84 mV for 1.3 nm EOT.

$In_{0.53}Ga_{0.47}As$ MOSFETs were fabricated on 300 nm p-type $In_{0.53}Ga_{0.47}As$ (Be-doped, $5 \times 10^{16}\,cm^{-3}$) epitaxially grown on p-InP substrate with ring-type structure. The surface oxides of InGaAs were removed with diluted HF cleaning, and then 100 Å Al_2O_3 (dummy capping layer) was deposited by ALD. After 35 keV, $2 \times 10^{14}\,cm^{-2}$ Si ion implantation at S/D region, S/D activation annealing was performed at 700 °C for 10 s. The Al_2O_3 was removed using BOE. Different gate oxides were then deposited by ALD including Al_2O_3 with thicknesses varying from 9 to 4.2 nm, HfO_2 from 7.8 to 4.5 nm, and $LaAlO_3$ from 5.9 to 3.6 nm. After 500 °C postdeposition annealing, TaN gate electrode was deposited by reactive sputter, AuGe/Ni/Au was deposited by e-beam evaporation for source and drain ohmic contact, while Cr/Au for back contact.

Figure 14.10a shows the EOT versus physical thicknesses for different gate dielectrics. The EOT value was obtained at the inversion region from split-*CV* of MOSFETs. HfO_2 shows the highest dielectric constant (κ) value of 17.0 and the thinnest EOT of 1 nm with 4.5 nm physical thickness. $LaAlO_3$ obtains κ-value of 12.1 and Al_2O_3 shows κ-value of 8.1. HfO_2 is demonstrated to have the best EOT scalability. The electron barrier height between Al_2O_3 and $In_{0.53}Ga_{0.47}As$ is usually larger than that between HfO_2 or $LaAlO_3$ and $In_{0.53}Ga_{0.47}As$ [60–62]. Figure 14.10b compares the gate leakage current density at $V_g = 1\,V$ for different gate dielectrics. For similar EOT of about 2.2 nm, Al_2O_3 has larger gate leakage current than HfO_2 and $LaAlO_3$. The gate leakage current density is about 0.80 A/cm^2 for EOT of 1 nm using

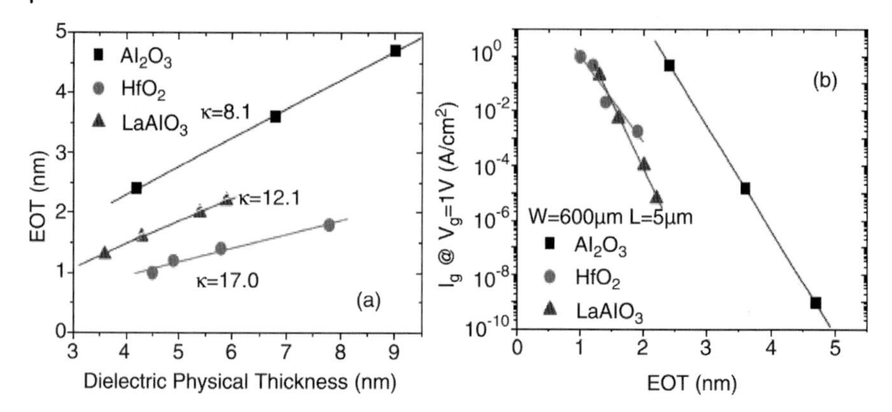

Figure 14.10 (a) EOT versus physical thicknesses for different gate dielectrics including HfO_2, $LaAlO_3$, and Al_2O_3. (b) Gate leakage current density at $V_g = 1\,V$ and $V_d = 50\,mV$ for MOSFETs using different gate dielectrics with various thicknesses ($W = 600\,\mu m$, $L = 5\,\mu m$).

HfO_2, 0.2 A/cm^2 for EOT of 1.3 nm using $LaAlO_3$, and 0.48 A/cm^2 for EOT of 2.4 nm using Al_2O_3.

The threshold voltages for various gate dielectrics with different thicknesses were measured and shown in Figure 14.11. For Al_2O_3 and $LaAlO_3$, the threshold voltage increases with reduced EOT, which is believed to be due to the positive fixed charges in the dielectrics. The positive fixed charges may come from oxygen vacancies. They may also exist in HfO_2, but not as many as they are in Al_2O_3 or $LaAlO_3$.

Figure 14.12a and b compares the drive current capability ($V_g - V_{th} = 2.5\,V$ and $V_d = 2.5\,V$) and maximum extrinsic transconductance G_{mmax} ($V_d = 0.05\,V$) for different gate dielectrics including Al_2O_3, HfO_2, and $LaAlO_3$. From the figure, for similar EOT of about 2.2 nm, Al_2O_3 has the highest drive current density and transconductance, which indicates its best interface quality with InGaAs substrate among these three kinds of dielectrics. This is demonstrated by D_{it} value

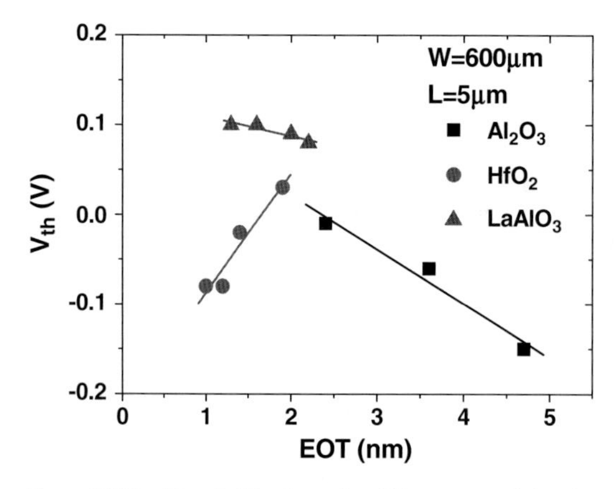

Figure 14.11 Threshold voltage for different gate dielectrics with various thicknesses.

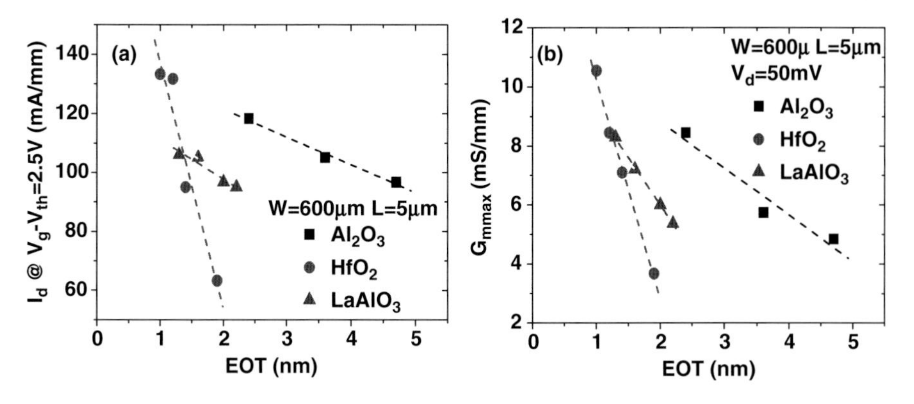

Figure 14.12 (a) Drive current density at $V_g - V_{th} = 2.5$ V and $V_d = 2.5$ V for MOSFETs using different gate dielectrics including Al_2O_3, HfO_2, and $LaAlO_3$ with various thicknesses ($W = 600\,\mu m$, $L = 5\,\mu m$). (b) Maximum extrinsic transconductance for different gate dielectrics ($W = 600\,\mu m$, $L = 5\,\mu m$, $V_d = 50$ mV).

(Figure 14.13) measured using full-conductance method at room temperature with frequency range from 100 Hz to 1 MHz. Full-conductance method on MOSFETs with S/D and bulk shorted provides a reliable solution to extract D_{it} with minority carrier responses for small bandgap materials [63]. From Figure 14.13, we can see that Al_2O_3 has the best interface quality (minimum $D_{it} = 1.17 \times 10^{12}\,cm^{-2}\,eV^{-1}$) with InGaAs substrate, while HfO_2 has minimum D_{it} of $4.41 \times 10^{12}\,cm^{-2}\,eV^{-1}$. In Figure 14.13, when $V_g - V_{th} = 0$, the position of D_{it} is close to conduction band edge (surface Fermi level close to conduction band). D_{it} first decreases and then increases as surface Fermi level moves toward valence band ($V_g - V_{th} < 0$). Thus, Figure 14.13 illustrates an asymmetric D_{it} distribution along bandgap at high-κ dielectric/InGaAs interface, and higher D_{it} near valence band is indicated. Due to the comparably high interface

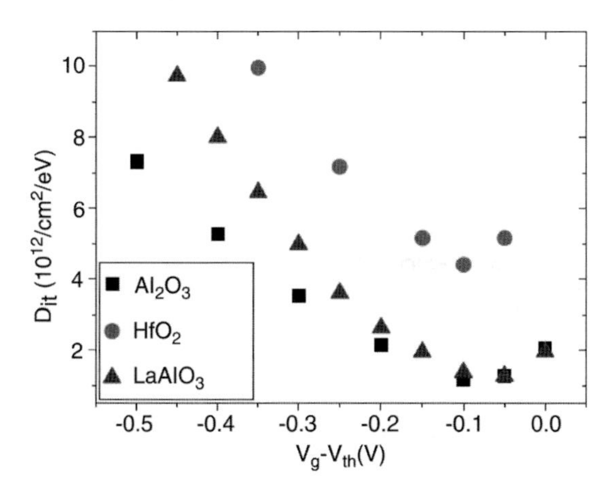

Figure 14.13 D_{it} distribution for MOSFETs with similar EOT of 2.2 nm using different gate dielectrics (Al_2O_3, HfO_2, and $LaAlO_3$).

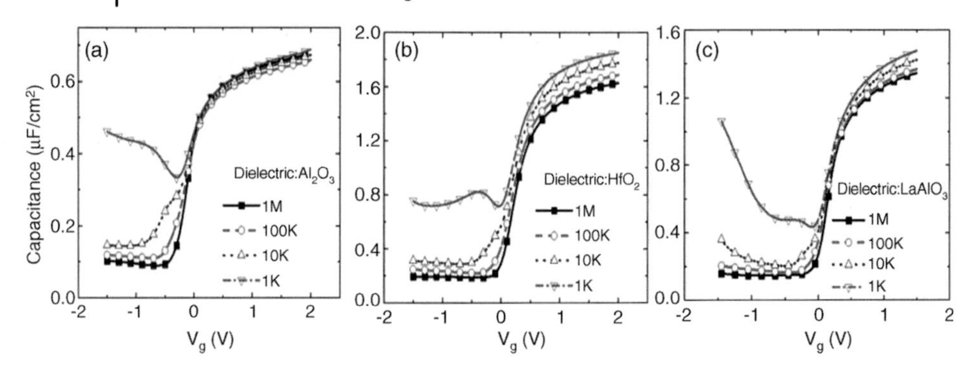

Figure 14.14 Split-*CV* at various frequencies from 1 kHz to 1 MHz for MOSFETs with 9 nm Al_2O_3 (a), 7.8 nm HfO_2 (b), and 5.9 nm $LaAlO_3$ (c) gate dielectrics.

trap density and inadequate data for capture cross section of trap states, it is difficult to locate the position of D_{it} in bandgap accurately. $V_g - V_{th}$ is used here to roughly indicate the location of D_{it}. Figure 14.14 shows the split-*CV* of MOSFETs (gate to channel capacitance) with different dielectrics at various frequencies from 1 kHz to 1 MHz. Al_2O_3 has smaller frequency dispersion value than HfO_2 and $LaAlO_3$, also indicating its better interface quality with InGaAs substrate.

Figure 14.15 shows the maximum effective channel mobility calculated from split-*CV* method for MOSFETs with different gate dielectrics. Long gate length of 20 μm was used to minimize the effect of source/drain contact resistance. As one can see, Al_2O_3 has the highest electron mobility that is believed to be due to its best interface quality with InGaAs substrate. There is no dependence of dielectric thicknesses on effective channel mobility from Figure 14.15. The differences in mobility among different EOTs are believed to be due to sample variation and device nonuniformity.

Figure 14.16 illustrates the subthreshold swing for different gate dielectrics at $V_d = 50$ mV. HfO_2 has larger D_{it} value than Al_2O_3 and $LaAlO_3$ and thus larger

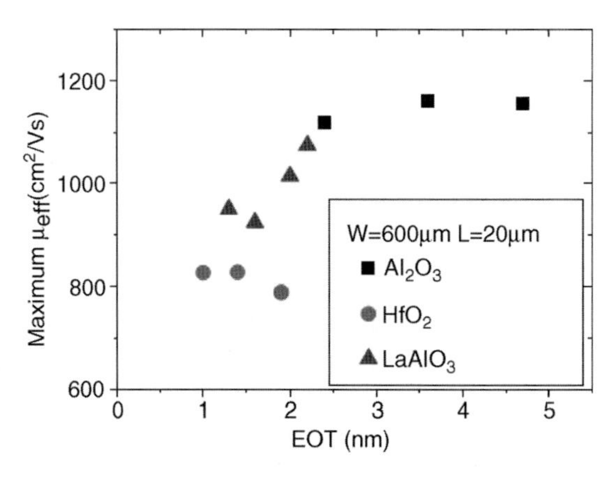

Figure 14.15 Maximum effective channel mobility for MOSFETs using different gate dielectrics including Al_2O_3, HfO_2, and $LaAlO_3$ with various thicknesses ($W = 600$ μm, $L = 20$ μm).

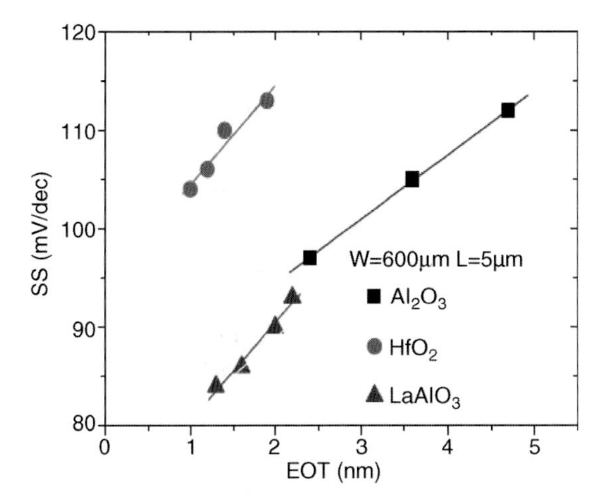

Figure 14.16 Subthreshold swing at $V_d = 50$ mV for MOSFETs using different gate dielectrics with various thicknesses ($W = 600$ μm, $L = 5$ μm).

subthreshold swing. $LaAlO_3$ can achieve smaller EOT than Al_2O_3 (larger C_{ox}) and thus smaller subthreshold swing. The minimum subthreshold swing of 84 mV/dec was obtained by $LaAlO_3$ with EOT of 1.3 nm. For $V_d = 1$ V, the subthreshold swing increases by 10–20 mV/dec due to source and drain junction leakage [64].

Figure 14.17a and b shows the characteristics of InGaAs n-MOSFETs with EOT of 1 nm using HfO_2 gate dielectric. Figure 14.17a shows the drive current I_d, gate leakage current I_g, and extrinsic transconductance g_m as a function of V_g at $V_d = 50$ mV for 5 μm gate length. The transistor has subthreshold swing of 104 mV/dec. The maximum extrinsic transconductance is 10.5 mS/mm at $V_d = 50$ mV and 67.5 mS/mm at $V_d = 1$ V. Figure 14.17b shows the I_d–V_d curves at $V_g = V_{th}$ to $V_g = V_{th} + 2.5$ V.

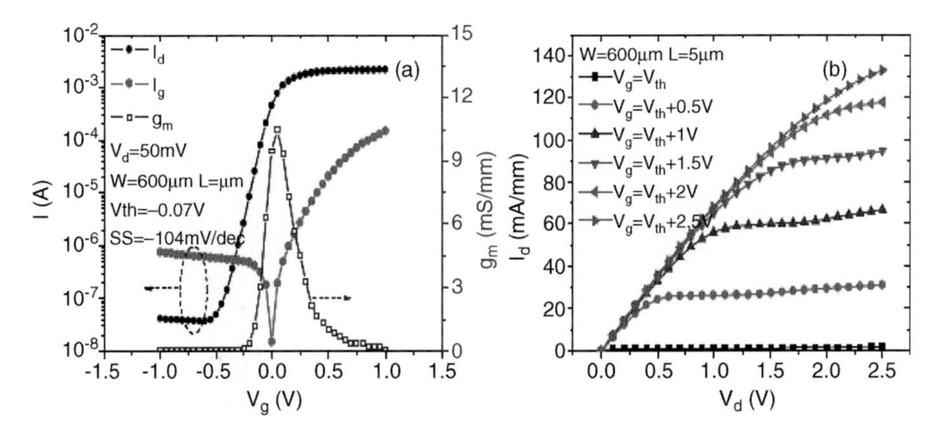

Figure 14.17 (a) I_d, I_g, and extrinsic transconductance g_m as a function of V_g for MOSFETs with HfO_2 gate dielectric (EOT = 1 nm) at $V_d = 50$ mV ($W = 600$ μm, $L = 5$ μm). (b) I_d–V_d curves from $V_g = V_{th}$ to $V_g = V_{th} + 2.5$ V with a step of 0.5 V for the same device.

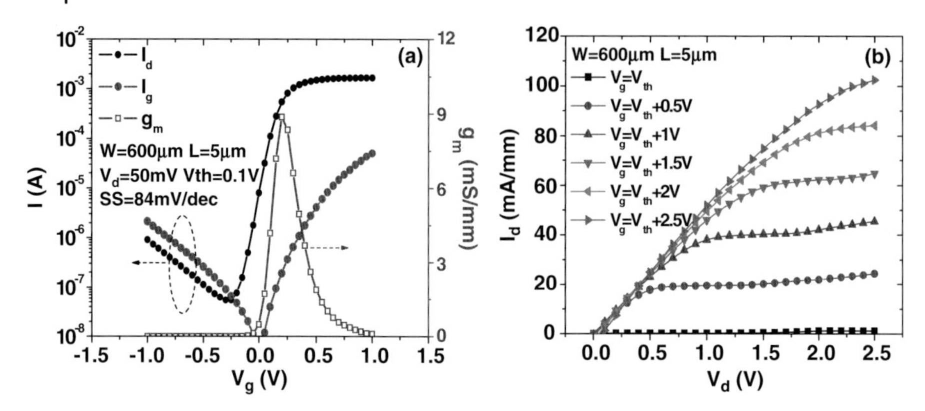

Figure 14.18 (a) Drive current I_d, gate leakage current I_g, and extrinsic transconductance g_m as a function of V_g for MOSFETs with $LaAlO_3$ gate dielectric (EOT = 1.3 nm) at $V_d = 50$ mV ($W = 600$ μm, $L = 5$ μm). (b) I_d–V_d curves at various V_g values for the same device.

High drive current density of 133.3 mA/mm at $V_g - V_{th} = 2.5$ V was obtained for 5 μm gate length. Figure 14.18 illustrates the characteristics of InGaAs n-MOSFETs with EOTof 1.3 nm using $LaAlO_3$ gate dielectric. Figure 14.18a shows I_d, I_g, and g_m as a function of V_g, while Figure 14.18b shows the I_d–V_d curve at various V_g values. The minimum subthreshold swing of 84 mV/dec was obtained. The maximum extrinsic transconductance is 8.8 mS/mm at $V_d = 50$ mV and 53.8 mS/mm at $V_d = 1$ V.

In summary, the performances for $In_{0.53}Ga_{0.47}As$ n-MOSFETs were compared among different ALD gate dielectrics including Al_2O_3, HfO_2, and $LaAlO_3$. HfO_2 shows the highest κ-value and the smallest EOT, while Al_2O_3 has the best interface quality with InGaAs. $LaAlO_3$ has higher κ-value than Al_2O_3 and better interface quality than HfO_2, and it obtains subthreshold swing of 84 mV/dec with EOT of 1.3 nm. High drive current of 133.3 mA/mm and maximum extrinsic transconductance of 67.5 mS/mm were achieved using 4.5 nm HfO_2 gate dielectric ($L = 5$ μm, EOT = 1 nm).

14.3 Buried Channel InGaAs MOSFETs

14.3.1 High-Performance $In_{0.7}Ga_{0.3}As$ MOSFETs with Mobility >4400 $cm^2/(V\,s)$ Using InP Barrier Layer

Even though surface channel InGaAs MOSFETs with ALD Al_2O_3, HfO_2, and ZrO_2 dielectrics [19–22], MBE $Ga_2O_3(Gd_2O_3)$ dielectrics [57], and Si IPL and high-κ gate stacks [26] show promising results on MOSFETs with high drive current capability, the reported effective channel mobility μ_{eff} of the surface channel devices is still relatively low compared to the bulk mobility of InGaAs (e.g., $\mu_{eff} \sim 1000-1700$ $cm^2/(V\,s)$ [19–22, 27]). On the other hand, buried channel InGaAs MOSFETs with MBE

InAlAs barrier layer and Si IPL [36], flatband InGaAs MOSFETs with GaAs/AlGaAs barrier layer and Si δ-doping using MBE GaGdO gate oxide [40], or MOS-HEMTs [42] can achieve much higher electron mobility (e.g., $3810\,cm^2/(V\,s)$ with Si IPL, $1280\,cm^2/(V\,s)$ without Si IPL [65], $5500\,cm^2/(V\,s)$ [23], and $4250\,cm^2/(V\,s)$ [42]).

The gate leakage current density of buried channel InGaAs MOSFETs or MOS-HEMTs can be several orders of magnitude lower than that of HEMTs [40, 43, 65]. Flatband InGaAs MOSFETs and MOS-HEMTs devices in general require a Si δ-doped layer, a spacer layer, and a barrier layer [23, 41], while buried channel MOSFETs only need a barrier layer [65]. Compared to MBE dielectrics, *ex situ* ALD gate dielectrics are preferable due to their potential manufacturability. Furthermore, as a barrier layer, InAlAs usually has the problem of excessive aluminum oxidation for *ex situ* process, while InP shows better interface quality with *ex situ* ALD dielectrics than GaAs and InAlAs [66, 67]. On the other hand, the conduction band offset between InP and InGaAs is smaller than that between InAlAs and InGaAs (0.26 eV for $InP/In_{0.53}Ga_{0.47}As$ compared to 0.47 eV for $In_{0.52}Al_{0.48}As/In_{0.53}Ga_{0.47}As$ [68]). In this paper, we have investigated and compared device performance for buried channel $In_{0.7}Ga_{0.3}As$ and $In_{0.53}Ga_{0.47}As$ MOSFETs with InP barrier layer and surface channel $In_{0.7}Ga_{0.3}As$ and $In_{0.53}Ga_{0.47}As$ MOSFETs without InP barrier layer. High device performance including drive current of 98 mA/mm ($L = 20\,\mu m$), SS of 106 mV/dec, and effective channel mobility of $4402\,cm^2/(V\,s)$ has been achieved for $In_{0.7}Ga_{0.3}As$ MOSFETs using 4 nm InP barrier layer and 5.5 nm ALD Al_2O_3 gate oxide.

$In_{0.7}Ga_{0.3}As$ and $In_{0.53}Ga_{0.47}As$ MOSFETs were fabricated on undoped InGaAs epitaxially grown on SI-InP substrate with a ring-type structure. Figure 14.19a shows the cross-sectional view of the substrate for $In_{0.7}Ga_{0.3}As$ MOSFETs. For $In_{0.7}Ga_{0.3}As$ MOSFETs, the InP barrier layer is 4 nm thick and $In_{0.7}Ga_{0.3}As$ channel layer is 10 nm thick, while for $In_{0.53}Ga_{0.47}As$ MOSFETs, the InP barrier layer is 6 nm thick and $In_{0.53}Ga_{0.47}As$ channel layer is 30 nm thick. The n^+ InGaAs contact layer at the channel region was selectively removed by citric acid-based solution. For some samples, the InP barrier layer was selectively etched by diluted hydrochloric acid.

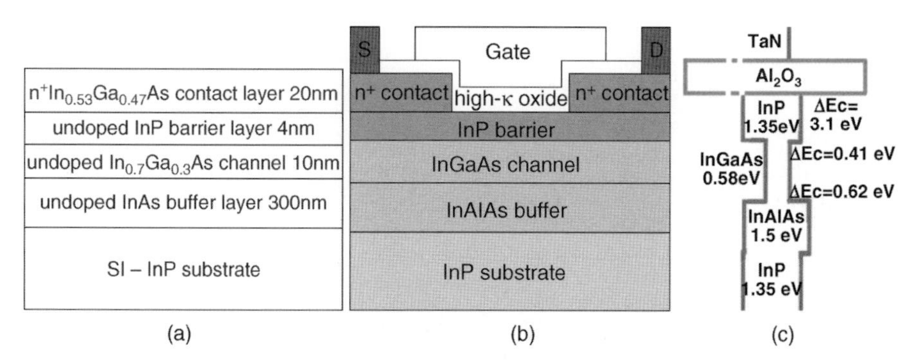

Figure 14.19 (a) Cross-sectional view of substrate structure for $In_{0.7}Ga_{0.3}As$ MOSFETs. (b) Cross-sectional view of $In_{0.7}Ga_{0.3}As$ and $In_{0.53}Ga_{0.47}As$ MOSFETs with InP barrier layer. (c) Energy band diagram for $In_{0.7}Ga_{0.3}As$ MOSFETs with InP barrier layer.

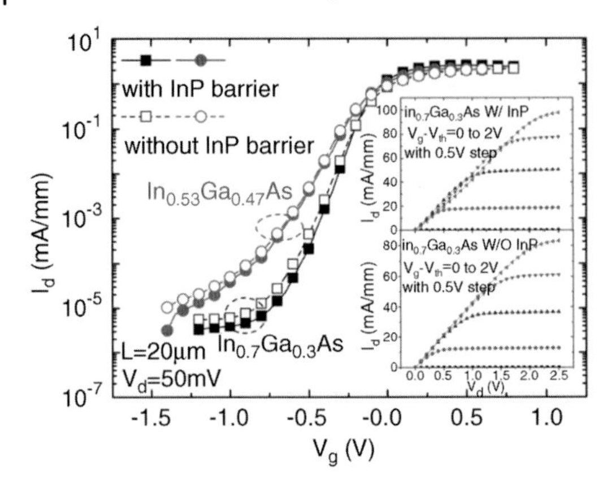

Figure 14.20 Log-scale I_d–V_g at $V_d = 50$ mV for $In_{0.7}Ga_{0.3}As$ and $In_{0.53}Ga_{0.47}As$ MOSFETs with and without InP barrier layer. Inset shows I_d–V_d at $V_g - V_{th}$ from 0 to 2 V with 0.5 V steps for $In_{0.7}Ga_{0.3}As$ MOSFETs with and without InP barrier layer. The gate length is 20 μm.

Gate oxide (5.5 nm Al_2O_3) was then deposited by ALD (EOT = 3.4 nm). After that, TaN gate electrode was deposited by PVD and AuGe/Ni/Au source and drain ohmic contact was deposited by e-beam evaporation. Figure 14.19b shows the cross-sectional view of the InGaAs MOSFETs with InP barrier layer. Figure 14.19c shows the energy band diagram for $In_{0.7}Ga_{0.3}As$ device with InP barrier layer.

Figure 14.20 illustrates the log-scale I_d–V_g characteristics of $In_{0.7}Ga_{0.3}As$ and $In_{0.53}Ga_{0.47}As$ MOSFETs with and without InP barrier layer. The gate leakage current density is less than 4×10^{-9} A/cm^2 at $V_g - V_{th} = 1$ V for all samples (data not shown). Compared to $In_{0.53}Ga_{0.47}As$ MOSFETs, $In_{0.7}Ga_{0.3}As$ MOSFETs show much lower subthreshold swing (106 mV/dec versus 154 mV/dec (Figure 14.21)). In addition to the shorter gate-to-channel distance due to thinner InP barrier layer for $In_{0.7}Ga_{0.3}As$

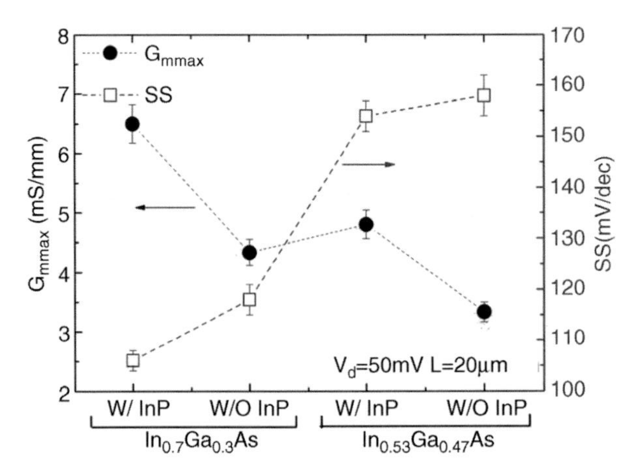

Figure 14.21 Maximum transconductance and subthreshold swing at $V_d = 50$ mV for $In_{0.7}Ga_{0.3}As$ and $In_{0.53}Ga_{0.47}As$ MOSFETs with and without InP barrier layer. The gate length is 20 μm.

MOSFETs, the thinner $In_{0.7}Ga_{0.3}As$ channel layer that is easier to be depleted by gate bias is another important reason for the lower subthreshold swing. Although InP barrier layer increases gate-to-channel distance compared to MOSFETs without InP layer, subthreshold swing for $In_{0.7}Ga_{0.3}As$ MOSFETs with InP barrier layer is actually decreased. There are two interfaces from oxide to channel in $In_{0.7}Ga_{0.3}As$ MOSFETs with InP barrier (Al_2O_3/InP interface and InP/$In_{0.7}Ga_{0.3}As$ interface (see Figure 14.19)) compared to $In_{0.7}Ga_{0.3}As$ MOSFETs without InP barrier that have only one oxide to channel interface (Al_2O_3/$In_{0.7}Ga_{0.3}As$). Although in $In_{0.7}Ga_{0.3}As$ MOSFETs with InP barrier, the ALD Al_2O_3/InP interface quality may not be as good as ALD Al_2O_3/InGaAs interface [67, 69], the MBE-grown InP/InGaAs interface is closer to channel, and we believe its excellent interface quality plays a more important role for the enhanced device characteristics. Therefore, the improved subthreshold swing is believed to be due to the better MBE-grown InP/InGaAs interface than ALD Al_2O_3/InGaAs interface. Inset in Figure 14.20 shows I_d–V_d curves at $V_g - V_{th}$ from 0 to 2 V with 0.5 V steps for $In_{0.7}Ga_{0.3}As$ MOSFETs with and without InP barrier layer. For $In_{0.7}Ga_{0.3}As$ MOSFETs with InP barrier at high V_g (1.5–2 V), some channel electrons spill over into the lower mobility InP layer. This results in the crossover characteristics of the I_d–V_d curves. Note that the $In_{0.7}Ga_{0.3}As$ MOSFETs with InP barrier layer exhibit high drive current of 98 mA/mm for gate length $L = 20\,\mu m$.

Figures 14.21 and 14.22 compare the device performance of $In_{0.7}Ga_{0.3}As$ and $In_{0.53}Ga_{0.47}As$ MOSFETs with and without InP barrier layer including maximum extrinsic transconductance G_{mmax}, subthreshold swing, and drive current at different gate voltages. InP barrier provides MOSFETs with higher transconductance due to better InP/InGaAs interface close to the channel. $In_{0.7}Ga_{0.3}As$ MOSFETs with InP barrier layer show much higher transconductance and much lower subthreshold swing than all other devices. This is believed to be due to higher mobility channel

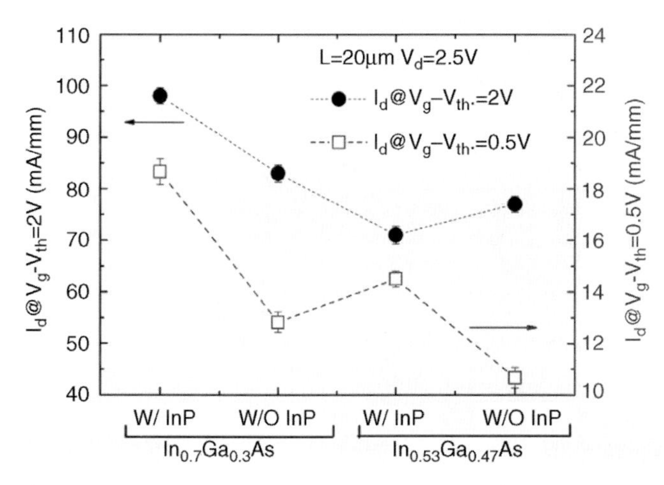

Figure 14.22 Drive current I_d at $V_g - V_{th} = 0.5$ and 2 V for $In_{0.7}Ga_{0.3}As$ and $In_{0.53}Ga_{0.47}As$ MOSFETs with and without InP barrier layer. The gate length is 20 μm and V_d is 2.5 V.

material and better gate-to-channel control. In Figure 14.22, for $V_g - V_{th} = 0.5$ V, $In_{0.53}Ga_{0.47}As$ MOSFETs with InP layer show higher drive current than MOSFETs without InP layer. However, at $V_g - V_{th} = 2$ V, the drive current for $In_{0.53}Ga_{0.47}As$ MOSFETs with InP barrier is smaller than those for $In_{0.7}Ga_{0.3}As$ and $In_{0.53}Ga_{0.47}As$ MOSFETs without InP layer. This is again because some channel electrons enter the lower mobility InP layer. For $In_{0.7}Ga_{0.3}As$ MOSFETs with larger conduction band offset between InP and $In_{0.7}Ga_{0.3}As$ (0.41 eV versus 0.26 eV for InP/$In_{0.53}Ga_{0.47}As$), less electrons can enter InP layer at high V_g. Consequently, $In_{0.7}Ga_{0.3}As$ MOSFETs with InP barrier layer show 46 and 18% enhancement in drive current compared to those without InP layer at $V_g - V_{th} = 0.5$ and 2 V, respectively. For $In_{0.7}Ga_{0.3}As$ MOSFETs with InP barrier, the subthreshold swing is 106 mV/dec compared to 118 mV/dec for $In_{0.7}Ga_{0.3}As$ without InP barrier. The maximum transconductance is 50% higher (Figure 14.21). These results illustrate that InP is an excellent barrier layer to enhance device performance, especially for $In_{0.7}Ga_{0.3}As$ MOSFETs.

We have measured the effective channel mobility of $In_{0.7}Ga_{0.3}As$ and $In_{0.53}Ga_{0.47}As$ MOSFETs with and without InP barrier layer using split-*CV* method and plotted it in Figure 14.23. The peak effective channel mobility for $In_{0.7}Ga_{0.3}As$ MOSFETs with InP barrier is 4402 $cm^2/(V\,s)$, which is much higher than that for $In_{0.7}Ga_{0.3}As$ with InAlAs barrier (e.g., 1280 $cm^2/(V\,s)$ without Si IPL [65]). Inset shows split-*CV* frequency dispersion and hysteresis for $In_{0.7}Ga_{0.3}As$ MOSFETs with and without InP barrier layer. They exhibit a small frequency dispersion and hysteresis.

In summary, we have fabricated $In_{0.7}Ga_{0.3}As$ and $In_{0.53}Ga_{0.47}As$ MOSFETs with and without InP barrier layer and compared their device performance. InP is an

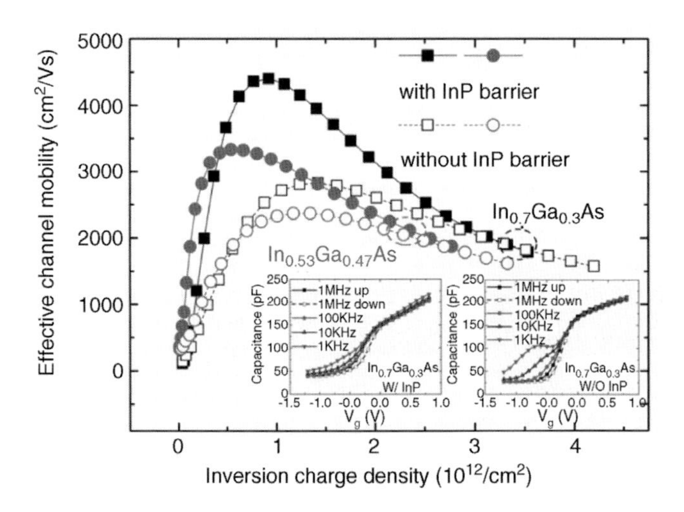

Figure 14.23 Effective channel mobility versus inversion charge density for $In_{0.7}Ga_{0.3}As$ and $In_{0.53}Ga_{0.47}As$ MOSFETs with and without InP barrier layer. Inset shows split-*CV* frequency dispersion from 1 MHz to 1 kHz and hysteresis at 1 MHz (up trace: V_g starts from −1 V; down trace: V_g starts from 1 V) for $In_{0.7}Ga_{0.3}As$ MOSFETs with and without InP barrier layer.

excellent barrier and passivation layer to enhance device current driving capability, especially for $In_{0.7}Ga_{0.3}As$ MOSFETs due to good conduction band offset. $In_{0.7}Ga_{0.3}As$ MOSFETs with InP barrier layer show much higher transconductance and lower subthreshold swing than other MOSFETs, and exhibit high drive current of 98 mA/mm ($L = 20\,\mu m$), subthreshold swing of 106 mV/dec, and effective channel mobility of 4402 $cm^2/(V\,s)$.

14.3.2
Effects of Barrier Layers on Device Performance of High-Mobility $In_{0.7}$ $Ga_{0.3}$ As MOSFETs

Due to the electron spillover into the barrier layer with lower mobility, improvement of electron mobility at high field of buried channel devices with InP barrier degrades as shown in the previous section. In this section, double-barrier (InP/InAlAs) structures were used to significantly improve peak mobility as well as high-field mobility. Single InP barrier with different thicknesses and no barrier MOSFETs have also been studied for comparison. MOSFETs with InP/InAlAs barrier achieve high μ_{eff} of 4889 $cm^2/(V\,s)$ using Al_2O_3 gate oxide and 3722 $cm^2/(V\,s)$ using HfO_2 gate oxide, which are among the highest μ_{eff} reported.

Ring-type $In_{0.7}Ga_{0.3}As$ MOSFETs were fabricated on 10 nm undoped $In_{0.7}Ga_{0.3}As$ channel. Various MBE barrier layers were applied including undoped 3 nm InP, 5 nm InP, and InP/InAlAs double barrier with 2 nm InP on the top and 3 nm $In_{0.52}Al_{0.48}As$ at the bottom (Figure 14.24). Various gate oxides were deposited by ALD including 4–8 nm Al_2O_3 (EOT = 2–4.1 nm), 5 nm HfO_2 (EOT = 1.2 nm), and 1 nm Al_2O_3 (bottom)/4 nm HfO_2 (top) (EOT = 1.4 nm). Sharp interface between InP barrier and Al_2O_3 is observed by TEM (Figure 14.25).

Using $In_{0.52}Al_{0.48}As$ barrier with higher conduction band offset to $In_{0.7}Ga_{0.3}As$ than InP can suppress the electron spilling over effect and improve high-field

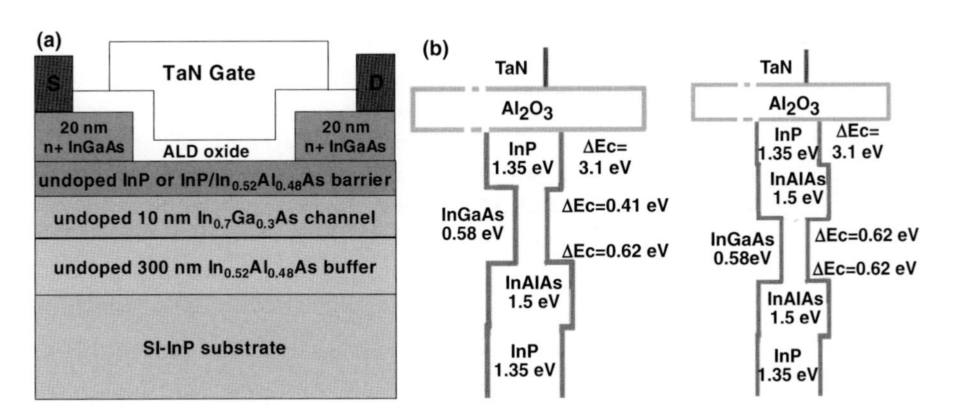

Figure 14.24 (a) Cross-sectional view and (b) band diagram of buried channel $In_{0.7}Ga_{0.3}As$ MOSFETs with single InP barrier (3 or 5 nm) or 2 nm InP (top)/3 nm $In_{0.52}Al_{0.48}As$ (bottom) double barrier. InAlAs/$In_{0.7}$GaAs shows larger ΔE_c (0.62 eV) than InP/$In_{0.7}$GaAs (0.41 eV).

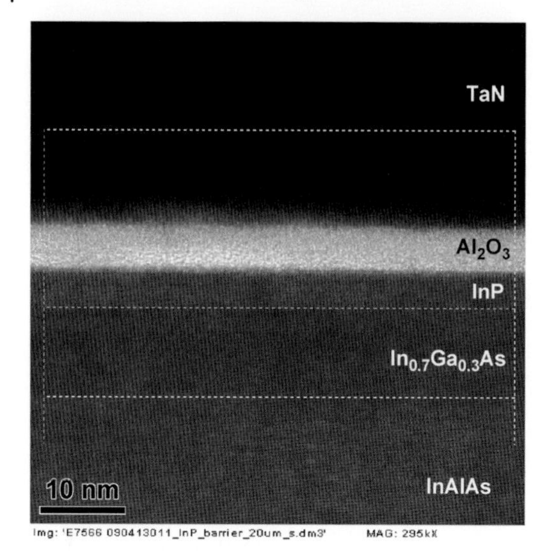

Figure 14.25 Cross-sectional high-resolution TEM for $In_{0.7}Ga_{0.3}As$ MOSFETs with 5 nm InP barrier. Sharp Al_2O_3/InP interface was observed.

mobility (Figure 14.24b). However, InAlAs is easy to be oxidized; thus, InP/InAlAs double barrier with 2 nm InP on the top and 3 nm $In_{0.52}Al_{0.48}As$ at the bottom was used to protect InAlAs from oxidation. Figure 14.26 illustrates I_d–V_g and extrinsic G_m–V_g characteristics for MOSFETs with 3 nm InP, 5 nm InP, and 2 nm InP/3 nm InAlAs barriers using Al_2O_3 gate oxide. The gate leakage current for MOSFETs with barriers and 4 nm Al_2O_3 (EOT = 2 nm) is less than 1×10^{-4} A/cm^2 (data not shown). Devices with 5 nm InP show 20% maximum transconductance (G_{mmax}) increase compared to those with 3 nm InP barrier due to better passivation from oxide/III–V interface by a using thicker barrier layer. Although MOSFETs using InP/InAlAs barrier only show slightly larger G_{mmax} than those with 5 nm InP at $V_d = 50$ mV, the better channel electron confinement using InP/InAlAs barrier results in 17% increase of G_{mmax} at $V_d = 1$ V (EOT of 2 nm in Figure 14.26b) and absence of I_d–V_d crossover (Figure 14.27b). Double-barrier MOSFETs also show 39% I_d increase at $V_g = V_{th} + 2$ V compared to single 5 nm InP barrier MOSFETs (Figure 14.27).

MOSFETs with InP/InAlAs barrier show larger subthreshold swing than those with single InP barrier (Figure 14.28). This might be because 2 nm InP is still not sufficient to passivate InAlAs layer, and there is small amount of InAlAs oxidation, which generates donor-like defects in InAlAs layer. This does not affect I_{on} characteristics but will degrade SS [70]. Small SS of ~95 mV/dec was achieved by MOSFETs using single InP barrier and Al_2O_3 with 2 nm EOT.

Devices with 3 nm InP and 5 nm InP achieve 23 and 56% peak mobility enhancement compared to those with no barrier, while MOSFETs with InP/InAlAs double barrier achieve both 68% peak mobility enhancement and 55% high field mobility ($Q_{inv} = 4 \times 10^{12}$ cm^{-2}) enhancement compared to devices with no barrier (Figures 14.29 and 14.30).

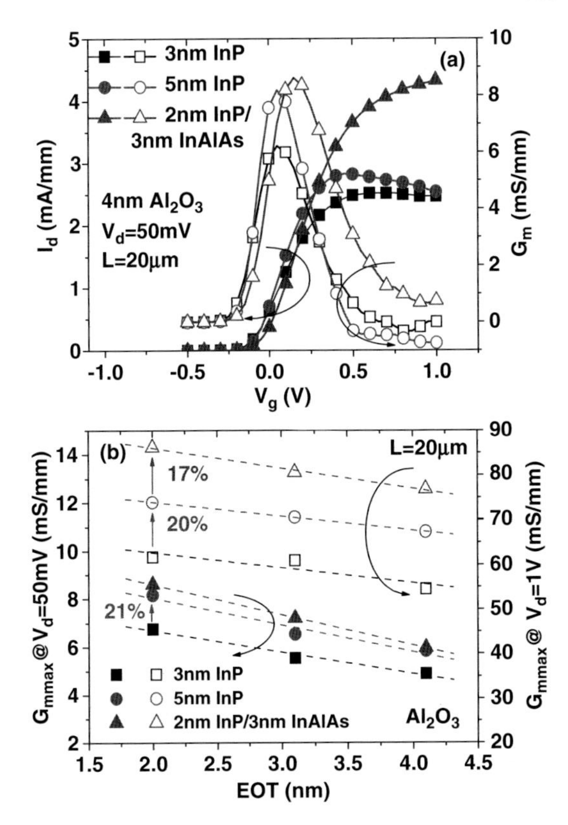

Figure 14.26 (a) MOSFETs with InP/InAlAs show much higher I_d and G_m at high V_g than those with InP. (b) Devices with 5 nm InP show 20% higher G_{mmax} than those with 3 nm InP, and MOSFETs with InP/InAlAs show 17% higher G_{mmax} than those with 5 nm InP using Al_2O_3 ($V_d = 1$ V).

To further scale down EOT, 5 nm HfO_2 (EOT = 1.2 nm) was applied to devices with InP and InP/InAlAs barriers. The gate leakage current for MOSFETs with barriers and 5 nm HfO_2 is less than 8×10^{-5} A/cm^2 (data not shown). MOSFETs using HfO_2 with InP/InAlAs barrier show significant high-field I_d and μ_{eff} improvement (Figures 14.31–14.34), as well as lower SS (135 mV/dec versus 152 mV/dec in Figure 14.32) and 16% peak mobility enhancement (Figure 14.34a), compared to those with 5 nm InP barrier. HfO_2/InP interface shows about one order of magnitude higher D_{it} than Al_2O_3/InP interface (Figures 14.35 and 14.36). Thus, it is even more critical to reduce scattering from HfO_2/InP interface to channel electrons. The good channel electron confinement by using InP/InAlAs barrier keeps electrons far from HfO_2/InP; therefore, it improves both low-field (peak μ_{eff}) and high-field characteristics compared to single InP barrier. MOSFETs with HfO_2 (EOT = 1.2 nm) show smaller G_m and higher SS than those with Al_2O_3 (EOT = 2 nm) due to higher D_{it} at HfO_2/InP interface than Al_2O_3/InP interface (Figures 14.35 and 14.36). Using Al_2O_3 (bottom)/HfO_2 (top) bilayer oxides helps to optimize oxide/InP interface and thus improves G_m (Figure 14.31), SS (Figure 14.32), I_d (10% increase in Figure 14.33),

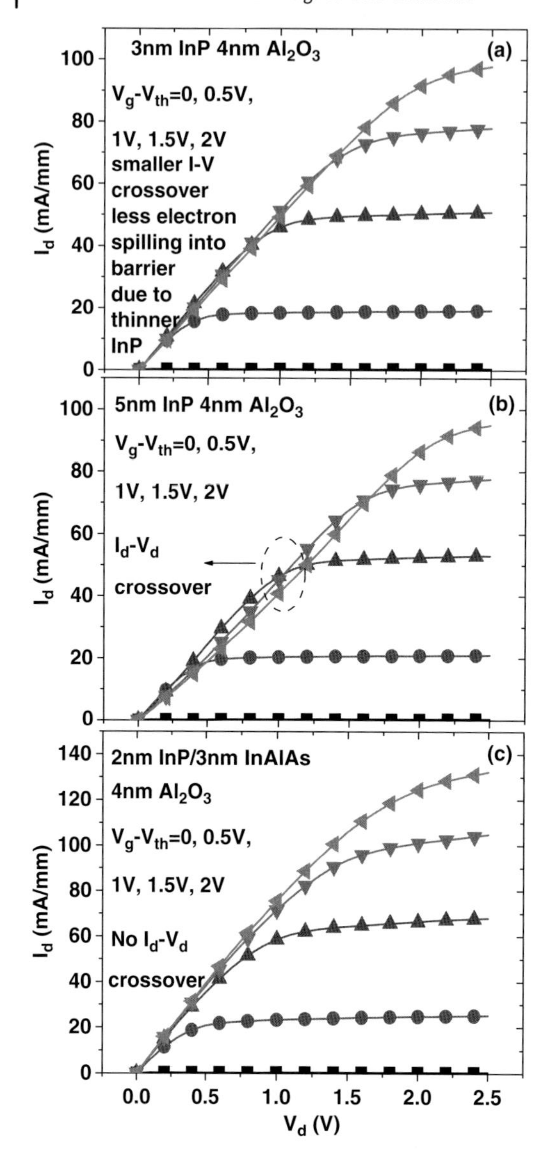

Figure 14.27 No I_d–V_d crossover and 39% I_d increase ($V_g = V_{th} + 2$ V) was obtained for MOSFETs with InP/InAlAs barrier (c) compared to those with 5 nm InP barrier (b).

and μ_{eff} (17% increase in Figure 14.34) even with slightly thicker EOT compared to single HfO_2 dielectric.

SS and on/off current were significantly improved at 115 K for MOSFETs with InP/InAlAs barrier and HfO_2, highlighting the presence of scattering due to interface traps (Figure 14.37). Effective channel mobility was measured by split-*CV* method at

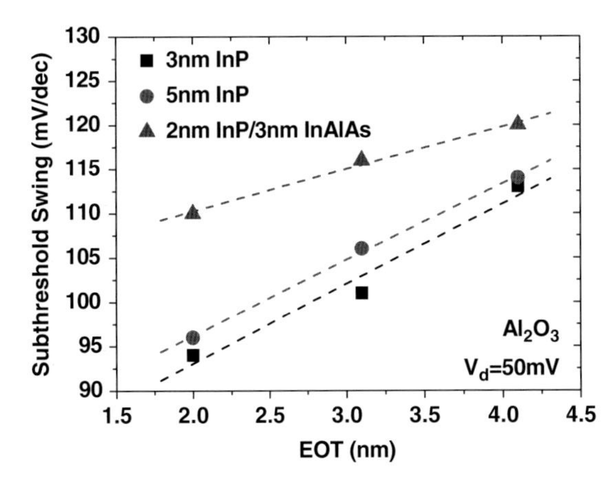

Figure 14.28 Small SS of ~95 mV/dec was achieved by MOSFETs with 3 or 5 nm InP barrier and Al_2O_3 with 2 nm EOT.

various temperatures from 115 to 433 K on $In_{0.7}Ga_{0.3}As$ MOSFETs with InP/InAlAs barrier and 9 nm HfO_2 or 9 nm Al_2O_3 to investigate the scattering mechanisms (Figure 14.38). Coulombic scattering and phonon scattering are the main factors affecting mobility at low electron density. Coulombic scattering (from interface and oxide charges) dominates at low temperature and optical phonon scattering (from both substrate and high-κ oxides) dominates at high temperature. At high Q_n, surface roughness scattering also plays a role. Compared to Al_2O_3, lower mobility of MOSFETS with HfO_2 results from high interface charges and high high-κ phonon scattering.

In conclusion, we have investigated the characteristics of $In_{0.7}Ga_{0.3}As$ MOSFETs with InP or InP/InAlAs barrier and various ALD gate dielectrics. Adding InAlAs into barrier significantly improves device performance at high field. Devices with thicker

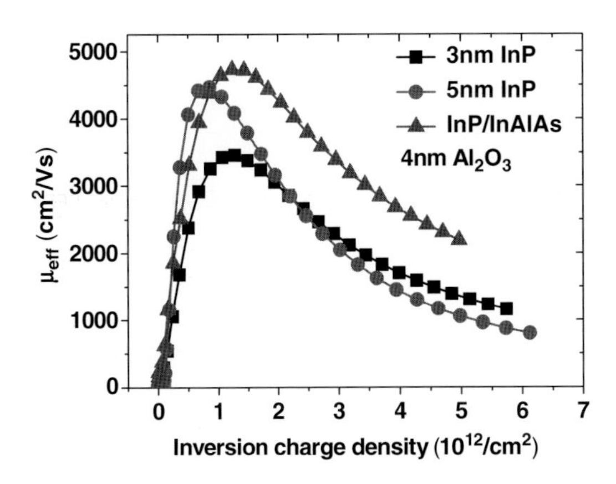

Figure 14.29 Thicker InP improves peak μ_{eff} and InP/InAlAs increases high-field μ_{eff} significantly.

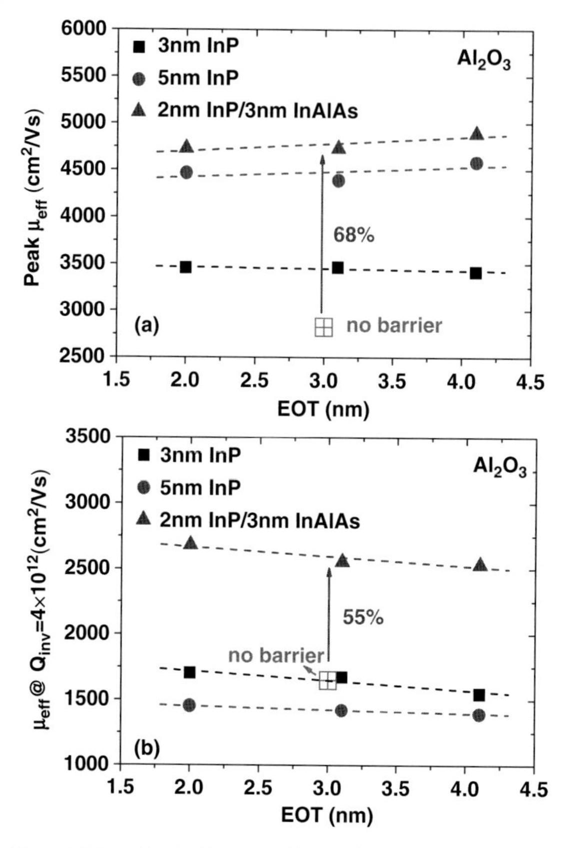

Figure 14.30 MOSFETs using Al_2O_3 with InP/InAlAs barrier show 68% higher peak μ_{eff} (a) and 55% higher high-field μ_{eff} (b) than those without barrier. Single InP barrier only improves peak μ_{eff} but not high-field μ_{eff}.

InP barrier exhibit higher μ_{eff} than those with thinner InP. MOSFETs with Al_2O_3 exhibit better interface quality than those with HfO_2, and Al_2O_3/HfO_2 bilayer improves transistor performance compared to single HfO_2. High μ_{eff} with low gate leakage has been demonstrated owing to our novel III–V MOSFET structure and fabrication process.

14.4 Summary

With the end of CMOS roadmap looming, there has been tremendous research in order to identify promising technologies to continue the historical trend of performance scaling. This chapter mainly explored the device characteristics of III–V MOSFETs with various substrate structures and gate oxides, aiming to realize high-performance III–V devices by improving the device structures and interfaces.

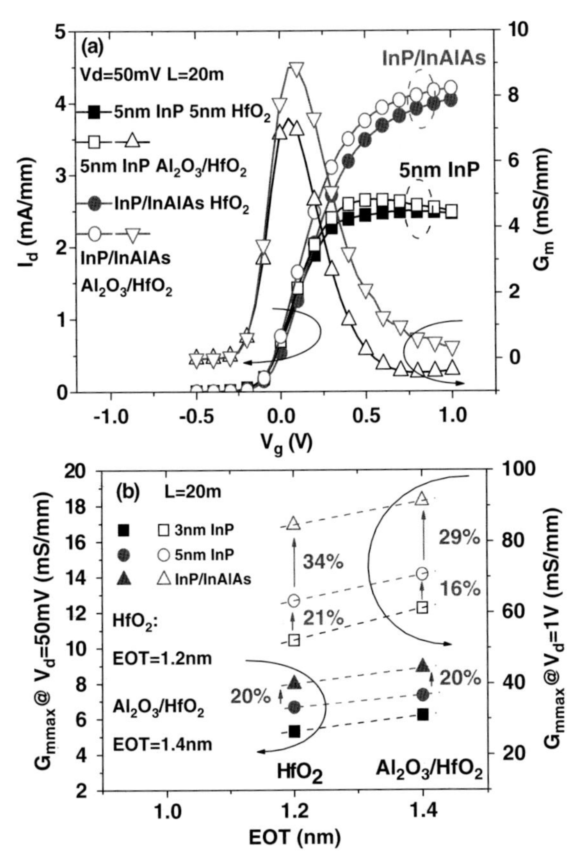

Figure 14.31 MOSFETs using HfO_2 show both 20% G_{mmax} increase at $V_d = 50$ mV and 34% G_{mmax} increase at $V_d = 1$ V with InP/InAlAs compared to those with 5 nm InP. Al_2O_3 (bottom)/HfO_2 (top) bilayer improves G_m and I_d compared to single HfO_2.

The proper fabrication process for $In_{0.53}Ga_{0.47}As$ MOSFETs with ALD oxides has been identified by comparing device characteristics from gate-first and gate-last processes. It has been found that applying the gate-last process provides significant smaller interface trap density compared to the gate-first process. This is due to the less interface oxide growth from the gate-last process in comparison to the gate-first process. By investigating the dependence of device performance on the channel doping concentration and the channel thickness for $In_{0.53}Ga_{0.47}As$ MOSFETs with ALD Al_2O_3 dielectrics, a p-type InGaAs substrate with proper doping concentration and sufficient thickness is identified to be the optimum substrate for both high drive current and small subthreshold swing. High drive current of 133.3 mA/mm and EOT of ~1 nm were achieved for $In_{0.53}Ga_{0.47}As$ MOSFETs using 4.5 nm HfO_2 gate dielectric ($L = 5\,\mu m$) due to good EOT scalability of ALD HfO_2.

The buried channel InGaAs MOSFETs using single InP barrier layer with different thicknesses and InP/$In_{0.52}Al_{0.48}As$ double-barrier layer have been investigated to

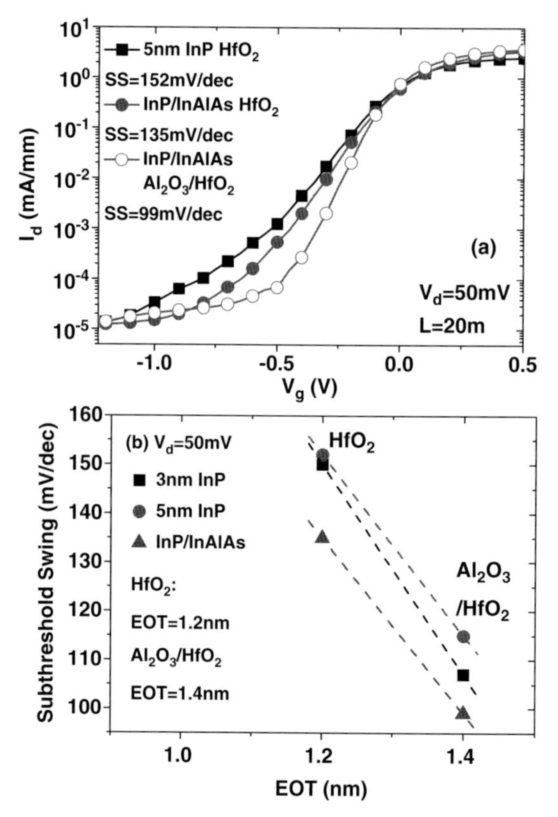

Figure 14.32 Devices using HfO_2 show smaller SS with InP/InAlAs barrier than those with 5 nm InP. MOSFETs with Al_2O_3/HfO_2 bilayer achieve smaller SS than those with single HfO_2. Small SS of 99 mV/dec was obtained by devices with InP/InAlAs barrier and Al_2O_3/HfO_2.

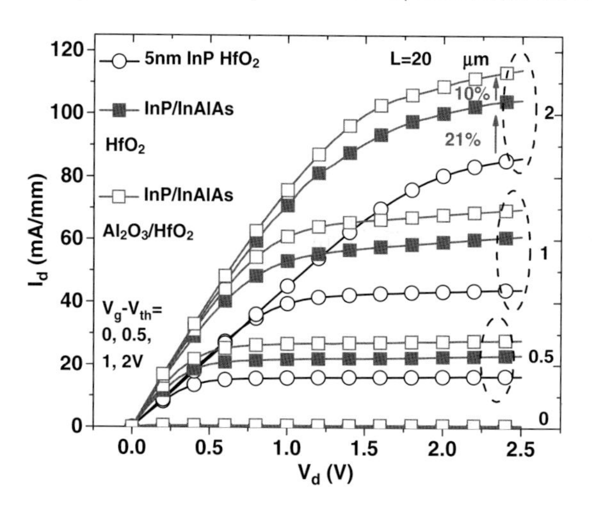

Figure 14.33 MOSFETs using HfO_2 show 21% I_d increase ($V_g = V_{th} + 2$ V) with InP/InAlAs barrier compared to those with 5 nm InP barrier. MOSFETs with InP/InAlAs barrier show 10% I_d increase using Al_2O_3/HfO_2 compared to those with HfO_2.

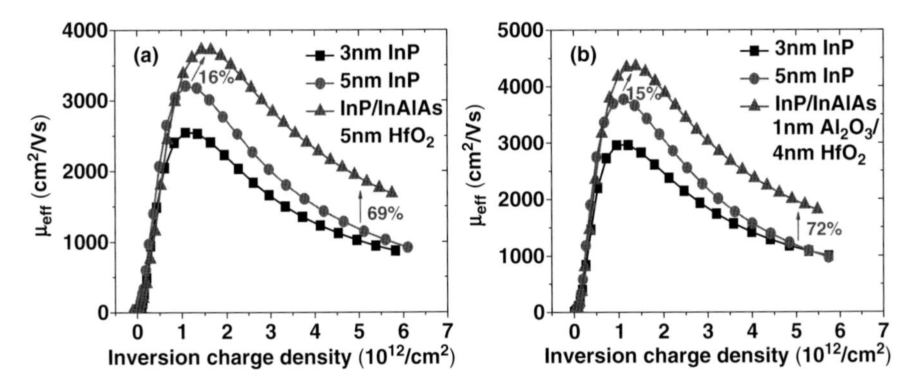

Figure 14.34 (a) MOSFETs using HfO_2 show 16% higher peak μ_{eff} and 69% higher high-field μ_{eff} ($Q_{inv} = 5 \times 10^{12}\,cm^{-2}$) with InP/InAlAs barrier than those with 5 nm InP barrier. (b) MOSFETs using Al_2O_3/HfO_2 show 15% higher peak μ_{eff} and 72% higher high-field μ_{eff} with InP/InAlAs barrier than those with 5 nm InP barrier. MOSFETs with InP/InAlAs barrier show 17% higher peak μ_{eff} using Al_2O_3/HfO_2 than those using HfO_2.

increase the channel mobility and drive current. InP barrier layer was found to be an effective barrier to improve the low-field mobility of $In_{0.7}Ga_{0.3}As$ MOSFETs. The InP/$In_{0.52}Al_{0.48}As$ double-barrier architecture significantly improves high-field mobility compared to same thickness single InP barrier and provides >50% improvement compared to surface channel devices.

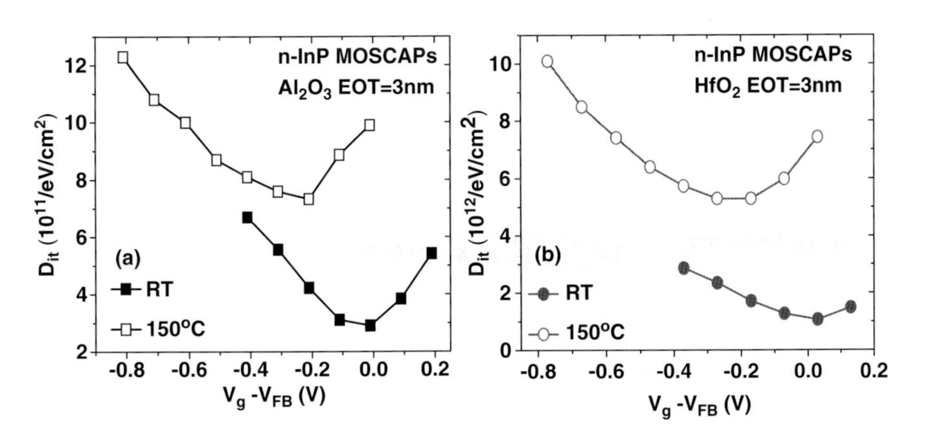

Figure 14.35 D_{it} for n-InP MOSCAPs with (a) Al_2O_3 or (b) HfO_2 gate dielectrics (same 3 nm EOT) at RT and 150 °C. 150 °C helps to detect D_{it} closer to mid-gap and results indicate larger D_{it} at mid-gap than closer to conduction band. MOSCAPs with HfO_2 show about one order of magnitude higher D_{it} than those with Al_2O_3.

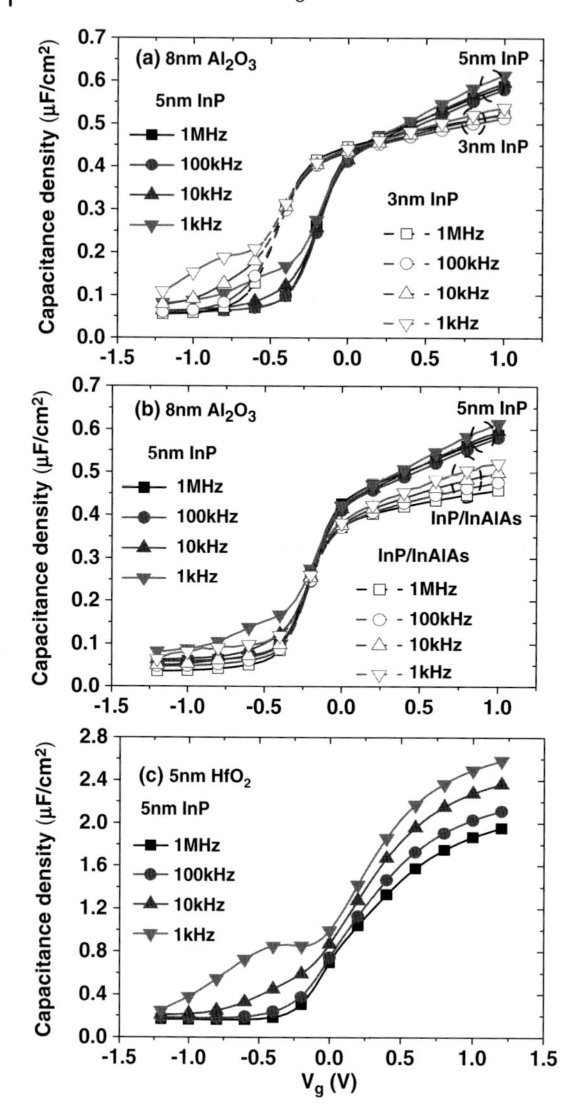

Figure 14.36 Frequency dispersion from split-*CV* for MOSFETs with (a) 3 or 5 nm InP barrier and 8 nm Al_2O_3, (b) 5 nm InP barrier or InP/InAlAs barrier and 8 nm Al_2O_3, and (c) 5 nm InP barrier and 5 nm HfO_2. Smaller capacitance at $V_g = 1$ V (thicker T_{inv}) for MOSFETs with 3 nm InP indicates reduced electrons spilling over into barrier than those with 5 nm InP (a). Similarly, thicker T_{inv} for MOSFETs with 2 nm InP/3 nm InAlAs barrier than those with 5 nm InP shows less electrons spilling over into barrier (b). Larger frequency dispersion for HfO_2 indicates higher D_{it} at HfO_2/InP interface than that for Al_2O_3 (c).

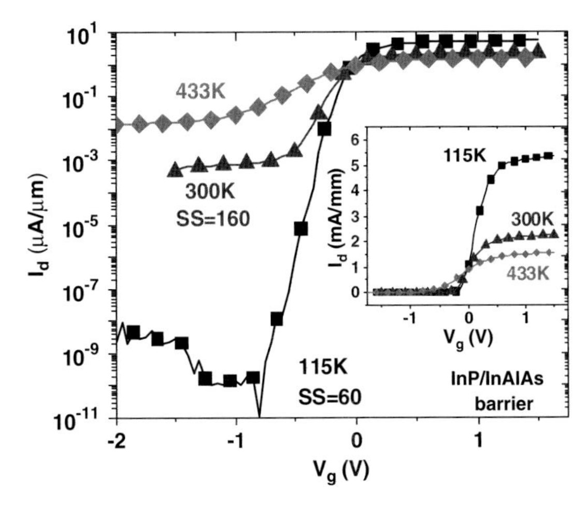

Figure 14.37 SS and I_d on/off current were significantly improved at 115 K for QW MOSFETs with InP/InAlAs barrier and HfO_2, showing effect of interface traps and importance of III–V/high-κ interface ($V_d = 50$ mV).

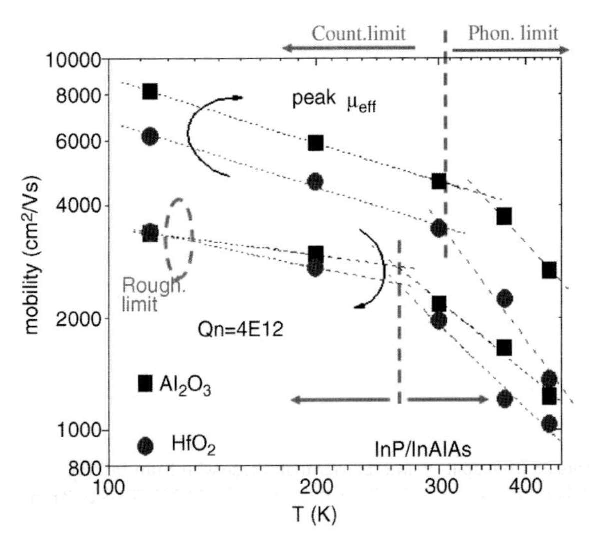

Figure 14.38 Temperature-dependent mobility of QW MOSFETs with HfO_2 and Al_2O_3 dielectrics at low N_{inv} (peak μ_{eff}) and high N_{inv} ($4 \times 10^{12}\,cm^{-2}$). Compared to Al_2O_3, lower mobility with HfO_2 results from high interface charges and high-κ phonon scattering.

References

1 Moore, G.E. (1965) MOS transistor as an individual device and in integrated arrays. *IEEE Spectrum*, **2**, 49.

2 Dennard, R.H. (1984) Evolution of the MOSFET dynamic RAM: a personal view. *IEEE Trans. Electron Devices*, **31**, 1549.

3 ITRS (2008) International Technology Roadmap for Semiconductors, 2008 edn, http://www.itrs.net/reports.html.

4 Wilk, G., Wallace, R., and Anthony, J. (2001) High-*k* gate dielectrics: current status and materials properties considerations. *J. Appl. Phys.*, **89**, 5243.

5 Henson, W., Ahmed, K., Vogel, E., Hauser, J., Wortman, J., Venables, R., Xu, M., and Venables, D. (1999) Estimating oxide thickness of tunnel oxides down to 1.4nm using conventional capacitance–voltage measurements on MOS capacitors. *IEEE Electron Device Lett.*, **20**, 179.

6 ITRS (2007) International Technology Roadmap for Semiconductors, 2007 edn, http://www.itrs.net/reports.html.

7 Ghani, T., Armstrong, M., Auth, C., Bost, M., Charvat, P., Glass, G., Hoffmann, T., Johnson;, K., Kenyon, C., Klaus, J., McIntyre, B., Mistry, K., Murthy, A., Sandford, J., Silberstein, M., Sivakumar, S., Smith, P., Zawadzki, K., Thompson, S., and Bohr, M. (2003) A 90nm high volume manufacturing logic technology featuring novel 45nm gate length strained silicon CMOS transistors. IEEE International Electron Devices Meeting, Technical Digest, p. 11.6.1.

8 Welser, J., Hoyt, J., and Gibbons, J. (1994) Electron mobility enhancement in strained-Si n-type metal–oxide–semiconductor field-effect transistors. *IEEE Electron Device Lett.*, **15**, 100.

9 Houssa, M. (ed.) (2003) *High-k Gate Dielectrics*, Institute of Physics, Bristol.

10 Cheng, B., Cao, M., Rao, R., Inani, A., Voorde, P., Greene, W., Stork, J., Yu, Z., Zeitzoff, P., and Woo, J. (1999) The impact of high-*k* gate dielectrics and metal gate electrodes on sub-100nm MOSFETs. *IEEE Trans. Electron Devices*, **46**, 1537.

11 Gusev, E., Buchanan, D., Cartier, E., Kumar, A., Dimaria, D., Guha, S., Callegari, A., Zafar, S., Jamison, P., Neumayer, D., Copel, M., Gribelyuk, M., Schmidt, H., Emic, C., Kozlowshi, P., Chan, K., Bojarczuk, N., Ragnarsson, L., Ronsheim, P., Fleming, T., Mocuta, A., and Ajmera, A. (2001) Ultrathin high-*k* gate stacks for advanced CMOS devices. IEEE International Electron Devices Meeting, Technical Digest, p. 451.

12 Datta, S., Dewey, G., Doczy, M., Doyle, B., Jin, B., Kavalieros, J., Kotlyar, R., Metz, M., Zelick, N., and Chau, R. (2003) High mobility Si/SiGe strained channel MOS transistors with HfO_2/TiN gate stack. IEEE International Electron Devices Meeting, Technical Digest, p. 653.

13 Chui, C., Kim, H., Chi, D., Triplett, B., Mcintyre, P., and Saraswat, K. (2002) A sub-400 °C germanium MOSFET technology with high-*k* dielectric and metal gate. IEEE International Electron Devices Meeting, Technical Digest, p. 437.

14 Ye, P. (2008) Main determinants for III–V metal–oxide–semiconductor field-effect transistors (invited). *J. Vac. Sci. Technol. A*, **26**, 697.

15 Roy, K., Mukhopadhyay, S., and Meimand, H. (2003) Leakage current mechanisms and leakage reduction techniques in deep-submicrometer CMOS circuits. *Proc. IEEE*, **91**, 305.

16 Yang, N., Henson, W., and Wortman, J. (2000) A comparative study of gate direct tunneling and drain leakage currents in n-MOSFET's with sub-2nm gate oxides. *IEEE Trans. Electron Devices*, **47**, 1636.

17 Robertson, J. (2002) Band offsets of high dielectric constant gate oxides on silicon. *J. Non-Cryst. Solids*, **303**, 94.

18 Hubbard, K. and Schlom, D. (1996) Thermodynamic stability of binary oxides in contact with silicon. *J. Mater. Res.*, **11**, 2757.

19 Xuan, Y., Wu, Y., and Ye, P. (2008) High-performance inversion-type enhancement-mode InGaAs MOSFET with maximum drain current exceeding 1 A/mm. *IEEE Electron Device Lett.*, **29**, 294.

20 Zhao, H., Chen, Y., Yum, J., Wang, Y., Goel, N., Koveshnikov, S., Tsai, W., and Lee, J.C. (2009) HfO_2-based $In_{0.53}Ga_{0.47}As$

MOSFETs (EOT ≈ 10 Å) using various interfacial dielectric layers. IEEE Device Research Conference.

21 Goel, N., Heh, D., Koveshnikov, S., Ok, I., Oktyabrsky, S., Tokranov, V., Kambhampati, R., Yakimov, M., Sun, Y., Pianetta, P., Gaspe, C., Santos, M., Lee, J., Datta, S., Majhi, P., and Tsai, W. (2008) Addressing the gate stack challenge for high mobility $In_xGa_{1-x}As$ channels for NFETs. IEEE International Electron Devices Meeting, Technical Digest, p. 363.

22 Zhao, H., Yum, J., Chen, Y., and Lee, J. (2009) $In_{0.53}Ga_{0.47}As$ n-MOSFETs with ALD Al_2O_3, HfO_2 and $LaAlO_3$ gate dielectrics. *J. Vac. Sci. Technol. B*, **27**, 2024.

23 Passlack, M., Zurcher, P., Rajagopalan, K., Droopad, R., Abrokwah, J., Tutt, M., Park, Y., Johnson, E., Hartin, O., Zlotnicka, A., and Fejes, P. (2007) High mobility III–V MOSFETs for RF and digital applications. IEEE International Electron Devices Meeting, Technical Digest, p. 621.

24 Hill, R., Moran, D., Li, X., Zhou, H., Macintyre, D., Thoms, S., Asenov, A., Zurcher, P., Pajagopalan, K., Abrokwah, J., Droopad, R., Passlack, M., and Thayne, I. (2007) Enhancement-mode GaAs MOSFETs with an $In_{0.3}Ga_{0.7}As$ channel, a mobility of over $5000 cm^2/Vs$, and transconductance of over 475 μS/μm. *IEEE Electron Device Lett.*, **28**, 1080.

25 Hong, M., Kwo, J., Tsai, P., Chang, Y., Huang, M., Chen, C., and Lin, T. (2007) III–V metal–oxide–semiconductor field-effect transistors with high-κ dielectrics. *Jpn J. Appl. Phys.*, **46**, 3167.

26 Koveshnikov, S., Tsai, W., Ok, I., Lee, J., Torkanov, V., Yakimov, M., and Oktyabrsky, S. (2006) Metal–oxide–semiconductor capacitors on GaAs with high-*k* gate oxide and amorphous silicon interface passivation layer. *Appl. Phys. Lett.*, **88**, 022106.

27 Ok, I., Kim, H., Zhang, M., Lee, T., Zhu, F., Yu, L., Koveshnikov, S., Tsai, W., Tokranov, V., Yakimov, M., Oktyabrsky, S., and Lee, J. (2006) Self-aligned n- and p-channel GaAs MOSFETs on undoped and p-type substrates using HfO_2 and silicon interface passivation layer. IEEE International Electron Devices Meeting, Technical Digest.

28 Kim, H., Ok, I., Zhang, M., Zhu, F., Park, S., Yum, J., Zhao, H., and Lee, J. (2008) Inversion-type enhancement-mode HfO_2 based GaAs metal–oxide–semiconductor field effect transistors with a thin Ge layer. *Appl. Phys. Lett.*, **92**, 032907.

29 Chin, H., Zhu, M., Lee, Z., Liu, X., Tan, K., Lee, H., Shi, L., Tang, L., Tung, C., Lo, G., Tan, L., and Yeo, Y. (2008) A new silane–ammonia surface passivation technology for realizing inversion-type surface channel GaAs N-MOSFET with 160nm gate length and high-quality metal-gate/high-κ dielectric stack. IEEE International Electron Devices Meeting, Technical Digest, p. 383.

30 Zhao, H., Kim, H., Zhu, F., Zhang, M., Ok, I., Park, S., Yum, J., and Lee, J. (2007) Metal–oxide–semiconductor capacitors on GaAs with germanium nitride passivation layer. *Appl. Phys. Lett.*, **91**, 172101.

31 Shahrjerdi, D., Tutuc, E., and Banerjee, S. (2007) Impact of surface chemical treatment on capacitance–voltage characteristics of GaAs metal–oxide–semiconductor capacitors with Al_2O_3 gate dielectric. *Appl. Phys. Lett.*, **91**, 063501.

32 Wu, Y., Xu, M., Wang, R., Koybasi, O., and Ye, P. (2009) High performance deep-submicron inversion-mode InGaAs MOSFETs with maximum G_m exceeding 1.1 mS/μm: new HBr pretreatment and channel engineering. IEEE International Electron Devices Meeting, Technical Digest, p. 323.

33 Oh, H., Lin, J., Suleiman, A., Lo, G., Kwong, D., Chi, D., and Lee, S. (2009) Thermally robust phosphorous nitride interface passivation for InGaAs self-aligned gate-first n-MOSFET integrated with high-κ dielectric. IEEE International Electron Devices Meeting, Technical Digest, p. 339.

34 Chen, Y., Zhao, H., Wang, Y., Xue, F., Zhou, F., and Lee, J. (2010) Effects of fluorine incorporation into HfO_2 gate dielectrics on InP and $In_{0.53}Ga_{0.47}As$ metal–oxide–semiconductor field-effect-transistors. *Appl. Phys. Lett.*, **96**, 253502.

35 Chen, Y., Zhao, H., Wang, Y., Xue, F., and Zhou, F. (2010) Fluorinated HfO_2 gate dielectric engineering on $In_{0.53}Ga_{0.47}As$ metal–oxide–semiconductor field-effect-transistors. *Appl. Phys. Lett.*, **96**, 103506.

36 Sun, Y., Koester, S., Kiewra, E., Fogel, K., Sadana, D., Webb, D., Fompeyrine, J., Locquet, J., Sousa, M., and Germann, R. (2006) Buried-channel $In_{0.7}Ga_{0.3}As$/ $In_{0.52}Al_{0.48}As$ MOS capacitors and transistors with HfO_2 gate dielectrics. Proceedings of the 64th IEEE Device Research Conference, p. 49.

37 Sun, Y., Kiewra, E., Souze, J., Bucchignano, J., Fogel, K., Sadana, D., and Shahidi, G. (2008) Scaling of $In_{0.7}Ga_{0.3}As$ buried-channel MOSFETs. IEEE International Electron Devices Meeting, Technical Digest, p. 367.

38 Zhao, H., Chen, Y., Yum, J., Wang, Y., Goel, N., and Lee, J. (2009) High performance $In_{0.7}Ga_{0.3}As$ metal–oxide–semiconductor transistors with mobility $>4400\,cm^2/V\,s$ using InP barrier layer. *Appl. Phys. Lett.*, **94**, 193502.

39 Zhao, H., Chen, Y., Yum, J., Wang, Y., Zhou, F., Xue, F., and Lee, J. (2010) Effects of barrier layers on device performance of high mobility $In_{0.7}Ga_{0.3}As$ metal–oxide–semiconductor field-effect-transistors. *Appl. Phys. Lett.*, **96**, 102101.

40 Passlack, M., Rajagopalan, K., Abrokwah, J., and Droopad, R. (2006) Implant-free high-mobility flatband MOSFET: principles of operation. *IEEE Trans. Electron Devices*, **53**, 2454.

41 Rajagopalan, K., Droopad, R., Abrokwah, J., Zurcher, P., Fejes, P., and Passlack, M. (2007) 1-μm enhancement mode GaAs n-channel MOSFETs with transconductance exceeding 250 mS/mm. *IEEE Electron Device Lett.*, **28**, 100.

42 Lin, H., Yang, T., Sharifi, H., Kim, S., Xuan, Y., Shen, T., Mohammadi, S., and Ye, P. (2007) Enhancement-mode GaAs metal–oxide–semiconductor high-electron-mobility transistors with atomic layer deposited Al_2O_3 as gate dielectric. *Appl. Phys. Lett.*, **91**, 212101.

43 Kim, D., Alamo, J., Lee, J., and Seo, K. (2007) Logic suitability of 50-nm $In_{0.7}Ga_{0.3}As$ HEMTs for beyond-CMOS applications. *IEEE Trans. Electron Devices*, **54**, 2606.

44 Kim, D. and Alamo, J. (2008) 30 nm E-mode InAs PHEMTs for THz and future logic applications. IEEE International Electron Devices Meeting, Technical Digest, p. 719.

45 Ma, T. and Lubow, A. (2008) Future CMOS technologies based on high-mobility channel materials. 5th International Symposium on Advanced Gate Stack Technology.

46 Chudzik, M., Doris, B., Mo, R., Sleight, J., Cartier, E., Dewan, C., Park, D., Bu, H., Natzle, W., Yan, W., Ouyang, C., Henson, K., Boyd, D., Callegari, S., Carter, R., Casaroto, D., Gribelyuk, M., Hargrove, M., He, W., Kim, Y., Linder, B., Moumen, N., Paruchuri, V., Stathis, J., Steen, M., Vayshenker, A., Wang, X., Zafar, S., Ando, T., Iijima, R., Takayanagi, M., Narayanan, V., Wise, R., Zhang, Y., Divakaruni, R., Khare, M., and Chen, T. (2007) High-performance high-κ/metal gates for 45nm CMOS and beyond with gate-first processing. IEEE Symposium on VLSI Technology, p. 194.

47 Zhao, H., Shahrjerdi, D., Zhu, F., Kim, H., Ok, I., Zhang, M., Yum, J., Banerjee, S., and Lee, J. (2008) Inversion-type indium phosphide metal–oxide–semiconductor field-effect transistors with equivalent oxide thickness of 12 Å using stacked $HfAlO_x/HfO_2$ gate dielectric. *Appl. Phys. Lett.*, **92**, 253506.

48 Brammertz, G., Lin, H., Matens, K., Mercier, D., Sioncke, S., Delabie, A., Wang, W., Caymax, M., Meuris, M., and Heyns, M. (2008) Capacitance–voltage characterization of GaAs–Al_2O_3 interfaces. *Appl. Phys. Lett.*, **93**, 183504.

49 Brammertz, G., Lin, H., Matens, K., Alian, A., Merckling, C., Penaud, J., Kohen, D., Wang, W., Sioncke, S., Delabie, A., Meuris, M., Caymax, M., and Heyns, M. (2009) Electrical properties of III–V/oxide interfaces. *ECS Trans.*, **19**, 375.

50 Hinkle, C., Milojevic, M., Brennan, B., Sonnet, A., Tostado, F., Hughes, G., Vogel, E., and Wallace, R. (2009) Detection of Ga suboxides and their impact on III–V

passivation and Fermi-level pinning. *Appl. Phys. Lett.*, **94**, 162101.

51 Milojevic, M., Hinkle, C., Tostado, F., Kim, H., Vogel, E., Kim, J., and Wallace, R. (2008) Half-cycle atomic layer deposition reaction studies of Al_2O_3 on $(NH_4)_2S$ passivated GaAs (100) surfaces. *Appl. Phys. Lett.*, **93**, 252905.

52 Kobayashi, M., Chen, P., Sun, Y., Goel, N., Majhi, P., Garner, M., Tsai, W., Pianetta, P., and Nishi, Y. (2008) Synchrotron radiation photoemission spectroscopic study of band offsets and interface self-cleaning by atomic layer deposited HfO_2 on $In_{0.53}Ga_{0.47}As$ and $In_{0.52}Al_{0.48}As$. *Appl. Phys. Lett.*, **93**, 182103.

53 Chang, C., Chiou, Y., Chang, Y., Lee, K., Lin, T., Wu, T., and Hong, M. (2006) Interfacial self-cleaning in atomic layer deposition of HfO_2 gate dielectric on $In_{0.15}Ga_{0.85}As$. *Appl. Phys. Lett.*, **89**, 242911.

54 Hinkle, C., Sonnet, A., Vogel, E., McDonnell, S., Hughes, G., Milojevic, M., Lee, B., Aguirre-Tostado, F., Choi, K., Kim, H., Kim, J., and Wallace, R. (2008) GaAs interfacial self-cleaning by atomic layer deposition. *Appl. Phys. Lett.*, **92**, 071901.

55 hinkle, C., Milojevic, M., Vogel, E., and Wallace, R. (2009) The significance of core-level electron binding energies on the proper analysis of InGaAs interfacial bonding. *Appl. Phys. Lett.*, **95**, 151905.

56 Xuan, Y., Wu, Y., Shen, T., Yang, T., and Ye, P. (2007) High performance submicron inversion-type enhancement-mode InGaAs MOSFETs with ALD Al_2O_3, HfO_2 and HfAlO as gate dielectrics. IEEE International Electron Devices Meeting, Technical Digest, p. 637.

57 Lin, T., Chen, C., Chiu, H., Chang, P., Lin, C., Hong, M., Kwo, J., and Tsai, W. (2008) Self-aligned inversion-channel and D-mode InGaAs MOSFET using Al_2O_3/$Ga_2O_3(Gd_2O_3)$ as gate dielectrics. 66th IEEE Device Research Conference Digest, p. 39.

58 Tyagi, S., Auth, C., Bai, P., Curello, G., Deshpande, H., Gannabaram, S., Golonzka, O., Heussneer, R., James, R. *et al.* (2005) An advanced low power, high performance, strained channel 65 nm technology. IEEE International Electron Devices Meeting, Technical Digest, p. 245.

59 Ok, I., Kim, H., Zhang, M., Zhu, F., Zhao, H., Park, S., Yum, J., Garcia, D., Majhi, P., Geol, N., Tsai, W., Gaspe, X., Santos, M., and Lee, J. (2008) Self-aligned n-channel MOSFET on InP and $In_{0.53}Ga_{0.47}As$ using physical vapor deposition HfO_2 and silicon interface passivation layer. 66th IEEE Device Research Conference Digest, p. 91.

60 Robertson, J., and Falabretti, B. (2008) Band offset of high-κ gate oxides on III–V semiconductors. *J. Appl. Phys.*, **100**, 014111.

61 Huang, M., Chang, Y., Chang, C., Lin, T., Kwo, J., Wu, T., and Hong, M. (2006) Energy-band parameters of atomic-layer-deposition Al_2O_3/InGaAs heterostructures. *Appl. Phys. Lett.*, **89**, 012903.

62 Goel, N., Majhi, P., Tsai, W., Warusawithana, M., Schlom, D., Santos, M., Harris, J., and Nishi, Y. (2007) High-indium-content InGaAs metal–oxide–semiconductor capacitor with amorphous $LaAlO_3$ gate dielectric. *Appl. Phys. Lett.*, **91**, 093509.

63 Martens, K., Chui, C., Brammertz, G., Jaeger, B., Kuzum, D., Meuris, M., Heyns, M., Krishnamohan, T., Saraswat, K., Maes, H., and Groeseneken, G. (2008) On the correct extraction of interface trap density of MOS devices with high-mobility semiconductor substrates. *IEEE Trans. Electron Devices*, **55**, 547.

64 Ye, P., Xuan, Y., Wu, Y., Shen, T., Pal, H., Varghese, D., Alam, M., and Lundstrom, M. (2008) Subthreshold characteristics of high-performance inversion-type enhancement-mode InGaAs NMOSFETs with ALD Al_2O_3 as gate dielectric. IEEE Device Research Conferences Digest, p. 93.

65 Sun, Y., Kiewra, E., Souza, J., Bucchignano, J., Fogel, K., Sadana, D., and Shahidi, G. (2008) High performance long- and short-channel $In_{0.7}Ga_{0.3}As$-channel MOSFETs. Proceedings of the 66th IEEE Device Research Conference, p. 41.

66 Yasuda, T., Miyata, N., Ishii, H., Itatani, T., Ichikawa, O., Fukuhara, N., Hata, M., Ohtake, A., Haimoto, T., Hoshii, T., Takenaka, M., and Takagi, S. (2008) Impact of cation composition and surface orientation on electrical properties of ALD-Al_2O_3/III–V interfaces. 39th IEEE Semiconductor Interface Specialists Conference.

67 Zhao, H., Shahrjerdi, D., Zhu, F., Zhang, M., Kim, H., Ok, I., Yum, J., Park, S., Banerjee, S., and Lee, J. (2008) Gate-first inversion-type InP metal–oxide–semiconductor field-effect transistors with atomic-layer-deposited Al_2O_3 gate dielectric. *Appl. Phys. Lett.*, **92**, 233508.

68 Frensley, W.R. and Einspruch, N.G. (1994) Chapter 1, in *Heterostructures and Quantum Devices, A Volume of VLSI Electronics: Microstructure Science*, Academic Press, San Diego, CA, p. 11.

69 Zhao, H., Zhu, F., Chen, Y., Yum, J., Wang, Y., and Lee, J. (2009) Effect of doping concentration and thickness on device performance for $In_{0.53}Ga_{0.47}As$ metal–oxide–semiconductor transistors with atomic-layer-deposited Al_2O_3 dielectrics. *Appl. Phys. Lett.*, **94**, 093505.

70 Varghese, D., Xuan, Y., Wu, Y., Shen, T., Ye, P., and Alam, M. (2008) Multi-probe interface characterization of $In_{0.65}Ga_{0.35}As$ MOSFET. IEEE International Electron Devices Meeting, Technical Digest, p. 379.

Part Five
High-k Application in Novel Devices

High-k Gate Dielectrics for CMOS Technology, First Edition. Edited by Gang He and Zhaoqi Sun.

15
High-k Dielectrics in Ferroelectric Gate Field Effect Transistors for Nonvolatile Memory Applications

Xubing Lu

15.1
Introduction

Current memory technologies, such as dynamic random access memory (DRAM) and NAND flash memory, are approaching very difficult issues related to their continued scaling to and beyond the 16 nm technology generation [1]. Fortunately, research over the past decades has led to discovery of several emerging memory technologies. These new memories include ferroelectric gate field effect transistor (FeFET) [2], spin-transfer torque magnetoresistive RAM [3], phase-change RAM [4], resistive-change RAM [5], and organic memory [6, 7]. Among these, FeFET has the advantages of high-speed, low-voltage operation, low power consumption, high endurance, and nondestructive readout [1, 8]. In the past two decades, FeFET has been extensively studied.

The cell structure of FeFET is composed of one metal–oxide–silicon (MOS) transistor, in which the gate dielectric is replaced by a ferroelectric film. The drain current can be controlled by the polarization direction of the ferroelectric film, through which the "on" and "off" states of the transistor are realized. Figure 15.1 is a demonstration of the device configuration and the basic working mechanism of the FeFETs. Since it is composed of only one metal–ferroelectric–Si (MFS) transistor, the structure of FeFET is suitable for high-density integration. Unlike DRAM and capacitor-type ferroelectric random access memory (FeRAM), it does not require destructive readout.

The principle of FeFET was proposed in 1950s [9] and many experimental studies were carried out in 1960s and 1970s. However, since it is very difficult to form a ferroelectric–semiconductor interface with good electrical properties, no commercially available devices have been fabricated yet. To solve this interface problem, one insulating buffer layer is often inserted between a ferroelectric film and Si substrate to construct the metal–ferroelectric–insulator–Si (MFIS) structure, which is shown in Figure 15.2. The insulator layer can prevent the reaction and interdiffusion between the ferroelectric layer and the silicon substrate as well as provide a potential barrier for charge injection from Si substrate to ferroelectric film [2]. Therefore, the electrical properties, especially the data retention time, can be significantly improved.

High-k Gate Dielectrics for CMOS Technology, First Edition. Edited by Gang He and Zhaoqi Sun.

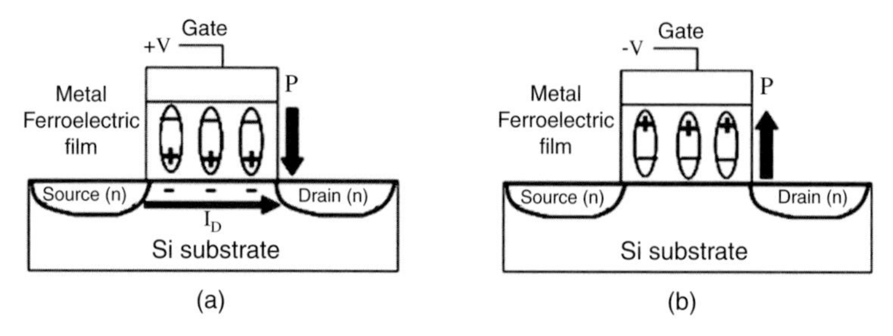

Figure 15.1 A schematic diagram of FeFET: (a) "on" state; (b) "off" state.

Figure 15.3 is a demonstration of the band alignment of the MFIS diode and the detailed working principle for the ferroelectric polarization to control the on/off state of the devices [10]. For an n-type substrate, when a positive gate bias is applied, the ferroelectric film is polarized and electrons are accumulated at the Si surface (Figure 15.3a). Due to the existence of the potential barrier, electrons are difficult to inject from Si to ferroelectric film. Otherwise, the electrons can be easily injected to the ferroelectric film or their interface. When the gate voltage is reduced to near zero, the remnant polarization in the ferroelectric film will keep inducing a smaller amount of negative charge in Si. Consequently, the Si surface remains in a weaker state of accumulation, which corresponds to the "off" state of the device. When a negative gate voltage is applied, the negative charge on the gate causes the ferroelectric polarization to switch direction, which induces a positive charge at the Si surface, bending the Si surface band upward. For an n-type semiconductor, this case corresponds to either the depletion or inversion condition. For sufficiently large polarization, the silicon surface will be in inversion (Figure 15.3c). When the gate voltage is reduced from this point to near zero, the remnant polarization in the ferroelectric film will cause the n-type Si to remain in either depletion or inversion. Figure 15.3d illustrates the case where a sufficiently large remnant polarization in the

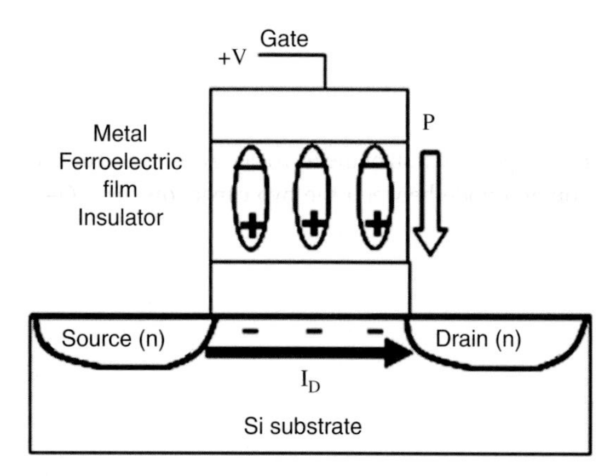

Figure 15.2 A schematic diagram of the MFIS structure.

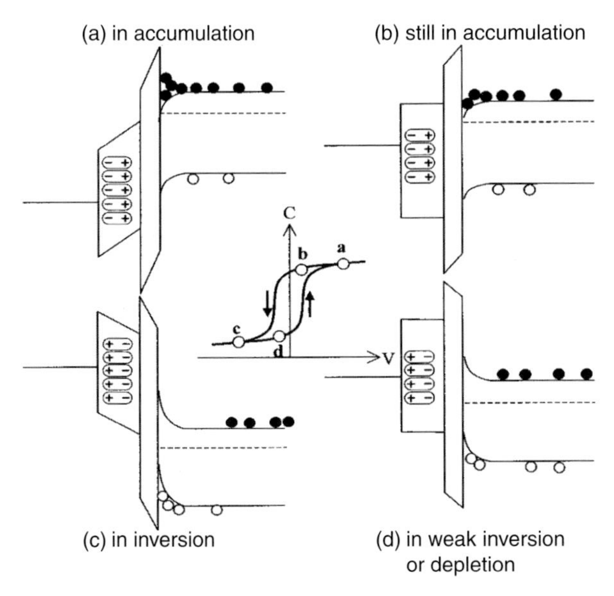

Figure 15.3 The band diagrams for MFIS structure under various bias conditions. Si surface is (a) in accumulation, (b) still in accumulation, (c) in inversion, and (d) in either depletion or inversion. Reprinted with permission from Ref. [10]. Copyright 2004, American Institute of Physics.

ferroelectric film causes the semiconductor surface to be in inversion, which corresponds to the "on" state of the device. It should be mentioned that Figure 15.3 only demonstrates a very ideal situation. In the real situation, the charge injection will inevitably exist, and its amount depends on the thickness of the buffer layer, interface defects, and other factors.

Although the retention time of the MFIS structure can be much improved compared to that of MFS structure, the MFIS structure still suffers from the following two major issues:

1) **Depolarization field.** When the power supply is off and the gate terminal of the FET is grounded, the top and bottom electrodes of the two capacitors are short-circuited. At the same time, electric charges $\pm Q$ appear on the electrodes of the both capacitors due to the remnant polarization of the ferroelectric film and due to the charge neutrality condition at a node between the two capacitors. The Q–V (charge versus voltage) relation in the dielectric capacitor is $Q = CV$ (where C is the capacitance), and thus the relation in the ferroelectric capacitor becomes $Q = -CV$ under the short-circuited condition. This relation means that the direction of the electric field (E_F) in the ferroelectric film is opposite to that of the polarization (P), as shown in Figure 15.4. This depolarization field reduces the data retention time significantly. In the early stage for the study of MFIS structure, SiO_2 was often used as the insulator buffer layer. The data retention time of MFIS devices with SiO_2 buffer layer is very short. In order to make the depolarization field low, C must be as large as possible. That is, a thin buffer layer with a high

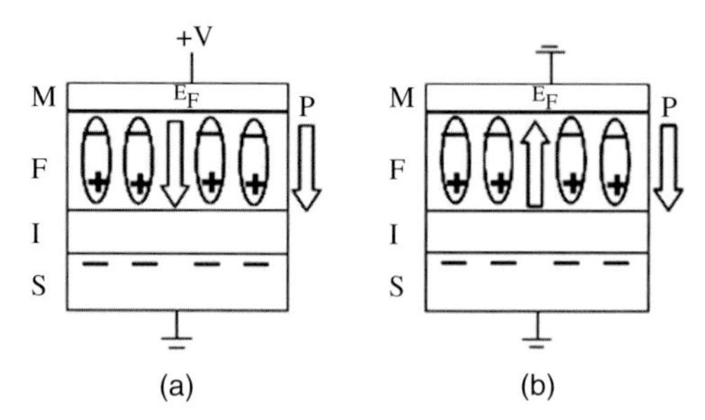

Figure 15.4 (a) During writing process, the direction of E_F is same as that of the polarization P. (b) During retention, the direction of E_F is opposite to that of P.

dielectric constant is desirable. Figure 15.5 is a theoretical simulation for the effects of the buffer layer dielectric constant on the data retention characteristics of the MFIS capacitors [11]. It can be clearly seen that the MFIS structure has a very short retention time when the dielectric constant of the buffer layer is small, while the retention time can be greatly increased with the increase in the dielectric constant of the buffer layer.

2) **Voltage drop across the insulator layer.** During the program/erase process, the bias voltage applied on the gate V_G will be divided into the voltage across the ferroelectric film (V_F) and the voltage across the buffer layer (V_I). That is, $V_G = V_F + V_I$. It is also known that $V_F C_F = V_I C_I$ according to charge matching. C_F and C_I are the ferroelectric capacitance and insulator capacitance, respectively. The remnant polarization values of typical ferroelectric films such as $SrBi_2Ta_2O_9$ (SBT) and $Pb(Zr_x, Ti_{1-x})O_3$ (PZT) are about 10 and 40 $\mu C/cm^2$, respectively. These values are generally much larger than the maximum charge density induced by a

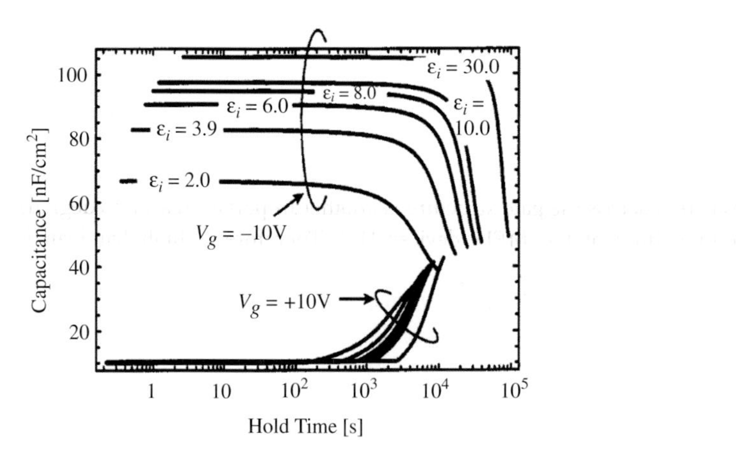

Figure 15.5 Calculated retention characteristics of MFIS structures as a parameter of dielectric constant of insulator layer. Reprinted with permission from Ref. [11]. Copyright 2003, American Institute of Physics.

dielectric film. For example, the maximum charge density induced on the surface of SiO_2 is 3.5 μC/cm^2 for an electric field of 10 MV/cm. This means that C_F is much larger than C_I. Consequently, V_F is much smaller than V_I. The gate voltage is mainly applied on the insulator layer. To minimize the voltage drop across the insulator layer, high dielectric constant insulators are desirable to make C_I as large as possible.

Other motivation to use high-*k* buffer layer is that the high dielectric constant of the insulator layer makes it possible to use a substantially thicker (physical thickness) dielectric for reduced gate leakage and raised retention time of the device, while the coupling between the ferroelectric polarization and the surface charge on Si substrate will not be reduced. That is to say, the retention time can be increased and the program/erase efficiency is not reduced, which is very important for the shrinking of the cell size.

15.2 Overview of High-*k* Dielectric Studies for FeFET Applications

15.2.1 Materials Requirements for High-*k* Buffer Layers

High-*k* dielectrics have been widely studied for their application to replace SiO_2 as the gate insulator for next-generation MOSFETs. Wilk *et al.* have given a comprehensive review on the material properties for different kinds of high-*k* materials [12]. Although the MFIS devices are much different from MOS devices from the viewpoints of working principle and application directions, the basic material selection criterion used in high-*k* dielectrics for MOSFETs can still be applied for MFIS structures. The selection of the buffer layer materials generally obeys the following criteria [13]:

1) **High dielectric constant**. Considering for depolarization field effect, voltage drop across the insulator layer, and larger physical thickness, a high dielectric constant is the basic and essential requirement for the buffer layer.
2) **High bandgap or band offset**. In addition to the depolarization field effect, the leakage current across the gate structure is another important reason to degrade the retention time of the MFIS devices [11]. Therefore, a high bandgap is necessary for the buffer layer to get small gate leakage current. At the same time, a high band offset with the conduction band or valence band of silicon is required to prevent the charge injection from Si to the ferroelectric film and its interface.
3) **Good thermal stability with silicon**. Generally, the ferroelectric film needs to be crystallized at a high temperature to obtain enough remnant polarization. This crystallization temperature is often higher than the fabrication temperature of the high-*k* buffer layer. Therefore, one interfacial layer (IL) with low dielectric constant can hardly be avoided after the fabrication of the total MFIS structure. A good

thermal stability of the high-*k* buffer layer with silicon will be desirable to reduce the IL thickness as small as possible.

4) **Amorphous or epitaxial structure on silicon**. Similar to high-*k* dielectric in MOS device, the amorphous or epitaxial structure of the buffer layer will be helpful in reducing the interface defects, which act as the leakage path or charge trapping sites.

15.2.2 Research Progress of High-*k* in the MFIS Devices

Starting from around 1994, different kinds of high-*k* buffer layer materials have been fabricated by various methods and integrated with ferroelectric films to construct the MFIS structure [14–30]. According to the available literature, there are more than 20 kinds of buffer layers studied for MFIS applications. For ferroelectric gate materials, $LiNbO_3$ [31], $PbTiO_3$ [32], (Ba, Sr)TiO_3 [33], $YMnO_3$ [34], and so on were investigated before 2000. With further investigation on ferroelectric films, the ferroelectric gate materials finally focused on $SrBi_2Ta_2O_9$ [11, 18], (Bi, La)$_4Ti_3O_9$ (BLT) [15, 35], Pb(Zr, Ti)O_3 [20, 21], and so on due to their excellent fatigue and leakage characteristics. Recently, $BiFeO_3$ (BFO) has been tremendously studied owing to its unique ferroelectric and ferromagnetic properties. At the same time, attention was also paid for its application as the ferroelectric gate material for FeFET applications due to its low crystallization temperature and high P_r value [19, 36].

Before 2000, the efforts on searching suitable buffer layers focused on the materials such as CeO_2 [27], TiO_2 [37], ZrO_2 [33], $SrTiO_3$ [38], $YMnO_3$ [39], and so on. The electrical properties of the MFIS devices were improved in comparison to devices with SiO_2 and without any buffer layer. For example, Hirai *et al.* reported a long retention time of 1.0×10^5 s for an Al/$PbTiO_3$ (81 nm)/CeO_2 (18 nm)/Si structure [27], in which 18 nm CeO_2 deposited by electron beam evaporation was used as the buffer layer. The retention time is much longer than that of 3.0×10^3 s for MFIS structure using SiO_2 buffer layer [14]. $YMnO_3$ has a high dielectric constant (about 30 for bulk ceramics) and high thermal stability, and especially yttrium can easily deoxidize SiO_2 formed on Si. Choi *et al.* investigated $YMnO_3$ as the buffer layer using a Pt/SBT (200 nm)/$YMnO_3$ (25 nm)/Si gate structure [39] and demonstrated a memory window of 1.5 V under 6 V bias sweeping. Furthermore, $YMnO_3$ exhibited better ability to prevent the interdiffusion between ferroelectric film and Si than CeO_2 does. Unfortunately, the retention data are not available in Choi's work. For these early studied high-*k* buffer layer materials, although some encouraging results were reported, the overall electrical properties of the MFIS devices are still far from their commercial applications. This is mostly due to the fact that they suffer from some of the essential material deficiencies such as small band offset with silicon [37, 38], easy crystallization [33, 37], thermal stability [27, 33, 37, 38], and so on.

Starting from 2000, efforts for high-*k* buffer layer materials shifted to materials such as Y_2O_3 [34], Ta_2O_5 [40], Si_3N_4 [15], SiON [16], Al_2O_3 [17], La_2O_3 [41], $LaAlO_3$ [24], PrO_x [11], and so on. New encouraging progresses were continuously made. Choi *et al.* studied Al_2O_3 thin films as buffer layers in the MFIS capacitors [17].

They obtained a maximum memory window of 1.52 V at 5 V bias voltage for Pt/$SrBi_2Nb_2O_9$ (SBN) (220 nm)/Al_2O_3 (11.4 nm)/Si structures, which is believed to be sufficient for the practical applications of MFIS-type FeFET operation at low voltage. Won *et al.* also demonstrated large memory window in an Al/$PbTiO_3$ (360 nm)/La_2O_3 (28 nm)/Si configuration [41]. The memory window increased from 0.3 to 2.6 V with increasing bias from 2 to 10 V. Although large memory window values were reported in Choi and Won's work, unfortunately, no retention results were reported in their work. Park and Ishiwara reported $LaAlO_3$ as the buffer layer in the Pt/$Sr_{0.8}Bi_{2.2}Ta_2O_9$ (210 nm)/$LaAlO_3$ (25 nm)/Si devices [24]. A large memory window of 3.0 V was observed under the bias voltage of 10 V. In addition, a capacitance retention time of 12 h was reported. Noda *et al.* reported excellent leakage current properties of PrO_x thin films formed by pulsed laser deposition [11]. The retention time of the corresponding MFIS structures was longer than 1×10^4 s; unfortunately, the observed memory window was only 0.3 V at a bias voltage of 12 V.

Through developing various kinds of high-*k* buffer layers, the above-mentioned work pushed forward the developing of MFIS devices from various aspects. However, none of these above-mentioned high-*k* buffer layers can really push the overall electrical properties, especially the retention time of the MFIS devices, to a commercial production level. It was not until 2004 that breakthrough for the retention characteristics of the MFIS devices was achieved, which made FeFET a real candidate for practical nonvolatile memory applications. The key work in contributing for this retention breakthrough is the developing of the Hf-based oxide as the insulator buffer layer in the MFIS-type devices. In 2004, Sakai and Ilangovan reported excellent device performance for a Pt/$SrBi_2Ta_2O_9$/HfAlO/Si MFIS FeFET [42]. The observed memory window was 1.6 V at 6 V bias voltage sweeping, as shown in Figure 15.6. Most importantly, the on/off ratio of the drain current can be kept larger than 10^6 even after a retention time of 12 days, as shown in Figure 15.7. In 2004, Ishiwara and coworkers also reported excellent retention properties for a MFIS FeFET based on HfO_2 buffer layer [43]. In their work, the Pt/$SrBi_2Ta_2O_9$/HfO_2/Si FeFET exhibited the drain current on/off ratio of approximately 10^5 even after 15.9 days had elapsed at

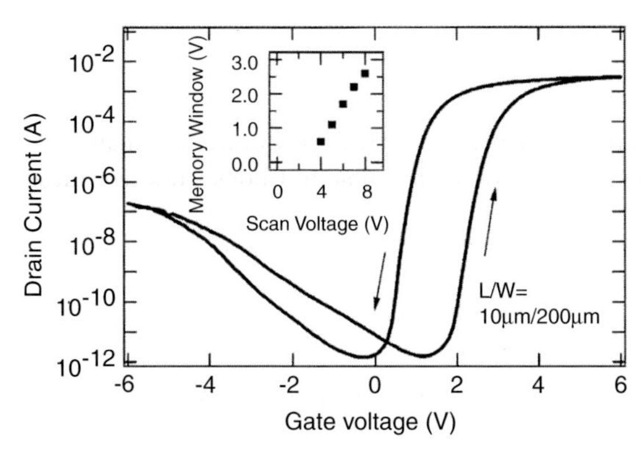

Figure 15.6 Drain current versus gate voltage of a Pt–$SrBi_2Ta_2O_9$–HfAlO–Si FET. Reproduced from Ref. [42]. Copyright 2004, IEEE.

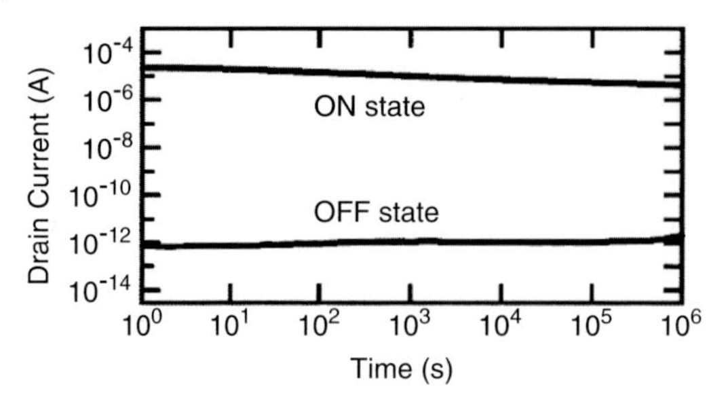

Figure 15.7 Data retention characteristic of the Pt–$SrBi_2Ta_2O_9$–HfAlO–Si FET. $V_{Keep} = 1.7$ V and $V_d = 0.1$ V. Reproduced from Ref. [42]. Copyright 2004, IEEE.

room temperature. It was also found that a write pulse width as short as 20 ns was enough to obtain a significant drain current on/off ratio. The corresponding results obtained by Ishiwara and coworkers are shown in Figures 15.8 and 15.9, respectively. The origin of the excellent device performance of the HfO_2- and HfAlO-based MFIS FeFETs was that the HfO_2-based films have high thermal stability, a high dielectric constant, and excellent interface properties. Therefore, the trapped charge density in the buffer layer and the interface is small. The ferroelectric polarization was not significantly screened, and can induce free carriers near the semiconductor surface. In addition, the intrinsic leakage current across the ferroelectric/HfO_2 (HfAlO) was very small. These two advantages directly improved the retention characteristics greatly.

By using a self-aligned gate process, Takahashi and Sakai successfully obtained a retention time of more than 33 days with an on- and off-state I_d ratio of over 10^5 for FeFET using HfAlO buffer layer [23]. Further work done by Ishiwara and

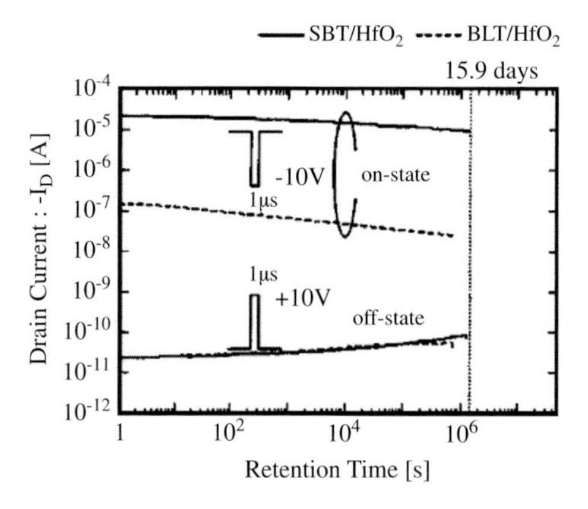

Figure 15.8 Typical retention characteristics of the fabricated MFIS FETs with SBT/HfO_2 and BLT/HfO_2 structures. Reprinted with permission from Ref. [43]. Copyright 2004, American Institute of Physics.

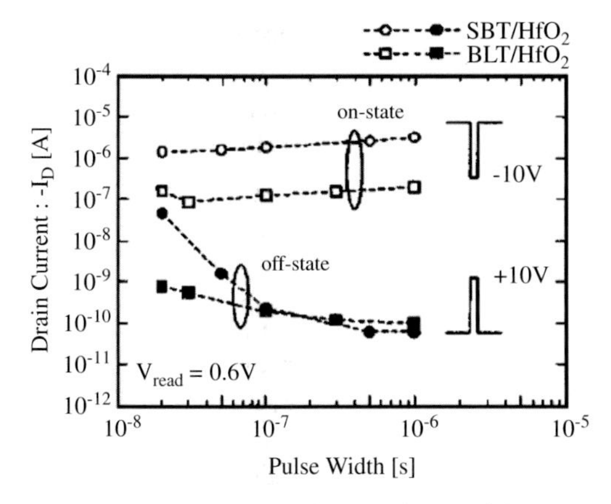

Figure 15.9 Write pulse width dependences of the drain currents in the SBT/HfO_2 and BLT/HfO_2 samples. Reprinted with permission from Ref. [43]. Copyright 2004, American Institute of Physics.

coworkers also demonstrated a retention time of 30 days for a Pt/$SrBi_2Ta_2O_9$/HfO_2/Si FeFET [26]. Although several new buffer layers such as Dy_2O_3 [20], $DyScO_3$ [29], and $LaZrO_x$ [44] were studied in recent years, the reported results were much worse than those of FeFET using Hf-based oxides. Up to now it is believed that the HfO_2-based dielectrics serve as the best buffer layer for MFIS applications. Table 15.1 provides a summary of the typical high-*k* buffer layers investigated in the past two decades and their typical results for MFIS devices. It was clearly shown that the dielectric constant of the buffer layer is one of the critical factors to affect the MFIS device performance. It should be mentioned that the mechanisms to affect the overall device performance are very complex. They are not only related to the material properties of high-*k* buffer layer, but also closely related to material properties of ferroelectric film and device configuration. Using different ferroelectric gate materials or different thickness ratios of ferroelectric/insulator [18, 45], the overall device properties such as memory window and retention time will vary in a wide range.

15.2.3
Issues for High-*k* Dielectric Integration into MFIS Devices

15.2.3.1 High-*k*/Si Interfacial Reaction

High-*k* materials such as HfO_2 and HfAlO are expected to have good thermal stability with silicon [12, 46], and the interfacial reaction between high-*k* dielectric and Si substrate can be well controlled for some of the high-*k* dielectrics in the MOS structure. For example, $LaAlO_3$ [47] and La_2O_3 [48] have been deposited on Si without interfacial layers by optimization of the process parameters. For the MFIS structure, however, the ferroelectric film often needs to be crystallized at high temperature for a long time to obtain enough P_r value. For example, the SBT films always receive a crystallization process at 750 °C for 30 min [18]. This long-time and high-temperature

Table 15.1 A summary of the typically studied high-*k* buffer layers and their typical electrical characteristics of the corresponding MFIS devices.

Buffer layer	k	Ferroelectric gate/buffer layer	M_W (V)	V_{sweep} (V)	Retention time (s)	Reference
SiO_2	3.9	SBT (400 nm)/SiO_2 (27 nm)	2.7	−6 to +6	3.0×10^3	[14]
Si_3N_4	7	BLT (100 nm)/Si_3N_4 (3 nm)	1.2	−5 to +5	$\sim 1.0 \times 10^4$	[15]
SiON	~7	SBT (400 nm)/SiON (20 nm)	0.3	−2 to +4	6.0×10^5 (7 days)	[16]
Al_2O_3	9	SBN (220 nm)/Al_2O_3 (11.4 nm)	1.5	−5 to +5	NA	[17]
HfSiON	~11	SBT (300 nm)/HfSiON (2 nm)	0.8	−4 to +4	8.6×10^4 (1 day)	[18]
PrO_x	12	SBT (400 nm)/PrO_x (32 nm)	0.3	−6 to +6	1.0×10^4	[11]
TiO_2	12	BFO (250 nm)/TiO_2 (150 nm)	1.1	−6 to +6	NA	[19]
Dy_2O_3	14	PZT (250 nm)/Dy_2O_3 (20 nm)	0.6	−6 to +6	1×10^4	[20]
Y_2O_3	12–18	PZT (280 nm)/Y_2O_3	1.5	−8 to +8	3.0×10^3	[21]
ZrO_2	20	SBT (210 nm)/ZrO_2 (28 nm)	2.6	−10 to +10	NA	[22]
HfAlO	20	SBT (420 nm)/HfAlO (12 nm)	1.6	−6 to +8	2.9×10^6 (33 days)	[23]
$LaAlO_3$	21–25	SBT (210 nm)/$LaAlO_3$ (25 nm)	3.0	−10 to +10	4.3×10^4	[24]
Ta_2O_5	22	P(VDF, TrFE) (100 nm)/Ta_2O_5 (3 nm)	4.6	−6 to +6	1.0×10^6 (11.6 days)	[25]
HfO_2	25	SBT (400 nm)/HfO_2 (8 nm)	1.0	−5 to +5	2.6×10^6 (30 days)	[26]
CeO_2	26	$PbTiO_3$ (81 nm)/CeO_2 (18 nm)	2.4	−4 to +4	1.0×10^5	[27]
La_2O_3	~25	PZT (160 nm)/La_2O_3 (16 nm)	0.7	−8 to +8	NA	[28]
$DyScO_3$	27	BNT (500 nm)/$DyScO_3$ (5 nm)	4.0	−12 to +12	1×10^3	[29]
$SrTiO_3$	~300	SBT (300 nm)/$SrTiO_3$ (23 nm)	1.1	−7 to +7	8.6×10^4 (1 day)	[30]

annealing process will inevitably induce interfacial reaction between high-*k* materials and Si substrate. Even for HfO_2 and HfAlO, there exists one clear IL between HfO_2/Si and HfAlO/Si in their MFIS structures [26]. Figure 15.10 shows the cross-sectional transmission electron microscopy image of BLT/HfO_2/Si structure, in which one 5 nm thick IL formed between 8 nm HfO_2 and Si substrate. This IL was mainly composed of SiO_2 identified by energy-dispersive X-ray spectroscopy (EDX). In Takahashi et al.'s work [49], clear IL was found between HfAlO and Si in the Pt/SBT/HfAlO/Si FeFETs, and the thickness of IL varies with the nitrogen ambient pressure during HfAlO deposition. The results were shown in Figure 15.11. According to the available reported results, one low dielectric constant IL between high-*k* buffer layer and Si inevitably exists for all of the reported MFIS structures. This low dielectric constant IL will greatly reduce the overall dielectric constant of the total insulator buffer layer, resulting in further reduction in the voltage drop across the ferroelectric film. Consequently, a high bias voltage or longer pulse width was needed to get enough polarization to control the on/off state of the channel. For example, the reported write/erase voltage is ±10 V, and the corresponding write/erase pulse width

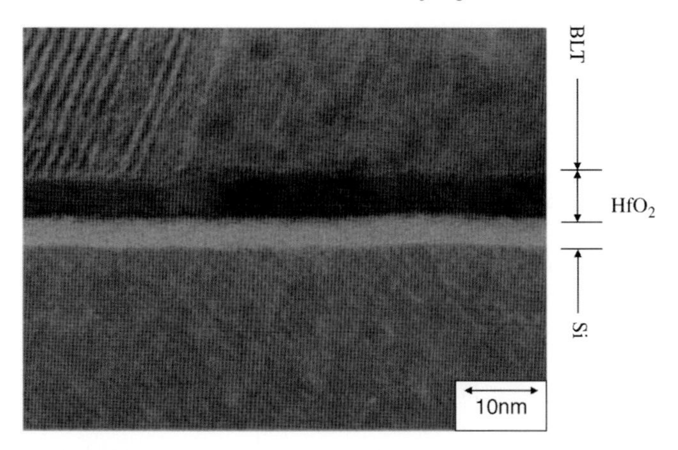

Figure 15.10 Cross-sectional TEM image of BLT/HfO_2/Si structure. Reprinted with permission from Ref. [26]. Copyright 2005, The Japan Society of Applied Physics.

is 1 μs for the HfO_2 FeFET reported by Ishiwara and coworkers [26]. The write/erase voltage is ±8 V, and the corresponding write/erase pulse width is around 10 μs for the HfAlO FeFET reported by Takahashi and Sakai [23]. It is known that the theoretical ferroelectric polarization switching speed is only ~100 ns and the operation voltage can be as small as 1–2 V for ferroelectric films [50]. The present results are not good enough for high-speed, low-voltage, and low-power operations for the next-generation nonvolatile memory scaling. How to control the formation of the interfacial low dielectric constant IL is still one of the challenges in the future FeFET studies.

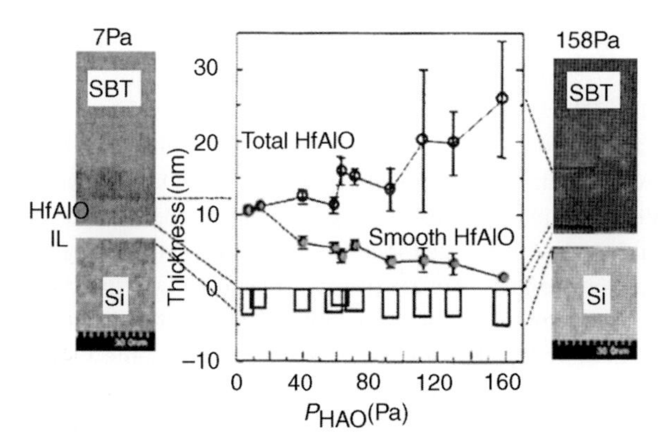

Figure 15.11 Estimated thicknesses of HfAlO and IL in Pt/SBT/HfAlO/Si FETs at various nitrogen ambient pressures during HfAlO deposition (P_{HAO}). Total HfAlO means a summation of rough and smooth part thicknesses. Reprinted with permission from Ref. [49]. Copyright 2008, American Vacuum Society.

15.2.3.2 **Crystallinity and Interdiffusion with Ferroelectric Film**

It is believed that an amorphous or epitaxial structure is necessary to obtain good high-*k*/Si interface not only for MOS structure but also for MFIS structure, because polycrystalline structure will form grain boundaries as the paths for fast oxygen diffusion and leakage current [12]. In addition, it is also the important origin of the interface charge traps. In MFIS structure, the ferroelectric films are always deposited and postannealed after the deposition of the high-*k* buffer layer. Therefore, the high-*k* buffer layer will also receive the long-time and high-temperature annealing during the deposition and crystallization process of ferroelectric film. Some high-*k* materials such as HfO_2 and Ta_2O_5 have low crystalline temperature. For example, the crystallization temperature of HfO_2 is lower than 500 °C. However, most of the oxide ferroelectric films have a crystallization temperature higher than 500 °C. As a result, buffer layers such as HfO_2 will be crystallized easily after the fabrication of the MFIS devices, which will degrade the overall electrical characteristics of the FeFET. As shown in Figure 15.10, the HfO_2 buffer layer exhibits crystalline structure after the fabrication of BLT layer. Although some high-*k* materials such as Al_2O_3 and HfAlO have crystallization temperature higher than 800 °C, the amorphous structure can remain even after the fabrication of the MFIS structure. The interdiffusion was found between high-*k* materials and ferroelectric film, which reduces the thickness of the buffer layer and degrades the interface quality between ferroelectric film and Si. As shown in Figure 15.11, the HfAlO buffer layer remains amorphous after SBT layer fabrication. However, the interdiffusion was clearly observed between SBT and HfAlO, and the degree of interdiffusion changes with the nitrogen ambient pressure during HfAlO deposition [49]. According to the available literature, this kind of high-*k*/ferroelectric interdiffusion often occurs for most of the MFIS structures, which is also one of the issues needed to be overcome in future FeFET studies.

15.2.3.3 **Possible Solutions**

To control the interfacial low-*k* oxide layer, efforts such as silicon surface passivation were introduced in the work by Horiuchi *et al.* [51] and Hirakawa *et al.* [52]. However, the passivation layer such as Si_3N_4 or SiON only has a low dielectric constant of ~7.0. Therefore, the overall electrical thickness still cannot be effectively reduced. In recent years, new ferroelectric gate materials such as organic ferroelectric film and oxide ferroelectric film with low crystallization temperature such as $BiFeO_3$ are investigated to replace the presently studied ferroelectric gate materials for MFIS applications. For example, Fujisaki *et al.* [25] reported that the Au/PMMA–P(VDF–TrFE)/Ta_2O_5/Si MFIS diode can operate at a very small gate voltage (~1.5 V memory window only by 2 V voltage sweeping). In addition, this diode exhibits excellent retention time of 11.6 days. In this chapter, detailed introduction on this topic will not be given. Another promising solution is to develop new high-*k* buffer layer materials, which is still believed to be one of the effective methods to overcome the issues existing in the studied high-*k* buffer layers. In this chapter, our recent work on developing new buffer layer materials for MFIS FeFET applications will be introduced. Details on the fabrication and characterization of the HfTaO buffer layer, MFIS diodes, and FeFET are demonstrated.

15.3
Developing of HfTaO Buffer Layers for FeFET Applications

15.3.1
Introduction

Recently, a novel gate dielectric material of HfTaO formed by incorporation of Ta into HfO_2 has been studied [53, 54]. Material studies indicate that the crystallization temperature of HfO_2 is significantly increased by adding Ta [53]. Simultaneously, no obvious lowering of dielectric constant in HfTaO [53] is observed due to the high dielectric constant of Ta_2O_5 (~25). Moreover, the interfacial state density and charge trapping are decreased considerably. Considering the requirements on crystallization temperature, dielectric constant, and electrical stability for high-*k* buffer layers, HfTaO seems to be very promising as the buffer layer for future MFIS device applications. In this work, we carry out a systematical evaluation on the microstructure and electrical properties of HfTaO films. The MFIS diodes and FeFETs were fabricated and their electrical properties were investigated [55].

15.3.2
Experimental Procedure

p-type Si(100) wafers were cleaned by wet chemical solution, dipped in a diluted HF solution (1%) to remove the surface oxide, and loaded into a vacuum chamber with a base pressure of $\sim 1.5 \times 10^{-9}$ Torr. For the deposition of HfTaO films, a sintered $HfO_2-Ta_2O_5$ pellet with the nominal atomic ratio of Hf:Ta = 0.6 : 0.4 was evaporated using an electron gun. The pressure during the deposition was around 10^{-7} Torr and the Si wafers were kept at room temperature (RT). The film thickness was *in situ* observed by a crystal thickness monitor attached in the chamber and a typical growth rate was around 0.01 nm/s. The deposited films were annealed at 700, 800, and 900 °C for 1 min in O_2 atmosphere.

SBT films were fabricated on the HfTaO/Si structure by chemical solution deposition. The deposited film was dried at 240 °C for 5 min in air in order to remove organic materials and successively fired by using a RTA (rapid thermal annealing) furnace at 750 °C for 1 min in O_2 flow. These processes were repeated for several times until the total thickness was approximately 300 nm. Then, the SBT films were crystallized by using a RTA furnace at 750 °C for 30 min in O_2 flow. For measurements of the electrical properties, dot-shaped Al and Pt top electrodes with an area of 3.14×10^{-4} cm^2 were deposited on the surface of the samples using a shadow mask by vacuum evaporation. Finally, backside contact was formed by thermal evaporation of a layer of Al with ~100 nm thickness and subsequent annealing at 400 °C in N_2 for 5 min to decrease the contact resistance. X-ray diffraction (XRD) analysis and reflection high-energy electron diffraction (RHEED) were used to analyze the crystallinity of HfTaO and SBT films. Cross-sectional TEM (transmission electron microscopy) images were also taken to investigate the interfacial structure of MFIS diodes. Capacitance–voltage (C–V)

and current density–voltage (*J*–*V*) characteristics were measured using a LCR meter (Toyo Corp.) and a high-precision semiconductor parameter analyzer (Agilent 4156C).

In the fabrication of MFIS FETs, a p-type Si wafer prepared for complementary MOS-type ferroelectric memories was used, in which source/drain p^+-regions were formed in n-wells and the FET regions were surrounded by field oxide layers. The deposition processes for HfTaO and SBT films were same as those described above. Pt gate electrodes were formed by etching a uniform film using a reactive ion etching apparatus. Via holes for the source and drain contacts were opened by wet and dry etching. Al electrodes were patterned by the liftoff process and Al backside contact was formed by vacuum evaporation. Finally, all the samples were sintered at 400 °C for 5 min in N_2 flow, aiming for a decrease in contact resistance. Typical channel width (*W*)/length (*L*) values are 20/2 and 50/2 μm, respectively. The electrical characteristics of the MFIS FeFETs were investigated by a high-precision semiconductor parameter analyzer (Agilent 4156C).

15.3.3 Crystallization Characteristics of HfTaO Films

The crystallization characteristics of the HfTaO films were studied by XRD measurement. A 6 nm thick HfTaO film deposited at RT was annealed at 800 °C for 1 min in O_2 and analyzed by XRD. However, no diffraction peaks related to HfTaO film can be seen, except for the peaks from the Si substrate. Then, RHEED observations were carried out to further check the crystallization characteristics of HfTaO films. Figure 15.12 shows the RHEED patterns of the Si substrate and the postannealed HfTaO film. Figure 15.12a exhibits a typical 1×1 structure of the hydrogen-terminated Si surface. After deposition of a 6 nm thick HfTaO film onto the Si substrate at RT and RTA at 800 °C for 1 min in O_2, the RHEED pattern changed from the 1×1 structure to a halo pattern, as shown in Figure 15.12b. This result clearly shows that the HfTaO film is amorphous even after being annealed at 800 °C for 1 min.

15.3.4 Electrical Properties of HfTaO Films on Si Substrates

Figure 15.13a shows typical *C*–*V* characteristics of the as-deposited and postannealed HfTaO films. The HfTaO films with 6 nm physical thickness were deposited at RT by electron beam evaporation. To reduce the leakage current in the films, postannealing was carried out at high temperatures of 700, 800, and 900 °C in O_2 for 1 min. As can be seen from the figure, the following results can be derived: (1) A large hysteresis loop can be seen in the *C*–*V* curve of the as-deposited film, indicating the existence of a large amount of fixed charges or charge traps. The bend down of the *C*–*V* curve in the large negative gate voltage also demonstrates a large leakage current in the film. (2) After annealing, the hysteresis loop width clearly decreases and good saturation characteristics of the *C*–*V* curves can be observed even under a large negative gate

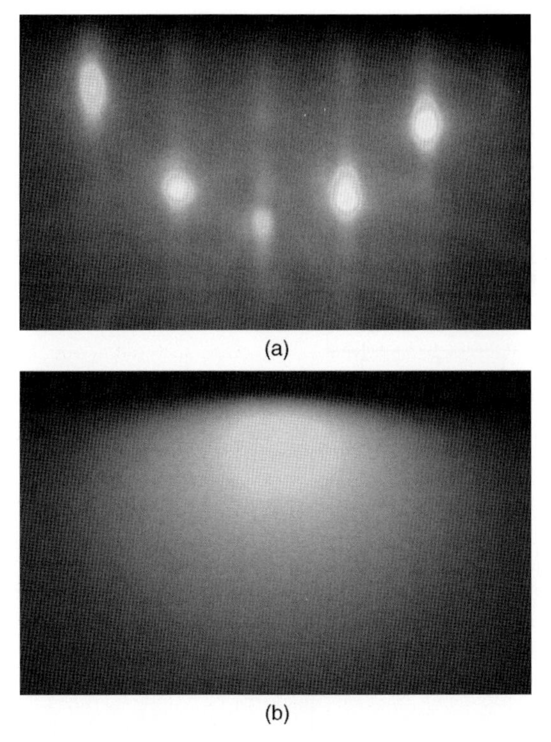

Figure 15.12 RHEED patterns of H-terminated Si(100) surface (a) and an HfTaO film annealed at 800 °C for 1 min in O_2 (b). Reprinted with permission from Ref. [55]. Copyright 2008, American Institute of Physics.

voltage. This demonstrates that both the fixed charges and leakage current in the HfTaO film have been effectively reduced. (3) The accumulation capacitance decreases with the increase in the annealing temperature. We assume that a thicker SiO_2 layer has been formed under higher annealing temperature. (4) A large hysteresis loop of 100 mV width is found in the HfTaO film annealed at 900 °C, which is much wider than that (15 mV) in the film annealed at 800 °C. This may be due to crystallization of the HfTaO film annealed at 900 °C.

According to the high-resolution transmission electron microscopy analysis results by Kiguchi, (Kiguchi, T., private communication). HfTaO films with Hf:Ta of 6 : 4 have been crystallized after annealing at 900 °C. Although XRD results by Yu *et al.* have demonstrated that an HfTaO film annealed at 900 °C is still amorphous [53], we assume that the HfTaO film annealed at 900 °C has been already crystallized in our experiment and thus more defects exist in the film. Figure 15.13b shows the leakage current versus gate voltage (*J*–*V*) characteristics of the as-deposited and postannealed HfTaO films. It can be seen from this figure that the leakage current has been effectively reduced after annealing. It should be noted that the leakage current in the HfTaO film with 900 °C annealing is larger than that of the film annealed at 800 °C when the applied electric field is below 4 MV/cm. This result also

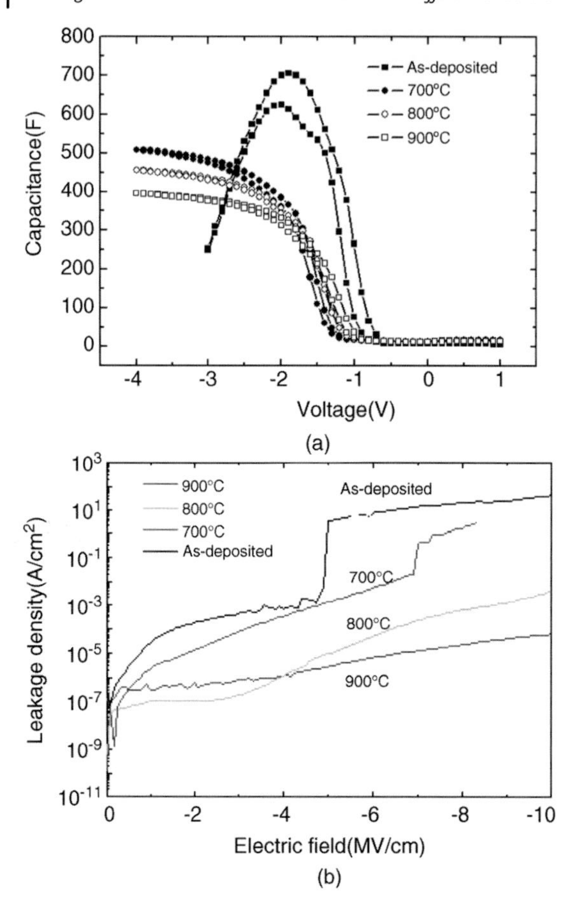

Figure 15.13 *C–V* characteristics (a) and *J–V* characteristics (b) of an as-deposited HfTaO film and annealed HfTaO films at 700, 800, and 900 °C in O_2 for 1 min. Reprinted with permission from Ref. [55]. Copyright 2008, American Institute of Physics.

suggests the crystallization of the HfTaO film at 900 °C. In crystalline HfTaO films, there may be more defect levels, which are close to the conductance band and act as the path for Poole–Frenkel emission current under a low electric field. However, the conduction mechanism is dominated by direct tunneling or Fowler–Nordheim tunneling under a high electric field, which is mainly affected by the barrier height and film thickness. Because of this mechanism, the leakage current in the film annealed at 800 °C becomes larger in the high electric field region.

Figure 15.14a shows typical *C–V* characteristics of 2, 4, and 6 nm thick HfTaO films deposited at RT, all of which were annealed at 800 °C in O_2 for 1 min at the same time. From the accumulation capacitance of the *C–V* curves, the effective oxide thicknesses (EOTs) of the 2, 4, and 6 nm thick HfTaO films can be calculated (with quantum mechanical correction) to be 1.2, 1.6, and 2.2 nm, respectively. A small

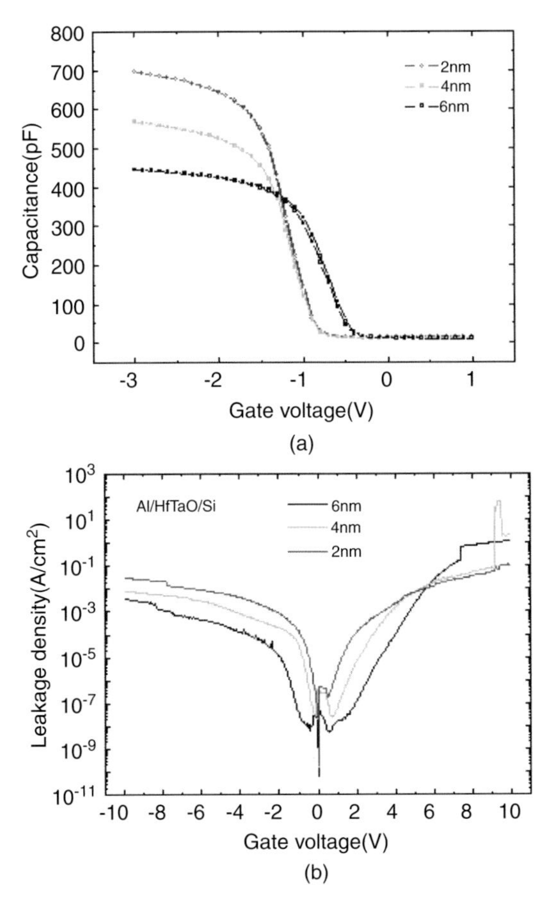

Figure 15.14 *C–V* characteristics (a) and *J–V* characteristics (b) of HfTaO films with 2, 4, and 6 nm physical thicknesses, all of which were annealed at 800 °C in O_2 for 1 min. Reprinted with permission from Ref. [55]. Copyright 2008, American Institute of Physics.

hysteresis loop of 15 mV width can be seen in the 6 nm thick HfTaO film. On the contrary, almost no hysteresis loop can be observed in the 2 and 4 nm thick HfTaO films. Figure 15.14b shows *J–V* characteristics of the same samples illustrated in Figure 15.14a. The leakage current increases with the decrease in the HfTaO film thickness. Some small current peaks can be seen around the gate voltage of 0 V, which may be due to some defect energy levels still existing in the films.

Summarizing the *C–V* and *J–V* characteristics in Figures 15.13 and 15.14, the leakage current versus EOT characteristics of the HfTaO films deposited in our work are shown in Figure 15.15. In this figure, the leakage current measured at a voltage shifted by 1 V from the flatband voltage ($V_{fb} - 1$ V) is plotted. As can be seen from the figure, the leakage current of the HfTaO film is significantly reduced, compared to that in SiO_2 films with the same EOT. In an HfTaO film with 1.6 nm EOT, at least

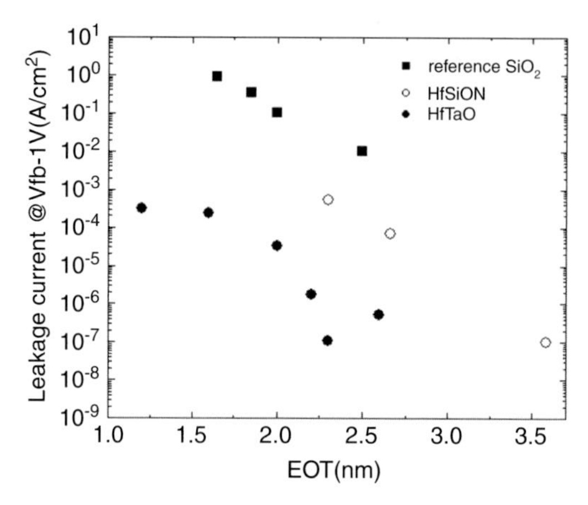

Figure 15.15 Relationship between leakage current and EOT in various HfTaO films. The leakage current is defined as a current at a voltage shifted by 1 V from the flatband voltage. Reprinted with permission from Ref. [55]. Copyright 2008, American Institute of Physics.

three-orders-of-magnitude reduction in leakage current can be seen, compared to the current in a 1.6 nm thick SiO_2 film. Moreover, it is much lower than that in HfSiON films deposited by electron beam evaporation in our work, and comparable to the leakage current in amorphous high-*k* rare-earth dysprosium scandate films annealed at 800 °C [56]. These results demonstrate excellent film quality of HfTaO deposited by electron beam evaporation.

If we assume that an interfacial SiO_2 is formed during postdeposition annealing, the relationship between EOT and the physical thickness of the HfTaO film can be expressed by a simple equation, as shown below:

$$\mathrm{EOT} = t_{\mathrm{SiO_2}} + 3.9/\varepsilon_r \times t_{\mathrm{HfTaO}}, \tag{15.1}$$

where t_{HfTaO} and $t_{\mathrm{SiO_2}}$ are the physical thicknesses of HfTaO and interfacial SiO_2, respectively, and ε_r is the relative dielectric constant of HfTaO films. The result is shown in Figure 15.16. From the above-mentioned equation, the interfacial SiO_2 thickness and the dielectric constant of HfTaO are determined to be 0.7 nm and 17, respectively. This dielectric constant value is a little lower than that reported by Yu *et al.* [53]. This difference may be due to the differences in the film composition and structure by different deposition methods. This value (17) is still much higher than that of HfSiO (~10) and HfSiON (12) [57].

15.3.5
Electrical Characteristics of Pt/SBT/HfTaO/Si Diodes

The Pt/SBT/HfTaO/Si MFIS diodes were fabricated, in which the thicknesses of the SBT and HfTaO were 300 and 4 nm, respectively. Figure 15.17 shows a cross-sectional TEM image in the vicinity of FIS structure, which shows that a SiO_2 layer of approximately 5 nm thickness has been formed between the HfTaO film and Si

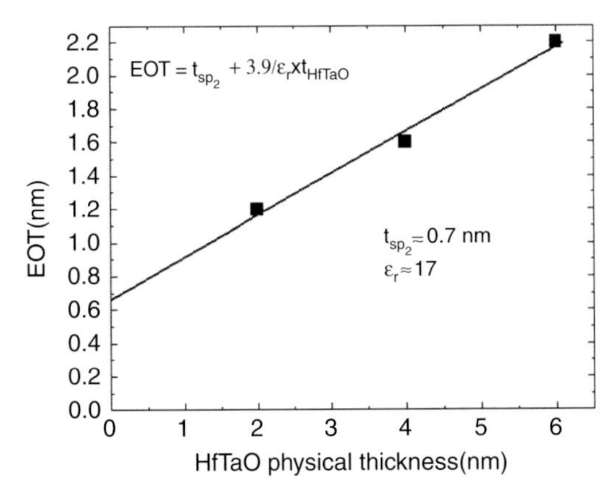

Figure 15.16 Relationship between EOT and HfTaO physical thickness. Reprinted with permission from Ref. [55]. Copyright 2008, American Institute of Physics.

substrate. Since the thickness of the interfacial SiO_2 layer derived from Figure 15.16 is 0.7 nm, we assume that the additional SiO_2 layer was formed during the deposition and crystallization processes of SBT films. Although the actual structure is Pt/SBT (300 nm)/HfTaO (4 nm)/SiO_2 (5 nm)/Si, we still use the expression of Pt/SBT/HfTaO/Si in the following discussions.

Figure 15.18 shows 1 MHz *C–V* characteristics of a typical MFIS diode. As can be seen from the figure, a memory window of 0.65 V was obtained during the back-and-forth sweeping between +4 and −4 V gate voltages. The clockwise hysteresis loop direction proves that the hysteresis characteristic is due to the ferroelectricity of the SBT film and not due to the trapped charges. The memory window width does not change under different voltage sweeping speeds in the range from 0.01 to 1 V/s, which also excludes the possibility of mobile ion effects on the hysteresis loop. In the subthreshold gate voltage region in a MOSFET, it takes about 0.1 V to change the drain current by a factor of 10. Thus, the current on/off ratio on the order of 10^6 can be expected from the obtained memory window width of 0.65 V, if the MFIS interface is as good as the MOS interface. Assuming this interface quality, the obtained memory

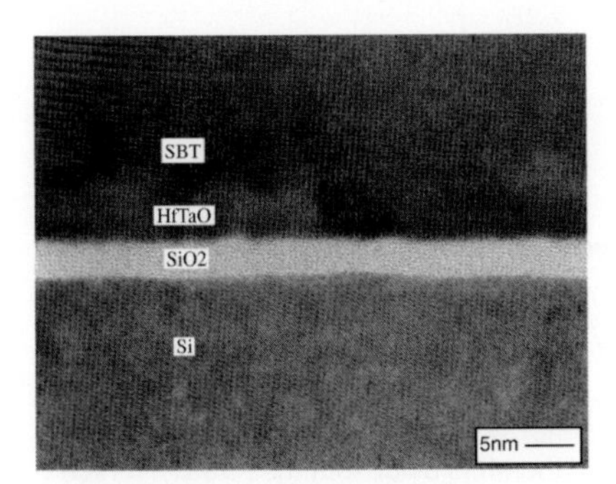

Figure 15.17 Cross-sectional TEM image for a SBT (300 nm)/HfTaO (4 nm)/Si (100) structure. Reprinted with permission from Ref. [55]. Copyright 2008, American Institute of Physics.

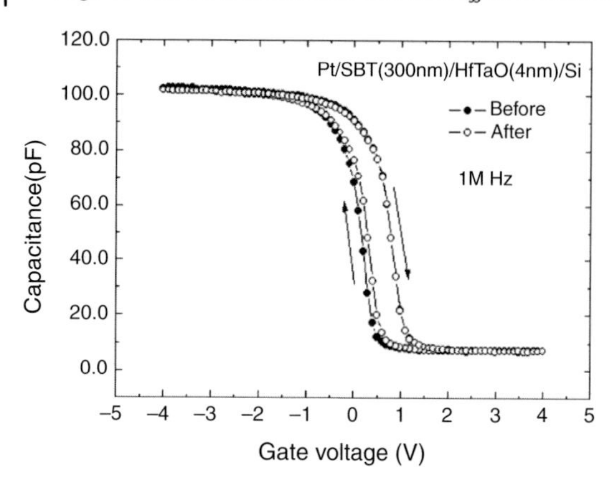

Figure 15.18 *C–V* characteristics of a MFIS diode with the Pt/SBT/HfTaO/Si gate structure. Closed and open circles represent the *C–V* curves before and after 24 h retention measurements, respectively. Reprinted with permission from Ref. [55]. Copyright 2008, American Institute of Physics.

window width and the relatively low writing voltage (4 V) are good enough for practical applications.

The long-term retention properties of the Pt/SBT (300 nm)/HfTaO (4 nm)/Si diode are shown in Figure 15.19. In this measurement, the "write" voltage of ± 5 V height and 100 ms width was applied to the sample. The time dependences of the high and low capacitance values were measured at a bias voltage of 0.5 V, at which a maximum difference between the high and low capacitances was observed in the *C–V* curve shown in Figure 15.18. (In practical applications, the threshold voltage of MFIS FETs is so adjusted that the center of the hysteresis loop is located at 0 V.) From Figure 15.19, it can be seen that the high and low capacitance values are clearly

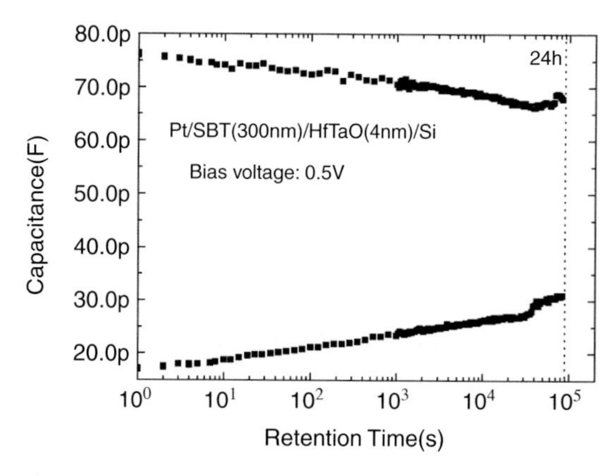

Figure 15.19 Long-term retention characteristics of the MFIS diode with the Pt/SBT/HfTaO/Si gate structure. Reprinted with permission from Ref. [55]. Copyright 2008, American Institute of Physics.

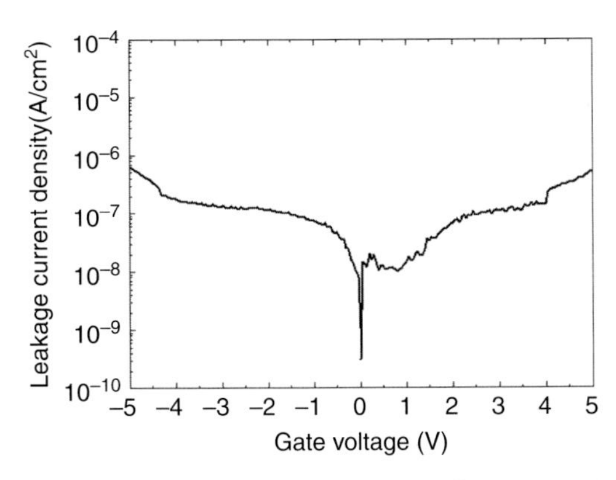

Figure 15.20 Leakage current properties of a Pt/SBT (300 nm)/HfTaO (4 nm)/Si MFIS diode. Reprinted with permission from Ref. [55]. Copyright 2008, American Institute of Physics.

distinguishable even after the retention measurement for 24 h. At the same time, only a slight decrease in the memory window was observed after the retention measurement, as also shown in Figure 15.18 with open circles. The long-term retention characteristics and *C*–*V* characteristics before and after the retention measurement demonstrate that good ferroelectric properties have been obtained for the MFIS diodes with HfTaO buffer layers. Figure 15.20 shows *J*–*V* characteristics of the same MFIS diode as illustrated in Figure 15.19, in which leakage current density lower than 10^{-6} A/cm^2 has been observed within the ±5 V measurement voltages. A small current peak in the positive sweeping voltage direction is considered to be due to the ferroelectricity of the SBT film.

15.3.6 Electrical Properties of MFIS FeFETs with HfTaO Buffer Layers

Finally, p-channel MFIS FET characteristics with the Pt/SBT/HfTaO/Si gate structure were investigated [58]. Figure 15.21a shows a schematic of the p-channel MFIS FET used in our study. The fabricated transistors have the same gate structure as that of the MFIS diodes. Figure 15.21b shows a typical drain current–gate voltage (I_{ds}–V_{gs}) characteristic of a device with *W*/*L* of 20/2 μm. The I_{ds}–V_{gs} characteristic exhibits a clockwise hysteresis loop, which is believed to be due to the polarization reversal of the SBT film. A memory window width of around 0.6 V and a drain current on/off ratio as high as 10^6 can be observed during the forward-and-reverse voltage sweeping from +4 to −4 V. The subthreshold slope derived from the figure is about 200 mV/decade.

Figure 15.22 shows the variation of drain on and off current with the writing pulse width for the same device illustrated in Figure 15.21. During the writing process, various pulse widths ranging from 10^{-7} to 10^{-3} s were applied to the sample while the

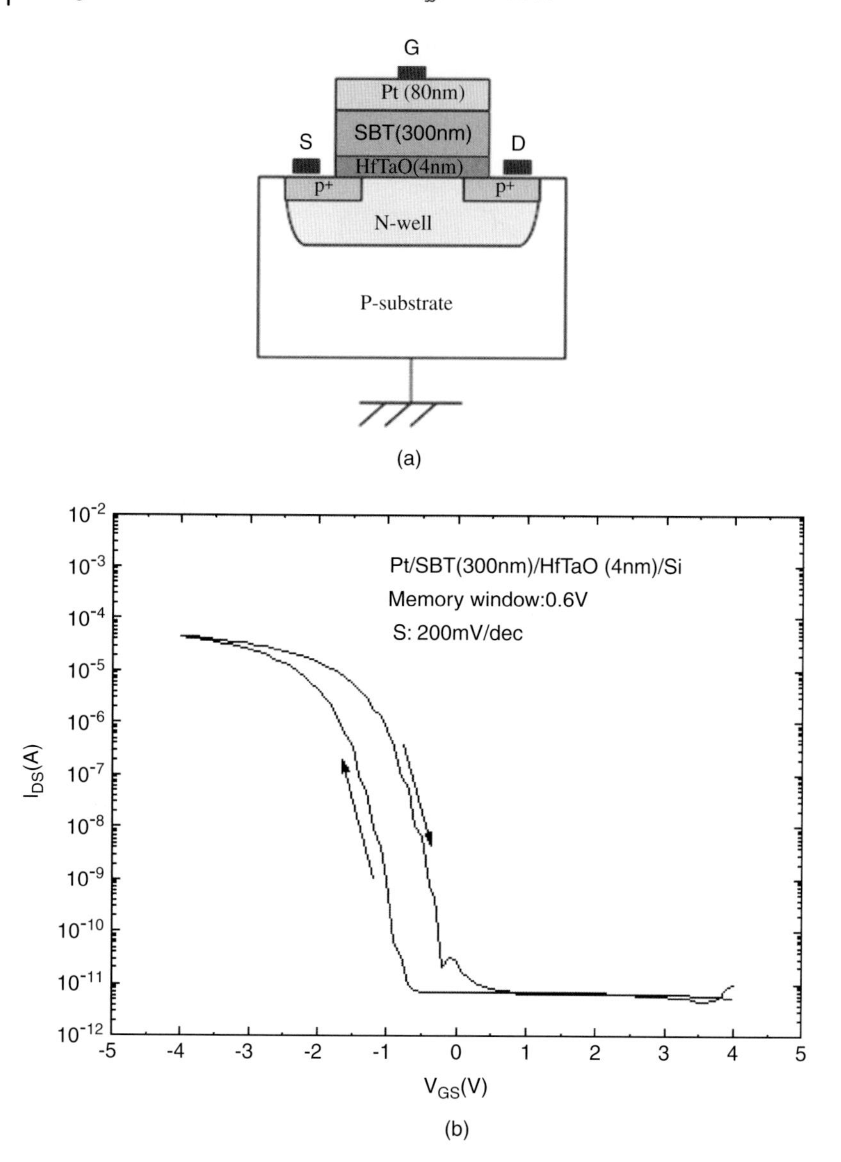

Figure 15.21 (a) A schematic diagram of a fabricated p-channel MFIS FET using HfTaO as the buffer layer. The channel width and length were 20 and 2 μm, respectively. (b) Typical I_{ds}–V_{gs} characteristic of a p-channel MFIS FET with the Pt/SBT (300 nm)/HfTaO (4 nm)/Si gate structure. Reproduced from Ref. [58]. Copyright 2008, IOP Publishing Ltd.

pulse amplitude was kept at +8 or −8 V. Then, the drain current was measured at a bias gate voltage of −0.5 V at a time of 30 s after the writing operation. In this measurement, the applied voltage was changed from −8 V (or +8 V) during the writing operation to −0.5 V during the data storage operation without returning to

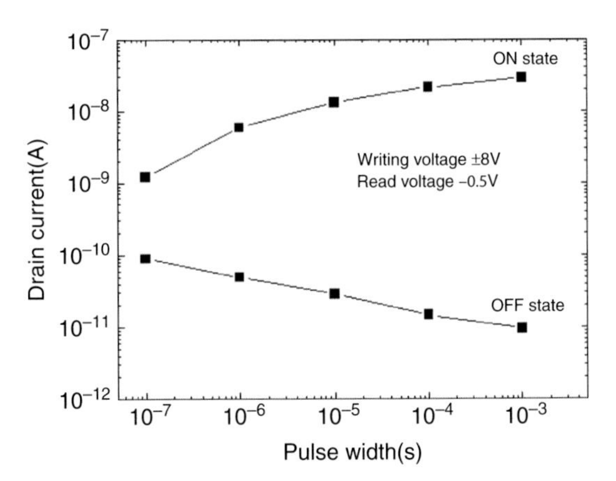

Figure 15.22 Writing pulse width dependences of the drain currents in the fabricated p-channel MFIS FET. Voltage pulses of +8 or −8 V in amplitude were applied to the sample. Reproduced from Ref. [58]. Copyright 2008, IOP Publishing Ltd.

0 V. As shown in the figure, the drain current on/off ratio increases with the increase in the pulse width, which implies that no significant charge injection occurs even after application of voltage pulses of ±8 V amplitude and up to at least 1 ms width. The maximum on/off ratio of 3000 was observed for pulses with ±8 V and 1 ms. Figure 15.23 shows retention characteristics of typical MFIS FeFETs on the same wafer as that in Figure 15.21a, in which a drain current was continuously monitored after applying a single writing pulse with 100 ms width and 8 V amplitude to the gate terminal. The applied gate voltage during retention measurement was fixed at −0.7 V. The expression of "on state" and "off state" exhibits the variation of the drain current after applying write pulses of −8 and +8 V, respectively. As can be seen from the figure, the reduction in the drain current on/off ratio is very small, even after over

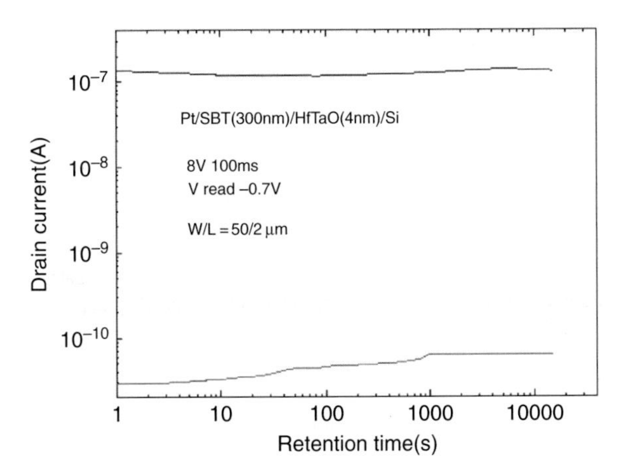

Figure 15.23 Retention characteristics of a Pt/SBT (300 nm)/HfTaO (4 nm)/Si MFIS FET. Reproduced from Ref. [58]. Copyright 2008, IOP Publishing Ltd.

10^5 s have elapsed at room temperature. It is believed from these results that the high film quality of HfTaO is responsible for the large drain current on/off ratio of MFIS FETs with HfTaO buffer layers. Considering the comparatively high subthreshold slope of 200 mV/decade of the present transistors, these properties will be further improved by optimizing the process parameters.

15.4
Summary

FeFET has the advantages of low-voltage, high-speed operation, high endurance, and nondestructive readout, and is one of the emerging memories to replace the conventional NAND flash memory for scaling to and below the 16 nm generation. High-*k* dielectric is necessary to be used in the MFIS structure as the insulator buffer layer to reduce the depolarization field effect, charge matching with ferroelectric film, and electrical thickness. In the past decades, various kinds of high-*k* materials have been studied as the buffer layers in the MFIS FeFETs, among which HfO_2-based high-*k* materials proved to be best buffer layer ever reported in the MFIS FeFETs. Although excellent retention properties were reported for HfO_2 and HfAlO FeFETs, the write/erase voltage is still high, and the operation speed is much slower than the theoretical predicted values. This is because the presently studied high-*k* buffer layers still suffer from some issues such as IL formation between Si and high-*k* buffer layer and interdiffusion between ferroelectric film and buffer layer, which limit the operation voltage and speed to a large degree. Therefore, developing new high-*k* buffer layers is still necessary to further improve the electrical properties of FeFETs.

In our work, ultrathin HfTaO films have been fabricated by electron beam evaporation-based molecular beam epitaxy. The high-quality HfTaO films demonstrated significant reduction in the leakage current compared to that in SiO_2 with the same EOT. The interfacial SiO_2 thickness and dielectric constant of the e-beam fabricated HfTaO films were determined to be 0.7 nm and 17, respectively. Small leakage current and long-term retention properties were observed for the MFIS diodes with Pt/SBT (300 nm)/HfTaO (4 nm)/Si gate structures. For the HfTaO FeFET, a reasonable memory window of 0.6 V and a drain current on/off ratio of 10^6 were observed for the voltage sweeping from +4 to −4 V, as well as their retention characteristics were excellent. The present results demonstrate that HfTaO is one of the most promising candidates used as the buffer layer in future ferroelectric gate FETs.

Acknowledgment

The author is very grateful to Professor Hiroshi Ishiwara at Tokyo Institute of Technology for his great support in the writing of this chapter. The author acknowledges the support from the University Talent Program of Guangdong Province, P. R. China(Grant No. C10314).

References

1 ITRS (2010) International Technology Roadmap for Semiconductors, http://www.itrs.net/reports.html.

2 Arimoto, Y. and Ishiwara, H. (2004) Current status of ferroelectric random-access memory. *MRS Bull.*, **29**, 823.

3 Miura, K., Kawahara, T., Takemura, R., Hayakawa, J., Ikeda, S., Sasaki, R., Takahashi, H., Matsuoka, H., and Ohno, H. (2007) A novel SPRAM (spin-transfer torque RAM) with a synthetic ferromagnetic free layer for higher immunity to read disturbance and reducing write-current dispersion. Proceedings of the International Symposium on VLSI Technology, p. 234.

4 Wuttig, M. and Yamada, N. (2007) Phase-change materials for rewriteable data storage. *Nat. Mater.*, **6**, 824.

5 Waser, R. and Aono, M. (2007) Nanoionics-based resistive switching memories. *Nat. Mater.*, **6**, 833.

6 Ouyang, J.Y., Chu, C.W., Szmanda, C.R., Ma, L.P., and Yang, Y. (2004) Programmable polymer thin film and non-volatile memory device. *Nat. Mater.*, **3**, 918.

7 Müller, R., Genoe, J., and Heremans, P. (2006) Nonvolatile Cu/CuTCNQ/Al memory prepared by current controlled oxidation of a Cu anode in LiTCNQ saturated acetonitrile. *Appl. Phys. Lett.*, **88**, 242105.

8 Ishiwara, H. (2009) Current status of ferroelectric-gate Si transistors and challenge to ferroelectric-gate CNT transistors. *Curr. Appl. Phys.*, **9**, s2.

9 Ross, I.M. (1957) Semiconductive translating device. U.S. Patent No. 2,791,760.

10 Han, J.P., Koo, S.M., Richter, C.A., and Vogel, E.M. (2004) Influence of buffer layer thickness on memory effects of $SrBi_2Ta_2O_9$/SiN/Si structures. *Appl. Phys. Lett.*, **85**, 1439.

11 Noda, M., Kodama, K., Kitai, S., Takahashi, M., Kanashima, T., and Okuyama, M. (2003) Basic characteristics of metal–ferroelectric–insulator–semiconductor structure using a high-*k* PrO_x insulator layer. *J. Appl. Phys.*, **93**, 4137.

12 Wilk, G.D., Wallace, R.M., and Anthony, J.M. (2001) High-*k* gate dielectrics: current status and materials properties considerations. *J. Appl. Phys.*, **89**, 5243.

13 Sakai, S. and Takahashi, M. (2010) Recent progress of ferroelectric-gate field-effect transistors and applications to nonvolatile logic and FeNAND flash memory. *Materials*, **3**, 4950.

14 Okuyama, M., Noda, M., and Yamashita, K. (1999) A low-temperature preparation of ferroelectric $Sr_xBi_{2+y}Ta_2O_9$ thin film and its application to metal–ferroelectric–insulator–semiconductor structure. *Mater. Sci. Semicond. Process.*, **2**, 239.

15 Kijima, T., Fujisaki, Y., and Ishiwara, H. (2001) Fabrication and characterization of Pt/(Bi, La)$_4$Ti$_3$O$_{12}$/Si$_3$N$_4$/Si metal ferroelectric insulator semiconductor structure for FET-type ferroelectric memory applications. *Jpn. J. Appl. Phys.*, **40**, 2977.

16 Noda, M., Kodama, K., Ikeguchi, I., Takahashi, M., and Okuyama, M. (2003) A significant improvement in memory retention of metal–ferroelectric–insulator–semiconductor structure for one transistor-type ferroelectric memory by rapid thermal annealing. *Jpn. J. Appl. Phys.*, **42**, 2055.

17 Choi, H.S., Lim, G.-S., Lee, J.-H., Kim, Y.T., Kim, S., Yoo, D.C., Lee, J.Y., and Choi, I.-H. (2003) Improvement of electrical properties of ferroelectric gate oxide structure by using Al_2O_3 thin films as buffer insulator. *Thin Solid Films*, **444**, 276.

18 Lu, X.B. and Ishiwara, H. (2008) Fabrication and characterization of metal–ferroelectric–insulator–Si diodes and transistors with different HfSiON buffer layer thickness. *J. Mater. Res.*, **23**, 2727.

19 Xie, D., Han, X., Li, R., Ren, T., Liu, L., and Zhao, Y. (2010) Characteristics of Pt/$BiFeO_3$/TiO_2/Si capacitors with TiO_2 layer formed by liquid-delivery metal organic chemical vapor deposition. *Appl. Phys. Lett.*, **97**, 172901.

20 Juan, T.P.-C., Chang, C.-Y., and Lee, J.Y.-M. (2006) A new metal– ferroelectric

($PbZr_{0.53}Ti_{0.47}O_3$)–insulator (Dy_2O_3)–semiconductor (MFIS) FET for nonvolatile memory applications. *IEEE Electron Device Lett.*, **27**, 217.

21 Shih, W.-C., Juan, P.-C., and Lee, J.Y.-M. (2008) Fabrication and characterization of metal–ferroelectric ($PbZr_{0.53}Ti_{0.47}O_3$)–insulator (Y_2O_3)–semiconductor field effect transistors for nonvolatile memory applications. *J. Appl. Phys.*, **103**, 094110.

22 Choi, H.S., Kim, E.H., Choi, I.H., Kim, Y.T., Choi, J.H., and Lee, J.Y. (2001) The effect of ZrO_2 buffer layer on electrical properties in Pt/$SrBi_2Ta_2O_9$/ZrO_2/Si ferroelectric gate oxide structure. *Thin Solid Films*, **388**, 226.

23 Takahashi, M. and Sakai, S. (2005) Self-aligned-gate metal/ferroelectric/insulator/semiconductor field-effect transistors with long memory retention. *Jpn. J. Appl. Phys.*, **44**, L800.

24 Park, B.E. and Ishiwara, H. (2001) Electrical properties of $LaAlO_3$/Si and $Sr_{0.8}Bi_{2.2}Ta_2O_9$/$LaAlO_3$/Si structures. *Appl. Phys. Lett.*, **79**, 806.

25 Fujisaki, S., Ishiwara, H., and Fujisaki, Y. (2008) Organic ferroelectric diodes with long retention characteristics suitable for non-volatile memory applications. *Appl. Phys. Express*, **1**, 081801.

26 Takahashi, K., Aizawa, K., Park, B.-E., and Ishiwara, H. (2005) Thirty-day-long data retention in ferroelectric-gate field-effect transistors with HfO_2 buffer layers. *Jpn. J. Appl. Phys.*, **44**, 6218.

27 Hirai, T., Teramoto, K., Nagashima, K., Koike, H., and Tarui, Y. (1995) Characterization of metal/ferroelectric/insulator/semiconductor structure with CeO_2 buffer layer. *Jpn. J. Appl. Phys.*, **34**, 4163.

28 Juan, T.P.-C., Lin, C.-L., Shih, W.-C., Yang, C.-C., Lee, J.Y.-M., Shye, D.-C., and Lu, J.-H. (2009) Fabrication and characterization of metal–ferroelectric ($PbZr_{0.6}Ti_{0.4}O_3$)–insulator (La_2O_3)–semiconductor capacitors for nonvolatile memory applications. *J. Appl. Phys.*, **105**, 061625.

29 Thomas, R., Melgarejo, R.E., Murari, N.M., Pavunny, S.P., and Katiyar, R.S. (2009) Metalorganic chemical vapor deposited $DyScO_3$ buffer layer in Pt/$Bi_{3.25}Nd_{0.75}Ti_3O_{12}$/$DyScO_3$/Si metal–ferroelectric–insulator–semiconductor diodes. *Solid State Commun.*, **149**, 2013.

30 Lu, X.B., Ishiwara, H., Gu, X., Lubyshev, D., Fastenau, J., and Pelzel, R. (2009) Characteristics of metal–ferroelectric–insulator–semiconductor diodes composed of Pt electrodes and epitaxial $Sr_{0.8}Bi_{2.2}Ta_2O_9$(001)/$SrTiO_3$(100)/Si(100) structures. *J. Appl. Phys.*, **105**, 024111.

31 Chen, J., Ho, K.S., Lin, J., and Rabson, T.A. (1995) Switching of ferroelectric films. *Integr. Ferroelectr.*, **10**, 215.

32 Hirai, T., Teramoto, K., Nishi, T., Goto, T., and Tarui, Y. (1994) Formation of metal/ferroelectric/insulator/semiconductor with a CeO_2 buffer layer. *Jpn. J. Appl. Phys.*, **33**, 5219.

33 Lim, M. and Kalkur, T.S. (1997) Electrical characteristics of Pt–bismuth strontium tantalate (BST)–p-Si with zirconium oxide buffer layer. *Integr. Ferroelectr.*, **14**, 247.

34 Yoshimura, T., Fujimura, N., Ito, D., and Ito, T. (2000) Characterization of ferroelectricity in metal/ferroelectric/insulator/semiconductor structure by pulsed *C–V* measurement: ferroelectricity in $YMnO_3$/Y_2O_3/Si structure. *J. Appl. Phys.*, **87**, 3444.

35 Park, B.E., Takahashi, K., and Ishiwara, H. (2004) Five-day-long ferroelectric memory effect in Pt/(Bi, La)$_4$$Ti_3O_{12}$/$HfO_2$/Si structures. *Appl. Phys. Lett.*, **85**, 4448.

36 Chiang, Y.W. and Wu, J.M. (2007) Characterization of metal–ferroelectric ($BiFeO_3$)–insulator (ZrO_2)–silicon capacitors for nonvolatile memory applications. *Appl. Phys. Lett.*, **91**, 142103.

37 Byun, C., Kim, Y.I., Lee, W.J., and Lee, B.W. (1997) Effect of a TiO_2 buffer layer on the CV properties of Pt/$PbTiO_3$/TiO_2/Si structure. *Jpn. J. Appl. Phys.*, **36**, 5588.

38 Tokumitsu, E., Nakamura, R., and Ishiwara, H. (1997) Nonvolatile memory operations of metal–ferroelectric–insulator–semiconductor (MFIS) FET's using PLZT/STO/Si(100) structures. *IEEE Electron Device Lett.*, **18**, 160.

39 Choi, K.J., Shin, W.C., Yang, J.H., and Yoon, S.G. (1999) Metal/ferroelectric/insulator/semiconductor structure of

$Pt/SrBi_2Ta_2O_9/YMnO_3/Si$ using $YMnO_3$ as the buffer layer. *Appl. Phys. Lett.*, **75**, 722.

40 Sze, C.Y. and Lee, J.Y.M. (2000) Electrical characteristics of metal–ferroelectric ($PbZr_xTi_{1-x}O_3$)–insulator (Ta_2O_5)–silicon structure for nonvolatile memory applications. *J. Vac. Sci. Technol. B*, **18**, 2848.

41 Won, D.J., Wang, C.H., and Choi, D.J. (2001) Characteristics of metal/ferroelectric/insulator/semiconductor using La_2O_3 thin film as an insulator. *Jpn. J. Appl. Phys.*, **40**, L1235.

42 Sakai, S. and Ilangovan, R. (2004) Metal–ferroelectric–insulator–semiconductor memory FET with long retention and high endurance. *IEEE Electron Device Lett.*, **25**, 369.

43 Aizawa, K., Park, B.E., Kawashima, Y., Takahashi, K., and Ishiwara, H. (2004) Impact of HfO_2 buffer layers on data retention characteristics of ferroelectric-gate field-effect transistors. *Appl. Phys. Lett.*, **85**, 3199.

44 Im, J.H., Jeon, H.S., Kim, J.N., Kim, J.H., Lee, G.G., Park, B.E., and Kim, C.J. (2009) Fabrication and characterization of Au/SBT/LZO/Si MFIS structure. *J. Electroceram.*, **23**, 284.

45 Tang, M.H., Sun, Z.H., Zhou, Y.C., Sugiyama, Y., and Ishiwara, H. (2009) Capacitance–voltage and retention characteristics of $Pt/SrBi_2Ta_2O_9/HfO_2/Si$ structures with various buffer layer thickness. *Appl. Phys. Lett.*, **94**, 212907.

46 Hubbard, K.J. and Schlom, D.G. (1996) Thermodynamic stability of binary oxides in contact with silicon. *J. Mater. Res.*, **11**, 2757.

47 Lu, X.B., Liu, Z.G., Wang, Y.P., Yang, Y., Wang, X.P., Zhou, H.W., and Nguyen, B.Y. (2003) Structure and dielectric properties of amorphous $LaAlO_3$ and $LaAlO_xN_y$ films as alternative gate dielectric materials. *J. Appl. Phys.*, **94**, 1229.

48 Kakushima, K., Tachi, K., Ahmet, P., Tsutsui, K., Sugii, N., Hattori, T., and Iwai, H. (2010) Advantage of further scaling in gate dielectrics below 0.5nm of equivalent oxide thickness with La_2O_3 gate dielectrics. *Microelectron. Reliab.*, **50**, 790.

49 Takahashi, M., Horiuchi, T., Wang, S., Li, Q.-H., and Sakai, S. (2008) Optimum ambient N_2 pressure during HfAlO pulsed-laser deposition in Pt/SBT/HfAlO/Si field effect transistors. *J. Vac. Sci. Technol. B*, **26**, 1585.

50 Fujisaki, Y. (2010) Current status of nonvolatile semiconductor memory technology. *Jpn. J. Appl. Phys.*, **49**, 100001.

51 Horiuchi, T., Takahashi, M., Li, Q.H., Wang, S.Y., and Sakai, S. (2010) Lowered operation voltage in $Pt/SBi_2Ta_2O_9/HfO_2/Si$ ferroelectric-gate field-effect transistors by oxynitriding Si. *Semicond. Sci. Technol.*, **25**, 055005.

52 Hirakawa, M., Hirooka, G., Noda, M., Okuyama, M., Honda, K., Masuda, A., and Matsumura, H. (2004) Nitridation of ultrathin SiO_2 layers in metal–ferroelectric–insulator–semiconductor structures. *Integr. Ferroelectr.*, **68**, 29.

53 Yu, X.F., Zhu, C.X., Yu, M.B., and Kwong, D.-L. (2004) Improvements on surface carrier mobility and electrical stability of MOSFETs using HfTaO gate dielectric. *IEEE Trans. Electron Devices*, **51**, 2154.

54 Yu, X.F., Zhu, C.X., Li, M.F., Chin, A., Du, A.Y., Wang, W.D., and Kwong, D.-L. (2004) Electrical characteristics and suppressed boron penetration behavior of thermally stable HfTaO gate dielectrics with polycrystalline-silicon gate. *Appl. Phys. Lett.*, **85**, 2893.

55 Lu, X.B., Maruyama, K., and Ishiwara, H. (2008) Characterization of HfTaO films for gate oxide and metal–ferroelectric–insulator–silicon device applications. *J. Appl. Phys.*, **103**, 044105.

56 Thomas, R., Ehrhart, P., Luysberg, M., Boese, M., Waser, R., Roeckerath, M., Rije, E., Schubert, J., Elshocht, S.V., and Caymax, M. (2006) Dysprosium scandate thin films as an alternate amorphous gate oxide prepared by metal-organic chemical vapor deposition. *Appl. Phys. Lett.*, **89**, 232902.

57 Visokay, M.R., Chambers, J.J., Rotondaro, A.L.P., Shanware, A., and Colombo, L. (2002) Application of HfSiON as a gate dielectric material. *Appl. Phys. Lett.*, **80**, 3183.

58 Lu, X.B., Maruyama, K., and Ishiwara, H. (2008) Metal–ferroelectric–insulator–Si devices using HfTaO buffer layers. *Semicond. Sci. Tech.*, **23**, 045002.

16
Rare-Earth Oxides as High-*k* Gate Dielectrics for Advanced Device Architectures

Pooi See Lee, Mei Yin Chan, and Peter Damarwan

16.1
Introduction

Aggressive scaling of sub-0.1 μm CMOS (complementary metal–oxide–semiconductor) devices poses significant leakage current and reliability concerns for SiO_2 in sub-20 Å thickness regime [1]. High-*k* dielectric materials ($k > 3.9$) are being intensively investigated to alleviate the scaling limitations. Different high-*k* materials have been studied as a potential alternative gate dielectric candidate, including HfO_2, ZrO_2, and Al_2O_3 [2, 3]. Recent works by research groups and chip manufacturers worldwide have demonstrated the use of hafnium-based high-*k* dielectric as a promising candidate as the first-generation high-*k* dielectric material. However, several material property considerations are critical to permit successful integration of high-*k* dielectrics with CMOS processing, including the dielectric permittivity and barrier height, interface property, thermal stability, carrier mobility, and process compatibility [4, 5]. Due to an improved thermal stability on Si expected from lanthanide-based high-*k* dielectrics, rare-earth oxides such as La_2O_3, Lu_2O_3, and Gd_2O_3 are also recently considered as the next-generation high-*k* materials.

Rare-earth oxides, also known as lanthanide oxides, have shown potential in various fields of technology. One of the unique characteristics of rare-earth elements is that their $6s^2$ shell is always occupied, with $5d^1$ configuration noted in La, Ce, Gd, and Lu. The 4f shell is filled gradually with increasing atomic number, where the extent to which the 4f shell is filled will give rare-earth oxides their unique characteristics [6]. $LaAlO_3$ has been shown as a promising candidate for its high dielectric constant, wide bandgap, and good thermal stability when in contact with Si [7, 8].

Rare-earth elements with half or fully filled 4f shell, in particular Gd and Lu, respectively, are expected to be stable. Another important characteristic of rare-earth elements is that all of these elements have the $+3$ oxidation state in the solid state. However, some of these elements such as Pr and Tm are also stable in the $+4$ and $+2$ oxidation states, respectively. The clear advantage of having the $+3$ oxidation state is that the rare-earth elements are likely to form unique stoichiometry, which

High-k Gate Dielectrics for CMOS Technology, First Edition. Edited by Gang He and Zhaoqi Sun.

leads to a simpler band structure as opposed to elements with more than one oxidation state that can lead to multiple stable stoichiometries.

16.2
Key Challenges for High-*k* Dielectrics

16.2.1
Interfaces Properties

One key challenge of high-*k* dielectric technology is the control of interface properties between the dielectric and Si substrate. Many high-*k* dielectric materials have unstable interfaces with Si, which forms an undesirable poor quality interfacial layer [1]. This introduces more defects at the substrate–dielectric interface, which act as additional charge trapping sites. Several simple oxides such as HfO_2 and ZrO_2 have been reported to have high oxygen diffusivities [9], resulting in uncontrolled interfacial layer formation due to rapid oxygen diffusion during the annealing treatment. Different interface engineering schemes have been studied, including the formation of an ultrathin silicon dioxide or oxynitride buffer layer as a barrier to minimize reaction between the high-*k* material and underlying silicon. However, the formation of series capacitance sets a limit on the minimum achievable EOT value, which poses challenges for scalability to later technology nodes. The requirement of high-quality oxide with minimal charge trapping at the interface between stacked dielectrics adds to the process complexity for future alternative gate dielectrics. This leads to the study of different material systems, including pseudo-binary oxides and lanthanide-based high-*k* dielectrics.

16.2.2
Thermal Stability

Many high-*k* dielectric materials crystallize to form polycrystalline films at relatively low temperature, resulting in detrimental leakage properties with the presence of grain boundaries. Studies on HfO_2 and ZrO_2 indicate the occurrence of crystallization upon moderate annealing treatment at $\sim$500 °C [9]. Hence, efforts to suppress the crystallization of high-*k* dielectrics were implemented, with the introduction of pseudo-binary material system such as $(HfO_2)_x(SiO_2)_{1-x}$ and $(ZrO_2)_x(SiO_2)_{1-x}$ as high-*k* dielectrics in field effect transistors [10]. The incorporation of SiO_2 into the high-*k* metal oxides provides improved thermal stability of hafnium and zirconium silicates, despite a trade-off on the high permittivity value. The addition of a capping layer on the high-*k* dielectric film also allows an increased crystallization temperature of the dielectric film stack [11]. To achieve a good compromise between the dielectric permittivity and thermal stability, recent works have also focused on the study of lanthanide-based high-*k* dielectrics that exhibit favorable thermal stability, including La_2O_3 [12, 13], Lu_2O_3 [14], Gd_2O_3 [15, 16], $LaAlO_3$ [17], and $LaLuO_3$ [18].

16.2.3
Fermi-Level Pinning

Other intrinsic issues for the integration of high-*k* dielectrics include Fermi-level pinning (FLP) at the high-*k*/poly transistors [19, 20], which causes threshold voltage instability and channel mobility degradation [21, 22] that slows down the device switching speed. The interfacial problems are associated with the presence of defects and high number of surface states (usually more than 10^{12} cm^{-2} eV^{-1}). At the top interface, interaction between the interfacial polysilicon and high-*k* metal bonds creates dipoles that pin the Fermi level of the gate electrode, resulting in substantial V_t shifts [20]. At the bottom interface, coupling of low-energy surface optical (SO) phonon modes arising from the polarization of the high-*k* dielectric to the channel charge carriers results in surface phonon scattering that severely degrades the channel mobility. The effect of high-*k* dipoles on FLP and mobility degradation can be alleviated with the introduction of metal gate electrodes [23]. The increased density of electrons allows screening of the surface phonons for improved channel mobility and reduced FLP effect. However, several challenges remain in the integration scheme for high-*k*/metal gate transistors [24].

16.2.4
Device Integration

Metal gate electrodes are desirable for the integration of high-*k* dielectrics due to better chemical stability and screening of surface phonons, reduced sheet resistance, and elimination of poly depletion effects. The key challenges for a prospective high-*k* metal gate (HKMG) scheme are to attain a good combination of high-*k* and metal gate for both PMOS and NMOS transistors, as well as thermal budget concerns. To achieve an optimal work function of metal gate electrodes for both NMOS and PMOS performance, different approaches include the adoption of a single mid-gap metal electrode and dual band edge work function metal gate electrodes [24]. The first approach affords a simpler CMOS processing scheme, but challenges remain in the search for gate electrode materials with the right work function and tolerance to high-temperature processing. In this context, Intel has taken the lead in introducing the second approach for its 45 nm generation HKMG technology, using two different metals with work functions matched to p-type and n-type substrates. To circumvent the thermal budget concerns of sustaining the source/drain activation anneals, a gate-last approach was adopted by depositing the gate electrode materials after the source/drain formation. The transistors feature high-*k* dielectric stack of ~10 Å EOT comprising 1.5–2.0 nm HfO/ZrO with a thin underlying SiO_2 interface. As shown in Figure 16.1, a replacement gate process was implemented, which involves the formation of high-*k*/TiN/polysilicon dummy transistors followed by polysilicon removal [24]. The gate trench was first filled with PMOS work function metal comprising tantalum and TiN. NMOS work function metal comprising TiAl is

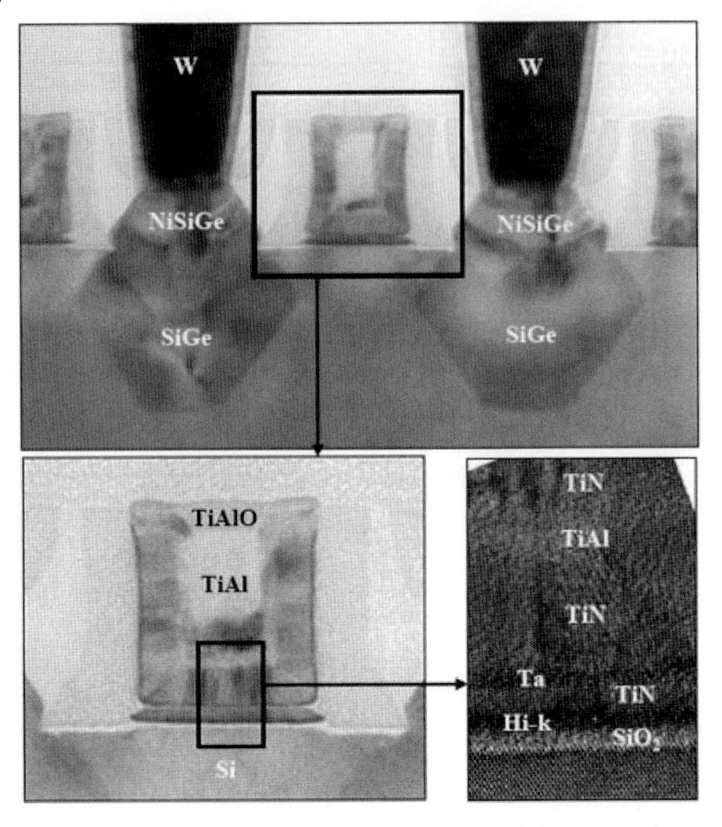

Figure 16.1 Intel 45 nm technology transistor device, which features a replacement high-*k* metal gate process. © 2008 IEEE. Reprinted, with permission, from Advanced Semiconductor Manufacturing Conference, IEEE/SEMI.

deposited after the removal of tantalum from NMOS areas, which provides favorable work function control after the thermal treatment.

On the other hand, IBM adopted a gate-first HKMG technology [25], featuring low-power devices with EOT of ~10–14 Å. The gate-first approach was developed by Sematech and the IBM-led Fishkill Alliance. It relies on very thin capping layers – Al_2O_3 for the PMOS and LaO_x for the NMOS transistors – to create dipoles that set the threshold voltage of the device [26]. By depositing different capping layers on the hafnium-based high-*k* dielectrics, V_t adjustment can be achieved via the interaction between La and Al atoms with the high-*k* dielectrics during the annealing treatment, as shown in Figure 16.2. However, as the device scaling continues at aggressive EOT, thermal instabilities can lead to significant threshold voltage shifts [27]. This poses challenges in achieving the optimum work function for lower threshold voltage, V_{th}, especially for PMOS devices, as the thermal budget gets tighter. For low power and memory technology, for example, DRAM applications, gate-first approach provides a cost-effective solution with better trade-off, since V_{th} and EOT requirements are more relaxed [28, 29]. However, as the technology proceeds beyond 28 nm, the need for low

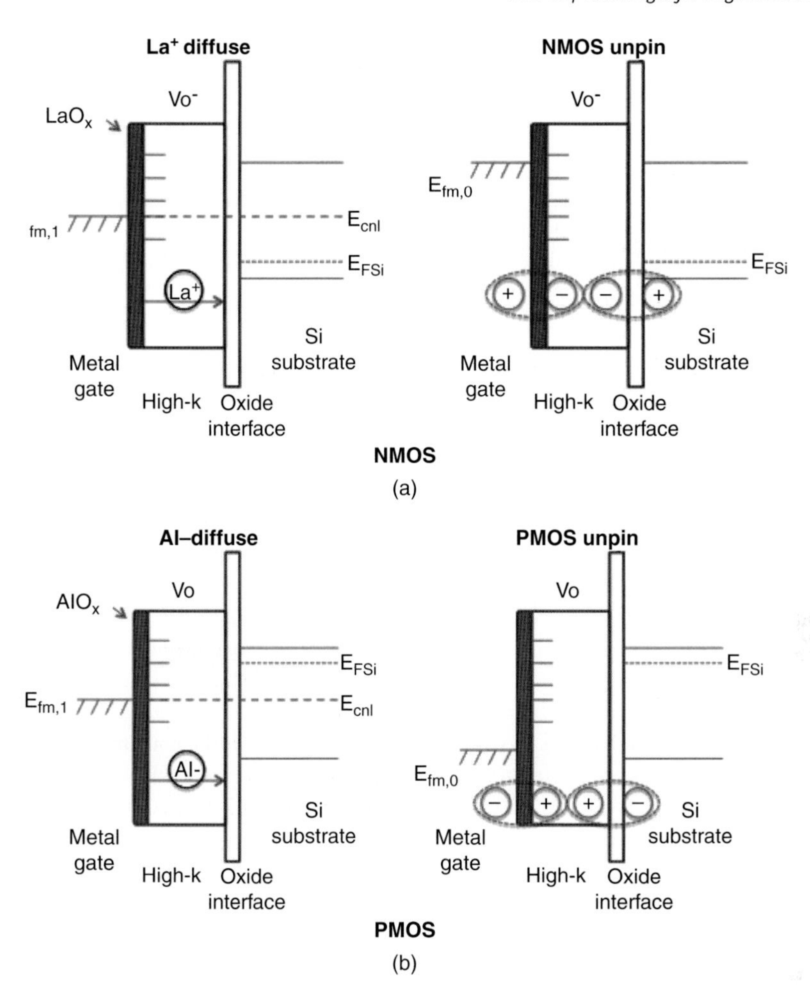

Figure 16.2 The effect of Fermi-level pinning with the metal work function, $E_{fm,1}$, pinned near the charge neutrality level, E_{cnl}, can be eliminated with different capping layers via dipole creation associated with the diffusion of (a) La atoms from LaO_x capping layer in NMOS device and (b) Al atoms from AlO_x capping layer in PMO device during the high-*k* stack anneal.

V_{th} and extra performance advantage makes gate-last approach a more viable option for both high-performance and low-power applications due to a better ability for work function control. Nevertheless, efforts to achieve better control of pFET variability have been carried out to allow for high-performance applications using the gate-first approach. Performance boost of pFET devices has been reported using epitaxial growth methods to create a thin layer of selectively grown SiGe for PMOS devices [30]. However, this adds to cost considerations of gate-first approach, despite the advantage over gate-last approach in terms of process complexity.

16.3
Rare-Earth Oxides as High-*k* Dielectrics

In the lanthanide group, the oxides in bulk form have been reported to exhibit a systematic decrease in relative permittivities from 14.8 to 12.4 for the cubic lanthanide oxides when moving from La_2O_3 to Lu_2O_3 [31]. However, similar trend was not observed in many different reports, mostly attributed to ultrathin films, differences in the impurities [32], method of deposition, deposition temperatures, or small thickness variations.

Among the key challenges for the application of rare-earth oxides as gate dielectrics, it is reckoned that rare-earth oxides easily break down the oxygen found in air or supplied during postannealing processes into its atomic oxygen. This leads to the formation of SiO_x or Si-based compounds [6, 33]. It has also been proven theoretically and experimentally that the SiO_2 or SiO_x compound formed would readily react with rare-earth oxides to form rare-earth oxide silicates. This is confirmed in the work of Marsella and Fiorentini in which they have calculated the formation enthalpies to determine the thermodynamical stability of some rare-earth oxides in contact with Si [34]. Their work concludes that La and Lu have been shown to be stable against the degradation into silica and metal component as well as formation of silicide. However, for La in particular, it is not stable against the formation of silicate, and this conclusion is very likely to apply to Lu as well, although it was not studied in the report [34]. The formation enthalpies are summarized in Table 16.1.

The high-*k* dielectric silicate may have been formed through the consumption of the SiO_2 layer. In the work of Stemmer, it has been shown that rare-earth oxides form silicates readily [35]. From the Gibbs free energy comparison for ZrO_2 and Y_2O_3, it was reported that the conversion into its silicate form by consuming a SiO_2 interfacial layer is more spontaneous at −135 kJ/mol for Y_2O_3 as compared to ZrO_2 at −2.96 kJ/mol [36]. In the case of Y_2O_3, it has been shown that Y_2O_3 and SiO_2 are reactive to form silicate layer and the silicate layer may react further with SiO_2 as shown in Equations 16.1 and 16.2 [35]:

$$Y_2O_3 + SiO_2 \rightarrow Y_2SiO_5 \tag{16.1}$$

$$Y_2SiO_5 + SiO_2 \rightarrow Y_2Si_2O_7 \tag{16.2}$$

Another issue to consider is that silicate layer is amorphous in nature, although the rare-earth oxide film has crystallized before the actual reaction [37, 38]. The formation

Table 16.1 Formation enthalpies (eV per formula unit) of oxides, silicides, and silicates.

Cation	Oxide	Silicide	Silicate
Y	−19.41	−2.96	−29.41
La	−18.84	−3.87	−29.74
Lu	−20.18	−2.30	—

The experimental formation enthalpy of silica is −8 eV [34].

of the amorphous layer has been attributed to the significant mobility differences of atoms in the film [39]. For example, in the case of Y_2O_3 and SiO_2, if Si in Y_2O_3 is able to diffuse freely as compared to Y in SiO_2 and/or Si, it would have caused Y_2O_3 to be supersaturated with Si, and subsequently, Y silicate should start to nucleate. However, it forms an amorphous silicate instead, if the nucleation barrier of crystalline silicate is high and also if the formation of the amorphous silicate leads to a large change in free energy [35]. In the following section, the main factors to be considered for the materials selection of rare-earth oxides as high-k dielectrics will be illustrated using the examples of lutetium oxide (Lu_2O_3) and gadolinium oxide (Gd_2O_3).

16.3.1 Lutetium Oxides as High-*k* Dielectrics

One of the most important considerations for selecting a high-k material is its thermal stability. Ideally, there should not be a formation of SiO_2 or SiO_x at the interface, which may lower the overall κ-value of the gate stack. A common method to determine the thermal stability is to subject the deposited high-k film to various rapid thermal annealing (RTA) temperatures. For example, 5 nm thick Lu_2O_3 films deposited by pulsed laser deposition (PLD) were subjected to RTA at temperatures ranging from 500 up to 900 °C with 60 s holding time in oxygen ambient. Various characterization tools such as the atomic force microscopy (AFM) and X-ray diffraction (XRD) were then employed to study the thermal stability of Lu_2O_3 on the Si substrate.

The AFM revealed a smooth topology across various annealing temperatures. The root mean square (rms) roughness value showed a downward trend with increasing annealing temperature as shown in Figure 16.3. At higher annealing temperature, the atoms are likely to be more mobile, thus causing a more efficient migration of the atoms as it attempts to alleviate defects leading to an enhancement in the film densification.

A rough interface either at the high-k/Si interface or at the high-k/metal gate interface is undesirable. Reported studies have shown that a rough surface at high-k/Si interface could lead to formation of scattering centers that will lead to reduction in device mobility [40]. Furthermore, a more recent study has shown that an increase in the high-k/metal gate interface roughness has led to a reduction in the dielectric constant as well as an increase in the leakage current density of the metal–oxide–semiconductor (MOS) stack structure [41].

The rms roughness of the 900 °C annealed sample was 0.24 nm, which is comparable to a thermally deposited SiO_2 film [42]. This result indicates that the deposited film is both smooth and uniform, without any distinct grain structure. Crystallizations in gate dielectric should be avoided as grain boundaries may form leakage path, which will be deleterious to the device as it not only increases the leakage current density of the device, but also causes instability in the device [43]. Lu_2O_3 has the largest lattice energy among the rare-earth oxides and was reported to have the tendency to form rigid crystals that lead to agglomeration and/or crystallization [43]. However, it was observed in this work that Lu_2O_3 maintains its

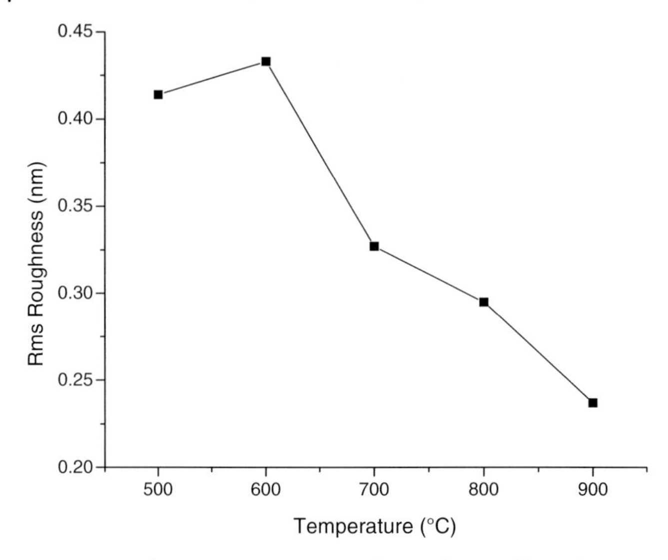

Figure 16.3 The root mean square roughness of Lu_2O_3 film with increasing annealing condition [14].

amorphous state and produces a relatively smooth surface even after RTA process at high temperature as seen in the high-resolution transmission electron microscope (HRTEM) micrograph shown in Figure 16.4. It was observed in the HRTEM micrograph that the interface between Si and Lu_2O_3 was abrupt, with little or no observable interfacial layer in between. This indicates that Lu_2O_3 can be a good candidate for high-*k* gate dielectric material as the film can remain thermally stable at

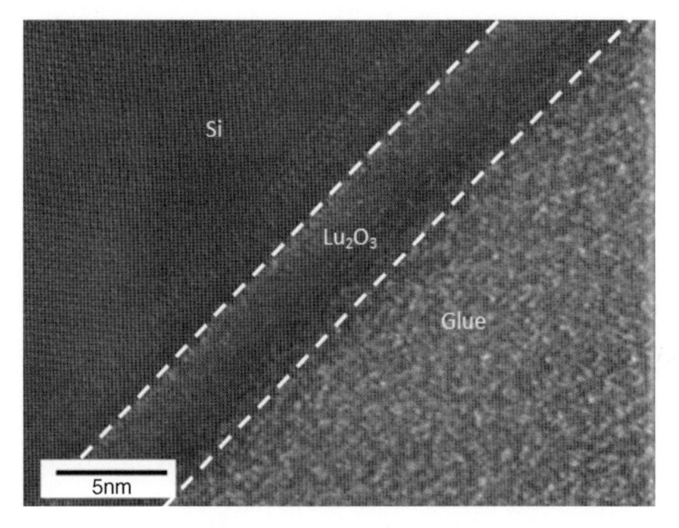

Figure 16.4 Cross-sectional HRTEM image of Lu_2O_3 films on n-type Si(100) revealing an amorphous structure after annealing at 900 °C in oxygen ambient [14].

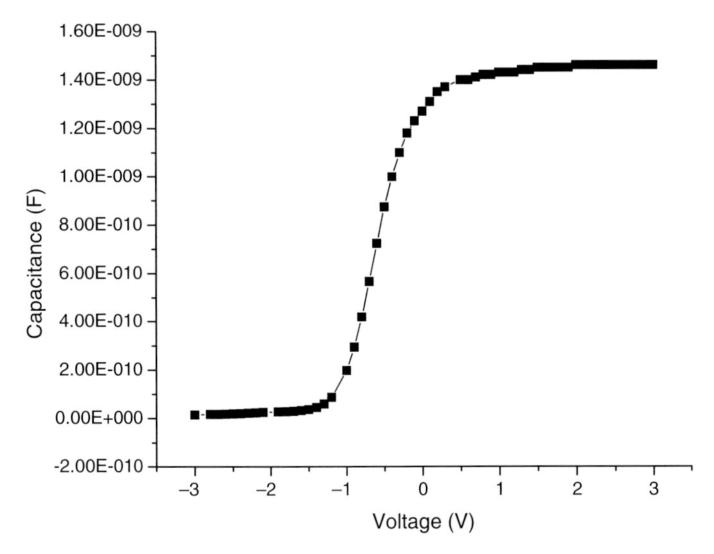

Figure 16.5 Capacitance–voltage characteristic of approximately 5 nm thick Lu_2O_3 subjected to rapid thermal annealing at 900 °C in oxygen ambient [44].

900 °C, which is near the processing temperature of CMOS device fabrication thermal annealing step.

In order to ascertain the suitability of Lu_2O_3 as a high-*k* dielectric film, a *C–V* measurement was performed on the Lu_2O_3 MOS capacitor. The Lu_2O_3 MOS capacitor was prepared by depositing ~5 nm thick Lu_2O_3 film onto an n-type Si(100) substrate using the PLD. This film was then subjected to RTA at 900 °C for about 1 min holding time in oxygen ambient. Following the RTA process, a 100 nm thick gold electrode (of area 7.07×10^{-4} cm^2) was sputtered on the post-annealed film. The *C–V* measurement on the Lu_2O_3 capacitor was conducted on a HP4284 LCR meter. In this measurement, a relatively high frequency of 100 kHz was used, along with a voltage sweep of −3 to 3 V bias in order to switch the capacitor from inversion to accumulation. The measured *C–V* curve for Lu_2O_3 MOS capacitor is shown in Figure 16.5. From the measured *C–V* curve, the dielectric constant of the deposited Lu_2O_3 film could be evaluated from the obtained accumulation capacitance value, as both thickness and the capacitor area are known.

Without taking into account the quantum mechanical tunneling, the dielectric constant κ could be calculated by applying Equation 16.3:

$$C_{ox} = \frac{\kappa \varepsilon_0 A}{t_{ox}}, \tag{16.3}$$

where C_{ox} refers to the maximum capacitance at accumulation derived from the high-frequency *C–V* measurement curve, κ is the relative dielectric constant of the high-*k* thin film, ε_0 refers to the relative permittivity of free space, A is the area of the capacitor, and t_{ox} pertains to the thickness of the high-*k* thin film. Therefore, solving Equation 16.3 for κ, a dielectric constant of 11.59 was obtained. This obtained experimental value is comparable to the value previously reported for Lu_2O_3 by other research groups (G. Scarel *et al.*) using ALD [45].

The EOT, which gives an indication of the equivalent SiO_2 thickness required in order to achieve the same capacitance level as the high-*k* dielectric, could also be calculated from the κ-value obtained earlier. The equation used to calculate the EOT is shown in Equation 16.4. In the equation, t_{eq} refers to the EOT that we are interested in, κ_{ox} refers to the standard SiO_2 dielectric constant, which in this case is approximately 3.9, $t_{high\text{-}k}$ refers to the thickness of Lu_2O_3 thin film, and $\kappa_{high\text{-}k}$ is the Lu_2O_3 dielectric constant that was calculated earlier:

$$\frac{t_{eq}}{\kappa_{ox}} = \frac{t_{high-k}}{\kappa_{high-k}}. \tag{16.4}$$

Therefore, inputting the parameters into Equation 16.4, an EOT of 1.68 nm was obtained for the 5 nm thick film. The calculated EOT in this experiment was lower than that previously obtained by another work on Lu_2O_3 with similar thickness [43], but annealed at lower annealing temperature of 600 °C. The result suggested that the film deposition method and postdeposition process chosen in this experiment seemed to provide good insulating properties for Lu_2O_3. Having a low EOT is not sufficient to qualify a material as a suitable high-*k* gate dielectric. Other factors such as the leakage current density and interface trap density must also be considered.

Another important information that we can extract from the *C–V* data is the flatband voltage (V_{FB}). The flatband voltage (V_{FB}) refers to the voltage that is required to bring the Fermi level of the electrodes and substrate, which is aligned by an energy difference into alignment. According to Wilk *et al.*, substantial high-*k* films reported in the literature showed a flatband shift away from the ideal *C–V* curve, which could cause significant issues for CMOS applications. The observed shift is undesirable and must be minimized given the scaling limitation on applied voltages due to power consumption. The gold electrode has an accepted work function of 5.31 eV, while the work function of n-type silicon with doping concentration of $\sim 1 \times 10^{16}\ cm^{-2}$ is reported to be 5.13 eV [46], which gives a work function difference of 0.18 eV. The amount of the fixed charge present in a high-*k* dielectric film can be related to the measured V_{FB} value as shown in Equation 16.5, where ϕ_{ms} refers to the work function difference between the metal gate and the semiconductor, Q_f refers to the fixed charges in the high-*k* dielectric film, and C_{acc} refers to the accumulation capacitance:

$$V_{FB} = \phi_{ms} \pm \frac{Q_f}{C_{acc}}. \tag{16.5}$$

If there is no fixed oxide charge present in the film, the flatband voltage of the oxide should be equal to the work function difference. Assuming an ideal flatband condition at $V = 0$, presence of positive fixed charges ($+Q_f$) in the high-*k* film would cause a negative flatband voltage shift, while presence of negative fixed charges ($-Q_f$) in the film would cause the flatband voltage to shift in the positive direction. The V_{FB} can be estimated from the *C–V* curve by first evaluating the estimated flatband capacitance C_{FB} using the equation proposed by Motorola as shown in Equation 16.6 [47]:

$$C_{FB} = C_{min} + 0.66(C_{max} - C_{min}), \tag{16.6}$$

where C_{min} refers to the minimum capacitance in the inversion region of the C–V curve and C_{max} is the maximum capacitance in the accumulation region.

By applying Equation 16.4 on the 900 °C annealed Lu_2O_3 sample, a flatband capacitance of 9.69×10^{-10} F is obtained. Relating these data on the C–V curve, the corresponding flatband voltage was found to be −400 mV. This shift is away from the ideal V_{FB}, which indicates presence of positive fixed charges (Q_F) in the Lu_2O_3 film, which resulted in a required applied voltage of −400 mV to achieve a flatband condition. The shift, although normally interpreted as presence of fixed charge in the film, could also arise from oxide damage associated with gate electrode deposition, as well as other forms of deposition treatment. However, considering that large ΔV_{FB} values have been measured by many independent groups using different processing conditions and electrodes, it is currently attributed to fixed charge within the film [40]. The origin of the fixed charge, although not always positive, is believed to arise from excess Si (trivalent Si) and/or the loss of electron from excess oxygen center (nonbridging oxygen) near the Si/high-k dielectric interface [40]. Therefore, due to the presence of defects at the Si interface, there are some localized states within the forbidden energy gap of Si because the silicon surface is where the properties of the bulk Si associated with its periodicity terminate. These surface interfaces trap charges causing the surface potential to bend down below the Fermi level, resulting in the V_{FB} shift [40].

The fixed charge is undesirable because large fixed charges in the dielectric film could cause the threshold voltage V_{th} to become too large in order to be compensated with dopant implants. In addition, although it is still debatable whether the observed V_{FB} arises from fixed charges within the high-k dielectric or some other phenomenon, a large amount of fixed charge could have an enormous influence on the viability of their insertion into a CMOS processing. If these high-k dielectrics have indeed been proven to contain significant amount of fixed charges, then the sign of the charge would become very important as it could affect the device performance [40]. Studies are still on going to determine if indeed there are substantial fixed charges on these films and if so, how to remove them or make them more manageable.

In relation to the V_{FB}, the interface trap density (D_{it}) could be calculated out together with the information obtained from the conductance (G) curve, which was also obtained simultaneously during the C–V measurement. The D_{it} is evaluated by the density of the charge states located at the mid-gap of the forbidden energy band. The trapped charges would affect not only the C–V characteristics, but also the device. By trapping electrons and holes, the surface states are able to reduce the drive current in MOSFETs, in which the trapped electrons and holes located at the interface will act as charge scattering centers that would lower the mobility of the mobile carriers traveling in the surface channel. In addition, interface states can behave like localized generation–recombination centers that may give rise to generation–recombination leakage currents and affect the gate dielectric reliability. The D_{it} near the mid-gap energy can be determined by taking into account the conductance–voltage characteristics of the MIS capacitor. The interface trap density can then be evaluated using the equation as proposed by Hill and Coleman, shown in Equation 16.7 [48]:

$$D_{it} = \frac{(2/qA)(G_m/\omega)}{(G_m/\omega C_{max})^2 + [1-(C_m/C_{max})]^2}, \tag{16.7}$$

where q corresponds to electronic charge, A refers to the area of the capacitor, G_m is the maximum conductance, C_m refers to the corresponding capacitance at the maximum conductance, C_{max} is the maximum capacitance at the accumulation region, and ω is the angular frequency.

Therefore, applying the equation as proposed by Hill and Coleman, a D_{it} of 2.66×10^{13} cm^{-2} eV^{-1} was obtained for the 5 nm Lu_2O_3 on n-type Si(100) after annealing at 900 °C in oxygen ambient. The calculated value of the interface trap density is about an order of magnitude higher than the reported value for Lu_2O_3 deposited by atomic layer deposition (ALD), which typically has a D_{it} value of $0.9–1.0 \times 10^{12}$ cm^{-2} eV^{-1}. Comparing to other deposition methods, a thermally evaporated Al_2O_3 was reported to produce a D_{it} of $>10^{11}$ eV^{-1} cm^{-2} [49]. Y_2O_3/Gd_2O_3 film deposited using molecular beam epitaxy (MBE) produces a $D_{it} < 10^{12}$ eV^{-1} cm^{-2}, while a sputtered Zr–Al–Si film produced a D_{it} in the range of $1–5 \times 10^{11}$ eV^{-1} cm^{-2} [50, 51]. The higher D_{it} value reported in this work was probably due to the inherent lower quality of the film deposited by the laser pulsed deposition as compared to the ALD, for example, in which layer by layer atomic deposition could be controlled precisely, thereby eliminating or significantly reducing possible defects and/or vacancy at the interface that could result in the interface trap contribution. The calculated D_{it} for the 600 °C falls in the range of 10^{13} cm^{-2} eV^{-1}.

Charge trapping issue is also an important issue for consideration. This is because in a gate stack we are likely to use different material interfaces, which have been known as charge trapping sites [52]. Unlike SiO_2 gate dielectric that is typically thermally grown, high-k films are deposited on the substrate and therefore have a higher tendency for trapping sites, even more so if crystallization occurs as well. Therefore, in addition to the κ-value calculation, the hysteresis of Lu_2O_3 was also investigated. One of the simplest ways to indicate charge trapping in a MOS capacitor structure is the hysteresis measurement from the $C–V$ curve by using trace–retrace voltage sweep during the measurement [52]. In the Lu_2O_3 hysteresis measurement in this work, the measurement was made starting with negative voltage on the gate (inversion mode) sweeping to positive voltage (accumulation mode) while recording the forward $C–V$ trace and proceeding by sweeping the trace in the opposite direction to record the backward trace.

Figure 16.6 showed the hysteresis $C–V$ sweep of Lu_2O_3 thin film sample annealed at 900 °C in oxygen ambient. A fairly small hysteresis of 48 mV was obtained for the 900 °C annealed sample in the oxygen ambient. It is observed that the backward sweep is more negative as compared to the forward sweep. This indicates a

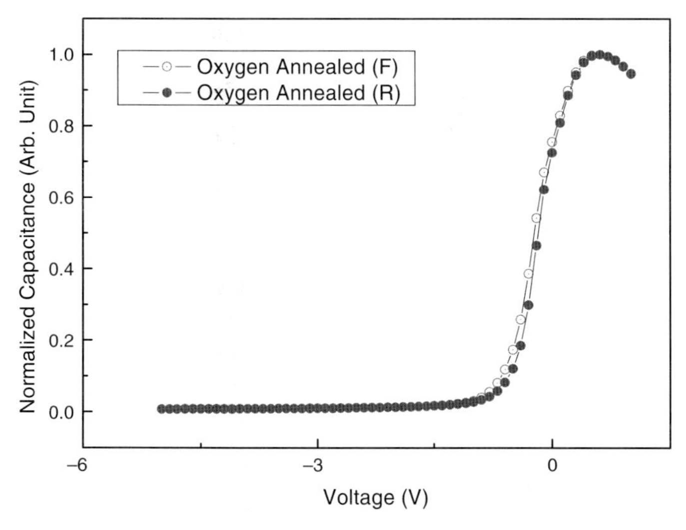

Figure 16.6 Normalized *C–V* curves for Lu_2O_3 samples annealed at 900 °C in oxygen ambient. (F) denotes forward voltage sweep, while (R) denotes reverse voltage sweep.

counterclockwise hysteresis, in which it is associated with the trapping of positive charge in the area near the high-*k* oxide and silicon interface.

In addition to the *C–V* measurement, current density–voltage (*J–V*) measurement is also typically conducted in the high-*k* material research. The leakage current density–voltage plot of the Lu_2O_3film (~5 nm thick) is shown in Figure 16.7. The leakage current density at the off state of n-type Si substrate at +1 V bias was found to

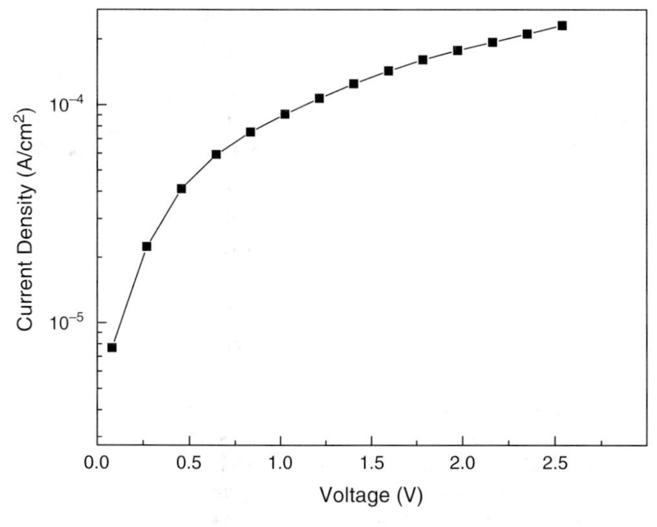

Figure 16.7 *J–V* characteristics of ~5 nm thick Lu_2O_3 film subjected to rapid thermal annealing at 900 °C in oxygen ambient [14].

be 1.0×10^{-4} A/cm^2. This is lower than the value reported by other research group with similar Lu_2O_3 thickness [43]. The lower leakage current density reported in this work is attributed to the better defect alleviation and film densification process resulting from the rapid thermal anneal along with an overall smoother surface, thereby reducing the leakage current paths from both metal/high-*k* dielectric interface and high-*k*/Si interface. In addition, as the sample was annealed in oxygen ambient, the processed Lu_2O_3 film is likely to have a much reduced oxygen vacancy, thus contributing to the lower leakage current density. Furthermore, as mentioned in the previous section, in the presence of O_2, there is a high chance of the formation of silicate layer. This silicate layer would have contributed to the reduction in the leakage current density as it acts as a barrier layer. Note that the leakage current density reported in this work is several orders of magnitude lower as compared to thermally grown SiO_2 with similar EOT, which is highly desirable for high-*k* gate dielectric application.

16.3.2
Gd_2O_3 as High-*k* Dielectric

Gd_2O_3 is thermodynamically stable in contact with Si, it has a considerably larger bandgap of $E_g = 5.9$ eV, which is favorable for low leakage current, and the *k*-value is reported to be about 14 when grown using ultrahigh vacuum vapor deposition [53]. Crystalline and amorphous Gd_2O_3 films have been deposited using different methods [54–58]. It was found that $Ga_{2-x}Gd_xO_3$ mixed oxides ($k = 12$) grown by ultrahigh vacuum deposition formed an excellent dielectric with low interfacial state density on GaAs surfaces [59, 60]. The material properties of Gd_2O_3 thin films formed using pulsed laser deposition coupled with an *in situ* substrate heating during deposition were discussed here.

The cross-sectional TEM image of the *in situ* heated sample at 800 °C is shown in Figure 16.8a. Two amorphous interlayers (layer marked "2" and "3") were found with different contrasts shown in the TEM between the Si substrate and the top crystalline Gd_2O_3 film. The thickness of the top Gd_2O_3 layer is about 6 nm. The thicknesses of the two interfacial layers (marked as "2" and "3") are 2.5 and 2 nm, respectively. The secondary ion mass spectroscopy (SIMS) analysis is shown in Figure 16.8b. It shows that the first 6 nm of the film corresponds to Gd_2O_3; below the Gd_2O_3 film, the intensity hump of the Gd and Si signal shows the mixing of the Gd_2O_3 and SiO_x interlayers. Similar structure has also been found by others for the Gd_2O_3 film deposited on Si wafer after oxygen annealing [54, 55, 61].

Figure 16.9 shows the XPS spectrum of the *in situ* heated 800 °C samples. The profile of the spectrum shows the multiplet splitting of the 4d hole with the 4f valence electrons forming the ^7D and ^9D ionic states [62] with the ^9D Gd deconvoluted into five peaks and the ^7D Gd fitted using one Gaussian peak. The spectrum presents a shift of the peak components toward higher binding energies corresponding to oxide phase. O 1s can be deconvoluted into three peaks at 530.9, 532.1, and 533.7 eV, which correspond to Gd–O bond, Gd–O–Si bond, and Si–O bond, respectively. This shows

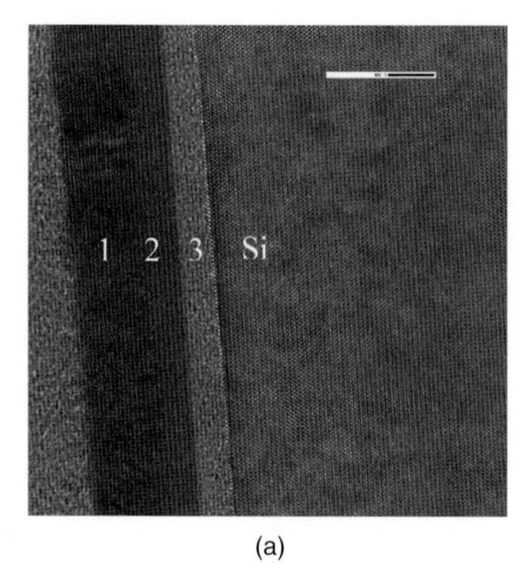

(a)

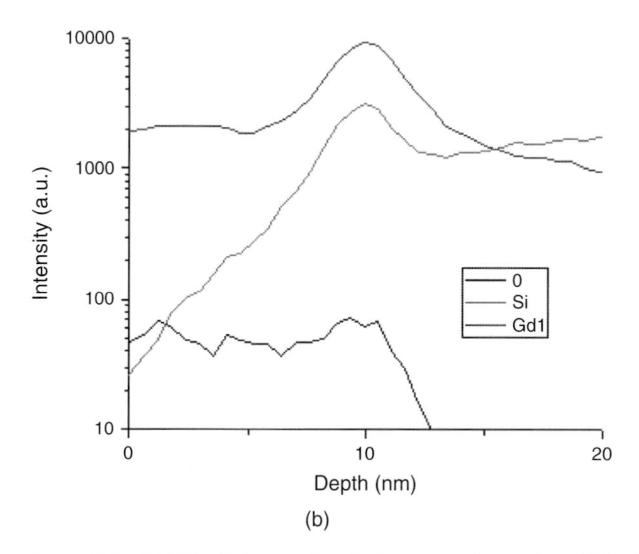

(b)

Figure 16.8 (a) HRTEM image of the *in situ* annealed samples at 800 °C. (b) SIMS analysis of the sample in part (a).

the coexistence of Gd_2O_3, SiO_2, and the silicate-like structure in these films. Detailed studies on the presence of hydroxide phase for rare-earth oxides can be conducted by using EELS [63].

The interfacial layer formation tends to reduce the dielectric constant due to the intermix layer silicates and oxides found between the crystalline Gd_2O_3 layer and Si substrate. Silicate has almost similar k-value to Gd_2O_3, and the SiO_2 layer will dominate the C–V measurement [62]. Having a postannealing RTP step instead of an *in situ* substrate heating step has been found to minimize the interfacial layer formation that improves the capacitance density as shown in Figure 16.10.

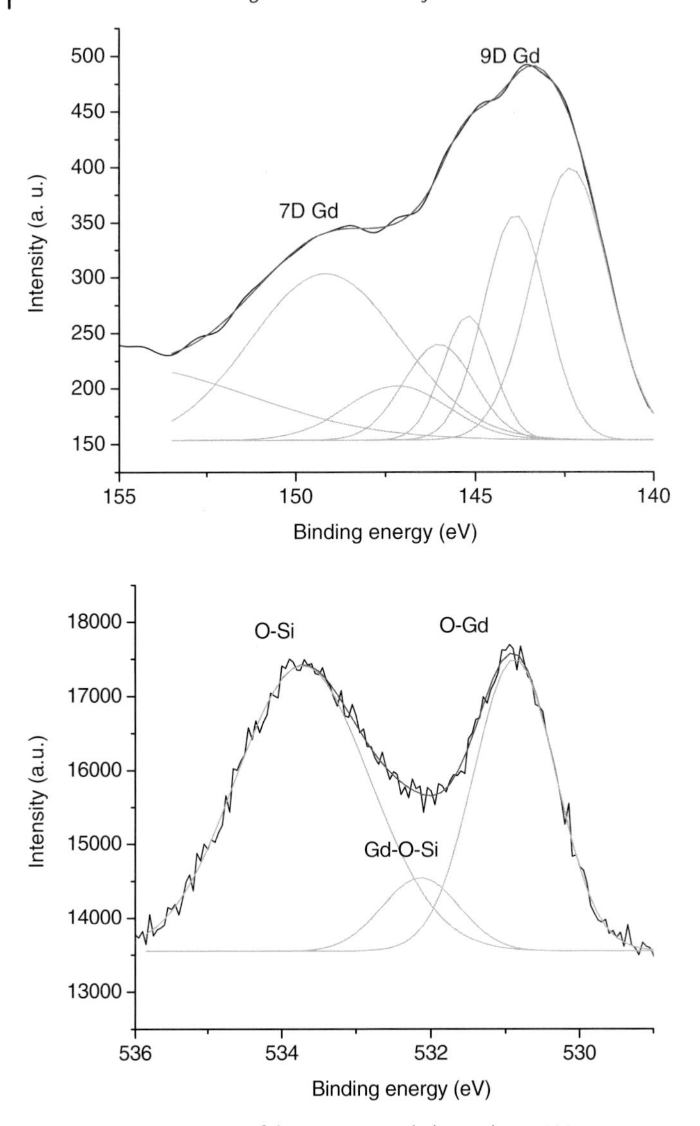

Figure 16.9 XPS spectra of the *in situ* annealed samples at 800 °C.

16.3.3
Summary

There are many ways to prevent the degradation of the layers after rare-earth oxide deposition, including capping with amorphous Si prior to postdeposition annealing. Molecular beam epitaxy growths of rare-earth oxides are able to provide abrupt interfaces as a model system to elucidate the critical processing integration issues; notably, the formation of a-Si/Gd_2O_3/c-Si and a-Si/Y_2O_3/c-Si gate stacks has been achieved [55].

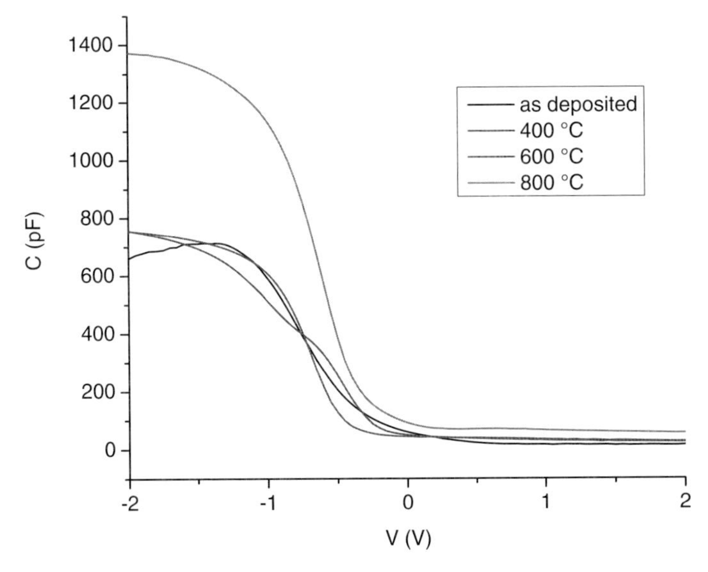

Figure 16.10 *C–V* curves of the rapid thermal annealed samples at various temperatures.

Table 16.2 Chemical composition and electronic band structure of rare-earth oxides.

Rare-earth oxide	Bandgap (eV)	Offset (eV)	
		C.B.	V.B.
La_2O_3	6.4	2.3	3.0
Gd_2O_3	6.4	3.1	2.2
Lu_2O_3	6.0	1.9	3.0

The composition and chemical structures have been analyzed in detail in the work by Hattori *et al.*, where the energy band discontinuities at rare-earth oxide/Si interfaces were determined by measuring the O 1s photoelectron energy loss and valence band spectra [56]. La_2O_3, Gd_2O_3, and Lu_2O_3 were found to be promising due to their relatively large energy barriers for electrons in Si substrates as shown in Table 16.2.

16.4
High-*k* Dielectrics in Advanced Device Architecture

16.4.1
HfO_2 Alloy with Rare-Earth and Bilayer Stacks

HfO_2 is considered as the most promising candidate as the first-generation high-*k* dielectric material due to the high dielectric constant (25–30), large energy bandgap

(~5.6 eV), high electron offset (1.5 eV), and high breakdown field (15–20 MV/cm). However, HfO_2 suffers from mobility degradation and threshold voltage instabilities resulting from oxygen vacancies [29, 30]. Another drawback of the material is the low crystallization temperature of the film of ~500–600 °C, which could lead to reliability issues due to dopant penetration through the film associated with grain boundary formation [64]. This leads to the study of new candidate materials for next-generation high-k dielectrics, with lanthanide-based rare-earth oxides, for example, La_2O_3 [13, 65], Lu_2O_3 [14], and Gd_2O_3 [16], being considered due to their high permittivity values ($k = 20$–27) and higher crystallization temperature than HfO_2. However, each individual material also exhibits undesirable properties, for example, hygroscopic properties of La_2O_3 that readily reacts with Si during high-temperature annealing [66, 67]. Thus, rather than switching to new materials, different pseudo-binary alloy systems have been considered as the most promising candidate dielectrics, which combine the desirable properties of different individual materials. Pseudo-binary oxides, for example, $(HfO_2)_x(SiO_2)_{1-x}$ and $(ZrO_2)_x(SiO_2)_{1-x}$, are being considered as potential high-k candidate materials as a compromise between high permittivity (k) and thermal stability during processing.

In order to provide substantial room for ultimate EOT scaling, much work has focused on the attempt to tweak HfO_2 with proprietary additives to achieve a zero interface layer technology [68]. Different issues need to be considered for the HfO_2 alloy system, including the dielectric permittivity value, thermal stability, threshold voltage control, and effect on carrier mobility. Recently, doping of transitional and rare-earth metal oxides, for example, La and Al into HfO_2, was demonstrated as a promising technique for future microelectronic devices [69, 70].

To enable V_{th} adjustment of PMOS transistors, the use of Al capping layer, for example, $TiN/Al_2O_3/HfO_2$ high-k/metal gate stack, has been studied by various groups [26, 71]. The Al-induced dipole layers resulting from the substitution of the Hf sites in the dielectric (Al_{Hf}) effectively increase the effective work function, which is favorable for meeting the PMOS threshold voltage requirements [71]. Despite having the advantages of improved thermal stability and reduction of interfacial layer growth with the incorporation of Al, several issues remain in achieving an optimum device performance. The Al-induced dipole and fixed oxide charges result in additional mobility degradation, which needs to be taken care for today's high-k/metal gate technology [72].

On the other hand, La-induced dipole layer has been introduced in today's high-k/metal gate technology for effective threshold voltage control of NMOS transistors [69, 73]. Gate dielectric stacks comprising La_2O_3 capping layer of <1 nm on top of HfO_2 dielectric exhibit potential for additional EOT scaling associated with high dielectric permittivity ($k \sim 30$) of La_2O_3. Upon annealing at high temperature, the La atoms diffuse into the high-k bulk, resulting in the formation of charge orientation, or dipole is formed near the substrate interface, which mitigates the Fermi-level pinning effect for V_{th} adjustment [74–77]. Besides that, passivation of oxygen vacancies by La dopants provides plausible improvement of PBTI reliability in NMOS devices [78–80]. The introduction of La_2O_3 also provides a promising approach to eliminate the interfacial layer in Hf-based high-k dielectrics for continued scaling beyond 22 nm node that requires EOT < 0.6 nm. This is associated with the high oxygen affinity properties of rare-earth elements, which act as an effective oxygen scavenging species from the interfacial layer [81, 82]. This enables aggressive EOT scaling of La-doped high-*k* dielectrics down to 0.42 nm, with low leakage current, adequate V_{th} control, and minimized mobility degradation [72, 82].

16.4.2 Advanced Device Architecture with High-*k* Dielectrics

16.4.2.1 High-*k* Dielectrics for Advanced CNT and Nanowire Devices

As fundamental physical limitations place conventional scaling limits on Si technology, different alternative device technologies have been studied for future electronic devices. One promising approach is the replacement of Si channel by alternative channel materials for mobility improvement, including Ge for pFET [83] and III–V materials such as InGaAs for nFET applications [84]. Recent studies have also demonstrated promising potential of one-dimensional channel devices, such as nanowire- or nanotube-based field effect transistors [85, 86] and graphene nanoribbon transistor [87] for the realization of high-speed devices with increased carrier mobility attributed to their unique transport properties [88].

In order to push the performance limits of electronic devices, advanced carbon nanotube (CNT) FET devices have been demonstrated with the integration of high-*k* dielectrics for improved electrostatic gate control [89–92]. The utilization of high-*k* dielectrics on CNTFET devices allows for reduced operation voltage, high on current with low leakage properties, and subthreshold swing close to the room temperature limit of 60 mV/decade [89, 90]. This is associated with the chemical inertness and lack of surface dangling bonds [93], which alleviates the mobility degradation effect of conventional transistor device as a result of carrier scattering at the high-*k*/Si channel interface. However, several challenges remain in the integration of high-*k* dielectrics on CNTFET devices. In terms of device architecture, top-gate [94] and bottom-gate approaches [95] have been employed for CNTFET devices.

Although the top-gate approach allows for high-performance device with low operating voltage, challenges remain in producing a uniform and conformal coating of dielectric layer on the nanotube. Atomic layer deposition of high-*k* dielectrics, including HfO_2 [93] and ZrO_2 [96], on CNT has been demonstrated using a relatively low-temperature process that does not destroy the electrical properties of the nanotubes with high composition uniformity. However, due to the chemical inertness of CNT, the ALD process is benign to nanotubes and poses limitations in producing a uniform and conformal coating of dielectric layer. Existing techniques rely on a substrate-assisted growth, which requires chemisorption between ALD precursor molecules and OH-terminated SiO_2 surface to achieve a uniform film surrounding the nanotube [93]. Relatively thick dielectric films (>8 nm) are needed to ensure sufficient coverage on the CNT for low leakage properties. This leads to studies on different functionalization techniques, including covalent functionalization [97] and noncovalent functionalization [90] of CNTs to achieve conformal ALD coating. The first technique has demonstrated conformal coating of ~10 nm on the nitro groups formed on the sp^3-modified CNT sidewalls, followed by annealing to recover the sp^2 structure and electrical properties of the CNT. The latter technique has next demonstrated conformal ALD of HfO_2 with a thickness down to 2 nm by utilizing noncovalent DNA functionalization of CNTs. This enables the formation of ultrathin dielectrics approaching the scaling limit with low subthreshold swing for advanced nanotube electronics [90].

Although CNTFET devices have been reported with unique carrier transport close to the ballistic limits [98], several issues remain to achieve reproducible properties for large-scale integration of nanotubes in CMOS technology [99, 100]. Nanowire FET (NWFET) device is viewed as a promising alternative, with significant work focused on radial core/shell nanowires for enhanced device performance [101–103]. Recent studies have reported high-performance NWFET devices comparable to similar length CNTFET devices, attributed to good interface properties with the passivation of surface states by controlled growth of shells surrounding the NW core [103, 104]. Ge/Si core/shell [102, 103] and InAs/InAs core/shell [101] NW heterostructures have been demonstrated for the realization of high-performance p-channel and n-channel NWFET devices, respectively, with enhanced quantum-confined transport properties. The devices were integrated with high-*k* dielectrics with minimal carrier scattering from the core/shell interface, including ALD of HfO_2 and Al_2O_3 with a thickness of ~4 nm. The conformal top-gate configuration with integral high-*k* dielectrics provides optimum electrostatic control for future high-speed, low-power nanoelectronic devices.

16.4.2.2 High-*k* Dielectrics for DRAM and Flash Memory Devices

Before the implementation in logic applications, high-*k* materials comprising metal–insulator–semiconductor (MIS) Ta_2O_5 [105] and Al_2O_3 [106] have been adopted in DRAM production. As the DRAM technology scales below 100 nm, there is a need to search for higher-*k* dielectrics to meet the storage capacitance requirements as the use of extreme topology structures for increased area is reaching a limit. High-quality Al_2O_3 films have been grown using atomic layer deposition as the first-generation DRAM capacitors with enhanced capacitance and low leakage behavior [106–108]. 1 Gbit DRAM was demonstrated with the films uniformly coated on the cylinder-type poly-Si to form a MIS structure with TiN top electrode [106].

In search for higher-*k* ($\varepsilon > 100$) dielectric films, perovskites, for example, $SrTiO_3$ (STO) [109] and (Ba, Sr)TiO_3 (BST) [109, 110], have been studied for DRAM capacitor applications. However, in order to achieve high dielectric permittivity, epitaxial growth of the dielectric requires submonolayer control of the interface [111]. Instead of the conventional MIS structure, metal–insulator–metal (MIM) structure is required to eliminate interfacial oxide formation. However, several issues remain including thermal stability and possible deoxidation of the MIM capacitor structure.

To extend the DRAM capacitor technology beyond 70 nm node, there is a need for new dielectrics with higher *k*-values as compared to today's Ta_2O_5 or Al_2O_3 capacitor, which meets the capacitor scaling requirements, with simultaneous reduction in leakage current. MIM capacitors compatible with DRAM flow with low EOT of 0.4 nm and low leakage properties have been reported using stack capacitors with ALD Sr-rich STO high-k dielectric and Ru bottom electrode. This was enabled using controlled ultrathin Ru oxidation and highly conformal ALD process with low thermal budget, which demonstrates the potential for DRAM scalability to the 3X nm node [112].

Recent works have also shown nanolaminates of HfO_x and other materials, for example, Al_2O_3–HfO_2 (AHO) [113], as a promising candidate, which combines the advantages of different dielectrics to give low leakage property and relatively thin EOT. Due to its process controllability and logic-friendly low-temperature process, this technology enables the formation of MIM capacitor cell for embedded DRAM (eDRAM) application beyond 70 nm generation. Recent progress has proven the formation of high-performance eDRAM coupled with 32 nm high-performance logic device based on HKMG constituting HfO_2 dielectric and TiN metal gate on SOI technology [114].

Significant progress in high-k dielectric studies has also provided new perspectives for flash memory technology. Besides short channel effects (SCEs), the diminishing number of electrons poses significant restrictions on the device lifetime, thus resulting in reduced charge loss tolerance to less than 10 electrons in 32 nm technology node [115]. Another significant scaling challenge involves maintaining adequate coupling of the control gate to the floating gate when the floating gate height reduced as the space between adjacent poly-Si gates shrink. Potential solutions to mitigate tunnel oxide scaling and coupling issue include the introduction of high-k dielectrics [116] and extension of the floating gate memory by charge trap memory devices utilizing nitride layer or nanocrystals as charge storage elements [117–119].

The most viable alternative for today's floating gate technology is the silicon–oxide–nitride–oxide–silicon (SONOS) charge trap flash (CTF) memory with discrete charges stored in a nitride layer sandwiched between the dielectric materials. The discrete charge storage mitigates the vulnerability of charge loss through localized oxide defects, thus improving the device retention characteristics. In recent years, much work is focused on the utilization of high-k dielectrics to achieve a low operation voltage without sacrificing the retention behavior. Due to a smaller EOT and larger physical thickness, a high program/erase (P/E) tunneling current can be obtained at low gate voltages. On the other hand, the off-state leakage current is reduced to lower levels under the retention mode [120, 121]. With increased field

Figure 16.11 Cross-sectional image of TANOS cell comprising SiO_2 (40 Å)/SiN (60 Å)/Al_2O_3 (150 Å) dielectric composite and 150 Å TaN electrode [122].

sensitivity, the P/E speeds can be improved by using a high-*k* tunnel barrier and high work function gate electrode. The nonvolatility of the memory device can also be improved by utilizing combined structures of high-*k* dielectric layers as the blocking oxide. The introduction of TANOS (TaN–Al_2O_3–nitride–oxide–silicon) architecture by Samsung represents the first application of HKMG into NAND flash memory device. Figure 16.11 shows a 40 nm TANOS cell structure fabricated for 32 Gb multilevel NAND flash memory application (Samsung Semiconductor) [122, 123].

With recent development on new high-*k* dielectric materials, different dielectric materials have also been demonstrated for memory device applications, including HfO_2 [120, 124], ZrO_2 [121, 125], Al_2O_3 [125], Lu_2O_3 [126], and $LaAlO_3$ [127]. Due to an improved thermal stability on Si expected from lanthanide-based high-*k* dielectrics, dielectrics, for example, La_2O_3, Lu_2O_3, and Gd_2O_3, are considered as promising candidates for the next-generation high-*k* materials. The dielectric stack can be further engineered with a sandwich of different high-*k* materials in order to achieve a favorable band alignment for improved P/E and retention trade-off. Different band-engineered dielectric stacks comprising advanced high-*k* material, for example, TaN/HfLaON/$Hf_{0.3}N_{0.2}O_{0.5}$/SiO_2/Si, provide further potential for improved charge storage and retention behavior, which represent an important avenue for future flash memory device [128].

References

1 Wilk, G.D. and Wallace, R.M. (1999) Electrical properties of hafnium silicate gate dielectrics deposited directly on silicon. *Appl. Phys. Lett.*, **74**, 2854.

2 Clark, R.D., Consiglio, S., Wajda, C.S., Leusink, G.J., Sugawara, T., Nakabayashi, H., Jagannathan, H., Edge, L.F., Jamison, P., Paruchuri, V.K., Iijima, R., Takayanagi, M., Linder, B.P., Bruley, J., Copel, M., and Narayanan, V. (2008) High-*K* gate dielectric structures by atomic layer deposition for the 32nm and beyond nodes, in *Atomic Layer Deposition Applications 4*, vol. 291 (eds A. Londergan, S.F. Bent, S. De Gendt, J.W. Elam, S.B. Kang, and O. van der Straten), Electrochemical Society, Inc., Pennington, NJ.

3 Wallace, R.M. and Wilk, G.D. (2003) High-kappa dielectric materials for microelectronics. *Crit. Rev. Solid State Mater. Sci.*, **28**, 231.

4 Rotondaro, A.L.P., Visokay, M.R., Shanware, A., Chambers, J.J., and Colombo, L. (2002) Carrier mobility in MOSFETs fabricated with Hf–Si–O–N gate dielectric, polysilicon gate electrode, and self-aligned source and drain. *IEEE Electron Device Lett.*, **23**, 603.

5 Wilk, G.D., Wallace, R.M., and Anthony, J.M. (2001) High-kappa gate dielectrics: current status and materials properties considerations. *J. Appl. Phys.*, **89**, 5243.

6 Scarel, G., Svane, A., and Fanciulli, M. (2007) *Rare Earth Oxide Thin Films*, Springer, Berlin.

7 Först, C.J., Schwarz, K., and Blöchl, P.E. (2005) Structural and electronic properties of the interface between the high-*k* oxide $LaAlO_3$ and Si(001). *Phys. Rev. Lett.*, **95**, 137602.

8 Edge, L.F., Schlom, D.G., Chambers, S.A., Cicerrella, E., Freeouf, J.L., Hollander, B., and Schubert, J. (2004) Measurement of the band offsets between amorphous $LaAlO_3$ and silicon. *Appl. Phys. Lett.*, **84**, 726.

9 Plummer, J.D. and Griffin, P.B. (2001) Material and process limits in silicon VLSI technology. *Proc. IEEE*, **89**, 240–258.

10 Kumar, A., Rajdev, D., and Douglass, D.L. (1972) Effect of oxide defect structure on electrical properties of ZrO_2. *J. Am. Ceram. Soc.*, **55**, 439.

11 Monaghan, S., Greer, J.C., and Elliott, S.D. (2005) Thermal decomposition mechanisms of hafnium and zirconium silicates at the atomic scale. *J. Appl. Phys.*, **97**, 114911.

12 Chin, A., Wu, Y.H., Chen, S.B., Liao, C.C., and Chen, W.J. (2000) High quality La_2O_3 and Al_2O_3 gate dielectrics with equivalent oxide thickness 5–10 Å. 2000 Symposium on VLSI Technology, Digest of Technical Papers.

13 Schamm, S., Coulon, P.E., Miao, S., Volkos, S.N., Lu, L.H., Lamagna, L., Wiemer, C., Tsoutsou, D., Scarel, G., and Fanciulli, M. (2009) Chemical/structural nanocharacterization and electrical properties of ALD-grown La_2O_3/Si interfaces for advanced gate stacks. *J. Electrochem. Soc.*, **156**, H1.

14 Darmawan, P., Lee, P.S., Setiawan, Y., Lai, J.C., and Yang, P. (2007) Thermal stability of rare-earth based ultrathin Lu_2O_3 for high-k dielectrics. *J. Vac. Sci. Technol. B*, **25**, 1203.

15 Gottlob, H.D.B., Schmidt, M., Stefani, A., Lemme, M.C., Kurz, H., Mitrovic, I.Z., Davey, W.M., Hall, S., Werner, M., Chalker, P.R., Cherkaoui, K., Hurley, P.K., Piscator, J., Engstrom, O., and Newcomb, S.B. (2009) Scaling potential and MOSFET integration of thermally stable Gd silicate dielectrics. *Microelectron. Eng.*, **86**, 1642.

16 Gottlob, H.D.B., Echtermeyer, T., Mollenhauer, T., Efavi, J.K., Schmidt, M., Wahlbrink, T., Lemme, M.C., and Kurz, H. (2006) Investigation of high-K gate stacks with epitaxial Gd_2O_3 and FUSI NiSi metal gates down to CET=0.86nm. *Mater. Sci. Semicond. Proc.*, **9**, 904.

17 Lu, X.B., Zhang, X., Huang, R., Lu, H.B., Chen, Z.H., Xiang, W.F., He, M., Cheng, B.L., Zhou, H.W., Wang, X.P., Wang, C.Z., and Nguyen, B.Y. (2004) Thermal stability of $LaAlO_3$/Si deposited by laser molecular-beam epitaxy. *Appl. Phys. Lett.*, **84**, 2620.

18 Triyoso, D.H., Gilmer, D.C., Jiang, J., and Droopad, R. (2008) Characteristics of thin lanthanum lutetium oxide high-k dielectrics. *Microelectron. Eng.*, **85**, 1732.

19 Hobbs, C., Dip, L., Reid, K., Gilmer, D., Hegde, R., Ma, T., Taylor, B., Cheng, B., Samavedam, S., Tseng, H., Weddington, D., Huang, F., Farber, D., Schippers, M., Rendon, M., Prabhu, L., Rai, R., Bagchi, S., Conner, J., Backer, S., Dumbuya, F., Locke, J., Workman, D., and Tobin, P. (2001) Sub-quarter micron Si-gate CMOS with ZrO_2 gate dielectric. 2001 Symposium on VLSI Technology.

20 Misra, V., Lucovsky, G., and Parsons, G.N. (2002) Issues in high-kappa gate stack interfaces. *MRS Bull.*, **27** (3), 212–216.

21 Gusev, E.P., Cartier, E., Buchanan, D.A., Gribelyuk, M., Copel, M., Okorn-Schmidt, H., and D'Emic, C. (2001) Ultrathin high-K metal oxides on silicon: processing, characterization and integration issues. *MRS Bull.*, **27**, 212.

22 Gusev, E.P., Narayanan, V., and Frank, M.M. (2006) Advanced high-kappa dielectric stacks with polySi and metal gates: recent progress and current challenges. *IBM J. Res. Dev.*, **50**, 387.

23 Lee, J., Suh, Y.S., Lazar, H., Jha, R., Gurganus, J., Lin, Y.X., and Misra, V. (2003) Compatibility of dual metal gate electrodes with high-k dielectrics for CMOS. 2003 IEEE International Electron Devices Meeting, Technical Digest.

24 Chau, R., Datta, S., Doczy, M., Doyle, B., Kavalieros, J., and Metz, M. (2004) High-kappa/metal-gate stack and its MOSFET characteristics. *IEEE Electron Device Lett.*, **25**, 408.

25 Mistry, K., Allen, C., Auth, C., Beattie, B., Bergstrom, D., Bost, M., Brazier, M., Buehler, M., Cappellani, A., Chau, R., Choi, C.H., Ding, G., Fischer, K., Ghani, T., Grover, R., Han, W., Hanken, D., Hatttendorf, M., He, J., Hicks, J., Huessner, R., Ingerly, D., Jain, P., James, R., Jong, L., Joshi, S., Kenyon, C., Kuhn, K., Lee, K., Liu, H., Maiz, J., McIntyre, B., Moon, P., Neirynck, J., Pei, S., Parker, C., Parsons, D., Prasad, C., Pipes, L., Prince, M., Ranade, P., Reynolds, T.,

Sandford, J., Schifren, L., Sebastian, J., Seiple, J., Simon, D., Sivakumar, S., Smith, P., Thomas, C., Troeger, T., Vandervoorn, P., Williams, S., and Zawadzki, K. (2007) A 45nm logic technology with high-*k* plus metal gate transistors, strained silicon, 9 Cu interconnect layers, 193nm dry patterning, and 100% Pb-free packaging. 2007 IEEE International Electron Devices Meeting, vols. 1–2, p. 247.

26 Lammers, D. (2010) GlobalFoundries adds Qualcomm, supports gate-first technology at 28nm generation. Semiconductor.net.

27 Arnaud, F., Liu, J., Lee, Y.M., Lim, K.Y., Kohler, S., Chen, J., Moon, B.K., Lai, C.W., Lipinski, M., Sang, L., Guarin, F., Hobbs, C., Ferreira, P., Ohuchi, K., Li, J., Zhuang, H., Mora, P., Zhang, Q., Nair, D.R., Lee, D.H., Chan, K.K., Satadru, S., Yang, S., Koshy, J., Hayter, W., Zaleski, M., Coolbaugh, D.V., Kirn, H.W., Ee, Y.C., Sudijono, J., Thean, A., Sherony, M., Samavedam, S., Khare, M., Goldberg, C., and Steegen, A. (2008) 32nm general purpose bulk CMOS technology for high performance applications at low voltage. 2008 IEEE International Electron Devices Meeting, Technical Digest, p. 633.

28 Tomimatsu, T., Goto, Y., Kato, H., Amma, M., Igarashi, M., Kusakabe, Y., Takeuchi, M., Ohbayashi, S., Sakashita, S., Kawahara, T., Mizutani, M., Inoue, M., Sawada, M., Kawasaki, Y., Yamanari, S., Miyagawa, Y., Takeshima, Y., Yamamoto, Y., Endo, S., Hayashi, T., Nishida, Y., Horita, K., Yamashita, T., Oda, H., Tsukamoto, K., Inoue, Y., Fujimoto, H., Sato, Y., Yamashita, K., Mitsuhashi, R., Matsuyama, S., Moriyama, Y., Nakanishi, K., Noda, T., Sahara, Y., Koike, N., Hirase, J., Yamada, T., Ogawa, H., and Ogura, M. (2009) Cost-effective 28-nm LSTP CMOS using gate-first metal gate/high-*k* technology. 2009 Symposium on VLSI Technology, Digest of Technical Papers.

29 Iwai, H., Ohmi, S., Akama, S., Ohshima, C., Kikuchi, A., Kashiwagi, I., Taguchi, J., Yamamoto, H., Tonotani, J., Kim, Y., Ueda, I., Kuriyama, A., and Yoshihara, Y. (2002) Advanced gate dielectric materials for sub-100nm CMOS. 2002 IEEE International Electron Devices Meeting, Technical Digest.

30 Capron, N., Broqvist, P., and Pasquarello, A. (2007) Migration of oxygen vacancy in HfO_2 and across the HfO_2/SiO_2 interface: a first-principles investigation. *Appl. Phys. Lett.*, **91**, 192905.

31 Xue, D., Betzler, K., and Hesse, H. (2000) Dielectric constants of binary rare-earth compounds. *J. Phys.: Condens. Matter*, **12**, 3113.

32 Paivasaari, J., Putkonen, M., and Niinisto, L. (2005) A comparative study on lanthanide oxide thin films grown by atomic layer deposition. *Solid Films*, **472**, 275.

33 Narayan, V., Guha, S., Copel, M., Bojarczuk, N.A., Flaitz, P.L., and Gribelyuk, M. (2002) Interfacial oxide formation and oxygen diffusion in rare earth oxide–silicon epitaxial heterostructures. *Appl. Phys. Lett.*, **81**, 4183.

34 Marsella, L. and Fiorentini, V. (2004) Structure and stability of rare-earth and transition-metal oxides. *Phys. Rev. B*, **69**, 172103.

35 Stemmer, S. (2004) Thermodynamic considerations in the stability of binary oxides for alternative gate dielectrics in complementary metal–oxide–semiconductors. *J. Vac. Sci. Technol. B*, **22**, 791.

36 Barin, I. (1995) *Thermochemical Data of Pure Substances*, vols. I–II, 3rd edn, VCH, Weinheim.

37 Stemmer, S., Klenov, D.O., Chen, Z.Q., Niu, D., Ashcraft, R.W., and Parsons, G.N. (2002) Reactions of Y_2O_3 films with (001) Si substrates and with polycrystalline Si capping layers. *Appl. Phys. Lett.*, **81**, 712.

38 Niu, D., Ashcraft, R.W., Chen, Z., Stemmer, S., and Parsons, G.N. (2002) Electron energy-loss spectroscopy analysis of interface structure of yttrium oxide gate dielectrics on silicon. *Appl. Phys. Lett.*, **81**, 676.

39 Cheng, Y.-T., Johnson, W.L., and Nicolet, M.-A. (1985) Dominant moving species in the formation of amorphous NiZr by solid-state reaction. *Appl. Phys. Lett.*, **47**, 800.

40 Wilk, G.D., Wallace, R.M., and Anthony, J.M. (2001) High-*k* gate dielectrics: current status and materials properties considerations. *J. Appl. Phys.*, **89**, 5243.

41 Son, J.Y., Maeng, W.J., Kim, W.-H., Shin, Y.H., and Kim, H. (2009) *Thin Solid Films*. doi: 10.1016/j.tsf.2009.1001.1117.

42 Klaus, J.W., Sneh, O., and George, S.M. (1997) Growth of SiO_2 at room temperature with the use of catalyzed sequential half-reactions. *Science*, **278**, 1934.

43 Ohmi, S., Takeda, H., Ishiwara, H., and Iwai, H. (2004) Electrical characteristics for Lu_2O_3 thin films fabricated by e-beam deposition method. *J. Electrochem. Soc.*, **151**, G279.

44 Darmawan, P., Chia, P.S., and Lee, P.S. (2007) Rare-earth based ultra-thin Lu_2O_3 for high-*k* dielectrics. *J. Phys.: Conf. Ser.*, **61**, 229.

45 Reynaert, J., Arkhipov, V.I., Borghs, G., and Heremans, P. (2004) Current–voltage characteristics of a tetracene crystal: space charge or injection limited conductivity? *Appl. Phys. Lett.*, **85**, 630.

46 Skriver, H.L. and Rosengaard, M.N. (1992) Surface energy and work function of elemental metals. *Phys. Rev. B*, **46**, 7157.

47 Zhu, M., Zhu, J., Liu, J.M., and Liu, Z.G. (2005) Pulsed laser deposition of zirconium silicate thin films as candidate gate dielectrics. *Appl. Phys. A*, **80**, 135.

48 Hill, W.A. and Coleman, C.C. (1980) A single-frequency approximation for interface-state density determination. *Solid State Electron.*, **23**, 987.

49 Chin, A., Liao, C.C., Liu, C.H., Chen, W.J., and Tsai, C. (1999) Device and reliability of high-*k* Al_2O_3 gate dielectric with good mobility and low D_{it}. Symposium on VLSI Technology, Digest of Technical Papers, p. 135.

50 Kwo, J., Hong, M., Kortan, A.R., Queeney, K.L., Chabal, Y.J., Mannaerts, J.P., Boone, T., Krajewski, J.J., Sergent, A.M., and Rosamilia, J.M. (2000) High ε gate dielectrics Gd_2O_3 and Y_2O_3 for silicon. *Appl. Phys. Lett.*, **77**, 130.

51 Manchanda, L., Green, M.L., Dover, R.B.V., Morris, M.D., Kerber, A., Hu, Y., Han, J.P., Silverman, P.J., Sorsch, T.W., Weber, G., Donnelly, V., Pelhos, K., Klemens, F., Ciampa, N.A., Kornblit, A., Kim, Y.O., Bower, J.E., Barr, D., Ferry, E., Jacobson, D., Eng, J., Busch, B., and Schulte, H. (2000) Si-doped aluminates for high temperature metal-gate CMOS: Zr–Al–Si–O, a novel gate dielectric for low power applications. 2000 IEEE International Electron Devices Meeting, Technical Digest, p. 23.

52 Brown, G.A. (2004) *High Dielectric Constant Materials: VLSI MOSFET Applications*, Springer, Berlin.

53 Kwo, J., Hong, M., Kortan, A.R., Queeney, K.L., Chabal, Y.J., Opila, R.L., Muller, D.A., Chu, S.N.G., Sapjeta, B.J., Lay, T.S., Mannaerts, J.P., Boone, T., Krautter, H.W., Krajewski, J.J., Sergnt, A.M., and Rosamilia, J.M. (2001) Properties of high kappa gate dielectrics Gd_2O_3 and Y_2O_3 for Si. *J. Appl. Phys.*, **89**, 3920.

54 Czernohorsky, M., Tetzlaff, D., Bugiel, E., Dargis, R., Osten, H.J., Gottlob, H.D.B., Schmidt, M., Lemme, M.C., and Kurz, H. (2008) Stability of crystalline Gd_2O_3 thin films on silicon during rapid thermal annealing. *Semicond. Sci. Technol.*, **23**, 035010.

55 Kwo, J., Hong, M., Busch, B., Muller, D.A., Chabal, Y.J., Kortan, A.R., Mannaerts, J.P., Yang, B., Ye, P., Gossmann, H., Sergent, A.M., Ng, K.K., Bude, J., Schulte, W.H., Garfunkel, E., and Gustafsson, T. (2003) Advances in high kappa gate dielectrics for Si and III–V semiconductors. *J. Cryst. Growth*, **251**, 645.

56 Hattori, T., Yoshida, T., Shiraishi, T., Takahashi, K., Nohira, H., Joumori, S., Nakajima, K., Suzuki, M., Kimura, K., Kashiwagi, I., Ohshima, C., Ohmi, S., and Iwai, H. (2004) Composition, chemical structure, and electronic band structure of rare earth oxide/Si(100) interfacial transition layer. *Microelectron. Eng.*, **72**, 283.

57 Jones, A.C., Aspinall, H.C., Chalker, P.R., Potter, R.J., Kukli, K., Rahtu, A., Ritala, M., and Leskela, M. (2004) Some recent developments in the MOCVD and ALD of high-kappa dielectric oxides. *J. Mater. Chem.*, **14**, 3101.

58 Kosola, A., Päiväsaari, J., Putkonen, M., and Niinistö, L. (2005) Neodymium oxide and neodymium aluminate thin films by atomic layer deposition. *Thin Solid Films*, **479**, 152.

59 Passlack, M., Hong, M., Mannaerts, J.P., Opila, R.L., Chu, S.N.G., Moriya, N., Ren, F., and Kwo, J.R. (1997) Low D_{it}, thermodynamically stable Ga_2O_3–GaAs interfaces: fabrication, characterization, and modeling. *IEEE Trans. Electron Devices*, **44**, 214.

60 Hong, M., Mannaerts, J.P., Bower, J.E., Kwo, J., Passlack, M., Hwang, W.Y., and Tu, L.W. (1997) Novel $Ga_2O_3(Gd_2O_3)$ passivation techniques to produce low D_{it} oxide–GaAs interfaces. *J. Cryst. Growth*, **175**, 422.

61 Landheer, D., Gupta, J.A., Sproule, G.I., McCaffrey, J.P., Graham, M.J., Yang, K.C., Lu, Z.H., and Lennard, W.N. (2001) Characterization of Gd_2O_3 films deposited on Si(100) by electron-beam evaporation. *J. Electrochem. Soc.*, **148**, G29.

62 Gupta, J.A., Landheer, D., McCaffrey, J.P., and Sproule, G.I. (2001) Gadolinium silicate gate dielectric films with sub-1.5nm equivalent oxide thickness. *Appl. Phys. Lett.*, **78**, 1718.

63 Kwo, J., Hong, M., Busch, B., Muller, D.A., Chabal, Y.J., Kortan, A.R., Mannaerts, J.P., Yang, B., Ye, P., Gossmann, H., Sergent, A.M., Ng, K.K., Bude, J., Schulte, W.H., Garfunkel, E., and Gustafsson, T. (2003) Advances in high κ gate dielectrics for Si and III–V semiconductors. *J. Cryst. Growth*, **251**, 645.

64 Tseng, H.H., Grant, J.M., Hobbs, C., Tobin, P.J., Luo, Z., Ma, T.P., Hebert, L., Ramon, M., Kalpat, S., Wang, F., Triyoso, D., Gilmer, D.C., White, B.E., Abramowitz, P., and Moosa, M. (2005) Mechanism of G_m degradation and comparison of V_t instability and reliability of HfO_2, $HfSiO_x$, and $HfAlO_x$ gate dielectrics with 80nm poly-Si gate CMOS. 2005 Symposium on VLSI Technology.

65 Kim, Y., Miyauchi, K., Ohmi, S., Tsutsui, K., and Iwai, H. (2005) Electrical properties of vacuum annealed La_2O_3 thin films grown by E-beam evaporation. *Microelectron. J.*, **36**, 41.

66 Yamamoto, Y., Kita, K., Kyuno, K., and Toriumi, A. (2007) Study of La-induced flat band voltage shift in metal/$HfLaO_x$/SiO_2/Si capacitors. *Jpn. J. Appl. Phys. 1*, **46**, 7251.

67 Yamada, H., Shimizu, T., and Suzuki, E. (2002) Interface reaction of a silicon substrate and lanthanum oxide films deposited by metalorganic chemical vapor deposition. *Jpn. J. Appl. Phys. 2*, **41**, L368.

68 Huang, J., Heh, D., Sivasubramani, P., Kirsch, P.D., Bersuker, G., Gilmer, D.C., Quevedo-Lopez, M.A., Hussain, M.M., Majhi, P., Lysaght, P., Park, H., Goel, N., Young, C., Park, C.S., Park, C., Cruz, M., Diaz, V., Hung, P.Y., Price, J., Tseng, H.H., and Jammy, R. (2009) Gate first high-*k*/metal gate stacks with zero SiO_x interface achieving EOT=0.59nm for 16nm application. 2009 Symposium on VLSI Technology, Digest of Technical Papers, p. 34.

69 Park, C.S., Yang, J.W., Hussain, M.M., Kang, C.Y., Huang, J., Sivasubramani, P., Park, C., Tateiwa, K., Harada, Y., Barnett, J., Melvin, C., Bersuker, G., Kirsch, P.D., Lee, B.H., Tseng, H.H., and Jammy, R. (2009) La-doped metal/high-*K* nMOSFET for sub-32nm HP and LSTP application. 2009 Symposium on VLSI Technology.

70 Lee, P.F., Dai, J.Y., Wong, K.H., Chan, H.L.W., and Choy, C.L. (2003) Growth and characterization of Hf-aluminate high-*k* gate dielectric ultrathin films with equivalent oxide thickness less than 10 Å. *J. Appl. Phys.*, **93**, 3665.

71 Xiong, K., Robertson, J., Pourtois, G., Petry, J., and Muller, M. (2008) Impact of incorporated Al on the TiN/HfO_2 interface effective work function. *J. Appl. Phys.*, **104**, 074501.

72 Ando, T., Copel, M., Bruley, J., Frank, M.M., Watanabe, H., and

Narayanan, V. (2010) Physical origins of mobility degradation in extremely scaled SiO_2/HfO_2 gate stacks with La and Al induced dipoles. *Appl. Phys. Lett.*, **96**, 132904.

73 Kamiyama, S., Ishikawa, D., Kurosawa, E., Nakata, H., Kitajima, M., Ootuka, M., Aoyama, T., Nara, Y., and Ohji, Y. (2008) Systematic study of V_{th} controllability using ALD-Y_2O_3, La_2O_3, and MgO_2 layers with HfSiON/metal gate first n-MOSFETs for hp 32nm bulk devices. 2008 IEEE International Electron Devices Meeting, Technical Digest, p. 41.

74 Ramanathan, S., Karthikeyan, A., Govindarajan, S.A., and Kirsh, P.D. (2008) Synthesis of nitrogen passivated rare-earth doped hafnia thin films and high temperature electrochemical conduction studies. *J. Vac. Sci. Technol. B*, **26**, L33; Erratum: **27**, 198 (2009).

75 Copel, M., Guha, S., Bojarczuk, N., Cartier, E., Narayanan, V., and Paruchuri, V. (2009) Interaction of La_2O_3 capping layers with HfO_2 gate dielectrics. *Appl. Phys. Lett.*, **95**, 212903.

76 Sivasubramani, P., Boscke, T.S., Huang, J., Young, C.D., Kirsch, P.D., Krishnan, S.A., Quevedo-Lopez, M.A., Govindarajan, S., Ju, B.S., Harris, H.R., Lichtenwalner, D.J., Jur, J.S., Kingon, A.I., Kim, J., Gnade, B.E., Wallace, R.M., Bersuker, G., Lee, B.H., and Jammy, R. (2007) Dipole moment model explaining nFET V_t tuning utilizing La, Sc, Er, and Sr doped HfSiON dielectrics. 2007 Symposium on VLSI Technology.

77 Kirsch, P.D., Sivasubramani, P., Huang, J., Young, C.D., Park, C.S., Freeman, K., Hussain, M.M., Bersuker, G., Harris, H.R., Majhi, P., Lysaght, P., Tseng, H.H., Lee, B.H., and Jammy, R. (2009) Dipole model explaining high-k/metal gate threshold voltage tuning, in *Advanced Gate Stack, Source/Drain, and Channel Engineering for Si-Based CMOS 5: New Materials, Processes, and Equipment*, vol. 269 (eds V. Narayanan, F. Roozeboom, D.L. Kwong, H. Iwai, E.P. Gusev, and P.J. Timans), Electrochemical Society, Inc., Pennington, NJ.

78 Sato, M., Umezawa, N., Shimokawa, J., Arimura, H., Sugino, S., Tachibana, A., Nakamura, M., Mise, N., Kamiyama, S., Morooka, T., Eimori, T., Shiraishi, K., Yamabe, K., Watanabe, H., Yamada, K., Aoyama, T., Nabatame, T., Nara, Y., and Ohji, Y. (2008) Physical model of the PBTI and TDDB of La incorporated HfSiON gate dielectrics with pre-existing and stress-induced defects. 2008 IEEE International Electron Devices Meeting, Technical Digest, p. 119.

79 Liu, D. and Robertson, J. (2009) Passivation of oxygen vacancy states and suppression of Fermi pinning in HfO_2 by La addition. *Appl. Phys. Lett.*, **94**, 042904.

80 Sato, M., Kamiyama, S., Sugita, Y., Matsuki, T., Morooka, T., Suzuki, T., Shiraishi, K., Yamabe, K., Ohmori, K., Yamada, K., Yugami, J., Ikeda, K., and Ohji, Y. (2009) Negatively charged deep level defects generated by yttrium and lanthanum incorporation into HfO_2 for V_{th} adjustment, and the impact on TDDB, PBTI and $1/f$ noise. 2009 IEEE International Electron Devices Meeting, p. 111.

81 Lide, D.R. (ed.) (2003) *CRC Handbook of Chemistry and Physics*, CRC Press, Boca Raton, FL.

82 Choi, K., Jagannathan, H., Choi, C., Edge, L., Ando, T., Frank, M., Jamison, P., Wang, M., Cartier, E., Zafar, S., Bruley, J., Kerber, A., Linder, B., Callegari, A., Yang, Q., Brown, S., Stathis, J., Iacoponi, J., Paruchuri, V., and Narayanan, V. (2009) Extremely scaled gate-first high-k/metal gate stack with EOT of 0.55nm using novel interfacial layer scavenging techniques for 22nm technology node and beyond. 2009 Symposium on VLSI Technology.

83 Harris, H.R., Majhi, P., Kirsch, P., Sivasubramani, P., Oh, J.W., and Song, S.C. (2009) The challenges in introducing PMOS dual channel in CMOS processing, in *Advanced Gate Stack, Source/Drain, and Channel Engineering for Si-Based CMOS 5: New Materials, Processes, and Equipment*, vol. **223** (eds V. Narayanan, F. Roozeboom, D.L. Kwong, H. Iwai, E.P. Gusev, and

P.J. Timans), Electrochemical Society, Inc., Pennington, NJ.

84 Lin, D., Brammertz, G., Sioncke, S., Fleischmann, C., Delabie, A., Martens, K., Bender, H., Conard, T., Tseng, W.H., Lin, J.C., Wang, W.E., Temst, K., Vatomme, A., Mitard, J., Caymax, M., Meuris, M., Heyns, M., and Hoffmann, T. (2009) Enabling the high-performance InGaAs/Ge CMOS: a common gate stack solution. 2009 IEEE International Electron Devices Meeting, p. 300.

85 Avouris, P. and Chen, J. (2006) Nanotube electronics and optoelectronics. *Mater. Today*, **9**, 46.

86 Lieber, C.M. (2003) Nanoscale science and technology: building a big future from small things. *MRS Bull.*, **28**, 486.

87 Wang, X.R., Ouyang, Y.J., Li, X.L., Wang, H.L., Guo, J., and Dai, H.J. (2008) Room-temperature all-semiconducting sub-10-nm graphene nanoribbon field-effect transistors. *Phys. Rev. Lett.*, **100**, 206803.

88 Avouris, P., Chen, Z.H., and Perebeinos, V. (2007) Carbon-based electronics. *Nat. Nanotechnol.*, **2**, 605.

89 Weitz, R.T., Zschieschang, U., Forment-Aliaga, A., Kalblein, D., Burghard, M., Kern, K., and Klauk, H. (2009) Highly reliable carbon nanotube transistors with patterned gates and molecular gate dielectric. *Nano Lett.*, **9**, 1335.

90 Lu, Y.R., Bangsaruntip, S., Wang, X.R., Zhang, L., Nishi, Y., and Dai, H.J. (2006) DNA functionalization of carbon nanotubes for ultrathin atomic layer deposition of high kappa dielectrics for nanotube transistors with 60mV/decade switching. *J. Am. Chem. Soc.*, **128**, 3518.

91 Javey, A., Guo, J., Farmer, D.B., Wang, Q., Wang, D.W., Gordon, R.G., Lundstrom, M., and Dai, H.J. (2004) Carbon nanotube field-effect transistors with integrated ohmic contacts and high-*k* gate dielectrics. *Nano Lett.*, **4**, 447.

92 Yang, M.H., Teo, K.B.K., Gangloff, L., Milne, W.I., Hasko, D.G., Robert, Y., and Legagneux, P. (2006) Advantages of top-gate, high-*k* dielectric carbon nanotube field-effect transistors. *Appl. Phys. Lett.*, **88**, 113507.

93 Javey, A., Guo, J., Farmer, D.B., Wang, Q., Yenilmez, E., Gordon, R.G., Lundstrom, M., and Dai, H.J. (2004) Self-aligned ballistic molecular transistors and electrically parallel nanotube arrays. *Nano Lett.*, **4**, 1319.

94 Tans, S.J., Verschueren, A.R.M., and Dekker, C. (1998) Room-temperature transistor based on a single carbon nanotube. *Nature*, **393**, 49.

95 Wind, S.J., Appenzeller, J., Martel, R., Derycke, V., and Avouris, P. (2002) Vertical scaling of carbon nanotube field-effect transistors using top gate electrodes. *Appl. Phys. Lett.*, **80**, 3817; Erratum: **81**, 1359 (2002).

96 Perkins, C.M., Triplett, B.B., McIntyre, P.C., Saraswat, K.C., and Shero, E. (2002) Thermal stability of polycrystalline silicon electrodes on ZrO_2 gate dielectrics. *Appl. Phys. Lett.*, **81**, 1417.

97 Farmer, D.B. and Gordon, R.G. (2005) ALD of high-kappa dielectrics on suspended functionalized SWNTs. *Electrochem. Solid State Lett.*, **8**, G89.

98 Javey, A., Guo, J., Wang, Q., Lundstrom, M., and Dai, H.J. (2003) Ballistic carbon nanotube field-effect transistors. *Nature*, **424**, 654.

99 Friedman, R.S., McAlpine, M.C., Ricketts, D.S., Ham, D., and Lieber, C.M. (2005) High-speed integrated nanowire circuits. *Nature*, **434**, 1085.

100 Jin, S., Whang, D.M., McAlpine, M.C., Friedman, R.S., Wu, Y., and Lieber, C.M. (2004) Scalable interconnection and integration of nanowire devices without registration. *Nano Lett.*, **4**, 915.

101 Jiang, X.C., Xiong, Q.H., Nam, S., Qian, F., Li, Y., and Lieber, C.M. (2007) InAs/InP radial nanowire heterostructures as high electron mobility devices. *Nano Lett.*, **7**, 3214.

102 Zhang, L., Tu, R., and Dai, H.J. (2006) Parallel core–shell metal–dielectric–semiconductor germanium nanowires for high-current surround-gate field-effect transistors. *Nano Lett.*, **6**, 2785.

103 Xiang, J., Lu, W., Hu, Y.J., Wu, Y., Yan, H., and Lieber, C.M. (2006) Ge/Si nanowire

heterostructures as high-performance field-effect transistors. *Nature*, **441**, 489.

104 Lauhon, L.J., Gudiksen, M.S., Wang, C.L., and Lieber, C.M. (2002) Epitaxial core–shell and core–multishell nanowire heterostructures. *Nature*, **420**, 57.

105 Jeong, G.T., Lee, K.C., Ha, D.W., Lee, K.H., Kim, K.H., Kim, I.G., Kim, D.H., and Kim, K. (1997) A high performance 16 Mb DRAM using giga-bit technologies. *IEEE Trans. Electron Devices*, **44**, 2064.

106 Park, I.S., Lee, B.T., Choi, S.J., Im, J.S., Lee, S.H., Park, K.Y., Lee, J.W., Hyung, Y.W., Kim, Y.K., Park, H.S., Park, Y.W., Lee, S.I., and Lee, M.Y. (2000) Novel MIS Al_2O_3 capacitor as a prospective technology for Gbit DRAMs. 2000 Symposium on VLSI Technology, Digest of Technical Papers.

107 Kim, Y.K., Lee, S.H., Choi, S.J., Park, H.B., Seo, Y.D., Chin, K.H., Kim, D., Lim, J.S., Kim, W.D., Nam, K.J., Cho, M.H., Hwang, K.H., Kim, Y.S., Kim, S.S., Park, Y.W., Moon, J.T., Lee, S.L., and Lee, M.Y. (2000) Novel capacitor technology for high density stand-alone and embedded DRAMs. 2000 IEEE International Electron Devices Meeting, Technical Digest.

108 Lutzen, J., Birner, A., Goldbach, M., Gutsche, M., Hecht, T., Jakschik, S., Orth, A., Sanger, A., Schroder, U., Seidl, H., Sell, B., and Schumann, D. (2002) Integration of capacitor for sub-100-nm DRAM trench technology. 2002 Symposium on VLSI Technology, Digest of Technical Papers.

109 Londergan, A.R., Ramanathan, S., Vu, K., Rassiga, S., Hiznay, R., Winkler, J., Velasco, H., Matthysse, L., Seidel, T.E., Ang, C.H., Yu, H.Y., and Li, M.F. (2002) Process optimization in atomic layer deposition of high-k oxides for advanced gate stack engineering, in *Rapid Thermal and Other Short-Time Processing Technologies III*, vol. 163 (eds P. Timans, E. Gusev, F. Roozeboom, M.C. Ozturk, and D.L. Kwong), Electrochemical Society, Inc., Pennington, NJ.

110 Kingon, A.I., Maria, J.P., and Streiffer, S.K. (2000) Alternative dielectrics to silicon dioxide for memory and logic devices. *Nature*, **406**, 1032.

111 Eisenbeiser, K., Finder, J.M., Yu, Z., Ramdani, J., Curless, J.A., Hallmark, J.A., Droopad, R., Ooms, W.J., Salem, L., Bradshaw, S., and Overgaard, C.D. (2000) Field effect transistors with $SrTiO_3$ gate dielectric on Si. *Appl. Phys. Lett.*, **76**, 1324.

112 Pawlak, N., Popovici, M., Swerts, J., Tomida, K., Kim, M.-S., Kaczer, B., Opsomer, K., Schaekers, M., Favia, P., and Bender, H. (2010) Enabling 3*X* nm DRAM: record low leakage 0.4nm EOT MIM capacitors with novel stack engineering. 2010 IEEE International Electron Devices Meeting, Technical Digest, p. 929.

113 Lee, J.H., Kim, J.P., Kim, Y.S., Jung, H.S., Lee, N.I., Kang, H.K., Suh, K.P., Jeong, M.M., Hyun, K.T., Baik, H.S., Chung, Y.S., Liu, X.Y., Ramanathan, S., Seidel, T., Winkler, J., Londergan, A., Kim, H.Y., Ha, J.M., and Lee, N.K. (2002) Mass production worthy HfO_2–Al_2O_3 laminate capacitor technology using Hf liquid precursor for sub-100nm DRAMs. 2002 IEEE International Electron Devices Meeting, Technical Digest.

114 Wang, G., Anand, D., Butt, N., Cestero, A., Chudzik, M., Ervin, J., Fang, S., Freeman, G., Ho, H., Khan, B., Kim, B., Kong, W., Krishnan, R., Krishnan, S., Kwon, O., Liu, J., McStay, K., Nelson, E., Nummy, K., Parries, P., Sim, J., Takalkar, R., Tessier, A., Todi, R.M., Malik, R., Stiffler, S., and Iyer, S.S. (2009) Scaling deep trench based eDRAM on SOI to 32nm and beyond. 2009 IEEE International Electron Devices Meeting, p. 236.

115 Spicer, R., Hawthorne, J., Peters, D., and Newcomb, R. (2009) Controlling process-induced charging heightens productivity. *Semiconducting International*, March 2009.

116 Lai, S.K. (2008) Flash memories: successes and challenges. *IBM J. Res. Dev.*, **52**, 529.

117 Buskrik, M.C. (2006) MirrorBit technology, past, present, and future: the on-going scaling of nitride-based flash memory. IEEE Non-Volatile Semiconductor Memory Workshop.

118 Prinz, E.J., Yater, J., Steimle, R., Sadd, M., Swift, C., and Chang, K.-M. (2006) A 90nm embedded 2-bit per cell nanocrystal flash EEPROM. Non-Volatile Semiconductor Memory Workshop, p. 62.

119 De Blauwe, J. (2002) Nanocrystal nonvolatile memory devices. *IEEE Trans. Nanotechnol.*, **1**, 72.

120 Lee, J.J., Wang, X., Bai, W, Lu, N., Liu, and Kwong, D.L. (2003) Theoretical and experimental investigation of Si nanocrystal memory device with HfO_2 high-k tunneling dielectric. Symposium on VLSI Technology, Digest of Technical Papers. p. 33–34.

121 Kim, D.W., Kim, T., and Banerjee, S.K. (2003) Memory characterization of SiGe quantum dot flash memories with HfO_2 and SiO_2 tunneling dielectrics. *IEEE Trans. Electron Devices*, **50**, 1823.

122 Kang, C., Choi, J., Sim, J., Lee, C., Shin, Y., Park, J., Sel, J., Jeon, S., Park, Y., and Kim, K. (2007) Effects of lateral charge spreading on the reliability of TANOS (TaN/AlO/SiN/oxide/Si) NAND flash memory. Proceedings of the 45th Annual IEEE International Reliability Physics Symposium, p. 167.

123 Park, Y., Choi, J., Kang, C., Lee, C., Shin, Y., Choi, B., Kim, J., Jeon, S., Sel, J., Park, J., Choi, K., Yoo, T., Sim, J., and Kim, K. (2006) Highly manufacturable 32 Gb multi-level NAND flash memory with 0.0098 μm^2 cell size using TANOS (Si–oxide–Al_2O_3–TaN) cell technology. 2006 International Electron Devices Meeting, vols. 1–2, p. 495.

124 Das, S., Das, K., Singha, R.K., Dhar, A., and Ray, S.K. (2007) Improved charge injection characteristics of Ge nanocrystals embedded in hafnium oxide for floating gate devices. *Appl. Phys. Lett.*, **91**, 233118.

125 Wan, Q., Lin, C.L., Liu, W.L., and Wang, T.H. (2003) Structural and electrical characteristics of Ge nanoclusters embedded in Al_2O_3 gate dielectric. *Appl. Phys. Lett.*, **82**, 4708.

126 Chan, M.Y., Lee, P.S., Ho, V., and Seng, H.L. (2007) Ge nanocrystals in lanthanide-based Lu_2O_3 high-*k* dielectric for nonvolatile memory applications. *J. Appl. Phys.*, **102**, 094307.

127 Lu, X.B., Lee, P.F., and Dai, J.Y. (2005) Synthesis and memory effect study of Ge nanocrystals embedded in $LaAlO_3$ high-*k* dielectrics. *Appl. Phys. Lett.*, **86**, 203111.

128 Yang, H.J., Cheng, C.F., Chen, W.B., Lin, S.H., Yeh, F.S., McAlister, S.P., and Chin, A. (2008) Comparison of MONOS memory device integrity when using $Hf_{1-x-y}N_xO_y$ trapping layers with different N compositions. *IEEE Trans. Electron Devices.*, **55**, 1417.

Part Six
Challenge and Future Directions

High-k Gate Dielectrics for CMOS Technology, First Edition. Edited by Gang He and Zhaoqi Sun.

17
The Interaction Challenges with Novel Materials in Developing High-Performance and Low-Leakage High-κ/Metal Gate CMOS Transistors

Michael Chudzik, Siddarth Krishnan, Unoh Kwon, Mukesh Khare, Vijay Narayanan, Takashi Ando, Ed Cartier, Huiming Bu, and Vamsi Paruchuri

17.1
Introduction

The basic principles for CMOS (complementary metal–oxide–semiconductor) scaling were described by Dennard in 1974 [1], in which he describes the effects of reducing all dimensions of a transistor. The resulting dimensional reduction results in a number of advantages, of which, arguably the most important are, an increase in speed and an increase in device density. The keystone for this scaling is the ability to increase the gate capacitance (gate scaling) of the device, a fact the industry has leveraged for decades. Since the 90 nm technology node, the gate dielectrics are now thin enough that quantum mechanical tunneling is causing the gate leakage and the subthreshold leakage to become significant fractions of the total leakage current [2]. Figure 17.1 shows IBM's historic SiO_2 and SiON (silicon oxynitride) gate dielectric thickness scaling over the past several decades from several hundred Å to values near 11 Å. The saturation in thickness scaling beyond CMOS 10S (90 nm technology node) marks the end of SiON scaling and the beginning of the need for a gate dielectric with a high dielectric constant, κ.

The advantage of high-κ materials is that their larger permittivity ($\kappa > 4.0$) enables physically thicker gate dielectrics at the same effective capacitance. This increased physical thickening reduces the quantum mechanical tunneling and thus in turn reduces the gate leakage and enables physical gate length scaling without an increase in subthreshold leakage [3]. High-κ and metal gate (HKMG) materials have re-enabled gate length scaling and were first introduced in the industry at the 45 nm node [4].

A HKMG stack for CMOS logic must also include two other very important components in order to deliver a viable solution for high-performance CMOS. One of these components includes a bottom interface layer of SiO_2 that sits between the high-κ dielectric and the silicon channel. The importance of the bottom interfacial layer is to improve mobility by reducing the impact of soft optical phonons [5] and secondly to maintain the high standards of reliability required by the industry [6].

High-k Gate Dielectrics for CMOS Technology, First Edition. Edited by Gang He and Zhaoqi Sun.

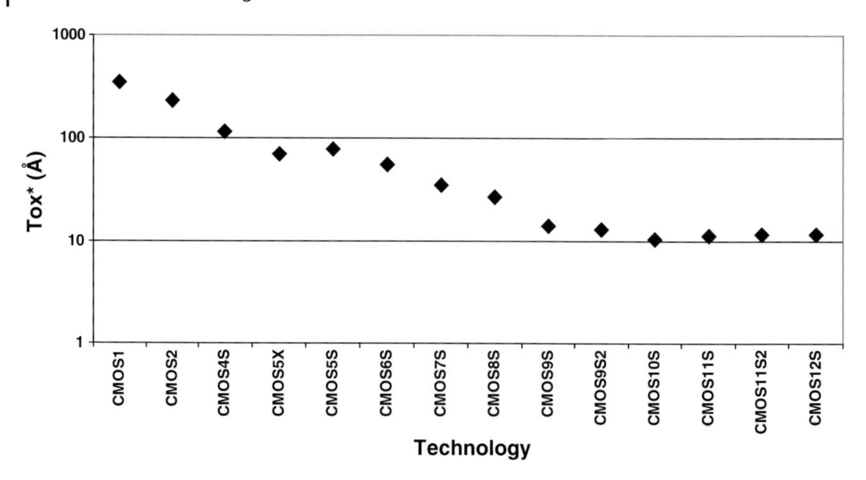

Figure 17.1 Gate oxide scaling in IBM technologies over several decades in time.

The second critical component is the metal gate electrode that sits atop these gate dielectrics. The metals for NFET and PFET must have the correct effective work function in order to have the required threshold voltages for proper device functionality; therefore, there are effectively two separate metal gates unique to each device type. The metal gates also serve to improve the capacitance of the total stack by removing the depletion capacitance of polycrystalline silicon gates. The development of stable metal gate electrodes was the last major issue overcome prior to the introduction of HKMG stacks in the industry. As discussed in detail, stabilizing the work functions during typical CMOS processing that require high thermal budgets ($T > 1000\,°C$) for device junction formation is challenging [7, 8].

In order to demonstrate appreciable performance and short channel benefits over SiON/poly-Si devices, high-performance logic using HKMG stacks requires dual work functions near 110 meV to the band edge (BE) values of silicon and inversion thicknesses (T_{inv}) of less than 14 Å (equivalent oxide thickness (EOT) $<$ 10 Å), as shown in the simulations in Figure 17.2a and b [4]. The stabilization of these work functions and the requirement of low T_{inv} for functional transistors have led to two divergent process flows by which HKMG stacks are integrated into CMOS transistors and are called gate-first and gate-last. These schemes will be described in detail in the next section.

17.2 Traditional CMOS Integration Processes

The standard or traditional CMOS process can be divided into three distinct and sequential groupings: front end of line (FEOL), middle of line (MOL), and back end of line (BEOL). Traditionally, transistor devices are fabricated in FEOL that includes the majority of the high thermal budget processes. In CMOS generations prior to HKMG, the gate dielectric was formed in FEOL with an oxidation and subsequent

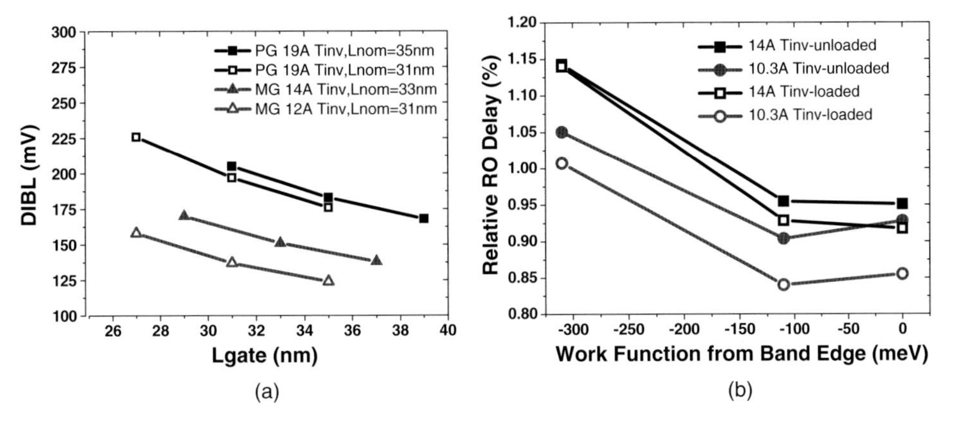

Figure 17.2 (a) Simulations showing DIBL response for HKMG and SiON/poly-Si devices as a function of L_{gate} at various T_{inv} values. (b) Simulations showing relative ring oscillator (RO) delay versus work function from BE for HKMG. Reproduced from Ref. [4].

nitridation and postanneal to enable a $\kappa > 4$. Gate dielectric and the contacting polycrystalline silicon are etched to form the gates using reactive ion etching (RIE) processes that are tuned to etch these materials in a vertical fashion.

The source and drain linkages are used for short channel control and threshold voltage adjustment implants are then performed by creating self-aligned spacers from the deposition of oxide and nitride materials. The implanted dopants from all or most of the halo, LDD, and S/D implanted regions are activated using rapid thermal anneal (RTA) at temperatures in excess of 1000 °C, often using abrupt thermal profiles called "spike" anneals that minimize time but maximize temperature. These activations serve to both tailor the dopant profile and activate the implanted species and there is a great deal of leveraging of this time and temperature space to perfect the device performance. In newer technologies (65 nm), these RTAs are followed by one or more millisecond anneals (MSAs) that are performed with either lasers or flashlamps to provide increased activation at temperatures in excess of 1100 °C with even shorter dwell times compared to spike anneals. Similarly to the well anneals, these activation anneals are performed primarily in inert ambients of N_2 and Ar but often contain a few percent oxygen.

After the final source/drain activation anneal, contacts to the source and drain are formed in the MOL by the deposition and reaction of metals (typically Ni) to form self-aligned silicide (salicide). The MOL is where the transistors are contacted with NiSi and then connected to the back-end wiring levels with contact vias. The reaction temperatures for NiSi are in the range of 400 °C and typically done in a single-wafer rapid thermal annealing chamber. An interlayer (IL) dielectric of oxide is then deposited and planarized to enable lithography of the contacts that will connect the source drains and gate contacts to the BEOL metallization layers. The final stage of CMOS logic processing is to wire the transistors into circuits using vias and wiring levels and is called the back end of line and is where the transistors are wired together with copper to form circuits and eventually connected to the packing chip.

17.3 High-κ/Metal Gate Integration Processes

The implementation of HKMG materials generally follows one of the two major integration paths to realize logic devices. One approach is to simply switch out the SiON gate dielectrics used in older technologies and replace them with the high-κ and work function-specific metal gates for the NFET and PFET. This processing is known as the gate-first approach and leverages the existing knowledge base on CMOS processing as much as possible to minimize disruption. The HKMG stack is formed in the FEOL. The drawbacks to gate-first are that the high-κ and metal gates now have to go through the high-temperature steps of source/drain activation that typically exceed 1000 °C. This approach requires a thermally stable HKMG stack and an added patterning step to separately form the NFET- and PFET-specific electrodes. A generalized processing flow for gate-first device integration is schematically depicted in Figure 17.3a.

Conversely, the CMOS integration flow can be altered to reduce the time/temperature impacts to HKMG stacks by implementing the HKMG as late as possible in the MOL just before the BEOL metallization, where the temperatures are restricted to <400 °C. This process uses dummy gate stacks of SiO_2 and polysilicon in the FEOL, and they are replaced by the high-κ and appropriate work function electrodes later in the flow in an attempt to circumvent as much of the high thermal budgets as possible. This integration flow is called gate-last or replacement gate flow. Figure 17.3b shows the generalized schematic flow for the gate-last scheme.

Despite dramatic differences in the integration schemes between gate-first and gate-last, they share the same common structure and both have a clear interfacial oxide layer, a high-κ layer, and the appropriate work function electrodes. The schematics in Figure 17.4 compare close-ups of a typical HKMG stack for the two integration schemes. Figure 17.5 shows representative transmission electron micrographs for the two different HKMG stacks. The importance of the interfacial layer, the high-κ dielectric, and the metal gate electrodes as well as how their performance and functionality is impacted by the thermal budgets and ambient gases used in modern CMOS processing is the focus of this chapter. Understanding these dependences is paramount in delivering viable gate stacks as their mobility, leakage, and work function properties can be enhanced or adversely impacted by structure, processing temperatures, and processing gases.

17.4 Mobility

New scattering mechanisms exist in HKMG-based technologies that are more or less evident at all fields. HfO_2, owing to its higher dielectric polarizability [5], has phonons within the dielectric that can couple with carriers in the channel. This coupling leads to carrier mobility degradation and is known as "remote" phonon scattering or soft optical phonon scattering (SOPS). In addition, HfO_2 is also known to have electrically

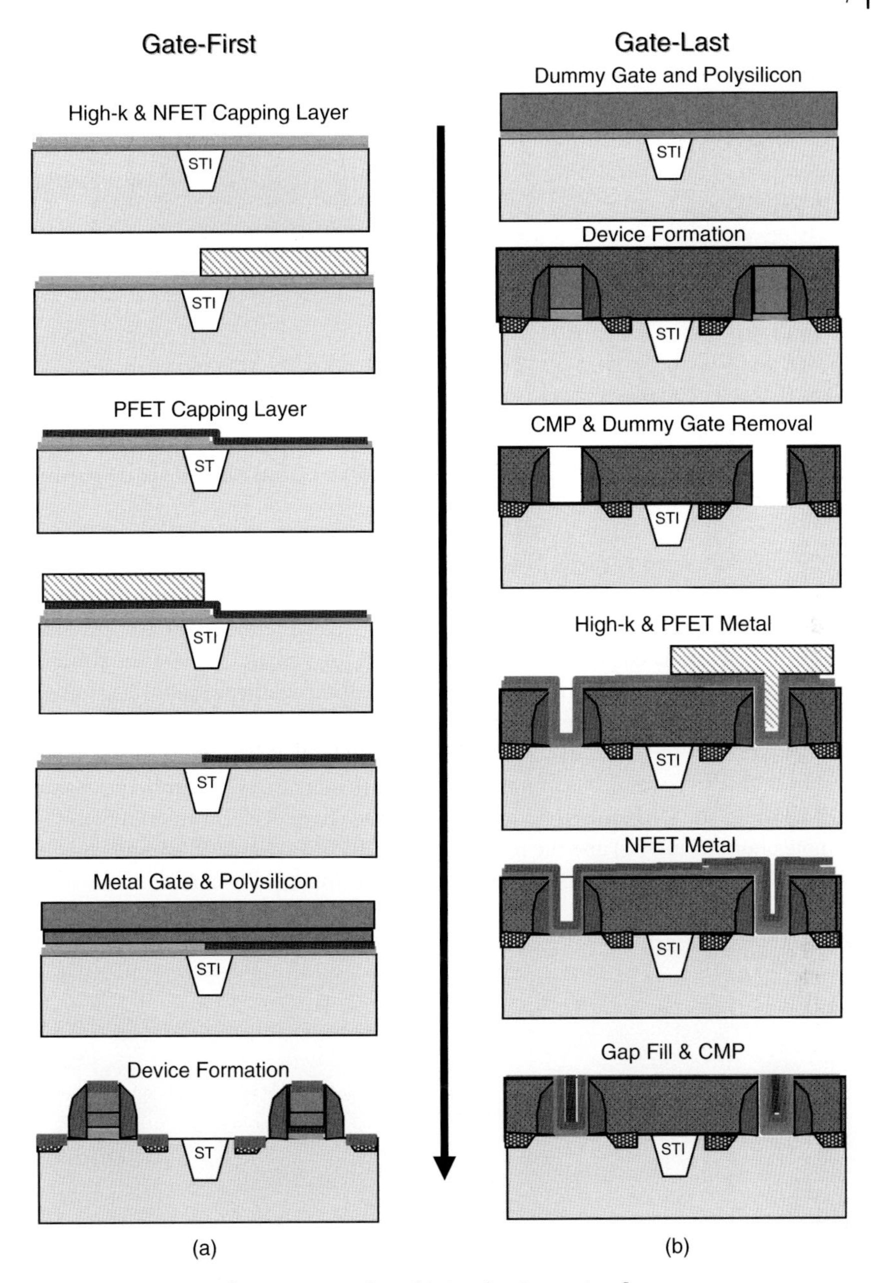

Figure 17.3 (a) Gate-first integration flow. (b) Gate-last integration flow.

charged defects that are denoted by the term "fixed" charge. The fixed charge in the HfO_2 also causes carrier mobility degradation through "remote" Coulomb scattering (RCS) [9]. Understanding the structural and processing impact on mobility is of key importance in creating useful technologies with HKMG.

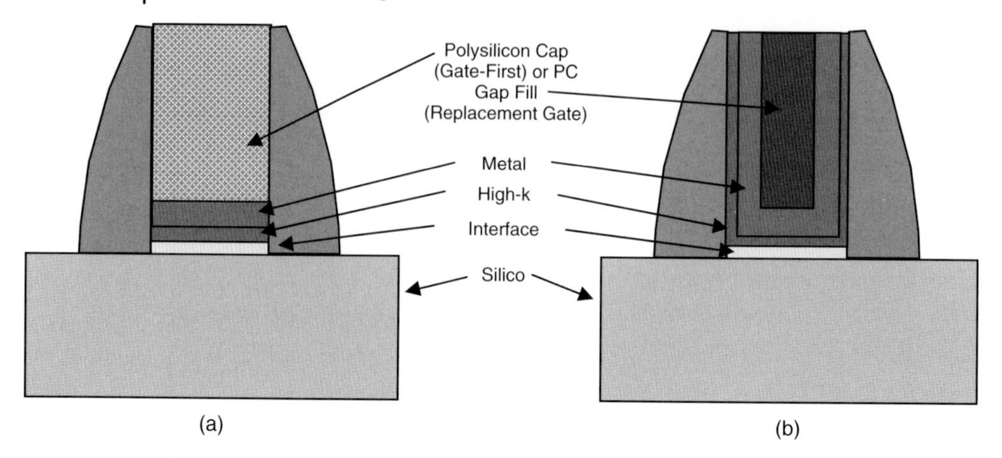

Figure 17.4 Schematic diagrams of the gate stack components for (a) gate-first and (b) gate-last HKMG stacks.

The interface layer between HfO_2 and Si is usually a SiO_2 or SiON layer that is grown intentionally to provide a high-quality, thermodynamically stable interface. This interface is critical for attaining high mobility. The interface layer, compared to bulk HfO_2, has reduced phonon scattering and less fixed charge. It therefore acts as a barrier layer, removing the HfO_2 phonons and the fixed charge from the Si channel, and mitigating their negative impact.

In Figure 17.6, the dependence of the electron mobility on the interface thickness is shown. As can be seen, electron mobility tends to increase rapidly with increasing IL thickness. The IL dampens the scattering of the channel electrons with both the remote phonons (SOPS) and the remote charge (RCS). However, this improvement in mobility due to IL thickness increase is also accompanied by a rapid increase in

Figure 17.5 Transmission electron micrograph of (a) gate-first (reproduced with permission from Ref. [4], copyright 2007 IEEE) and (b) gate-last HKMG stacks.

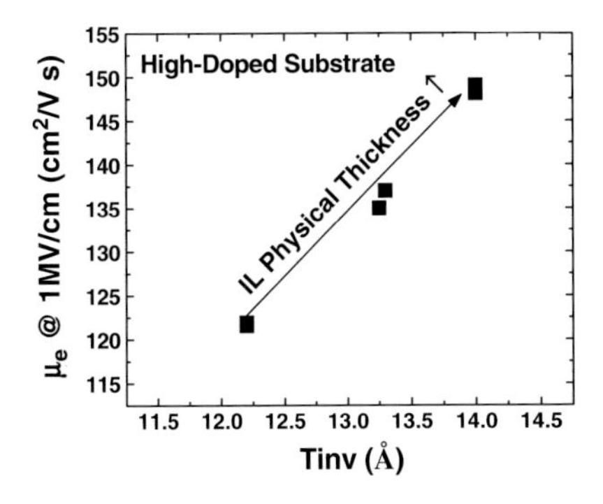

Figure 17.6 Increasing the interface layer thickness rapidly increases carrier mobility, at the expense of T_{inv}.

T_{inv}, quickly going beyond technologically useful numbers. The increase in T_{inv} is due to the low dielectric constant of the IL (usually, $k \sim 4$–5 as compared to $k \sim 20$ for HfO_2). Adding nitrogen to the IL increases the IL dielectric constant, so that the T_{inv} penalty is mitigated, however, nitrogen is a known coulombic scattering center and may require additional thermal treatments to mitigate mobility degradation.

Increasing the high-κ thickness generally decreases both electron [10] and hole mobility, due to a corresponding increase in the number of scattering centers (remote phonons and fixed charge), as illustrated in Figure 17.7, where the T_{inv} increase is achieved only though high-κ physical thickness increase.

In practical gate stacks, capping layers are also used to set the work function, as discussed in the next section. To set the work function, the material in the capping layer needs to diffuse through the high-κ layer to react with the IL. This reaction with the IL leads to the formation of an interfacial silicate, resulting in reduced T_{inv}. The mobility is concomitantly reduced along a mobility/T_{inv} trend line [11]. The choice of the high-κ layer thickness for manufacturing is, due to the reasons mentioned above,

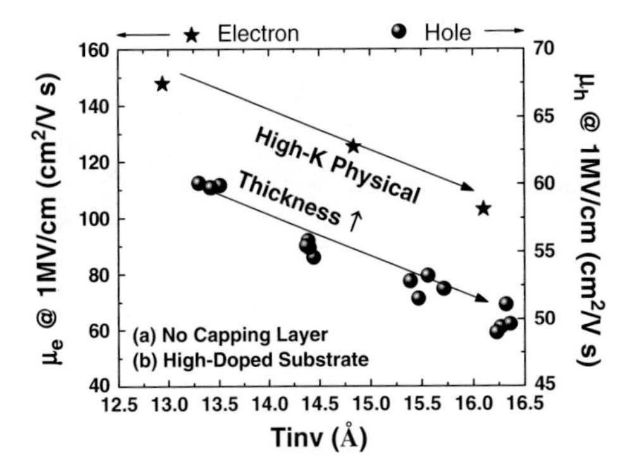

Figure 17.7 Carrier mobility reduction due to high-κ thickness attributed to an increase in remote charges and soft optical phonons.

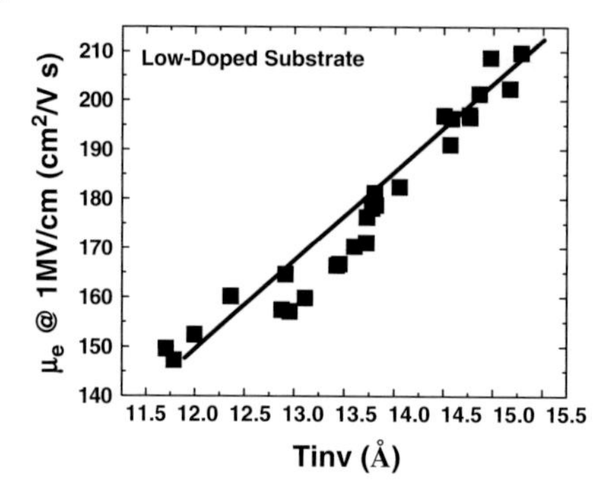

Figure 17.8 Universal trend of electron mobility in HKMG NFETs as a function of T_{inv}.

crucial in maintaining NFET and PFET mobility and T_{inv}. For the gate-first integration scheme, in net, changing the various constitutional elements such as HfO_2 thickness, interface thickness, interface nitrogen, and capping layer thickness proves to change the mobility along a universal trend line, as a function of T_{inv} (Figure 17.8).

The gate-last integration necessitates a low thermal budget after the deposition of the metal gates. Many of the mobility degradation mechanisms discussed in the previous sections are prevalent in the gate-last scenario. The higher thermal budgets usually serve to harden the interface, by reflowing the Si–O bonds and reducing the interface state density. In the absence of the high thermal budgets, interface state density tends to be high and degrade mobility in addition to remote phonon and fixed charge scattering. This degradation in mobility can be improved (Figure 17.9) with

1) increased thermal budget by implementing high-temperature post-HfO_2 deposition anneals in an attempt to mimic the thermal budgets seen in a gate-first scheme;
2) improved passivation of the interface.

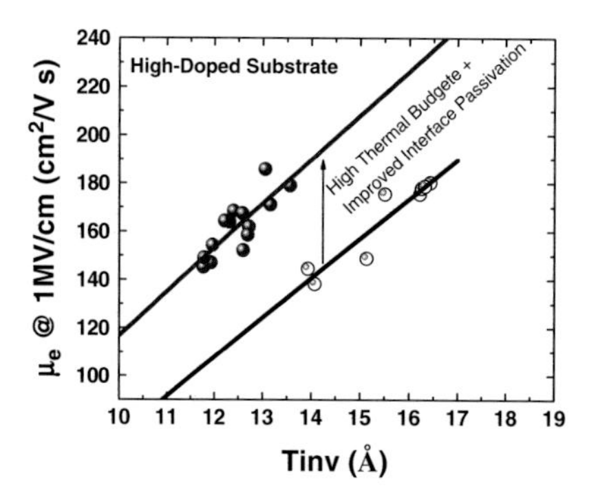

Figure 17.9 Gate-last carrier mobility improvements by high-temperature postdeposition anneals and improved interface passivation.

17.5 Metal Electrodes and Effective Work Function

Gate-first and gate-last integration schemes have very different requirements when it comes to the structural and thermal stability of the metal electrode in contact with the high-κ dielectric. However, in both cases, electrodes are required that satisfy the 110 meV from band edge requirements for high-performance logic. This is schematically described at flatband in Figure 17.10.

The choice of appropriate metal alloy gates in gate-first integration such as TiN, TaN, or TaC [12–14] was shown to be critical since these types of electrodes had distinct advantages over other elemental metals that either were thermally unstable (such as Al, Ti, or V) [7] or tended to act as oxygen reservoirs (such as W and Re) [8], which resulted in thick and technologically irrelevant equivalent oxide thicknesses. Although TiN and TaN are thermally stable, with low EOTs and high electron mobility, they tend to have near mid-gap work functions and thus unacceptably high V_t. This problem can be solved by inserting nanoscale capping layers containing group IIA and IIIB elements into proven high-mobility mid-gap HfO_2/TiN gate stacks, thereby moving the effective work function (EWF) to NFET BE [15, 16]. Using the oxides or nitrides of La or Mg within the capping films, one can shift the V_t toward NFET band edge and concomitantly provide significant T_{inv} scaling (Figure 17.11a). Detailed experiments have shown that the mixing of the La_2O_3 and interfacial SiO(N) results in the formation of an interfacial dipole, inducing a corresponding negative V_t shift appropriate for NFETs [17]. Positive V_t shifts appropriate for PFETs have been observed by using less electronegative elements such as the oxides of Al [18]. In addition, it has been shown that the V_{fb} shifts observed in MgO, La_2O_3 (NFET capping layers), and Al_2O_3 (a PFET capping layer) are strongly modulated by activation temperature that determines the diffusion of Mg, La, or Al through the HfO_2 into the SiO(N) IL [18, 19]. The impact of annealing temperature on La oxide-capped HfO_2/TiN MOSCAPs is shown in Figure 17.11b.

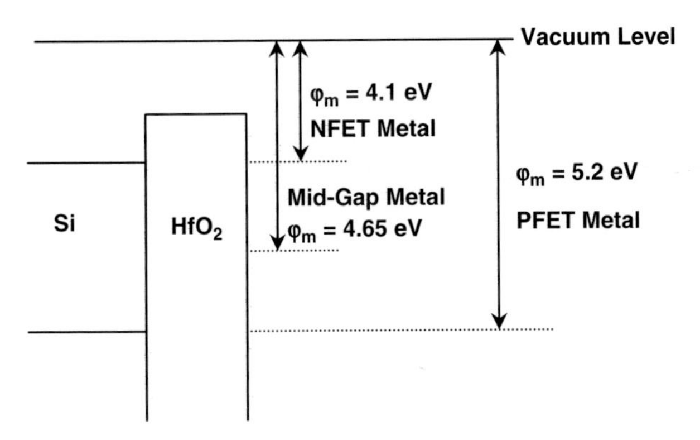

Figure 17.10 Schematic representation of work function requirements for NFET and PFET relative to Si conduction and valence band edges, respectively.

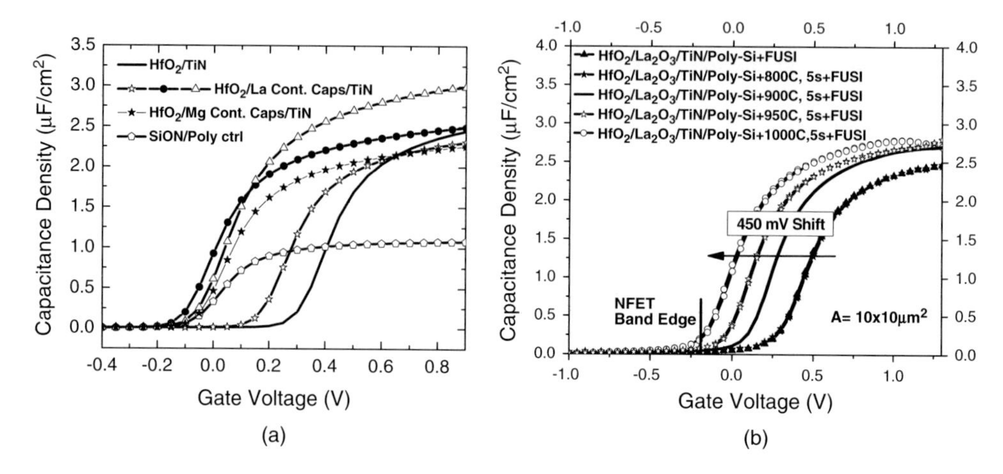

Figure 17.11 (a) Impact of La- and Mg-containing capping layers on V_t and T_{inv} of NFET device. Reproduced with permission from Ref. [15], copyright 2006 IEEE. (b) Impact of process temperature on V_t and T_{inv} of an NFET device containing a La_2O_3 capping layer inserted between HfO_2 and TiN/poly-Si followed by full silicidation (FUSI) of the poly-Si electrode.

For gate-last integration, work functions near the silicon conduction band for NFET and the valence band for PFET are achieved by utilizing the intrinsic vacuum work functions of gate electrode materials. Figure 17.12 shows various choices of metals having work functions extending from NFET band edge to PFET band edge [20]. However, in order to prevent work functions from being shifted toward mid-gap, it is required to limit thermal budget on gate electrode materials as much as possible. Although gate-last integration provides higher flexibility in the choice of metal gate electrode materials to achieve work functions near the NFET or PFET band edge, the choice of work function materials is often limited by the compatibility with a CMOS integration scheme including its ability to deposit in the nonplanar structures created from the dummy gate removal and only limited numbers of materials become industry viable candidates. Aluminum or Al-containing alloys are widely

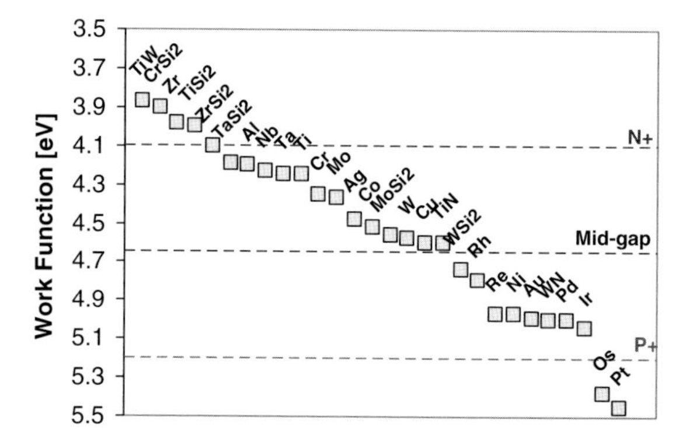

Figure 17.12 Metal gate electrode materials with various work functions. After Ref. [20].

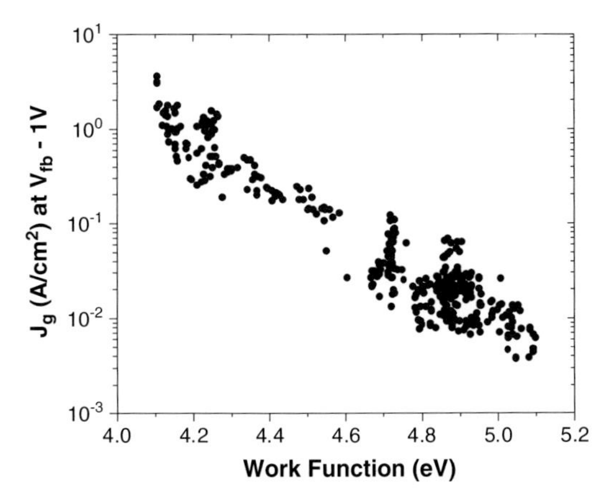

Figure 17.13 Work function tuning in TiN and Al (or Al alloy) electrodes.

used as NFET work function metal [21], while TiN is often utilized as PFET work function metal. Figure 17.13 shows examples of work function tuning in low thermal budget replacement metal gate systems that contain Al (or Al-containing alloys) and TiN for NFET and PFET, respectively. By proper process optimization, work function modulation is possible down from 4.1 eV, which is close to NFET band edge, up to 5.1 eV, which is near PFET band edge.

Despite the capability of controlling work function in a wide range with the Al-containing alloy–TiN system, this system still has various challenges and concerns. As indicated in Figure 17.14, work function control toward the NFET band edge is often coupled with increased gate leakage current that is likely due to lack of thermal stability of Al (or Al-containing alloys) causing Al spiking and interaction with the high-κ dielectric to create excessive oxygen vacancies. More details of this work

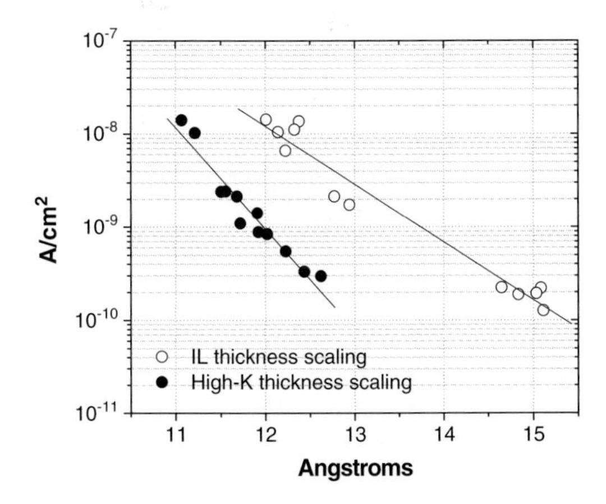

Figure 17.14 T_{inv}–leakage current density (J_g) trend with respect to HK layer and IL thickness scaling.

function–gate leakage interaction will be discussed in a later section. Alternative NFET metal electrodes, such as TaN, TaC [22, 23], or HfSi [24], have been reported to be stable in contact with Hf-based high-κ dielectrics and have a reduced impact on gate leakage current.

Another challenge of the Al-containing alloy–TiN system is to obtain PFET band edge work functions. Since the TiN work function typically saturates near 4.95 eV, a further increase in work function toward the PFET band edge requires careful optimization of the stack with additional innovative treatments. Using materials that have intrinsically high vacuum work function, such as Pt, Mo, Ru, or their alloys, can be an alternative solution to achieve near or beyond PFET band edge [25–27].

17.6 T_{inv} Scaling and Impacts on Gate Leakage and Effective Work Function

As described earlier, all HKMG structures contain a SiO_2-based IL, a high-κ dielectric layer, and a metal gate. A conventional approach to scale T_{inv} in this system is to simply scale the thickness of either the IL or the high-κ dielectric. Figure 17.14 shows T_{inv} versus J_g trend with high-κ dielectrics and ILs at various thicknesses. The scalability by using this conventional approach, however, is often limited by gate leakage current increase as well as mobility degradation [28].

One way to scale T_{inv}, while maintaining low gate leakage current, is to increase the dielectric constant of the IL (which is typically SiO_2, with $k = 3.9$) by incorporating nitrogen or other higher-κ elements. Nitride ILs (SiO_xN_y) can be achieved either by nitriding the IL directly or through the high-κ dielectric using plasma-assisted or thermal nitridation processes.

Difficulties of further T_{inv} scaling arise also from unintended regrowth of the interface layer by oxygen ingress during device fabrication, which often requires high thermal budget. Therefore, it is critical to control ambient to prevent IL regrowth downstream in the integration flow, which will be discussed in detail in the following section. Nitrided interface layer (SiO_xN_y) could also be helpful to suppress the regrowth. The metal gate itself can either act as an oxygen reservoir that can release oxygen during high-temperature process, resulting in IL regrowth [12], or serve as an oxygen scavenging (from IL) layer [29]. Optimization of process conditions of metal gate deposition to reduce amount of oxygen in the metal itself is, therefore, another key method to suppress regrowth and enable T_{inv} scaling.

Extreme T_{inv} scaling for future technology nodes (beyond 32/28 nm) may require one or all of the following: further optimization of HfO_2 through process improvements, new higher-κ materials, and innovative interface engineering to obtain ultrathin IL or higher-κ IL. Replacing HfO_2 with materials having κ-values greater than 20 is a long-term solution that can support multiple technology nodes to come. In pursuit of higher-κ materials, the trade-off between κ-value and bandgap needs to be taken into account. It is generally known that bandgap values have a roughly inverse dependence on κ-values ($E_g \sim \kappa^{-0.65}$) [30]. Therefore, materials with too high a κ-value typically result in excessive direct tunneling currents and most work showing better EOT–J_g characteristics than those of HfO_2 has been achieved with κ-value ranging from 20 to 30. Various groups have reported k-value increase in this range by stabilizing the higher-κ phase (tetragonal or cubic) of HfO_2 via doping of elements such as zirconium [31], yttrium [32], and silicon [33]. This is a practical method to attain a modest κ-value increase; however, controllability of crystallinity in an ultrathin HfO_2 thickness regime (<20 Å) has yet to be demonstrated for implementation in the state-of-the-art CMOS devices. Other groups have suggested La-based higher-κ materials such as $La_xAl_{1-x}O_3$ or $La_xLu_{1-x}O_3$ [34]. These materials have recently been demonstrated to show a lot of promise to outperform Hf-based high-κ dielectrics on a device level [35, 36]. However, most lanthanide-based higher-κ gate dielectrics show very low effective work function near the Si conduction band edge due to formation of interface dipoles, which makes application of these materials to the PFET extremely difficult. Development of CMOS-compatible higher-κ materials is one of the biggest challenges to overcome.

An alternative approach to scaling is to reduce the IL thickness by remote IL scavenging that can be performed by doping a TiN gate electrode with oxygen scavenging elements at certain isolation from the high-κ/TiN interface [11]. This enables uniform T_{inv} scaling for NFETs and PFETs to $T_{inv} < 10$ Å. However, there are many challenges associated with IL scaling, including mobility degradation (covered in Section 17.4) and degraded reliability (covered in Section 17.8). There is also a clear trade-off between PFET V_t and T_{inv}, which is a potential showstopper for aggressive T_{inv} scaling. It has been reported that oxygen flow from a high-κ layer to an IL is promoted with a thinner IL and positively charged oxygen vacancies are generated, resulting in the PFET V_t/V_{fb}–T_{inv} roll-off behavior [37]. It has been shown that a gate-last process can mitigate this roll-off behavior to some extent via recovery of oxygen vacancies [38]. Accurate control of oxygen vacancy concentration in the high-κ layer is indispensable to achieve low PFET V_t in the scaled T_{inv}/EOT regime.

17.7
Ambients and Oxygen Vacancy-Induced Modulation of Threshold Voltage

High-κ dielectrics are significantly more prone to oxygen vacancy formation than SiO_2 and SiON dielectrics, and the newly introduced metal electrodes, such as TiN, have typically higher oxygen solubility and oxidize more easily than Si. For these reasons, much greater control over oxidizing impurities during thermal processes is required throughout HKMG CMOS integration. Small amounts of O_2, H_2O, or OH^- in processing gases (N_2, Ar, or N_2/H_2 used for PDA, PMA, or FGA) can alter the final V_t, greatly impacting the performance of HKMG transistors. The intentional modulation of the oxygen vacancy concentration in the HfO_2 layer itself was shown to cause an immediate modulation of the threshold voltage [39–43]. The V_t values of both NFETs and PFETs can be impacted by the subtle interplay between metal oxidation, IL thickness/regrowth [44], and vacancy passivation/formation in the high-κ layer. When looking at the whole body of experimental data in the listed references, it becomes quite obvious that oxygen redistribution either intentionally or unintentionally is a key aspect of V_t tuning in CMOS processing. If these oxygen effects are understood in detail, it is indeed possible to take advantage of metal oxidation and/or HfO_2 vacancy formation to tune the threshold voltage of FETs in a continuous manner.

TiN oxidation is observed at temperatures as low as 400 °C [45–51]. Even in an inert ambient with only ppm level of oxidizing impurities, TiN tends to incorporate oxygen, forming TiON compounds (possibly with enhanced resistance) before the TiN layer is converted into insulating TiO_2. Oxygen incorporation in the TiN can directly impact V_t by modulating the metal work function. TiON films have strongly enhanced work functions over stoichiometric TiN [41, 42]. In addition, oxidation on refractory and noble metals also results in higher effective work functions [52–54].

Residual O_2, OH, or H_2O in processing tools can lead to IL regrowth [12] and V_t modulation. For IL regrowth to occur, the oxygen has to diffuse to the Si substrate in the channel region. Several known ingress pathways for oxygen have been identified. Oxidizing species can penetrate the gate stack through the metal electrodes [41, 43], likely diffusing along grain boundaries without fully oxidizing the electrode and conserving its metallic nature, or over the device perimeters such as the spacer edges [42] or the STI edges, where the structure may be more open toward the ambient. Oxygen vacancy modulation through the metal, referred to as "top-down" oxygenation [41, 43], and oxygen vacancy modulation over the S/D edge of PFETs, referred to as "lateral" oxygenation [42], have been proposed as methods to control the PFET V_t directly and without excessive IL growth. The temperature–time space for V_t tuning via "top-down" oxygenation is summarized in Figure 17.15. V_t tuning by lateral oxygenation is summarized in Figure 17.16. In these studies, very low partial

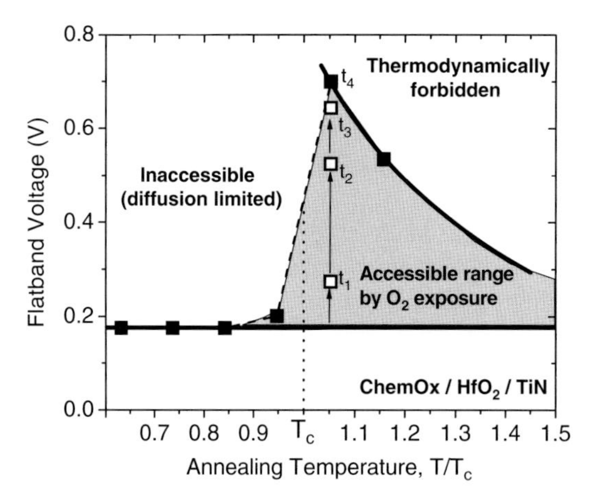

Figure 17.15 Effect of oxygenation on the flatband voltage of TiN-gated MOS capacitors as a function of temperature. Three regimes are identified: inaccessible, oxygen cannot diffuse through the TiN; thermodynamically forbidden, the vacancy concentration cannot be increased because of IL regrowth [56]; and accessible, the flatband voltage can be set at any value by varying oxygenation temperature, time, and pressure. After Ref. [41].

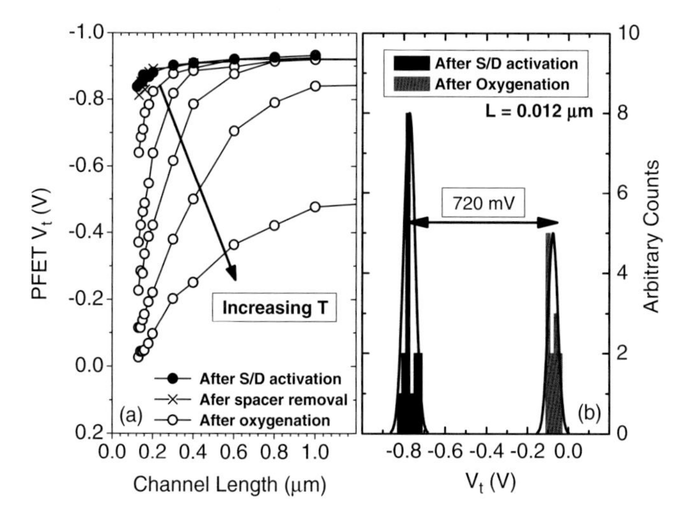

Figure 17.16 (a) PFET V_t shift versus channel length induced by oxygenation from the gate stack sidewall at different temperatures. The *L*-dependences for various oxygenation times are shown. (b) V_t distribution across an 8″ wafer before and after oxygenation. Reproduced with permission from Ref. [42], copyright 2009 IEEE.

pressure of oxygen was used, showing that only small amounts of oxygen can induce very large V_t variations. The HfO_2 layer tends to laterally redistribute the oxygen that penetrates the device structure (from the device edges or along grain boundaries) and IL regrowth tends to be uniform over the device area, suppressing the formation of bird peaks common to conventional CMOS processing. Since oxygen vacancy formation is believed to be promoted by Si oxidation [55], the IL thickness can have a direct impact on the threshold voltage. The much discussed roll-off effect of V_t with decreasing IL thickness [44] is a well-documented example of the impact of the IL thickness on V_t. If size-dependent IL regrowth would occur, a corresponding V_t modulation should be expected because of the IL thickness-dependent oxygen exchange between the HfO_2 and the Si substrate [56].

17.8 Reliability

The requirement for ever thinner gate dielectrics to enable channel length scaling in advanced technology nodes has not only led to unacceptably larger gate leakage currents (the primary reason for the transition to HKMG), but also placed ever higher constraints on reliability margins [57]. With the introduction of HKMG stacks, time-dependent dielectric breakdown (TDDB), negative bias temperature instability (NBTI) [58], and degradation due to hot carrier injection (HCI) continue to be critical reliability limiters, requiring a continuous feedback loop of reliability estimation and gate stack optimization. Positive bias temperature instability (PBTI), a minimal concern in SiON dielectrics, poses a new reliability threat in HKMG

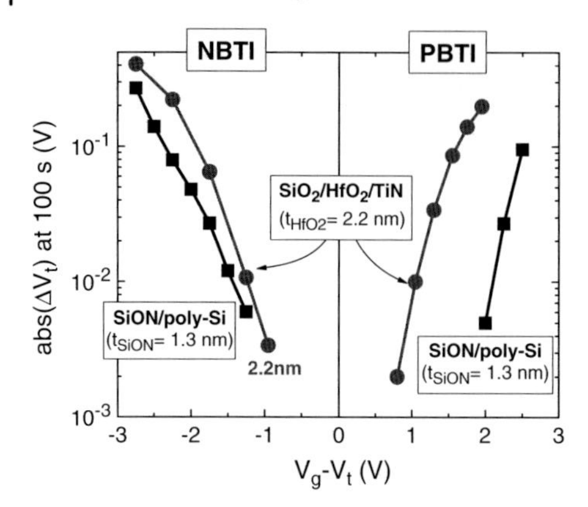

Figure 17.17 Voltage dependence of the NBTI and the PBTI in NFETs and PFETs. FETs with conventional SiON/poly-Si and HfO_2/TiN gate stacks are compared. A conventional "stress and sense" measurement procedure with delay times of ~0.2 s between stress and sense was used here. Reproduced with permission from Ref. [60], copyright 2009 IEEE.

stacks [59, 60]. This is illustrated in Figure 17.17. As can be seen, while NBTIs are comparable, a substantial enhancement of the PBTI degradation is observed. Closely related to PBTI is the stress-induced leakage current phenomenon in HfO_2-gated NFETs [61–63].

- **TDDB**: The primary difference, from a reliability point of view, between HKMG and poly/SiON is the dual-layer nature of the dielectric in HKMG gate stacks. The percolation theory [57, 64], used to explain breakdown and wearout in SiON, has been extended to the dual-layer structure to successfully explain the increase in Weibull slopes at low failure percentiles (Figure 17.18) [65]. This improvement in Weibull slope is a key element to passing reliability specifications for TDDB.

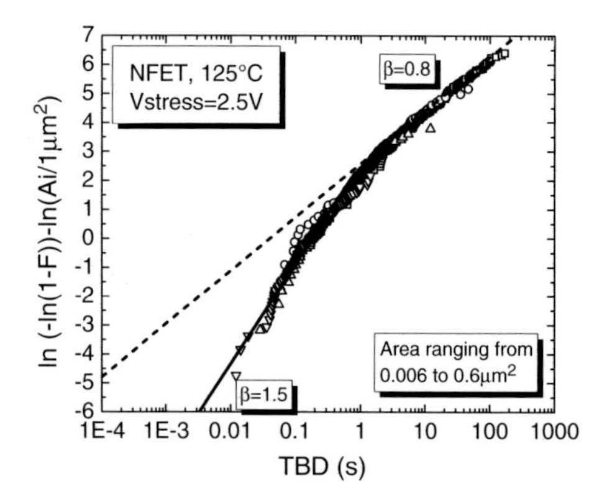

Figure 17.18 Failure distribution for various NFET HKMG devices based on vertical Poisson scaling. Note the significant increase in Weibull slope from $\beta = 0.8$ to $\beta = 1.5$ for low Weibull numbers (larger areas and lower failure percentile). Reproduced with permission from Ref. [60], copyright 2009 IEEE.

Further, ramp voltage stress tests [66] on HfO_2-based gate stacks show that the breakdown voltage monotonically decreases with increasing gate leakage, with a constant slope [6], suggesting that T_{inv} scaling might become limited by TDDB reliability requirements.

- **NBTI**: For conventional SiON gate stacks, NBTI has been shown to be a Si/IL interface phenomenon that is controlled by the oxide field. Two physical processes contribute to NBTI: hole trapping at/near the Si/SiON interface and interface state generation [58]. Since a SiO(N)-like IL is used in HKMG stacks, the learning from SiON gate stacks can be directly applied. The high-κ layer to first order acts only as a field modulator. As for SiON, the hole trapping is found to be transient, leading to the much studied reversible component [67]. Generated interface states are significantly more stable [68] and the V_t shift due to interface state generation is therefore referred to as the permanent component. For lifetime projections, depending on the circuit functionality, the reversible component may not limit the lifetime [69]. One of the primary modulators of NBTI, apart from the oxide electric field, is the nitrogen content and the nitrogen distribution in the IL. In general, as for SiON, enhanced nitrogen content tends to reduce NBTI lifetime. Ramp voltage NBTI test [70] shows that heavily nitrided interfaces lead to reduced NBTI lifetime, as illustrated in Figure 17.19.

  In HKMG stacks, nitrogen may be incorporated not only during intentional nitridation steps, but also during metal nitride deposition and subsequent PDAs. Typically, higher annealing temperatures tend to harden the IL with respect to NBTI. This IL hardening is at least in part related to nitrogen redistribution. Great care needs to be taken to carefully account for CET/T_{inv} differences when studying the impact of annealing on NBTI. Any increase in T_{inv} will reduce the stress field for a constant voltage test and indicate an apparent interface improvement with respect to NBTI.
- **PBTI**: The PBTI results in a threshold voltage increase in the NFET under positive gate bias. This new phenomenon is usually attributed to electron trapping in preexisting defects (perhaps due to oxygen vacancies [71]). Two process

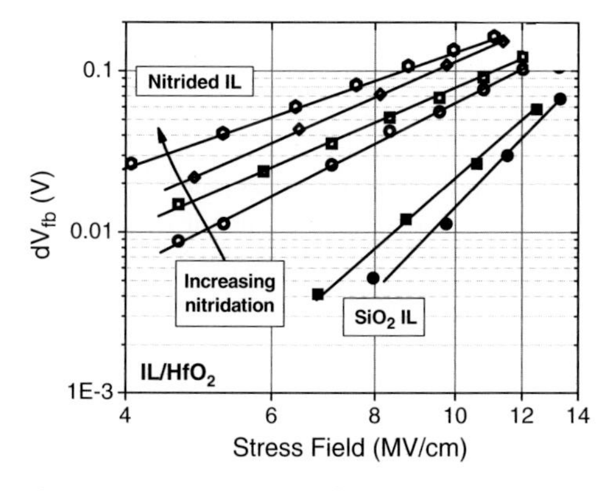

Figure 17.19 Comparison of NBTI response (ramp voltage test) for HKMG transistors with thermal SiO_2 interlayers and with nitrided interlayers with various degrees of nitridation.

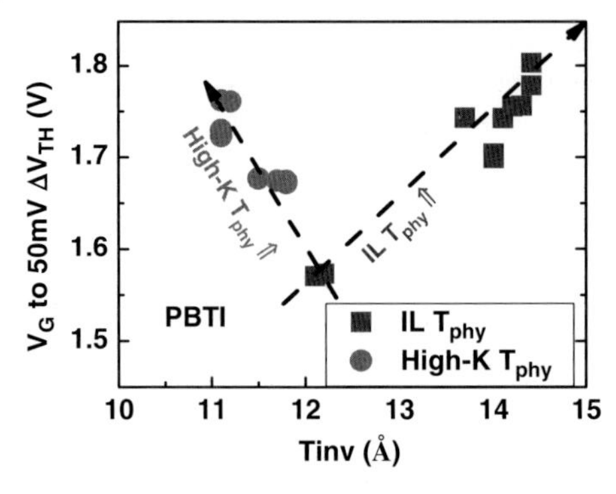

Figure 17.20 Voltage at which a PBTI shift of 50 mV is observed in a ramp voltage test for NFETs with various HKMG stacks. Decreasing the IL thickness leads to enhanced gate currents and enhanced trapping, reducing the voltage at which a 50 mV shift is observed.

parameters have been shown to effectively reduce PBTI-induced voltage shifts. As can be seen from Figure 17.20, both reduction in high-κ thickness and increase in the IL thickness reduce PBTI. Reducing the high-κ thickness reduces the electron trapping in HfO_2 bulk defects while increasing the interface thickness makes the trapping less likely simply because the gate leakage current density is reduced. The minimal high-κ thickness is determined by manufacturability, gate leakage, and TDDB requirements and cannot be reduced at will. The IL thickness is dictated by performance and T_{inv} requirements. The PBTI impact, while impossible to eliminate altogether for high-performance applications, is also reduced due to the fact that it shows a large recoverable component. It is to be noted that recovery during AC stress causes the PBTI impacts to be significantly lower than those under DC stress [72, 73].

17.9 Conclusions

High-κ/metal gate devices have re-enabled classical gate length scaling due to their ability to reduce subthreshold leakage, reduce gate leakage, and improve capacitance over prior technologies made from SiON/polysilicon gates. The use of metal gates and high-κ dielectrics brings with them numerous materials and integration challenges that must be overcome in order to realize the benefits of these stacks in CMOS devices. Modern CMOS integration methods and practices refined over generations pose challenges for the stability of these stacks and have been shown to cause changes in effective work function, regrowth of the IL causing T_{inv} increase, reductions in mobility, and deleterious impacts to reliability. In fact, the materials and integration methodology are so interlinked that one impacts the other. Understanding how to coax the maximum performance out of HKMG materials is paramount to enable continued scaling of HKMG technologies for the next CMOS generations.

References

1 Dennard, R.H., Gaensslen, F.H., Yu, H.-N., Rideout, V.L., Bassous, E., and LeBlanc, A. (1974) Design of ion-implanted MOSFET's with very small physical dimensions. *IEEE J. Solid-State Circuits*, **9**, 256.

2 Haensch, W., Nowak, E.J., Dennard, R.H., Solomon, P.M., Bryant, A., Dokumaci, O.H., Kumar, A., Wang, X., Johnson, J.B., and Fischetti, M.V. (2006) Silicon CMOS devices beyond scaling. *IBM J. Res. Dev.*, **50**, 339.

3 Chudzik, M., Doris, B., Mo, R., Sleight, J., Cartier, E., Dewan, C., Park, D., Bu, H., Natzle, W., Yan, W., Ouyang, C., Henson, K., Boyd, D., Callegari, A., Carter, R., Casarotto, D., Gribelyuk, M., Hargrove, M., He, W., Kim, Y., Linder, B., Moumen, N., Paruchuri, V.K., Stathis, J., Steen, M., Vayshenker, A., Wang, X., Zafar, S., Ando, T., Iijima, R., Takayanagi, M., Narayanan, V., Wise, R., Zhang, Y., Divakaruni, R., Khare, M., and Chen, T.C. (2007) High-performance high-k/metal gates for 45 nm CMOS and beyond with gate-first processing. Symposium on VLSI Technology, Digest of Technical Papers, p. 194.

4 Mistry, K., Allen, C., Auth, C., Beattie, B., Bergstrom, D., Bost, M., Brazier, M., Buehler, M., Cappellani, A., Chau, R., Choi, C.-H., Ding, G., Fischer, K., Ghani, T., Grover, R., Han, W., Hanken, D., Hattendorf, M., He, J., Hicks, J., Huessner, R., Ingerly, D., Jain, P., James, R., Jong, L., Joshi, S., Kenyon, C., Kuhn, K., Lee, K., Liu, H., Maiz, J., Mclntyre, B., Moon, P., Neirynck, J., Pae, S., Parker, C., Parsons, D., Prasad, C., Pipes, L., Prince, M., Ranade, P., Reynolds, T., Sandford, J., Shifren, L., Sebastian, J., Seiple, J., Simon, D., Sivakumar, S., Smith, P., Thomas, C., Troeger, T., Vandervoorn, P., Williams, S., and Zawadzki, K. (2007) A 45 nm logic technology with high-κ/metal gate transistors, strained silicon, 9 Cu interconnect layers, 193 nm dry patterning, and 100% Pb-free packaging. IEDM Technical Digest, p. 247.

5 Fischetti, M.V., Neumayer, D.A., and Cartier, E.A. (2001) Effective electron mobility in Si inversion layers in metal–oxide–semiconductor systems with a high-κ insulator: the role of remote phonon scattering. *J. Appl. Phys.*, **90**, 4587.

6 Linder, B.P., Kerber, A., Cartier, E., Krishnan, S., and Stathis, J.H. (2009) The effect of interface thickness of high-κ/metal gate stacks on NFET dielectric reliability. IEEE International Reliability Physics Symposium, p. 510.

7 Gusev, E.P., Narayanan, V., and Frank, M.M. (2006) Advanced high-κ dielectric stacks with poly-Si and metal gates: recent progress and current challenges. *IBM J. Res. Dev.*, **50**, 387.

8 Guha, S. and Narayanan, V. (2009) High-κ/metal gate science and technology. *Annu. Rev. Mater. Res.*, **39**, 181.

9 Saito, S., Hisamoto, D., Kimura, S., and Hiratani, M. (2003) Unified mobility model for high-κ gate stacks [MISFETs]. IEDM Technical Digest, p. 797.

10 Kirsch, P.D., Quevedo-Lopez, M.A., Li, H.-J., Senzaki, Y., Peterson, J.J., Song, S.C., Krishnan, S.A., Moumen, N., Barnett, J., Bersuker, G., Hung, P.Y., Lee, B.H., Lafford, T., Wang, Q., Gay, D., and Ekerdt, J.G. (2006) Nucleation and growth study of atomic layer deposited HfO_2 gate dielectrics resulting in improved scaling and electron mobility. *J. Appl. Phys.*, **99**, 023508.

11 Ando, T., Frank, M.M., Choi, K., Choi, C., Bruley, J., Hopstaken, M., Copel, M., Cartier, E., Kerber, A., Callegari, A., Lacey, D., Brown, S., Yang, Q., and Narayanan, V. (2009) Understanding mobility mechanisms in extremely scaled HfO_2 (EOT 0.42 nm) using remote interfacial layer scavenging technique and V_t-tuning dipoles with gate-first process. IEDM Technical Digest, p. 423.

12 Narayanan, V., Maitra, K., Linder, B.P., Paruchuri, V.K., Gusev, E.P., Jamison, P., Frank, M.M., Steen, M.L., La Tulipe, D., Arnold, J., Carruthers, R., Lacey, D.L., and Cartier, E. (2006) Process optimization for high electron mobility in nMOSFETs with aggressively scaled

HfO_2/metal stacks. *IEEE Electron Device Lett.*, **27**, 591.

13 Datta, S., Dewey, G., Doczy, M., Doyle, B.S., Jin, B., Kavalieros, J., Kotlyar, R., Metz, M., Zelick, N., and Chau, R. (2003) High mobility Si/SiGe strained channel MOS transistors with HfO_2/TiN gate stack. IEDM Technical Digest, p. 653.

14 Schaeffer, J.K., Capasso, C., Fonseca, L.R.C., Samavedam, S., Gilmer, D.C., Liang, Y., Kalpat, S., Adetutu, B., Tseng, H.-H., Shiho, Y., Demkov, A., Hegde, R., Taylor, W.J., Gregory, R., Jiang, J., Luckowski, E., Raymond, M.V., Moore, K., Triyoso, D., Roan, D., White Jr, B.E., and Tobin, P.J. (2004) Challenges for the integration of metal gate electrodes. IEDM Technical Digest, p. 287.

15 Narayanan, V., Paruchuri, V.K., Bojarczuk, N.A., Linder, B.P., Doris, B., Kim, Y.H., Zafar, S., Stathis, J., Brown, S., Arnold, J., Copel, M., Steen, M., Cartier, E., Callegari, A., Jamison, P., Locquet, J.-P., Lacey, D.L., Wang, Y., Batson, P.E., Ronsheim, P., Jammy, R., Chudzik, M.P., Ieong, M., Guha, S., Shahidi, G., and Chen, T.C. (2006) Band-edge high-performance high-κ/metal gate n-MOSFETs using cap layers containing group IIA and IIIB elements with gate-first processing for 45 nm and beyond. Symposium on VLSI Technology, Digest of Technical Papers, p. 224.

16 Guha, S., Paruchuri, V.K., Copel, M., Narayanan, V., and Wang, Y.Y. (2007) Examination of flatband and threshold voltage tuning of HfO_2/TiN field effect transistors by dielectric cap layers. *Appl. Phys. Lett.*, **90**, 092902.

17 Kamimuta, Y., Iwamoto, K., Nunoshige, Y., Hirano, A., Mizubayashi, W., Watanabe, Y., Migita, S., Ogawa, A., Ota, H., Nabatame, T., and Toriumi, A. (2007) Comprehensive study of V_{fb} shift in high-κ CMOS: dipole formation, Fermi-level pinning and oxygen vacancy effect. IEDM Technical Digest, p. 341.

18 Jagannathan, H., Narayanan, V., and Brown., S.L. (2008) Engineering high dielectric constant materials for band-edge CMOS applications. *ECS Trans.*, **16**, 19.

19 Copel, M., Guha, S., Bojarczuk, N., Cartier, E., Narayanan, V., and Paruchuri, V.K. (2009) Interaction of La_2O_3 capping layers with HfO_2 gate dielectrics. *Appl. Phys. Lett.*, **95**, 212903.

20 Skotnicki, T. (2004) Transistor scaling to the end of the roadmap. Symposium on VLSI Technology, Short Course.

21 Hinkle, C.L., Galatage, R.V., Chapman, R.A., Vogel, E.M., Alshareef, H.N., Freeman, C., Wimmer, E., Niimi, H., Li-Fatou, A., Shaw, J.B., and Chambers, J.J. (2010) Dipole controlled metal gate with hybrid low resistivity cladding for gate-last CMOS with low V_t. Symposium on VLSI Technology, Digest of Technical Papers, p. 183.

22 Pan, J., Woo, C., Yang, C.Y., Bhandary, U., Guggilla, S., Krishna, N., Chung, H., Hui, A., Yu, B., Xiang, Q., and Lin, M.-R. (2003) Replacement metal-gate NMOSFETs with ALD TaN/EP-Cu, PVD Ta, and PVD TaN electrode. *IEEE Electron Device Lett.*, **24**, 304.

23 Mizubayashi, W., Akiyama, K., Wang, W., Ikeda, M., Iwamoto, K., Kamimuta, Y., Hirano, A., Ota, H., Nabatame, T., and Toriumi, A. (2008) Novel V_{th} tuning process for HfO_2 CMOS with oxygen-doped TaC_x. Symposium on VLSI Technology, Digest of Technical Papers, p. 42.

24 Ando, T., Hirano, T., Tai, K., Yamaguchi, S., Yoshida, S., Iwamoto, H., Kadomura, S., and Watanabe, H. (2010) Low threshold voltage and high mobility N-channel metal–oxide–semiconductor field-effect transistor using Hf–Si/HfO_2 gate stack fabricated by gate-last process. *Jpn. J. Appl. Phys.*, **49**, 016502.

25 Gu, D., Dey, S.K., and Majhi, P. (2006) Effective work function of Pt, Pd, and Re on atomic layer deposited HfO_2. *Appl. Phys. Lett.*, **89**, 082097.

26 Qiang, L., Lin, R., Ranade, P., Yeo, Y.C., Meng, X., Takeuchi, H., King, T.-J., Hu, C., Luan, H., Lee, S., Bai, W., Lee, C.-H., Kwong, D.-L., Guo, X., Wang, X., and Ma, T.-P. (2000) Molybdenum metal gate

MOS technology for post-SiO gate dielectrics. IEDM Technical Digest, p. 641.

27 Todi, R.M., Warren, A.P., Sundaram, K.B., Barmak, K., and Coffey, K.R. (2006) Characterization of Pt–Ru binary alloy thin films for work function tuning. *IEEE Electron Device Lett.*, **27**, 542.

28 Chau, R. (2004) Advanced metal gate/high-κ dielectric stacks for high-performance CMOS transistors. 5th AVS International Conference on Microelectronics and Interfaces, p. 3.

29 Wu, L., Yu, H.Y., Li, X., Pey, K.L., Hsu, K.Y., Tao., H.J., Chiu, Y.S., Lin, C.T., Xu, J.H., and Wan, H.J. (2010) Investigation of ALD or PVD (Ti-rich vs. N-rich) TiN metal gate thermal stability on HfO_2 high-κ. Symposium on VLSI Technology, Digest of Technical Papers, p. 90.

30 Robertson, J. (2008) Maximizing performance for higher K gate dielectrics. *J. Appl. Phys.*, **104**, 124111.

31 Hegde, R.I., Triyoso, D.H., Tobin, P.J., Kalpat, S., Ramon, M.E., Tseng, H.-H., Schaeffer, J.K., Luckowski, E., Taylor, W.J., Capasso, C.C., Gilmer, D.C., Moosa, M., Haggag, A., Raymond, M., Roan, D., Nguyen, J., La, L.B., Hebert, E., Cotton, R., Wang, X.-D., Zollner, S., Gregory, R., Werho, D., Rai, R.S., Fonseca, L., Stoker, M., Tracy, C., Chan, B.W., Chiu, Y.H., and White, B.E., Jr. (2005) Microstructure modified HfO_2 using Zr addition with Ta_xC_y gate for improved device performance and reliability. IEDM Technical Digest, p. 35.

32 Kita, K., Kyuno, K., and Toriumi, A. (2005) Permittivity increase of yttrium-doped HfO_2 through structural phase transformation. *Appl. Phys. Lett.*, **86**, 102906.

33 Tomida, K., Kita, K., and Toriumi, A. (2006) Dielectric constant enhancement due to Si incorporation into HfO_2. *Appl. Phys. Lett.*, **89**, 142902.

34 Schlom, D.G. and Haenni, J.H. (2002) A thermodynamic approach to selecting alternative gate dielectrics. *MRS Bull.*, **27**, 198.

35 Arimura, H., Brown, S.L., Callegari, A., Kellock, A., Bruley, J., Copel, M., Watanabe, H., Narayanan, V., and Ando, T. (2011) Maximized benefit of La–Al–O higher-κ gate dielectrics by optimizing the La/Al atomic ratio. *IEEE Electron Device Lett.*, **32**, 288.

36 Edge, L.F., Vo, T., Paruchuri, V.K., Iijima, R., Bruley, J., Jordan-Sweet, J., Linder, B.P., Kellock, A.J., Tsunoda, T., and Shinde, S.R. (2011) Materials and electrical characterization of physical vapor deposited $La_xLu_{1-x}O_3$ thin films on 300 mm silicon. *Appl. Phys. Lett.*, **98**, 122905.

37 Song, S.C., Park, C.S., Price, J., Burham, C., Choi, R., Wen, H.C., Choi, K., Tseng, H.H., Lee, B.H., and Jammy, R. (2007) Mechanism of V_{fb} roll-off with high work function metal gate and low temperature oxygen incorporation to achieve PMOS band edge work function. IEDM Technical Digest, p. 337.

38 Ragnarsson, L.-Å., Li, Z., Tseng, J., Schram, T., Rohr, E., Cho, M.J., Kauerauf, T., Conard, T., Okuno, Y., Parvais, B., Absil, P., Biesemans, S., and Hoffmann, T.Y. (2009) Ultra low-EOT (5 Å) gate-first and gate-last high performance CMOS achieved by gate-electrode optimization. IEDM Technical Digest, p. 663.

39 Schaeffer, J.K., Fonseca, L.R.C., Samavedam, S.B., Liang, Y., Tobin, P.J., and White, B.E. (2004) Contributions to the effective work function of platinum on hafnium dioxide. *Appl. Phys. Lett.*, **85**, 1826.

40 Cartier, E., McFeely, F.R., Narayanan, V., Jamison, P., Linder, B.P., Copel, M., Paruchuri, V.K., Basker, V., Haight, R., Lim, D., Carruthers, R., Shaw, T., Steen, M., Sleight, J., Rubino, J., Deligianni, H., Guha, S., Jammy, R., and Shahidi, G. (2005) Role of oxygen vacancies in V_{FB}/V_t stability of pFET metals on HfO_2. Symposium on VLSI Technology, Digest of Technical Papers.

41 Cartier, E., Hopstaken, M., and Copel, M. (2009) Oxygen passivation of vacancy defects in metal-nitride gated $HfO_2/SiO_2/Si$ devices. *Appl. Phys. Lett.*, **95**, 042901.

42 Cartier, E., Steen, M., Linder, B.P., Ando, T., Iijima, R., Frank, M.,

Newbury, J.S., Kim, J.S., McFeely, F.R., Copel, M., Haight, R., Choi, C., Callegari, A., Paruchuri, V.K., and Narayanen, V. (2009) p-FET V_T control with HfO_2/TiN/poly-Si gate stack using a lateral oxygenation process. Symposium on VLSI Technology, Digest of Technical Papers, p. 42.

43 Cartier, E. (2010) The role of oxygen in the development of Hf-base high-κ/metal gate stacks for CMOS technologies. *ECS Trans.*, **33**, 83.

44 Bersuker, G., Park, C.S., Wen, H.-C., Choi, K., Price, J., Lysaght, P., Tseng, H.-H., Sharia, O., Demkov, A., Ryan, J.T., and Lenahan, P. (2010) Origin of the flatband-voltage roll-off phenomenon in metal/high-κ gate stacks. *IEEE Trans. Electron Devices*, **57**, 2047.

45 Ernsberger, C., Nickerson, J., Smith, T., Miller, A.E., and Banks, D. (1986) Low temperature oxidation behavior of reactively sputtered TiN by X-ray photoelectron spectroscopy and contact resistance measurements. *J. Vac. Sci. Technol. A*, **4**, 2784.

46 Tompkins, H.G. (1991) Oxidation of titanium nitride in room air and in dry O_2. *J. Appl. Phys.*, **70**, 3876.

47 Chen, H.Y. and Lu, F.-H. (2005) Oxidation behavior of titanium nitride films. *J. Vac. Sci. Technol. A*, **23**, 1006.

48 Hinodea, K., Homma, Y., Horiuchi, M., and Takahashi, T. (1997) Morphology-dependent oxidation behavior of reactively sputtered titanium–nitride films. *J. Vac. Sci. Technol. A*, **15**, 2017.

49 Wittmer, M., Noser, J., and Melchior, H. (1981) Oxidation kinetics of TiN thin films. *J. Appl. Phys.*, **52**, 6659.

50 Soriano, L., Abbate, M., Fuggle, J.C., Prieto, P., Jimenez, C., Sanz, J.M., Galan, L., and Hofmann, S. (1993) Thermal oxidation of TiN studied by means of soft X-ray absorption spectroscopy. *J. Vac. Sci. Technol. A*, **11**, 47.

51 Naresh, C., Tompkins, S., and Tompkins, G.T. (1992) Titanium nitride oxidation chemistry: an X-ray photoelectron spectroscopy study. *J. Appl. Phys.*, **72**, 3072.

52 Lim, D. and Haight, R. (2005) *In situ* photovoltage measurements using femtosecond pump-probe photoelectron spectroscopy and its application to metal–HfO_2–Si structures. *J. Vac. Sci. Technol. A*, **23**, 1698.

53 Lim, D. and Haight, R. (2005) Temperature dependent defect formation and charging in hafnium oxides and silicates. *J. Vac. Sci. Technol. B*, **23**, 201.

54 Lim, D., Haight, R., Copel, M., and Cartier, E. (2005) Oxygen defects and Fermi level location in metal–hafnium oxide–silicon structures. *Appl. Phys. Lett.*, **87**, 72902.

55 Shiraishi, K., Yamada, K., Torii, K., Akasaka, Y., Nakajima, K., Konno, M., Chikyow, T., Kitajima, H., and Arikado, T. (2004) Physics in Fermi level pinning at the polySi/Hf-based high-κ oxide interface. *Jpn. J. Appl. Phys. 2*, **43**, L1413.

56 Guha, S. and Narayanan, V. (2007) Oxygen vacancies in high dielectric constant oxide–semiconductor films. *Phys. Rev. Lett.*, **98**, 196101.

57 Lombardo, S., Stathis, J.H., Linder, B.P., Pey, K.L., Palumbo, F., and Thung, C.H. (2005) Dielectric breakdown mechanisms in gate oxides. *Appl. Phys. Rev.*, **98**, 121302.

58 Stathis, J.H. and Zafar, S. (2006) The negative bias temperature instability in MOS devices: a review. *Microelectron. Reliab.*, **46**, 270.

59 Kerber, A., Cartier, E., Pantisano, L., Degraeve, R., Kauerauf, T., Kim, Y., Hou, A., Groeseneken, G., Maes, H.E., and Schwalke, U. (2003) Origin of the threshold voltage instability in SiO_2/HfO_2 dual layer gate dielectrics. *IEEE Electron Device Lett.*, **24**, 87.

60 Kerber, A. and Cartier, E. (2009) Reliability challenges for CMOS technology qualification with hafnium oxide/titanium nitride gate stacks. *IEEE Trans. Dev. Mater. Reliab.*, **9**, 147.

61 Cartier, E. and Kerber, A. (2009) Stress induced leakage current and defect generation in nFETs with HfO_2/TiN gate stacks during positive bias temperature stress. IEEE International Reliability Physics Symposium, p. 486.

62 Crupi, F., Degraeve, R., Kerber, A., Kwak, D.H., and Groeseneken, G. (2003) Correlation between stress-induced leakage current (SILC) and the HfO_2 bulk trap density in a SiO_2/HfO_2 stack. IEEE International Reliability Physics Symposium, p. 181.

63 Chen, C.-Y., Ran, Q., Cho, H.J., Kerber, A., Liu, Y., and Dutton, R. (2011) Correlation of Id- and Ig-random telegraph noise to positive bias temperature instability in scaled high-κ/metal gate n-type MOSFETs, IEEE International Reliability Physics Symposium, p. 3A.2.1 - 3A.2.6.

64 Degraeve, R., Groeseneken, G., Bellens, R., Ogier, J.L., Depas, M., Roussel, P.J., and Maes, H.E. (1998) New insights in the relation between electron trap generation and the statistical properties of oxide breakdown. *IEEE Trans. Electron Devices*, **45**, 904.

65 Nigam, T., Kerber, A., and Peumans, P. (2009) Accurate model for time dependent dielectric breakdown of high-κ metal gate stacks. IEEE International Reliability Physics Symposium, p. 523.

66 Kerber, A., Pantisano, L., Veloso, A., Groeseneken, G., and Kerber, M. (2007) Reliability screening of high-κ dielectrics based on voltage ramp stress. *Microelectron. Reliab.*, **47**, 513.

67 Rangan, S., Mielke, N., and Yeh, E.C.C. (2003) Universal recovery behavior of negative bias temperature instability. IEDM Technical Digest, p. 341.

68 Huard, V. (2010) Two independent components modeling for negative bias temperature instability. IEEE International Reliability Physics Symposium, p. 33.

69 Nigam, T. and Harris, E.B. (2006) Lifetime enhancement under high frequency NBTI measured on ring oscillators. IEEE International Reliability Physics Symposium, p. 289.

70 Kerber, A., Krishnan, S.A., and Cartier, E. (2009) Voltage ramp stress for bias temperature instability testing of metal gate/high-κ stacks. *IEEE Electron Device Lett.*, **30**, 1347.

71 Cartier, E., Linder, B.P., Narayanan, V., and Paruchuri, V.K. (2006) Fundamental understanding and optimization of PBTI in nFETs with SiO_2/HfO_2 gate stack. IEDM Technical Digest, p. 321.

72 Kerber, A., Maitra, K., Majumdar, R., Hargrove, M., Carter, R.J., and Cartier, E. (2008) Characterization of fast relaxation during BTI stress in conventional and advanced CMOS devices with HfO_2/TiN gate stacks. *IEEE Trans. Electron Devices*, **55**, 3175.

73 Zhao, K., Stathis, J.H., Kerber, A., and Cartier, E. (2010) PBTI relaxation dynamics after AC vs. DC stress in high-κ/metal gate stacks. IEEE International Reliability Physics Symposium, p. 50.

Index

High-k Gate Dielectrics for CMOS Technology, First Edition. Edited by Gang He and Zhaoqi Sun.